Electrical Engineering

Concepts and Applications

Electrical Engineering
Concepts and Applications

S. A. Reza Zekavat

Michigan Technological University

PEARSON

Upper Saddle River Boston Columbus San Franciso New York
Indianapolis London Toronto Sydney Singapore Tokyo Montreal
Dubai Madrid Hong Kong Mexico City Munich Paris Amsterdam Cape Town

Vice President and Editorial Director, ECS: *Marcia J. Horton*
Executive Editor: *Andrew Gilfillan*
Editorial Assistant: *William Opaluch*
Vice President, Production: *Vince O'Brien*
Senior Managing Editor: *Scott Disanno*
Production Liaison: *Irwin Zucker*
Production Editor: *Pavithra Jayapaul, Jouve India*
Operations Specialist: *Lisa McDowell*
Executive Marketing Manager: *Tim Galligan*
Marketing Assistant: *Jon Bryant*
Art Editor: *Greg Dulles*
Art Director: *Jayne Conte*
Cover Image: *Photo of wireless sensor used on the Golden Gate Bridge in San Francisco. Courtesy of Shamim Pakzad, Lehigh University.*
Composition/Full-Service Project Management: *Jouve India*

MATLAB is a registered trademark of The Math Works, Inc., 3 Apple Hill Drive, Natick, MA 01760-2098

OrCAD and PSPICE content reprinted with permission of Cadence Design Systems, Inc. All rights reserved.

The author and publisher of this book have used their best efforts in preparing this book. These efforts include the development, research, and testing of the theories and programs to determine their effectiveness. The author and publisher make no warranty of any kind, expressed or implied, with regard to these programs or the documentation contained in this book. The author and publisher shall not be liable in any event for incidental or consequential damages in connection with, or arising out of, the furnishing, performance, or use of these programs.

Library of Congress Cataloging-in-Publication Data

Zekavat, Seyed A.
  Electrical engineering: concepts and applications / Seyed A. (Reza) Zekavat.—1st ed.
    p. cm.
  ISBN-13: 978-0-13-253918-0
  ISBN-10: 0-13-253918-7
   1. Electrical engineering—Textbooks.   I. Title.
  TK165.Z45 2012
  621.3—dc23

                                                                        2011029582

7 2023

PEARSON

ISBN 10: 0-13-253918-7
ISBN 13: 978-0-13-253918-0

Dedication

To my father, Seyed Hassan, and mother Azardokht

CONTENTS

PREFACE

A multi-disciplinary effort was initiated at Michigan Technological University, with a support from the U.S. National Science Foundation's Engineering Education division. The goal was to create a curriculum that (1) encourages students to pursue the life-long learning necessary to keep pace with the rapidly-evolving engineering industry and emerging interdisciplinary technologies, (2) maintains sufficient connection between the students' chosen engineering fields and class content; and (3) motivates and excite the students about the importance of EE concepts to their discipline and career.

Seven faculty members across different departments contributed to this process. Participating departments included: electrical engineering, chemical engineering, civil and environmental engineering, mechanical engineering, biomedical engineering, and the education division of the cognitive and learning science department. The group's curriculum reform efforts were informed by a nationwide survey of engineering schools. The survey outcomes were analyzed to fine tune different curriculum options for this course for different engineering disciplines. Then, those options were integrated to create the final draft of the curriculum. The final draft of the curriculum was used as a layout to create a new textbook for this course.

Although no single text can *perfectly* meet the needs of every institution, diverse topics have been included to address the mixed survey response and allow this book to address the needs of lecturers in different institutions worldwide. The resulting textbook creates a prototype curriculum available to electrical engineering departments that are charged with providing an introduction to electrical engineering for non-EE majors. The goals of this new curriculum are to be attractive, motivational, and relevant to students by creating many application-based problems; and provide the optimal level of both range and depth of coverage of EE topics in a curriculum package.

The book features:

a. *Application-based examples:* A large number of application-based examples were selected from different engineering fields and are included in each chapter. They aim to bridge EE and diverse non-EE areas. These examples help to address the question: "why I should take this course?" Non-EE students will better understand: (1) why they should learn how to solve circuits; and; (2) what are the applications of solving circuits in mechanical, chemical, and civil engineering areas.

b. *PSpice lectures, examples, and problems:* The text offers a distributed approach for learning PSpice. A PSpice component is integrated in many chapters. Chapter 2 provides an initial tutorial, and new skills are added in Chapters 3–11. This part includes lectures that teach students how to use PSpice and can be considered as an embedded PC-based lab for the course. In addition, many PSpice-specific examples have been developed, which help students better understand the process of building a circuit and getting the desired results. There are also many end-of-chapter PSpice problems.

c. *Innovative chapters:* Based on our nationwide survey, the topics in these chapters have been highlighted by many professionals as important topics for this course. It should be noted that each instructor has the liberty to include or exclude some of these topics from his/her curriculum. Some topics include:

- *Chapter 1—Case Study:* This chapter presents the applications of electrical engineering components in mechanical engineering, chemical engineering, and civil engineering through real life scenarios. A bridge across these case studies and the topics that will be covered later in the book is maintained. The goal is to better motivate students by placing the concepts of electrical engineering in the context of their chosen fields of study. Each section of this chapter was been prepared by a different member of the faculty at Michigan Tech who contributed to the NSF project.

- *Chapter 7—Frequency Response with MATLAB and PSpice:* This chapter discusses the frequency response of circuits and introduces different types of filters and uses MATLAB and PSpice examples and end-of chapter problems. This chapter creates an opportunity for students to learn some features of MATLAB software. In other words, this chapter promotes an integrated study using both PSpice and MATLAB.
- *Power Coverage: Chapters 9, 12, 13*—Based on our nationwide survey, and motivated by concerns about global warming and the need for clean energy, industry respondents requested a more thorough treatment of power. Thus, power coverage is supported by three chapters. Chapter 9 introduces the concept of three-phase systems, transmission lines, their equivalent circuits, and power transfer. Chapter 12 studies another important topic of energy transfer—transformers. Finally Chapter 13 studies the topic of motors and generators. This chapter offers the concept of motors and generators in a clear and concise approach. The chapter introduces applications of motors and generators and introduces many applications of both.
- *Chapter 15—Electrical Safety:* This unique chapter discusses interesting electric safety topics useful in the daily life of consumers or engineers working in the field.

d. *Examples and sorted end-of-chapter problems:* The book comes with more than 1100 examples and end-of-chapter problems (solutions included). End-of-chapter problems are sorted to help instructors select basic, average, and difficult problems.

e. *A complete solution manual:* A complete solutions manual for all problems will be available via download for all adopting professors.

ACKNOWLEDGMENTS

Professor William Bulleit (Civil and Environmental Engineering Department, Michigan Tech), Professor Tony Rogers (Chemical Engineering Department, Michigan Tech) and Professor Harold Evensen (Mechanical Engineering, Engineering Mechanics Department, Michigan Tech) are the authors of chapter one. The research on this National Science Foundation project was conducted with the support of many faculty members. Here, in addition to Professor Bulleit, Professor Rogers and Professor Evensen, I should acknowledge the efforts of Professor Kedmon Hungwe (Education Department, Michigan Tech), Mr. Glen Archer (Electrical and Computer Engineering Department, Michigan Tech), Professor Corina Sandu (Mechanical Engineering Department, Virginia Tech), Professor David Nelson (Mechanical Engineering Department, University of South Alabama), Professor Sheryl Sorby (Mechanical Engineering, Engineering Mechanics Department, Michigan Tech), and Professor Valorie Troesch (Institute for Interdisciplinary Studies, Michigan Tech). The preparation of the book was not possible without the support of many graduate students that include Luke Mounsey, Xiukui Li, Taha Abdelhakim, Shu G. Ting, Wenjie Xu, Zhonghai Wang, Babak Bastaami, Manaas Majumdar, Abdelhaseeb Ahmed, Daw Don Cheam, Jafar Pourrostam and Greg Price. I would like to thank all of them. Moreover, I should thank the support of the book's grand reviewer Mr. Peter A. Larsen (Sponsored Programs, Michigan Tech) which improved the quality of its presentation. In addition, I should acknowledge many colleagues whose names are listed below, who reviewed the book and provided me with invaluable comments and feedback.

Paul Crilly—*University of Tennessee*
Timothy Peck—*University of Illinois*
George Shoane—*Rutgers University*
Ziqian Liu—*SUNY Maritime College*
Ralph Tanner—*Western Michigan University*
Douglas P. Looze—*University of Massachusetts, Amherst*
Jaime Ramos-Salas—*University of Texas, Pan American*
Dale Dolan—*California Polytechnic State University, San Luis Obispo*
Munther Hassouneh—*University of Maryland*
Jacob Klapper—*New Jersey Institute of Technology*
Thomas M. Sullivan—*Carnegie Mellon University*
Vijayakumar Bhagavatula—*Carnegie Mellon University*
S. Hossein Mousavinezhad—*Idaho State University*
Alan J. Michaels—*Harris Corporation*
Sandra Soto-Caban—*Muskingum University*
Wei Pan—*Idaho State University*

Finally, I should acknowledge the support of late Professor Derek Lile, the former department head of Electrical and Computer Engineering of Colorado State University, while I was creating the ideas of this project while I was a Ph.D. candidate at Colorado State University, Ft. Collins, CO.

S. A. Reza Zekavat
Michigan Technological University

Electrical Engineering

Concepts and Applications

Why Electrical Engineering?

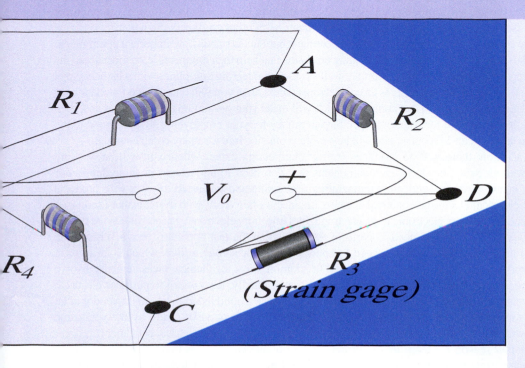

1.1 INTRODUCTION

If you are reading these words, then you are probably an engineering student who is about to take a course in electrical engineering (EE), or possibly an engineer who wants to learn about EE. In either case, it is safe to say that you are *not* majoring in EE nor are you already an electrical engineer. So, why are you doing this to yourself?

As an engineering student, there are two likely possible reasons: (1) You are being forced to because it is required for your major, and/or (2) you believe that it will help you pass the Fundamentals of Engineering (FE) examination that you will take before you graduate or shortly thereafter. If you are already an engineer, then you are likely reading this book because you need to learn EE for the FE exam that you put off until after graduation, or you need to learn EE to perform your job better. Studying EE because it is required for you to graduate or because you want all the help you can get to pass the FE exam are both laudable reasons. But, the second possible reason mentioned earlier for our hypothetical practicing engineer needs to be considered further. Can learning EE help you in your engineering career? The short answer is, yes. The long answer follows.

1.2 ELECTRICAL ENGINEERING AND A SUCCESSFUL CAREER

As a practicing engineer, you will work on projects that require a wide range of different engineers and engineering disciplines. Communication among those engineers will be vital to the successful completion of the project. You will be in a better position to communicate with the engineers working on electrical systems of all sorts if you have a basic background in EE. Certainly— through this course alone—you will not be able to design complicated electrical systems, but you will be able to get a feel of how the system works and be better able to discuss the implications of areas where the non-EE system you are designing and the electrical system overlap. For example, mechanical engineers often design packages for electronic systems where heat dissipation due to electronic components can be a major problem. In this instance, the non-EE engineer should be able to help the EE with component placement for optimum heat dissipation. In short, no engineer works in isolation and the more you can communicate with other engineers the better.

The company that hires you out of engineering school understands how important communication is. Thus, they will most likely have training programs that help their engineers learn more about the specific engineering that they will perform as well as other engineering disciplines with which they will be associated. If you have taken EE as an engineering student, then you will have a good foundation for learning EE topics specific to that company, which will make your on-the-job training easier and, consequently, less expensive for your employer. Saving money for your employer is always a good thing. So, by having taken an EE course, you will be a more promising hire for many companies.

In addition, there will be instances in your engineering career where you will be working directly with electrical or electronic components that you need to understand in some depth. For example, many engineers work in manufacturing processing and will need to work with products that have electrical/electronic content. Likewise, engineers often work with systems that used to be mechanical, but are now electronic (e.g., electronic fuel injection, electronic gas pedals). In the course of your work, you may also need to perform tests in which the test apparatus uses a Wheatstone bridge. If that is the case, then you need to know how a Wheatstone bridge, which is an electric circuit, works to use the equipment adequately. In addition, most mechanical measurements involve converting the mechanical quantity to an electrical signal. Finally, if you need to purchase electrical components and equipment you will need a fundamental background in EE to talk to the vendor in an intelligent manner and get the type of equipment that your company needs. Thus, by knowing some EE, you will be better able to obtain and use electrical components and electrical equipment.

Another reason for learning the principles and practices of EE is that you may be able to make connections between your engineering discipline and EE that lead to creative problem solutions or even inventions. For instance, maybe your job will require you to monitor a system on a regular basis that requires you to perform a significant number of tedious by-hand techniques. Your familiarity with the monitoring process, combined with your background in EE, might allow you to teach yourself enough in-depth EE to design and build a prototype monitoring system that is faster and less hands-on. This type of invention could lead to a patent or could lead to a significant savings in monitoring costs for your company. In this scenario, you would have been able to do the work yourself and would thus gain ownership of your work and ideas, that is, the design and fabrication of a prototype monitoring system. Learning EE (as well as other engineering fundamentals outside your specific discipline) may allow you to make connections that could lead to creative solutions to certain types of engineering problems.

In conclusion, studying EE will not only help you pass the FE exam, but it will make you more marketable, give you capabilities that will enhance your engineering career, and increase your self-confidence, all of which may allow you to solve problems in ways you cannot now imagine.

1.3 WHAT DO YOU NEED TO KNOW ABOUT EE?

Electric circuits are an integral part of nearly every product on the market. In any engineering career, you will need a working knowledge of circuits and the various elements that make up a circuit, including resistors, capacitors, transistors, power supplies, switches, and others. You need to know circuit analysis techniques and by learning these gain an understanding of how voltage, current, and power interact. You will likely be required to purchase equipment during your career, so you will need to learn how to determine technical specifications for that equipment. Working with electrical equipment exposes you to certain hazards with which you must be aware. Thus, you will need to learn to respect electrical systems and work with them safely.

Although engineers in all disciplines are expected to understand and use electrical systems, power sources, and circuits in many job assignments, expert knowledge is not required. For example, many non-EEs are plant managers and are called upon to manage heating and air conditioning (HVAC) systems. While the engineer will not be asked to design the system or its components, a basic EE knowledge is useful for day-to-day management. The practicing engineer must be able to design and analyze simple circuits and be able to convey technical requirements to vendors, electricians, and electrical and computer engineers.

A typical on-the-job application is data acquisition from temperature, pressure, and flow sensors that monitor process or experimental equipment. Process control and monitoring situations in plants and refineries require working knowledge of data acquisition and logging, signal processing, analog-to-digital (A/D) conversion, and interfacing valves and other control devices with controllers. In-line, real-time chemical analysis is sometimes necessary, as well as monitoring process temperatures, pressures, and flow rates with in-line sensors.

Familiarity with power generation and general knowledge of generators, electric motors, and power grids is also beneficial to engineers. A major job objective is frequently to reduce utility expenses, primarily electricity costs. For example, EE knowledge is necessary to design and operate cogeneration systems for simultaneous production of heat and electricity. In such situations, high-pressure steam can be throttled through a turbine-generator system for power production, and the lower-pressure exhaust steam is available for plant use. Alternatively, natural gas can be combusted in a gas turbine to generate electricity, and the hot gas exhaust can make steam in a boiler. One issue is how to operate the system to match electricity use patterns in the plant.

Process engineers also need to recapture energy (as electricity) from process streams possessing high thermodynamic availability, that is, streams at high pressure and/or temperature. Often, this can be done by putting the process stream through an isentropic expander (turbine) and using the resulting shaft work to operate an electric generator.

Electrochemistry involves knowledge and use of potentiostats, battery testing equipment, cyclic voltammetry measurements, electrode selection, and electrochemical cells. Electroplating operations are also of interest to chemical engineers. A background in EE will help you understand these and other related processes.

1.4 REAL CAREER SUCCESS STORIES

The bottom line for engineers is frequently the economic consequence of operating a process. The profit motive is paramount, with safety and environmental considerations providing constraints in operation. Electrical engineering principles often directly affect a process's profitability and operability. Learning and applying the concepts in this textbook may help you get noticed (favorably) in a future job by saving money for your company.

Consider the case of a chemical engineering graduate who began work a few years ago in a major refinery that had recently implemented a cogeneration system that produced steam and electric power simultaneously. Generation of high-pressure steam in a natural gas fired boiler, followed by expansion of the steam through a turbine, produced shaft work that was used to operate a generator. For internal plant use, this generated electricity was valued at the retail electricity price. (External sale of excess electricity is regulated by the Public Utility Resource Power Act (PUPA), and the price is the cost the utility company incurs to make incremental electricity, i.e., the utility company's "avoided cost.") The new employee did an economic analysis, looking at the trade-off between the equipment capital investments and operating costs versus the anticipated electricity savings. The bottom line is that the employee's recommendation was to "Turn it off!" since the cogeneration scheme was losing money. This result was not popular with the refinery's management (at first), but the employee was right on target with the analysis and recommendation. Saving money for the refinery provided a jump-start to a very successful career. Working knowledge of power generation cycles and equipment was necessary to do this critical analysis.

1.5 TYPICAL SITUATIONS ENCOUNTERED ON THE JOB

The case studies in this section are intended to illustrate, with discipline-relevant projects, how the general principles presented in this textbook may be applied in a real-life job or research setting. After reading the descriptions, you should better understand where, how, and why electrical engineering fits into an engineer's job. Until you have completed this course, some of the terms in these case studies may be unfamiliar. This illustrates the importance of completing this course prior to encountering these situations in real, on-the-job situations. For more detailed information on the equipment and processes described in the case studies, the interested reader can view the PowerPoint® presentations that accompany this textbook.

1.5.1 On-the-Job Situation 1: Active Structural Control

Imagine that you are a young civil engineer in Charleston, SC, who is working on the design and construction of a 15-story building that will have motion-sensitive equipment in it and must be able to withstand both hurricane winds and earthquakes. (Charleston has a history of strong earthquake events.) Since the building motion must be controlled during both moderate wind and earthquake events as well as during hurricanes and a major earthquake, active structural control is required. Passive structural control systems may be used in conjunction with active control systems, but the scenario described here pertains to an active control system.

You have been chosen by your boss to be the liaison between the company that will design and install the active control system and your company. You will need to give them information about the building that will allow them to design the control system, and you will need to feed information from them back to engineers in your company who are designing the building. If the building is being designed in the manner described, then the design process is a simple feedback loop: the initial building design affects the initial design of the control system, which in turn affects the building design, which in turn affects the control system design, and so on until both designs are compatible. So it is clear that you must have a reasonable understanding of the building design (your field) and a reasonable understanding of the active control system (not your field—the EE concepts are found here).

In general, structural control is the control of dynamic behavior of structures such as building and bridges. This type of control becomes important when the structure is relatively flexible, such as tall buildings and long-span bridges, or is sensitive to damage, such as historic buildings in earthquake regions. Engineers want to control structures: (1) to prevent damage and/or occupant discomfort during typical events, relatively high wind, and small earthquakes, and (2) to prevent collapse of the structure during large events, for example, major earthquakes. In your case, the building contents are vibration sensitive and the building is susceptible to damage from hurricanes and earthquakes.

Structural control is performed in one of three ways: passive control, active control, or hybrid control. *Passive control* is accomplished using the mass, stiffness, and damping built into the system. A passive control system cannot be readily altered. *Active control*, which is the method we will examine in this case study, is accomplished using control actuators, which require external energy, to modify the dynamic behavior of the system. An active control system can be adaptive. *Hybrid control* is simply a combination of passive and active control.

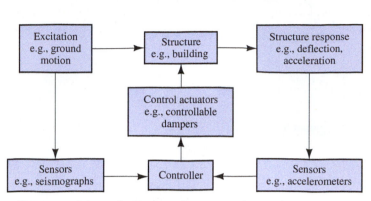

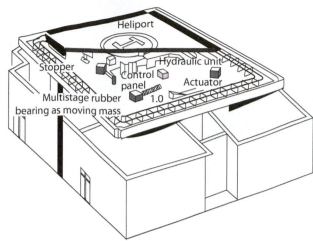

FIGURE 1.1 Schematic diagram of a structural control system. (Adapted from Spencer and Sain 1997.)

FIGURE 1.2 An application of an active mass damper (AMD) control system. AMD using a rooftop heliport.

The first time you arrive at the company who will design the control system, the engineer there shows you a diagram of an active control system. That diagram is shown in Figure 1.1. The first thing you notice is that the total system exhibits feedback, a concept discussed in EE and other areas. For the design of your building, the excitation will be wind force time histories and earthquake ground acceleration records. For the actual building, the excitation will be the real wind and earthquakes.

Your job will be to work in conjunction with the control system designer to develop a mathematical model of the building with the control system so that the building with the control system can be analyzed. From this design, the building and the control system will be built.

The control system could consist of controllable dampers, a mass damper, some other type of system, or a combination of more than one. A damper is a mechanical device that absorbs shocks or vibrations and prevents structural damage. The control system consultant is recommending an active mass damper (AMD) system since such systems have been used in Japan. He or she shows you the example given in Figure 1.2. The objective of the AMD system in Applause Tower was the suppression of building vibrations in strong winds and small-to-medium earthquakes. Your building will need to meet that objective as well as have a system that will minimize damage during a hurricane or a major earthquake. The Japanese building used the heliport on top of the building as the mass for the AMD, which is great because your building will have a heliport on top as well.

With this background on the proposed building, you need to get a better feel of how a building would be controlled using an AMD. The control system engineer shows you a simple example of a two-story building with an AMD controller with the components included (Figure 1.3).

The structure in Figure 1.3 consists of two rigid masses (the floors): m_1 and m_2, connected to the building's columns. The type of control actuator used is an AMD in which the mass used for control, m_a, is moved back and forth by signals from the computer controller. Refer to Figure 1.1.

The computer controller receives information on the structural response, accelerations a_1 and a_2 at each floor level, and masses m_1 and m_2, respectively. These accelerations are measured with accelerometers placed at each floor. The relative position of the mass in the mass damper, m_a, with respect to the structure is measured with a potentiometer. The structure shown is excited with a base acceleration, a_g, that is representative of the ground acceleration in an earthquake. The structure could also be excited by wind forces. Note that in Figure 1.1, the excitation is also monitored with sensors and fed to the controller. In the simple building example, the excitation is not monitored, but monitoring both the structural response and the excitation input is possible, particularly for earthquakes. In this simple building, one or more accelerometers could have been placed below the foundation to measure the base acceleration.

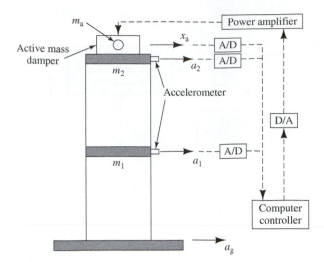

FIGURE 1.3 Simple building structure and AMD control system.

As you will note from Figure 1.3 and its description, there are a number of components that you will not understand without a basic EE knowledge. For example: What do the terms D/A and A/D mean? How does an accelerometer work? Why do you need an amplifier? How does the computer controller do its job? What is a potentiometer? Without an introduction to EE—like the one provided in this text—you will have difficulty in a job setting like this, and your learning curve will be much steeper, making your life more difficult. Note: If you would like to know the explanation now to the topics introduced in this example, see Chapter 11 of this book.

1.5.2 On-the-Job Situation 2: Chemical Process Control

In this scenario, imagine that you are a chemical engineering graduate who has just taken a job with a petroleum company in Baton Rouge, LA. Dwindling reserves of easy-to-access petroleum, as well as government subsidies for alternative fuel ventures, has led your company to diversify into fuel-grade ethanol production.

The process of interest at your plant (see Figure 1.4) is the application of corn (grain) fermentation to produce a dilute aqueous solution of ethyl alcohol that is further purified by distillation. Corn, sugar, yeast, water, and nutrients are continuously fed into a fermenter, which produces CO_2, spent yeast and grain, and a dilute (~8%) ethanol–water solution. This slurry is filtered and clarified to remove the yeast and grain before being sent to a preheater (see Figure 1.5). The heated ethanol–water solution is then fed to a distillation column that produces a 95% to 96% pure ethanol product as its overheads. Due to the large amount of energy required to separate the ethanol and water, it is desired to keep the temperature of the dilute feed elevated. Plant data, however, reveal that the feed is currently well below its boiling temperature.

Your plant manager intends to make the process profitable, so your first major assignment is a blanket imperative to "Reduce costs!" wherever possible. Since this is your first week on the job, you want to do a superlative job without panicking (or damaging any equipment). Thinking about the problem, you remember from school that it is more economical to preheat the distillation column feed than to introduce it into the column cold. You therefore decide to further preheat the temperature of the feed stream to its boiling point.

Preheating of the dilute ethanol–water feed in a heat exchanger (using Dowtherm®, an industrial heat transfer fluid produced by Dow Chemical) would significantly reduce the amount of energy needed for the separation, reducing the cost of operation (i.e., reducing the reboiler steam heating duty). The easiest approach would be to heat the feed at a constant rate. However, several

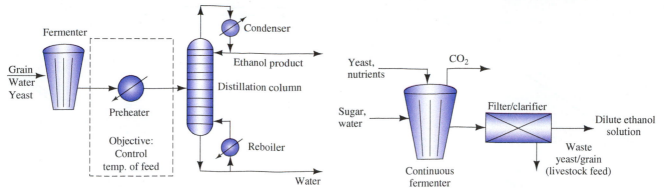

FIGURE 1.4   Ethanol distillation flow diagram.

FIGURE 1.5   Filtering of fermenter products.

key variables of the feed stream, including temperature, flow rate, and ethanol concentration are subject to change, requiring a variable preheating rate. The rate at which the feed is preheated is also affected by the temperature of the Dowtherm®, which may also vary with time. All of these factors make the heat load on the preheater vary with time.

Knowing these considerations, you decide to carefully control the rate at which the feed is preheated to lower operating costs (i.e., lower the reboiler duty and Dowtherm® consumption). Applying a feedback control strategy is your choice of an efficient way to regulate the column's feed temperature.

A basic feedback system is seen in Figure 1.6. A variable of a stream exiting a process is measured and sent to a controller. The controller compares the signal with a predetermined set point and determines the appropriate action to take. A signal is then sent from the controller to a piece of equipment that changes a variable affecting the process, resulting in a measured variable that is closer to the set point. For our case study, the temperature exiting the preheater is monitored and compared with a predetermined set point. A computer then determines what action must be taken, and adjusts the flow of Dowtherm® to the preheater, raising the temperature of the column feed as needed. Figure 1.7 shows the basic setup for the preheater.

The control scheme you wind up choosing for the process consists of five major components:

1. A thermocouple that measures the temperature of the feed exiting the preheater and produces an analog signal (TC) (Figure 1.8)
2. An analog-to-digital signal converter (A/D)
3. A proportional controller (CPU) that compares the feed temperature with a predetermined set point

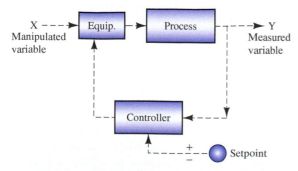

FIGURE 1.6   Basic feedback control strategy.

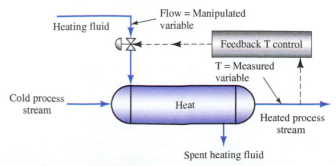

FIGURE 1.7   Schematic diagram of a generalized preheater.

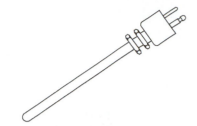

FIGURE 1.8   A thermocouple.

FIGURE 1.9   Pneumatic control valve (© Alexander Malyshev/Alamy).

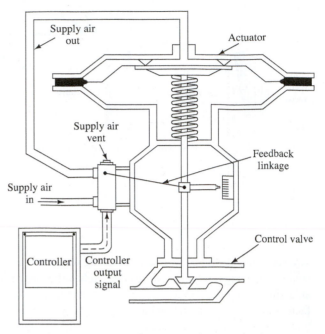

FIGURE 1.10   Diagram of a pneumatic control valve.

4. A digital-to-analog signal converter (D/A)

5. A pneumatic diaphragm control valve (Figures 1.9 and 1.10)

The dilute column feed to the ethanol separator is fed to the preheater where its temperature is increased by thermal contact with the Dowtherm®. As the feed exits the preheater, the thermocouple (TC) measures its temperature and emits an analog voltage signal (see Figure 1.11). The A/D converts the analog signal to a digital signal, and sends it to the CPU where the temperature is compared to a predetermined set-point temperature.

The CPU produces a digital signal proportional to the error (difference in temperatures), which is converted back to an analog signal by the D/A. This analog signal is interpreted as an analog air signal, typically scaled to 3 to 15 psig. The air pressure signal adjusts the valve stem position on the pneumatic diaphragm control valve, manipulating the amount of Dowtherm® flowing in the preheater. By keeping the measured temperature error below a specified tolerance, the temperature of the feed to the column is kept at its optimum and costs are kept down.

Net profit goes up as a result, and you get a big pat on the back—and maybe a big raise!

Clearly, to do this on-the-job assignment, you must be able to apply EE skills and knowledge in a plant environment. As this scenario illustrates, you must often define the problem, select a solution strategy, and pick equipment to implement it. The interested reader is encouraged to examine this textbook's related topics (see Table 1.1). Chapter 11 explains many fundamentals required for understanding the basics of a PC-controlled system.

1.5.3 On-the-Job Situation 3: Performance of an Off-Road vehicle prototype

A mechanical engineer working on the design and development of an off-road vehicle, SUV, or snowmobile is likely to be assigned the task of monitoring the field performance of a prototype during prescribed maneuvers, either to confirm that it is operating within design specifications, or to identify and troubleshoot malfunctions. A sample of vehicle performance features is given below:

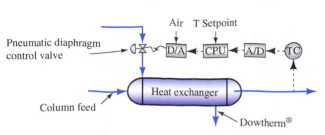

FIGURE 1.11   Detailed control scheme for situation 2.

TABLE 1.1	EE Connections to Process Control	
Case Study	**Textbook Topic**
Pneumatic Valve[a]	Analog Actuator
Thermocouple[a]	Sensors Data Acquisition
CPU	Microprocessor Control Feedback Control
Digital and Analog Signal	A/D and D/A Conversion

[a] PowerPoint presentations for these topics accompany this textbook.

Engine

- Dynamic temperature in piston, connecting rod, or cylinder wall
- Dynamic pressure in combustion chamber
- Dynamic stress in piston, connecting rod, or cylinder wall
- Temperature fluctuations in water or oil
- Output torque fluctuations
- Bearing clearance (related to lateral load)

Suspension and Drive Train

- Stress in suspension elements
- Deflections and clearances in suspension elements
- Windup and backlash in gear train
- Temperature during braking

Vehicle Interior

- Vibration of steering wheel, dash, mirrors, and floor
- Sound pressure level at driver's ear or passenger's ear
- Vibration at passenger's seat
- Driver's head–neck vibration

Even though your contribution to the design itself may be "purely" mechanical (a rarity!), as a mechanical engineer you are expected to interface strongly with other disciplines to *evaluate the actual performance of that design*. The mechanical engineer would be expected to specify the appropriate measurements, and even to select the sensors that will perform reliably in this demanding environment. Because each sensor has its own particular electrical requirements, you must work closely with a technical staff of instrumentation, computer, and electronic professionals capable of advising you, describing your options, and implementing your decisions on these options. This will require that you have a reasonable understanding not only of the vehicle's design but also the terminology and electrical features of the instrumentation used by your staff.

In this situation, you are part of the engineering team engaged in testing an SUV prototype during field tests, to ensure that the interactions between the shock tower and the shock absorber are within specifications.

The shock tower provides the attachment between the shock absorber (as seen at its threaded end in Figure 1.12) and the frame supporting the shock tower. In your particular assignment, it might be necessary to measure: (a) the stress history in the shock tower (fatigue life); (b) the history of the load transmitted through the shock absorber (attachment life); or (c) the history of the relative displacement between the shock absorber shaft and the shock tower (blowout of the rubber washer).

In each case, you are trying to create a voltage analog—a voltage signal proportional to the measured phenomenon—of the stress/load/displacement history. The voltage form is usually most desirable because it can be easily sampled and stored in files for computer-aided analysis.

FIGURE 1.12   Front shock tower. (Photo courtesy of Rob Robinette.)

STRESS IN THE SHOCK TOWER

The force exerted on the shock tower by the shock absorber causes the shock tower to deform slightly, in proportion to that load. If, at a key location on the tower, the strain associated with that deformation can be sensed, it could be used to produce a voltage proportional to that strain, or to the stress at that location. This is accomplished using a device called a strain gage, a fine metal wire, or foil, which is glued to the chosen point on the surface of the shock tower (see Figure 1.13).

As the surface beneath the strain gage stretches or contracts, the resistance of the wire increases or decreases in proportion to that strain. This very small change of resistance is measured using a Wheatstone bridge that is especially adapted to measure extremely small *changes* in resistance while ignoring the relatively large basic resistance of the gage. Figure 1.14 represents the Wheatstone bridge. A Wheatstone bridge (discussed in Chapter 11) works like a balance scale in that an unknown quantity is measured by comparing it with a known quantity.

The output of the Wheatstone bridge is a voltage proportional to the change of resistance of the deforming strain gage. This strain is, in turn, proportional to the stress at the attachment point of the gage.

LOAD ON THE SHOCK TOWER

This can be measured through a specially designed load cell attached to the shock tower, with one side fastened to the shock absorber shaft and the other fastened to the tower structure. Any load exerted by the shock absorber on the shock tower must pass through the load cell, which produces a voltage proportional to that load. This can be accomplished using a transducer containing a piezoelectric crystal, a crystal that develops a charge proportional to its deformation under load (see Figure 1.15). Chapter 11 discusses sensors that use different properties including piezoelectric.

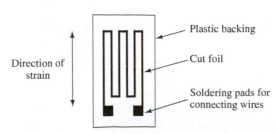

FIGURE 1.13   Electrical resistance strain gage.

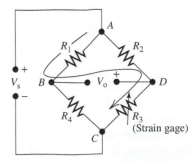

FIGURE 1.14   Wheatstone bridge circuit incorporating a single strain gage.

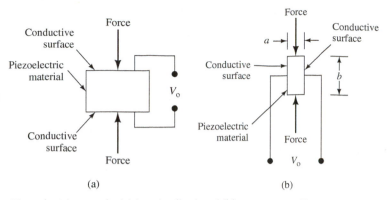

FIGURE 1.15 Piezoelectric crystals: (a) longitudinal and (b) transverse effect.

Because the charge developed during deformation is small, it is difficult to measure this charge without dissipating the accumulated electrons through the measurement device. A device with very high "input impedance" is necessary to accomplish this measurement without degrading the signal itself. This small charge is measured with the aid of an operational amplifier—configured with extremely high input impedance to prevent drainage of the signal, and extremely high output gain to produce a strong output voltage proportional to that charge.

The operational amplifier (Figure 1.16) is configured in a so-called charge amplifier circuit to produce a voltage proportional to the small charge signal, which is, in turn, proportional to the deformation of the piezoelectric crystal under load. The principles of the operational amplifier and the charge amplifier will be explained in Chapter 8.

RELATIVE DISPLACEMENT AT AN ATTACHMENT POINT

This can be measured through a capacitive load cell, one plate attached to the end of the shock absorber, the other plate attached to the shock tower. As the shock tower and the shock absorber move relative to each other, the gap between the capacitors changes and the net capacitance of the capacitor changes in inverse proportion to the gap distance (see Figure 1.17).

This capacitor is incorporated into a capacitive load cell, consisting of the capacitor and a stiff nonconducting elastic medium between the plates. The load applied by the shock absorber is passed through the elastic medium into the shock tower.

The capacitive load cell is connected into the feedback arm of a simple operational amplifier circuit, which produces a voltage proportional to the dynamic component of the displacement signal, while tending to ignore the static component (see Figure 1.18). Chapter 8 explains the fundamentals of operational amplifiers. This separation can be enhanced by low-pass filtering. Chapter 7 explains the principles of low-pass filters.

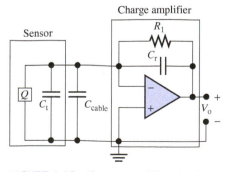

FIGURE 1.16 Charge amplifier circuit, incorporating a piezoelectric sensor and operational amplifier.

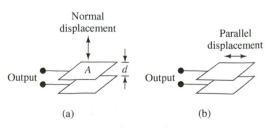

FIGURE 1.17 Capacitive transducers, sensitive to (a) normal and (b) parallel displacements.

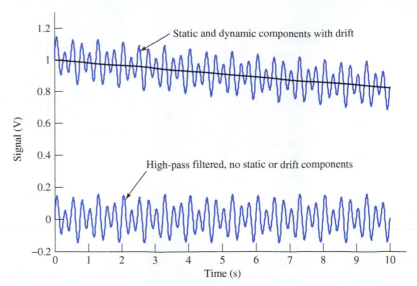

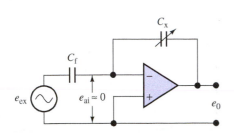

FIGURE 1.18 Operational amplifier with capacitive transducer in feedback arm.

FIGURE 1.19   Separation of dynamic and static components.

The separation of the static and dynamic components allows the engineer to observe the actual dynamic component, which is usually superimposed on a fairly large, but already known, static component (see Figure 1.19).

It should be clear that a mechanical engineer with no understanding of electrical transducers will have to place his or her success in the hands of the staff, who will be obliged to select the appropriate transducers, assure that they respond to only the desired phenomenon, produce signals in the desired format, and are correctly interpreted. Nevertheless, it is the engineer, not the staff, who will be held responsible for the acquisition and interpretation of that data.

Further Reading

Battaini, M., Yang, G., and Spencer, B. F., Jr. (2000) "Bench-Scale Experiment for Structural Control," *Journal of Engineering Mechanics,* ASCE, 126(2), pp. 140–148. (Available at www. uiuc.edu/sstl/default.html)

Spencer, B. F., Jr. and Sain, M. K. (1997) "Controlling Buildings: A New Frontier in Feedback," *IEEE Control Systems*, vol. 17, no. 6, pp. 19–35.

CHAPTER 2

Fundamentals of Electric Circuits

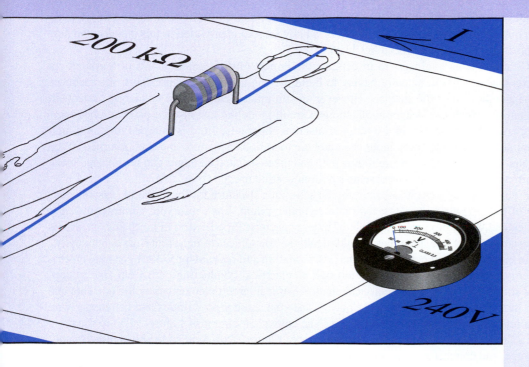

2.1 INTRODUCTION

This chapter introduces the main variables and tools needed to analyze an electric circuit. You may wonder: How can an understanding of circuit theory help me better understand and solve real-world problems in my field of engineering?

As you learned through the case studies in Chapter 1, there is a clear link between many engineering disciplines and electrical engineering. In Chapter 2, we will examine more examples that will clarify the applications of electric circuits in your engineering field. After completing this chapter, you should better understand the link between engineering fields and circuit theory. In addition, through many application-based examples, the chapter will clarify how circuit theory can be applied to solve problems in your field of interest. Please note that the application-based examples you will encounter in this chapter are not the only application-based examples. Many more application-based examples and problems are provided in other chapters.

Circuits consist of individual elements that together form a model structure that can be used to simplify the process of analyzing and interpreting the behavior of complex engineering structures. The two main variables needed to analyze electric circuits are (1) charge flow or current, and (2) voltage. Current is measured by placing an Ammeter (ampere meter) in series, and voltage is measured by placing a voltmeter across elements in the circuit. The three main tools needed to analyze circuits are (1) Kirchhoff's current law (KCL), (2) Kirchhoff's voltage law (KVL), and (3) Ohm's law. Analysis of an electric circuit includes computation of the voltage across an element or the current going through that element. The following sections discuss the aforementioned laws, and detail how each is used to compute current and voltage.

Circuit models enable engineers to analyze the impact of different individual elements of interest. For instance, this chapter discusses problem solving and the generation of waveforms using the advanced computer software package PSpice. You will learn about and run PSpice applications to study the effect of each variable within a circuit. Please note that we give the PSpice tutorial in a distributive format. Thus, a basic tutorial is provided in this chapter, and new concepts in PSpice are introduced in Chapters 3–13.

Circuits enable engineers to investigate the impact of different elements of a system via software, that is, without actually having to build the engineering structure. This often reduces the time and the cost of the analysis. However, circuit models can be used to analyze systems in more than just electrical system applications. Circuit modeling can also be used in other engineering areas. For example, in a hydraulic machine, hydraulic fluid is pumped to gain a high pressure and transmitted throughout the machine to various actuators and then returned to the reservoirs. The hydraulic fluid circulation path and the associated elements can be modeled using a circuit model. Figure 2.1(a) represents a hydraulic structure.

Figure 2.1(b) represents a circuit model schematic structure for the hydraulic system, consisting of the following elements: hydraulic cylinder, pump, filter, reservoir, control valve, and retract/extend. Here, pipes play the role of interconnectors. Pipe friction in mechanics measures the resistance. Therefore, the pipes should ideally be frictionless to represent high conductivity.

In electrochemical cells (e.g., batteries), oxidation and reaction processes generate electrical energy. The energy is transferred through an electrical conducting path from zinc to copper. As shown in Figure 2.2(a), classical electrochemical circuit elements may include zinc and copper metals and a lamp to show the power. The zinc and copper sulfides and the copper wire are the interconnecting parts or conductors. The equivalent circuit of Figure 2.2(a) is shown in Figure 2.2(b). This circuit simplifies the structure of Figure 2.2(a) to include only a resistor, voltage source, and connectors (e.g., copper wires).

In the case of electric circuits, a circuit model includes various types of circuit elements connected in a path by conductors. Therefore, in an electric circuit, the interconnecting materials are conductors. Copper wires are usually used as conductors. These interconnectors may be

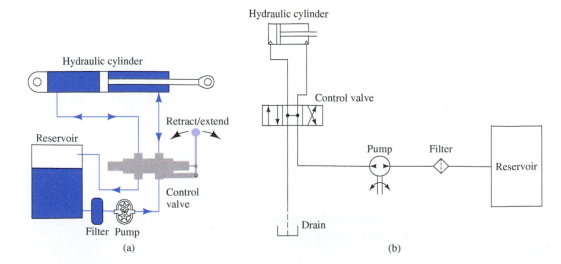

FIGURE 2.1

(a) Hydraulic circuit structure and (b) its equivalent schematic structure.

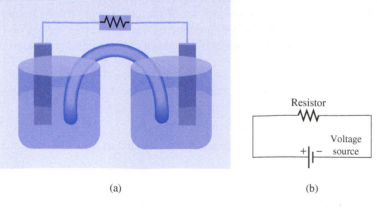

(a)   (b)

FIGURE 2.2
(a) Electrochemical
circuit and (b) its
equivalent schematic
circuit.

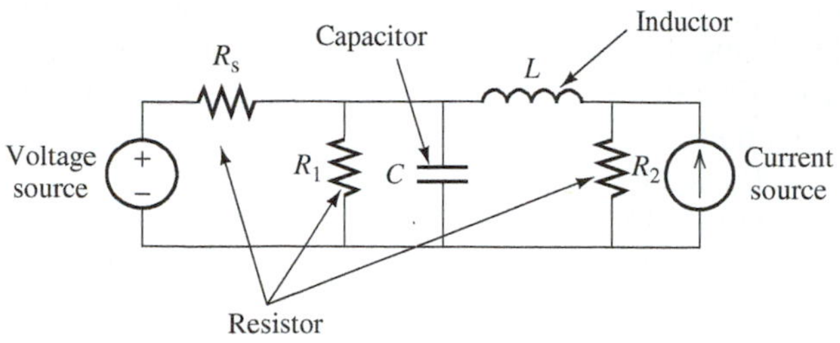

R_s Capacitor   Inductor

                                  L

Voltage   +   Current
source   − R_1 C R_2 source

Resistor

FIGURE 2.3 An electric
circuit.

as small as one micrometer (e.g., in electronic circuits) or as large as hundreds of kilometers (e.g., in power transmission lines that may connect generated hydroelectric power to cities; see Chapter 9). Voltage source, resistance, capacitance, and inductance are examples of circuit elements. An example of an electric circuit is shown in Figure 2.3.

2.2 CHARGE AND CURRENT

Fundamental particles are composed of positive and negative electric charges. Niels Bohr (1885–1962) introduced an atom model [Figure 2.4(a)] in which electrons move in orbits around a nucleus containing neutrons and protons. Protons have a positive charge, electrons have a negative charge, and neutrons are particles without any charge. Particles with the same charge repel each other, while opposite charges attract each other [Figure 2.4(b)]. In normal circumstances, the atom is neutral, that is, the number of protons and electrons are equivalent.

Charge is measured in *coulombs*. Electrons are the smallest particles and each possesses a negative charge equal to $e = -1.6 \times 10^{-19}$ C. Equivalently, the charge of each proton is $+1.6 \times 10^{-19}$ C.

Electric charge (q), measured by the discrete number of electrons, corresponds to:

$$q = ne \qquad\qquad (2.1)$$

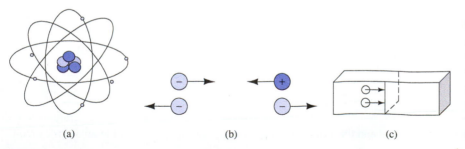

(a)   (b)   (c)

FIGURE 2.4 (a) Bohr
atom model,
(b) interaction between
the same and the
opposite charges and
(c) charges are passing
through a cross section
at a constant rate.

where n is a positive or negative integer. *Electric current* is the rate at which charges flow through a conductor; specifically, the amount of charges moving from a point per unit of time.

In Figure 2.4(c), charges are moving through a cross section of a conductor with a constant current, i. During time t, the total amount of charges passing through the cross section is:

$$q = i \times t \qquad (2.2a)$$

Then the constant current, i, can be expressed as:

$$i = \frac{q}{t} \qquad (2.2b)$$

When the current is time-varying, denoted as a function of time t, Equation (2.2a) can be expressed in integration form:

$$q(t) = \int_{t_0}^{t} i(\tau)d\tau \qquad (2.3a)$$

In Equation (2.3a), $q(t)$ is the total amount of charge flowing through the cross section within the time t_0 to t ($t-t_0$), t_0 is the initial time, and t is the time at which we intend to measure the charge. Equation (2.3a) states that the total charge flowing through a cross section at any given time period is the integration of the current passing through that cross section from the initial time t_0 up to the time t.

Equation (2.2b) shows that the current is the rate of variation of charge within a specific time period. Therefore, the relationship between the time-varying current, the charge, and the time corresponds to:

$$i(t) = \frac{dq(t)}{dt} \qquad (2.3b)$$

where $dq(t)$ represents the variation of charges and i shows the current measured in amperes. In fact, 1 *ampere* (A) = 1 *coulomb/second* (C/s). Microamperes ($\mu A = 10^{-6}$ A), milliamperes ($\mu A = 10^{-3}$ A), and kiloamperes ($kA = 10^{+3}$ A) are also widely used for measuring current. For instance, kA is the unit commonly used in power distribution systems, while μA is used regularly in digital microelectronic systems.

Current has both value and direction. In general, the direction of the movement of *positive* charges is defined as the conventional direction of the current (Figure 2.5). However, in an electric circuit with various elements, for the purpose of analysis of a circuit, we may assign arbitrary directions to the current that flows through an element. These arbitrarily selected directions for current determine the direction of voltage (see details in Section 2.3). Figure 2.6 shows an example of how current direction can be assigned. Here, the current variables i_A, i_B, i_C, i_E are assigned to different circuit elements. Note that we do not need to know the real direction of current; rather, arbitrary directions can be assigned to each branch. Real directions are determined after performing the calculations: if the calculated voltage or current is positive, we can deduce that we have selected the actual direction; if the calculated voltage or current is negative, the actual direction would be the inverse of the selected one.

If the magnitude and the direction of the current are constant over time, it is referred to as *direct current* and it is usually denoted by I; if the amount of the current or its direction changes

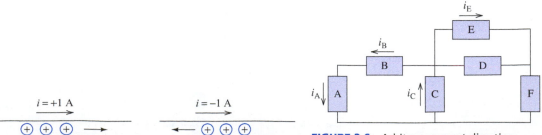

FIGURE 2.5 Current and the movement of positive charges.

FIGURE 2.6 Arbitrary current direction assignment in an electric circuit.

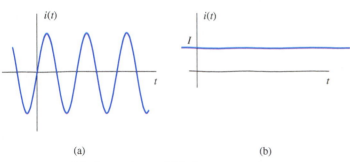

(a)   (b)

FIGURE 2.7   (a) Alternating and (b) direct currents.

FIGURE 2.8   Water gravitational potential energy at the top of the waterfall is converted to kinetic energy.

over time, it is called *time-varying current* and it is usually denoted by $i(t)$. Alternating currents are an example of time-varying currents. In alternating currents, both the current amplitude and current direction periodically change with time. Direct and alternating currents are abbreviated as DC and AC, respectively. Figure 2.7 represents examples of DC and AC.

2.3 VOLTAGE

To better explain the phenomenon of voltage, let us examine an analogy based on potential energy. Potential energy is the energy that is able to do work if it is converted to another type of energy. For example, the water at the top of a waterfall has a potential energy with respect to the bottom of the waterfall due to gravity forces. As water moves down, this potential energy is gradually converted to kinetic energy. As a result, at the bottom of the waterfall, the stored energy that we observed at the top of the waterfall shows itself as full kinetic energy (Figure 2.8). Therefore, the potential energy *defined between the two points* (the top and the bottom of the waterfall) is converted to kinetic energy (resulting in the current of the water flow). This represents a real-life example: dams, in fact, use this technique to create electricity (hydroelectric power). In hydroelectric dams, potential energy is converted to kinetic energy that is applied to turbines and electric generators to create electrical energy.

Similarly, the difference in voltage between two points in a circuit forces or *motivates electrons* (charges) to travel. This is why voltage is referred to as *electromotive force* (EMF). In an electric circuit, differences in potential energy at different locations (called voltage) force electrons to move. Therefore, electric current is the result of a difference in the amount of potential energy at two points in a circuit. Like the example of the dammed water outlined earlier, where potential energy was converted to the movement of water, here, the voltage makes the charges move. Voltage is measured in volts (V).

One joule (1 J) of energy is needed to move one coulomb (1 C) of charge through one volt (1 V) of potential difference. Therefore:

$$1 \text{ Volt (V)} = 1 \text{ Joule/Coulomb (J/C)}.$$

A DC voltage is a constant voltage, which is always either negative or positive. In contrast, an AC voltage alternates its size and sign with time. In Figure 2.9, v_A represents the voltage across element A; v_B represents the voltage across element B; and v_F represents the voltage across

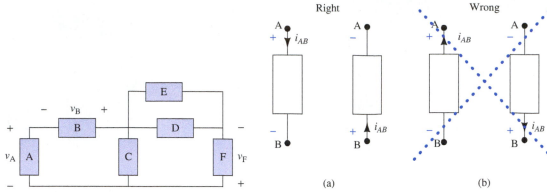

FIGURE 2.9 Arbitrary voltage polarity assignment in an electric circuit.

FIGURE 2.10 Current and voltage respective direction: (a) correct; (b) incorrect.

element F. The same indexing approach can be used to represent the voltage across any given element. In Section 2.2, we explained that the current flow direction within an element can be assigned arbitrarily. However, if we select the current direction, the voltage direction is determined based on the selected current direction. Similarly, the voltage polarity across an element can be determined arbitrarily as shown in Figure 2.9. However, if we select the voltage direction across an element, the corresponding current is determined relative to the voltage. This point is discussed further in the next section.

2.4 RESPECTIVE DIRECTION OF VOLTAGE AND CURRENT

In the prior two sections, we explained that the current flow direction within an element and the voltage polarity across an element might be determined arbitrarily. Here, we will clarify this concept: for an element that consumes power, the current flows into the side of the element that has the greatest amount of positive voltage and out of the side with the greatest amount of negative voltage. Once the direction of the current has been assigned, the assignment of voltage polarity must follow. The alternative case is also true: once the polarity of an element has been assigned, the direction of the current flow must follow from the known polarity.

Please note that the selected current flow and the voltage polarity may not be consistent with real (actual) current flow and voltage direction. If, upon calculation, the current flow is computed to be a positive number, then the directions assigned were correct. On the other hand, if the current flow is calculated to be a negative number, then the directions assigned were the reverse of the actual current flow and voltage direction. Figure 2.10 summarizes the rule for the selection of voltage polarity as it relates to current direction in an element that consumes power. Figure 2.10(a) represents correct selection. Here, the current enters from the positive sign of the voltage across the element and leaves from the negative sign of the voltage across the element. Figure 2.10(b) shows an incorrect selection.

2.5 KIRCHHOFF'S CURRENT LAW

Circuit analysis refers to characterizing the current flowing through and voltage across every circuit element within a given circuit. Some general rules apply when analyzing any circuit with any number of elements. However, before discussing these rules, we need to define other terms that are commonly used in circuit analysis literature: a *node* and a *branch*.

A **node** is the connecting point of two (or more) elements of a circuit.

A **branch** represents a circuit element that is located between any two nodes in a circuit.

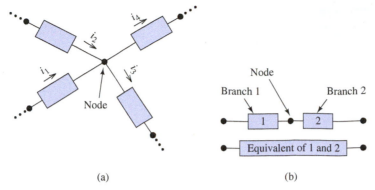

FIGURE 2.11   Definition of a node and branch in an electric circuit.

Therefore, nodes are the starting and ending points of branches. Several branches may meet at a single node. In Figure 2.11(a), four circuit elements in four branches are connected to each other at a common node. In some special cases, two elements might be connected back to back as shown in Figure 2.11(b). In that case, we may consider each one as a branch, and their connecting point as a node. We may also replace them with an equivalent element. In that case, we consider the element that is the equivalent of both as one branch.

Kirchhoff's current and voltage laws are the two fundamental laws in electric circuits. Russian scientist Gustav Robert Kirchhoff (1824–1887) introduced the two laws that now bear his name. These laws allow the calculation of currents and voltages in electric circuits with multiple loops using simple algebraic equations.

Kirchhoff's current law (KCL) states that the net current entering a node in a circuit is zero. Some currents enter into a node and some leave the node. Thus, based on this law, the sum of the currents entering a node is equal to the sum of the currents leaving that node. KCL results from the law of the conservation of charge:

(Used with permission from © INTERFOTO/ Alamy.)

> **Law of the conservation of charge:** Charges are always moving in a circuit and cannot be stored at a point; charge can neither be destroyed nor created.

EXAMPLE 2.1   Series Elements: Currents in the "Same Direction"

In Figure 2.12, Node A is between two elements 1 and 2, and the current entering Node A (i_1) equals the current leaving Node A (i_2). In the scenario shown in Figure 2.12, Node A connects only two branches. In this case, we say element 1 is in series with element 2. Therefore, currents passing through the two elements in series are always the same in size and in direction.

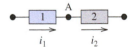

FIGURE 2.12
Kirchhoff's current law for two series elements with similar current direction assignment.

EXAMPLE 2.2   Series Elements: Currents in "Opposite Directions"

As discussed earlier, we can assign arbitrary directions for the branch currents. For example, in the simple circuit of Figure 2.13, we have assigned the current directions different from Figure 2.12. In this case, the algebraic sum of the currents entering the node equals $i_1 + i_2$. Based on KCL, if we set this sum to zero, then $i_2 = -i_1$. This confirms the same notion that the currents through two series elements are equal in size and direction.

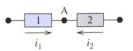

FIGURE 2.13
Kirchhoff's current law for two series elements with opposite current direction assignment.

EXAMPLE 2.3   **KCL**

KCL can also be applied to a node with more than two branches. Consider the scenario of applying KCL equations to the node represented in Figure 2.11. In this case:

$$i_1 + i_2 = i_3 + i_4 \qquad (2.4)$$

or

$$i_1 + i_2 - i_3 - i_4 = 0 \qquad (2.5)$$

Knowing this, we can resketch Figure 2.11 as the one shown in Figure 2.14. Applying Kirchhoff's law to the node shown in this figure leads to:

$$i_1 + i_2 + (-i_3) + (-i_4) = 0 \qquad (2.6)$$

This equation is equivalent to Equation (2.5).

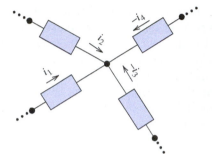

FIGURE 2.14 Kirchhoff's current law for arbitrary current assignment.

In general, KCL can be applied to every closed surface (e.g., closed sphere) within any circuit.

Based on the Kirchhoff's current law:

The net current entering any closed surface is zero.

A node is indeed a special case of a closed surface when the area of that closed surface tends to be zero. Using Figure 2.15, a generalized form of KCL for a node containing N branches can be shown as:

$$\sum_{k=1}^{n} i_k = \sum_{k=n+1}^{N} i_k \qquad (2.7)$$

In Equation (2.7), $i_k, k \in \{1, 2, ..., n\}$ are currents entering Node A, while $i_k, k \in \{n+1, ..., N\}$ are currents leaving Node A.

For instance, if we apply KCL to the closed surface shown by dotted line in Figure 2.16, the equation will be:

$$i_1 + i_2 + i_3 + i_4 + i_5 = 0 \qquad (2.8)$$

The closed surface can be considered as a black box. The black box covers a portion of circuit that includes multiple branches and circuit elements.

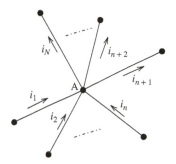

FIGURE 2.15 The generalized Kirchhoff's current law.

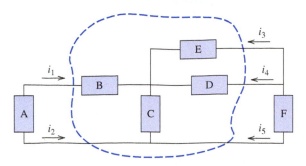

FIGURE 2.16 Generalized KCL for a closed surface.

In conclusion, by applying KCL, only four steps are needed to find an unknown current entering or leaving a node or a closed surface. The steps include the following:

1. Determine currents entering the node (or closed surface).
2. Determine currents leaving the node (or closed surface).
3. Apply KCL to the nodes.
4. Use the created set of equations to find the unknown current.

EXAMPLE 2.4   **KCL**

Find i_C in Figure 2.17.

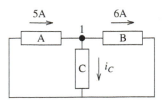

FIGURE 2.17 Circuit diagram for Example 2.4.

SOLUTION

• The current from only one branch enters Node 1—the one that passes through element A. Therefore, if we show the current entering Node 1 by i_A, we can express this as:

$$i_A = 5A$$

• The currents passing through elements B and C leave Node 1. Therefore, if we indicate the current leaving Node 1 by i_B, we see that:

$$i_A = 5 = i_B + i_C = 6A + i_C$$

Using this equation, we find that:

$$i_C = -1\,A$$

Therefore, the i_C current size is 1 A with a direction opposite to the one shown in Figure 2.17.

EXAMPLE 2.5   **KCL**

Find the currents i_B, i_C, and i_D in Figure 2.18.

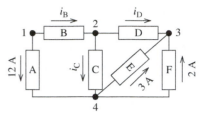

FIGURE 2.18   Circuit diagram for Example 2.5.

SOLUTION

To find the desired currents, apply KCL to Nodes 1, 2, and 3.

1. Currents entering Nodes 1, 2, and 3 are, respectively:

$$i_{i1} = 0\,\text{A}$$
$$i_{i2} = i_B$$
$$i_{i3} = i_D + 3\,\text{A} + 2\,\text{A} = i_D + 5\,\text{A}$$

2. Currents leaving Nodes 1, 2, and 3 are, respectively:

$$i_{o1} = i_B + 12\,\text{A}$$
$$i_{o2} = i_C + i_D$$
$$i_{o3} = 0\,\text{A}$$

3. Applying KCL, we show that $i_{i1} = i_{o1}$, $i_{i2} = i_{o2}$, and $i_{i3} = i_{o3}$, therefore,

$$i_B = -12\,\text{A}$$
$$i_D = -5\,\text{A}$$
$$i_C = -7\,\text{A}$$

The signs of the currents we calculated (negative) show that the direction of each is opposite to the initially assigned direction.

EXERCISE 2.1

Can all unknown currents in Figure 2.18 be found if we assume that element D is shorted? If yes, find all currents for this scenario. Note that in this case we say elements E and C are in parallel.

2.6 KIRCHHOFF'S VOLTAGE LAW

Before discussing the second fundamental law used to analyze electric circuits, we must first introduce another variable in circuits, called a **loop**. A loop is a closed path that starts at a node, proceeds through some circuit elements, and returns to the starting node. In other words, a loop is formed by tracing a closed path in a circuit without passing through any node more than once.

For instance, in Figure 2.19, Path 1 is a loop, because its starting and ending points are the same and it passes by each node in its path only once. However, Path 2 is not a loop. Although Path 2 starts and ends at the same node, it is not a loop because it passes by Node 1 twice.

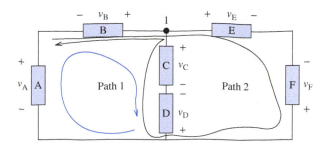

FIGURE 2.19 Loop definition—Path 1 is a loop but Path 2 is not.

Kirchhoff's voltage law states that:

The algebraic sum of the voltages equals zero for any loop in an electrical circuit.

Each defined loop in a circuit can have an arbitrary direction. Using the KVL equation, we can assign positive and negative signs, respectively, to the rising and dropping voltages. For simplicity, from now on, we will consider the polarity of voltages across each element to follow the direction of the currents through that element, as explained in detail in Figure 2.10. When voltages rise, the polarity changes from negative to positive; when voltages drop, the polarity changes from positive to negative. To apply KVL, the algebraic sum of the raised and dropped voltages in the direction of the loop is set equal to zero. Then, we can determine the unknown voltage in the loop.

Figure 2.20 outlines the procedure for writing KVL equations. When following the loop, we consider the change of the voltage polarity of an element and assign the relevant sign for the voltage in the KVL equation. For Figure 2.20, the equation is:

$$-v_A + v_B - v_C = 0 \qquad (2.9)$$

We can also move the negative voltages to the other side of the equation; therefore, Equation (2.9) can also be written as:

$$v_B = v_A + v_C$$

This is really another way of writing the KVL: it equates the sum of the absolute values of the dropping voltages to that of the rising voltages.

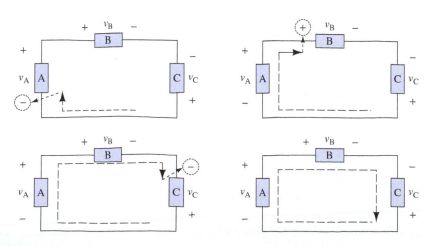

FIGURE 2.20 KVL procedure.

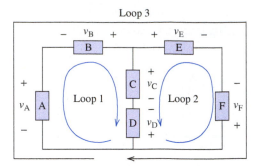

FIGURE 2.21 KVL example.

Figure 2.21 shows three loops; KVL can be applied to each. If we do this, the generated set of KVL equations applied to loops 1, 2, and 3, respectively, corresponds to:

$$\text{Loop 1:} \qquad -v_A - v_B + v_C - v_D = 0$$
$$\text{Loop 2:} \qquad v_F - v_E + v_C - v_D = 0$$
$$\text{Loop 3:} \qquad -v_A - v_B + v_E - v_F = 0$$

Assuming one of these voltages (e.g., v_D) is known, other unknown voltages can be easily extracted via this set of algebraic equations.

EXAMPLE 2.6 **KVL**

Find v_B in Figure 2.22.

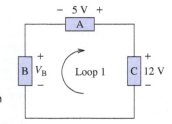

FIGURE 2.22 Circuit diagram for Example 2.6.

SOLUTION

For this circuit and the considered loop, the equation can be written:

$$-v_B - 5 \text{ V} + 12 \text{ V} = 0$$

Using this equation, we find:

$$v_B = 7 \text{ V}$$

EXAMPLE 2.7 **KVL**

Find the unknown voltages v_C and v_D in Figure 2.23.

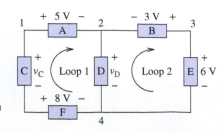

FIGURE 2.23 Circuit diagram for Example 2.7.

SOLUTION

In this circuit, there are two unknown voltages. Therefore, at least two KVL equations are needed to solve both unknowns. Two loops are shown in Figure 2.23. For Loop 1 the KVL equation can be written as:

$$-v_C + 5\text{ V} + v_D - 8\text{ V} = 0$$

For Loop 2, KVL leads to:

$$-v_D - 3\text{ V} + 6\text{ V} = 0$$

Solving these two equations, the following values are obtained for the unknown voltages:

$$v_D = 3\text{ V}$$
$$v_C = 0\text{ V}$$

APPLICATION EXAMPLE 2.8   Chemical Reaction

For this example, assume that the temperature of a chemical reaction needs to be monitored for the entire duration of an experiment. To monitor this temperature, an electric temperature sensor is connected to a computer that records the data (see Figure 2.24). The temperature sensor used is basically a resistor; its resistance changes with temperature (this type of sensor is known as a thermistor; see Chapter 11, Section 11.2 for a review on thermistors).

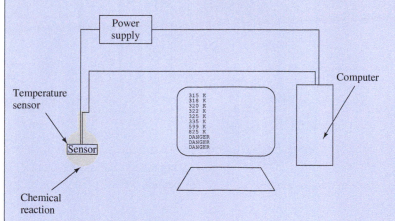

FIGURE 2.24 Chemical reaction system.

The computer can be modeled as an amplifier, which in turn can be modeled as a resistance. The entire system can be modeled as shown in Figure 2.25. The elements presented in this diagram are explained more fully later in this chapter. Find the voltage at the computer terminals.

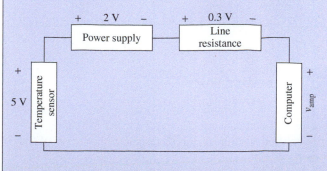

FIGURE 2.25 Equivalent circuit for Example 2.8.

(continued)

APPLICATION EXAMPLE 2.8   Continued

SOLUTION

Applying KVL, we see that:

$$0 = 2.0\,\text{V} + 0.3\,\text{V} + v_{\text{amp}} - 5\,\text{V}$$

Next, we can calculate the voltage by solving the equation, resulting in:

$$v_{\text{amp}} = 2.7\,\text{V}$$

APPLICATION EXAMPLE 2.9   Cooling Fan

An automobile cooling fan is connected to a 12 V car battery (Figure 2.26). The car battery supplies voltage to the car motor as well. The voltage drop across the fan motor consists of the voltage drop due to the internal resistance of the armature's windings. Back electromotive force (EMF) functions as a source. The equivalent circuit is depicted in Figure 2.27. Line resistance represents the resistance of the wire connecting the fan to the battery. What is the voltage drop over the line resistance, v_{R}?

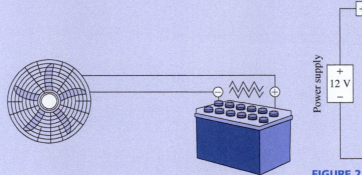

FIGURE 2.26   Cooling fan connected to car battery.

FIGURE 2.27   Equivalent circuit for Example 2.9.

SOLUTION

Applying KVL to the equivalent circuit shown in Figure 2.27, we see that:

$$-12\,\text{V} + v_{\text{R}} - 3\,\text{V} + 10\,\text{V} = 0$$

Solving for the voltage using this equation, we see that:

$$v_{\text{R}} = 5\,\text{V}$$

APPLICATION EXAMPLE 2.10   Loudspeaker

A basic loudspeaker is shown in Figure 2.28. The variable current flowing through the electromagnetic coil pushes and pulls on the magnet, causing the speaker cone to vibrate, producing sound. Figure 2.29 shows a simplified version of the electrical components of the speaker.

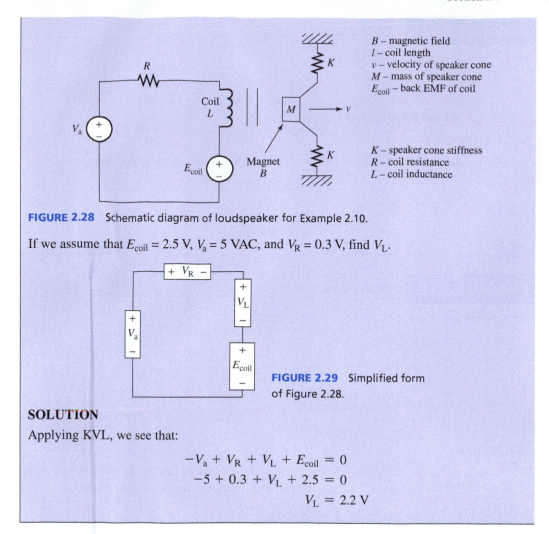

FIGURE 2.28 Schematic diagram of loudspeaker for Example 2.10.

If we assume that $E_{coil} = 2.5$ V, $V_a = 5$ VAC, and $V_R = 0.3$ V, find V_L.

FIGURE 2.29 Simplified form of Figure 2.28.

SOLUTION

Applying KVL, we see that:

$$-V_a + V_R + V_L + E_{coil} = 0$$
$$-5 + 0.3 + V_L + 2.5 = 0$$
$$V_L = 2.2 \text{ V}$$

2.7 OHM'S LAW AND RESISTORS

So far, we have introduced the circuit element variables, current, and voltage. The next important question to address is: what is the relationship between the current and the voltage of a passive element in a circuit?

The answer to this basic question depends on the properties of the element. The simplest passive element in an electric circuit is a resistor. In fact, every conducting material is a resistor. The voltage across the terminals of an ideal resistor is directly proportional to the current passing through that terminal. If we measure the ratio of the voltage and the current associated with an ideal resistor, we observe a constant value (at any given time). Accordingly, the relationship of the voltage and the current of a resistor can be stated as:

$$v = i \times R \qquad (2.10)$$

where the constant R is called the resistance and its unit is Ohm (Ω). Another important fact is that, 1 Ohm (Ω) = 1 volt/ampere (V/A). The Ohm is named after George Simon Ohm, who was probably the first physicist to measure the voltage and the current of a conductor and describe the constant voltage to current ratio, that is, the resistance.

Figure 2.30 shows the symbol commonly used to signify a resistor in an electric circuit. The voltage polarity and the direction of the current satisfy the sign conventions. The voltage–current or $v-i$ curve of an ideal (linear) resistor is a straight line as plotted in Figure 2.31. This curve is called the characteristic of a resistor.

FIGURE 2.30 Ideal resistor symbol.

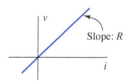

FIGURE 2.31 The linearity of an ideal resistor.

EXAMPLE 2.11   **Ohm's Law**

Find R in Figure 2.32.

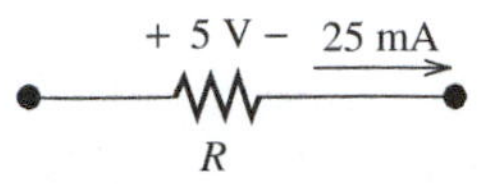

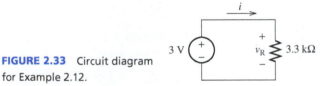

FIGURE 2.32   For Example 2.11.

SOLUTION

As shown in Figure 2.32, $v = 5$ V and $i = 0.025$ A. According to the Ohm's law:

$$5 \text{ V} = R \times 0.025 \text{ A}$$

Therefore:

$$R = 200 \text{ } \Omega$$

EXAMPLE 2.12   **Ohm's Law**

Find the current, i, in the circuit shown in Figure 2.33.

FIGURE 2.33   Circuit diagram
for Example 2.12.

SOLUTION

The voltage across the terminals of the resistor equals the voltage across the voltage source, in this case, 3 V. Therefore, $v_R = 3$ V. Now, using Ohm's law, we can develop the equation:

$$3 \text{ V} = 3300 \times i$$

Calculating i from this equation, we find:

$$i = 0.91 \text{ mA}$$

EXERCISE 2.2

Assume that another 3.3 kΩ resistor is added to the circuit in series in Example 2.12. Calculate the current based on this revised circuit.

APPLICATION EXAMPLE 2.13   **Electric Shock**

An unfortunate bald utility worker climbs an aluminum ladder with bare feet and comes into contact with an overhead power line with the maximum voltage of 30 kV (Figure 2.34). The worker's body resistance is 315 kΩ. In this situation, his body can be modeled as a resistor connected between the terminals of a power supply (the wire) (Figure 2.35). How much current flows through his body? If the minimum current needed for a shock to be fatal is 100 mA, does the worker survive?

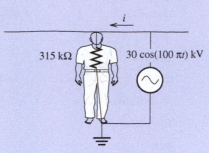

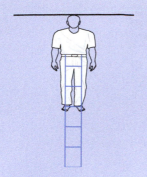

FIGURE 2.34 Application problem—electric shock.

FIGURE 2.35 The equivalent circuit for Example 2.13.

SOLUTION

The equivalent circuit has been shown in Figure 2.35. Using Ohm's law, the maximum current through the worker's body is:

$$i_{max} = \frac{30,000 \text{ V}}{315 \times 10^3 \text{ } \Omega} = 95.2 \text{ mA}$$

Because the current passing through the worker is less than 100 mA, the worker survives, but barely! Note: As discussed in Chapter 15, to reduce the amount of current flow through the worker's heart, one hand should have been placed in a pocket. This creates a shortcut around the heart and reduces the current that goes through the heart.

2.7.1 Resistivity of a Resistor

The value of an ideal resistor depends on its physical characteristics. These physical characteristics include the conducting performance of the material and its physical and geometrical shape. A sample conductor, in the shape of a cylinder, is shown in Figure 2.36. The relationship, which expresses the resistance, R, of a conductor in terms of its physical parameters, corresponds to:

$$R = \rho \frac{L}{A} \tag{2.11}$$

where ρ is the resistivity of the material, expressed in units called ohm meters (Ωm). L and A, respectively, are the length and the area of the cylinder as shown in Figure 2.36.

Materials are usually categorized as conductors, semiconductors, and insulators. Conductors have the smallest resistivity and insulators have the largest resistivity. Semiconductors have values of resistivity that fall between conductors and insulators. Therefore:

$$\rho \text{ (conductors)} < \rho \text{ (semiconductors)} < \rho \text{ (insulators)}$$

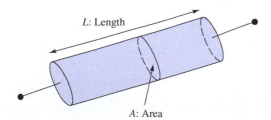

L: Length

A: Area

FIGURE 2.36 The geometry of a cylindrical conductor.

TABLE 2.1 The Resistivity Value of Some Materials

Material	Resistivity (Ωm)
Silver	1.59×10^{-8}
Copper	1.72×10^{-8}
Gold	2.44×10^{-8}
Aluminum	2.82×10^{-8}
Tungsten	5.6×10^{-8}
Iron	1×10^{-7}
Brass	0.8×10^{-7}
Silicon	6.4×10^{2}
Glass	10^{10} to 10^{12}

It should be noted that the value of the resistivity of a material may vary with time. For example, while aluminum is a conductor, aluminum contacts turn into aluminum oxide and aluminum oxide functions almost like an insulator. This is one reason why aluminum wire is discouraged for most home wiring settings. See Sections 9.2 and 9.3 for details.

The values of the resistivity at normal temperatures for selected materials are shown in Table 2.1. Materials such as silver and copper are good conductors. Silicon is a semiconductor and glass is an insulator.

APPLICATION EXAMPLE 2.14   **Length of Tissue**

Suppose that surgeons remove a cylindrical-shaped piece of tissue of unknown length from a patient. Their measurement for a piece of tissue shows that $\rho = 250\ \Omega$m and the cross section is 6.875×10^{-9} m^2. When a voltage of 30 V is applied across the entire length of the tissue, the current flowing through it is 5.65 µA (see Figure 2.37). Find the length of the tissue.

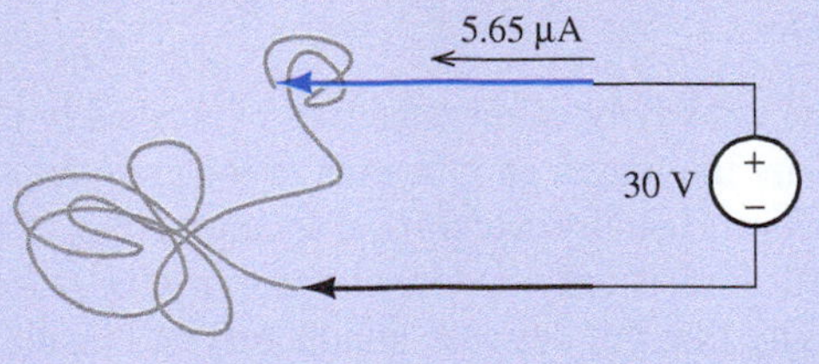

FIGURE 2.37 Application problem (length of tissue).

SOLUTION

Assume R is the resistance of the tissue. Using Ohm's law, we can state that:

$$30 = R \times 0.00000565$$

Next, we can calculate that R is:

$$R = 5310000\ \Omega = 5.31\ \text{M}\Omega$$

According to Equation (2.11),

$$R = \frac{\rho L}{A} \Rightarrow 5.31\ \text{M}\Omega = \frac{250\ \Omega\text{m} \times L}{6.875 \times 10^{-9}\,\text{m}^2}$$

Using this equation, we can find the length $L = 146 \, \mu$m. The equivalent circuit is shown in Figure 2.38.

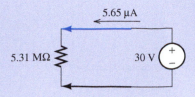

FIGURE 2.38   The equivalent circuit for Example 2.14.

EXAMPLE 2.15   **Computing Resistance**

Water conducts electricity. Electricity and water can be a fatal combination, as will be detailed in later chapters. As is also the case with other conductors, water resists the flow of electric current. Pure water has a resistivity of $\rho = 182 \, k\Omega$-m. However, most water has some additive minerals or salt, and has a resistivity of about $\rho = 175 \, k\Omega$-m.

In the water tank of Figure 2.39, find the resistance between (1) points A and B and (2) points X and Y, given $\rho = 175 \, k\Omega$-m. Assume that the sides of the tank do not conduct current.

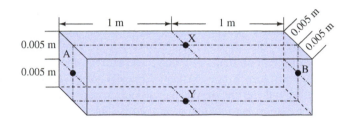

FIGURE 2.39   Water tank for Example 2.15.

SOLUTION

Using Equation (2.11), we know that:

$$R = \frac{\rho \times L}{A}$$

Next, we can apply the parameters outlined earlier to this equation.

a.   When measuring the resistance between points A and B (i.e., the right and the left sides of the pool), we see that the length of the pool is 2 m and the cross section is 0.1 times 0.1. Thus:

$$R_{AB} = \frac{175,000 \times 2}{0.01 \times 0.01} = 3500 \, M\Omega$$

b.   When measuring the resistance between points X and Y, that is, the top and the bottom of the pool, the length is 0.01 and the cross section is 0.1 by 2 m. Thus:

$$R_{XY} = \frac{175,000 \times 0.01}{2 \times 0.01} = 87.5 \, k\Omega$$

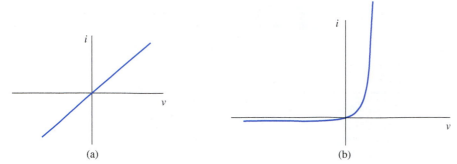

FIGURE 2.40   The *v–i* characteristics of a (a) linear and (b) nonlinear (diode) resistor.

2.7.2 Nonlinear Resistors

The linear relationship between voltage and current defined in Equation (2.10) is valid for linear resistors. In other words, as shown in Figure 2.40(a), for a linear resistor, if you sketch the voltage in terms of current or vice versa, the slope of this curve will not change with current and voltage. Therefore, for all currents and voltages, the resistance of a linear resistor is constant and remains unchanged.

However, note that in general, resistors are not linear. Therefore, the voltage–current relationship of the resistor is not linear. In other words, the resistance of a resistor may not be independent of the current that flows through it or the voltage across it. As a result, the slope of the characteristic curve of nonlinear resistors changes with current and/or voltage. Diodes are an example of nonlinear resistors (see Chapter 8).

The *v–i* characteristics of a diode are sketched in Figure 2.40(b). A diode is a device that allows the current to flow in only one direction in a circuit. Diodes are used in many applications, including voltage regulation. Chapter 8 discusses the details of how diodes are constructed. In Figure 2.40(b), we observe that for different amounts of voltage and current, the slope of the curve is different. Thus, the equivalent resistance is not constant. Typically, when the voltage is high enough, the slope tends toward infinity; therefore, the resistance tends toward zero. Inversely, when the voltage is negative, the slope is almost zero. In that case, the resistance is infinity.

2.7.3 Time-Varying Resistors

The magnitude of resistors may change with time due to environmental factors, such as temperature. As an example, resistivity of resistors changes with temperature. Because of this, in general, the resistance of a resistor varies with time due to the changing environmental factors. As a result, the characteristic curve of a time-varying resistor will not have the same slope at all times. In other words, the slope of the line in Figure 2.40(a) may change with time. In general, in this book, we consider only linear time-invariant resistors and we simply call them resistors.

2.8 POWER AND ENERGY

Electric circuits are used to generate energy for diverse applications. The electrical energy in power distribution systems is used to provide light, heat, and so on. The power and energy associated with a circuit element can be determined directly from the element's current, *i*, and voltage, *V*.

From the definition of voltage, we know that voltage is in fact the potential energy per charge and we know that current is the rate of charge passing through a circuit element. In other words, voltage and current correspond to:

$$\text{Voltage} = \frac{\text{Energy}}{\text{Charge}} \quad \text{Current} = \frac{\text{Charge}}{\text{Time}} \tag{2.12}$$

TABLE 2.2 Notations for Voltage and Current

	DC	AC	Total: AC + DC	Amplitude of AC
Current	I_A	i_a	i_A	I_a
Voltage	V_A	v_a	v_A	V_a

Multiplying voltage and current results in the rate of energy transfer, also called power. Therefore, if power is denoted by p, the following equation can be developed:

$$p = v \cdot i \tag{2.13}$$

where the corresponding units for power are:

$$\text{Volt} \times \text{Ampere} = \frac{\text{Joule}}{\text{Coulomb}} \times \frac{\text{Coulomb}}{\text{Second}} = \frac{\text{Joule}}{\text{Second}} = \text{Watt} \tag{2.14}$$

Units of power are called watts (W), where 1 watt = 1 joule/second (J/s) = volts × amperes.

Note that lower case v is a general notation for a voltage that can be constant or time varying. If we know that the voltage is constant, usually, capital V is used. If we know that voltage is time varying then the notation $v(t)$ is used. The same notation is applied to current, i, i.e., i is a general notation for any current and $i(t)$ is for time-varying current. In addition, throughout this book, we consider two general types of currents and voltages: DC and AC (see Figure 2.7). The notations for AC and DC currents and voltages are listed in Table 2.2. It should be noted that in Table 2.2, "A" refers to a branch for current and a node for a voltage. That is, i_A is the total current through branch "A," and v_A is the total voltage at Node "A."

In Table 2.2, the total current (voltage) refers to the DC plus AC current (voltage) (see the third column of Table 2.2). These notations are used, because in general, a circuit includes both DC and AC sources. Similarly, upper case P is used for DC power (or for phasors in Chapter 6). In addition, lower case p is used for time-varying power. For example, in a radio (e.g., your cell phone), the DC source is the battery, and the AC source is the signal induced over the radio antenna. As detailed in Chapter 8, the DC source is used to bias the transistors of a radio system to allow the amplification of the weak AC signal induced on the antenna. Note that DC analysis of circuits is discussed in Chapter 3, while AC analysis of circuits is discussed in Chapter 6. As discussed in Chapter 6, an AC signal is represented using a sinusoid. Figure 2.7(a) represents a sinusoid. As shown in Figure 2.7, a sinusoid includes an amplitude. The last column in Table 2.2 represents the notation for the amplitude of the sinusoid. For details, refer to Chapter 6.

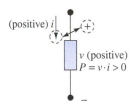

FIGURE 2.41 Passive element.

An electric circuit consists of one or more sources. A source supplies power and energy to other elements. A source is also called an **active element**. All power supplied by active elements in a circuit is consumed, dissipated, or converted by other elements in the circuit.

Elements that absorb energy are called **passive elements**. Passive and active elements in a circuit can be identified by considering the element's voltage polarity relative to the element's current direction. When the current direction is the same with the voltage drop direction (Figure 2.41), the element absorbs energy and it is passive. In this case, the power, $P = v \times i$, is a positive quantity. In other words, the element consumes positive power.

On the other hand, if the current direction is not the same with the voltage drop direction (Figure 2.42), the element supplies energy and is considered an active element. In this case, the element consumes negative power (delivers positive power). It should be noted that in Figures 2.41 and 2.42, we consider the real (positive) directions/polarities of current/voltage. In contrast, the directions discussed in Figure 2.10 represent the relationship of current/voltage directions/polarities required for circuit analysis.

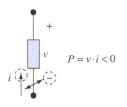

FIGURE 2.42 Active element.

In general, if the current multiplied by the voltage for an element is negative, that element is an active element; otherwise it is a passive element (see Figures 2.41 and 2.42).

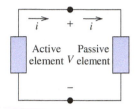

FIGURE 2.43 Passive and active elements connected to each other.

If active and passive elements are connected together as shown in Figure 2.43, the active element will supply energy to the passive element. As a result, the active element delivers energy and the passive element absorbs energy. The voltage polarity for both the elements is the same, but the current direction relative to the voltage polarity is different for each.

In general, if the element's current and voltage vary with time, the power of the element is a function of time as well. In this case, we can develop the following equation:

$$p(t) = v(t) \cdot i(t) \tag{2.15}$$

The total amount of energy supplied by a source element and consumed by a passive element can be determined by integrating the power time-dependent function over a specified time interval. As a result, the transferred energy, w, between time instants t_1 and t_2 corresponds to:

$$w = \int_{t_1}^{t_2} p(t) \mathrm{d}t \tag{2.16}$$

EXAMPLE 2.16   **Voltage Direction Versus Current Direction**

Assuming the directions of current and the polarities of voltages shown in Figure 2.44 refer to the true (positive) values, do the following voltage/current labels conform to how passive elements should be referenced?

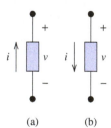

FIGURE 2.44   Configuration for Example 2.16.

(a)   (b)

SOLUTION

In passive elements, the current enters the positive polarity of the voltage; therefore, elements a and b in Figure 2.44 are active and passive, respectively.

APPLICATION EXAMPLE 2.17   Robotic Arm

A robotic arm—5 m long—is designed to load containers onto a cargo ship (Figure 2.45). The maximum mass that can be lifted by the robotic arm is 500 kg at a constant angular velocity of 0.2 rad/s. The power required to initiate the lifting is five times the power required for the lifting process. Assume that the voltage source for the robotic arm is 50 kV. Calculate the current flow through the circuit at the initial time, $t = 0$, and the current during the loading process. Assume the gravitational acceleration is $g = 10 \text{ ms}^{-2}$.

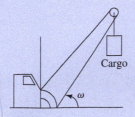

FIGURE 2.45 An example of a robotic arm used to load and unload a ship's cargo.

The equations needed to solve this problem include the following:

1. The relationship between the torque, distance (L), and force (F) is: $\tau = L \times F$.
2. The minimum force required to lift an object with mass m is: $F = m \times g$.
3. The relationship of power and torque is: $P = \tau \times \omega$.

SOLUTION

The torque required to lift up a 500 kg container is:

$$
\begin{aligned}
\tau = L \times F &= L \times m \times g \\
&= 5 \text{ m} \times 500 \text{ kg} \times 10 \text{ ms}^{-2} \\
&= 25{,}000 \text{ Nm}
\end{aligned}
$$

The angular velocity is $\omega = 0.2$ rad/s; thus, the required power for the lifting process is:

$$
P_{\text{process}} = \tau \times \omega = 25{,}000 \times 0.2 = 5000 \text{ W}
$$

The power required to initiate the lifting is five times the power required for the lifting process, thus:

$$
P_{\text{initial}} = 5P_{\text{process}} = 25{,}000 \text{ W}
$$

Given voltage, V, is 500,000 V, the initial current is:

$$
I_{\text{initial}} = \frac{P_{\text{initial}}}{V} = 0.5 \text{ A}
$$

The current during the lifting process is:

$$
I_{\text{process}} = \frac{P_{\text{process}}}{V} = 0.1 \text{ A}
$$

APPLICATION EXAMPLE 2.18   Voltage Adapter

Many electronic devices are connected to their power source through an adapter. Because the "standard" voltage across different countries is not the same, universal device adapters have become popular. One example of this is adapters for laptop computers that enable the capability to connect to both 110 V and 220 V voltage sources without an additional step-down transformer.

Figure 2.46 represents a block diagram of a universal adapter. The power factor control (PFC) consists of a variable resistor that changes the resistance so that the power input to the transformer remains unchanged, regardless of the input voltage. Transformers will be discussed in more detail in Chapters 8 and 12.

FIGURE 2.46   Block diagram of a universal adapter.

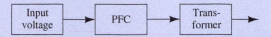

Assume that the desired adapter output is 15 V and 6 A, with 100% power efficiency. What is the value of the PFC resistor when:

a. input voltage is 110 V
b. input voltage is 220 V

a. First, calculate the output power:

$$P = IV$$
$$P = 6 \times 15$$
$$P = 90\,\text{W}$$

For a 100% power efficiency transformer, the power input would be the same as output power. In addition, we know that the input voltage is 110 V. Therefore, the required input current is:

$$I_{\text{input}} = \frac{P}{V_{\text{input}}} \rightarrow I_{\text{input}} = \frac{90}{110} \rightarrow I_{\text{input}} = 0.8182\,\text{A}$$

The required resistance for 110 V input voltage for the PFC is therefore:

$$R_{110} = \frac{V_{\text{input}}}{I_{\text{input}}} \rightarrow R_{110} = 134.44\,\Omega$$

b. We know that the input and output powers are the same in this problem. We have already calculated the power as 90 W. Thus, the resistance required for a 220 V input voltage can be determined directly using the definition below:

$$P = \frac{V^2}{R}$$

Where V is the input voltage and R is the required resistance. Thus, the required resistance is:

$$R_{220} = \frac{V^2}{P} \rightarrow R_{220} = \frac{220^2}{90} \rightarrow R_{220} = 537.78\,\Omega$$

2.8.1 Resistor-Consumed Power

Because resistors are passive elements, they always absorb energy. The absorbed electrical energy is converted to other types of energies such as heat or light. The power, P, absorbed by a resistor can be found by multiplying the current, I, passing through it and voltage, V, across its terminals (Section 2.8 discusses the power and energy in detail). Because in a resistor the ratio of voltage to current is expressed by the constant resistance value, R, we can state that:

$$P = V \times I = \frac{V^2}{R} = R \times I^2 \qquad (2.17)$$

Because R is always a positive value, Equation (2.17) confirms that the power of a resistor (similar to the power of all other passive elements) is positive. As has been discussed, the value of the resistance of an ideal resistor or conductor is completely independent of the voltage across its terminals and the current passing through it.

EXAMPLE 2.19   **Power Dissipation**

Figure 2.47 represents a circuit that includes a dependent current source as detailed in Section 2.9. Here, the current of the current source is a function of the current in the 3 kΩ resistor. How much power is dissipated in the 33 kΩ resistor in Figure 2.47?

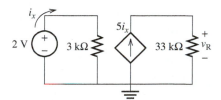

FIGURE 2.47   Circuit diagram for Example 2.19.

SOLUTION

The current source generates a current in the 33 kΩ resistor that depends on the current passing through the 3 kΩ resistor. To compute the voltage across the 33 kΩ resistor and then evaluate its power, we first find the value of i_x. Using Ohm's law, we can state:

$$i_x = \frac{2\ \text{V}}{3\ \text{k}\Omega} = 667\ \mu\text{A}$$

Therefore, $5i_x = 3.33$ mA. Then:

$$v_R = 3.33\ \text{mA} \times 33\ \text{k}\Omega = 110\ \text{V}$$

Next, we are able to find the value of the power dissipated in the 33 kΩ resistor. Using Equation (2.17), we see that:

$$P_R = 110 \times 3.33 \times 10^{-3} = 366.3\ \text{mW}$$

EXERCISE 2.3

Find the power dissipated in the 33 kΩ resistor if the dependent current source in Figure 2.47 is replaced by a dependent voltage source with its voltage a function of the voltage across the 3 kΩ resistor. (See Section 2.9 for a more detailed investigation of dependent and independent sources.)

APPLICATION EXAMPLE 2.20   Door Opener

A driveway gate shown in Figure 2.48 is driven by an electric motor that is connected to a voltage source of 200 V. In order to move the driveway gate, a force of 500 N or more is required. Assume that the motor drives the driveway gate at a constant speed of 0.2 ms^{-1} and that the gate path is 4 m long. Also, assume that the power required to start the motor is the same as the power required to move the gate. Calculate the total power that will be required for this process, and the possible resistance in the motor.

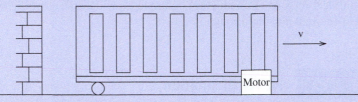

FIGURE 2.48   A simple illustration of a driveway gate driven by an electric motor.

SOLUTION

Here, force, $F = 500$ N, velocity, $V_e = 0.2$ ms^{-1}, and distance, $d = 4$ m. Thus, the total time of the process is $t = d/V_e = 20$ s. In addition, we can calculate the power needed as:

$$P = \frac{Fd}{t} = 100 \text{ W}$$

To calculate the resistance, we can use the following equation:

$$P = \frac{V^2}{R}$$

Given that $V = 200$ V, and the result of the power calculation ($P = 100$ W), we see that the resistance corresponds to $R = 400 \, \Omega$.

2.9 INDEPENDENT AND DEPENDENT SOURCES

An *independent voltage source* is a voltage source (e.g., a battery or utility-connected electrical outlet) applied across a circuit that generates a current through the circuit. A voltage source maintains a constant or time-varying voltage across its terminal nodes.

An independent voltage source generates a predetermined voltage that is independent of the current flowing through and/or voltage across any given element of the circuit.

Figure 2.49(a) shows an ideal independent voltage source. The current passing through this element is determined by other circuit elements.

The voltage across the terminals of an independent DC voltage source (e.g., a battery) is constant and does not change with time. For example, the constant DC voltage in Figure 2.49(b) is 5 V.

EXERCISE 2.4

Sketch the voltage (*y*-axis) versus current (*x*-axis) diagram of an independent voltage source. Is it a line parallel to the (*x*- axis)? This diagram is called a *v–i* diagram. Resketch the voltage source *v–i* diagram when the voltage is zero. This represents the *v–i* diagram of a short circuit. In other words, a zero voltage source represents a short circuit.

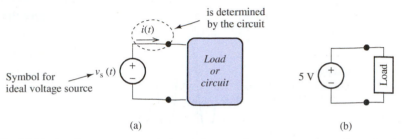

FIGURE 2.49   (a) Ideal voltage source symbol and (b) a 5 V DC source connected to a load.

In AC voltage sources, the voltage varies with time. An example of an AC voltage source is the signal that can be obtained from standard electrical outlets in a home. In Figure 2.50, the AC voltage is in the form of a sinusoid where the voltage changes periodically with the time. The symbol for an independent voltage source is a circle (see Figure 2.49). The sinusoidal waveform in the circle in Figure 2.50 represents the varying nature of the voltage. Note: standards for residential electric power vary from country to country. In this figure, the voltage source frequency is 50 Hz $(50 = 100 \, \pi/2\pi)$.

A *dependent or controlled voltage source* creates a voltage across its terminals. An automotive alternator is an example of a dependent voltage source—voltage is dependent on the revolutions per minute (RPM) of the engine. A dependent voltage is a function of the voltage across or current through a circuit element. In Figure 2.51, the voltage across the terminals of the voltage source is controlled by the voltage across the terminals of the element B. In this circuit, at a given time, the voltage delivered to the circuit by the source is four times the voltage across the element B at that time.

The voltage supplied by a dependent voltage source can also be controlled by a current passing through an element. Figure 2.52 shows a dependent voltage source. Here, the voltage

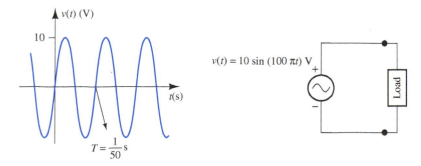

FIGURE 2.50   AC voltage source.

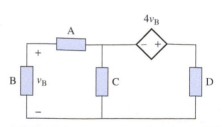

FIGURE 2.51   Voltage controlled voltage source.

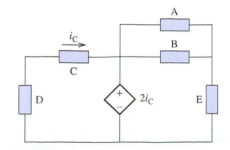

FIGURE 2.52   Current voltage source.

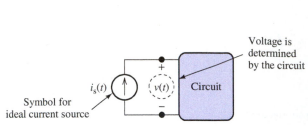

FIGURE 2.53 symbol labels: Symbol for ideal current source, $i_s(t)$, $v(t)$, Circuit, Voltage is determined by the circuit

FIGURE 2.53   An ideal current source.

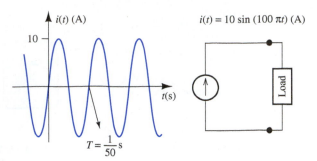

$i(t) = 10 \sin(100\,\pi t)$ (A)

$T = \dfrac{1}{50}$ s

FIGURE 2.54   AC source.

provided by the source is two times the current passing through element C. A diamond is the common symbol for showing a current or voltage-controlled source in a circuit. The energy might be delivered to a circuit via a voltage source or a current source.

An *independent current source* is a current source that supplies energy to a circuit in the form of current. A current source maintains a specific current flow through its branch. Unlike voltage sources (e.g., batteries that can be observed physically) current sources are usually a model for electronic systems. Figure 2.53 shows an ideal independent current source that generates a current independent of the connected circuit.

EXERCISE 2.5

Sketch the voltage (y-axis) versus current (x-axis) diagram for an independent current source. Is it a line parallel to the (y-axis)? Resketch the current source v–i diagram when the current is zero. This represents the v–i diagram of an open circuit. As described, a zero-current source represents an open circuit.

The symbol commonly used for an independent current source is a circle with an arrow inside. The direction of the arrow represents the true direction of the flow of current. The DC and AC sources, respectively, force constant and time-varying currents in a circuit. Figure 2.54 represents an example of AC supply.

Dependent or controlled current sources generate currents that are determined by a current through or voltage across a circuit element. Voltage-dependent current sources establish a current in a circuit that is controlled by a voltage across the terminals of an element in the circuit. Current-dependent current sources create a current that is controlled by a current passing through an element. Figures 2.55 and 2.56 show some examples of current- and voltage-controlled current sources. A diamond with an arrow inside is the common symbol for dependent current sources.

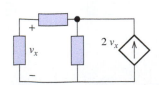

FIGURE 2.55   Voltage-controlled current source.

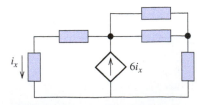

FIGURE 2.56   Current-controlled current source.

EXAMPLE 2.21   **Dependent Source**

Find the current, I_x, that is shown in Figure 2.57.

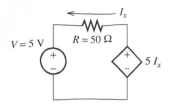

FIGURE 2.57   Circuit diagram for Example 2.21.

SOLUTION

Here, the current direction is arbitrarily selected to be counterclockwise, which is the opposite direction of the voltage source, 5 V. Using the values from Figure 2.57, the equation is:

$$5I_x = 50I_x + 5$$

Therefore, the current, I_x, is:

$$-45I_x = 5$$
$$I_x = -0.11 \text{ A}$$

The negative sign of the current shows that the correct current direction is clockwise instead of counterclockwise, as shown in Figure 2.57.

EXAMPLE 2.22   **Dependent Source**

Determine the voltage, V_A, across the 50-Ω resistor in Figure 2.58.

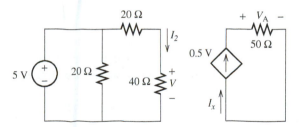

FIGURE 2.58   Circuit diagram for Example 2.22.

SOLUTION

V_2 is the voltage that corresponds to the current, I_2, which is also the voltage across both 20-Ω and 40-Ω resistors. Then, using the amounts from Figure 2.58, we can develop the equation:

$$I_2 = \frac{5}{40 + 20} = 0.0833 \text{ A}$$

The voltage across 40-Ω resistor, V, is:

$$V = 40I_2 = 3.33 \text{ V}$$

Then, the current, I_x, is:

$$I_x = 0.5 V = 1.667 \text{ A}$$

The voltage, V_A, can then be calculated as:

$$V_A = 50I_x = 83.33 \text{ V}$$

2.10 ANALYSIS OF CIRCUITS USING PSPICE

This section provides an overview of the capabilities of PSpice software for circuit analysis. You will learn how to set up a PSpice circuit step by step. In order to analyze a circuit using PSpice, three stages are required:

a. Creating a new project file
b. Drawing circuit in the PSPICE schematic window
c. Running simulation analysis

This section explains the details of these stages.

a. Creating a new project file
1. Click on the PSpice icon and open the main page. Figure 2.59 shows a screenshot of the main PSpice page:
2. Go to "File" > "New" and choose Project.
3. A window (as shown in Figure 2.60) will pop up. Type a filename of your choice, for example, "Test01." Next, you will choose from among the different categories of projects allowed by PSpice, including the following:
    i. *Analog or Mixed A/D:* Analog or digital circuits can be set up on the PSpice layout, and simulations can be run for circuit analysis.
    ii. *The PC Board:* Used for setting up a PC Board layout.
    iii. *The Programmable Logic Wizard:* Allows the users to prepare their design with the aid of a complex programmable logic device (CPLD) or a field programmable gate array (FPGA) tool. The libraries of the project are configured based on the manufacturer's type/part that is chosen.
    iv. *The Schematic:* This option creates a project that contains only a design file.

    In this book, we mainly deal with analog or mixed A/D circuits. As such, choose the option "Analog or Mixed A/D project," as shown in Figure 2.60, and press OK. The file can be saved in a location of your choice.

4. As soon as "OK" is selected, the next window (shown in Figure 2.61) will pop up to ask whether to use a project created based on an existing project for the file. At this stage, you may use an existing project and modify the parameters or you may create a new one. In most cases, we will create a new project. Thus, we choose "Create a blank project" and press OK.
5. Next, the main PSpice schematic circuit board window will appear (as shown in Figure 2.62). Circuits can be easily created in this board.

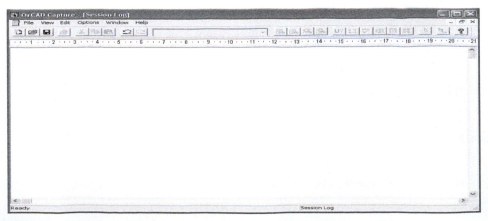

FIGURE 2.59 Main page of PSpice.

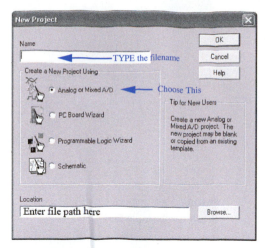

FIGURE 2.60   New project.

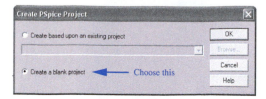

FIGURE 2.61   Create a PSPICE project.

FIGURE 2.62   PSpice schematic window.

b. Drawing a circuit in the PSpice schematic window

PSpice consists of many tools to facilitate drawing circuits. Each category of these tools is allocated a specific icon located on the right-hand side of the schematic window. The following are some important examples of these icons:

- *Place Part icon:*   This contains the various analog and digital parts that are needed to set up an electronic circuit.
- *Wire icon:*   This connects the parts on the circuit board to create a complete circuit.
- *Ground icon:*   All the analog circuits should be connected to a Ground to form a complete circuit.

Now, considering the circuit of Example 2.12 as an example; the basic PSpice schematic and simulation steps are shown below:

1. Select "Place" > "Part" or click on the icon ⊃ . A window (as shown in Figure 2.63) will pop up.
2. The electronic parts are stored in a specific library file. Therefore, the library files are required to load into the Parts Database and search and place the particular electronic parts on the PSpice circuit board. To place the parts, choose Add Library (STEP A in Figure 2.63) and highlight all of the files or "Ctrl+A" to add all of the *.olb/*.lib files into the library.

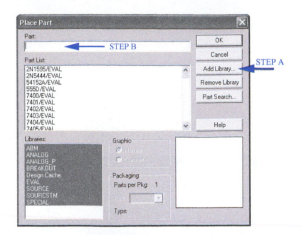

FIGURE 2.63 Place Part and Add Library.

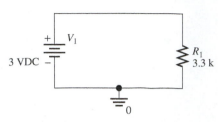

FIGURE 2.64 PSpice schematic circuit for Example 2.12.

3. Continuing with Example 2.12, to add a DC voltage source, type "Vdc" in the Part column (STEP B in Figure 2.63) and then click OK. Place the "Vdc" icon as shown in Figure 2.64, and then right click and choose "End Mode." *Press "r" to change the orientation or direction of the voltage source.*

4. To set the voltage source value, double-click on the "0 V DC" and type "3 V DC," where DC stands for direct current.

5. To place a resistor on the circuit board, go to "Place" > "Part" and type "R" in the Part column, then click OK. (Note: be sure to use R/analog and not R/discrete.) Place the resistor, "R1" as shown in Figure 2.64. *Press "r" to change the orientation of the resistor.*

6. To change the resistance of resistor, double click the "1 k" and change it into "3.3 k."

7. To connect the voltage and resistor, click "Place" > "Wire," or the icon ⌐ on the vertical bar to the right. Left-click inside the square box of the voltage source (shown in Figure 2.65) to create the connection between the voltage source and the wire. For the other end of the wire, click on the empty square box of the resistor to create a connection between the voltage source and the resistor. A colored solid circle will appear during the connection if it is connected properly (shown in blue in Figure 2.66). Right-click and choose "End Wire" after connecting the wire to the resistor.

8. The direction of the wire can be changed by left-clicking once on the circuit board.

9. Most of the time, the parts and their square boxes appear small on the screen. To ease the process of wiring the electronic parts in the circuit, you can zoom in so that the electronic parts are shown larger on the screen. You may also highlight the electronic parts as shown in Figure 2.67. Note: hold the Ctrl" button down on the computer to select multiple parts simultaneously. We can change the orientation of the resistor by pressing "r."

10. Next, press the "zoom in" (solid arrow) icon as shown in Figure 2.68. If you wish to zoom out, press the "zoom out" (dashed arrow) icon in Figure 2.68.

11. Every circuit in PSpice is required to be connected to a ground to become a complete circuit. Click "Place" > "Ground" or the icon ⏚ on the vertical bar to the right.

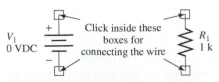

FIGURE 2.65 Empty square box on voltage source and resistor.

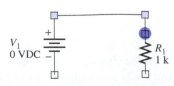

FIGURE 2.66 Colored circle for connection confirmation.

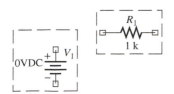

FIGURE 2.67 Highlight the electronic parts.

FIGURE 2.68 Zoom in and zoom out buttons.

12. A window that is similar to Place Part (Figure 2.63) will pop up. Choose add library, and choose source.olb/source.lib to add the ground into the library. Next, choose "0/Source" and place it as shown in Figure 2.64. Note: Always either press "Esc" or right click and choose "End" to complete the insertion.

13. You should be cautious not to choose and place any parts in the list of Figure 2.63 that have the suffix "Design Cache," that is, called "*partname*/Design Cache" in the PSpice circuit board. PSpice is unable to run any simulation analysis with those components.

c. Running simulation analysis

PSpice can run four types of simulation analyses, including the following:

i. *Time Domain (Transient):* The analysis of voltage and current over time.

ii. *DC Sweep:* The analysis of the circuit by changing voltage in linear or logarithmic steps.

iii. *AC Sweep:* Usually used when the voltage or current are time varying and also for frequency response analysis (e.g., Chapter 7).

iv. *Bias Point:* The analysis of voltage and current for a given (constant voltage or current), or when voltage or current varies around a bias voltage or current, for example, for the analysis of current and voltage in transistors and diodes (Chapter 8).

In this section, we will only look into Time Domain and Bias Point analysis. AC Sweep analysis will be discussed in Chapter 7.

Bias Point Analysis

1. First, go to "PSpice" > "New simulation profile" or click the bottom left icon that the **colored** arrow points to in Figure 2.69.

2. A window (shown in Figure 2.70) will pop up to prompt you to insert a simulation name. Type a preferred simulation name, for example,, "Bias Point" and then click "Create."

3. Another window, "Simulation Setting" will pop up, as shown in Figure 2.71. Go to the "Analysis" Tab, and choose the "Analysis Type" to be "Bias Point," and then click OK.

4. Go to "PSpice" > "Run" or click the "Play" icon (see the dotted arrow in Figure 2.69). The results are shown in Figure 2.72.

5. If either the voltage or current does not show up after running the simulation, click the "V" or "I" icon (for location, see the dashed arrow in Figure 2.69).

FIGURE 2.69 Create new simulation profile.

FIGURE 2.70 New simulation.

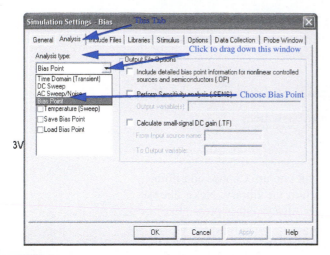

FIGURE 2.71   Simulation results for Example 2.12.

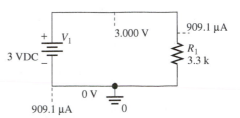

FIGURE 2.72   Simulation setting.

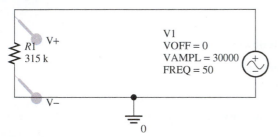

FIGURE 2.73   PSpice schematic circuit for Example 2.13.

FIGURE 2.74   Voltage differential markers.

Time Domain (Transient) Analysis

Consider the circuit of Example 2.13 that is shown in Figure 2.73. Construct this circuit in PSpice using techniques similar to those shown earlier. It is clear that there is a set of "Voltage Differential Markers" at the resistance, R_1. The Voltage Differential Markers will allow the simulation to plot the desired observation automatically whenever the simulation is run. To place the Voltage Differential Markers on the circuit, click the "Voltage Differential Markers" (solid arrow) on the toolbars as shown in Figure 2.74. The first click (or odd number) of placing onto the circuit will be a positive Voltage Differential Marker, while the second click (or even number) of placing will be a negative Voltage Differential Marker.

Next, follow these steps to set up the time domain analysis simulation:

1. First, go to "PSpice" > "New simulation profile" or click the bottom left icon that is indicated by the solid arrow shown in Figure 2.69.
2. Insert a simulation name in the window when prompted. Type a preferred simulation name, for example, "Time Analysis" and then click "Create."
3. Another window, "Simulation Setting" will pop up in Figure 2.75. Go to the "Analysis" Tab, and choose the "Analysis Type" to be "Time Domain (Transient)."
4. Set the "Run to time" to 50 ms (milliseconds) and press OK.
5. Go to "PSpice" > "Run" or click the "Play" icon to run the simulation.
6. Notice that the Simulation Schematic Window will plot the voltage plot of the circuit, which is shown in Figure 2.76.
7. The current plot can be added using the "Add Trace" function. However, the amplitude of the current plot is much smaller than the voltage plot in most cases; therefore, we need to delete the voltage plot and replace it with the current plot. To do so, click the

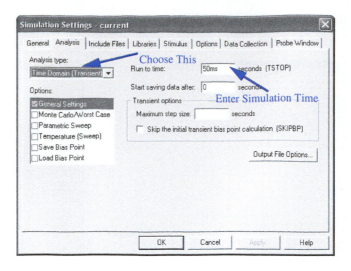

FIGURE 2.75 Set up a time domain (transient) simulation.

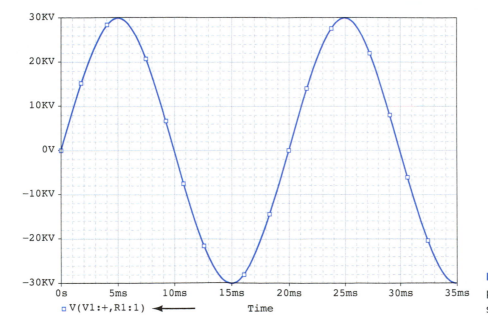

FIGURE 2.76 Voltage plot for time domain simulation.

"V(V1:+,R1:1)" that is located near the arrow in Figure 2.76, and press the "Delete" key on the keyboard.

8. Next, to add the current plot, go to "Trace" > "Add Trace."

9. A window will pop up as shown in Figure 2.77. The "add trace" window allows the user to perform mathematic summation, subtraction, multiplication, and so on on the "Simulation Output Variables" in the "Trace Expression" column for various plotting. However, in this section, we only intend to observe the current flow through the resistor R1; therefore, click "I(R1)" and choose OK, or double-click "I(R1)" in Figure 2.77.

10. The output of the plot for the current plot is shown in Figure 2.78.

Copy the Simulation Plot to the Clipboard to Submit Electronically

Sometimes, you may be requested to submit your work electronically. Unfortunately, the PSpice student/demo version does not allow the user to save the simulation plot in any type of file.

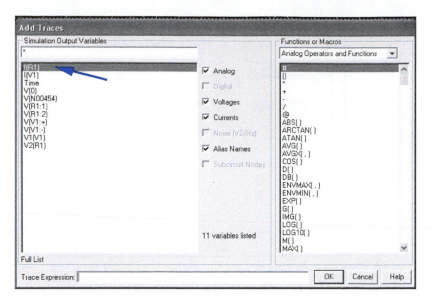

FIGURE 2.77 Add traces.

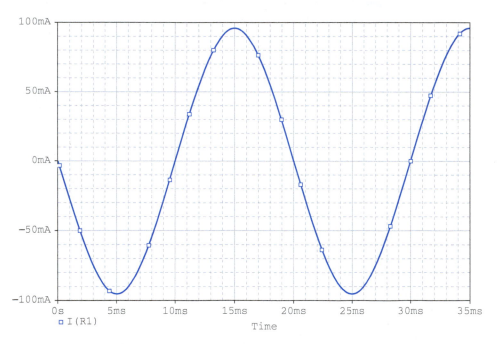

FIGURE 2.78 Current plot for time domain simulation.

Therefore, several steps are required to submit your work electronically. There are two ways to submit a PSpice plot electronically:

a. Use the Microsoft Windows *Print Screen* function, and paste it into image editor software, such as *Microsoft Paint*, *Photoshop*, and so on. However, this method requires an extra image editing process.

b. Use the "copy to clipboard" function and paste it into *Microsoft Paint*. This is a more straightforward method.

The following steps explain how to use the "copy to clipboard" function in PSpice. For these instructions, we will use the simulation plot for Example 2.13 that is shown in Figure 2.76.

a. In the schematic plot window, go to "Window" and choose "Copy to Clipboard" as shown in Figure 2.79.

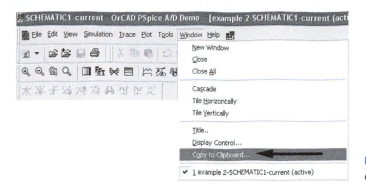

FIGURE 2.79 Copy to clipboard.

b. A copy to clipboard option window will pop up as shown in Figure 2.80. Choose either the option "change white to black" or "change all colors to black," then press OK.

FIGURE 2.80 Copy to clipboard option.

c. Next, go to the Microsoft Windows Start Menu > All Programs > Accessories and click "Paint." Press "CTRL+V" or go to the "edit" menu and choose "Paste" to paste the plot.

d. Go to file > save to save the file with a specific file type, for example, JPG or JPEG.

e. The plot will be saved as shown in Figure 2.81.

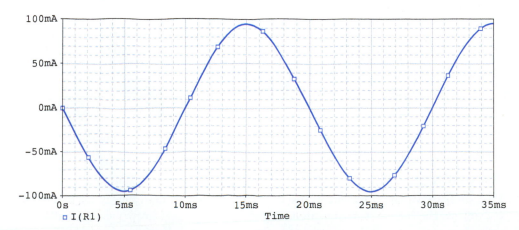

FIGURE 2.81 Simulation plot using copy to clipboard.

EXAMPLE 2.23 **PSpice Example**

Use PSpice to find the current, I, in the circuit of Example 2.12 (shown in Figure 2.82).

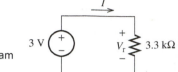

FIGURE 2.82 Circuit diagram
for Example 2.23.

SOLUTION

Follow the steps in the PSpice tutorial to set up the PSpice circuit for Example 2.12, shown in Figure 2.82. The PSpice schematic solution is shown in Figure 2.83.

Next, set the simulation type to be "Bias Point" and run the simulation. The PSpice result is shown in Figure 2.84.

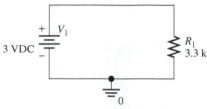

FIGURE 2.83 PSpice schematic for the
circuit in Example 2.23.

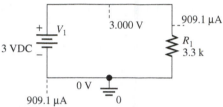

FIGURE 2.84 Simulation results for
Example 2.23.

APPLICATION EXAMPLE 2.24 **Electric Shock**

Consider the unfortunate bald utility worker of Example 2.13 who climbs an aluminum ladder with bare feet and comes into contact with an overhead power line bearing 30 cos(100 πt) kV (Figure 2.86). The worker's body resistance is 315 kΩ. This situation can be modeled as a resistor (his body) connected between the terminals of a power supply (the wire) (Figure 2.85). How much current flows through his body? If the minimum current needed for a shock to be fatal is 100 mA, does the worker survive?

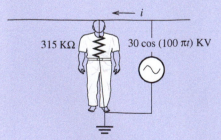

FIGURE 2.85 The equivalent circuit for
Example 2.24.

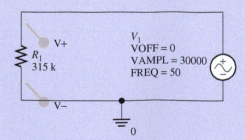

FIGURE 2.86 PSpice schematic circuit for
Example 2.13.

SOLUTION

First, given the voltage source of 30 cos(100 πt) kV, we need to determine of the frequency of the voltage. The voltage source function is $V_S = V \cos(\omega t)$, and the frequency, f, is defined as

$f = \omega/2\pi$. This clearly shows that: $\omega = 100\pi$, and the frequency of the circuit is: $f = 100\pi/2\pi$ or $f = 50\,\text{Hz}$.

PSpice Schematic Setup:

1. Instead of using the Direct Current Voltage Source, a sinusoidal voltage source has been chosen. The sinusoidal voltage source for PSpice can be obtained by typing "VSIN" in the "Place Part" window; its amplitude and frequency are set as shown in Figure 2.86.
2. Follow the steps in the Tutorial section to set up a Time Domain Analysis Simulation. The current plot across the human body is shown in Figure 2.87.
3. Figure 2.87 shows the maximum (also known as amplitude) of the current is less than 100 mA. Therefore, the worker is able to survive the electric shock.

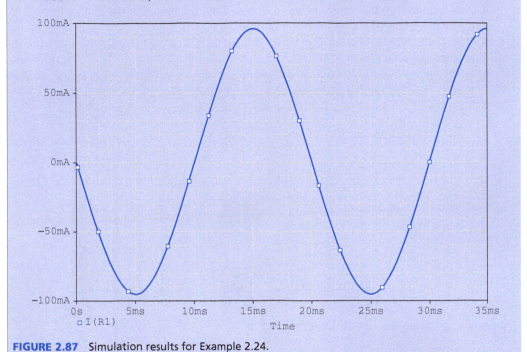

FIGURE 2.87 Simulation results for Example 2.24.

EXAMPLE 2.25   **PSpice Example**

Use PSpice to calculate the dissipated power in Example 2.16 (see Figure 2.88).

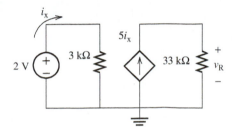

FIGURE 2.88 Circuit diagram for Example 2.25.

SOLUTION

You have already learned how to set up the PSpice Schematic. You will discover some other useful PSpice functions in other chapters. In this example, you will learn how to use the current-controlled current source (CCCS) gain in the PSpice software. After gaining familiarity with CCCS, you will

(continued)

EXAMPLE 2.25   Continued

be able to easily apply the same procedure to voltage-controlled voltage source (VCVS), voltage-controlled current source (VCCS), and current-controlled voltage source (CCVS).

1. First, set up the PSpice Schematic as shown in Figure 2.89.
2. Notice that there is an additional resistor with a very high resistance of 100 MΩ in the circuit. The reason for this is that an error will occur if the PSpice circuit is open. A high-resistance resistor is required and will act as a "bridge" to connect both sides of the circuit, while preventing any additional current flow through it.
3. To use the CCCS gain function in PSpice, go to "Place Part" and type "F." Choose "F/Analog."
4. After connecting the wires shown in Figure 2.89, double-click the F icon to open the schematic properties window shown in Figure 2.90.
5. Change the "GAIN" value from "1" to "5," and then click the "Display…" that is located near the arrow in Figure 2.90.
6. A "Display Properties" window will pop up as shown in Figure 2.91. In the "Display Format" section, choose to display "Name and Value," then click OK.
7. Then the CCCS gain block will show "GAIN = 5" as shown in Figure 2.92.

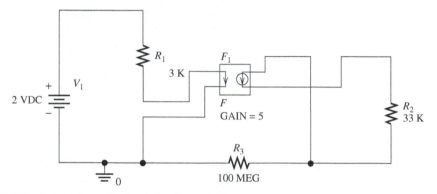

FIGURE 2.89   Spice schematic circuit for Figure 2.88.

FIGURE 2.90   Schematic properties.

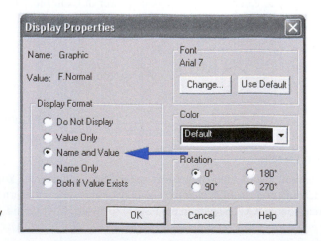

FIGURE 2.91   Display properties.

8. Set up the simulation profile with "Analysis Type" as "Bias Point" and run the simulation. The simulation result is shown in Figure 2.92.

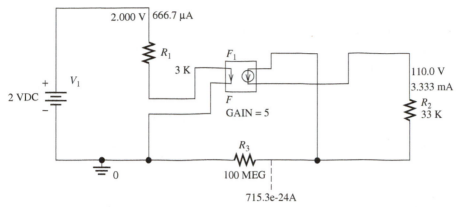

FIGURE 2.92   Simulation results.

2.11 WHAT DID YOU LEARN?

- The relationship between current and charges is expressed as [Equation (2.3b)]:

$$i(t) = \frac{dq(t)}{dt}$$

- Current direction is defined as the direction of the movement of *positive* charges.
- The voltage between two points in a circuit indicates the energy that is needed to move one coulomb (1 C) of charge and one volt (1 V); it is defined as:

$$1 \text{ volt (V)} = 1 \text{ joule/coulomb (J/C)}$$

- *Respective direction* of voltage and current (as shown in Figure 2.10) means that the direction of the current is from the positive sign of the voltage to the negative sign of the voltage.
- Kirchhoff's current law states that the currents entering a node are equal to the currents leaving the node [Equation (2.7)]:

$$\sum_{k=1}^{n} i_k = \sum_{k=n+1}^{N} i_k$$

- Kirchhoff's voltage law states that the algebraic sum of the voltages equals zero for any loop in an electric circuit.
- Ohm's law [Equation (2.10)] defines the relationship between voltage, v, current, i, and resistance, R, and can be written as:

$$v = i \times R$$

- The resistance, R, of a conductor is determined by its resistivity, ρ, length, L, and cross-sectional area, A, as shown in Equation (2.11):

$$R = \rho \frac{L}{A}$$

- The power consumed by an electric element is defined by [Equation (2.13)]:

$$p = v \times i$$

If $p > 0$, the electric element absorbs energy and is called a *passive element*.
If $p < 0$, the electric element supplies energy and is called an *active element*.
- Energy suppliers in a circuit are divided into dependent and independent sources.
- PSpice can be used to analyze simple circuits using the techniques outlined in the chapter.

Problems

*B refers to Basic, A refers to Average, H refers to Hard, and * refers to problems with answers.*

SECTION 2.1 INTRODUCTION

2.1 (B) Identify the device in Figure P2.1:

FIGURE P2.1 Image for Problem 2.1.

2.2 (B) Identify the device in Figure P2.2:

FIGURE P2.2 Image for Problem 2.2.

2.3 (B) Identify the device in Figure P2.3

FIGURE P2.3 Image for Problem 2.3.

2.4 (B) Identify the device in Figure P2.4:

FIGURE P2.4 Image for Problem 2.4.

2.5 (B) Identify the device in Figure P2.5:

FIGURE P2.5 Image for Problem 2.5.

SECTION 2.2 CHARGE AND CURRENT

2.6 (B) What is the definition of current direction?

2.7 (A) Find the charge as a function of time if:

$$i(t) = 2 \sin (10\pi t)$$

2.8 (A) Find the number of electrons, n, as a function of time if:

$$i(t) = 12t$$

2.9 (A)* Direct current in steady state does not vary with time. If a DC current of 5 A flows through a wire, find the charge through the wire with respect to time. The charge at $t = 0$ is 0.

2.10 (A)* In Problem 2.9, how many charges flow in 5 s?

2.11 (H)* Given that the number of charges measured at t_1 is $n_1 = 2 \times 10^{19}$, and the number of charges measured at t_2 is, $n_2 = 5.75 \times 10^{19}$. Find the current flow if the time interval between t_1 and t_2 is 2 s, with the assumption that the current, $i(t)$ is a linear function of charges, $q(t)$.

2.12 (H) Application: electroplating
A depth of 0.15 mm nickel is required to be coated on a metal surface by electroplating. The nickel density $\rho = 8.8 \times 10^3$ kg/m^3, the electrochemical equivalent $k = 0.304 \times 10^{-6}$ kg/C (which is the weight of the nickel produced by electrolysis during the flow of a quantity of electricity equal to 1 C), and the current per square meter $\rho = 35$ A/m^2. How long does it take to achieve the desired coating depth? (*Hint: Find the mass of the nickel.*)

SECTION 2.3 VOLTAGE

2.13 (B) The voltage across the device in Figure P2.13 is 1 V. What is the voltage polarity of Nodes a and b if the potential energy is reduced when a positive charge moves from a to b.

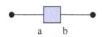

FIGURE P2.13 Image for Problem 2.13.

2.14 (A)* If a current of 3 A flows from a 5 V DC source for 1 s, how much energy, in Joules, is used?

2.15 (H) If 5×10^{16} electrons are pushed by the energy of 15 Joules, what is the voltage?

2.16 (H)* If an alternating current of $i(t) = 2 \sin (3/2 \pi t)$ flows from a voltage source of $v(t) = 5$, find the energy used in 1 s.

2.17 (H) 20 J is used from a 2 V source in 4 s. How much current, i, is flowing during this time?

SECTION 2.4 RESPECTIVE DIRECTION OF VOLTAGE AND CURRENT

2.18 (B) True/false: The notation in Figure P2.18 conforms to passive reference.

FIGURE P2.18 Image for Problem 2.18.

2.19 (B) True/false: The notation in Figure P2.19 conforms to passive reference.

FIGURE P2.19 Image for Problem 2.19.

2.20 (B) True/false: The notation in Figure P2.20 conforms to passive reference.

FIGURE P2.20   Image for Problem 2.20.

SECTION 2.5 KIRCHHOFF'S CURRENT LAW

2.21 (B) Use KCL to find i_2 in Figure P2.21.

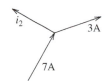

FIGURE P2.21   Image for Problem 2.21.

2.22 (B)* Use KCL to find i_4, i_5, and i_6 in Figure P2.22.

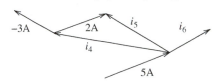

FIGURE P2.22   Image for Problem 2.22.

2.23 (A)* Find v in the circuit given in Figure P2.23.

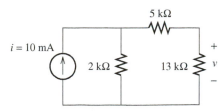

FIGURE P2.23   Circuit for Problem 2.23.

2.24 (A) Use KCL to show that i_1 is equal to i_4 in Figure P2.24.

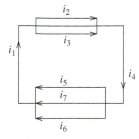

FIGURE P2.24   Circuit for Problem 2.24.

2.25 (H) In Figure 2.11, explain how KCL results from the law of conservation of charge.

2.26 (H)* Find the current, i, in the circuit shown in Figure P2.26.

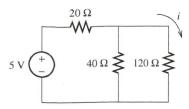

FIGURE P2.26   Circuit for Problem 2.26.

2.27 (H) Find I_x in this Figure P2.27.

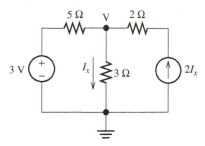

FIGURE P2.27   Circuit for Problem 2.27.

2.28 (H)* Use KCL to find the equivalent resistance between Node A and B, that is, $R_{AB} = V_{AB}/I_{AB}$.

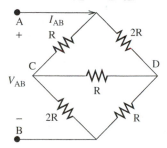

FIGURE P2.28   Circuit for Problem 2.28.

SECTION 2.6 KIRCHHOFF'S VOLTAGE LAW

2.29 (B) Find v_A in Figure P2.29:

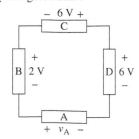

FIGURE P2.29   Circuit for Problem 2.29.

2.30 (B)* (Application) An automobile cooling fan is connected to the car's 12 V battery. The fan motor can be modeled as two voltage sources in series, a negative back EMF voltage, and a positive armature voltage, as shown in Figure P2.30.

Find the current through the line resistance, if the line resistance is 20 Ω.

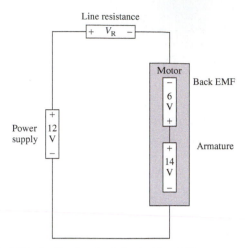

FIGURE P2.30   Cooling fan circuit for Problem 2.30.

2.31 (A) Find v_C and v_D in Figure P2.31:

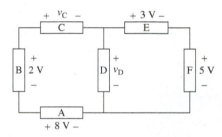

FIGURE P2.31   Circuit for Problem 2.31.

2.32 (A) Find v_C and v_E in Figure P2.32:

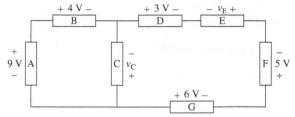

FIGURE P2.32   Circuit for Problem 2.32.

2.33 (A) Find v_C, v_D, and v_E in Figure P2.33:

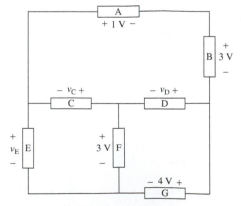

FIGURE P2.33   Circuit for Problem 2.33.

2.34 (H) Find v_A, v_B, v_D, and v_E in Figure P2.34:

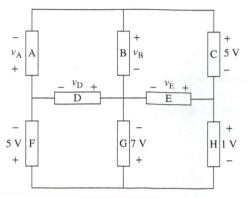

FIGURE P2.34   Circuit for Problem 2.34.

2.35 (H) Find V_C, V_D, V_H, and V_I in Figure P2.35:

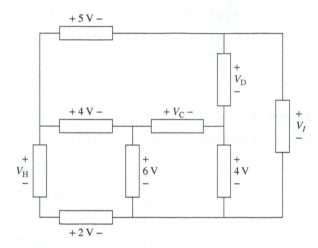

FIGURE P2.35   Circuit for Problem 2.35.

2.36 (H)* Shown in Figure P2.36 is a non-series-parallel connection known as a *Wheatstone bridge circuit*.
   a. Given that $R_1 = 6\ \Omega$, $R_2 = 3\ \Omega$, find i.
   b. When the current $I = 0$, we say the bridge is *balanced*. Under what condition (find an expression relating R_1 and R_2) will this bridge be balanced?

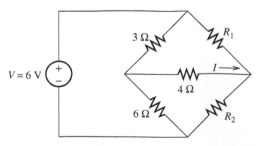

FIGURE P2.36   Bridge circuit.

2.37 (H)* Use KVL to find the equivalent resistance between Nodes A and B, $R_{AB} = V_{AB}/I_{AB}$.

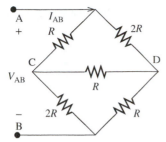

FIGURE P2.37   Circuit for Problem 2.37.

SECTION 2.7 OHM'S LAW AND RESISTORS

2.38 (B) Find V in the circuit of Figure P2.38.

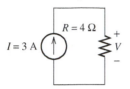

FIGURE P2.38   Circuit for Problem 2.38.

2.39 (B) Find R in the circuit shown in Figure P2.39.

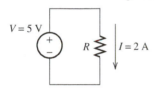

FIGURE P2.39   Circuit for Problem 2.39.

2.40 (B) Find I in the circuit shown in Figure P2.40.

FIGURE P2.40   Circuit for Problem 2.40.

2.41 (A)* A 50-cm-long copper wire with a cross-sectional area $1.2286 \times 10^{-10}\ \text{m}^2$ is used to connect a voltage source and a device. What is the maximum voltage if the current is required to be not bigger than 1 A?

2.42 (A) (Application Problem)

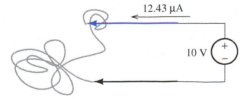

FIGURE P2.42   Circuit for Problem 2.42.

Surgeons remove a cylindrical-shaped piece of tissue of unknown length from a patient. They find that the piece of tissue has a length of 15.94 m and a cross-sectional area of $3.468 \times 10^{-3}\ \text{m}^2$. When a voltage of 10 V is applied

across the entire length of the tissue as shown in Figure P2.42, a current of 12.43 μA flows through it. Find the resistivity of the tissue.

2.43 (H) Find an expression for i in Figure P2.43.

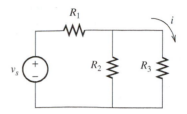

FIGURE P2.43   Circuit for Problem 2.43.

2.44 (H) Find V_A, V_B, V_C, and V_S in Figure P2.44.

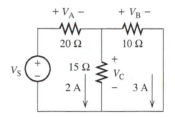

FIGURE P2.44   Circuit for Problem 2.44.

2.45 (A)* (Application Problem)

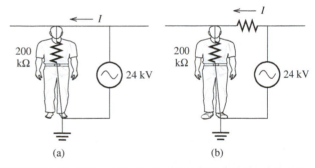

FIGURE P2.45   (a) Circuit for Problem 2.44. (b) Safety shoe included.

A careless utility worker climbs an aluminum ladder and comes into contact with an overhead power line bearing 24 kV, as shown in Figure P2.45a. His body resistance is 200 kΩ. How much current flows through his body? The worker wears a good safety shoe. The safety shoe can be modeled as a resistance that is located in series with the 240 V voltages (see Figure P2.45b). In this situation, the current drops to 40 mA. Calculate the resistance of the shoe.

2.46 (H) Find the equivalent resistance of the short circuit and the open circuit shown in Figure P2.46.

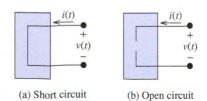

FIGURE P2.46   Circuit for Problem 2.46.

2.47 (H) Application: temperature meter

The circuit shown in Figure P2.47 is a bridge circuit. R_S is a resistive temperature sensor that varies with temperature. This circuit can be used as a thermometer by measuring the voltage V_{out}. To enable a high sensitivity in reading the temperature, the circuit should be designed to ensure that a small change in R_S would lead to a large change in V_{out}.

Determine the conditions for which the output voltage changes the most for a given change in the resistor R_S, that is, the derivative of V_{out} with respect to R_S, (dV_{out}/dR_S) would be maximum. (*Hint: Properly set R_C.*)

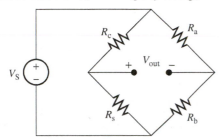

FIGURE P2.47   Circuit for Problem 2.47.

2.48 (H)* A thermistor is a temperature-dependent resistor. When the temperature changes, the resistance of the thermistor changes in a predictable way. The relationship of the absolute temperature and the thermistor's resistance is:

$$1/T = A + B \ln(R) + C \, [\ln(R)]^3$$

Here are some data points for a typical thermistor.

T (K)	R (Ω)
273	16,330
298	5000
323	1801

Find the coefficients A, B, and C.

SECTION 2.8 POWER AND ENERGY

2.49 (A) Identify the passive and active devices in Figure P2.49, and compute the power absorbed or supplied.

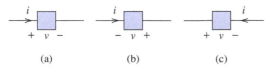

FIGURE P2.49   Circuit for Problem 2.49.

    a. $v = -3$ V, $i = 1$ A
    b. $v = -2$ V, $i = 2$ A
    c. $v = 2$ V, $i = 3$ A

2.50 (H) Repeat Problem 2.49 with the following new assumptions.
    a. $v = 20$ V, $i = 2.5e^{-2t}$ mA
    b. $v = 20$ V, $i = 2 \sin t$ mA

2.51 (A) Find the value of the resistance, R, in Figure P2.51.

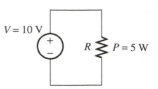

FIGURE P2.51   Circuit for Problem 2.51.

2.52 (A)* Find the power, P, dissipated by the 20-Ω resistor in Figure P2.52.

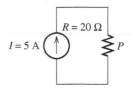

FIGURE P2.52   Circuit for Problem 2.52.

2.53 (H) Find the power P_1, P_2, and P_3 dissipated by all resistors in the circuit shown in Figure P2.53.

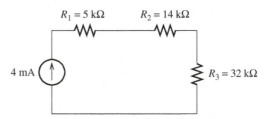

FIGURE P2.53   Circuit for Problem 2.53.

2.54 (H) Find v in the circuit shown in Figure P2.54.

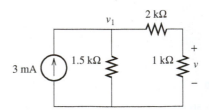

FIGURE P2.54   Circuit for Problem 2.54.

2.55 (H) Find the power dissipated in all the resistors of Figure P2.54.

2.56 (A) Show that the power absorbed in each resistor is a non-negative number.

2.57 (H) The voltage and the current across an electrical device are shown in Figure P2.57. The voltage and the current are associated in their direction. Draw the waveform of the power absorbed by the device and compute the energy consumed during the time interval [0, 2].

2.58 (A)* For the circuit shown in Figure P2.58:
    a. Find the power provided by the −5 V voltage source.
    b. In order to zero the power mentioned earlier, what will the current of the 4 A current source be changed to?

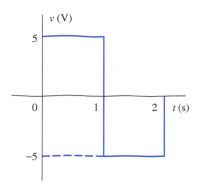

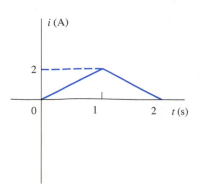

FIGURE P2.57 Voltage and current waveforms.

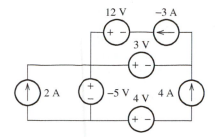

FIGURE P2.58 Circuit for Problem 2.58.

SECTION 2.9 INDEPENDENT AND DEPENDENT SOURCES

2.59 (B) Plot voltage versus current curves for ideal voltage source and current source.

2.60 (B) In Figure P2.60, what is the value that *i* can take?

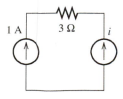

FIGURE P2.60 Circuit for Problem 2.60.

2.61 (A) In Figure P2.61, compute the voltage across the 3Ω resistor, the voltage across the current source, and the power of the current source.

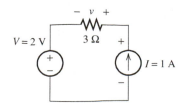

FIGURE P2.61 Circuit for Problem 2.61.

2.62 (A)* Find V_x in the circuit shown in Figure P2.62:

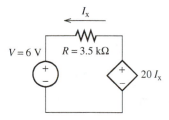

FIGURE P2.62 Circuit for Problem 2.62.

2.63 (A) Find the value of I_x in Figure P2.63.

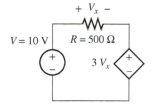

FIGURE P2.63 Circuit for Problem P2.63.

2.64 (A) In the circuit given in Figure P2.64, find the current of the dependent source.

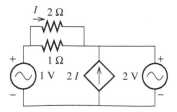

FIGURE P2.64 Circuit for Problem 2.64.

2.65 (H) Find the value of v_o with respect to v_S in Figure P2.65.

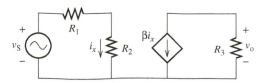

FIGURE P2.65 Circuit for Problem 2.65.

2.66 (H)* A current controlled current source (CCCS) circuit is given in Figure P2.66.

Find v_S and the power supplied by the dependent source.

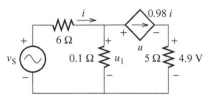

FIGURE P2.66   Circuit for Problem 2.66.

SECTION 2.10 ANALYSIS OF CIRCUITS USING PSPICE

2.67 (B) Use PSpice and find the current, i, that shown in Figure P2.66.

2.68 (A) In Figure P2.68, the current source A is 10 A, and resistors R_1, R_2, and R_3 are equal to 4 Ω, 3 Ω, and 2 Ω, respectively. Use PSpice and find the current i_2 and i_3. In addition, assume we replace the current source with a voltage source. Assuming i_1, i_2, and i_3 remain unchanged, find the magnitude of the voltage source.

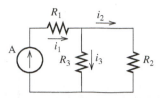

FIGURE P2.68   Circuit diagram for PSpice Problem 2.68.

2.69 (H) Given the circuit shown in Figure P2.69, assume a 10 V voltage source. Find the values of the resistor R_6 and the current i_5, assuming: $i_2 = 1$ A, $R_1 = R_5 = 2$ Ω, $R_2 = 4$ Ω, $R_3 = R_4 = 1$ Ω, $R_7 = 3$ Ω. Verify the answer through PSpice software using the value of R_6 obtained from calculation.

2.70 (H) Revise the PSpice Problem 2.70 by adding two new voltage sources v_2 and v_3 to the circuit as shown in Figure P2.70. Also, a new resistor R_8 has been added. Also, note that the values of the resistors R_1 through R_8 have been changed. Use PSpice to find the current at i_4 and i_6 with the resistance of resistors are given as below: $R_1 = 3$ Ω, $R_2 = R_3 = R_7 = 2$ Ω, $R_4 = 1$ Ω, $R_5 = 20$ Ω, $R_6 = 1.6$ Ω, and $R_8 = 8$ Ω.

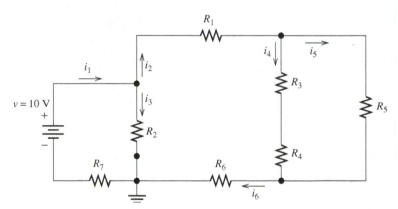

FIGURE P2.69   Circuit diagram for Problem 2.69.

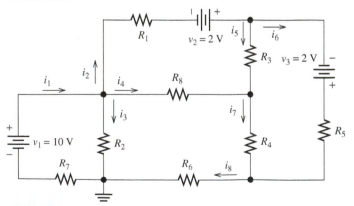

FIGURE P2.70   Circuit diagram for PSpice Problem 2.70.

Resistive Circuits

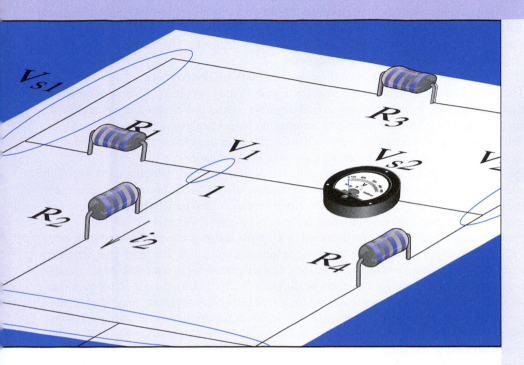

3.1 INTRODUCTION

Chapter 2 focused on the main variables of electric circuits: voltage, current, and power. Three elements of electric circuits were introduced: the voltage source, the current source, and resistors. This chapter outlines how to analyze complex electric circuits and calculate the voltage and current of circuit elements. This chapter focuses on circuits that consist of resistors and independent or dependent current and voltage sources. The strategies outlined here will enable you to find unknown current, voltage, power, and resistance values in terms of known circuit values. These techniques are applicable to any electric circuit, regardless of its complexity.

A resistive circuit consists of only resistors and sources. For example, the circuits shown in Figure 3.1 are resistive. The tools and techniques used for analyzing resistive circuits include Kirchhoff's voltage laws (KVL), Kirchhoff's current law (KCL), and Ohm's law, which were introduced in Chapter 2. To analyze a resistive circuit network with a large number of sources, resistors, nodes, and branches may seem extremely difficult; however, by applying the techniques outlined in this

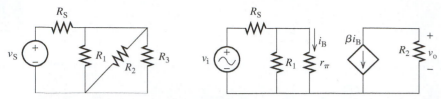

FIGURE 3.1 Resistive circuits.

chapter, you will learn to decrease the number of nodes and branches of a resistive circuit and therefore ease the process of resistive circuit analysis.

The first section of this chapter introduces *series* and *parallel* arrangements of resistors. In either case, an equivalent resistor can be developed. A network of resistors is usually a combination of parallel and series resistors. Therefore, by using an equivalent resistor for parallel and series configurations, several resistors in a network can be replaced by a single one. This technique reduces the complexity of a given circuit, simplifying the analysis.

This chapter also examines how voltages and currents are divided between the series and parallel resistors. Finding an equivalent circuit for a network may not always be easy or possible. This chapter outlines the techniques used to find unknown values in a circuit that are applicable to all circuits, regardless of their structural complexity. Nodal analysis will be discussed as a good tool for this goal. Nodal analysis applies Kirchhoff's rules to different nodes and loops of a circuit and finds unknown values by solving a series of linear equations.

Finally, this chapter summarizes Thévenin and Norton methods which are other useful techniques for finding an equivalent circuit for a resistive circuit in the presence of dependent or independent sources. Each of the techniques listed here are based on Kirchhoff's current and voltage laws. This chapter explains how these basic rules are applied to analyze circuits.

Many electric circuits can be modeled using a simple resistive circuit. For example, consider a 120-V power supply connected to a lamp, a fan, or a computer. The lamp and computer draw specific currents. Thus, a resistor can be used as a simple model to represent these circuits. Resistive circuit modeling allows engineers in different disciplines to use the techniques presented in this chapter to analyze systems. For instance, mechanical engineers may need to analyze resistive circuits to understand automotive system power requirements and ignition systems. Civil engineers may need to analyze resistive circuits formed by resistive strain gages. Resistive circuits are also vital to the study of building wiring, electric shock, and various other applications as detailed throughout the many applied examples in this chapter.

This chapter also offers examples to explain how circuit currents and voltages can be computed and analyzed using PSpice.

3.2 RESISTORS IN PARALLEL AND SERIES AND EQUIVALENT RESISTANCE

As discussed in Table 2.2, different notations for current and voltage might be used. In this chapter, lower case i/v (that is a general notation for current/voltage) and upper case I/V (that is used for DC current and voltage) are used interchangeably. The series elements in a circuit refer to those arranged in a chain. In other words, if any two branches share only one node, the elements are said to be arranged *in series*. Consider the two series elements in Figure 3.2. If KCL is applied to node a:

$$i_1 - i_2 = 0 \implies i_1 = i_2$$

As a result, currents that flow through two series elements are equal. Current flowing through two series elements has only one path to take. In practice, there might be more than two series elements in a circuit (see Figure 3.3 for example). For the current flowing through the series elements of Figure 3.3, by applying KCL to each node between the elements we arrive at the following equation:

$$i_1 = i_2 = i_3 = i_4 = \cdots = i \tag{3.1}$$

As shown, the same current flows through all of the series elements. Next, because this chapter studies resistive circuits, consider an example where the circuit element in Figure 3.3 is replaced

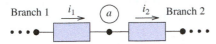

FIGURE 3.2 Two series elements.

FIGURE 3.3 A circuit with several series elements.

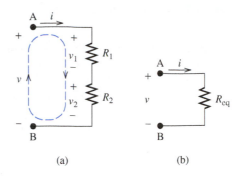

(a)   (b)

FIGURE 3.4   (a) Two series resistors. (b) The equivalent resistor.

by an ideal resistor. In Figure 3.4(a), a voltage, v, is applied to two series resistors. Applying KVL to the loop specified in Figure 3.4(a):

$$-v + v_1 + v_2 = 0 \tag{3.2}$$

Using Ohm's law for the two resistors and knowing that the currents passing through the two series elements are equal, that is, $i_1 = i_2 = i$, results in:

$$v_1 = iR_1 \tag{3.3}$$

In addition:

$$v_2 = iR_2 \tag{3.4}$$

Substituting Equations (3.3) and (3.4) into (3.2) yields:

$$v = v_1 + v_2 \Rightarrow v = iR_1 + iR_2 \Rightarrow v = i(R_1 + R_2)$$

Thus, the equivalent resistance through the terminals corresponds to:

$$R_{eq} = \frac{v}{i} = R_1 + R_2$$

Therefore, if the voltage, V, is applied to the equivalent resistor, R_{eq}, there will be no change in the current flowing through the circuit. The same approach can be applied to situations where multiple resistors are in series. In general, the equivalent resistance of N resistors in series is:

$$R_{eq} = \sum_{n=1}^{N} R_n \tag{3.5}$$

The current flowing through series resistors in a circuit is the same as the current flowing through their equivalent resistor. Therefore, the equivalent resistance (instead of series elements) can be used to analyze the circuit.

The other well-known configuration for circuit elements is the *parallel connection*. When two elements in a circuit are in parallel, their terminals are connected to each other. Parallel elements are arranged with their two sides (called *heads* and *tails*) connected together. In general, the voltages across two parallel elements are the same. Figure 3.5 shows two parallel elements.

Applying KVL to loop a in Figure 3.5 results in:

$$-v_1 + v_2 = 0 \Rightarrow v_1 = v_2$$

As shown above, the voltage across the two parallel elements is the same. In general, N circuit elements will be in parallel if they are connected at each end to the same point with no circuit elements in between (see Figure 3.6). KVL can be applied to each of the N loops shown in Figure 3.6. The voltage across all parallel elements is equal, that is, $v = v_1 = v_2 = v_3 = v_4 = \cdots$

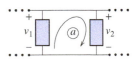

FIGURE 3.5 Two parallel elements.

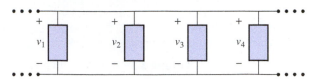

FIGURE 3.6 Parallel elements.

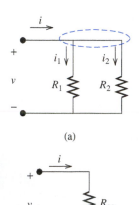

(a)

(b)

FIGURE 3.7   (a) Two
parallel resistors.
(b) The equivalent
resistor.

Next, consider the situation where resistors are the circuit elements. Figure 3.7(a) shows two parallel resistors. A voltage, v, is applied to the two parallel resistors. The current of the circuit is broken up at the initial point of the two branches. Specifically, the current at the beginning of the source is divided between the two resistors. The point at which the two parallel resistors are connected possesses the same potential or voltage. This point, called a *node*, is represented by the dashed line in Figure 3.7(a). If KCL is applied to the node specified in the figure:

$$i - i_1 - i_2 = 0 \Rightarrow i = i_1 + i_2 \qquad (3.6)$$

Next, Ohm's law can be used to define the currents of resistors in terms of their voltages and resistances. Because the voltages across parallel elements are the same:

$$i_1 = \frac{v}{R_1} \quad i_2 = \frac{v}{R_2} \qquad (3.7)$$

If the equivalent resistor, R_{eq}, is attached to the voltage, V [Figure 3.7(b)], the same current is expected to flow through the circuit. This can be expressed as:

$$R_{eq} = \frac{v}{i} \Rightarrow i = \frac{v}{R_{eq}} \qquad (3.8)$$

Now, substituting i_1 and i_2 in Equations (3.7) and (3.8) into (3.6):

$$i = i_1 + i_2 \Rightarrow \frac{v}{R_{eq}} = \frac{v}{R_1} + \frac{v}{R_2}$$

Therefore:

$$\frac{1}{R_{eq}} = \frac{1}{R_1} + \frac{1}{R_2} \qquad (3.9)$$

Sometimes, the notation of two parallel lines ∥ is used to show that two resistors are in parallel:

$$R_{eq} = R_1 \| R_2 = \frac{R_1 R_2}{R_1 + R_2}$$

If R_{eq} is used instead of the two parallel resistors in a circuit, there will be no change in the voltage or the current of the circuit. Using the same approach, it is possible to find an equivalent resistance for N parallel resistors:

$$R_{eq} = \frac{1}{\dfrac{1}{R_1} + \dfrac{1}{R_2} + \cdots \dfrac{1}{R_N}} \qquad (3.10)$$

Note that if $R_1 = R_2 = \cdots = R_N = R$, then:

$$R_{eq} = \frac{R}{N} \qquad (3.11)$$

Sometimes, the process of circuit analysis can be simplified by reducing the number of circuit resistors. This can be done by using an *equivalent resistor*. Replacing several resistors in a resistive network with an equivalent resistor simplifies analysis but does not change the voltage or the current of the circuit.

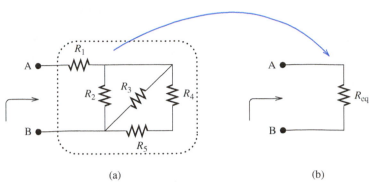

(a)   (b)

FIGURE 3.8 The equivalent resistor for five resistors in a circuit.

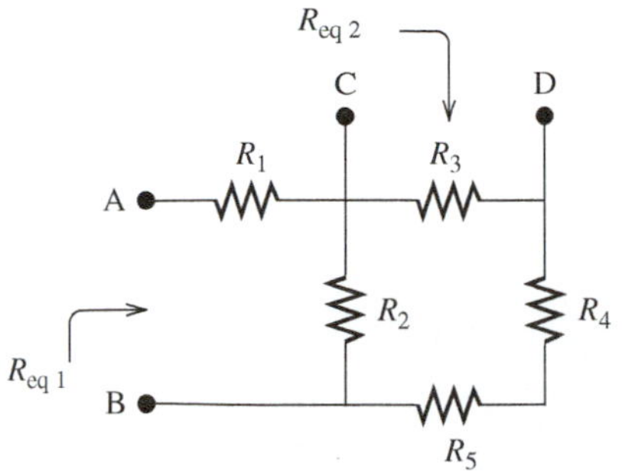

FIGURE 3.9 The equivalent resistors observed from two different locations in a circuit are not necessarily equal.

For example, in Figure 3.8, resistors R_1–R_5 can be replaced with the equivalent resistor R_{eq}. Now, by applying the same voltage across the two terminals A and B, the same current will flow in those terminals of the circuit.

> Therefore, the equivalent resistance observed across any two points in a circuit network can be found by applying a voltage across the two points, measuring the current flowing through the circuit, and computing the voltage to current ratio, which is the equivalent resistance.

Note that in general the equivalent resistances observed from two different locations in a resistive network are not the same. For instance, in Figure 3.9, R_{eq1} and R_{eq2} are not necessarily equal because they are seen through two different sets of terminals A–B and C–D.

Resistive circuits are usually a combination of parallel and series resistors. The equivalent resistance of a group of resistors can be found, step by step, by finding the equivalent resistances for series and parallel configurations. This approach of finding the equivalent resistance is illustrated in Figure 3.10.

Figure 3.10 shows the step-by-step process to find the equivalent resistance of the circuit in Figure 3.8(a). Each step shows a portion of the circuit being replaced by the equivalents of series and parallel resistor arrangements. First, observe that the resistors R_4 and R_5 are in series. Thus, they can be replaced with their equivalent resistance:

$$R_{54} = R_5 + R_4$$

Now, the resistance R_{54} is in parallel to the resistance R_3 (i.e., $R_{543} = R_3 \| R_{54}$), and the resistance:

$$R_{543} = R_3 \| R_{54} = \frac{R_{54}R_3}{R_{54} + R_3}$$

is equivalent. In the next step, R_{543} and i_2 are parallel and their equivalent resistance is:

$$R_{5432} = R_{543} \| R_2 = \frac{R_{543}R_2}{R_{543} + R_2}$$

Finally, R_{5432} and R_1 are in series; therefore, their equivalent resistance is $R_{eq} = R_{54321} = R_{5432} + R_1$. This step-by-step process is reviewed in Figure 3.10.

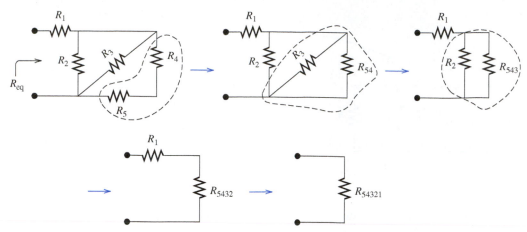

FIGURE 3.10 Step-by-step process to find an equivalent resistance.

EXAMPLE 3.1 **Resistor Tolerance**

In practice, the resistance of a given resistor varies within a given range. The accuracy of the stated value of a resistor is described as *resistor tolerance*. Resistor tolerance is depicted by colored strips on the resistor. Figure 3.11(a) represents different types of resistors. In the circuit shown in Figure 3.11(b), the resistances are given with a tolerance, $R_1 = 2.5$ k$\Omega \pm 10\%$ and $R_2 = 7.5$ k$\Omega \pm 5\%$. Find the nominal, minimum, and maximum equivalent resistances.

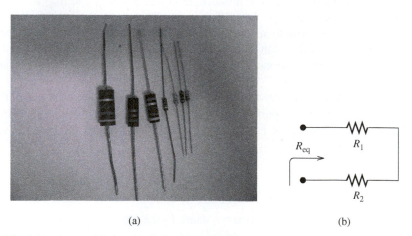

(a)   (b)

FIGURE 3.11 (a) Resistors; (b) the circuit for Example 3.1.

SOLUTION

The minimum and maximum values of R_1 and R_2 are:

$$R_1(\text{min}) = 2.25\,\text{k}\Omega \quad R_1(\text{max}) = 2.75\,\text{k}\Omega$$
$$R_2(\text{min}) = 7.125\,\text{k}\Omega \quad R_2(\text{max}) = 7.875\,\text{k}\Omega$$

The two resistors are in series; therefore, their equivalent resistances are as follows:

$$R_{eq}(\text{nominal}) = R_1 + R_2 = 10.0\,\text{k}\Omega$$
$$R_{eq}(\text{min}) = R_1(\text{min}) + R_2(\text{min}) = 9.38\,\text{k}\Omega$$
$$R_{eq}(\text{max}) = R_1(\text{max}) + R_2(\text{max}) = 10.63\,\text{k}\Omega$$

EXAMPLE 3.2   **Series and Parallel Resistors**

Which resistors in Figure 3.12 are in series, and which are in parallel?

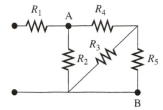

FIGURE 3.12   The circuit for Example 3.2.

SOLUTION

From Figure 3.12, it is clear that R_3 and R_5 are in parallel. The combination of these two is in series with R_4. R_2 is in parallel to the combination of R_3, R_4, and R_5. Finally, R_1 is in series with R_2–R_5.

EXCERCISE 3.1

In Example 3.2, if another resistor, R_6, is connected between nodes A and B, determine which resistors will be in parallel and which will be in series with others. Note: It may be that in some situations resistors are not in parallel and/or series with each other.

EXAMPLE 3.3   **Equivalent Resistance**

Find the equivalent resistance in Figure 3.13.

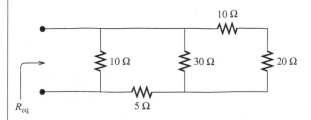

FIGURE 3.13   The circuit for Example 3.3.

SOLUTION

The equivalent resistance can be found using the step-by-step process of replacing the series and parallel resistor configurations with their equivalents. The steps are shown in Figure 3.14.

(continued)

| EXAMPLE 3.3 | Continued |

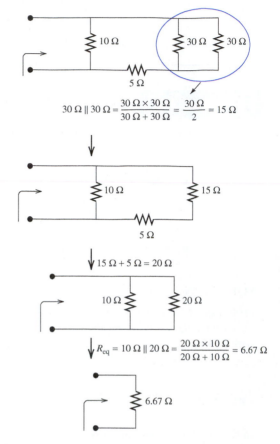

FIGURE 3.14 Steps for finding the equivalent resistance for Example 3.3.

Thus, the equivalent resistance is 6.67 Ω.

APPLICATION EXAMPLE 3.4   Is Building Wiring Parallel or Series?

Figure 3.15 shows a basic electric circuit with a power supply driving a single load. Two configurations are provided for study, series (Figure 3.16) and parallel (Figure 3.17). Assuming that $R_1 = 10\ \Omega$, $R_2 = 20\ \Omega$, and $R_3 = 30\ \Omega$:

a. Using the circuit in Figure 3.15, find V_1 and I_1.
b. For the circuit in Figure 3.16, find V_1, I_1, V_2, I_2, V_3, and I_3.
c. For the circuit in Figure 3.17, find V_1, I_1, V_2, I_2, V_3, and I_3.

Based on the results of this example, assume that in a building multiple devices, such as a computer, a television, and a lamp, are connected to the same circuit. Should these devices be connected in series or in parallel? Why?

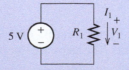

FIGURE 3.15 The basic electric circuit for Example 3.4.

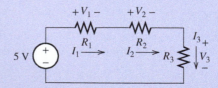

FIGURE 3.16 The circuit wired in series.

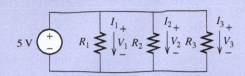

FIGURE 3.17 The circuit wired in parallel.

SOLUTION

(a)	$I_1 = 500$ mA	$V_1 = 5.000$ V	
(b)	$I_1 = 83.3$ mA	$V_1 = 0.833$ V	
	$I_2 = 83.3$ mA	$V_2 = 1.667$ V	
	$I_3 = 83.3$ mA	$V_3 = 2.500$ V	
(c)	$I_1 = 500$ mA	$V_1 = 5.000$ V	
	$I_2 = 250$ mA	$V_2 = 5.000$ V	
	$I_3 = 167$ mA	$V_3 = 5.000$ V	

When the circuit is wired in series, the current through and voltage over the elements changes as elements are added or removed. This poses several problems: (a) to use any single piece of equipment, all equipment must be turned on; (b) equipment must be designed to be powered by widely varying voltage levels; and (c) if an element burns out (i.e., is open circuited) all other elements will turn off. However, if the circuit is wired in parallel, the voltage across and the current through each element does not change as elements are added or removed. Thus, building wires are generally connected in parallel.

APPLICATION EXAMPLE 3.5   Shock and Water

Suppose that on a very rainy day a careless bald worker, wearing a hat and shoes, retrieves a dry wooden ladder from the garage. He climbs the ladder and begins working. Without warning, the old wooden ladder breaks and collapses below him as shown in Figure 3.18. Instinctively, he grabs the power line with his bare, wet hands. He grabs the soaked wooden power pole with his other hand, and feels a sharp tingle.

    Due to the water on his hands and legs, his bodily resistance is lowered to 200 kΩ. The resistance of the wet telephone pole is 600 kΩ. The maximum power line voltage is 30 kV. What is the maximum current flow through his body?

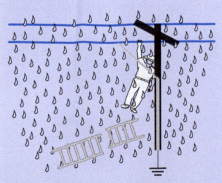

FIGURE 3.18 Application problem: shock and water.

(continued)

APPLICATION EXAMPLE 3.5   **Continued**

SOLUTION

The situation shown in Figure 3.18 can be modeled as a series circuit as shown in Figure 3.19. Because it is a series circuit, current I_{max} is the same at all points in the circuit. Therefore, in order to find the current through the worker's body, compute the total current in the circuit. In order to do this, first compute the equivalent resistance of the circuit. For this simple series circuit, the equivalent resistance is 200 kΩ + 600 kΩ = 800 kΩ. Now, using Ohm's law, the maximum current is:

$$I_{max} = \frac{30 \text{ kV}}{800 \text{ k}\Omega} = 37.5 \text{ mA}$$

While this is not a fatal level of current, the worker may still fall hard to the ground.

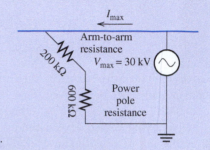

FIGURE 3.19   The circuit model for Example 3.5.

APPLICATION EXAMPLE 3.6   **Solar Cells**

The sunlight can be directly converted to electricity using solar cells. Solar cells are also called photovoltaic (PV) cells. Solar cells were first developed in the 1950s for the U.S. space satellites. Recently, they have become more widely used in daily life. As shown in Figure 3.20, solar cells can be used in homes for lights and appliances. The energy captured by these sources can be stored in batteries to power a lamp or an emergency roadside wireless telephone when no telephone wires are around (see Figure 3.20). They can also be used in vehicles to directly power electric motors.

Solar cells are usually arranged in a PV module and modules are wired together in a PV array to produce the required output power, as shown in Figure 3.21(a). The current produced by a solar cell is proportional to the solar illumination level and the produced voltage is almost invariable to the solar illumination level.

Assume that there are two panels of 10 cells each. One panel has 10 cells connected in series and the other has parallel connection. Each cell creates 0.50 V and 1 A; thus, the output power of each cell is 0.5 W. Because each panel consists of 10 cells, the total power output is 5.0 W (10 × 0.50 × 1).

 a. If a single cell is completely shaded in both panels, what is the power output of each panel?
 b. Which type of cell connection [series-connected and parallel-connected; see Figure 3.21(b)] is more robust ?

SOLUTION

 a. In a series-connected panel, if one cell is completely shaded, the current produced by this cell drops to zero. Therefore, the output current will be limited by the current of the shaded cell and it also drops to zero. In this case, the output power is:

$$P_s = V_{out} \times I_{out} = 0$$

(a)   (b)

FIGURE 3.20 Sample applications of solar cells. (Used with permission from © Martin Shields/Alamy and (b) Sebastien Burel/Shutterstock.com.)

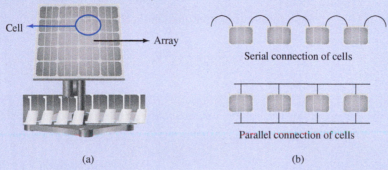

(a)   (b)

FIGURE 3.21 Solar cell module and array.

In the parallel-connected panels, if the current of one cell drops to zero, then the output current becomes:

$$I_{out} = 9 \times 1\,A = 9\,A$$

and the output voltage remains. Therefore the power output is:

$$P_{out} = V_{out}I_{out} = 0.5\,V \times 9\,A = 4.5\,W$$

b. From the computation shown in part (a), we can see that the parallel-connected panel is more robust. In a series network of solar cells, no output power is created when any single cell fails; however, in a parallel network, the failure of one cell only results in a partial decrease in power.

In practice, cells are usually put in series to meet minimum system voltage requirements, then multiple strings of series-connected cells are paralleled to attain the required system current.

3.3 VOLTAGE AND CURRENT DIVISION/DIVIDER RULES

3.3.1 Voltage Division

In order to determine the current and voltage of the parallel- and series-configured resistors in a circuit, it must first be determined how the voltage or current is divided across series and parallel resistors. The current passing through the two series resistors in Figure 3.22 is the same but the voltage across each of them is a fraction of the source voltage, V_s.

Applying KVL to the loop shown in the Figure 3.22:

$$-V_s + V_1 + V_2 = 0 \tag{3.12}$$

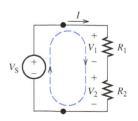

FIGURE 3.22 Voltage divider.

Next, Ohm's law can be used to maintain the relationship of current in the loop and voltage across the two resistors. Because $V_1 = R_1 \cdot I$ and $V_2 = R_2 \cdot I$:

$$I = \frac{V_1}{R_1} = \frac{V_2}{R_2}$$

This equation can be used to maintain the relationship of the two voltages V_1, and V_2, which is:

$$V_2 = \frac{R_2}{R_1}V_1 \qquad (3.13)$$

Substituting V_2 from Equation (3.13) into Equation (3.12):

$$V_1 + \frac{R_2}{R_1}V_1 = V_s$$

Factorizing V_1:

$$V_1 = \frac{R_1}{R_1 + R_2}V_s \qquad (3.14)$$

Equivalently, if $V_1 = \frac{R_1}{R_2}V_2$ is substituted into Equation (3.12):

$$V_2 = \frac{R_2}{R_1 + R_2}V_s \qquad (3.15)$$

Similarly, the voltage across each of the N series-configured resistors is obtained using:

$$V_k = \frac{R_k}{R_1 + R_2 + \cdots + R_N}V_s; \quad k = 1, 2, \ldots N \qquad (3.16)$$

Therefore, the voltage of each resistor in a series combination is a fraction of the total voltage. This fraction is the ratio of the resistor's individual resistance to the total series resistance. Larger resistance leads to larger voltage drop. Equation (3.16) is called *voltage division rule*.

EXCERCISE 3.2

Can we use voltage division rule depicted in Equation (3.16) to find the value of an unknown resistance? How?

Do you feel that this method of measuring an unknown resistance involves measurement errors?

Note that a voltage source such as a battery usually has an internal resistance. This resistance increases as the battery ages. Thus, at a given time, the resistance of a battery source is not constant. A better method for measuring the resistance of a resistor is Wheatstone bridge as discussed in Example 3.22.

APPLICATION EXAMPLE 3.7   Rocket Launch Setup

A simple rocket launch setup is shown in Figure 3.23. A controller with a voltage source of 24 V is used to control the switch and the current flow through the igniter. Switch 1 is opened to serve as a safety backup while the igniter is set up. The indicator serves as a high-resistance controller so that it lowers the current flow through the igniter to avoid launching rocket when switch 2 is opened and switch 1 is closed. When both switch 1 and switch 2 are closed, the

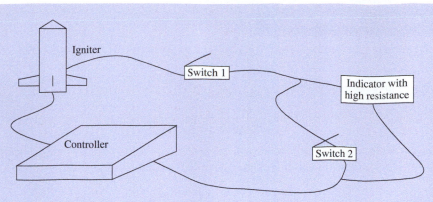

FIGURE 3.23   A simple rocket launch setup.

current flow through the igniter will become high enough to initiate the launch. Assume that a voltage source is in series with the controller, and it supplies 24 V. The resistances of the controller, igniter, and indicator are 500 Ω, 10 Ω, and 1 MΩ, respectively. Calculate the current flow through the igniter and the voltage across it when:

a. Switch 1 is closed and switch 2 is opened.
b. Both switch 1 and switch 2 are closed.

SOLUTION

The equivalent circuit for the rocket launch setup is shown in Figure 3.24.

a. If switch 1 is closed and switch 2 is opened, the total resistance is:

$$R_{total} = R_{controller} + R_{igniter} + R_{indicator}$$
$$R_{total} = 500 + 10 + 10^6 = 1000510 \ \Omega$$

The current flow through the igniter is:

$$I = \frac{V_{source}}{R_{total}}$$

$$I = \frac{24}{1,000,510} = 2.3988 \times 10^{-5} \ A$$

The voltage across the igniter is:

$$V_{igniter} = I \times R_{igniter} = 0.23988 \ mV$$

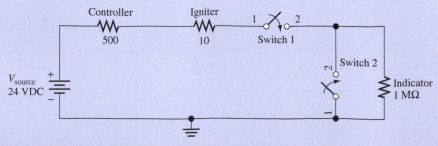

FIGURE 3.24   The equivalent circuit for the rocket launch setup.

(continued)

APPLICATION EXAMPLE 3.7   **Continued**

This problem can also be solved using voltage division rule of Equation (3.16):

$$V_{\text{igniter}} = \frac{R_{\text{igniter}}}{R_{\text{total}}} \times V_{\text{source}}$$

$$= \frac{10}{1{,}000{,}510} \times 24 = 0.23988 \text{ mV}$$

b. If both switch 1 and switch 2 are closed, a short circuit is created by switch 2. Therefore, the current does not flow through the indicator. The total resistance is:

$$R_{\text{total}} = R_{\text{controller}} + R_{\text{igniter}}$$
$$R_{\text{total}} = 500 + 10 = 510 \ \Omega$$

In this case, the current flow through the igniter is:

$$I = \frac{V_{\text{source}}}{R_{\text{total}}}$$

$$I = \frac{24}{510} = 0.0471 \text{ A}$$

The voltage across the igniter is:

$$V_{\text{igniter}} = I \times R_{\text{igniter}} = 0.471 \text{ V}$$

Again, using voltage division rule of Equation (3.16):

$$V_{\text{igniter}} = \frac{R_{\text{igniter}}}{R_{\text{total}}} \times V_{\text{source}}$$

$$= \frac{10}{510} \times 24 = 0.471 \text{ V}$$

3.3.2 Current Division

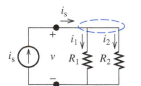

FIGURE 3.25 Current divider.

The current entering a configuration with parallel resistors is divided between the resistors. Therefore, a parallel resistor combination is also called a current divider. Consider Figure 3.25 as a current divider. Applying KCL:

$$i_s - i_1 - i_2 = 0 \tag{3.17}$$

Using Ohm's law:

$$v_1 = R_1 i_1 = R_2 i_2 \tag{3.18}$$

Equation (3.18) can be used to find the relationship of the currents in the two parallel branches:

$$i_2 = \frac{R_1}{R_2} i_1 \tag{3.19}$$

Substituting Equations (3.19) into (3.17):

$$i_1 + \frac{R_1}{R_2} i_1 = i_s$$

Factorizing i_1:

$$i_1 = \frac{R_2}{R_1 + R_2}i_s \qquad (3.20)$$

Similarly, using Equation (3.18) to write i_1 in terms of i_2 results in:

$$i_2 = \frac{R_1}{R_1 + R_2}i_s \qquad (3.21)$$

Equations (3.20) and (3.21) are *current division rules*.

EXERCISE 3.3

Prove Equation (3.21).

Each parallel resistor receives a fraction of the total current. For two resistors, this fraction equals the ratio between the other resistance in the circuit and the sum of the two resistances. In this case, a smaller amount of current flows through larger resistor. Recall that resistors resist the flow of current.

EXAMPLE 3.8   Voltage Division in Resistive Circuits

In Figure 3.26, find the resistor voltage drops for V_1, V_2, and V_3.

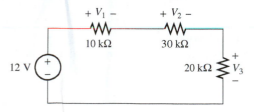

FIGURE 3.26   Figure for Example 3.8.

SOLUTION

According to the voltage division principle of Equation (3.16):

$$V_1 = 12\ \text{V} \times \frac{10\,\text{K}}{10\,\text{K} + 30\,\text{K} + 20\,\text{K}} = 2\ \text{V}$$

$$V_2 = 12\ \text{V} \times \frac{30\,\text{K}}{10\,\text{K} + 30\,\text{K} + 20\,\text{K}} = 6\ \text{V}$$

$$V_3 = 12\ \text{V} \times \frac{20\,\text{K}}{10\,\text{K} + 30\,\text{K} + 20\,\text{K}} = 4\ \text{V}$$

As expected from Kirchhoff's voltage law, the total resistance drop is equal to the applied voltage, that is, $2 + 4 + 6 = 12$ V.

EXAMPLE 3.9   Current Division in Resistive Circuits

Figure 3.27 is a current divider with a dependent source. Find I in this circuit.

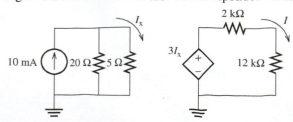

FIGURE 3.27   A current divider with a dependent source.

(continued)

EXAMPLE 3.9   **Continued**

SOLUTION

This circuit has a 10-mA independent current source and a dependent voltage source equal to $3I_x$. In the section of the circuit located in the right-hand side:

$$I = \frac{3I_x}{2\text{ k}\Omega + 12\text{ k}\Omega} = \frac{3I_x}{14\text{ k}\Omega}$$

To find I, first find I_x. According to the current division principle (see Equation (3.20)):

$$I_x = 10\text{ mA} \times \frac{20\,\Omega}{20\,\Omega + 5\,\Omega} = 8\text{ mA}$$

Now, replacing I_x with 8 mA:

$$I = \frac{3 \times 8\text{ mA}}{14\text{ k}\Omega} = 1.71\,\mu\text{A}$$

APPLICATION EXAMPLE 3.10   **Strain Gage**

A strain gage is a variable resistor whose resistance varies by the strain on the device to which it is attached. Strain gages can be used as sensors in "smart" buildings. Smart buildings are equipped with sensors that enable load specifications to be monitored and can inform authorities if there is any probability of building damage.

In this example, consider a strain gage attached to a beam (see Figure 3.28). This strain gage is connected to a computer that records the data over a given period of time. The computer has an equivalent resistance of 10 kΩ. The parameters of the strain gage are:

$R_{min} = 1.285$ kΩ
$R_{max} = 20.1$ kΩ
$I_{max} = 325\ \mu$A (maximum current the gage can handle)
$P_{max} = 2.9$ mW (maximum power the gage can dissipate).

This system can be modeled as shown in Figure 3.29. Given $I_{max} = 325\ \mu$A for the strain gage, determine if the current-limiting resistance meets the I_{max} and P_{max} requirements for the strain gage. If not, how can this problem be addressed?

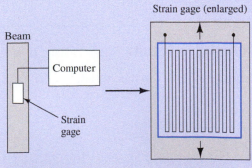

FIGURE 3.28 Strain gage.

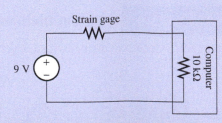

FIGURE 3.29 Equivalent circuit for Figure 3.28.

SOLUTION

No, the resistance does not meet the I_{max} and P_{max} requirements because at the minimum resistance, the current through the circuit is $9/(1.285\text{k}\Omega + 10\text{k}\Omega) = 797\ \mu\text{A}$, which is greater than I_{max} (325 μA).

In order to fix this problem, a resistor must be added in series with the gage to limit the current. Now, calculate the value of the required current-limiting resistor to ensure all of the strain gage specifications are met.

Because $I_{max} = 325\ \mu\text{A}$, the resistor of the circuit should be [see Figure 3.30(a)]:

$$\frac{9}{325\ \mu\text{A}} = 1.285\,\text{k}\Omega + R + 10\text{k}\Omega$$

$$\Rightarrow R = 16.4\,\text{k}\Omega$$

To allow for resistor tolerances, adjust its value to 17 kΩ as shown in Figure 3.30.

Including this resistor, the maximum voltage drop across the gage for the two situations of R_{max} and R_{min}, respectively, equals:

$$V_{R_{max}} = \frac{20.1\ \text{k}\Omega}{20.1\ \text{k}\Omega + 17\ \text{k}\Omega + 10\ \text{k}\Omega} \times 9\,\text{V} = 3.84\ \text{V}$$

$$V_{R_{min}} = \frac{1.285\ \text{k}\Omega}{1.285\ \text{k}\Omega + 17\ \text{k}\Omega + 10\ \text{k}\Omega} \times 9\,\text{V} = 0.409\ \text{V}$$

In addition, their corresponding currents are:

$$I_{R_{max}} = \frac{V_{R_{max}}}{20.1\,\text{k}\Omega} = 191\ \mu\text{A}$$

$$I_{R_{min}} = \frac{V_{R_{min}}}{1.285\,\text{k}\Omega} = 318\ \mu\text{A}$$

$$P_{max} = 9 \times 318 = 2.868\,\text{mW} < 2.9\,\text{mW}$$

Note that 2.9 mW is the maximum power allowable. Thus, this circuit with its current-limiting resistance meets the design requirements for the strain gage.

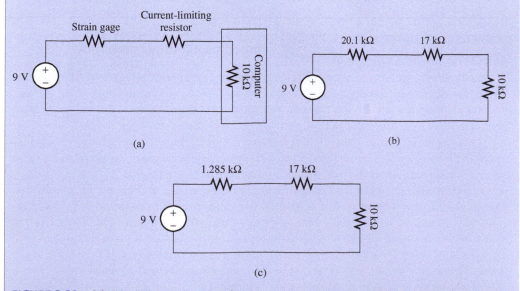

(a)

(b)

(c)

FIGURE 3.30   (a) Calculating the resistor that is needed to limit the current. (b) Equivalent circuit for R_{max}. (c) Equivalent circuit for R_{min}.

APPLICATION EXAMPLE 3.11   Automotive Power System

To design an automotive power supply appropriately, it is important to properly compute the amount of current that is required by its electrical system. The components of an automotive electrical system are connected to the power supply (battery–alternator combination) in parallel, as shown in Figure 3.31. Each additional component causes more current to be drawn from the supply. Find the total current i in Figure 3.31 given that $i_1 = 3$ A, $R_2 = 2\ \Omega$, $R_3 = 6\ \Omega$, $R_4 = 10\ \Omega$, and $i_5 = 1$ A.

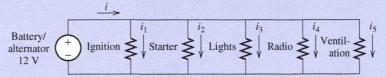

FIGURE 3.31   Circuit for Example 3.11.

SOLUTION

The voltage across all resistors is constant and is equal to 12 V; thus:

$$i_2 = \frac{12}{2} = 6\,\text{A} \quad i_3 = \frac{12}{6} = 2\,\text{A} \quad i_4 = \frac{12}{10} = 1.2\,\text{A}$$

Now, using KCL, and given the two known currents, i_1 and i_5:

$$-i + i_1 + i_2 + i_3 + i_4 + i_5 = 0$$

Replacing for the known currents results in:

$$-i + 3 + 6 + 2 + 1.2 + 1 = 0$$

$$i = 13.2\,\text{A}$$

EXERCISE 3.4

Why are the components of an automotive electric system connected in parallel?

APPLICATION EXAMPLE 3.12   Impedance Plethysmography

Biomedical engineers must measure biological features without invasive techniques. For example, measuring the volume of a heart chamber is not easily accomplished by conventional means, as it could require the removal of the heart from the body. Other less invasive techniques are therefore preferred.

One technique to measure the left ventricular (LV) volume is impedance plethysmography. This involves measuring the electrical resistance (impedance) of the blood in the heart chamber and using formulas to convert this measurement into a volume.

Electrodes are implanted along the length of the ventricle at an evenly spaced distances. A known voltage is applied across the two outer electrodes as shown in Figure 3.32. The current and the voltages between each pair of electrodes are measured. From this, the cross-sectional area of the cylinder between each electrode pair (as shown in the figure) can be calculated using the formula: $R = \rho L/A$, where ρ is the resistivity, L is the length of the material, and A is its cross-sectional area. Then, the volume of each cylinder can be easily found and added to get an approximate total volume.

Suppose five electrodes are inserted into the left ventricle with a spacing of 1 cm. $\rho = 150$ mΩ. The resistance between the electrodes A and E is found to be 7.5 kΩ. With

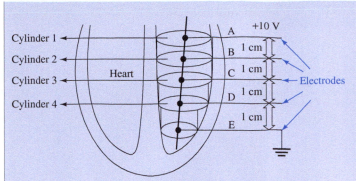

FIGURE 3.32 Impedance plethysmography.

a voltage of 10 V applied between A and E, the electrode voltages (with respect to the ground) are:

Electrode	Voltage
Electrode A	$V_A = 10$ V
Electrode B	$V_B = 8$ V
Electrode C	$V_C = 6.2$ V
Electrode D	$V_D = 3.8$ V
Electrode E	$V_E = 0$ V

Find the resistance of each cylinder, and from this, find the cross-sectional area and volume for each cylinder.

SOLUTION

This system can be modeled as the circuit shown in Figure 3.33. The resistance voltages can be calculated to be $V_{AB} = V_A - V_B = 2$ V, $V_{BC} = V_B - V_C = 1.8$ V, $V_{CD} = V_C - V_D = 2.4$ V and $V_{DE} = V_D - V_E = 3.8$ V. Using these values, the resistances can be calculated:

$$V_{AB} = 2\,\text{V} = \frac{10\,\text{V} \times R_1}{7.5\,\text{k}\Omega} \rightarrow R_1 = 1.5\text{k}\Omega$$

$$V_{BC} = 1.8\,\text{V} = \frac{10\,\text{V} \times R_2}{7.5\,\text{k}\Omega} \rightarrow R_2 = 1.35\,\text{k}\Omega$$

$$V_{CD} = 2.4\,\text{V} = \frac{10\,\text{V} \times R_3}{7.5\,\text{k}\Omega} \rightarrow R_3 = 1.8\,\text{k}\Omega$$

$$V_{DE} = 3.8\,\text{V} = \frac{10\,\text{V} \times R_4}{7.5\,\text{k}\Omega} \rightarrow R_4 = 2.85\,\text{k}\Omega$$

These resistances can then be used to calculate the cross-sectional area of the cylinders. The electrode spacing is 1 cm. Using $A = \dfrac{\rho L}{R}$:

$$A_1 = \frac{150\ \Omega\text{m} \times 0.01\ \text{m}}{1.5\ \text{k}\Omega} = 10\ \text{cm}^2$$

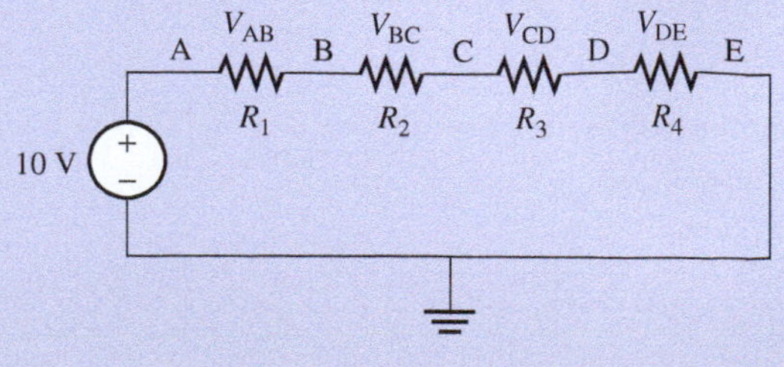

FIGURE 3.33 The circuit model for Figure 3.32.

(continued)

APPLICATION EXAMPLE 3.12   Continued

$$A_2 = \frac{150\ \Omega\text{m} \times 0.01\ \text{m}}{1.35\ \text{k}\Omega} = 11.1\ \text{cm}^2$$

$$A_3 = \frac{150\ \Omega\text{m} \times 0.01\ \text{m}}{1.8\ \text{k}\Omega} = 8.33\ \text{cm}^2$$

$$A_4 = \frac{150\ \Omega\text{m} \times 0.01\ \text{m}}{2.85\ \text{k}\Omega} = 5.26\ \text{cm}^2$$

Now, find the volume by adding the volume of each cylinder:

Total volume $= 10\,\text{cm}^2 \times 1\,\text{cm} + 11.1\,\text{cm}^2 \times 1\,\text{cm} + 8.33\,\text{cm}^2 \times 1\,\text{cm} + 5.26\,\text{cm}^2 \times 1\,\text{cm}$

$\qquad\qquad = 34.69\,\text{cm}^3$

APPLICATION EXAMPLE 3.13   Automobile Ignition System

The ignition system of an automobile generates a high voltage (>20,000 V) from the 12-V battery. Figure 3.34 represents a very simplified ignition system. When the switch, S (known as the "points") is opened, a high voltage is generated in the secondary coil, which is directed to the appropriate spark plug. The function of the ignition system will be detailed in future chapters.

    a. If the switch is closed, and $V_R = 5.2$ V, what is V_{SW}?

    b. What is V_L?

    c. If $R = 10\ \Omega$, what is the coil resistance R_L?

    d. What is the voltage V_{SW} when the switch opens?

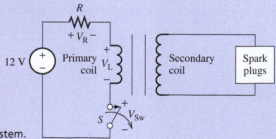

FIGURE 3.34   A simplified ignition system.

SOLUTION

    a. Because the switch is closed, the resistance across its contacts is close to zero. According to Ohm's law, $V = I \times R$. Using the equation, if the resistance is zero, the voltage will also be zero. Therefore:

$$V_{SW} = 0\,\text{V}$$

    b. According to KVL:

$$-12 + 5.2 + V_L + 0 = 0$$

This equation can be used to calculate the voltage, $V_L = 6.8$ V

    c. According to the voltage divider formula:

$$6.8 = 12 \times \frac{R_L}{10 + R_L + 0}$$

Now, the resistance can be calculated as $R_L = 13.08\ \Omega$.

d. When the switch is open, the resistance across its contacts is nearly infinite. According to the voltage divider formula:

$$V_{SW} = 12 \times \frac{R_{SW}}{R + R_L + R_{SW}}$$

As R_{SW} approaches infinity, V_{SW} approaches 12 V. Therefore:

$$V_{SW} = 12V$$

3.4 NODAL AND MESH ANALYSIS

In some circuits, equivalent resistance and voltage/current divider rules are not enough for resistive circuit analysis. For example, consider the circuit in Figure 3.35. None of the resistors in the circuit is in series or parallel to each other.

Therefore, a different analysis method is needed that works regardless of circuit configuration and size. Nodal voltage analysis and mesh current analysis are examples of such systems. In the following sections, nodal and mesh analysis are introduced.

3.4.1 Nodal Analysis

In the nodal analysis method, the nodes in the circuit are determined first. In a resistive circuit, a node is the connecting point of two or more circuit resistors. The nodes in Figure 3.35 are specified in Figure 3.36. Four nodes are observed in the figure.

In the next step, a node is picked from the specified nodes as the *reference node*. The voltages of the other nodes are then measured with respect to this node.

As shown in Figure 3.37, a voltage is assigned to each node. Any node in the circuit can be used as the reference node. However, the circuit analysis can be simplified by selecting one of the end points of a voltage source as the reference node. As can be observed in Figure 3.37, ground is the symbol used for the reference node. The reference node is assigned zero voltage. The voltage of some nodes in the circuit can be determined easily. For example, in Figure 3.37, there is only one voltage source between Node 1 and the reference node. As a result, the difference of the voltage between Node 1 and the zero voltage of the reference node is V_s and V_s is assigned to I_1.

The main step in node voltage analysis is to write KCL equations in terms of the node voltages. After writing the KCL equations corresponding to each node, solve the system of equations to find unknown voltages. For example, assume that Figure 3.38 is a part of the circuit in

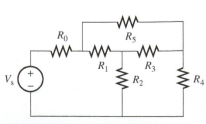

FIGURE 3.35 A resistive circuit.

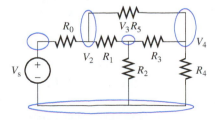

FIGURE 3.36 The nodes of the circuit of Figure 3.35.

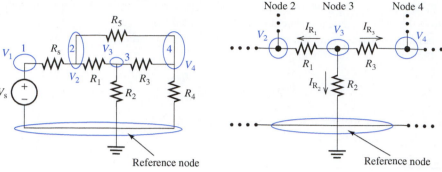

FIGURE 3.37 A section of Figure 3.35.

FIGURE 3.38 Different nodes in a resistive circuit.

Figure 3.37. Ohm's law states that the current of each resistor is the difference of the potentials of their terminals divided by their resistances, that is:

$$I_{R_1} = \frac{V_3 - V_2}{R_1}$$

$$I_{R_2} = \frac{V_3 - 0}{R_2}$$

$$I_{R_3} = \frac{V_3 - V_4}{R_3}$$

Actually, for each node, it is assumed that the current is leaving the node. Therefore, according to KCL, the sum of all the leaving currents is zero. Therefore, in Figure 3.38 (Node 3):

$$I_{R_1} + I_{R_2} + I_{R_3} = 0, \quad I_{R_1} = \frac{V_3 - V_2}{R_1}, \quad I_{R_2} = \frac{V_3}{R_2}, \quad I_{R_3} = \frac{V_3 - V_4}{R_3}$$

Similar equations can be generated for Nodes 2 and 4.

EXCERCISE 3.5

Write the equations for Nodes 1 and 2 of Figure 3.37.

Writing the leaving current from each node in terms of the node voltages and the resistance enables generation of KCL equations. Using KCL for the three nodes of Figure 3.37 and noting that $V_1 = V_s$:

KCL for Node 2:

$$\frac{V_2 - V_s}{R_s} + \frac{V_2 - V_3}{R_1} + \frac{V_2 - V_4}{R_5} = 0$$

KCL for Node 3:

$$\frac{V_3 - V_2}{R_1} + \frac{V_3}{R_2} + \frac{V_3 - V_4}{R_3} = 0$$

KCL for Node 4:

$$\frac{V_4 - V_3}{R_3} + \frac{V_4}{R_4} + \frac{V_4 - V_2}{R_5} = 0$$

Algebraic manipulation results in the following system of equations for the node voltages:

$$\begin{cases} \left(\dfrac{1}{R_s} + \dfrac{1}{R_1} + \dfrac{1}{R_5}\right)V_2 - \dfrac{V_3}{R_1} - \dfrac{V_4}{R_5} = \dfrac{V_s}{R_S} \\[2mm] -\dfrac{V_2}{R_1} + \left(\dfrac{1}{R_1} + \dfrac{1}{R_2} + \dfrac{1}{R_3}\right)V_3 - \dfrac{V_4}{R_3} = 0 \\[2mm] -\dfrac{V_2}{R_5} - \dfrac{V_3}{R_3} + \left(\dfrac{1}{R_3} + \dfrac{1}{R_4} + \dfrac{1}{R_5}\right)V_4 = 0 \end{cases} \qquad \textbf{(3.22)}$$

EXCERCISE 3.6

Show the details of generating the above set of equations.

The system of linear equations can also be shown in matrix form. The matrix form of the above node voltage equations is:

$$\begin{pmatrix} \dfrac{1}{R_5} + \dfrac{1}{R_1} + \dfrac{1}{R_5} & -\dfrac{1}{R_1} & -\dfrac{1}{R_5} \\[3mm] -\dfrac{1}{R_1} & \dfrac{1}{R_1} + \dfrac{1}{R_2} + \dfrac{1}{R_3} & -\dfrac{1}{R_3} \\[3mm] -\dfrac{1}{R_5} & -\dfrac{1}{R_3} & \dfrac{1}{R_3} + \dfrac{1}{R_4} + \dfrac{1}{R_5} \end{pmatrix} \begin{pmatrix} V_2 \\[2mm] V_3 \\[2mm] V_4 \end{pmatrix} = \begin{pmatrix} \dfrac{V_s}{R_S} \\[2mm] 0 \\[2mm] 0 \end{pmatrix}$$

Matrix form facilitates solving the problem and obtaining the voltages using a calculator or software such as MATLAB®. The technique used to solve these matrix equations and evaluate the unknown parameters is discussed in Appendix A.

EXAMPLE 3.14   Reference Node

Which node in Figure 3.39 is the best choice for a reference node?

SOLUTION

The negative terminal of the voltage source (in this case, node B) in the circuit is the best choice for the reference node.

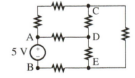

FIGURE 3.39   The circuit for Example 3.14.

EXAMPLE 3.15   Reference Node

Which node is the best choice for a reference node in Figure 3.40?

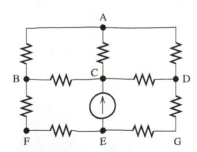

FIGURE 3.40   The circuit for Example 3.15.

(continued)

EXAMPLE 3.15   **Continued**

SOLUTION

There is no voltage source in the circuit. Therefore, any of the nodes in Figure 3.40 can be used as a reference node.

EXCERCISE 3.7

In Figure 3.40, if all resistors are equivalent to R, find the voltage of nodes B and D. Are these voltages equivalent? If two points (nodes) have the same voltage, they can be overlapped. This simplifies the topology of the circuit. Use this method to find the equivalent resistance across nodes C and E.

EXAMPLE 3.16   **Nodal Analysis**

Find V_1, V_2, and V_3 in Figure 3.41 using nodal equations.

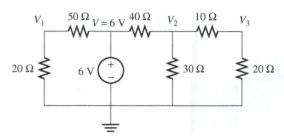

FIGURE 3.41   The circuit for Example 3.16.

SOLUTION

If the sum of the current leaving from each node is set to zero, the following set of linear equations can be obtained in terms of node voltages.

$$\begin{cases} \dfrac{V_1}{20} + \dfrac{V_1 - 6}{50} = 0 \\ \dfrac{V_2 - 6}{40} + \dfrac{V_2}{30} + \dfrac{V_2 - V_3}{10} = 0 \\ \dfrac{V_3 - V_2}{10} + \dfrac{V_3}{20} = 0 \end{cases} \longrightarrow \begin{cases} \left(\dfrac{1}{20} + \dfrac{1}{50} \right) V_1 = \dfrac{6}{50} \\ \left(\dfrac{1}{10} + \dfrac{1}{30} + \dfrac{1}{40} \right) V_2 - \dfrac{V_3}{10} = \dfrac{6}{40} \\ \left(\dfrac{-1}{10} \right) V_2 + \left(\dfrac{1}{10} + \dfrac{1}{20} \right) V_3 = 0 \end{cases}$$

Solving this system of equations:

$$V_1 = 1.71 \text{ V}$$
$$V_2 = 1.64 \text{ V}$$
$$V_3 = 1.09 \text{ V}$$

EXCERCISE 3.8

Use the voltage divider method to find the voltages V_1, V_2, and V_3 in Example 3.16.

| APPLICATION EXAMPLE 3.17 | Automobile Acceleration |

The inventor of a new sports car intends to verify if its initial acceleration from 0 to 100 mph is indeed better than any competitors. She has gathered the acceleration data from the leading competitors to compare to her own. To record her own data, she connects an accelerometer to her initial prototype. The accelerometer is connected to a PDA located in the car, which logs the data. The PDA has an equivalent resistance of 50 Ω which is considerably lower than the resistances of the resistors in the network. Therefore, the current through it would be large.

To avoid having the PDA overload the accelerometer's resistive network, it is isolated from the accelerometer using a buffer, modeled by the dependent voltage source. The accelerometer gives the acceleration as a varying output voltage, V_A. The voltage, V_A, ranges from 0 to $V_{max} = 5$ V. The circuit model is shown in Figure 3.42. To calibrate the accelerometer and the PDA, the inventor needs to find the circuit parameters as follows:

a. Find V_o if $V_A = 0.3$ V
b. Find I_A when $V_A = V_{A, max}$
c. How much power is dissipated by the PDA?

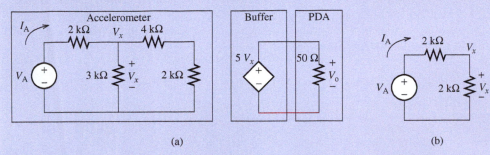

(a)   (b)

FIGURE 3.42 (a) The circuit model for the accelerometer and the PDA; (b) equivalent circuit for the accelerometer.

SOLUTION

a. Using KCL for the node with the voltage V_x:

$$\frac{V_x - V_A}{2\,k} + \frac{V_x}{3\,k} + \frac{V_x}{4\,k + 2\,k} = 0$$

If $V_A = 0.3$ V:

$$V_x = 0.15 \text{ V}$$

Therefore (see Buffer–PDA circuit):

$$V_o = 5V_x = 0.75 \text{ V}$$

b. The equivalent resistor of the accelerometer is 4 kΩ and $V_{max} = 5$ V. As a result:

$$I_A = \frac{5\,V}{4\,K} = 1.25\,mA$$

c. Figure 3.42(b) shows the equivalent circuit for the accelerator obtained by finding the equivalent resistance of 3||(2 + 4) = 2 kΩ. Using simple voltage division, and taking $V_A = V_{max}$:

$$V_x = \frac{1}{2}V_A = \frac{1}{2} \times 5 = 2.5 \text{ V}$$

(continued)

APPLICATION EXAMPLE 3.17   Continued

In addition, the voltage of the dependent voltage source of the buffer is $V_o = 2.5 \times 5 = 12.5\,\text{V}$ and the power dissipated at the load resistance will be:

$$P = \frac{V_o^2}{R} = \frac{(12.5)^2}{50} = 3.125\,\text{W}$$

APPLICATION EXAMPLE 3.18   Chemical Process

The temperature of a chemical process needs to be monitored in order to control the process (Figure 3.43). This requires a device that can electrically measure temperature to allow it to be monitored by an electric controller. There are two types of electric temperature sensors: thermocouples and thermistors. A thermistor is simply a resistor whose value changes with temperature. A thermistor is simpler to use than a thermocouple. The approximate temperature-to-resistance relationship of a thermistor is given by the following equation, known as the *Steinhart–Hart Equation*:

$$T = \frac{1}{a + b\ln(R_T) + c\,(\ln(R_T))^3}$$

where

T = temperature in Kelvin

a, b, c = parameters specific to the thermistor device

R_T = normalized resistance of the thermistor; the normalization is with respect to the resistance in ambient temperature.

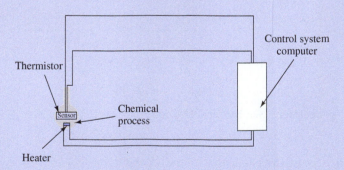

FIGURE 3.43 Chemical thermistor.

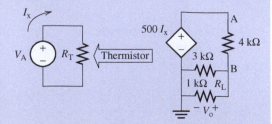

FIGURE 3.44 Circuit model for the controlled thermistor.

For this example, assume the associated ambient temperature is 1 Ω and that the thermistor has the following device parameters:

$$a = 0.0160717; \quad b = -0.00286422; \quad c = 0.0000193965$$

The thermistor is connected as shown in Figure 3.44. The output voltage, V_o, is measured to be 0.73 V. This is the voltage seen by the monitoring computer. The voltage, V_A, is used to convert the varying resistance of the thermistor to the varying current, I_x. This is isolated from the computer by the dependent source.

a. Find the voltage at node A.
b. What is the temperature at the thermistor, in kelvin?

SOLUTION

a. Based on Figure 3.44, the output voltage is given, therefore, $V_o = V_B = 0.73$ V. Now, using node voltage analysis applied to node B:

$$\frac{0.73 - V_A}{4\,\text{k}} + \frac{0.73}{3\,\text{k}} + \frac{0.73}{1\,\text{k}} = 0$$

Through calculation:

$$V_A = 4.623 \text{ V}$$

b. To solve part (b), start with the result from above:

$$V_A = 500\,I_x = 4.623 \text{ V}$$

Calculating the current:

$$I_x = 9.247 \text{ mA}$$

Now, see that:

$$R_T = \frac{V_A}{I_x} = \frac{5\text{ V}}{9.247\,\text{mA}} = 540.7\,\Omega$$

Using the *Steinhart–Hart* Equation:

$$T = \frac{1}{a + b\ln(R_T) + c(\ln(R_T))^3}$$

$$T = 347 \text{ K}$$

The node voltage analysis procedure can be summarized in the following steps:
1. Determine all nodes in the circuit.
2. Pick a node as a reference node.
3. Assign voltage variables to each node.
4. For each node, write KCL equations in terms of node voltages.
5. Solve the system of equations to find unknown voltages.

3.4.2 Mesh Analysis

In addition to nodal analysis, mesh analysis can be used to find the circuit parameters in Figure 3.35. The procedures are outlined as follows.

A mesh is defined as a closed loop of a circuit. The first step in mesh analysis is to identify the meshes in the given circuit. In the circuit of Figure 3.45, three meshes are identified and then the mesh current variables i_1, i_2, i_3 are assigned to them, respectively. To avoid confusion in writing circuit equations, unknown mesh currents are defined exclusively clockwise.

The major step is to write KVL equations for each mesh. Beginning with mesh 1, note that the voltages around the mesh have been assigned in Figure 3.45 *consistent with* the direction of the mesh current, i_1. Now it is important to observe that while mesh current i_1 is equal to the current flowing through resistor R_0, it is not equal to the currents through R_1 and R_2. The branch current through R_1 is the difference between the two mesh currents, that is, $i_1 - i_2$. Similarly, the current through R_2 is $i_1 - i_3$. Note that the current direction is determined consistent with the polarity of the voltage across R_2 which is from the positive pole toward the negative one. Here, the direction of i_1 is consistent with the polarity of v_2 while the direction of i_3 is not consistent with this polarity. Therefore, the current through R_2 would be $i_1 - i_3$.

According to the polarity of voltage v_1, the voltage, v_1, is given by:

$$v_1 = (i_1 - i_2)R_1$$

Similarly, the voltage, v_2, corresponds to:

$$v_2 = (i_1 - i_3)R_2$$

Therefore, the KVL equation for mesh 1 is:

$$v_s - i_1 R_0 - (i_1 - i_2) R_1 - (i_1 - i_3) R_2 = 0$$

The same approach can be applied to mesh 2. Figure 3.46 depicts the voltage assignment around mesh 2, which follows the clockwise direction mesh current i_2.

The mesh current i_2 is the branch current that passes through resistor R_5. However, the current that passes through resistor R_1 that is shared by mesh 1 and mesh 2 is equal to $i_2 - i_1$, and the current through resistor R_3 that is shared by mesh 2 and mesh 3, is equal to $i_2 - i_3$. Note that the current direction is determined consistent with the polarity of the voltage across R_3 which is from the positive pole toward the negative one. Here, the direction of i_2 is consistent with the polarity of v_3 while the direction of i_2 is not consistent with this polarity. Therefore, the current through R_3 would be $i_2 - i_3$.

Therefore, the voltages across R_1 and R_3 are:

$$v_1 = (i_2 - i_1)R_1$$
$$v_3 = (i_2 - i_3)R_3$$

Accordingly, the KVL equation for mesh 2 is:

$$i_2R_5 + (i_2 - i_3)R_3 + (i_2 - i_1)R_1 = 0$$

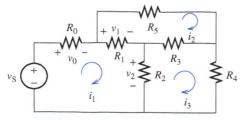

FIGURE 3.45 Assignment of currents and voltages around mesh 1 for the circuit in Figure 3.35.

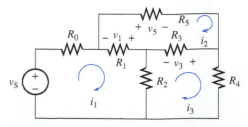

FIGURE 3.46 Assignment of currents and voltages around mesh 2.

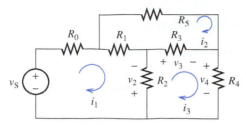

FIGURE 3.47   Assignment of currents and voltages around mesh 3.

Figure 3.47 depicts that the voltage assignment around mesh 3, which follows the clockwise direction for mesh current i_3.

Following the voltage polarity shown in Figure 3.47, voltages v_2 and v_3 can be expressed by:

$$v_2 = (i_3 - i_1)R_2$$
$$v_3 = (i_3 - i_2)R_3$$

Then, the KVL equation for mesh 3 is:

$$(i_3 - i_1)R_2 + (i_3 - i_2)R_3 + i_3 R_4 = 0$$

Combining the KVL equations for each mesh 1, 2, and 3, we obtain the following system of equations:

$$(R_0 + R_1 + R_2)i_1 - R_1 i_2 - R_2 i_3 = v_s$$
$$-R_1 i_1 + (R_1 + R_3 + R_5)i_2 - R_3 i_3 = 0$$
$$-R_2 i_1 - R_3 i_2 + (R_2 + R_3 + R_4)i_3 = 0$$

This system of equation can be solved to find the mesh currents i_1, i_2, and i_3.

The mesh current analysis procedure can be summarized in the following steps:

1. Define each mesh current (usually exclusively clockwise).
2. Assign current variables to each mesh.
3. To write the KVL for each mesh, determine the voltage across the components of that mesh based on the mesh current direction.
4. For each mesh, write the KVL equation.

EXAMPLE 3.19   Mesh Determination

Find the meshes in the circuit of Figure 3.48.

SOLUTION

By inspection there are three meshes in the circuit of Figure 3.48. The mesh currents are defined clockwise and denoted in Figure 3.49.

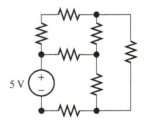

FIGURE 3.48   Find the meshes.

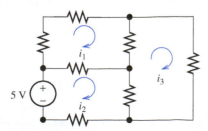

FIGURE 3.49   Find the meshes.

EXAMPLE 3.20   **Mesh Analysis**

Use mesh analysis to find the current, i, flowing through the 6-V voltage source in Figure 3.50.

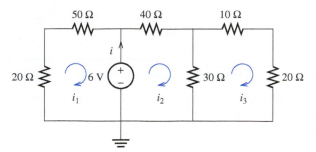

FIGURE 3.50 The circuit for Example 3.20.

SOLUTION

First, three meshes are identified in the circuit of Figure 3.50. The current flowing through the 6-V voltage source is:

$$i = i_2 - i_1$$

We need to find the mesh currents i_1 and i_2. Thus, we write KVL equations for each mesh to find i_1 and i_2.

Mesh 1:

$$20i_1 + 50i_1 + 6 = 0$$

Mesh 2:

$$40i_2 + 30(i_2 - i_3) - 6 = 0$$

Mesh 3:

$$30(i_3 - i_2) + 10i_3 + 20i_3 = 0$$

Rearranging the above three equations, we obtain the system of equations:

$$i_1 = -0.086$$
$$70i_2 - 30i_3 = 6$$
$$-30i_2 + 60i_3 = 0$$

Solving the system of equations:

$$i_1 = -0.086\text{A}, \quad i_2 = 0.109\text{A}, \quad i_3 = 0.055\text{A}$$

Therefore:

$$i = i_2 - i_1 = 0.195\text{A}$$

EXAMPLE 3.21   **Mesh Analysis with Current Sources**

Find the mesh currents in the circuit shown in Figure 3.51, when $i_s = 0.5$ A, $v_s = 6$ V, $R_1 = 3\ \Omega$, $R_2 = 8\ \Omega$, $R_3 = 2\ \Omega$.

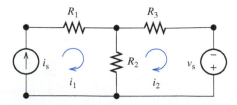

FIGURE 3.51 Mesh analysis with current source.

SOLUTION

For mesh 1, the mesh current is equal to the current source i, that is, $i_1 = i_s = 0.5$ A. Therefore, it is not necessary to write the KVL equation for mesh 1.

For mesh 2, the KVL equation is:

$$(i_2 - i_1)R_2 + i_2R_3 - 6 = 0$$

And i_2 can be expressed by:

$$i_2 = (6 + i_1R_2)/(R_2 + R_3)$$

Substituting i_1 and the resistor values, we can solve the mesh currents:

$$i_1 = 0.5\,\text{A}, \quad i_2 = 1\,\text{A}$$

EXAMPLE 3.22   **Wheatstone Bridge**

The circuit in Figure 3.52 is known as a Wheatstone bridge. This circuit is commonly used to identify the value of unknown resistor. The application of this bridge has already been introduced in Chapter 1. Chapter 14 also reveals the application of this bridge in measurement devices. Example 3.38 represents another application of Wheatstone bridge.

As shown in Figure 3.52, the circuit of this bridge consists of four elements (in this case, resistors). In the circuit, R_1 and R_2 are known resistors; resistor R_4 is known and adjustable; resistor R_3 is unknown and three meshes are identified.

a. List the KVL equations for all meshes.
b. In this bridge, the value of resistor R_4 can be adjusted to maintain zero current flow between the two nodes A and B. In this case, find the unknown resistor R_3 in terms of R_1, R_2, and R_4.

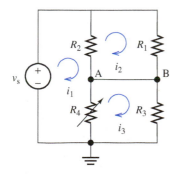

FIGURE 3.52   Wheatstone bridge.

SOLUTION

a. For mesh 1, the KVL equation is:

$$R_2(i_1 - i_2) + R_4(i_1 - i_3) - v_s = 0$$

For mesh 2, the KVL equation is:

$$R_1i_2 + R_2(i_2 - i_1) = 0$$

For mesh 3, the KVL equation is:

$$R_3i_3 + R_4(i_3 - i_1) = 0$$

b. When the current that flows from node A to node B is zero, $i_2 - i_3 = 0$. In this case, the KVL equation for mesh 2 becomes:

$$R_1/R_2 = (i_1 - i_2)/i_2 = (i_1 - i_3)/i_3$$

(continued)

EXAMPLE 3.22 **Continued**

In addition, the KVL equation for mesh 3 is:

$$R_3/R_4 = (i_1 - i_3)/i_3$$

Equating the left-hand sides of above two equations:

$$R_3 = (R_1 R_4)/R_2$$

Using this equation, it is clear that in the Wheatstone bridge the voltage source does not impact the measurement of the resistance R_3. Therefore, the weakness of the battery does not have any effect on the accuracy of the resistance measurement.

3.5 SPECIAL CONDITIONS: SUPER NODE

Recall again the circuit in Figure 3.37. Now, consider this question: what happens if a KCL equation is written for the reference node in addition to the three specified nodes?

The KCL equation for the reference node in this circuit is:

$$\frac{V_s - V_2}{R_s} + \frac{0 - V_3}{R_2} + \frac{0 - V_4}{R_4} = 0$$

Combining this with the system of Equations of (3.22) results in a system of four equations that corresponds to:

$$\begin{cases} \left(\dfrac{1}{R_s} + \dfrac{1}{R_1} + \dfrac{1}{R_5}\right)V_1 - \dfrac{V_2}{R_1} - \dfrac{V_3}{R_5} = \dfrac{V_s}{R_s} \\[2mm] -\dfrac{V_1}{R_1} + \left(\dfrac{1}{R_1} + \dfrac{1}{R_2} + \dfrac{1}{R_3}\right)V_2 - \dfrac{V_3}{R_3} = 0 \\[2mm] -\dfrac{V_1}{R_5} - \dfrac{V_2}{R_3} + \left(\dfrac{1}{R_3} + \dfrac{1}{R_4} + \dfrac{1}{R_5}\right)V_3 = 0 \\[2mm] \dfrac{V_2}{R_s} + \dfrac{V_3}{R_2} + \dfrac{V_4}{R_4} = \dfrac{V_s}{R_s} \end{cases}$$

Now, add up the first three equations. The result is the last equation. What conclusion can be drawn from this? We can conclude that these four equations are *linearly dependent*, that is, by combining three of them, the next one is found. As a result, only three equations are sufficient in order to find the unknown voltages. In contrast, any three equations in this set are *linearly independent* from each other; considering any three equations in the set, you cannot combine two of them in order to find the third one. Therefore, any linearly independent set of equations can be used to find the unknown variables. In general, for n nodes in a circuit, $n - 1$ independent equations are needed to find the unknown voltages.

Sometimes, the current in a circuit branch cannot be expressed in terms of its terminal node voltages. In such cases, the aforementioned nodal voltage analysis methods cannot be applied. For example, consider Figure 3.53. The two circuits in this figure are the same. There are two voltage sources. In (a) and (b), two different reference nodes are shown at the negative end of voltage sources. Applying KCL to Node 1 in the circuit of Figure 3.53(a) results in:

$$\frac{V_1 - V_{s1}}{R_1} + \frac{V_1}{R_2} + I_a = 0$$

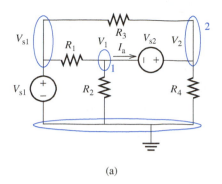
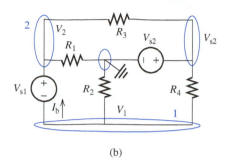

(a)

(b)

Accordingly, the current I_a cannot be expressed in the circuit shown in Figure 3.53(a) in terms of its node voltages, V_1 and V_2. Similarly, the current I_b in the circuit shown in Figure 3.53(b) can be found using:

$$\frac{V_1 - V_{s2}}{R_4} + \frac{V_1}{R_2} + I_b = 0$$

The same problem exists for another reference node as well: the current I_b cannot be written in terms of its node voltages.

In order to solve this problem, in these situations a *super node* is defined instead of a single node. A super node is a closed surface in a circuit that includes several nodes. As described in Chapter 2, KCL can be applied to any closed surface in a circuit. Recall that a node is actually a closed surface with a zero-limit area. Considering the circuit shown in Figure 3.54, the super node shown contains two single nodes.

If KCL is applied to the super node's closed surface:

$$I_1 + I_2 + I_3 + I_4 = 0$$

This can be written in terms of voltages as:

$$\frac{V_1 - V_{s1}}{R_1} + \frac{V_1}{R_2} + \frac{V_2 - V_{s1}}{R_3} + \frac{V_2}{R_4} = 0$$

In cases where super nodes are used instead of single nodes, a KVL equation must also be written for the super node's closed surface:

$$-V_1 - V_{s2} + V_2 = 0$$

This equation represents the KVL for the loop formed by the super node. Therefore, the following system of equations is derived for the circuit in Figure 3.54:

$$\begin{cases} \left(\frac{1}{R_1} + \frac{1}{R_2}\right)V_1 + \left(\frac{1}{R_3} + \frac{1}{R_4}\right)V_2 = \left(\frac{1}{R_1} + \frac{1}{R_3}\right)V_{s1} \\ -V_1 + V_2 = V_{s2} \end{cases}$$

Note that there is another approach for finding the voltage or current in circuits that include more than one independent source. This approach is called superposition and will be discussed in Section 3.7.

Nodal voltage analysis is also useful for cases where a dependent voltage or current sources exist in the circuit. In these situations, similar to the previous cases, a reference node is assigned to one of the end points (usually the negative point) of the independent sources. Figure 3.55 shows an example of a circuit with a dependent current source. In this example, the KCL equations at nodes 1 and 2, respectively, are:

$$\frac{V_1 - V_{s1}}{R_1} + \frac{V_1}{R_2} + \frac{V_1 - V_2}{R_3} = 0 \tag{3.23}$$

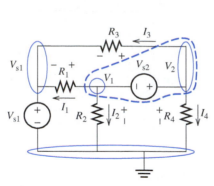

FIGURE 3.54   A super node in a circuit.

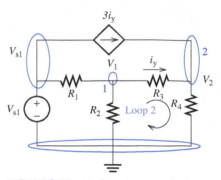

FIGURE 3.55   Node voltage analysis in a circuit with a dependent current source.

and

$$\frac{V_2 - V_1}{R_3} + \frac{V_2}{R_4} - 3i_y = 0 \tag{3.24}$$

The KCL equation at Node 2 is in terms of the current i_y as well. This term has appeared due to the dependent current source. Because i_y flows through the resistor R_3, the KVL in loop 2 shown in Figure 3.55 can be written as:

$$i_y = \frac{V_1 - V_2}{R_3} \tag{3.25}$$

Replacing i_y in Equation (3.24) with that of Equation (3.25), results in the following system of equations for this circuit:

$$\text{Equation (3.23)}: \left\{ \begin{array}{l} \dfrac{V_1 - V_{s1}}{R_1} + \dfrac{V_1}{R_2} + \dfrac{V_1 - V_2}{R_3} = 0 \\[3mm] \dfrac{V_2 - V_1}{R_3} + \dfrac{V_2}{R_4} - 3\dfrac{V_1 - V_2}{R_3} = 0 \end{array} \right.$$
$$\text{Modified Equation (3.24)}:$$

Now, consider Figure 3.56 as a circuit with a dependent voltage source. In this case, if node voltage analysis is applied with single nodes, it will not be possible to express the current passing through the dependent voltage source in terms of voltage. As a result, a super node is used which includes the nodes of the dependent voltage source. The KCL equations are:

$$\frac{V_1}{R_1} + \frac{V_1 - V_3}{R_2} - I_s + \frac{V_2 - V_3}{R_3} = 0$$

$$\frac{V_3 - V_1}{R_2} + \frac{V_3 - V_2}{R_3} + \frac{V_3}{R_4} = 0$$

KVL must also be applied to the super node's closed surface node:

$$-V_1 - 2V_x + V_2 = 0$$

where:

$$V_x = V_3$$

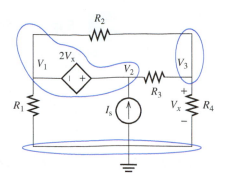

FIGURE 3.56 Voltage analysis in a circuit with a dependent voltage source.

Accordingly, the following set of node voltage equations is derived:

$$
\begin{cases}
\left(\dfrac{1}{R_1} + \dfrac{1}{R_2}\right)V_1 + \dfrac{V_2}{R_3} - \left(\dfrac{1}{R_2} + \dfrac{1}{R_3}\right)V_3 = I_s \\[2mm]
-\dfrac{V_1}{R_2} - \dfrac{V_2}{R_3} + \left(\dfrac{1}{R_2} + \dfrac{1}{R_3} + \dfrac{1}{R_4}\right)V_3 = 0 \\[2mm]
\qquad\qquad - V_1 + V_2 - 2V_3 = 0
\end{cases}
$$

EXAMPLE 3.23   **Node Analysis**

Find V_0 in the circuit shown in Figure 3.57.

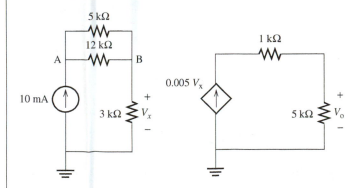

FIGURE 3.57 The circuit for Example 3.23.

SOLUTION

The following are the KCL equations for nodes A and B, respectively:

$$
-10\,\text{mA} + \frac{V_A - V_B}{12\,\text{k}} + \frac{V_A - V_B}{5\,\text{k}} = 0
\longrightarrow
\begin{cases}
\left(\dfrac{1}{12} + \dfrac{1}{5}\right)V_A - \left(\dfrac{1}{12} + \dfrac{1}{5}\right)V_B = 10 \\[3mm]
\end{cases}
$$

$$
\frac{V_B - V_A}{12\,\text{k}} + \frac{V_B - V_A}{5\,\text{k}} + \frac{V_B}{3\,\text{k}} = 0
\qquad
\begin{cases}
-\left(\dfrac{1}{12} + \dfrac{1}{5}\right)V_A + \left(\dfrac{1}{12} + \dfrac{1}{5} + \dfrac{1}{3}\right)V_B = 0
\end{cases}
$$

Solving this set of equations:

$$
V_B = 30\,\text{V}
$$

In addition, $V_x = V_B$. Using these voltages, the output voltage will be:

$$
V_0 = 0.005 \times V_B \times 5\,\text{k}\Omega = 750\,\text{V}
$$

EXAMPLE 3.24   Super Node

Find the nodal voltages V_1 and V_2 for the circuit shown in Figure 3.58.

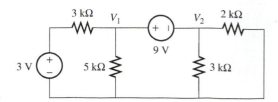

FIGURE 3.58 The circuit for Example 3.24.

SOLUTION

The current that flows through the 9-V source cannot be expressed in terms of its end voltages. Therefore, as shown in Figure 3.59, the 9-V source nodes are selected as a super node. Applying KCL at the super node results in:

$$\frac{V_1 - 3}{3\,k} + \frac{V_1}{5\,k} + \frac{V_2}{3\,k} + \frac{V_2}{2\,k} = 0$$

$$\left(\frac{1}{3} + \frac{1}{5}\right)V_1 + \left(\frac{1}{3} + \frac{1}{2}\right)V_2 = 1$$

Applying KVL to the loop formed by the super node loop, that is, the loop formed by the two voltages across the node, as shown in Figure 3.59 results in:

$$V_2 = V_1 - 9\text{ V}$$

These two equations are used to find V_1 and V_2:

$$V_1 = 6.22\text{ V}$$
$$V_2 = -2.78\text{ V}$$

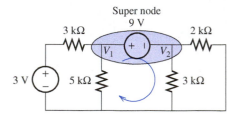

FIGURE 3.59 Circuit for Example 3.24.

EXAMPLE 3.25   Super Node

Find the nodal voltages in Figure 3.60.

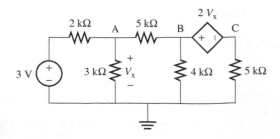

FIGURE 3.60 The circuit for Example 3.25.

SOLUTION

A current that flows through a dependent source cannot be expressed in terms of its end voltages. Therefore, as shown in Figure 3.61, the two nodes across the dependent source are selected as a super node.

Next, write two KCL equations for node A and super node "B, C" and one KVL equation for the super node loop. The system of equations corresponds to:

$$\frac{V_A - 3}{2k} + \frac{V_A}{3k} + \frac{V_A - V_B}{5k} = 0$$

$$\frac{V_B - V_A}{5k} + \frac{V_B}{4k} + \frac{V_C}{5k} = 0$$

$$V_B - V_C = 2V_x = 2V_A$$

The system of equations can be represented in matrix form as follows:

$$\begin{bmatrix} 31 & -6 & 0 \\ -4 & 9 & 4 \\ 2 & -1 & 1 \end{bmatrix} \begin{bmatrix} V_A \\ V_B \\ V_C \end{bmatrix} = \begin{bmatrix} 45 \\ 0 \\ 0 \end{bmatrix}$$

This form helps us to find the voltages using an inverse matrix, as discussed in the Appendix A:

$$\begin{bmatrix} V_A \\ V_B \\ V_C \end{bmatrix} = \begin{bmatrix} 1.767 \text{ V} \\ 1.631 \text{ V} \\ -1.903 \text{ V} \end{bmatrix}$$

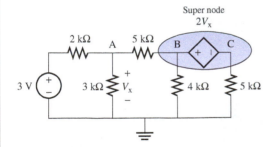

FIGURE 3.61   The circuit for Example 3.25.

APPLICATION EXAMPLE 3.26   **Sports Car Inventor/Accelerometer**

The inventor of the sports car discussed earlier has been informed by an anonymous source that connecting an accelerometer to a PDA is a good way to measure initial acceleration data. The 6-V source gives the signal a small amount of amplification. A circuit model for this example is shown in Figure 3.62. Remember, for this accelerometer, $0 \le V_A \le 5$ V.

  a. If the accelerometer produces its maximum output voltage, 5 V, what voltage is delivered to the PDA?
  b. In addition, if the maximum input voltage the PDA can handle is 5 V, will this circuit damage the PDA?

(*continued*)

APPLICATION EXAMPLE 3.26   **Continued**

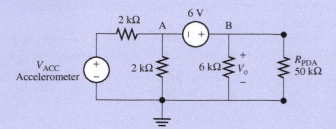

FIGURE 3.62   The circuit for Example 3.26.

SOLUTION

In Figure 3.63 the nodal voltage equations are:

$$\text{KCL for the super node } A,B: \frac{V_A - 5}{2\,k} + \frac{V_A}{2\,k} + \frac{V_B}{6\,k} + \frac{V_B}{50} = 0$$

$$\text{KVL for the super node loop: } V_B = V_A + 6$$

Solving these two equations results in $V_o = V_B = 0.402V$. Thus, this circuit is feasible, because the maximum voltage it can produce is not higher than the maximum voltage the PDA can handle.

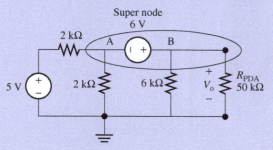

FIGURE 3.63   The circuit for Example 3.26.

APPLICATION EXAMPLE 3.27   **A Chemical Sensor**

A titration sensor is connected to measure the PH level of a solution. The solution is kept at a constant temperature by a heater connected to a thermostat. The sensor produces a voltage output and requires the circuit shown in Figure 3.64(b) to connect it to an LCD display. If the sensor voltage $V_s = 0.201$ V, find the voltage V_o seen by the LCD display.

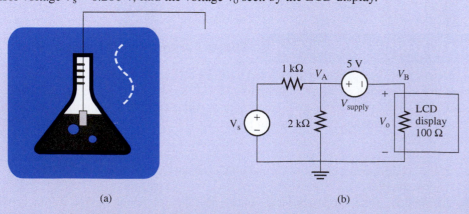

(a)   (b)

FIGURE 3.64   (a) A titration sensor; (b) its circuit model.

SOLUTION

First, write the KCL and KVL equations for the super node shown in Figure 3.65.

$$\frac{V_A}{2\,k} + \frac{V_A - V_s}{1\,k} + \frac{V_B}{0.1k} = 0$$

$$V_A - V_B = 5\text{ V}$$

This system of equations can be rearranged and written again in the matrix format and solved for the voltages to find $V_A = 4.3653$ V and $V_o = V_B = -0.6347$ V. The mathematical details of calculating these voltages are left to the student.

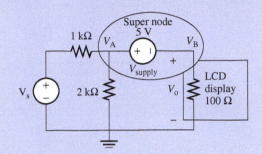

FIGURE 3.65 The circuit for Example 3.27.

3.6 THÉVENIN/NORTON EQUIVALENT CIRCUITS

This section introduces two theorems which facilitate the possibility of replacing a complex system or circuit containing resistors and sources with a simple equivalent circuit. This replacement eases the process of evaluating the current and voltage of a load when it is connected to the system. For example, consider a computer that is connected to a robot to control the movement of the robot. How much current and, accordingly, how much power is delivered to the robot. Is that power sufficient enough to control the robot's movement? If not, then what kind of amplifier (see Chapter 8 for more information about amplifiers) should be connected to the output of the computer to enable the desired task?

The first theorem, known as the Thévenin theorem, states that any resistive circuit regardless of its complexity can be represented by a voltage source in series with a resistance. The equivalent circuit is called the *Thévenin equivalent circuit*. In this equivalent circuit, the voltage source is referred to as the Thévenin voltage (V_{th}) while the resistance is called the Thévenin resistance (R_{th}). The second theorem, known as the Norton theorem, states that any resistive circuit regardless of its complexity can be represented by a current source in parallel with a resistance. The equivalent circuit is called a *Norton equivalent circuit*. In this equivalent circuit, the current source is referred to as the *Norton voltage* (V_n) while the resistance is called the *Norton resistance* (R_n).

Figure 3.66 shows an example of Thévenin and Norton equivalent circuits. If the resistance, R_L, is connected to the equivalent Thévenin and Norton circuits instead of the resistive circuit (shown at left), there will be no change in its current and voltage. Therefore, compared to the original circuit, the voltage (current) of R_L remains unchanged.

Consider the circuit in Figure 3.67 and its Thévenin equivalent where terminals A and B are open-circuited. The Thévenin equivalent circuit doesn't change the current through or the voltage across the terminals A and B when they are connected to a load. As a result, V_{oc1} and V_{oc2} will be equal. Thus, Thévenin and Norton equivalent circuits are used to represent the internal structure of complex circuits simply by using a source and a resistor independent of their complexity.

Now, if KVL is applied to the open-circuited Thévenin equivalent circuit sketched in Figure 3.67 (right):

$$-V_{th} + R_{th}i + V_{oc2} = 0$$

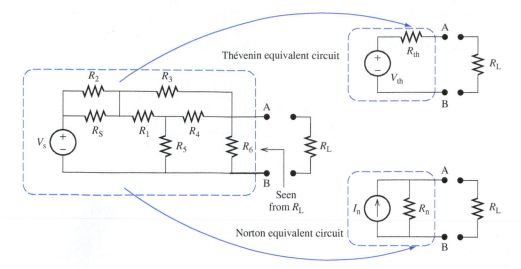

FIGURE 3.66
Thévenin and Norton
equivalent circuits.

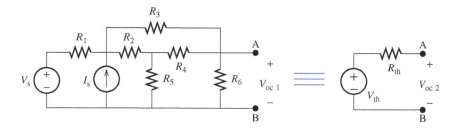

FIGURE 3.67
A Thévenin equivalent
circuit with open-
circuited terminals.

The circuit is open and therefore $i = 0$, so:

$$V_{\text{th}} = V_{\text{oc}\,2} = V_{\text{oc}\,1}$$

Accordingly, the Thévenin voltage will be equal to the open-circuit voltage of the original circuit, and in order to find it the terminals are opened, through which the Thévenin circuit can be found. Note that V_{th} is always proportional to V_{s}. Thus, if $V_{\text{s}} = 0$, V_{th} will be equal to zero. In that case, the resistor seen through the terminals A and B will be equal to R_{th}. This is an easy approach for finding R_{th}.

Now, consider the circuit shown in Figure 3.68 with its Norton equivalent. In this case, the terminals are short-circuited. According to the Norton theorem, $I_{\text{sc}\,1} = I_{\text{sc}\,2}$. In addition, by writing KCL for node A in the Norton equivalent short-circuit of Figure 3.68 (right) we find that:

$$-I_{\text{n}} + I + I_{\text{sc}\,2} = 0$$

Now, because $V = 0$, then $I = V/R_n = 0$. Thus:

$$I_{\text{n}} = I_{\text{sc}\,2} = I_{\text{sc}\,1}$$

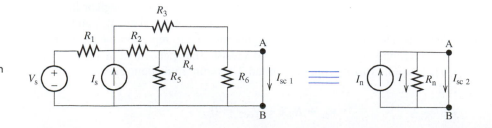

FIGURE 3.68 Norton
equivalent circuit
with short-circuited
terminals.

Therefore, the Norton current source is the same as the short-circuit current of the original circuit. In order to find the Thévenin and Norton resistances, consider the open-circuited Norton and short-circuited Thévenin equivalent circuits in Figures 3.69(a) and (b).

From Figure 3.69(a) and (b):

$$R_n = \frac{V_{oc}}{I_n}$$

and

$$R_{th} = \frac{V_{th}}{I_{sc}}$$

But, as shown in Figure 3.67, because no current flows through the circuit when it is open-circuited, $V_{th} = V_{oc}$. In addition, as seen in Figure 3.69(a), if terminals A and B are short-circuited, all I_n current will flow through these terminals, and $I_n = I_{sc}$; therefore:

$$R_{th} = R_n = \frac{V_{oc}}{I_{sc}} \tag{3.26}$$

Consequently, the Thévenin and Norton resistances are the same. In addition, in practice they can be found by first opening the circuit to find V_{oc} and then shortening the circuit to find I_{sc}. Then, by dividing V_{oc} and I_{sc} [Equation (3.26)] R_{th} can be found. Note that R_{th} or R_n is indeed the equivalent resistance seen through the same terminals.

Thus, Equation (3.26) represents another approach for finding the equivalent resistance of a circuit seen through any terminal: simply open-circuit the terminal and find the voltage seen through that open-circuited terminal; then, short-circuit that terminal and find the current through the short-circuited terminal; the division of the open-circuited voltage and short-circuited current is the equivalent resistance.

As stated earlier, $R_{th} = R_n$ can also be found by setting all independent sources in the main circuit to zero. This point is depicted in Figure 3.70.

Therefore, there are two ways to calculate R_{th}:

1. Using Equation (3.26), that is, finding V_{oc} and I_s.
2. Setting all sources to zero and finding the equivalent resistance through the terminals as discussed in Section 3.2.

Note that:

- An independent voltage source with zero voltage is equivalent to a short circuit.
- An independent current source with zero current is equivalent to an open circuit.

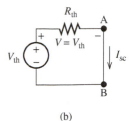

(a)

(b)

FIGURE 3.69 (a) Open-circuited Norton equivalent circuit. (b) Short-circuited Thévenin equivalent circuit.

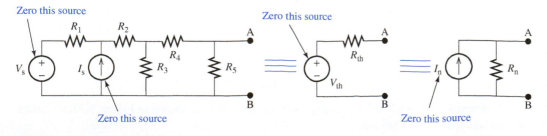

FIGURE 3.70 By zeroing the sources, the Thévenin and Norton theorems will still be true.

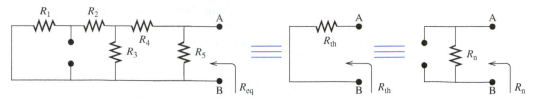

FIGURE 3.71 Zeroing the sources in order to find the Thévenin and Norton resistances.

Therefore, in order to zero the independent sources in a circuit, replace the voltage sources with a short circuit and the current sources with an open circuit. For example, consider Figure 3.70. If all the sources are zeroed, the circuit and its equivalents will correspond to Figure 3.71. After zeroing the sources, the equivalent resistance between the terminals will be the same as the Thévenin or Norton equivalent resistance. In this case, this equivalent resistance is:

$$R_{th} = R_n = R_{eq} = \{[(R_1 + R_2)\|R_3] + R_4\}\|R_5$$

Here, the notation $\|$ represents parallel resistors.

> It is *important to note* that the source zeroing technique can only be used for independent, *not dependent*, sources. For circuits that include only dependent sources, the first technique [i.e., Equation (3.26)] is the only method for finding the equivalent resistance of the circuit.

EXAMPLE 3.28 Thévenin Equivalent Circuit

Find the Thévenin equivalent of the circuit in Figure 3.72(a) as seen by R_L.

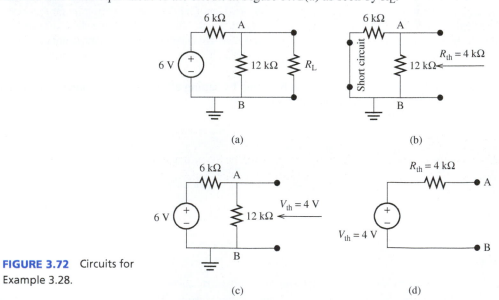

(a)   (b)

(c)   (d)

FIGURE 3.72 Circuits for Example 3.28.

SOLUTION

Equating the independent sources to zero, the 6-V voltage source is replaced by a short circuit. Recall that by equating a voltage source to zero, it will be replaced by a short circuit. In addition, equating a current source to zero, it will be replaced by an open circuit. The Thévenin resistance is the equivalent resistance between the A and B terminals when these terminals are open-circuited, which is shown in Figure 3.72(b):

$$R_{th} = R_{eq} = 6\,k\Omega\|12\,k\Omega = 4\,k\Omega$$

The Thévenin voltage is the open-circuit voltage of node B as shown in Figure 3.72(c). Using the voltage dividing technique:

$$V_{th} = \frac{12}{6 + 12} \times 6 = 4 \text{ V}$$

The Thévenin equivalent circuit of Figure 3.72(c) is sketched in Figure 3.72(d).

EXAMPLE 3.29   **Norton Equivalent Circuit**

Find the Norton equivalent of the circuit in Figure 3.73(a) as seen by R_L.

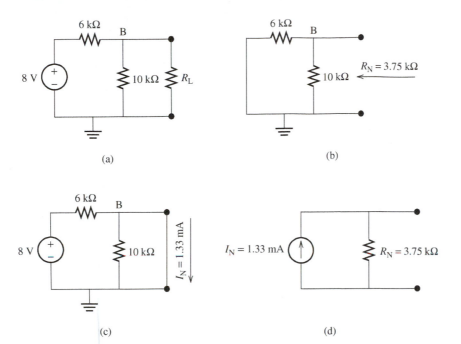

FIGURE 3.73   Circuits for Example 3.29.

SOLUTION

Equating the independent voltage sources to zero, the 8-V voltage source will be replaced by a short circuit. The Norton resistance is the equivalent resistance between the terminals in Figure 3.73(b), when they are open-circuited, which is:

$$R_n = R_{eq} = 10 \text{ k}\Omega \parallel 6 \text{ k}\Omega = 3.75 \text{ k}\Omega$$

The Norton current is the current passing from the short-circuited terminals in Figure 3.73(c). In this case, the current through the 10-kΩ resistance is zero. This can be verified simply by using the current division technique. As a result:

$$I_n = \frac{8V}{6 \text{ k}\Omega} = 1.33 \text{ mA}$$

The Norton equivalent circuit is sketched in Figure 3.73(d).

EXCERCISE 3.9

Consider current I is applied to two parallel resistances: one has a nonzero resistance of R and the other one has a zero resistance (i.e., short circuit). Calculate the current through each resistor.

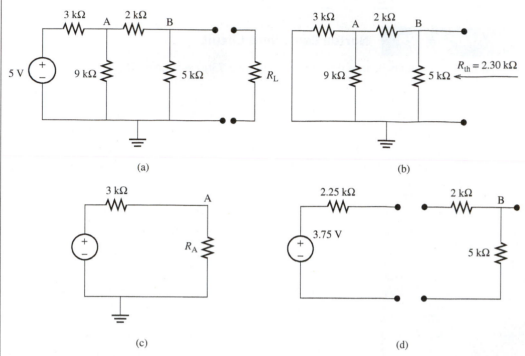

| EXAMPLE 3.30 | **Thévenin Equivalent Circuit through a Resistor** |

In Figure 3.74, find the Thévenin equivalent as seen by R_L.

FIGURE 3.74   Circuits for Example 3.30.

SOLUTION

Replace the 3-V voltage source with a short circuit. The Thévenin resistance is the equivalent resistance between the terminals in Figure 3.74(b), which is:

$$R_{th} = R_{eq} = 5\,k\Omega \parallel (2\,k\Omega + (9\,k\Omega \parallel 3\,k\Omega)) = 2.30\,k\Omega$$

The Thévenin voltage is the open-circuit voltage of node B as shown in Figure 3.74(a). To find this voltage, any of the following approaches can be selected:

Approach 1: Writing node voltage equations for nodes A and B results in:

$$\begin{cases} \dfrac{V_A - 5}{3\,K} + \dfrac{V_A}{9\,K} + \dfrac{V_A - V_B}{2\,K} = 0 \\[3mm] \dfrac{V_B - V_A}{2\,K} + \dfrac{V_B}{5\,K} = 0 \end{cases} \rightarrow \begin{cases} \left(\dfrac{1}{3} + \dfrac{1}{2} + \dfrac{1}{9}\right)V_A - \dfrac{1}{2}V_B = \dfrac{5}{3} \\[3mm] -\dfrac{1}{2}V_A + \left(\dfrac{1}{2} + \dfrac{1}{5}\right)V_B = 0 \end{cases}$$

Solving for V_A and V_B, the results are $V_A = 2.838$ V and $V_B = 2.027$ V. Therefore, $V_{th} = V_B = 2.027$ V.

Approach 2: The voltage dividing concept can be used. First, find the voltage of node A and then the voltage of node B. The resistance at node A, R_A corresponds to:

$$R_A = 9\,k\Omega \parallel (2\,k\Omega + 5\,k\Omega) = 3.9375\,k\Omega$$

Therefore, the equivalent circuit from the point of view of node A is the one sketched in Figure 3.74(c). Accordingly, the voltage at V_A is:

$$V_A = \frac{R_A}{R_A + 3} \times 5 = 2.838$$

Now, once more using voltage dividing concept:

$$V_{th} = V_B = \frac{5}{5+2} \times V_A = 2.027 \text{ V}$$

Approach 3: Find the Thévenin equivalent circuit seen through node A first by disconnecting the 2-kΩ and 5-kΩ resistors. Now the Thévenin equivalent circuit across terminal A can be easily calculated. Its voltage can be found using voltage division:

$$3.75 \text{ V} = \frac{9}{3+9} \times 5 \text{ V}$$

In addition, the equivalent resistance will be $2.25 \text{ k}\Omega = 3 \text{ k}\Omega \parallel 9 \text{ k}\Omega$. Thus, Figure 3.74(d) shows the equivalent Thévenin circuit of the circuit seen through terminal A. Now, by connecting it to the rest of the circuit [i.e., 2-kΩ and 5-kΩ resistors, as shown in Figure 3.74(d)], the equivalent circuit seen through terminal B is found. Now, using voltage division rule:

$$V_{th} = V_B = \frac{5}{5+2+2.25} \times 3.75 = 2.027 \text{ V}$$

EXCERCISE 3.10

In Example 3.30, use approach 3 to find the equivalent Thévenin resistance and compare it with the computed value.

EXAMPLE 3.31   Norton Resistance

Find the Norton resistance as shown in Figure 3.75 by equating the sources to zero.

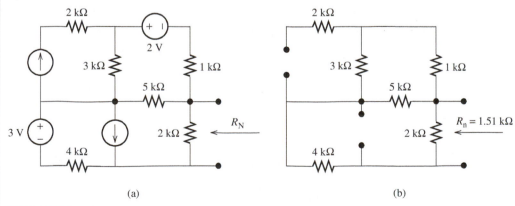

(a)   (b)

FIGURE 3.75 Circuits for Example 3.31.

SOLUTION

After equating the current and voltage sources to zero, the equivalent circuit between the terminals will be as shown in Figure 3.75(b). Accordingly, it can be verified that:

$$R_n = R_{eq} = [\{(1 \text{ k}\Omega + 3 \text{ k}\Omega \parallel (5 \text{ k}\Omega)\} + 4 \text{ k}\Omega] \parallel (2 \text{ k}\Omega) = 1.5135 \text{ }\Omega$$

| EXAMPLE 3.32 | **Norton Resistance** |

Find the Norton equivalent resistance as seen by R_L in Figure 3.76(a).

SOLUTION

The Norton equivalent resistance is $R_n = V_{oc}/I_{sc}$. It is important to determine both the open-circuit voltage and the short-circuit current of the terminals. Therefore, two separate circuits need to be analyzed. The open-circuit scenario is shown in Figure 3.76(b). Here, no current flows through the 2-kΩ resistor.

a. KCL in node A: Here, the current in the 3-kΩ resistor leaves the node. The voltage across this resistor is $V_A - 4$. Thus, the current in it will be $(V_A - 4)/3$ k. In addition, the current that enters this node is 0.004 V_x. Thus:

$$\frac{V_A - 4}{3\,\text{k}} - 0.004\,V_x = 0 \quad \rightarrow \quad 12\,V_x = V_A - 4$$

b. KVL in loop A:

$$V_x = V_A - 4$$

Now, observe that $12\,V_x = V_A - 4$ and $V_x = V_A - 4$, thus, $V_x = 0$ and:

$$V_A = V_{oc} = 4\ \text{V}$$

To find the short-circuit current, short-circuit the output terminal [see Figure 3.76(c)].

a. KCL in node A:

$$\frac{V_A - 4}{3\,\text{k}} - 0.004\,\text{V}_x + \frac{V_A}{2\,\text{k}} = 0 \quad \rightarrow \quad \left(\frac{1}{2} + \frac{1}{3}\right)V_A - 4\,\text{V}_x = \frac{4}{3}$$

b. KVL in loop A:

$$V_x = V_A - 4$$

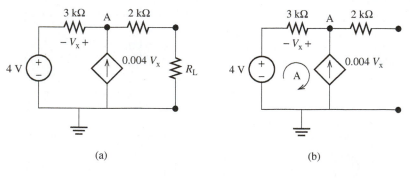

(a)   (b)

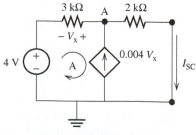

(c)

FIGURE 3.76   Circuits for Example 3.32.

Combining these two equations results in: $V_A = 4.63157\,\text{V}$ and $I_{sc} = \dfrac{V_A}{2\,\text{K}} = $ 2.3157mA $= I_n$. Therefore, the Thévenin/Norton resistance corresponds to:

$$R_n = \frac{V_{oc}}{I_{sc}} = \frac{4}{2.3157\ \text{mA}} = 1.727\ \text{k}\Omega$$

APPLICATION EXAMPLE 3.33   Aerial Fireworks

Aerial fireworks are used in various celebration events. In order to synchronize the firework ignition process, a computer is required. Figure 3.77(a) shows a simple setup of the computer and the igniter for an aerial fireworks show.

Assume a voltage source of 20 V is connected to the computer in series; the computer has a resistance of 500 Ω, and the igniter has a resistance of 50 Ω when it receives a signal from the computer to order it to launch. Assume the same ground for all igniters and the computer voltage source. The equivalent circuit for the computer and igniters is shown in Figure 3.77(b).

a. Sketch the Thévenin and Norton equivalent circuits of the computer.
b. Calculate the current flow through each igniter when it receives the order to ignite the firework. Use two methods: (i) calculate the total resistance and then the current through the computer first, and (ii) use the Norton equivalent circuit of computer and current division.

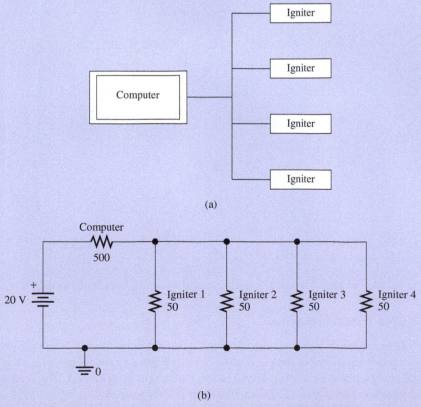

(a)

(b)

FIGURE 3.77   (a) A simple computer setup and igniters; (b) its equivalent circuit.

(continued)

APPLICATION EXAMPLE 3.33   Continued

SOLUTION

a. From Figure 3.77(b), the computer's equivalent circuit is a voltage source in series with a 500-Ω resistor. Indeed, this is the Thévenin equivalent circuit of the computer. The Norton equivalent circuit of the computer will be a current source in parallel to the same resistor. The Thévenin and Norton equivalent circuits are shown in Figure 3.78(a) and (b), respectively.

b. The total resistance from the igniters can be calculated as follows:

$$\frac{1}{R_{\text{parallel}}} = \frac{1}{R_1} + \frac{1}{R_2} + \frac{1}{R_3} + \frac{1}{R_4}$$

$$\frac{1}{R_{\text{parallel}}} = \frac{1}{50} + \frac{1}{50} + \frac{1}{50} + \frac{1}{50}$$

$$R_{\text{parallel}} = 12.5 \ \Omega$$

The total resistance is:

$$R_{\text{total}} = R_{\text{computer}} + R_{\text{parallel}}$$

$$R_{\text{total}} = 500 + 12.5 = 512.5 \ \Omega$$

Therefore, the current of the computer is:

$$I = \frac{V}{R_{\text{total}}}$$

$$I = \frac{20}{512.5} = 39 \text{ mA}$$

Because the resistance of each igniter is equivalent, the current flow in each igniter is:

$$I_{\text{igniter}} = \frac{1}{4} I$$

$$I_{\text{igniter}} = 9.75 \text{ mA}$$

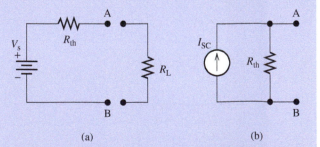

FIGURE 3.78   (a) Thévenin and (b) Norton equivalent circuit for the circuit of Figure 3.77.

(a)   (b)

3.6.1 Source Transformation

We have been exclusively working with ideal voltage and current sources which provide constant voltage or current regardless of the load resistor. A more accurate source model contains an internal resistance which accounts for observed reduction in voltage or current. The practical voltage and current sources are given in Figure 3.79(a) and (b), respectively.

We will see that the practical voltage and current sources may be interchangeable without affecting the remainder of the circuit. Such practical sources are defined as being equivalent if they produce identical values of I_L and V_L when they are connected to the identical load R_L, which is shown in Figure 3.79. Here, we intend to find the relationship of R_{th} and R_n, and the relationship of V_{th} and I_n.

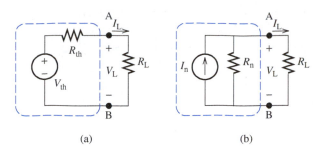

<table>
<tr><td align="center">(a)</td><td align="center">(b)</td><td></td></tr>
</table>

FIGURE 3.79   Equivalent sources.

When $R_L = 0$, the circuit between nodes A and B is short, that is, the total voltage of the source is across R_{th} in Figure 3.79(a), and no current flows through R_n in Figure 3.79(b). Therefore, I_L for Figure 3.79 (a) and (b) is:

$$I_L^a = V_{th}/R_{th}$$
$$I_L^b = I_n$$

When $R_L = \infty$, the circuit between nodes A and B is open, that is, there is no current through R_{th} in Figure 3.79(a) and the total current of the source flows through R_n in Figure 3.79(b).

Therefore, V_L for Figure 3.79(a) and (b) is:

$$V_L^a = V_{th}$$
$$V_L^b = I_n R_n$$

Because of equivalency, $V_L^a = V_L^b$ and $I_L^a = I_L^b$, then:

$$I_n = V_{th}/R_{th}$$
$$I_n = V_{th}/R_n$$

Therefore:

$$R_{th} = R_n$$
$$I_n = V_{th}/R_n$$

As a conclusion we can convert a Norton circuit to Thévenin and vice-versa. This approach can be used to calculate the equivalent Thévenin and Norton circuit for more complicated circuits (see Examples 3.34 and 3.35).

EXAMPLE 3.34   **Equivalent Thévenin and Norton Using Source Transformation**

The circuit of Example 3.28 has been re-sketched in Figure 3.80(a). Find the equivalent Thévenin and Norton circuit seen across the load R_L using source transformation.

(continued)

EXAMPLE 3.34   **Continued**

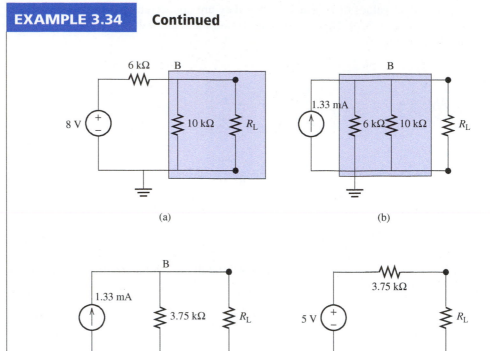

FIGURE 3.80   The figure of Example 3.34.

SOLUTION

Here, first, for simplicity, we assume everything beyond node B as the circuit load. Next, we find the equivalent Thévenin circuit of the 8-V source that is in series with the 6-kΩ resistor. The equivalent current source is:

$$I = \frac{8 \text{ V}}{6 \text{ k}\Omega} = 1.33 \text{ mA}$$

This equivalent circuit is sketched in Figure 3.80(b). Now, there are two parallel resistors of 6 kΩ, and 10 kΩ. The equivalent resistance of these two is 3.75 kΩ. This equivalent circuit has also been depicted in Figure 3.80(c). Now, the 1.33 mA in parallel to 3.75 kΩ form the equivalent Norton circuit that is seen through the terminals of the load resistor. We can equivalently find the corresponding Thévenin resistor seen through the terminals of the load resistor which is sketched in Figure 3.80(d).

EXAMPLE 3.35   **Equivalent Thévenin and Norton Using Source Transformation**

In the circuit shown in Figure 3.81, find the equivalent Norton and Thévenin circuit observed across the load resistor using source transformation.

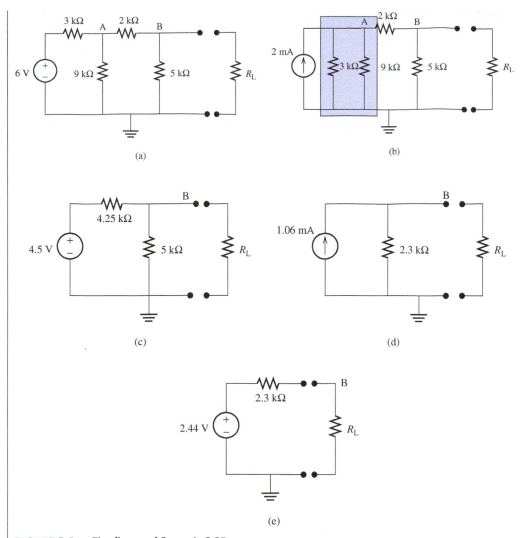

(a)

(b)

(c)

(d)

(e)

FIGURE 3.81 The figure of Example 3.35.

SOLUTION

The step-by-step process has been detailed in Figure 3.81(b–d). In Figure 3.81(b), we have converted the Thévenin circuit of 6 V voltage and the series resistor of 3 kΩ to its Norton equivalent circuit. Now there is another resistor of 9 kΩ. The equivalent resistance of 3 kΩ ∥ 9 kΩ is 2.5 kΩ. Thus, the 2-mA source would be indeed in parallel to the 2.5-kΩ resistor.

Now, the 2-mA current source and the parallel 2.5-kΩ resistance create a new Norton circuit. Converting this Norton circuit into Thévenin, we will have the 5-V source in series with the 2.5-kΩ resistor. In this condition, the 2.5-kΩ resistor will be in series with the 2-kΩ between the terminals A and B. Thus, the equivalent resistance will be 4.5 kΩ, as shown in Figure 3.81(c).

Now, again, the 5-V source is in series with the 4.5-kΩ resistor which forms another Thévenin circuit. Converting it to its Norton equivalent, the current source will be 4.5 V/4/25 kΩ = 1.06 mA. In addition, we will have two parallel resistances of 4.5 kΩ ∥ 5 kΩ, which is equivalent to 2.368 kΩ. Thus, the total Norton circuit across the load would be as shown in Figure 3.81(d). The Thévenin converted of this load is shown in Figure 3.81(d).

3.7 SUPERPOSITION PRINCIPLE

A circuit may consist of multiple sources. An example is a vehicle where the electrical systems might be supplied by both the car's battery and its generator. When multiple sources are available, it might seem difficult to find the voltage or current of the resistor. In a resistive circuit, the voltage or the current of a resistor can be considered a response to independent sources. The principle of superposition states that the total voltage (current) in any part of a linear circuit equals the algebraic sum of the voltages (currents) produced by each source, when other sources are set to zero.

It is *important to note* that superposition only applies to *linear* systems. The resistors, capacitors, and inductors described in this book are considered linear elements and the circuits they make up are considered linear systems. However, not all resistors, capacitors, or inductors are not linear systems. A linear system is the one, whose input, x, and output, $f(x)$, relationship can be maintained using a linear function. A function, $f(x)$, is called linear if it possesses the two conditions of additivity and homogeneity that are defined as:

Additivity: $f(x_1 + x_2) = f(x_1) + f(x_2)$

Homogeneity: $f(ax) = af(x)$

For example for a linear resistor, input–output relationship is defined by Ohm's law, that is, $v = R \cdot i$. It is clear that if we replace the current, i, with $i_1 + i_2$, we will have:

$$v = R \cdot (i_1 + i_2) = R \cdot i_1 + R \cdot i_2 = v_1 + v_2$$

In addition, if we replace i, with αi, we will have:

$$v = R \cdot \alpha i = \alpha(R \cdot i) = \alpha v$$

Thus, a linear resistor fulfills both additivity and homogeneity conditions and can be considered as a linear system. Note that, in general the current–voltage relationship might not be linear.

EXCERCISE 3.11

Show that if voltage–current relationship is represented by $v = Ri + \beta i^2$, the system would be nonlinear.

In order to apply the superposition principle in a circuit, first find the response (voltage or current) of a resistor to a source by setting other sources in the circuit to zero. Recall that the voltage and current sources can be equated to zero by replacing them with short and open circuits, respectively. Note that, in applying the superposition principle, the dependent sources are not set to zero.

For example, consider the circuit in Figure 3.82(a). The goal is to determine the current, I, flowing through the resistor R_2. By applying nodal voltage analysis, and writing KCL for node A in Figure 3.82(b):

$$\frac{V - V_s}{R_1} + \frac{V}{R_2} - I_s = 0$$

Rearranging to find the voltage, V:

$$V = \frac{R_1 R_2}{R_1 + R_2}\left(\frac{V_s}{R_1} + I_s\right) = \frac{R_1 R_2}{R_1 + R_2}\frac{V_s}{R_1} + \frac{R_1 R_2}{R_1 + R_2}I_s \qquad (3.27)$$

Now, the current through R_2, is equal to its voltage divided by the resistance, that is:

$$I = \frac{V}{R_2} = \frac{R_1}{R_1 + R_2}\left(\frac{V_s}{R_1} + I_s\right) = \frac{V_s}{R_1 + R_2} + \frac{R_1 I_s}{R_1 + R_2} \qquad (3.28)$$

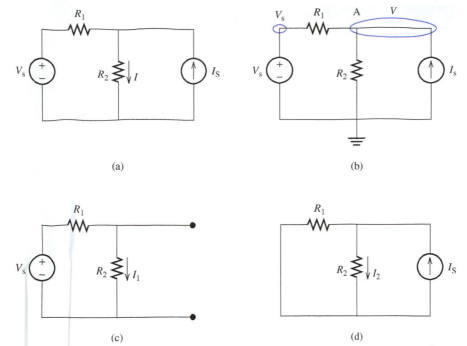

(a)

(b)

(c)

(d)

FIGURE 3.82
Superposition principles.

In Equations (3.27) and (3.28), the first term is the response to the voltage source (setting $I_s = 0$) and the second term is the response to the current source (setting $V_s = 0$). Thus, the superposition principle can also be used to find the current, I, by first setting $V_s = 0$ and finding V that is due to I_s, and then setting $I_s = 0$ and finding a second V that is due to V_s, and finally, adding these two voltages. Note that (Chapter 2, Exercises 2.4 and 2.5):

A zero-current source is equivalent to an open circuit.
A zero-voltage source is equivalent to a short circuit.

The current due to the voltage source and in the absence of the current source [Figure 3.82(c)] is:

$$I_1 = \frac{V_s}{R_1 + R_2}$$

The current due to the current source while the voltage source is zero [Figure 3.82(d)] will be:

$$I_2 = \frac{R_1 I_s}{R_1 + R_2}$$

Applying the superposition principle, the current i, in the original circuit is:

$$I = I_1 + I_2 = \frac{R_1}{R_1 + R_2}\left(\frac{V_s}{R_1} + I_s\right)$$

which is equivalent to Equation (3.28).

EXAMPLE 3.36   **Superposition**

Find the voltage V_1 in Figure 3.83 using the superposition technique.

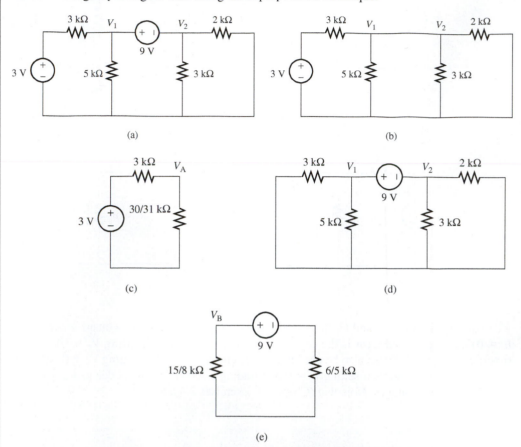

(a)   (b)

(c)   (d)

(e)

FIGURE 3.83   Circuits for Example 3.37. (a) The original circuit. (b,c) Setting 9 V to zero. (d, e) Setting 3 V to zero.

SOLUTION

First, equate the 9-V and 3-V voltage sources to zero, respectively. The resulting circuits and their equivalents are shown in Figure 3.83. According to the equivalent circuit in Figure 3.83(c), which is for the case when the 9-V voltage source is equated to zero [see Figure 3.83(b)], the node voltage due to the 3-V voltage source (V_A) is:

$$V_A = \frac{\dfrac{30}{31}}{3 + \dfrac{30}{31}} \times 3 = 0.7317$$

Similarly, according to the equivalent circuit in Figure 3.83(e), which is for the case when the 3-V voltage source is equated to zero [see Figure 3.83(d)] the node voltage due to the 9-V voltage source (V_B) is:

$$V_B = \frac{\dfrac{15}{8}}{\dfrac{15}{8} + \dfrac{6}{5}} \times 9 = 5.4878 \text{ V}$$

Applying the superposition principle to the node voltage V_1 in Figure 3.83(a) is the summation of the voltages V_A and V_B, that is:

$$V_1 = V_A + V_B = 6.22 \text{ V}$$

APPLICATION EXAMPLE 3.37   Solar-Powered Heater

Figure 3.84 shows a solar-powered shower heating system. In order to reduce the cost of energy, solar energy is produced during the daytime to heat the water that is used for shower purposes.

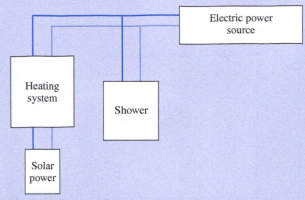

FIGURE 3.84   A solar-powered shower system.

Assume the solar power produces a voltage source of 100 V, the electric power supply is 110 V, the heater resistance is 100 Ω and the shower resistance is 80 Ω. Calculate the current flow through the shower.

SOLUTION

The equivalent circuit for the system in Figure 3.84 is shown in Figure 3.85.

Using the superposition concept, first, equate the electric voltage source to zero as shown in Figure 3.85(b). In this case, the shower is short-circuited. Thus, the current through the shower, I_1, is zero.

Next, consider the case when there is no solar power source [Figure 3.85(c)]. In this case, the heater and the shower become parallel resistors. The current flow through the shower, I_2, can be calculated as:

$$I_2 = \frac{V_{\text{electric}}}{R_{\text{shower}}}$$

$$I_2 = \frac{110}{80} = 1.375 \text{ A}$$

Therefore, the current flow through the shower is:

$$I_{\text{shower}} = I_1 + I_2$$

$$I_{\text{shower}} = 0 + 1.375 = 1.375 \text{ A}$$

So far superposition has been discussed for use across independent sources. However, superposition can also be applied to circuits that include both independent and dependent sources. The following examples clarify this point.

(continued)

APPLICATION EXAMPLE 3.37   **Continued**

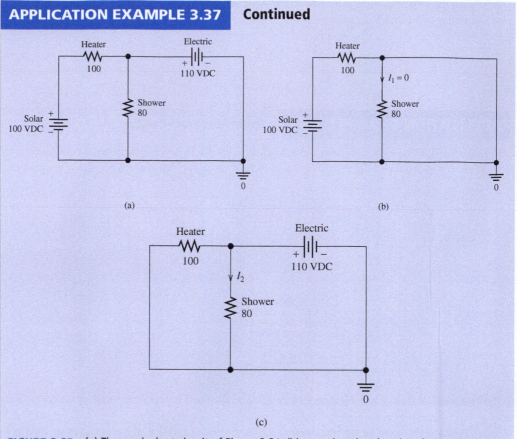

(a)

(b)

(c)

FIGURE 3.85   (a) The equivalent circuit of Figure 3.84; (b) equating the electric voltage source to zero; (c) equating the heater to zero.

EXAMPLE 3.38   **Superposition**

Find the current, I, in Figure 3.86 using the superposition technique.

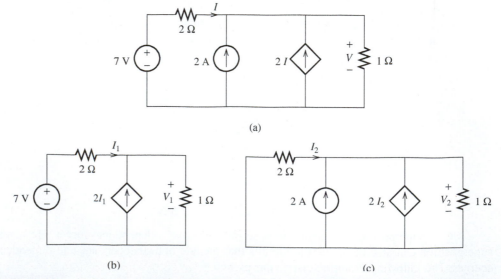

(a)

(b)

(c)

FIGURE 3.86   Circuits for Example 3.38. (a) The original circuit. (b) Setting 2 A to zero. (c) Setting 7 V to zero.

SOLUTION

Equate the 7-V voltage sources and the 2-A current sources to zero, respectively. The resulting circuits and their equivalents are shown in Figure 3.86.

According to the equivalent circuit in Figure 3.86(b), the node voltage due to the 7-V voltage source (V) is:

$$V_1 = 7 - 2I_1 = (I_1 + 2I_1) \times 1$$

$$I_1 = \frac{7}{5}A = 1.4\,A$$

Similarly, according to the equivalent circuit in Figure 3.86(c), the node voltage due to the 2-A current source (V) is:

$$V_2 = -2I_2 = (3I_2 + 2) \times 1$$

$$I_2 = -\frac{2}{5}A = -0.4\,A$$

According to the superposition principle, the current, I, in Figure 3.86(a) is the summation of the currents I_1 and I_2. Thus:

$$I = I_1 + I_2 = 1\,A$$

EXAMPLE 3.39   **Superposition**

Find the voltage, V, in Figure 3.87 using the superposition technique.

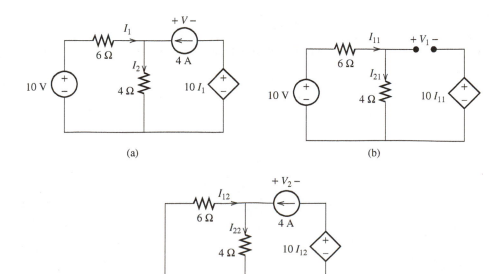

(a)

(b)

(c)

FIGURE 3.87   Circuits for Example 3.39. (a) The original circuit. (b) Setting 4 A to zero. (c) Setting 10 V to zero.

(*continued*)

EXAMPLE 3.39   **Continued**

SOLUTION

First, zero the 10-V voltage sources and 4-A current sources, respectively. The resulting circuits and their equivalents are shown in Figure 3.87 (b) and (c). According to the equivalent circuit in Figure 3.87(b), the node voltage due to the 10-A voltage source, I_{11}, is:

$$I_{11} = \frac{10}{6 + 4} = 1 \text{ A}$$

$$V_1 = -10I_{11} + 4I_{11} = -6 \text{ V}$$

Similarly, according to the equivalent circuit in Figure 3.87(c), the node voltage due to the 2-A voltage source, I_{12}, is:

$$I_{12} = -\frac{4}{6 + 4} \times 4 = -1.6 \text{ A}$$

$$I_{22} = 4 + I_{12} = 2.4 \text{ A}$$

$$V_2 = -10I_{12} + 4I_{22} = 25.6 \text{ V}$$

According to the superposition principle, the current I in Figure 3.87(a) is the summation of the voltages V_1 and V_2. Thus:

$$V = V_1 + V_2 = 19.6 \text{ V}$$

3.8 MAXIMUM POWER TRANSFER

As demonstrated by the Thévenin theorem, every two-terminal resistive circuit can be replaced by a Thévenin equivalent circuit that consists of a voltage source and a resistance. Suppose a load resistance, R_L, is connected to the terminals of the Thévenin equivalent circuit as shown in Figure 3.88.

In this section, the goal is to find the load resistance that absorbs the maximum power from the two-terminal resistive circuit. The current flowing through the load resistance is given by:

$$I_L = \frac{V_{th}}{R_L + R_{th}}$$

The power delivered to the load is:

$$P_L = R_L I_L^2$$

Therefore:

$$P_L = \frac{V_{th}^2 R_L}{(R_{th} + R_L)^2}$$

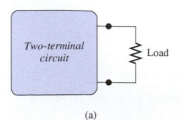
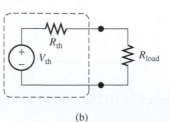

FIGURE 3.88 (a) A two-terminal circuit and (b) its Thévenin equivalent.

(a)

(b)

In order to determine the load resistance such that the maximum possible power is delivered to the load, the derivative of the power, P_L, with respect to R_L, is set to zero. This is a general concept in mathematics.

To find the maximum or minimum of a function, $f(x)$, with respect to the variable x, find the derivative of $f(x)$ with respect to x, that is, $f'(x)$, and set it to zero. Solving $f'(x) = 0$, for x, will result in the value of x, namely x_m, that maximizes or minimizes $f(x)$. In addition, from the concepts learned in mathematics, replacing x in the second derivative, that is, $f''(x_m)$, with x_m, $f''(x_m) < 0$, then x_m represents a maxima; otherwise, it represents a minima.

Solving $dP_L/dR_L = 0$ with respect to R_L, it can be determined that:

$$for\ \mathrm{P}_{max}:\ R_L = R_{th} \tag{3.29}$$

As a result, the load resistance that absorbs the maximum power is equal to the Thévenin equivalent resistance. In this case, the power transferred to the load will be:

$$P_L = \frac{V_{th}^2}{4R_{th}} \tag{3.30}$$

EXCERCISE 3.12

Work out the details of differentiation $dP_L/dR_L = 0$ and show that indeed $R_L = R_{th}$ leads to the maximum (and not the minimum) power transfer.

EXAMPLE 3.40   Maximum Power Transfer

In Figure 3.89, find the maximum power transferred to a speaker, when the resistor R is equal to the speaker resistance. Here, $V = 10$ V and $R_{th} = 100$ Ω.

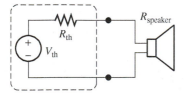

FIGURE 3.89 Figure for Example 3.40.

SOLUTION

Using Equation (3.30), and assuming $R_{speaker} = R_{th}$ we have:

$$P_L = \frac{V_{th}^2}{4R_{th}} = \frac{10^2}{4 \times 100} = 0.25\ \mathrm{W}$$

APPLICATION EXAMPLE 3.41   Strain Gage

Strain gage has already been introduced and discussed in Example 3.10. Resistive strain gages are often connected in groups of four, as shown in Figure 3.90(a), known as quad cells. These groupings are to eliminate the effect of the surrounding temperature on the strain reading. Only two of the gages are connected to the beam.

a. Find the Thévenin equivalent voltage and resistance between V_A and V_B;
b. Find the power dissipated in a 20.8-kΩ resistance installed between nodes A and B.

(*continued*)

APPLICATION EXAMPLE 3.41 **Continued**

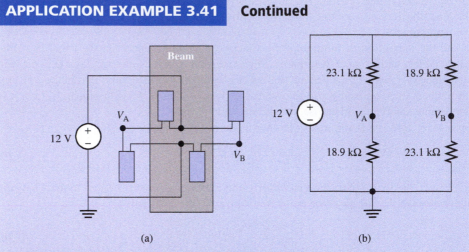

(a) (b)

FIGURE 3.90 (a) The strain gage and (b) its equivalent circuit.

SOLUTION

a. The equivalent circuit of Figure 3.90(a) is sketched in Figure 3.90(b). This structure is called a *Wheatstone bridge*. Here, the approach of $R_{th} = V_{oc}/I_{sc}$ is used. The open-circuit voltage V_{oc} is across the terminals A and B, that is, when A and B are detached (open-circuited), and I_{sc} is for the scenario that these two terminals are connected (short-circuited).

To find the short-circuit current between nodes A and B, short-circuit these two nodes as shown in Figure 3.91 and write the nodal voltage equations as:

$$\frac{V_A - 12}{23.1\,K} + \frac{V_A}{18.9\,K} + I_{sc} = 0$$
$$\frac{V_B - 12}{18.9\,K} + \frac{V_B}{23.1\,K} - I_{sc} = 0$$
$$V_A = V_B$$

Solving for I_{sc}, $I_{sc} = -57.72\ \mu A$.

Now, find the open-circuit voltage of the nodes $V_{oc} = V_A - V_B$. Based on the voltage division rule, the node voltages can be simply calculated:

$$V_A = 12\ V \times \frac{18.9\ k\Omega}{23.1\ k\Omega + 18.9\ k\Omega} = 5.4\ V$$

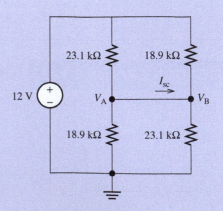

FIGURE 3.91 Circuit for Example 3.41.

$$V_{\mathrm{B}} = 12 \text{ V} \times \frac{23.1 \text{ k}\Omega}{23.1 \text{ k}\Omega + 18.9 \text{ k}\Omega} = 6.6 \text{ V}$$

Next, see that:

$$V_{\mathrm{th}} = V_{\mathrm{oc}} = V_{\mathrm{A}} - V_{\mathrm{B}} = -1.2 \text{ V}$$

Therefore, the Thévenin equivalent resistance will be:

$$R_{\mathrm{th}} = \frac{V_{\mathrm{oc}}}{I_{\mathrm{sc}}} = 20.8 \text{ k}\Omega$$

b. The Thévenin equivalent circuit was calculated seen through terminals A and B. It can be used to calculate the maximum power transfer for a load connected between terminals A and B. Incorporating Equation (3.30), it corresponds to:

$$P_{\mathrm{L}} = \frac{V_{\mathrm{th}}^2}{4 R_{\mathrm{th}}} = \frac{1.44}{4 \times 20.8 \text{ k}\Omega} = 17.3 \mu\text{W}$$

APPLICATION EXAMPLE 3.42   Sports Car Inventor

Back to our inventor and the hot rod acceleration test (see Example 3.17). Determine which of the circuits shown in Figure 3.92(a) or (b) match output resistance with the 50-Ω input resistance of the PDA. For this accelerometer, $0 \leq V_{\mathrm{ACC}} \leq 5$ V.

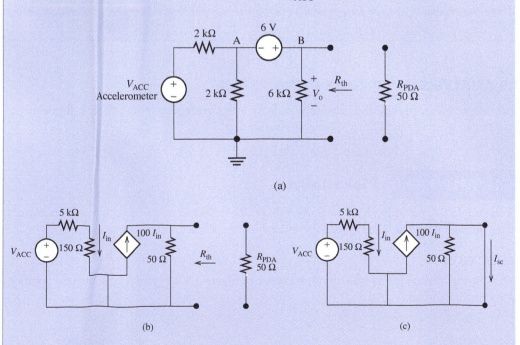

(a)

(b)   (c)

FIGURE 3.92   The circuits for Example 3.42.

SOLUTION

To find if the 50-Ω resistance has a good match with the circuit, first find the Thévenin equivalent circuit seen through terminal B (with respect to the ground), and check if $R_{\mathrm{th}} = R_{\mathrm{PDA}}$. For the circuit of Figure 3.92(a), in order to determine the Thévenin equivalent resistance,

(continued)

APPLICATION EXAMPLE 3.42 **Continued**

equate all sources to zero and find the equivalent resistance between the terminals. In this case, the equivalent resistance will be $R_{\text{th}} = 2\,\text{k}\Omega \,\|\, 2\,\text{k}\Omega \,\|\, 6\,\text{k}\Omega = 857\,\Omega$, but $R_{\text{PDA}} = 50\,\Omega$. Thus, $R_{\text{th}} \neq R_{\text{PDA}}$ and this circuit does not deliver maximum energy to the output (PDA).

For the second circuit in Figure 3.92(b), $V_{\text{ACC,max}} = 5\,\text{V}$ and $I_{\text{in}} = 5/(5000 + 150)$. The open-circuit voltage between the terminals is given by:

$$V_{\text{oc},max} = 100 I_{\text{in}} \times 50 = 100 \times \frac{5}{5150} \times 50 = 4.85\,\text{V}$$

As a result, the Thévenin voltage will be:

$$V_{\text{th},max} = V_{\text{oc,max}} = 4.85\,\text{V}$$

Moreover, the short-circuit current between the terminals [Figure 3.91(c)] will be obtained simply as:

$$I_{\text{sc},max} = 100 I_{\text{in}} = 100 \times \frac{5}{5150} = 97.1\,\text{mA}$$

Using the open-circuit voltage and the short-circuit current between the terminals:

$$R_{\text{th}} = \frac{V_{\text{oc},max}}{I_{\text{sc},max}} = \frac{4.85\,\text{V}}{97.1\,\text{mA}} = 49.94\,\Omega$$

Therefore, in this case, $R_{\text{th}} = R_{\text{PDA}}$, and the maximum power is absorbed by the PDA resistance.

3.9 ANALYSIS OF CIRCUITS USING PSPICE

Here, examples are provided to help students learn to solve problems using PSpice. Students are encouraged to review the PSpice tutorial in Section 2.9.

EXAMPLE 3.43 **PSpice Analysis**

In Example 3.4, assuming $R_1 = 10\,\Omega$, $R_2 = 20\,\Omega$, $R_3 = 30\,\Omega$, compute and compare the current flow through the circuits in Figure 3.16 and 3.17 using PSpice software.

SOLUTION

The PSpice solution for a series circuit is shown in Figure 3.93. Figure 3.94 shows the solution for a parallel circuit.

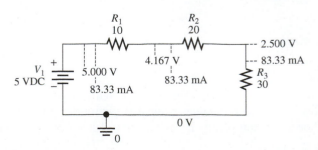

FIGURE 3.93 PSpice solution for Example 3.4, series circuit.

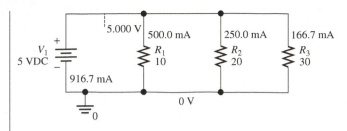

FIGURE 3.94 PSpice solution for Example 3.4, parallel circuit.

EXAMPLE 3.44   PSpice Analysis

Generate the results of Example 3.17 using PSpice.

SOLUTION

a. Find V_o if $V_A = 0.3$ V.
   1. To add a voltage-controlled voltage source into the PSpice schematic board go to "Parts" (see Figure 2.63 in Chapter 2) and type "E."
   2. Double-click the "E" icon to get into properties menu. Change the Gain to "5," and choose "display setting" while the "Gain" column is highlighted. Choose "Name and Value" for the display format and click OK.
   3. Run the simulation for Bias Point Analysis.

   The result is shown in Figure 3.95(a).

b. Find I_A when $V_A = V_{Amax}$. The result is shown in Figure 3.95(b).

c. How much power is dissipated by the PDA?

$$P = \frac{V_o^2}{R} = \frac{(12.5)^2}{50} = 3.125W$$

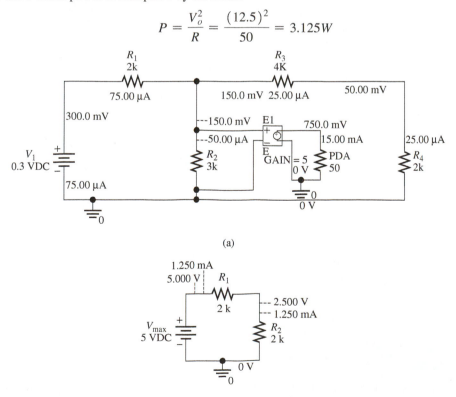

(a)

(b)

FIGURE 3.95 PSpice solution for Example 3.17 (a) Part (a); (b) Part (b).

EXAMPLE 3.45   PSpice Analysis

Use PSpice to find the nodal voltages V_1 and V_2 in Example 3.20.

SOLUTION

To study the voltage at a particular node, click and drag the voltage bar to the particular node as shown in Figure 3.96.

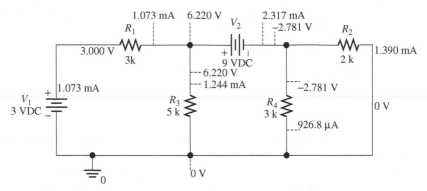

FIGURE 3.96   PSpice solution for Example 3.20.

EXAMPLE 3.46   PSpice Analysis

Use PSpice to find the Thévenin equivalent of the circuit in Example 3.28 [Figure 3.72(a)] seen by R_L.

SOLUTION

Because PSpice cannot run the simulation with an open circuit, the process instead re-defines the value of resistor R_L. In order to study the Thévenin equivalent of voltage, we set the value of R_L to be a very large resistance, which is shown in Figure 3.97.

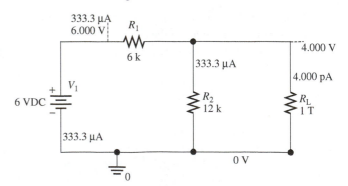

FIGURE 3.97   PSpice solution for Example 3.28.

EXAMPLE 3.47   PSpice Analysis

Find the Norton equivalent of the circuit in Example 3.29 as seen by R_L.

SOLUTION

To study the Norton equivalent current, the value of resistor, R_L needs to be selected as very small to represent a short circuit. Thus, the corresponding circuit corresponds to Figure 3.98.

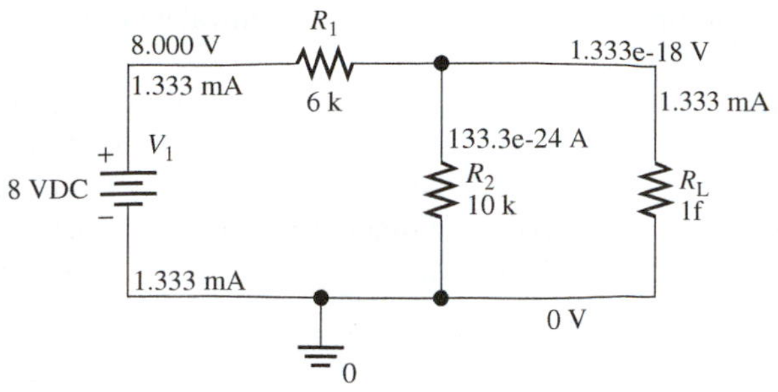

FIGURE 3.98 PSpice solution for Example 3.29.

3.10 WHAT DID YOU LEARN?

- The equivalent resistance of N resistors in series is shown in Equation (3.5) and is equal to:

$$R_{eq} = \sum_{n=1}^{N} R_n$$

- The equivalent resistance of N resistors in parallel is shown in Equation (3.10) and is equal to:

$$R_{eq} = \frac{1}{\dfrac{1}{R_1} + \dfrac{1}{R_2} + \cdots + \dfrac{1}{R_N}}$$

- The *voltage division rule* states that voltage is divided between N series resistors in direct proportion to their resistance [Equation (3.16)]:

$$v_k = \frac{R_k}{R_1 + R_2 + \cdots + R_N} v_s \ ; \ k = 1, 2, \ldots N$$

- The *current division rule* states that current is divided between two parallel resistors in inverse proportion to their resistance [see Equations (3.20) and (3.21)]:

$$i_1 = \frac{R_2}{R_1 + R_2} i_s$$

$$i_2 = \frac{R_1}{R_1 + R_2} i_s$$

- The nodal analysis procedure involves the following steps:
  1. Determine all nodes in the circuit.
  2. Pick a node as a reference node.
  3. Assign voltage variables to each node.
  4. For each node, write KCL equations in terms of node voltages.
  5. Solve the system of equations to find unknown voltages.
- The mesh analysis procedure involves the following steps:
  1. Define each mesh current (usually exclusively clockwise).
  2. Assign current variables to each mesh.
  3. To write the KVL for each mesh, determine the voltage across the components of that mesh based on the mesh current direction.
  4. For each mesh, write the KVL equation.
  5. Solve the system of equations to find unknown mesh currents.
- A *super node* is a closed surface in a circuit that includes several nodes. KCL can be applied to super nodes.
- Thévenin and Norton equivalent techniques can be used to simplify complex resistive circuits (see Figure 3.66).
  The *Thévenin theorem* states that any resistive circuit can be represented by a voltage source in series with a resistance.
  The *Norton theorem* states that any resistive circuit can be represented by a current source in parallel with a resistance.

Source transformation can be used to find the Thévenin or Norton equivalent of a complex circuit.

- The *principle of superposition* simplifies the solution of circuits with multiple sources. It states that the total voltage (or current) in any part of a linear circuit equals the algebraic sum of the voltages (or currents) produced by each source, when other sources are set to zero.
- *Maximum power transfer* means that the load resistance that absorbs the maximum power from a resistive network is equal to the Thévenin equivalent resistance of the resistive network [Equation (3.29)].

$$R_L = R_{th}$$

- PSpice can be used to analyze complex resistive circuits to find the voltage across and the current through all elements.

Problems

*B refers to Basic, A refers to Average, H refers to Hard, and * refers to problems with answers.*

SECTION 3.1 INTRODUCTION

3.1 (B) Which of the following is not resistive circuit?

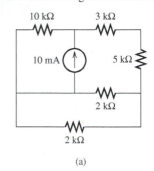

(a)

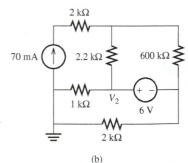

(b)

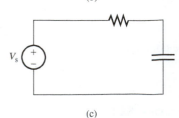

(c)

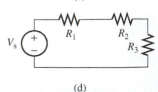

(d)

FIGURE P3.1 Circuit for Problem 3.1.

SECTION 3.2 RESISTORS IN PARALLEL AND SERIES AND EQUIVALENT RESISTANCE

3.2 (B) Identify which resistors are in parallel and which are in series in Figure P3.2.

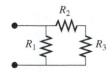

FIGURE P3.2 Circuit for Problem 3.2.

3.3 (B) Identify which resistors are in parallel and which are in series in Figure P3.3.

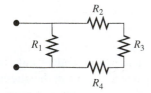

FIGURE P3.3 Circuit for Problem 3.3.

3.4 (B) Find R_{eq} in terms of R_1, R_2, R_3 in Figure P3.4.

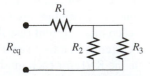

FIGURE P3.4 Circuit for Problem 3.4.

3.5 (A) Identify which resistors are in parallel and which are in series in Figure P3.5.

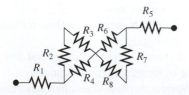

FIGURE P3.5 Circuit for Problem 3.5.

3.6 (A)* Find R_2 in Figure P3.6, given $R_{eq} = 17\ \Omega$, $R_1 = 5\ \Omega$, $R_3 = 15\ \Omega$.

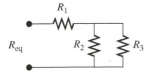

FIGURE P3.6   Circuit for Problem 3.6.

3.7 (A) Find R_{eq} in Figure P3.7, given

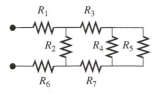

FIGURE P3.7   Circuit for Problem 3.7.

3.8 (A)* Find the R_{eq} in Figure P3.8, given that every resistor has an equal resistance, R (the center crossing is connected).

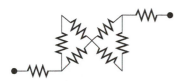

FIGURE P3.8   Circuit for Problem 3.8.

3.9 (H) Find R_{AB} in Figure P3.9:

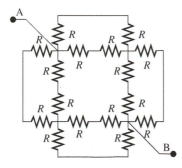

FIGURE P3.9   Circuit for Problem 3.9.

3.10 (H) Find R_{AB} and R_{CD} in Figure P3.10. Let node C and D open when calculating R_{AB} and let node A and B open when calculating R_{CD}.

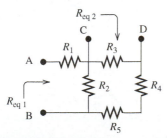

FIGURE P3.10   Circuit for Problem 3.10.

3.11 (H)* For Figure P3.11, find (a) the equivalent resistance, R_{AB} and (b) the equivalent resistance R_{CD}.

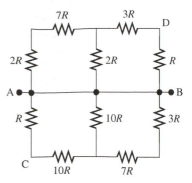

FIGURE P3.11   Resistor network for Problem 3.11.

3.12 (H) **Δ-Y transformation**

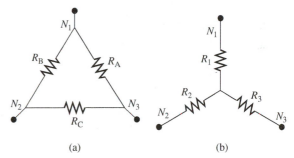

(a)   (b)

FIGURE P3.12   Δ-Y transformation.

The resistant network consisting of three resistors R_A, R_B, and R_C as shown in Figure P3.12 (a) forms a delta (Δ) connection and the resistors R_1, R_2, and R_3 form a Y connection shown in Figure P3.12 (b). Verify that the two circuits are equivalent provided that:

$$R_1 = R_A R_B/(R_A + R_B + R_C)$$
$$R_2 = R_B R_C/(R_A + R_B + R_C)$$
$$R_3 = R_C R_A/(R_A + R_B + R_C)$$

3.13 (H)* Find R_{AB} in Figure P3.13 using Δ-Y transformation shown in Problem 3.12.

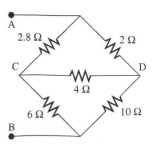

FIGURE P3.13   Bridge circuit.

3.14 (H) Find the resistance seen through the terminals A and B (R_{AB}) in Figure P3.14. The circuit in the box repeats infinitely.

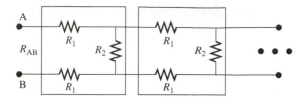

FIGURE P3.14 Circuit for Problem 3.14.

3.15 (H)* Each segment in Figure P3.15 represents a resistor R. Find R_{AB}, that is, the resistance between terminals A and B. (*Hint*: Let a current source I connect terminals A and B, and find the current through each resistor by symmetry of the circuit.)

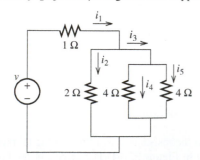

FIGURE P3.15 Circuit for Problem 3.15.

SECTION 3.3 VOLTAGE AND CURRENT DIVISION/DIVIDER RULES

3.16 (B) Find V_1, V_2, V_3 in Figure P3.16.

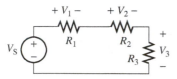

FIGURE P3.16 Circuit for Problem 3.16.

3.17 (A) Find V_s given V in Figure P3.17.

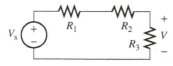

FIGURE P3.17 Circuit for Problem 3.17.

3.18 (A) Find I_1 and I_2 in Figure P3.18.

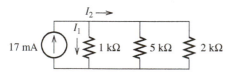

FIGURE P3.18 Circuit for Problem 3.18.

3.19 (A)* Find i_1, i_2, i_3, and i_4 in Figure P3.19 supposing $i_5 = 2$ A.

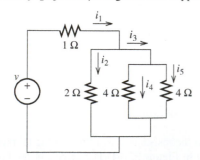

FIGURE P3.19 Circuit for Problem 3.19.

3.20 (H) Find i_2, i_3, i_4, and i_5 in Figure P3.19 assuming $i_1(t) = 4_{\cos t}$ A.

3.21 (H) **Current Division in Conductance**

Given a resistor R, we define its conductance G to be $G = 1/R$. Therefore in Figure P3.21, the conductances of R_1 and R_2 are $G_1 = 1/R_1$ and $G_2 = 1/R_2$, respectively. Prove that the current division formulas in terms of conductance are:

$$i_1 = \frac{G_1}{G_1 + G_2}i_s, \quad i_2 = \frac{G_2}{G_1 + G_2}i_s$$

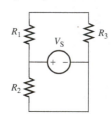

FIGURE P3.21 Circuit for Problem 3.21.

3.22 (A)* Using the current division formulas in Problem 3.21 to recompute the currents through 1 kΩ, 5 kΩ, and 2 kΩ in Figure P3.18.

3.23 (A) Find the voltage drop across all resistors in Figure P3.23.

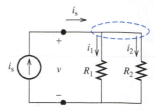

FIGURE P3.23 Circuit for Problem 3.23.

3.24 (A) Given V_o, find R_3 in Figure P3.24.

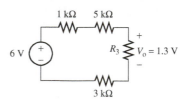

FIGURE P3.24 Circuit for Problem 3.24.

3.25 (H) Find I_1, I_2, and I_3 in Figure P3.25.

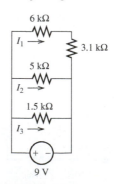

FIGURE P3.25 Circuit for Problem 3.25.

3.26 (H) Assume that you need to power a system from a V_s power supply, but the system requires a voltage of V_{sys}. Use voltage division as shown in Figure P3.26, express R_1 as a function of R_2, V_s, V_{sys}, and R_L. ($V_{sys} < V_s$)

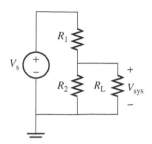

FIGURE P3.26   Circuit for Problem 3.26.

3.27 (A) Find I_1, I_2, I_3, and I_4 in Figure P3.27.

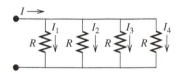

FIGURE P3.27   Circuit for Problem 3.27.

3.28 (H)* Find V_0 in Figure P3.28.

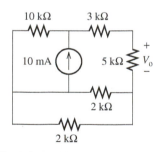

FIGURE P3.28   Circuit for Problem 3.28.

SECTION 3.4 NODAL AND MESH ANALYSIS

3.29 (B) Which node is the best choice for a reference node in the circuit in Figure P3.29?

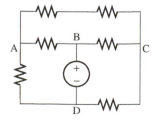

FIGURE P3.29   Circuit for Problem 3.29.

3.30 (A)* Which node is the best choice for a reference node in the circuit in Figure P3.30?

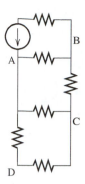

FIGURE P3.30   Circuit for Problem 3.30.

3.31 (A) Use MATLAB® to find V_1, V_2, and V_3 if $2V_1 + 2V_2 - 4V_3 = -2$, $3V_1 + 2V_2 = 1$, $-2V_1 - 3V_2 + 5V_3 = -3$.

3.32 (A) Use Cramer's rule to find V_1, V_2, and V_3 in Problem 3.31 (MATLAB® command "det" may be used to compute determinant).

3.33 (H) Determine the best choice of node as a reference node for the circuit in Figure P3.33. Then determine the current between B and D.

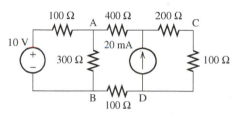

FIGURE P3.33   Circuit for Problem 3.33.

3.34 (B) Find V_1 and V_2 in Figure P3.34.

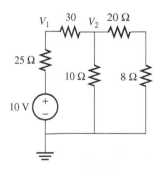

FIGURE P3.34   Circuit for Problem 3.34.

3.35 (B)* Find V_1 and V_2 in Figure P3.35.

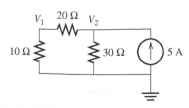

FIGURE P3.35   Circuit for Problem 3.35.

3.36 (A) Find V_1, V_2, and V_3 in Figure P3.36:

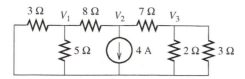

FIGURE P3.36   Circuit for Problem 3.36.

3.37 (A) Find V_1, V_2, and V_3 in Figure P3.37:

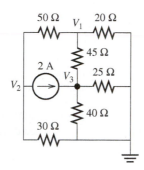

FIGURE P3.37   Circuit for Problem 3.37.

3.38 (A)* Find V_1, V_2, and V_3 in Figure P3.38:

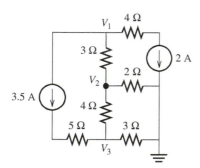

FIGURE P3.38   Circuit for Problem 3.38.

3.39 (H) Find I_x in Figure P3.39.

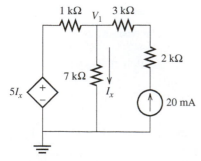

FIGURE P3.39   Circuit for Problem 3.39.

3.40 (H) Find I_x in Figure P3.40.

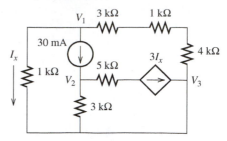

FIGURE P3.40   Circuit for Problem 3.40.

3.41 (H) The circuit shown in Figure P3.41 is a simple bipolar junction transistor amplifier. The portion of the circuit in the shaded box is an approximate T-model of a transistor in the common-emitter configuration.

Use nodal analysis to find the voltage gain V_2/V_1 when $R_B = 80$ kΩ, $R_C = R_E = 2$ kΩ.

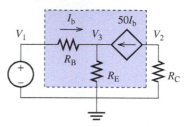

FIGURE P3.41   Circuit for Problem 3.41.

3.42 (B) Identify the meshes and mark the mesh currents in Figure P3.29.

3.43 (B) Identify the meshes and mark the mesh currents in Figure P3.30.

3.44 (A) Find the mesh currents in the circuit shown in Figure 3.44, when $I = 1$ A, $V = 8$ V, $R_1 = 3$ Ω, $R_2 = 4$ Ω, $R_3 = 20$ Ω.

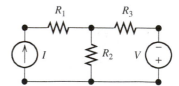

FIGURE P3.44   Circuit for Problem 3.44.

3.45 (A) Use mesh analysis to find the current through the 8-Ω resistor in Figure P3.34.

3.46 (H) Find the mesh currents in Figure P3.46.

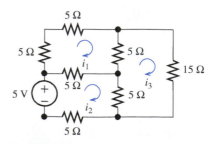

FIGURE P3.46   Circuit for Problem 3.46.

SECTION 3.5 SPECIAL CONDITIONS: SUPER NODE

3.47 (B) Which of the following circuits requires super node analysis?

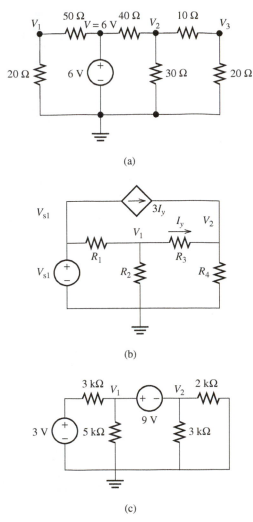

(a)

(b)

(c)

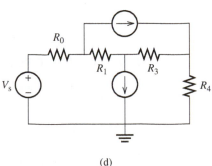

(d)

FIGURE P3.47 Circuit for Problem 3.47.

3.48 (A) Under what circumstances is a super node required in nodal analysis?

3.49 (H)* Find V_1, V_2, and V_3 in Figure P3.49:

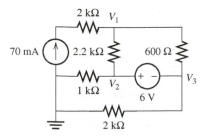

FIGURE P3.49 Circuit for Problem 3.49.

3.50 (H) Find V_1 and V_2 in Figure P3.50:

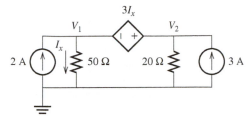

FIGURE P3.50 Circuit for Problem 3.50.

3.51 (H) Find I in Figure P3.51:

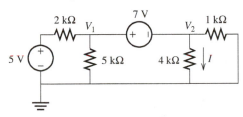

FIGURE P3.51 Circuit for Problem 3.51.

SECTION 3.6 THÉVENIN/NORTON EQUIVALENT CIRCUITS

3.52 (B) Find R_{th} in Figure P3.52:

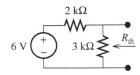

FIGURE P3.52 Circuit for Problem 3.52.

3.53 (B)* Find the Thévenin resistance for the circuit in Figure P3.53:

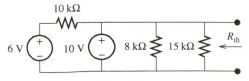

FIGURE P3.53 Circuit for Problem 3.53.

3.54 (A) Find the Thévenin equivalent resistance and voltage as seen by R_L for the circuit in Figure P3.54.

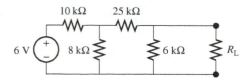

FIGURE P3.54 Circuit for Problem 3.54.

3.55 (A) Find the Norton equivalent as seen by R_L in the circuit in Figure P3.55.

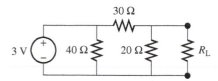

FIGURE P3.55 Circuit for Problem 3.55.

3.56 (A) Find the Thévenin equivalent circuit as seen by R_L in Figure P3.56.

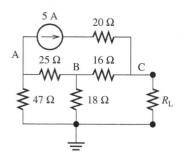

FIGURE P3.56 Circuit for Problem 3.56.

3.57 (A) Find the Thévenin/Norton equivalents for Figure P3.57.

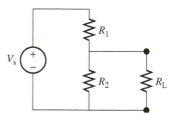

FIGURE P3.57 Circuit for Problem 3.57.

3.58 (H)* Find the Thévenin/Norton equivalents as seen by R_L in Figure P3.58.

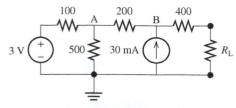

FIGURE P3.58 Circuit for Problem 3.58.

3.59 (H) **Source transformation**

In Figure P3.59(a) and (b) use the methods introduced in Examples 3.34 and 3.35 to calculate the total Thévenin

and Norton Circuit observed across the load terminals A and B.

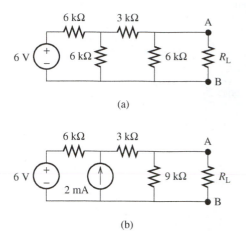

(a)

(b)

FIGURE P3.59 Circuits of Problem 3.59.

3.60 (A) Repeat Problem 3.52 using source transformation.

3.61 (H) Assume that there are 10 nonideal 1.5-V voltage sources. Each has an internal resistance of 1.5 Ω. Those voltage sources are divided into $10/n$ groups so that n voltage sources are connected in series and the $10/n$ groups are connected in parallel. A 5-Ω external resistor is driven by the voltage source combination. Find the maximum current through the 5-Ω resistors and determine the value of n.

3.62 (H)* In Figure 3.62, $\varepsilon = 100$ V, $\varepsilon' = 40$ V, $R = 10\ \Omega$, $R' = 30\ \Omega$. Find the current through R'.

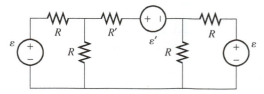

FIGURE P3.62 Circuit for Problem 3.62.

SECTION 3.7 SUPERPOSITION PRINCIPLE

3.63 (B) Find the current I_2 in Figure P3.63 using the superposition technique.

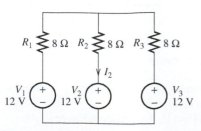

FIGURE P3.63 Circuit for Problem 3.63.

3.64 (A) Find I_x in Figure P3.64 using the principle of superposition.

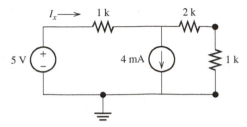

FIGURE P3.64   Circuit for Problem 3.64.

3.65 (A) Use the principle of superposition to find the power dissipated in the 2-kΩ resistor in Figure P3.64.

3.66 (H) If $R_L = 0$, which source in Figure P3.66 contributes the most to the power dissipated in the 500-Ω resistor? Which contributes the least? What is the power dissipated in the 500-Ω resistor? Does it equal the sum of power dissipation contributed by each source?

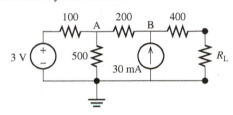

FIGURE P3.66   Circuit for Problem 3.66.

3.67 (A) Repeat Problem 3.33 using the superposition principle.
3.68 (A) Repeat Problem 3.51 using the superposition principle.
3.69 (A)* Find the current I in Figure P3.69 using the superposition principle.

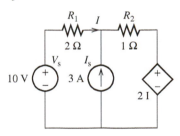

FIGURE P3.69   Circuit for Problem 3.69.

3.70 (A) Repeat Problem 3.50 using the superposition principle.
3.71 (H) When $V_s = 4$ V and $I_s = 2$ A, then $V_x = 8$ V; when $V_s = 7$ V and $I_s = 1$ A, then $V_x = 6$ V. Find the voltage V_x in Figure P3.71 using the superposition principle when $V_s = 2$ V and $I_s = 6$ A.

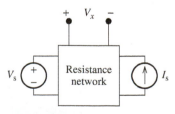

FIGURE P3.71   Circuit for Problem 3.71.

3.72 (H) Using the superposition principle, find the Thévenin equivalence seen by R_L in Figure P3.72.

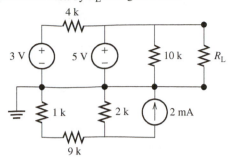

FIGURE P3.72   Circuit for Problem 3.72.

SECTION 3.8 MAXIMUM POWER TRANSFER

3.73 (B) Find the resistance R_1 in Figure P3.73 to transfer maximum power to R_2.

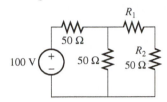

FIGURE P3.73   Circuit for Problem 3.73.

3.74 (A)* Find the value of R where R_L equals 50 Ω and draws the maximum power in Figure P3.74.

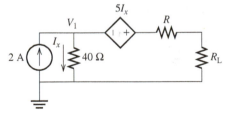

FIGURE P3.74   Circuit for Problem 3.74.

3.75 (A) Determine the maximum power transfer for the circuit shown in Figure P3.75.

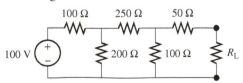

FIGURE P3.75   Circuit for Problem 3.75.

3.76 (A)* In Figure 3.76, find R_L which draws the maximum power from the source.

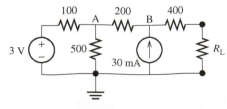

FIGURE P3.76   Circuit for Problem 3.76.

3.77 (H) For the circuit shown in Figure P3.77, draw a graph of the maximum output power of R_L when R_s changes from 1 Ω to 100 Ω. R_L is adjustable.

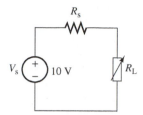

FIGURE P3.77   Circuit for Problem 3.77.

3.78 (H) The light bulb shown in Figure P3.78 has a resistance of $R_0 = 2$ Ω. The working voltage $V_0 = 4.5$ V. $V = 6$ V. Find the conditions under which the power efficiency η (defined as the ratio of the power consumed by the bulb and the power provided by the source) is maximum.

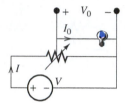

FIGURE P3.78   Circuit for Problem 3.78.

SECTION 3.9 ANALYSIS OF CIRCUITS USING PSPICE

3.79 (B)* Use PSpice to find I_1, I_2, and I_3 in Figure P3.25.
3.80 (B)* Find I_x in Figure P3.39 using PSpice.

3.81 (A) For the circuit shown in Figure 3.81, run the Bias Point simulation analysis for three cases in which the negative terminal of each voltage source is selected as a reference node. Assume $R_1 = 3$ Ω, $R_2 = R_3 = R_7 = 2$ Ω, $R_4 = 1$ Ω, $R_5 = 20$ Ω, $R_6 = 1.6$ Ω, $R_8 = 8$ Ω. (*Hint*: Apply a ground (GND) at the negative terminal of the voltage source that is selected as reference node.)

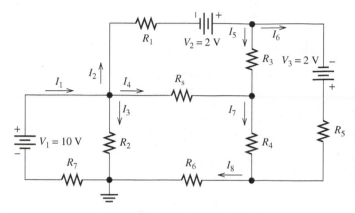

FIGURE P3.81   Circuit of PSpice Problem 3.81.

3.82 (A) Use PSpice to find the Thévenin equivalent resistance and voltage as seen by R_L for the circuit shown in Figure P3.54.

Capacitance and Inductance

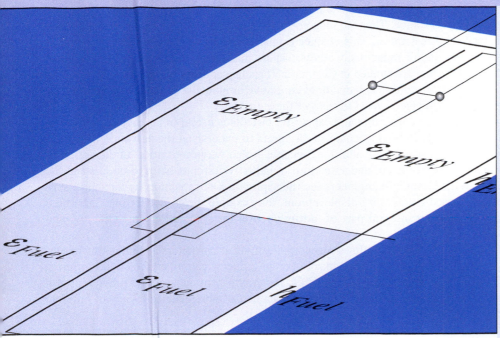

4.1 INTRODUCTION

In Chapter 3, we examined resistors and circuits. Resistors are elements that resist the flow of current. Resistance creates a voltage drop across the resistor which in turn leads to power consumption. Resistors are energy dissipaters. In this chapter, we consider energy storage elements. Resistors are a good model for many circuits but are not the ideal model for all circuits. Modeling of some circuits can be made more realistic by adding in other components. When investigating a system, it is important that the system is replaced with the best model.

For example, a light can be modeled by a resistor; however, the best model for a light is a combination of a resistor and an inductor. This combined model is more realistic because the inductor creates sparks when closing or opening a switch. These sparks are hazardous and should be prevented. (Note: Chapter 15 explains the details of preventing sparks and maintaining a safe environment.) Thus, a resistor alone is not the best circuit model for a light, because it does not account for the potential sparks caused by the switch.

Other circuits also benefit from models with additional elements. For example, a computer can be modeled using a combination of resistors and capacitors. Likewise, transmission lines can be modeled as a combination of capacitors, resistors, and inductors. Chapter 9 studies circuits made by transmission lines.

This chapter focuses on two popular circuit components, that is, capacitors and inductors, which store power and energy. Similar to resistors, these elements are passive; that is, they do not supply power. This chapter describes the relationship between voltage, current, power, and energy for inductors and capacitors. In addition, it explains the effects of connecting these components in series and in parallel.

A *capacitor* stores energy in its electric field. It consists of an insulating material, called dielectric, that is sandwiched between two conductors. An *inductor* stores energy in its magnetic field. Any piece of wire can create a magnetic field; thus, it can form an inductor. However, wires can be twisted to make inductances with higher values. Thus, an inductor is often in a spiral shape. It is interesting to note that because inductors are usually made using a twisted piece of wire, they also offer resistance. Thus, an inductor is most accurately modeled using a combination of inductance and resistance. This chapter primarily examines circuits that can be made using only capacitors and/or inductors. Circuits that are made using a combination of resistors and capacitors or inductors are studied in Chapter 5. This chapter also discusses the steps used to analyze the equivalent capacitor and inductor of capacitive or inductive circuits. Similar to previous chapters, PSpice will also be used to analyze these circuits.

Learning to analyze capacitors and capacitive circuits is useful in a variety of engineering disciplines. These skills help mechanical engineers to understand the operation of capacitive vibration or motion sensors. In addition, because coiled wires are critical in the design of electric motors and solenoids, an understanding of inductive circuits helps to understand the operation of these devices. Civil engineers see the application of capacitors in capacitive sensors that are often used to measure displacement, for example, beam deflection. Capacitors and inductors are also an integral part of tuning circuits for applications in radio, TV, and cellular phones.

Section 4.6 of this chapter outlines many applications of capacitors and inductors in sensors. This section also explains how a capacitor or an inductor can be used to build different types of sensors for use as measurement devices, for example, to measure the pressure, sense vibration, and/or measure the level of liquid in a container. A more detailed review of different types of sensors is provided in Chapter 11.

4.2 CAPACITORS

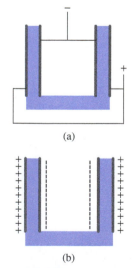

(a)

(b)

FIGURE 4.1 Leyden Jar (a) charging and (b) charged.

A capacitor stores energy in its electric field. It consists of an insulating material, the dielectric, sandwiched between two conductors. The basic function of a capacitor was famously illustrated by the Leyden jar in 1746. To demonstrate a capacitor using the Leyden jar approach, the interior and exterior of a glass jar are lined with a thin metal conductor [See Figure 4.1(a)]. The inner conductor is negatively charged, while the outer conductor is connected to a ground (zero charge), as shown in Figure 4.1(a). The glass material is between the two conductors. As a result of the glass material, no charge is able to reach the outer conductor. The negative charge on the inner conductor repels the negative charge on the outer to the ground, resulting in a net positive charge on the outer conductor. When the electrical connections are removed, the charges remain on the surface of the conductor. The glass and air are insulators, and they shield the charges. The net charges are equal in magnitude, and opposite in sign. The result is a stored electric charge on the jar, as shown in Figure 4.1(b).

Capacitors are used for storing electric charge in applications that require a quick discharge, for example, a camera flash or a defibrillator. Capacitors also play a major role in the tuning circuits of TVs and radios.

A basic capacitor is simply a dielectric placed between two conducting plates, as shown in Figure 4.2. A dielectric is an insulator. Examples include ceramic, vacuum, air, electrolyte, oil, paper, plastic, and mica. If the spacing between the two plates is filled with

TABLE 4.1	Permittivity of Some Typical Materials
Material	**Permittivity**
Polyethylene	2.1
Paper	3
Salt	3–15
Methanol	30
Sulfuric acid	83.6
Silicon	11.68
Rubber	7

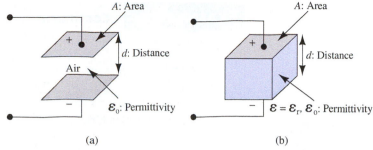

FIGURE 4.2 Parallel-plate capacitor with (a) an air dielectric and (b) a solid dielectric.

electrolyte, the capacitor will be a polarized one. It is very critical to connect polarized capacitors in the right order. If they are not connected in the right order, the capacitor could be exploded. Other dielectrics do not polarize the capacitor, and the capacitor can be connected in any direction. Electrical characteristic of dielectrics are usually determined by their permittivity.

The capacitance of a parallel-plate capacitor corresponds to:

$$C = \frac{\varepsilon_0 \varepsilon_r \times A}{d} \tag{4.1}$$

where

$\varepsilon_0 =$ permittivity of free space $\approx 8.85 \times 10^{-12}$ F/m

$\varepsilon_r =$ relative permittivity of the dielectric (that is relative to air) (see Table 4.1 for examples of relative permittivity)

$A =$ area of the plates (in square meters, m^2)

$C =$ capacitance, in farads (F)

$d =$ distance between the plates.

Capacitance is measured in *farads* (F), where 1 F = 1 coulomb/volt (C/V). The unit farad is named in honor of Michael Faraday. Faraday was the first to show the existence of electric and magnetic fields via experiments.

EXAMPLE 4.1   Capacitance

A parallel-plate capacitor has dimensions as shown in Figure 4.3. The dielectric is air, that is, $\varepsilon_r = 1$. Find the total capacitance.

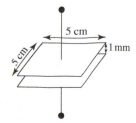

FIGURE 4.3 Capacitor for Example 4.1.

(*continued*)

| EXAMPLE 4.1 | Continued |

SOLUTION

Using Equation (4.1):

$$C = \frac{\varepsilon_0 \varepsilon_r \times A}{d}$$

Replacing for all parameters,

$$C = \frac{8.85 \times 10^{-12} \times 1 \times (0.05)^2}{0.001}$$

Therefore,

$$C = 22.13 \times 10^{-12}\,\text{F}$$

4.2.1 The Relationship Between Charge, Voltage, and Current

The symbols used to designate a capacitor are shown in Figure 4.4. The symbol in the left might be used for polarized capacitors. As mentioned, the spacing between the plates of these capacitors is filled with electrolyte.

The charge on a capacitor, $q(t)$, is the product of the capacitance and voltage:

$$q(t) = C \cdot v(t) \tag{4.2}$$

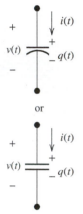

FIGURE 4.4 Capacitor symbols.

| EXAMPLE 4.2 | Stored Charge |

Find the stored charge in a 10-μF capacitor that is charged to 6 V.

SOLUTION

$$q = Cv$$
$$q = 10 \times 10^{-6} \times 6$$
$$q = 60\,\mu\text{C}$$

By definition, current is the derivative of charge with respect to time.

$$i(t) = \frac{dq(t)}{dt} \tag{4.3}$$

Substituting for $q(t)$ from Equation (4.2), the current through a capacitor corresponds to:

$$i(t) = C\frac{dv(t)}{dt} \tag{4.4}$$

| EXAMPLE 4.3 | Current of Capacitor |

In Figure 4.5(a), calculate the current through the capacitor assuming:

   a. $v(t) = 5\,\text{V}$
   b. $v(t) = 5\cos(2\pi \times 10^3 t)\,\text{V}$

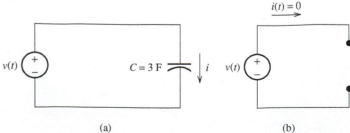

FIGURE 4.5 (a) The circuit for Example 4.3; (b) equivalent circuit of a capacitor when a constant current is applied to.

SOLUTION

Using Equation (4.4):

$$i(t) = 3 \times \frac{dv(t)}{dt}$$

a. For $v(t) = 5\,\text{V}$, $\frac{dv(t)}{dt} = 0$; therefore $i(t) = 0$. Thus, the capacitor would be equivalent to an open circuit as shown in Figure 4.3 (b).

b. For $v(t) = 5\cos(2\pi \times 10^3 t)$ V, $\frac{dv(t)}{dt} = -10\pi \times 10^3 \sin(2\pi \times 10^3 t)$; therefore:

$$i(t) = -30\pi \times 10^3 \sin(2\pi \times 10^3 t) \ \text{A}$$

Note: If a *constant* voltage is applied directly to a capacitor, the capacitor will ultimately (i.e., when it is fully charged) act as an *open circuit*, because the current through it will be zero. The equivalent circuit for this situation is shown in Figure 4.5(b). Note that the derivative of a constant voltage (see Equation (4.4)) is zero.

Based on Equation (4.3), the charge–current relationship is as follows:

$$q(t) = \int_{t_0}^{t} i(t)\,dt + q(t_0) \tag{4.5}$$

Here, $q(t_0)$ represent the initial charge of capacitor at $t = t_0$. Setting the right-hand side of Equations (4.2) and (4.5) equal, and solving for $v(t)$:

$$v(t) = \frac{1}{C}\int_{t_0}^{t} i(t)\,dt + \frac{q(t_0)}{C} \tag{4.6}$$

Because the initial voltage across the capacitor corresponds to:

$$v(t_0) = \frac{q(t_0)}{C} \tag{4.7}$$

Equation (4.6) may be rewritten as:

$$v(t) = \frac{1}{C}\int_{t_0}^{t} i(t)\,dt + v(t_0) \tag{4.8}$$

If the capacitor has zero charge at $t = t_0 = 0$, that is $q(t_0) = 0$, or a zero initial condition, then based on Equation (4.7), Equation (4.8) simplifies to:

$$v(t) = \frac{1}{C}\int_{0}^{t} i(t)\,dt \tag{4.9}$$

Comparing Equations (4.4) and (4.8), it is observed that current is a linear function of voltage [see Equation (4.4)]. Recall that linearity consists of two properties: additivity and homogeneity. Therefore, a function $f(x)$ is considered linear if it meets both conditions:

Additivity: $f(x_1 + x_2) = f(x_1) + f(x_2)$

Homogeneity: $f(\alpha x) = \alpha \cdot f(x)$

Additivity and homogeneity are summarized in the following equation:

$$f(\alpha_1 x_1 + \alpha_2 x_2) = \alpha_1 f(x_1) + \alpha_2 f(x_2) \tag{4.10}$$

EXERCISE 4.1

Show that Equation (4.8) represents a linear function with respect to $i(t)$ and $v(t_0)$. Show that Equation (4.9) is only linear with respect to $i(t)$.

4.2.2 Power

As mentioned in Chapter 2, instantaneous power equals voltage times current.

$$p(t) = v(t)i(t) \tag{4.11}$$

Using Equations (4.3) and (4.4), the stored delivered power by a capacitor corresponds to:

$$p(t) = v(t)\frac{dq(t)}{dt} = Cv(t)\frac{dv(t)}{dt} \tag{4.12}$$

EXAMPLE 4.4   Power Delivered to a Capacitor

A 1.5-μF capacitor shown in Figure 4.6 has been previously charged to 5 V. It is then discharged within 50 ms. What is the instantaneous power delivered?

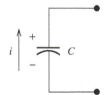

FIGURE 4.6   Circuit for Example 4.3.

SOLUTION

Using Equation (4.2):

$$q = Cv \rightarrow q = 1.5 \times 10^{-6} \times 5 = 7.5 \times 10^{-6}$$

$$\text{The rate of discharge} = \frac{7.5 \times 10^{-6}C}{50 \times 10^{-3}s} = 150 \frac{\mu C}{s} = 150 \ \mu A = i_{\text{capacitor}}$$

Referring to the sign of the voltage and current (as discussed in Chapter 2) as indicated in Figure 4.6:

$$p = -vi = 5 \times 150 \times 10^{-6} = -750 \ \mu W$$

The minus sign indicates that the power is delivered by the capacitor.

4.2.3 Energy

The power–energy relationship corresponds to:

$$p(t) = \frac{dE(t)}{dt} \tag{4.13}$$

Therefore:

$$E(t) = \int_{t_0}^{t} p(t)\, dt \tag{4.14}$$

Using Equation (4.12):

$$p(t) = Cv(t)\frac{dv(t)}{dt} = \frac{1}{2}C\frac{dv^2(t)}{dt} \tag{4.15}$$

The last equality holds since $v(t)\dfrac{dv(t)}{dt} = \dfrac{1}{2}\dfrac{dv^2(t)}{dt}$.

Comparing Equations (4.12) and (4.15), energy stored in or delivered by a capacitor is given by:

$$E(t) = \frac{1}{2}Cv^2(t) \tag{4.16}$$

EXAMPLE 4.5   **Voltage and Energy of Capacitor**

At time $t = 0$, a 2-mA current source is applied to an uncharged 1.2-µF capacitor, as shown in Figure 4.7 below. Determine the voltage, v_C, and the energy stored in the capacitor, E_C, at $t = 20$ ms assuming zero initial voltage across the capacitor prior to the closing of the switch, that is, a zero initial condition.

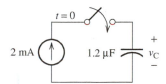

FIGURE 4.7   Circuit for Example 4.5.

SOLUTION

Given a constant current and zero initial charge, using Equation (4.5), the charge corresponds to:

$$q = i \cdot t = 2\text{ mA} \cdot 20\text{ ms} = 40\ \mu C$$

Now, using Equation (4.2):

$$q = Cv_C \rightarrow v_C = \frac{q}{C} = \frac{40\ \mu C}{1.2\ \mu F} = 33.3\ \frac{C}{C/v} = 33.3\text{ V}$$

Finally, applying Equation (4.16):

$$E_C = \frac{1}{2}Cv^2 = \frac{1}{2} \times 1.2 \times 10^{-6} \times 33.3^2 = 667\ \mu J$$

4.3 CAPACITORS IN SERIES AND PARALLEL

Capacitors, like resistors, can be arranged into series and parallel connections. In this section, the equivalent capacitances of simple series and parallel combinations are derived. The formulas derived can in turn be used to calculate the capacitance of more complex capacitor networks.

4.3.1 Series Capacitors

Figure 4.8(a) represents three capacitors, C_1, C_2, and C_3, connected in series.

Applying KVL to the structure of Figure 4.8(a):

$$v_C = v_1 + v_2 + v_3 \tag{4.17}$$

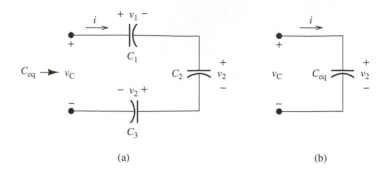

FIGURE 4.8 (a) Series capacitors; (b) an equivalent circuit.

Because the current $i(t)$ flowing through all capacitors is equivalent, the voltage across each capacitor corresponds to:

$$v_k = \frac{1}{C_k} \int i(t)\,dt + v_k(0), \quad k \in \{1,2,3\} \tag{4.18}$$

Assuming $v_k(0) = 0$ for all values of k (i.e., the initial voltage of all capacitors is zero), and substituting Equation (4.18) into Equation (4.17), the total voltage is:

$$v_C(t) = \frac{1}{C_1} \int i(t)\,dt + \frac{1}{C_2} \int i(t)\,dt + \frac{1}{C_3} \int i(t)\,dt \tag{4.19}$$

Factorizing the integral component that is the same across all terms above:

$$v_C(t) = \left(\frac{1}{C_1} + \frac{1}{C_2} + \frac{1}{C_3} \right) \cdot \int i(t)\,dt \tag{4.20}$$

Also, the relationship between voltage, v_C, and current, i, for the equivalent circuit of Figure 4.8(b) corresponds to:

$$v_C(t) = \left(\frac{1}{C_{eq}} \right) \cdot \int i(t)\,dt \tag{4.21}$$

Comparing Equations (4.20) and (4.21), the equivalent capacitance for three capacitors in series corresponds to:

$$C_{eq} = \left(\frac{1}{C_1} + \frac{1}{C_2} + \frac{1}{C_3} \right)^{-1} \tag{4.22}$$

For a general case with N capacitors in series:

$$C_{eq} = \left(\frac{1}{C_1} + \frac{1}{C_2} + \cdots + \frac{1}{C_N} \right)^{-1} \tag{4.23}$$

As a result, the method for calculating the equivalent capacitance of capacitors in series is equivalent to that of parallel resistors.

EXERCISE 4.2

Explain how Equation (4.22) can be generalized to Equation (4.23).

4.3.2 Parallel Capacitance

Next, consider the equivalent capacitance of parallel capacitors. As an example, consider the three capacitors connected in parallel in Figure 4.9(a).

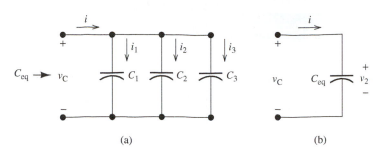

FIGURE 4.9 (a) Parallel capacitors; (b) their equivalent circuit.

Because the voltage across all capacitors is equivalent, the current–voltage relationship for each capacitor $R \in \{1,2,3\}$ corresponds to:

$$i_k(t) = C_k \frac{dv_C(t)}{dt}, k \in \{1,2,3\} \tag{4.24}$$

Applying KCL to Figure 4.9(a), the total current is:

$$i(t) = i_1(t) + i_2(t) + i_3(t) \tag{4.25}$$

Substituting Equation (4.24) into (4.25):

$$i(t) = C_1 \frac{dv_C(t)}{dt} + C_2 \frac{dv_C(t)}{dt} + C_3 \frac{dv_C(t)}{dt} \tag{4.26}$$

Rearranging the terms:

$$i(t) = (C_1 + C_2 + C_3) \frac{dv_C(t)}{dt} = (C_{eq}) \frac{dv_C(t)}{dt} \tag{4.27}$$

Also, the current–voltage relationship for the equivalent circuit of Figure 4.8(b) corresponds to:

$$i(t) = C_{eq} \frac{dv_C(t)}{dt} \tag{4.28}$$

Comparing Equations (4.27) and (4.28), the equivalent capacitance of the parallel capacitors in Figure 4.9 is:

$$C_{eq} = C_1 + C_2 + C_3 \tag{4.29}$$

For a general case with N capacitors:

$$C_{eq} = C_1 + C_2 + \cdots + C_N \tag{4.30}$$

Thus, the equivalent capacitance of *parallel* capacitors is computed in a way similar to the method used for *series* resistors. To demonstrate the calculation of the equivalent circuit of capacitor networks connected in more complex configurations, the following examples are presented.

EXAMPLE 4.6   **Equivalent Capacitance**

Find the equivalent capacitance in the circuit shown in Figure 4.10(a).

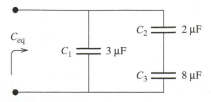

FIGURE 4.10(a)   The circuit for Example 4.6.

(*continued*)

EXAMPLE 4.6   **Continued**

SOLUTION

The total capacitance is found in a way that is similar to finding the total resistance of a resistive network. Here, there are two parallel branches. One branch includes two capacitors, C_2 and C_3, that are in series. Equation (4.21) is used to find the series capacitance.

$$C_{23} = \left(\frac{1}{2 \times 10^{-6}} + \frac{1}{8 \times 10^{-6}} \right)^{-1} = 1.6 \; \mu\text{F}$$

Now, in Figure 4.10(a), replace the series capacitors, C_2 and C_3, with the equivalent capacitor C_{23}, as shown in Figure 4.10(b).

Figure 4.10(b) represents two parallel capacitors, C_1 and C_{23}. As shown by Equation (4.30), the total equivalent capacitance corresponds to:

$$C_{eqv} = 3 \; \mu\text{F} + 1.6 \; \mu\text{F} = 4.6 \; \mu\text{F}$$

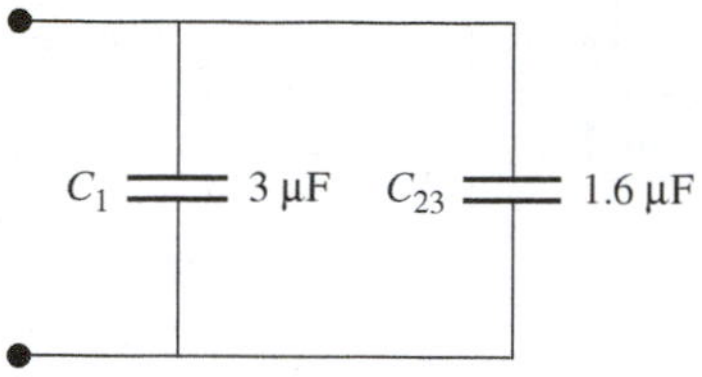

FIGURE 4.10(b)   Modification of Figure 4.10(a).

EXAMPLE 4.7   **Equivalent Capacitance**

Find the equivalent capacitance of the capacitor network shown in Figure 4.11(a) below.

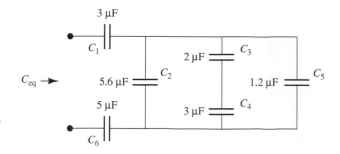

FIGURE 4.11(a)   The circuit for Example 4.7.

SOLUTION

Note that C_3 and C_4 are in series. Combine these capacitors using Equation (4.23).

$$C_{34} = \left(\frac{1}{2 \times 10^{-6}} + \frac{1}{3 \times 10^{-6}} \right)^{-1} = 1.2 \; \mu\text{F}$$

Next, replace series C_3 and C_4 capacitors with C_{34}, as shown in Figure 4.11(b).

In Figure 4.11(b), note that C_2, C_{34}, and C_5 are in parallel, and their equivalent capacitance corresponds to:

$$C_{2345} = C_2 + C_{34} + C_5 = 8 \; \mu\text{F}$$

Replacing these three capacitors with C_{2345}, results in the circuit shown in Figure 4.11(c).

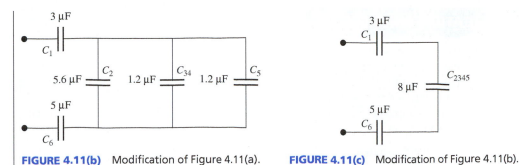

FIGURE 4.11(b) Modification of Figure 4.11(a).   **FIGURE 4.11(c)** Modification of Figure 4.11(b).

Finally, find the series combination of the three capacitors, C_1, C_{2345}, and C_6:

$$C_{eq} = \left(\frac{1}{3 \times 10^{-6}} + \frac{1}{8 \times 10^{-6}} + \frac{1}{5 \times 10^{-6}} \right)^{-1} = 1.52 \ \mu F$$

APPLICATION EXAMPLE 4.8   Tuning Circuits

The tuning circuit in a radio consists of several capacitors. The tuning knob of the radio adjusts the value of a variable capacitor. Figure 4.12 represents the capacitive portion of a radio tuning circuit. Find the minimum and maximum values for C_{eqv}.

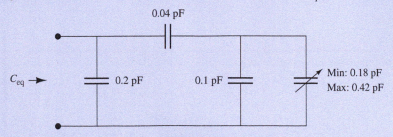

FIGURE 4.12   A radio tuner's capacitor network.

SOLUTION

To find the minimum and maximum, consider two scenarios. Considering the minimum capacitance for the tuning capacitor:

$$C_{eq} = \left(\frac{1}{0.18 \ \text{pF} + 0.1 \ \text{pF}} + \frac{1}{0.04 \ \text{pF}} \right)^{-1} + 0.2 \ \text{pF} = 0.2350 \ \text{pF}$$

Now, considering the maximum capacitance for the tuning capacitor:

$$C_{eq} = \left(\frac{1}{0.42 \ \text{pF} + 0.1 \ \text{pF}} + \frac{1}{0.04 \ \text{pF}} \right)^{-1} + 0.2 \ \text{pF} = 0.2371 \ \text{pF}$$

Therefore, the range of values for the circuit capacitance becomes:

$$C_{eq} \in \left[0.2350, 0.2371 \right] \text{pF}$$

APPLICATION EXAMPLE 4.9   A Capacitive Displacement Sensor

In civil engineering, it is important to know how far a beam or other structure made of a certain material deflects with a given force. Study of deflections and other force-related deformities is known as *solid mechanics*. A simple parallel-plate capacitor sensor can be used to determine this displacement.

(continued)

APPLICATION EXAMPLE 4.9   **Continued**

This capacitive sensor is shown in Figure 4.13(a). The sensor has three plates, two fixed and one movable, using air as the dielectric. Moving the center plate changes the capacitance across the terminals. The movable plate is connected to the deflecting beam and the fixed plate is connected to a fixed structure. Note that $C_2 = C_3$.

Consider a sample device where all the plates have dimensions of 4 cm × 4 cm. Spacing $S_1 = 2.5$ mm and $S_2 = 1$ mm. Air has a relative permittivity, $\varepsilon_r = 1$ and $\varepsilon_0 = 8.85 \times 10^{-12}$.

a. Find the total capacitance as a function of the displacement, d.
b. Find the displacement if the total capacitance is 6.181×10^{-12} F.

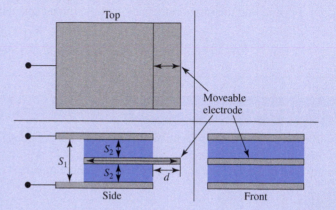

FIGURE 4.13(a)   Figure for Example 4.7.

SOLUTION

a. To find the total capacitance as a function of the displacement, d, use Equation (4.1).

$$C_1 = 8.85 \times 10^{-12} \frac{0.04 \cdot d}{0.0025} = 141.6 \times 10^{-12} \cdot d$$

$$C_2 = C_3 = 8.85 \times 10^{-12} \times \frac{0.04 \cdot (0.04 - d)}{0.001}$$

$$= 8.85 \times 10^{-9} \cdot (0.0016 - 0.04d)$$

$$= -354 \times 10^{-12} \cdot d + 14.16 \times 10^{-12}$$

Now, using the sensor network model of Figure 4.13(b) and knowing that $C_3 = C_2$:

$$C_{total} = C_1 + \left(\frac{1}{C_2} + \frac{1}{C_3} \right)^{-1} = C_1 + \left(\frac{2}{C_2} \right)^{-1} = C_1 + \frac{C_2}{2}$$

$$= 141.6 \times 10^{-12} \cdot d + \frac{-354 \times 10^{-12} \cdot d + 14.16 \times 10^{-12}}{2}$$

$$= 141.6 \times 10^{-12} \cdot d - 177 \times 10^{-12} \cdot d + 7.08 \times 10^{-12}$$

$$C_{total} = -35.4 \times 10^{-12} d + 7.08 \times 10^{-12} \text{F}$$

b. To find the displacement if the total capacitance is 6.181×10^{-12} F, solving for d in the equation above produces:

$$d = \frac{C_{total} - 7.08 \times 10^{-12}}{-35.4 \times 10^{-12}}$$

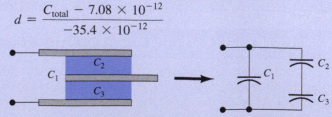

FIGURE 4.13(b)   Model of Figure 4.13(a).

$$d = \frac{6.181 \times 10^{-12} - 97.08 \times 10^{-12}}{-35.4 \times 10^{-12}}$$

$$d = 25.4 \times 10^{-3}\,\text{m} = 2.54\,\text{cm}$$

4.4 INDUCTORS

An inductor is a passive element that stores energy in its magnetic field. It consists of a *coil* of wire that is wrapped around a *core*. Current flowing through this coil produces a magnetic flux (see Figure 4.14). This flux, $\phi(t)$, is a function of the current, $i(t)$. Current flowing through any wire produces a slight magnetic field. Coiling the wire intensifies both this field and its related flux. A magnetic flux is the integral of the magnetic field. The concepts of magnetic field and flux will be introduced in Chapter 12.

A core of magnetic material is often used to direct the magnetic flux for a given current. Increasing the flux for a given current increases the inductance. Inductors, like capacitors, play an important role in tuning circuits (e.g., radios). Different types of inductors are shown in Figure 4.15.

The circuit schematic symbol for an inductor is shown in Figure 4.16.

Inductance, L, is measured in *henrys* (H):

$$1\,\text{H} = 1\,\text{V}\cdot\text{s/A} \tag{4.31}$$

The induced voltage across an inductor is the rate of change of the flux with time. Therefore, if the flux does not vary with time, the voltage will be zero. As a result, the induced voltage corresponds to:

$$v(t) = \frac{d\varphi}{dt}\frac{\text{weber}}{\text{s}} \tag{4.32}$$

4.4.1 The Relationship Between Voltage and Current

In an inductor, flux is proportional to current:

$$\phi(t) = Li(t) \tag{4.33}$$

Here, L is the inductance, and is defined as:

$$L = \frac{\phi}{i}\left(\frac{\text{weber}}{\text{ampere}} = \text{henry}\right) \tag{4.34}$$

Therefore, the induced voltage, $v(t)$, corresponds to:

$$v = \frac{d\varphi}{dt} = L\frac{di}{dt} \tag{4.35}$$

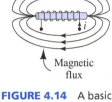

FIGURE 4.14   A basic inductor.

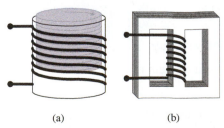

(a)   (b)

FIGURE 4.15   Different kinds of inductors. (a) Adjustable inductor via movable magnetic core; (b) laminated core transformer.

FIGURE 4.16   The symbol for an inductor.

Thus, for an inductor, voltage is a linear function of current. Rearranging the terms results in:

$$\mathrm{d}i = \frac{1}{L}v(t)\mathrm{d}t \qquad (4.36)$$

Integrating both sides to find the current:

$$i(t) = \frac{1}{L}\int_{t_0}^{t} v(t)\mathrm{d}t + i(t_0) \qquad (4.37)$$

Here, $i(t_0)$ is the initial current at time t_0.

EXAMPLE 4.10   **Inductor Voltage**

Calculate the voltage across the inductor in Figure 4.17(a), assuming:

  a. $i(t) = 5\,\mathrm{A}$
  b. $i(t) = 5\cos(2\pi \times 10^3 t)\,\mathrm{A}$

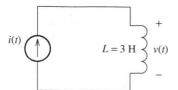

FIGURE 4.17(a)   Circuit for Example 4.10.

SOLUTION

Using Equation (4.35):

$$v(t) = 3 \times \frac{\mathrm{d}i(t)}{\mathrm{d}t}$$

  a. For $i(t) = 5$, $\dfrac{\mathrm{d}i(t)}{\mathrm{d}t} = 0$, therefore, $v(t) = 0\,\mathrm{V}$

  b. For $i(t) = 5\cos(2\pi \times 10^3 t)$, $\dfrac{\mathrm{d}i(t)}{\mathrm{d}t} = -10\pi \times 10^3 \sin(2\pi \times 10^3 t)$, therefore,

$$v(t) = -30\pi \times 10^3 \sin(2\pi \times 10^3 t)\,\mathrm{V}.$$

Note: If a constant current is applied directly to an inductor, the inductor will ultimately i.e., after a long time duration, act as a *closed circuit*. This is because the voltage across the inductor is zero. The equivalent circuit of the inductor is shown in Figure 4.17(b).

FIGURE 4.17(b)   Equivalent circuit of an inductor when constant current is applied to it.

4.4.2 Power and Stored Energy

The power delivered to a circuit element is the product of the voltage and current.

$$p(t) = v(t) \times i(t) \qquad (4.38)$$

Using Equation (4.35) to substitute for the voltage:

$$p(t) = Li(t) \times \frac{\mathrm{d}i}{\mathrm{d}t} \qquad (4.39)$$

The energy delivered to the inductor within the time period t_0 to t corresponds to the integration of the power from t_0 to t.

$$E(t) = L \int_{t_0}^{t} i(t) \times \frac{di}{dt} dt \qquad (4.40)$$

Canceling differential time and changing the limits of integration to the corresponding instantaneous currents result in:

$$E(t) = L \int_{i(t_0)}^{i(t)} i(t) \, di \qquad (4.41)$$

Integrating (4.41), assuming the initial current $i(t_0)$ is zero, leads to:

$$E(t) = \frac{1}{2} L i^2(t) \qquad (4.42)$$

EXAMPLE 4.11   Inductor Current

Given the circuit in Figure 4.18, find the current, i, as a function of time.

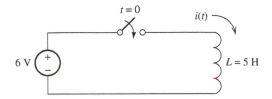

FIGURE 4.18   Circuit for Example 4.11.

SOLUTION

Find the current using Equation (4.37).

$$i(t) = \frac{1}{L} \int_{t_0}^{t} v(t) \, dt + i(t_0)$$

The switch closes at $t = 0$, and $t_0 = 0$. No current flows prior to closing the switch, that is, $i(t_0) = 0$. Therefore, the current after closing the switch is:

$$i(t) = \frac{1}{5} \int_0^t 6 \, dt$$

Integrating and evaluating results in:

$$i(t) = \frac{6}{5} t \text{ for } t > 0$$

Thus, applying a DC voltage to an inductor, the current linearly increases to infinity with time ($i(t) = (6/5)t$). As current increases beyond a limit, the connecting wires and the inductor will not be able to tolerate the heat created due to high values of current. As a result, the wires and/or inductor may burn, therefore opening the circuit. To avoid this situation, a resistor, limiter, or fuse can be used to limit the current in the circuit.

4.5 INDUCTORS IN SERIES AND PARALLEL

Inductors, like resistors and capacitors, can be arranged into networks. The equivalent inductance of an inductor network can be calculated by the same method used to calculate the equivalent resistance of a resistor network. **Series inductor combinations are simply added together.**

Parallel combinations are calculated by taking the reciprocal of the sum of the reciprocals of the inductance values.

4.5.1 Inductors in Series

Applying Kirchhoff's voltage law (KVL) around the loop in Figure 4.19, results in:

$$v = v_1 + v_2 + v_3 \tag{4.43}$$

Substituting Equation (4.35) into Equation (4.43) for the voltages yields:

$$v = L_1 \times \frac{di}{dt} + L_2 \times \frac{di}{dt} + L_3 \times \frac{di}{dt} \tag{4.44}$$

Separating the terms produces a result comparable to Equation (4.35), that is:

$$v = (L_1 + L_2 + L_3) \times \frac{di}{dt} = (L_{eq}) \times \frac{di}{dt} \tag{4.45}$$

Therefore, it can be seen that the equivalent inductance of the three series inductors is:

$$L_{eq} = L_1 + L_2 + L_3 \tag{4.46}$$

Similarly, for N inductors in series:

$$L_{eq} = L_1 + L_2 + \cdots + L_N \tag{4.47}$$

4.5.2 Inductors in Parallel

Applying Kirchhoff's current law (KCL) to the upper node of Figure 4.20 results in:

$$i = i_1 + i_2 + i_3 \tag{4.48}$$

Assuming that the initial current for all inductors is zero, Equation (4.37) can be substituted for i_1, i_2, i_3 in Equation (4.48), which leads to:

$$i = \frac{1}{L_1} \int_{t_0}^{t} v(t)\,dt + \frac{1}{L_2} \int_{t_0}^{t} v(t)\,dt + \frac{1}{L_3} \int_{t_0}^{t} v(t)\,dt \tag{4.49}$$

Equation (4.49) can be rewritten as:

$$i = \left(\frac{1}{L_1} + \frac{1}{L_2} + \frac{1}{L_3} \right) \int_{t_0}^{t} v(t)\,dt = \frac{1}{L_{eq}} \int_{t_0}^{t} v(t)\,dt \tag{4.50}$$

Therefore, the equivalent inductance corresponds to:

$$L_{eq} = \left(\frac{1}{L_1} + \frac{1}{L_2} + \frac{1}{L_3} \right)^{-1} \tag{4.51}$$

Therefore, the equivalent inductance for N inductors in parallel is:

$$L_{eq} = \left(\frac{1}{L_1} + \frac{1}{L_2} + \cdots + \frac{1}{L_N} \right)^{-1} \tag{4.52}$$

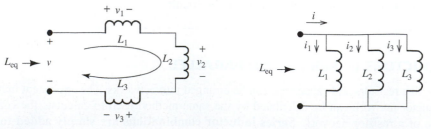

FIGURE 4.19 Inductors in series.　　　　**FIGURE 4.20** Inductors in parallel.

EXAMPLE 4.12   **Equivalent Inductance**

Find the equivalent inductance in Figure 4.21(a).

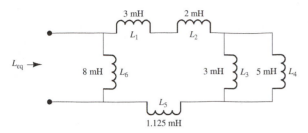

FIGURE 4.21(a)   Circuit for Example 4.12.

SOLUTION

L_3 and L_4 are in parallel. Thus, the circuit can be simplified to that shown in Figure 4.21(b).

$$L_{34} = \left(\frac{1}{L_3} + \frac{1}{L_4} \right)^{-1}$$

Because L_1, L_2, L_{34}, and L_5 are in series, the inductor network simplifies to Figure 4.21(c).

$$L_{12345} = L_1 + L_2 + L_{34} + L_5$$

Now L_6 and L_{12345} are in parallel. As a result, the equivalent inductance corresponds to:

$$L_{eq} = \left(\frac{1}{8 \text{ mH}} + \frac{1}{8 \text{ mH}} \right)^{-1} = 4 \text{ mH}$$

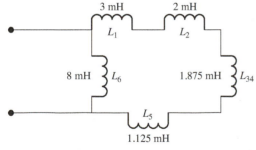

FIGURE 4.21(b)   Simplified circuit of Figure 4.21(a).

FIGURE 4.21(c)   Simplified circuit of Figure 4.21(b).

EXAMPLE 4.13   **Equivalent Inductance**

Find the equivalent inductance of the circuit shown in Figure 4.22(a).

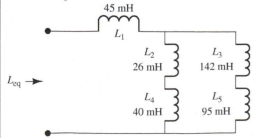

FIGURE 4.22(a)   Circuit of Example 4.13.

SOLUTION

Note that in Figure 4.22(a), there are two parallel branches, each consisting of two inductances in series: L_2 is in series with L_4, and L_3 is in series with L_5. Thus, the equivalent inductances of $L_{24} = L_2 + L_4$ and $L_{35} = L_3 + L_5$ and simplify Figure 4.22(a) to Figure 4.22(b).

(continued)

EXAMPLE 4.13   **Continued**

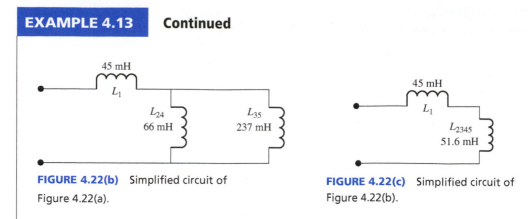

FIGURE 4.22(b)   Simplified circuit of Figure 4.22(a).

FIGURE 4.22(c)   Simplified circuit of Figure 4.22(b).

The parallel connection of L_{24} with L_{35} can be combined as shown in Figure 4.22(c).

$$L_{2345} = \left(\frac{1}{L_{24}} + \frac{1}{L_{35}} \right)^{-1}$$

Now, the network of Figure 4.22(c) is a simple series connection; therefore, the equivalent inductance will correspond to:

$$L_{eq} = 96.62 \text{ mH}$$

4.6 APPLICATIONS OF CAPACITORS AND INDUCTORS

Capacitors and inductors have many applications to different systems and technologies. This section explains some examples of these practical applications.

4.6.1 Fuel Sensors

Capacitors can be used as fuel gauges, for example, the fuel gauges used in airplanes. Automotive fuel gauges use a resistive sensor instead of capacitors. For a capacitive fuel gauge, low-voltage capacitors are stacked vertically in the fuel tank, as shown in Figure 4.23(a). In addition, one large parallel-plate capacitor is used, as shown in Figure 4.23(b).

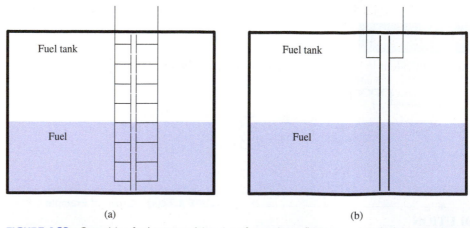

(a)   (b)

FIGURE 4.23   Capacitive fuel sensors: (a) series of capacitors; (b) large parallel-plate capacitor.

These capacitors are designed to allow the fuel to penetrate the area between the plates to contact the conductors. As the fuel level changes, the dielectric between the parallel plates changes and as a result its capacitance also changes. The magnitude of the capacitance is used to determine the fuel level.

The capacitance can be modeled as two capacitors in parallel as shown in Figure 4.24. The permittivity of the empty section of the tank (air) differs from the permittivity of the fuel. When the fuel level decreases, the capacitance associated with the air, C_{empty}, increases, while the capacitance associated with the fuel, C_{fuel}, decreases.

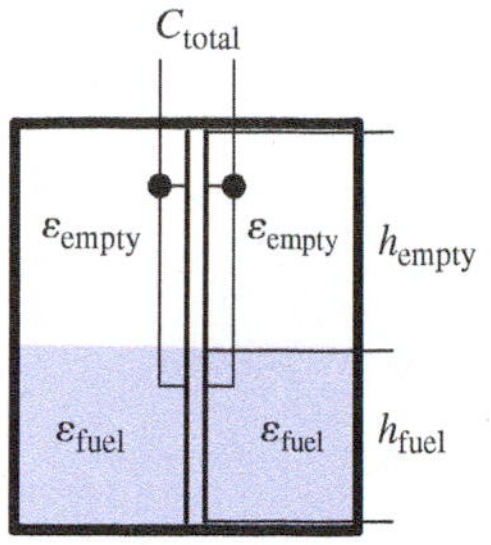

The total capacitance corresponds to:

$$C_{total} = C_{empty} + C_{fuel} \tag{4.53}$$

Applying the formula for the capacitance of a parallel-plate capacitor:

$$C = \varepsilon_o \varepsilon_r \times \frac{A}{d} \tag{4.54}$$

The total capacitance is:

$$C_{total} = \frac{w\left(\varepsilon_{empty}h_{empty} + \varepsilon_{full}h_{full}\right)}{d} \tag{4.55}$$

where:

w = width of the capacitor plates

h_{empty} = height of the empty section

h_{full} = height of the full section

ε_{empty} = permittivity of the empty section

ε_{full} = permittivity of the full section

d = displacement of the capacitor plates.

Equation (4.55) maintains a relationship (scale) between the capacitor and the fuel level. Accordingly, this design can function as an indicator for the fuel level.

FIGURE 4.24
Capacitance model for Figure 4.23.

4.6.2 Vibration Sensors

Capacitors and inductors can each be used to sense vibration and motion. These sensors have two components: a fixed component and a moving component. The moving component is attached to the moving object, while the fixed component is anchored to a fixed structure (see Figure 4.25). Vibrations in the moving object create pulsation in the moving component. This changes the inductance or the capacitance of the sensor. As shown in Figure 4.25, the sensor can be connected between two structures where the vibration needs to be measured.

Figure 4.25 details two capacitive vibration sensors. In Figure 4.25(a), the capacitance that changes the moving conducting plate changes position between the plates of a parallel-plate

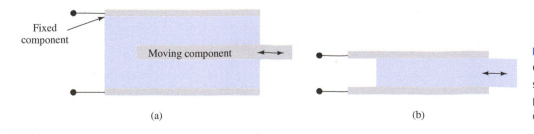

Fixed component

Moving component

(a)   (b)

FIGURE 4.25
Capacitive vibration sensors: (a) Moving plate. (b) Moving dielectric.

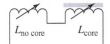

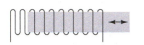

capacitor. The dielectric can be a fluid or air. In the sensor shown in Figure 4.25(b), the capacitance varies by moving a solid dielectric in and out between the conducting plates. Capacitive displacement sensors have been detailed in Example 4.9 of Section 4.3.

Figure 4.26 represents a simple inductive vibration sensor. Here, the inductance varies by moving a magnetic core in and out of the inductance coil.

This sensor can be modeled using two series-variable inductors as shown in Figure 4.27. As the core moves out of the coil, L_{core} decreases and L_{nocore} increases. Conversely, moving the core into the coil increases L_{core} and decreases L_{nocore}. As explained in Section 4.4, a magnetic core increases the overall inductance. Therefore, when L_{core} increases, the total inductance increases, and vice versa.

A linear variable differential transformer (LVDT) is a commonly used inductive sensor. A LVDT, as shown in Figure 4.28, is very similar to the sensor shown in Figure 4.26.

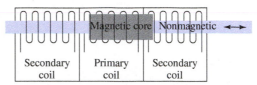

| Secondary coil | Primary coil | Secondary coil |

FIGURE 4.28   Linear variable differential transformer.

In a zero position, the magnetic core rests completely within the primary coil. When movement is applied to the core, it moves partially into one of the secondary coils. LVDT will be discussed in detail in Chapter 11.

APPLICATION EXAMPLE 4.14   Pressure Sensors

Capacitive sensors can also be used to measure the change of pressure. A change in pressure leads to a change of the distance between two capacitor plates. This changes the capacitance of the capacitor (Figure 4.29). If a reference pressure is given, with a known coefficient that represents pressure changes as a change in capacitance, the capacitance can be computed for any given pressure.

Assume that the change of capacitance due to the change of pressure is $\Delta C = 0.0085 \times \Delta P$, where P is the pressure measured in Pascal, Pa. If the pressure changes from 1000 Pa to 1050 Pa, with an initial capacitance of 3 F, what is the capacitance after the pressure changes?

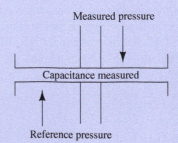

FIGURE 4.29   Capacitive pressure sensor.

SOLUTION

Change of pressure = 1050 − 1000 = 50 Pa, thus, the change in capacitance is:

$$\Delta C = 0.0085 \times \Delta P$$
$$\Delta C = 0.0085 \times 50 = 0.425 \text{ F}$$

The capacitance after the pressure change is:

$$C_{\text{final}} = C_{\text{initial}} + \Delta C$$
$$C_{\text{final}} = 3 + 0.425 = 3.425 \text{ F}$$

APPLICATION EXAMPLE 4.15   Traffic Light Detectors

In an automated traffic light system, like the one shown in Figure 4.30, the red traffic light stays ON longer if no vehicle stops at it; but it turns to green earlier when at least one vehicle is stopped at the light. This system operates as follows.

First, an inductor is placed underneath the ground in the lane where a vehicle will stop in the event of the red light. When there is a vehicle at the stop light, the size of the vehicle changes the inductance. A change in the inductance is detected by a specific device, which makes the traffic light to turn to green. Given the voltage source of 12 V, the inductance when there is not any vehicle is 2 H, and the sensitivity of the device to the change of inductance is 0.005 H.

Calculate the change of inductance if:

a. A car is stopped at the light, with the dimensions (mm) of: $4450 \times 1750 \times 1420$
b. A motorcycle is stopped at the light, with the dimensions (mm) of: $1945 \times 785 \times 1055$.

Assume the inductance changes with the respect to the size of vehicle as shown here:

$$L = 0.002 \times \text{volume}$$

Note:

$$\text{Volume} = \text{height} \times \text{width} \times \text{length, in m}^3.$$

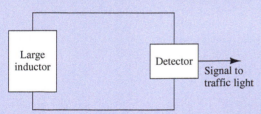

FIGURE 4.30   A traffic light detector.

SOLUTION

For part a:

The volume of the car is:

$$v = 4.45 \times 1.75 \times 1.42$$
$$v = 11.058 \text{ m}^3$$

The change in inductance is:

$$L = 0.002 \times 11.058$$
$$L = 0.0221 \text{ H} > 0.005 \text{ H (the device sensitivity)}$$

Thus, the device responds to the presence of the car.

For part b:

The volume of the motorcycle is:

$$v = 1.945 \times 0.785 \times 1.055$$
$$v = 1.611 \text{ m}^3$$

(continued)

APPLICATION EXAMPLE 4.15   **Continued**

The change of inductance is:

$$L = 0.002 \times 1.611$$
$$L = 0.0032\,\text{H} < 0.005\,\text{H (the device sensitivity)}$$

Thus, the detector is not able to detect the change of inductance when there is a motorcycle stopped at the light.

4.7 ANALYSIS OF CAPACITIVE AND INDUCTIVE CIRCUITS USING PSPICE

This section describes the process of analyzing capacitive and inductive circuits using PSpice. Here, we use examples to explain the approach.

EXAMPLE 4.16   **PSpice Analysis**

In the Figure used in Example 4.3, repeated as Figure 4.31, use PSpice to simulate the current through the capacitor, assuming $v(t) = 5 \cos(2\pi \times 10^3 t)$ V.

FIGURE 4.31   The circuit of Example 4.3.

SOLUTION

First, determine the frequency of the sinusoidal voltage source.

$$f = \frac{\omega}{2\pi}$$
$$f = \frac{2\pi \times 10^3}{2\pi}$$
$$f = 1000\,\text{Hz}$$

1. Choose VSIN as the voltage source, following the process shown in Example 2.24 in Chapter 2 to calculate and set up the parameters for VSIN.
2. To choose the capacitor, go to "Part" and type "C" in part name. Instead of using "C," choose "C_elect." If "C" is unavailable in "Part," add "Analog" into the library (see Figure 2.63 in Chapter 2).
3. To set the initial condition for the capacitor, right-click the capacitor and choose "Edit properties." Under the "IC" tab (solid arrow), insert the desired initial condition, "0" because $v_C(t = 0)$ is 0 V. The initial condition can optionally be displayed in the PSpice circuit by highlighting the "IC" tab then choosing "Display…" (dashed arrow). Choose "Name and Value" under the Display format.

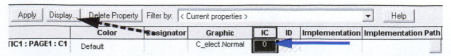

FIGURE 4.32   Setting the capacitor's initial conditions.

4. Set up the circuit as shown in Figure 4.33.

5. To run the simulation for the capacitor and inductor circuit, choose the "Analysis Type" to be time domain (see Figure 2.71 in Chapter 2). Set the "Run to time" as 10 ms. The result is shown in Figure 4.34. Use "Add Trace" in the Trace menu (Figure 2.77 in Chapter 2), to add the current flow through the capacitor into the plot.

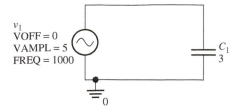

FIGURE 4.33 PSpice solution for Example 4.3.

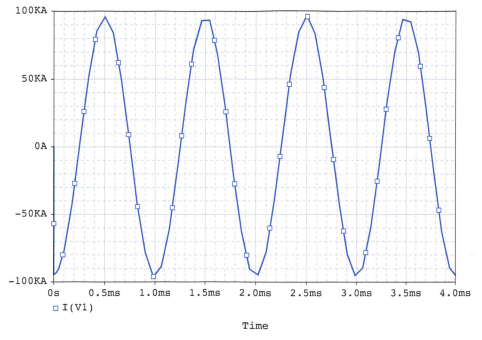

FIGURE 4.34 Results for Example 4.3.

EXAMPLE 4.17   PSpice Analysis

In Figure 4.17(a), repeated in Figure 4.35, use PSpice to simulate the voltage across the inductor, assuming $i(t) = 5 \cos(2\pi \times 10^3 t)$ A.

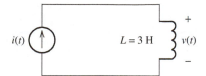

FIGURE 4.35 Circuit for Example 4.10.

SOLUTION

First, determine the frequency of the sinusoidal current source.

$$f = \frac{\omega}{2\pi} \quad f = \frac{2\pi \times 10^3}{2\pi} \quad f = 1000 \, \text{Hz}$$

(*continued*)

EXAMPLE 4.17   **Continued**

1. An "ISIN" sinusoidal current source has been chosen for this example. Go to "Place" > "Part" and type "ISIN" to obtain the sinusoidal current source part (see Figure 2.63 in Chapter 2). Follow the steps in Example 2.24 in Chapter 2 to set up the sinusoidal current source parameters.
2. To add an Inductor into the PSpice circuit, go to "Part" and type "L."
3. Define the initial condition, as shown in Example 4.16. Set IC to 0 for the case of I_L $(t = 0) = 0$ A.
4. Wire the circuit as shown in Figure 4.36 and run the simulation for 5 ms. The result is shown in Figure 4.37.

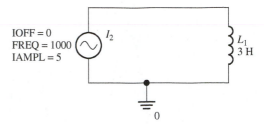

FIGURE 4.36   PSpice solution for Example 4.10.

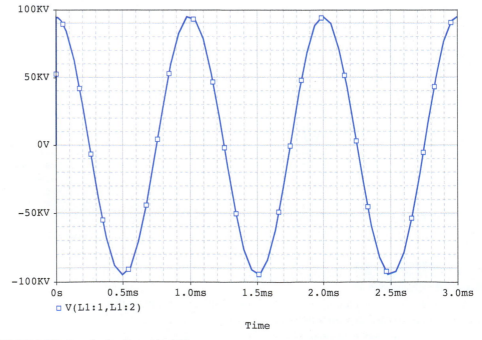

FIGURE 4.37   Results for Example 4.10.

4.8 WHAT DID YOU LEARN?

- The capacitance of a parallel-plate capacitor is proportional to the area of the plates and is inversely proportional to the distance between the plates [Equation (4.1)]:

$$C = \frac{\varepsilon_0 \varepsilon_r \times A}{d}$$

- The relationship between capacitors and inductors is expressed through the following equations:

Capacitor	Inductor
$q(t) = C \cdot v(t)$	$L = \dfrac{\phi}{i}$
$i(t) = C\dfrac{dv(t)}{dt}$	$di = \dfrac{1}{L}v(t)\,dt$
$v(t) = \dfrac{1}{C}\displaystyle\int_{t_0}^{t} i(t)\,dt + v(t_0)$	$i(t) = \dfrac{1}{L}\displaystyle\int_{t_0}^{t} v(t)\,dt + i(t_0)$
$P(t) = Cv(t)\dfrac{dv(t)}{dt}$	$P(t) = Li(t) \times \dfrac{di}{dt}$
$E(t) = \dfrac{1}{2}Cv^2(t)$	$E(t) = \dfrac{1}{2}Li^2(t)$

- Capacitors in series and parallel:
  Capacitors *in series* can be combined like resistors in parallel [Equation (4.23)]:

$$C_{eq} = \left(\frac{1}{C_1} + \frac{1}{C_2} + \cdots + \frac{1}{C_N}\right)^{-1}$$

  Capacitors *in parallel* can be combined like resistors in series [Equation (4.30)]:

$$C_{eq} = C_1 + C_2 + \cdots + C_N$$

- Inductors in series and parallel:
  Inductors *in series* can be combined like resistors in series [Equation (4.47)]:

$$L_{eq} = L_1 + L_2 + \cdots + L_N$$

  Inductors *in parallel* can be combined like resistors in parallel [Equation (4.52)]:

$$L_{eq} = \left(\frac{1}{L_1} + \frac{1}{L_2} + \cdots + \frac{1}{L_N}\right)^{-1}$$

Problems

*B refers to Basic, A refers to Average, H refers to Hard, and * refers to problems with answers.*

SECTION 4.1 INTRODUCTION

4.1 (B)* Which of the following devices is not passive?
  a. Resistor
  b. Capacitor
  c. Inductor
  d. Current-controlled current source.
4.2 (B) Which of the following devices store energy?
  a. Resistor
  b. Capacitor
  c. Voltage source
  d. Inductor.

SECTION 4.2 CAPACITORS

4.3 (B) A parallel plate capacitor has gasoline as the dielectric. The area of the plate is 0.25 m² and the distance between two plates is 5 cm. Find the capacitance given that the relative permittivity of gasoline is 2.
4.4 (B) In problem 4.3, how does the capacitance change if:
  a. the plate area is doubled

  b. the distance between plates is doubled
  c. the dielectric becomes air.
4.5 (B)* Find the stored charge in a 10-µF capacitor that is charged to 6 V.
4.6 (H) Application: How can one electron be detected?
  a. In the market, we can find many capacitors in the range of 1 µF. How many electrons should be available in a 1-µF capacitor to create a total voltage of 1 V?
  b. Suppose we want to see a change of 1 V in a capacitor before and after one electron is stored in it. What should the capacitance be?
  c. We can detect whether one electron is being stored in the capacitor by observing the change of the capacitor voltage. Why can we not detect one electron being stored in a 1-µF capacitor? (*Hint:* compute the capacitor voltage change while one electron being stored)
4.7 (H)* In Problem 4.6, if the smallest observable voltage is 0.1 V, compute the side-length of a square plate capacitor that can detect one electron, assuming the dielectric is air

with relative permittivity of 1 and the distance between plates is 10 nm (10×10^{-9} m).

4.8 (B) If a 1.5-μF capacitor, charged to 5 V, is discharged in a time of 50 ms, what is the average power provided?

4.9 (A) A 5-μF capacitor is charged with a steady current of 50 μA. How long does it take to charge to 12 V?

4.10 (B) Find the voltage, v, in Figure P4.10.

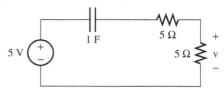

FIGURE P4.10 Circuit for Problem 4.10.

4.11 (A) The voltage across a 2.5-μF capacitor is $10 \sin(200\,\pi t)$ V as shown in Figure P4.11. Find an expression for (a) i_C through the capacitor and (b) the stored energy.

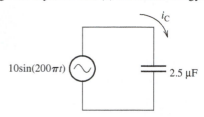

FIGURE P4.11 Circuit for Problem 4.11.

4.12 (H) Suppose $C = 2$ F in the circuit shown in Figure P4.12(a) and the voltage, $v(t)$, is shown in Figure P4.12(b).

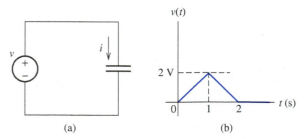

FIGURE P4.12 (a) A capacitor circuit; (b) waveform of $V(t)$.

Compute the current, $i(t)$, through the capacitor.

4.13 (A) In Problem 4.12:
 a. Compute the power absorbed by the capacitor
 b. Compute the energy stored in the capacitor.

4.14 (A)* At $t = 0$, a 2-mA source is applied to an uncharged 1.2-μF capacitor as shown in Figure P4.14. Determine (a) the capacitor voltage, v_C. at $t = 20$ ms and (b) the energy, w_C, stored in the capacitor at $t = 20$ ms.

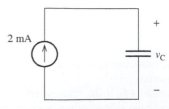

FIGURE P4.14 Circuit for Problem 4.14.

4.15 (H) At $t = 0$, a 1-mA source is connected to a 2-μF capacitor which has been previously charged to 3 V. Give (a) the expression for the capacitor voltage, v_C. as a function of time, and (b) v_C at $t = 10$ ms.

4.16 (H) For the circuit shown in Figure P4.16(a), $i(t)$ is given in Figure P4.16(b). Assume the initial value of v, $v(0) = 0$.

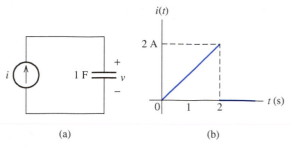

(a)   (b)

FIGURE P4.16 (a) Capacitor circuit; (b) waveform of $i(t)$.

Compute the voltage $v(t)$ across the capacitor.

4.17 (A) In Problem 4.16:
 a. Compute the power absorbed by the capacitor
 b. Compute the energy stored in the capacitor.

4.18 (H) Prove that the voltage across a capacitor cannot change instantaneously. (*Hint:* if the voltage at time a is $v(a) = 1/C \int_0^a i(t)\,dt$, compute the voltage at time $(a + \varepsilon)$, where ε is a positive number. When ε gets arbitrarily small, what will be the behavior of $v(a + \varepsilon)$?)

4.19 * Find ε_4 and the current through each resistor shown in Figure P4.19. $\varepsilon_1 = 4$ V, $\varepsilon_2 = 8$ V, $\varepsilon_3 = 12$ V, $C_1 = C_2 = C_3 = C_4 = 1$ μF. All resistors have the same resistance of 1Ω.

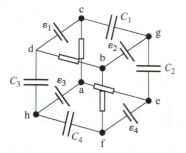

FIGURE P4.19 Circuit for Problem 4.19.

4.20 (A) Find the charge on each capacitor shown in Figure P4.19. Use the parameters in Problem 4.19.

4.21 (A) Find the total energy stored in capacitors shown in Figure P4.19. Use the parameters in Problem 4.19.

4.22 (H)* Connect nodes a and b in Figure P4.19, repeat Problem 4.19.

4.23 (A) Connect nodes a and b in Figure P4.19, repeat Problem 4.19.

4.24 (A)* Connect nodes a and b in Figure P4.19, repeat Problem 4.19.

SECTION 4.3 CAPACITORS IN SERIES AND PARALLEL

4.25 (B) Find the equivalent capacitance between points A and B in Figure P4.25.

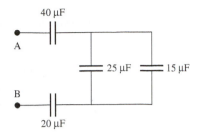

FIGURE P4.25 Capacitor network.

4.26 (B) For the capacitor network in Figure P4.26, what is the equivalent capacitance, C_{eq}?

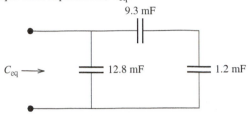

FIGURE P4.26 Circuit for Problem 4.26.

4.27 (A) Find the capacitance, C_A, needed to produce a network capacitance of $C_{eq} = 11.545$ μF.

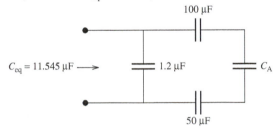

FIGURE P4.27 Capacitor network.

4.28 (A)* For the capacitor network in Figure P4.28, what is the equivalent capacitance, C_{eq}?

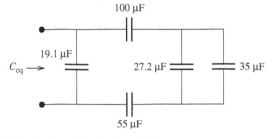

FIGURE P4.28 Circuit for Problem 4.28.

4.29 (A) Find the equivalent capacitance between points A and B in Figure P4.29.

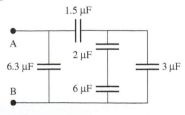

FIGURE P4.29 Capacitor network.

4.30 (A) Find the minimum and maximum values for C_{eq} as shown in Figure P4.30.

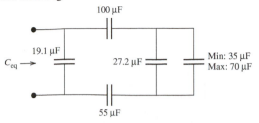

FIGURE P4.30 Capacitor network.

4.31 (H) Find the equivalent capacitance between points A and B in Figure P4.31.

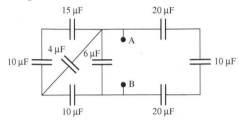

FIGURE P4.31 Capacitor network.

4.32 (H)* Look at the fuel sensor example in Section 4.6.1. The permittivity of air is 1 and the permittivity of gasoline is 2. Let the height of the gasoline level be h.

  a. Find the expression of the total capacitance C_{total} in terms of h, width of the capacitor w, distance between plates d, and the free space permittivity ε_0.

  b. When the tank is full, C_{total} is 12×10^{-12} F; when the tank is empty, C_{total} is 6×10^{-12} F. Suppose the capability of the tank is 15 gallon. How much gasoline in gallon is in the tank when C_{total} is read as 8×10^{-12} F?

SECTION 4.4 INDUCTORS

4.33 (B) Given the circuit in Figure P4.33, find the current, i, as a function of time. (*Note:* v_s is a sinusoid.)

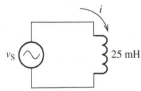

FIGURE P4.33 Inductor circuit.

4.34 (B) Find i_1 in the circuit as shown in Figure P4.34.

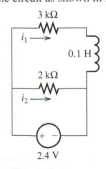

FIGURE P4.34 Circuit with an inductor.

4.35 (A)* Find v_o in the circuit given in Figure P4.35.

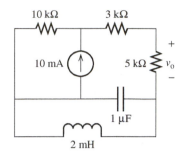

FIGURE P4.35   Circuit with capacitor and inductor.

4.36 (B) If a current of $i(t) = 20\cos(100\pi t)$ mA flows through an inductor of $L = 45$ mH, find the voltage induced across the inductor.

4.37 (A) Current through an inductor is turned on at time $t = 0$, as shown in Figure P4.37. $v_s = \cos(200\pi t)$. Calculate the energy delivered to the inductor at $t = 21$ ms.

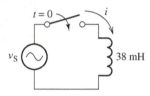

FIGURE P4.37   Circuit for Problem 4.37.

4.38 (H) For the circuit shown in Figure P4.38(a), $v(t)$ is given in Figure P4.38(b). Assume the initial value of i, $i(0) = 0$.

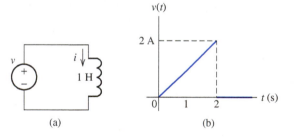

FIGURE P4.38   (a) Inductor circuit; (b) waveform of $v(t)$.

Compute the current, $i(t)$, through the inductor.

4.39 (A) In Problem 4.38:
   a. Compute the power absorbed by the inductor
   b. Compute the energy stored in the inductor.

4.40 (H) For the circuit shown in Figure P4.40(a), the current $i(t)$ is shown in Figure P4.40(b).

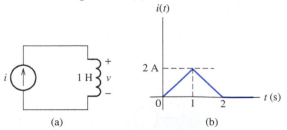

FIGURE P4.40   (a) Inductor circuit; (b) waveform of $i(t)$.

Compute the voltage, $v(t)$, across the inductor.

4.41 (A) In Problem 4.40:
   a. Compute the power absorbed by the inductor.
   b. Compute the energy stored in the inductor.

4.42 (H) Prove that the current through an inductor cannot change instantaneously. (*Hint:* refer to the hint of Problem 4.18.)

SECTION 4.5 INDUCTORS IN SERIES AND PARALLEL

4.43 (B)* For the inductor network in Figure P4.43, what is the equivalent inductance, L_{eq}?

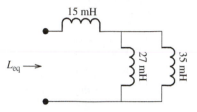

FIGURE P4.43   Circuit for Problem 4.43.

4.44 (B) For the inductor network in Figure P4.44, what is the equivalent inductance, L_{eq}?

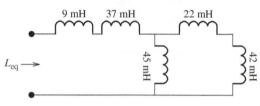

FIGURE P4.44   Circuit for Problem 4.44.

4.45 (A)* Find the equivalent inductance of the inductor network of Figure P4.45.

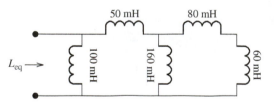

FIGURE P4.45   Inductor network.

4.46 (H) Find the equivalent inductance across points A and B in Figure P4.46.

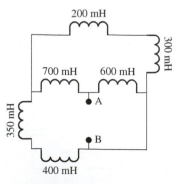

FIGURE P4.46   Inductor network.

4.47 (A) Find the equivalent inductance across points A and B in Figure P4.47.

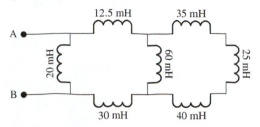

FIGURE P4.47 Inductor network.

4.48 (A)* (Application) oil pipeline, running from Mosul, Iraq to Beirut, Lebanon is being restored so it can be placed back into service. A new pump is needed to replace a broken one. A large electric motor is needed to drive this pump. The motor has the equivalent circuit as shown in Figure P4.48. Solve for the equivalent inductance.

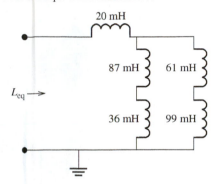

FIGURE P4.48 Circuit for Problem 4.48.

4.49 (H) Find the expression for the equivalent inductance across the points A and B in Figure P4.49.

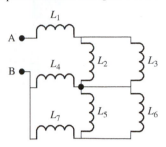

FIGURE P4.49 Inductor network.

SECTION 4.6 ANALYSIS OF THE CIRCUITS USING PSPICE

4.50 (B) The voltage across a 2.5-μF capacitor is 10 sin(200 πt) V, as shown in Figure P4.11. Use PSpice and plot the current, I_C, through the capacitor.

4.51 (A) Consider a circuit with three capacitor C_1, C_2, and C_3 connected in parallel to a voltage source of 10 sin(120πt) V. Use PSpice to plot the current flow through C_3 if C_1 and C_2 is 100 μF and C_3 is 200 μF.

4.52 (H) Given the voltage source is 5 sin(πt), current through an inductor is turned on at time, $t = 0$, as shown in Figure P4.37. Use PSpice to plot the voltage and current flow through the inductor for 25 ms.

CHAPTER 5

Transient Analysis

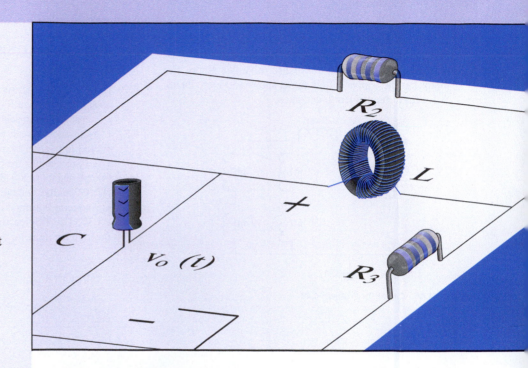

5.1 INTRODUCTION

Chapter 4 explained that capacitors and inductors are energy storage elements. Capacitors store energy within the electric field generated between the capacitor's parallel plates. Inductors store energy within the magnetic field created around the inductor. Chapter 4 also examined circuits that included only capacitors or inductors. This chapter extends the knowledge to describe strategies for analyzing circuits with diverse elements, namely, resistors and capacitors, or resistors and inductors.

When an electric circuit that contains a resistor, a capacitor, and/or an inductor is exposed to a sudden change in voltage or current from the source (i.e., a decrease or increase), the capacitor and/or the inductor begin to lose or gain energy. This process is called **charge** or **discharge**. Note that resistors do not exhibit transient behavior unless combined with an energy storage device. In this case, the average voltage across and the average current flowing through capacitors and inductors change with time. This variation of voltage and current over time is called **transient response**. Voltage or current reaches a steady state as the time moves toward infinity.

As discussed in Chapter 4, a resistor is not the best model for many electrical devices. A combination of resistors, inductors, and/or capacitors may better represent electrical devices. Electrical devices may exhibit a transient response when switched ON or OFF due to the presence of parasitic capacitance and inductance. For example, power lines have a high parasitic capacitance between the lines and the ground (see Chapter 9). Motors have a high inductance due to the multiple coils available within their rotor and stator. Both of these systems exhibit a noticeable transient when they are switched ON. It is important for engineers to acknowledge these issues.

For example, engineers must be aware of switch bounce. Examples 5.2 and 14.10, and Chapter 15 discuss switch bounce issue and its compensating methods. Electrical switches are mechanical devices and therefore have mechanical properties. When a switch is opened or closed, it does not immediately go from its first state to its second: it fluctuates for a brief period of time. Systems must be designed so that these fluctuations do not damage electrical devices or affect safety. Transient analysis can be used to design timers and auto-dimming lights as seen in newer automobiles. Biomedical engineers also utilize transient analysis to examine defibrillators, which use transients to store charge over time and deliver it instantaneously to restart patients' hearts.

In the analysis of transient circuits, engineers usually model the signal that is applied to the circuit after closing a switch as a step function (voltage) [see Figure 5.1(a)]. This figure shows that the step function is zero if $t < 0$ and is unity if $t > 0$. The step function is represented by $U(t)$. For this example, the voltage can be generated using a DC voltage source and a switch, as shown in Figure 5.1(b). The arrow in Figure 5.1(b) indicates that the voltage exists between the two terminals when $t > 0$ (i.e., when the switch is closed).

5.2 FIRST-ORDER CIRCUITS

A circuit is called **first order** if it contains one type of reactive element (i.e., a capacitor or an inductor) and other resistive elements. In general, the **order** of a circuit is determined by the number of types of reactive elements in the circuit and the circuit's topology. Accordingly, first-order circuits are categorized as RC (resistive–capacitive) or RL (resistive–inductive) circuits.

5.2.1 RC Circuits

In their final, reduced form, RC first-order circuits contain a single capacitor and resistors. An example of an RC circuit is shown in Figure 5.2(a). To analyze this circuit in a more simple way, it can be replaced with the circuit shown in Figure 5.2(b) using the Thévenin theorem as discussed in Chapter 3. In this example, the Thévenin voltage is the open circuit voltage between the capacitor terminals (i.e., A and B), and the Thévenin resistance is the equivalent resistance between the A and B terminals after short-circuiting the voltage source. Because all first-order RC circuits can be reduced using the Thévenin theorem, the circuit shown in Figure 5.2(b) can be used for analysis regardless of the original circuit.

5.2.1.1 CHARGING A CAPACITOR USING A DC VOLTAGE SOURCE THROUGH A RESISTANCE
Consider the circuit shown in Figure 5.3. The voltage source, v_S, is constant (i.e., it is a DC voltage). The source is connected to the RC circuit through a switch that is closed at time $t = 0$.

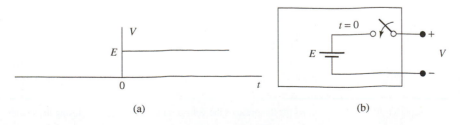

(a)

(b)

FIGURE 5.1 (a) Step voltage waveform; (b) circuit representation of step voltage.

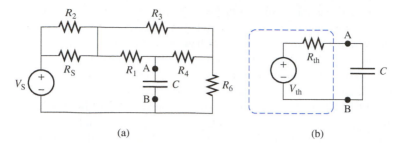

FIGURE 5.2
(a) First-order circuit;
(b) Thévenin equivalent circuit.

Assuming the initial voltage across the capacitor is zero (i.e., $v_C(0^-) = 0$),[*] KVL can be used to find the voltage across the capacitor for $t > 0$. Writing KVL:

$$V_S - Ri(t) - v_C(t) = 0 \tag{5.1}$$

Now, using the relationship between the capacitor's voltage and current (i.e., $i(t) = i_C(t)(dv_C(t)/dt)$ as discussed in Chapter 4), Equation (5.1) can be represented as:

$$V_S - RC\frac{dv_C(t)}{dt} - v_C(t) = 0 \tag{5.2}$$

A simple displacement of the terms in Equation (5.2) leads to:

$$RC\frac{dv_C(t)}{dt} = -(v_C(t) - V_S) \tag{5.3}$$

or:

$$\frac{dv_C(t)}{v_C(t) - V_S} = -\frac{1}{RC}\,dt \tag{5.4}$$

Integrating both sides of Equation (5.4):

$$\int_{0^+}^{v_C(t)} \frac{dv_C(\tau)}{(v_C(\tau) - V_S)} = -\int_{0^+}^{t} \frac{1}{RC}\,d\tau \tag{5.5}$$

The result corresponds to:

$$\ln\left(\frac{v_C(t) - V_S}{v_C(0^+) - V_S}\right) = -\frac{1}{RC}(t - 0^+) \tag{5.6}$$

To complete the solution and find $v_C(t)$, the value of the initial voltage, $v_C(0^+)$, must be known. Note that 0^- is the time just *before* the switch is closed, and 0^+ refers to the time right *after* the switch is closed, which is also the time immediately after the (voltage) source is applied to the RC circuit. In general, the voltage across the capacitor cannot change suddenly. This is because the current through the capacitor corresponds to $i_C(t) = C\,dv_C(t)/dt$; thus, $dv_C(t) = i_C(t)dt/C$. Now, a sudden change in voltage, $dv_C(t)$ is equivalent to an infinitely large change in the current, $i_C(t)$, because dt is infinitely small. But, because of the existence of the resistor, $R > 0$, the current $i_C(t)$

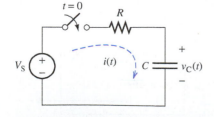

FIGURE 5.3 Capacitor charging through a resistance from a DC voltage source.

---

[*] The superscript "–" which follows the 0 represents the voltage immediately *before* closing the switch.

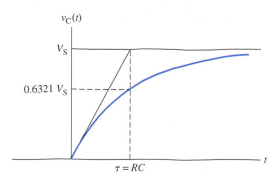

FIGURE 5.4 The charging transient for the circuit shown in Figure 5.3.

cannot increase to infinity, even if the capacitor is short-circuited. Thus, within a short period of time, from 0^- to 0^+, the voltage remains constant. This is expressed by the following equation:

$$v_C(0^+) = v_C(0^-) = 0 \tag{5.7}$$

Using Equation (5.7), Equation (5.6) is equivalent to:

$$\ln\left(1 - \frac{v_C(t)}{V_S}\right) = -\frac{1}{RC}t \tag{5.8}$$

Now, applying exponential function to both sides of the equation results in:

$$1 - \frac{v_C(t)}{V_S} = e^{-\frac{t}{RC}} \tag{5.9}$$

Solving for $v_C(t)$ in Equation (5.9) leads to:

$$v_C(t) = V_S\left(1 - e^{-\frac{t}{RC}}\right) \tag{5.10}$$

The plot of the capacitor's voltage is shown in Figure 5.4. It is clear that the voltage increases exponentially from zero to V_S. Equation (5.10) shows that at $t = 0$, $v_C(t) = 0$. That is, the capacitor is equivalent to a short circuit.

5.2.1.2 TIME CONSTANT

Time constant is a measure of a circuit's "sluggishness" in its ability to quickly respond to a change in input. The more resistance and/or capacitance, the more sluggish or slow the circuit will be. The value of the time constant in seconds is equal to the product of the circuit resistance (in ohms) and the circuit capacitance (in farads). In other words, the time constant corresponds to:

$$\tau = RC \tag{5.11}$$

According to Equation (5.11), the unit of time, *second*, is equal to: ohm × farad. Substituting Equation (5.11) into (5.10):

$$v_C(t) = V_S\left(1 - e^{-\frac{t}{\tau}}\right) \tag{5.12}$$

When $t = \tau$, the equation results in $v_C(\tau) = V_s(1 - e^{-1}) = 0.6321V_s$. Therefore, within the time constant $\tau = RC$, the capacitor voltage is charged up to 63% of its maximum value, v_S. After about five time constants, the difference between the voltage across the capacitor and the full charge voltage is negligible. In other words, when $t = 5\tau$, the capacitor voltage, $v_C(4\tau) = V_S(1 - e^{-5}) = 0.9933V_S$. After these five time constants (generally as $t \to \infty$), it is said that the capacitor is completely charged.

Referring to Figure 5.3 and Equation (5.12), if a capacitor is fully charged ($t \to \infty$), the voltage across the capacitor will be equal to the voltage source (i.e., $v_C(t) = V_S$). In this case, the voltage across the resistor is almost zero, and the current through the circuit corresponds to:

$$i(t) = \frac{V_S - v_C(t)}{R}$$

$$= \frac{V_S - V_S}{R} \tag{5.13}$$

$$= 0$$

In other words, the current flowing through the capacitor tends toward zero as the voltage across it tends toward the source voltage, V_S, or as $t \to \infty$, that is, the capacitor will be equivalent to an open circuit.

Note that the current flowing through the capacitor corresponds to:

$$i(t) = C\frac{dv_C(t)}{dt} \tag{5.14}$$

Now, replacing $v_C(t)$ in Equation (5.14) with that of Equation (5.12):

$$i(t) = \frac{V_S}{R}e^{-t/\tau} \tag{5.15}$$

The current flowing through the capacitor, $i(t)$, is shown in Figure 5.5. As shown, the current at $t = 0^+$ is V_S/R and as t tends to infinity, $i(t)$ tends to zero. This is consistent with Equation (5.13). Based on this explanation, at $t = 0^+$, the capacitor is equivalent to a short circuit [see Figure 5.6(a)]. Writing KVL for the circuit shown in Figure 5.6(a), it is clear that $i(t) = V_S/R$. However, as t tends to infinity, $i(t) = 0$. In this case, the voltage drop across the resistor will be zero, thus, the equivalent circuit of the capacitor will be an open circuit [see Figure 5.6(b)].

EXERCISE 5.1

If the capacitor in the equivalent circuit shown in Figure 5.6(a) has an initial voltage, v_0, explain why the short circuit will be replaced by a voltage source.

As a general rule, at $t = 0^+$, the capacitor is equivalent to a voltage source with the value of its initial voltage, v_0 (short circuit, if $v_0 = 0$). Thus, in Figure 5.6 (d), if $v_0 \neq 0$, the short circuit should be replaced by a voltage source of v_0. However, a fully charged (i.e., when $t \to \infty$) capacitor is equivalent to an open circuit [see Figure 5.6(b)].

A fully charged capacitor converts the complex RC circuit into a simple resistive circuit. This rule helps to easily study the behavior of RC circuits at $t = 0$ and at $t = \infty$ (infinity).

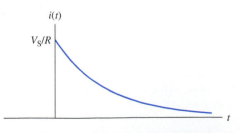

FIGURE 5.5   The current passing through the capacitor in the circuit of Figure 5.3.

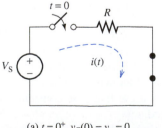

(a) $t = 0^+$, $v_C(0) = v_0 = 0$

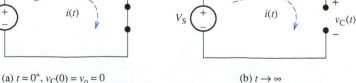

(b) $t \to \infty$

FIGURE 5.6   Equivalent circuit for the capacitor shown in Figure 5.3.

EXAMPLE 5.1	**Capacitor Voltage**

In the circuit shown in Figure 5.7, the switch is closed at $t = 0$, and the initial voltage across the capacitor $v_C(0^-) = 0$.

   a. Find an expression for the capacitor voltage, $v_C(t)$
   b. Find the voltage across the capacitor at $t = 5$ s
   c. Find the capacitor voltage at $t = \infty$

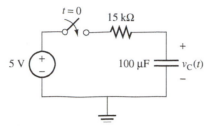

FIGURE 5.7 Circuit for Example 5.1.

SOLUTION

   a. The source voltage is: $V_S = 5$ V
       The time constant is: $\tau = RC = 15 \times 10^3 \times 100 \times 10^{-6} = 1.5$ s
   Substituting these values in Equation (5.12):

$$v_C(t) = 5\left(1 - e^{\frac{-t}{1.5}}\right)$$

   b. At $t = 5$ s

$$v_C(t) = 5\left(1 - e^{\frac{-5}{1.5}}\right) = 4.8216 \text{ V}$$

   c. At $t = \infty$

$$v_C(t) = 5\left(1 - e^{\frac{-\infty}{1.5}}\right) = 5(1 - 0) = 5 \text{ V}$$

   This value is called **steady-state** voltage, a term that will be explained and discussed later in this chapter.

APPLICATION EXAMPLE 5.2	**Switch Bounce**

When a switch is closed, the contacts (and thus the voltage) fluctuate for a time before settling into their final position. This is known as **switch bounce**. Figure 5.8(a) shows the ideal switch response, while Figure 5.8(b) shows a typical response. Chapter 15 discusses the corresponding safety issues of switch bounce. Switch bounce, in general, is not a major problem for basic electric circuits such as lights, radios, and motors. However, it does pose a problem for digital circuits such as computers and CD players.

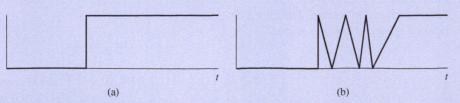

        (a)   (b)

FIGURE 5.8 The transient response of (a) an ideal switch; (b) a typical switch that experiences switch bounce.

(*continued*)

APPLICATION EXAMPLE 5.2   **Continued**

If a keyboard button is pushed and exhibits switch bounce, how can the computer know how many keystrokes to register? Digital circuits often require "debouncing" of the switch. This example discusses a debounce system for an RC-based circuit. Chapter 10 outlines how flip-flop systems can be used for debouncing.

Figure 5.9 shows a simple RC switch debouncing circuit. When the switch is first turned on, instead of fluctuating with the switch bounce, v_o slowly rises with each bounce as the capacitor charges. Note that in Figure 5.9, the equivalent circuit of the switch also includes a small resistance. When the switch is closed, the capacitor is charged through the switch as shown in Figure 5.10. When the switch is opened, the capacitor starts discharging through the load resistor R.

Based on Figure 5.9, when the switch is closed, the Thévenin equivalent resistance across the capacitor is the parallel combination of the equivalent resistor of the switch and the resistor, R. Because the switch resistance is very small, the equivalent resistance across the capacitor will be very small. Thus, the time constant of the charge is very small (ideally zero). The values for R and C are chosen such that the time constant of discharge is large enough so that the capacitor will barely be discharged when the switch is bounced off. This eliminates the bounce and makes the voltage across the resistor rise smoothly. The resulting switch response is sketched in Figure 5.10. Another switch bounce solution will be introduced in Chapter 10.

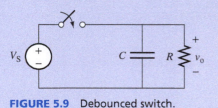

FIGURE 5.9 Debounced switch.

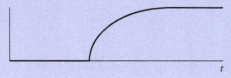

FIGURE 5.10 Transient response for the debounced switch.

EXERCISE 5.2

Assume that a voltage source is replaced by a voltage source and a resistance, R_S, placed in series. What should be the relationship between R and R_S to allow the debounce system introduced in Figure 5.9 to work in this setting?

EXAMPLE 5.3   **Capacitor Voltage**

The switch in the circuit shown in Figure 5.11 has been closed for a long time period and opens at $t = 0$.

   a. Find an expression for $v_C(t)$
   b. Find the capacitor voltage after 60 ms
   c. Find $v_C(t)$ when t tends to infinity

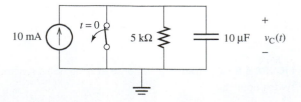

FIGURE 5.11 Circuit for Example 5.3.

SOLUTION

Because the switch was closed for a long time, the initial condition of the capacitor voltage will be zero, that is, $v_C(0) = 0$. Note that for simplicity the minus or plus sign for time zero has been removed because the capacitor voltage is the same for both cases.

a. This problem can be solved using two methods:
    Method 1: Applying KCL (note that there is only one non-reference node):

$$-10 \times 10^{-3} + \frac{v_C(t)}{5 \times 10^3} + 10 \times 10^{-6} \frac{dv_C(t)}{dt} = 0$$

Some mathematical manipulation leads to:

$$\frac{dv_C(t)}{dt} = \frac{-1}{5 \times 10^3 \times 10 \times 10^{-6}}(v_C(t) - 5 \times 10^3 \times 10 \times 10^{-3})$$

or:

$$\frac{dv_C(t)}{v_C(t) - 50} = \frac{-dt}{0.05}$$

Integrating both sides of this equation yields:

$$\ln \frac{v_C(t) - 50}{0 - 50} = \frac{-t}{0.05} \rightarrow 1 - \frac{v_C(t)}{50} = e^{-20t} \rightarrow v_C(t) = 50(1 - e^{-20t})$$

Method 2: Another approach is to find the Thévenin equivalent circuit for the given circuit. Figure 5.12(a) shows the desired circuit.

The circuit time constant corresponds to:

$$\tau = RC = 5 \times 10^3 \times 10 \times 10^{-6} = 0.05 \text{ s}$$

Substituting V_S and τ into Equation (5.12) yields:

$$v_C(t) = 50\left(1 - e^{\frac{-t}{0.05}}\right) = 50(1 - e^{-20t})$$

b. When $t = 60$ ms:

$$v_C(0.06) = 50(1 - e^{-20 \times 0.06}) = 50(1 - 0.3012) = 34.9403$$

c. When $t \rightarrow \infty$, the capacitor is fully charged. In this case, the capacitor will be equivalent to an open circuit as shown in Figure 5.12(b).

$$v_C(\infty) = 10 \text{ mA} \times 5 \text{ k}\Omega = 50 \text{ V}$$

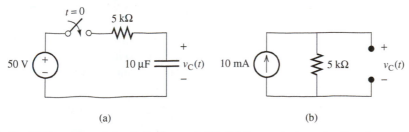

(a)   (b)

FIGURE 5.12   (a) The reduced circuit for Example 5.3; (b) the equivalent circuit as $t \rightarrow \infty$.

5.2.1.3 DISCHARGE OF A CAPACITOR THROUGH A RESISTANCE

Next, consider the circuit shown in Figure 5.13. When a voltage is applied to the capacitor for several time constants prior to $t = 0$, the capacitor is considered fully charged to an initial value of v_i (i.e., $v_C(0^-) = v_i$). Then, at time $t = 0$ the switch closes and the current flows through the resistor, discharging the capacitor.

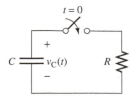

FIGURE 5.13
Discharging of the capacitor through a resistance.

Note that the current that flows through the capacitor is the opposite of the current that flows through the resistor, that is, $i_C(t) = -i_R(t)$. Thus, applying KCL to the circuit shown in Figure 5.13 yields:

$$C\frac{dv_C(t)}{dt} + \frac{v_C(t)}{R} = 0 \tag{5.16}$$

Multiplying both sides of Equation (5.16) by R:

$$RC\frac{dv_C(t)}{dt} + v_C(t) = 0 \tag{5.17}$$

Equation (5.17) can be rewritten as:

$$\frac{dv_C(t)}{v_C(t)} = -\frac{dt}{RC} \tag{5.18}$$

Integrating Equation (5.18) yields:

$$\ln\left(\frac{v_C(t)}{v_C(0^+)}\right) = -\frac{1}{RC}(t - 0) \tag{5.19}$$

As stated in the previous section, the voltage across the capacitor cannot change suddenly. Thus:

$$v_C(0^+) = v_C(0^-) = v_i \tag{5.20}$$

Substituting the value of $v_C(0^+)$ in Equation (5.19):

$$\ln\left(\frac{v_C(t)}{v_i}\right) = -\frac{t}{RC} \tag{5.21}$$

Applying exponential function to both sides of the equality:

$$v_C(t) = v_i e^{-\frac{t}{RC}} \tag{5.22}$$

Using $\tau = RC$:

$$v_C(t) = v_i e^{-\frac{t}{\tau}} \tag{5.23}$$

Note that at time $t = 0$, $v_C(t) = v_i$. Thus, at $t = 0$, the current through the resistor corresponds to $i(0) = v_i/R$. In other words, at $t = 0$, a fully charged capacitor will be equivalent to a voltage source of $v = v_i$. Previously, it was explained that at $t = 0$, a noncharged capacitor is equivalent to a short circuit. These two conclusions are consistent, because a zero-voltage source is equivalent to a short circuit.

A plot of $v_C(t)$ is shown in Figure 5.14. Notice that the voltage decays exponentially with time from V_S toward zero. The **time constant** corresponding to the discharge process is defined as the time required to discharge the capacitor to 36.79% from its initial voltage (i.e., $v_C(\tau) = e^{-1}v_i = 0.3679\,v_i$). This time constant is equal to RC as defined in Equation (5.11). The capacitor is said to be completely discharged after about five time constants. Here, 5τ is large enough to be

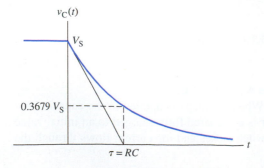

FIGURE 5.14

Discharging transient for Figure 5.13.

considered as $t \to \infty$, because after 5τ, the voltage that remains across the capacitor is negligible compared to the initial value. That is in Equation (5.23), replacing t with 5τ, we have

$$v_C(5\tau) = e^{-5}v_i = 0.0067\, v_i$$

which is a very small value when compared to v_i.

This exponential drop of voltage is due to the decay of the number of capacitor charges. This process is equivalent to a tank filled with water with a hole in the bottom. Here, the flow or the rate of loss of *volume of water* in the tank is analogous to the rate of loss of *electric charge* from the capacitor. As the width of the hole decreases, the resistance that water faces increases, and as a result, the water flow (discharge) will be slower. This process is shown in Figure 5.15.

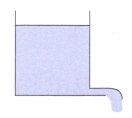

FIGURE 5.15 Water-flow analogy: discharging through a hole.

EXAMPLE 5.4   Capacitor Voltage

The switch in the circuit shown in Figure 5.16 has stayed in position 1 for a long time period. At time $t = 0$ the switch moves to position 2.

  a. Find $v_1(0^-)$, that is, $v_1(t)$ prior to the displacement of the switch.
  b. Find $v_2(0^+)$, that is, $v_2(t)$ right after the displacement of the switch.
  c. Find an expression for $v_C(t)$ for $t > 0$.
  d. Find the value of t at which the capacitor voltage reduces to 6 V.

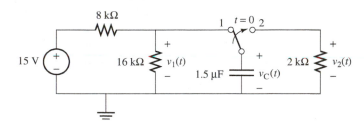

FIGURE 5.16 Circuit for Example 5.4.

SOLUTION

  a. After a long time period of the switch being in position 1, the capacitor is fully charged and can be replaced by an open circuit. Thus, using voltage divider rule, the voltage across the 16-kΩ resistor is:

$$v_1(0^-) = v_C(0^-) = 15\frac{16}{16 + 8} = 10 \text{ V}$$

In other words, the fully charged capacitor has been charged to 10 V.
  b. After the movement of the switch to position 2, $v_C(0^+) = v_C(0^-) = 10$ V. Thus,

$$v_2(0^+) = 10 \text{ V}$$

  c. At $t = 0$, when the switch is changed to position 2, the capacitor starts discharging through the 2-kΩ resistor. Using the same procedure as in the two previous examples, the voltage across the capacitor at time $t = 0$ corresponds to:

$$v_i = v_C(0^-) = v_C(0^+) = 10 \text{ V}$$

For $t > 0$, the time constant is:

$$\tau = RC = 2 \times 10^3 \times 1.5 \times 10^{-6} = 3 \times 10^{-3} \text{ s}$$

Substituting v_i and τ in Equation (5.23):

$$v_C(t) = 10e^{-\frac{t}{3\times10^{-3}}}$$

(continued)

EXAMPLE 5.4   **Continued**

   d. Substituting $v_C(t) = 6$ V:

$$6 = 10e^{-\frac{t}{3\times10^{-3}}} \rightarrow e^{-\frac{t}{3\times10^{-3}}} = 0.6 \rightarrow -\frac{t}{3\times10^{-3}} = \ln 0.6$$

   The value of t at which the capacitor voltage drops to 6 V is $t = 1.5325$ ms.

EXERCISE 5.3

In Example 5.4, if the switch is returned to position 1 after 2 ms, find the voltage v_1 (2 ms).

APPLICATION EXAMPLE 5.5   **Light Timer**

RC circuits can be used to build simple timers. Figure 5.17 shows a circuit that turns off a light after a given period of time. The transistor Q1 and the relay ensure that the light goes from fully ON to fully OFF. Transistor circuits are discussed in detail in Chapter 8.

   When the momentary contact switch SW1 is pushed, the capacitor charges and the light turns on. When the voltage between the base and the emitter of the transistor, $v_{BE}(t)$, is below 0.6 V, the transistor, and as a result the relay, turns off. This turns off the light. Given the simplified circuit shown in Figure 5.18, how long after pressing SW1 will the light turn off?

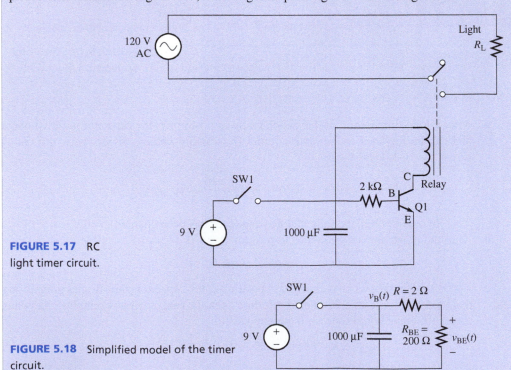

FIGURE 5.17 RC light timer circuit.

FIGURE 5.18 Simplified model of the timer circuit.

SOLUTION

In Figure 5.18, the switch has been closed for a long time and capacitor is fully charged to 9 V. Using KCL, the equation for $v_B(t)$, that is, the voltage across the capacitor after the switch SW1 is turned off, corresponds to:

$$0 = \frac{v_B(t)}{2200} + 1000 \ \mu F \frac{dv_B(t)}{dt}$$

Thus:

$$0 = \frac{v_B(t)}{2.2} + \frac{dv_B(t)}{dt} \rightarrow v_B(t) = Ke^{-t/2.2}$$

Because the initial charge of the capacitor is 9 V, in the initial conditions, $v_B(0^+) = 9$ V, accordingly:

$$9 = Ke^0 \rightarrow K = 9$$

Thus:

$$v_B(t) = 9e^{-t/2.2}$$

Now, using a simple voltage division rule, the voltage across the base and the emitter, $v_{BE}(t)$, and the voltage across the capacitor are related by:

$$v_{BE}(t) = \frac{200}{2000 + 200} \times v_B(t) = 0.09 v_B(t)$$

Therefore:

$$v_{BE}(t) = 0.818 e^{-t/2.2}$$

Solving for t:

$$0.6 = 0.818 e^{-t/2.2} \rightarrow \ln\left(\frac{0.6}{0.81}\right) = \frac{-t}{2.2} \rightarrow t = 0.682 \, s$$

Thus, the light turns off after 0.682 s. It is clear that the delay time can be increased if the 2 kΩ increases. For example, increasing it to 20 kΩ, the delay will increase to about 6 s.

EXERCISE 5.4

Select R in Example 5.5 (Figure 5.18) such that the delay is adjusted to 5 s.

APPLICATION EXAMPLE 5.6   Power Shock

Suppose that a careless worker wires a transformer improperly and uses the wrong chemicals for its coolant. As a result, the worker creates a massive explosion that destroys the transformer.

The worker is fired and goes out panhandling, limping on a broken leg and wearing a sign, "will fix downed power lines for food." The worker hitchhikes around the state looking for work until he finds the little white farmhouse in Figure 5.19.

"If you get those power lines out of my yard, I'll give you a hamburger," says the farmer. "And if you get them out of my fields, I'll give you a whole cow. The one that got struck by lightning, that one over there, is lying under the wreckage of the barn." The worker agrees, and goes to remove the lines from the farmer's yard. After testing the lines to make sure the power was indeed off, he takes off his thick rubber gloves, because it was "too hot for gloves."

Unfortunately, the careless worker forgot about the test signals that the power company occasionally sends through a disabled power line. Power companies sometimes send a

(continued)

APPLICATION EXAMPLE 5.6 **Continued**

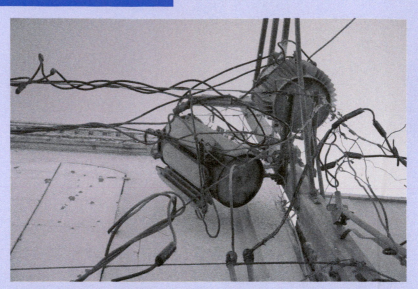

FIGURE 5.19 Destroyed transformer and transmission lines. (Used with permission from © Design Pics Inc. – RM Content/Alamy.)

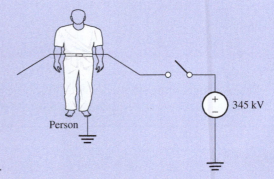

FIGURE 5.20 Circuit for Example 5.6.

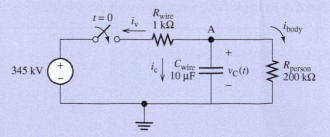

FIGURE 5.21 Model of the situation in Figure 5.20.

test current through a disconnected line to check if it could be reconnected safely. The test signal checks if the problem has been resolved before power for that line is turned back on. The power company, ignorant of the fact that the storm has twisted and toppled their towers beyond repair, sends a 345-kV test signal through the lines.

The scenario is shown in Figure 5.20. The scenario of Figure 5.20 can be modeled by the circuit shown in Figure 5.21. This is indeed a simplified version of the power line circuits. The details of the equivalent circuits for a power line will be discussed in Chapter 9.

Calculate the time, t, that it will take for the worker to feel a lethal shock. Lethal shock occurs when the body current reaches $i_{body} = 100$ mA. See Chapter 15 for details on electric safety.

SOLUTION

Method 1: Applying KCL to node A in Figure 5.21, $i_v + i_c + i_{body} = 0$. Now, because:

$$i_v = \frac{v_C(t) - 345 \text{ kV}}{1 \text{ k}\Omega}, \quad i_c = 10 \text{ μF} \frac{dv_C(t)}{dt}, \quad \text{and} \quad i_{body} = \frac{v_C(t)}{200 \text{ k}\Omega}$$

the equation becomes:

$$\frac{v_C(t) - 345 \text{ kV}}{1 \text{ k}\Omega} + 10 \text{ μF} \frac{dv_C(t)}{dt} + \frac{v_C(t)}{200 \text{ k}\Omega} = 0$$

After some mathematical manipulation:

$$\frac{dv_C(t)}{dt} + 100.5 \, v_C(t) = 34.5 \times 10^6$$

To find $v_C(t)$, rearrange the equation:

$$\frac{dv_C(t)}{dt} = -100.5 \left(v_C(t) - \frac{34.5 \times 10^6}{100.5} \right)$$

$$= -100.5 \, (v_C(t) - 3.4328 \times 10^5)$$

Solving for $v_C(t)$, and using the fact that the initial voltage is zero:

$$v_C(t) = 3.4328 \times 10^5 (1 - e^{-100.5t})$$

Note that $v_C(t) = R_{body} \, i_C(t)$. Next, find the time when the current through the body is 100 mA. In this case, $v_C(t) = 200 \text{ k}\Omega \times 100 \text{ mA} = 20 \text{ kV} = 20 \times 10^3$. Replacing for $v_C(t)$:

$$20 \times 10^3 = 3.4328 \times 10^5 (1 - e^{-100.5t})$$

Now, the time can be found:

$$t = \frac{1}{-100.5} \ln\left(1 - \frac{20 \times 10^3}{3.4328 \times 10^5} \right) \rightarrow t = 597 \, \mu s$$

Does he have enough time to drop the wire? No.

Method 2: The voltage, $v_C(t)$, can be found using the Thévenin equivalent:

$$v_{th} = 345 \times 10^3 \frac{200}{200 + 1} = 3.4328 \times 10^5 \text{ V}$$

$$R_{th} = \frac{200 \times 1}{200 + 1} = 0.9950 \text{ k}\Omega$$

The time constant is:

$$\tau = R_{th} C = 0.9950 \times 10^3 \times 10 \times 10^{-6} = \frac{1}{100.5}$$

The capacitor voltage can be written:

$$v_C(t) = v_{th}(1 - e^{\frac{-t}{\tau}}) = 3.4328 \times 10^5 (1 - e^{-100.5t})$$

and the solution can be completed as stated earlier.

APPLICATION EXAMPLE 5.7   Defibrillator

After being shocked by the power line, the careless worker from Example 5.6 lies motion-less on the ground. "Ma, his heart ain't beating. Get my defibrillator," yells the farmer, while attempting cardiopulmonary resuscitation (CPR) on him.

    A defibrillator contains a large capacitor that is charged to a high voltage. This voltage is discharged when the pads are pressed on the patient's chest. The scenario shown in Figure 5.22 can be modeled as that shown in Figure 5.23. Note that before $t = 0$, S1 and S2 have been closed and opened, respectively, for a long time period. Usually, defibrillators have an inductor in series with the patient, but it has been omitted to simplify this problem.

  a. The farmer shocks the worker across his chest with the defibrillator. What is the current through the worker's chest at $t = 0$?

  b. How long does it take until the voltage across his chest drops to $v_o(t) = \dfrac{1}{2} V_{\text{in}}$?

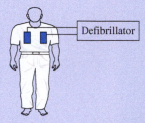

FIGURE 5.22   Model for Example 5.7.

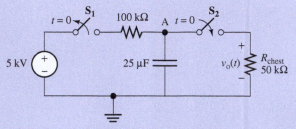

FIGURE 5.23   Circuit model for Example 5.7.

SOLUTION

  a. At $t = 0$, the capacitor has been fully charged to $v_0(0) = 5$ kV. Thus, $I = v_0(0)/R = 5000/50,000 = 0.1$ A

  b. Applying KCL to the parallel capacitor–resistor (i.e., Node A in Figure 5.23) for $t > 0$:

$$0 = \frac{v_0(t)}{50 \text{ k}\Omega} + 25 \text{ }\mu\text{F}\frac{dv_0(t)}{dt} \rightarrow 0 = \frac{v_0(t)}{1.25} + \frac{dv_0(t)}{dt}$$

or:

$$\frac{dv_0(t)}{v_0(t)} = \frac{-dt}{1.25}$$

Integrating both sides of this equation yields:

$$\ln\frac{v_0(t)}{5000} = \frac{-t}{1.25}$$

This leads to:

$$v_0(t) = 5000e^{-\frac{t}{1.25}}$$

Now, substituting $v_0(t) = 0.5 \times V_{\text{in}} = 0.5 \times 5000 = 2500$ V:

$$2500 = 5000e^{\frac{-t}{1.25}}$$

Next, find the corresponding time t required to drop to 2500 V, which is:

$$t = -1.25\ln 0.5 = 0.8664 \text{ s}$$

APPLICATION EXAMPLE 5.8	**Gyro Sensor**

The purpose of the gyro is to detect and determine the orientation of an object to which it is attached. These systems have applications in self-localization such as Inertial Navigation Systems (INS), used in aeronautical systems, boats, etc. This allows a vehicle (mobile) to find its position at any time instance with respect to a bearing point Figure 5.24(a) shows a gyro sensor while Figure 5.24(b) shows a simplified circuit for a gyro sensor setup. In practice, a transistor is required in the circuit to maintain a minimum amount of current flow through the gyro's sensor.

Use the simple equivalent circuit for the gyro shown in Figure 5.24(b) to find the expression for the voltage across gyroscope, v_R.

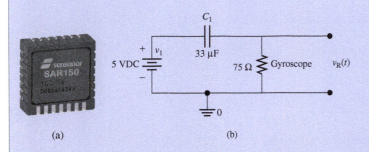

(a)　　　　　　　　　　　　　(b)

FIGURE 5.24 (a) Gyro sensor (Photo courtesy of Sensonor Technologies AS.); (b) a simple equivalent circuit.

SOLUTION

Using the definition of $v_C(t) = V_S(1 - e^{\frac{-t}{RC}})$, obtain the expression for the voltage across the capacitor, which is:

$$v_C(t) = 5\left(1 - e^{\frac{-t}{0.002475}}\right)$$

Applying KVL, the expression for the voltage across the gyroscope, V_R, is:

$$v_R(t) = V_S - v_C(t)$$

Replacing for $v_C(t)$:

$$v_R(t) = e^{\frac{-t}{0.002475}}$$

Notice that the voltage across the gyroscope drops to 0 V in a very short time period. In practice, a transistor is implemented to control and ensure a fixed amount of voltage and current flow through the gyroscope.

5.2.2 RL Circuits

A first-order RL circuit contains independent DC sources, resistors, and an inductor. Note that independent of the topology of the circuit and the number of resistors connected to the inductor, by using a Thévenin equivalent circuit and looking at the terminals of the inductor, all resistors can be replaced with a resistor and a voltage source. This is the same as what was discussed for capacitive circuits shown in Figure 5.2(a). Moreover, using the Norton theorem, a voltage source in series with a resistor is equivalent to a current source in parallel with the same resistor.

5.2.2.1 CHARGING AN INDUCTOR USING A DC SOURCE THROUGH A RESISTANCE

The circuit shown in Figure 5.25 can be analyzed as an example of first-order RL circuits. In this example, the DC source charges the inductor. Note that the term **charging** inductors refers to the

FIGURE 5.25 Inductor charging through a resistance from a DC source.

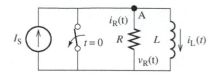

inductors gaining magnetic energy. Here, charge refers to **magnetic charge** versus the **electric charge** that was discussed for capacitors. This terminology is used for ease in presentation. Prior to $t = 0$, assume that the switch was closed for a long time period and the input current, i_s, flows through the short circuit. The source is connected to the RL circuit through a switch that opens at time $t = 0$. Assuming the initial current flowing through the inductor is zero (i.e., $i_L(0^-) = 0$), KCL can be applied to Node A in Figure 5.25 to find the current through the inductor, which corresponds to:

$$I_s = i_L(t) + i_R(t) = i_L(t) + \frac{1}{R}v_R(t) \tag{5.24}$$

The resistor and inductor are in parallel; thus, $v_L(t) = v_R(t)$.

$$I_s = i_L(t) + \frac{1}{R}v_L(t) \tag{5.25}$$

Now, using $v_L(t) = L\, di_L(t)/dt$:

$$I_s = i_L(t) + \frac{L}{R}\frac{di_L(t)}{dt} \tag{5.26}$$

Equation (5.26) can be rearranged, resulting in:

$$\frac{di_L(t)}{(i_L(t) - I_s)} = -\frac{R}{L}dt \tag{5.27}$$

Integrating the left- and right-hand sides of Equation (5.27) from 0^+ to time t, we obtain:

$$\int_{0^+}^{t} \frac{di_L(\tau)}{(i_L(\tau) - I_s)} = \int_{0^+}^{t} -\frac{R}{L}d\tau \tag{5.28a}$$

where the time argument, t, of the integrands is replaced by τ to avoid symbol clash.
Then:

$$\ln(i_L(t) - I_s) - \ln(i_L(0^+) - I_s) = -\frac{R}{L}(t - 0) \tag{5.28b}$$

Again, to complete the solution of Equation (5.28b), the value of $i_L(0^+)$ must be known. Using the same justification as for RC circuits, the current through the inductor cannot change suddenly; thus:

$$i_L(0^+) = i_L(0^-) = 0 \tag{5.29}$$

Substituting $i_L(0^+) = 0$ in Equation (5.28b) yields:

$$\ln\left(1 - \frac{i_L(t)}{I_s}\right) = -\frac{R}{L}t \tag{5.30}$$

Thus, the current, $i_L(t)$, corresponds to:

$$i_L(t) = I_s\left(1 - e^{-\frac{t}{L/R}}\right)$$
$$= I_s\left(1 - e^{-\frac{t}{\tau}}\right) \tag{5.31}$$

Similar to RC circuits, in this setting, the **time constant** is defined as the time required to charge the inductor to 63.21% of full charge. The time constant in seconds corresponds to:

$$\tau = \frac{L}{R} \tag{5.32}$$

According to Equation (5.32), the unit of time, *second*, is equal to: henry/ohm where L is in henries and R is in ohms. Based on this equation and Equation (5.11):

$$s = \frac{henry}{ohm} = ohm \times farad \rightarrow ohm^2 = \frac{henry}{farad} \rightarrow s^2 = henry \times farad$$

EXERCISE 5.5

Explain in detail why $s^2 = henry \times farad$ and why it can be concluded that:

$$Hz = \frac{1}{\sqrt{henry \times farad}}$$

This point will be verified using other methods later in this chapter.

A plot for the inductor charging transient is shown in Figure 5.26. Note that the current increases from zero (initial value) to i_S (final value) exponentially. Similar to when discussing an RC circuit, it can be said that the inductor is fully charged after five time constants. According to Equation (5.31), at time $t = 0$ (right after the displacement of the switch), the inductor current is zero. This means that the inductor will be equivalent to an open circuit. After a long time period, the inductor tends to be equivalent to a short circuit.

In general, a discharged inductor is equivalent to an open circuit. As a result, Figure 5.25 is equivalent to Figure 5.27(a) at $t = 0$, when the inductor is not charged. In addition, a fully charged (when $t \rightarrow \infty$) inductor is equivalent to a short circuit [see Figure 5.27(b)]. This is similar to the earlier discussion on the capacitor in this chapter (e.g., compare Figures 5.27 and 5.6). It can be observed that the capacitor is equivalent to a short circuit at time zero and is equivalent to an open circuit as $t \rightarrow \infty$. Each of these situations is the inverse of what happens to an inductor.

EXERCISE 5.6

Show that if the inductor's initial charge is i_0, the open circuit of Figure 5.27(a) will be replaced by a current source that is equivalent to i_0.

As a general rule, at $t = 0$, when the inductor has the initial charge of i_0, it is equivalent to a current source with the value of its initial current, i_0 (an open circuit, if $i_0 = 0$). But, a fully charged (i.e., when $t \rightarrow \infty$) inductor is equivalent to a short circuit.

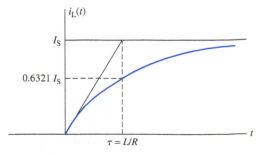

FIGURE 5.26   The charging transient for the circuit shown in Figure 5.25.

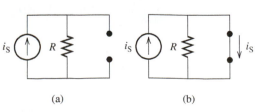

(a)   (b)

FIGURE 5.27   Equivalent circuit for Figure 5.25: (a) a discharged inductor; (b) a fully charged inductor.

| EXAMPLE 5.9 | **Inductor Current** |

Find an expression for $i_L(t)$ in the circuit shown in Figure 5.28. Assume the switch has been closed for a long time and opens at $t = 0$.

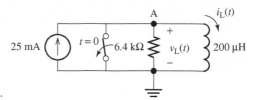

FIGURE 5.28 Circuit for Example 5.9.

SOLUTION

KCL for Node A corresponds to:

$$-i_s + i_R(t) + i_L(t) = 0$$

Because $v_R(t) = v_L(t)$:

$$i_R(t) = \frac{v_L(t)}{6.4 \text{ k}\Omega}$$

In addition:

$$v_L(t) = 200 \text{ μH} \frac{di_L(t)}{dt}$$

Replacing for $i_R(t)$ and $v_L(t)$ in the node A KCL equation:

$$-25 \times 10^{-3} + \frac{200 \times 10^{-6}}{6400} \frac{di_L(t)}{dt} + i_L(t) = 0$$

A simple mathematical manipulation leads to:

$$\frac{di_L(t)}{i_L(t) - 25 \times 10^{-3}} = -32 \times 10^6 dt$$

Integrating both sides of this equation yields (Note that the switch has been closed for a long time; thus, $i_L(\bar{0}) = 0$.):

$$\ln\left(\frac{i_L(t) - 25 \times 10^{-3}}{0 - 25 \times 10^{-3}}\right) = -32 \times 10^6 (t - 0)$$

Solving for $i_L(t)$:

$$i_L(t) = 25 \times 10^{-3}(1 - e^{-32 \times 10^6 t})$$

| EXAMPLE 5.10 | **Inductor Voltage** |

Find an expression for $v_L(t)$ in the circuit shown in Figure 5.29. Switch is opened at $t = 0$ after being closed for a long time.

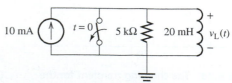

FIGURE 5.29 Circuit for Example 5.10.

SOLUTION

This problem can be solved in a way similar to the previous example using Equation (5.31). The time constant is:

$$\tau = \frac{L}{R} = \frac{20 \times 10^{-3}}{5 \times 10^3} = 4 \text{ μs}$$

$$i_L(t) = 0.01(1 - e^{-\frac{t}{4 \times 10^{-6}}}) = 0.01(1 - e^{-2.5 \times 10^5 t})$$

For the circuit shown in Figure 5.29:

$$v_L(t) = L\frac{di_L(t)}{dt}$$

$$= 20 \times 10^{-3} \times 0.01\frac{d(1 - e^{-2.5 \times 10^5 t})}{dt}$$

$$= 0.2 \times 10^{-3}(0 + 2.5 \times 10^5 e^{-2.5 \times 10^5 t})$$

$$= 50e^{-2.5 \times 10^5 t} \text{ V}$$

APPLICATION EXAMPLE 5.11   DC Motor

A motor, when switched on, does not immediately attain its final speed. This is due to the inductance of the motor coils. A DC motor is connected as shown in Figure 5.30(a) and its model is shown in Figure 5.30(b). Assume an inductor with zero initial charge (or assume the switch is open for a long time period). The motor reaches its steady-state speed when $v_{\text{motor}}(t) = V_S$. DC and AC motors have been discussed in detail in Chapter 13.

a. Find $v_{\text{motor}}(0^+)$ and $v_{\text{motor}}(\infty)$
b. Find an expression for $v_{\text{motor}}(t)$
c. Find the time at which $v_{\text{motor}}(t) = 95\%$ of V_S ($V_S = 120$ V)

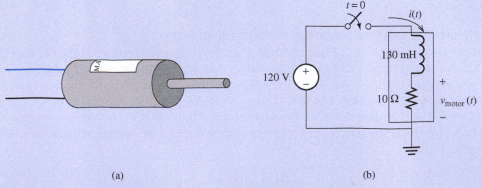

(a)   (b)

FIGURE 5.30   (a) Connection of a DC motor; (b) model of a DC motor's connection.

SOLUTION

a. The inductor is an open circuit at $t = 0$; thus, $v_{\text{motor}}(0) = 0$ V. The inductor is a short circuit at $t = \infty$; thus, $v_{\text{motor}}(\infty) = 120$ V.
b. Applying KVL for $t > 0$:

$$120 - v_L(t) - v_{\text{motor}}(t) = 0$$

(continued)

APPLICATION EXAMPLE 5.11 **Continued**

In addition:

$$v_{\text{motor}}(t) = 10i_{\text{L}}(t)$$

and:

$$v_{\text{L}}(t) = 0.13\frac{di_{\text{L}}(t)}{dt}$$

Thus:

$$120 - 0.13\frac{di_{\text{L}}(t)}{dt} - 10i_{\text{L}}(t) = 0$$

Solving for $i_{\text{L}}(t)$ yields:

$$i_{\text{L}}(t) = 12\left(1 - e^{\frac{-t}{1.3 \times 10^{-2}}}\right)$$

Accordingly:

$$v_{\text{motor}}(t) = 10 \times 12\left(1 - e^{\frac{-t}{1.3 \times 10^{-2}}}\right)$$

$$= 120\left(1 - e^{\frac{-t}{1.3 \times 10^{-2}}}\right)$$

c. $v_{\text{motor}}(t) = 0.95\,V_{\text{S}}$, thus:

$$120 \times 0.95 = 120\left(1 - e^{\frac{-t}{1.3 \times 10^{-2}}}\right)$$

Accordingly:

$$t = -1.3 \times 10^{-2} \times \ln 0.05 = 38.95 \text{ ms}$$

5.2.2.2 DISCHARGING AN INDUCTOR THROUGH A RESISTOR

Consider the circuit shown in Figure 5.31(a). Prior $t = 0$, the switch connects the DC source to the RL circuit and stays in this position for a long time period. As a result, the inductor is fully charged to i_{S} (i.e., $i_{\text{L}}(0^-) = i_{\text{S}}$). At $t = 0$ the RL circuit is detached from the current source by changing the switch position. The inductor begins to discharge through the resistor as shown in Figure 5.31(b). To find the inductor current for $t > 0$, write KVL for the circuit shown in Figure 5.31(b) (note the selected directions of current through the resistor and inductor shown in Figure 5.31).

$$v_{\text{L}}(t) + v_{\text{R}}(t) = 0 \tag{5.33}$$

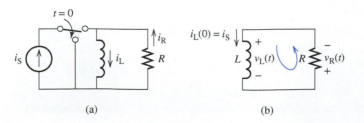

FIGURE 5.31
Discharging an inductor.

Replacing for $v_L(t) = L\,di_L(t)/dt$, and $v_R(t) = R\,i_R(t)$, and using the KCL equation of $i_L(t) = i_R(t)$:

$$L\frac{di_L(t)}{dt} + i_L(t)R = 0 \tag{5.34}$$

A simple mathematical manipulation leads to:

$$\frac{di_L(t)}{i_L(t)} = -\frac{R}{L}dt \tag{5.35}$$

Integrating Equation (5.35) yields:

$$\ln\left(\frac{i_L(t)}{i_L(0^+)}\right) = -\frac{R}{L}t \tag{5.36}$$

Because, after a long time, the inductor acts as a short circuit, the entire current will flow through it. Now, because the current through the inductor does not change immediately, that is,

$$i_L(0^+) = i_L(0^-) = I_s \tag{5.37}$$

As a result:

$$i_L(t) = I_s e^{-\frac{t}{L/R}}$$

$$= I_s e^{-\frac{t}{\tau}} \tag{5.38}$$

APPLICATION EXAMPLE 5.12   Explosive Atmosphere

When working in a hazardous, explosive atmosphere, workers must ensure that equipment does not produce any sparks that may lead to an explosion. At a minimum, any sparks must be isolated from the surrounding environment. Chapter 15 outlines how even a simple flashlight can create potentially dangerous sparks.

A standard flashlight circuit is shown in Figure 5.32. A flashlight bulb consists of both a resistance and an inductance. To illustrate the spark hazard, find the voltage across the switch at time $t = 0^+$.

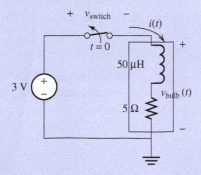

FIGURE 5.32 Circuit for Example 5.12.

SOLUTION

Before opening the switch (i.e., $t < 0$), the circuit can be modeled as shown in Figure 5.33(a). Because the switch has been closed for a long time period, the inductor functions similar to a short circuit. Replacing the inductor with a short circuit in Figure 5.33(a), the current flowing through the resistor corresponds to:

$$i(t) = \frac{3}{5} = 0.6$$

As stated earlier, the current through the inductor cannot change suddenly. Therefore:

$$i_L(0^+) = i_L(0^-) = 0.6\,\text{A}$$

(*continued*)

APPLICATION EXAMPLE 5.12 **Continued**

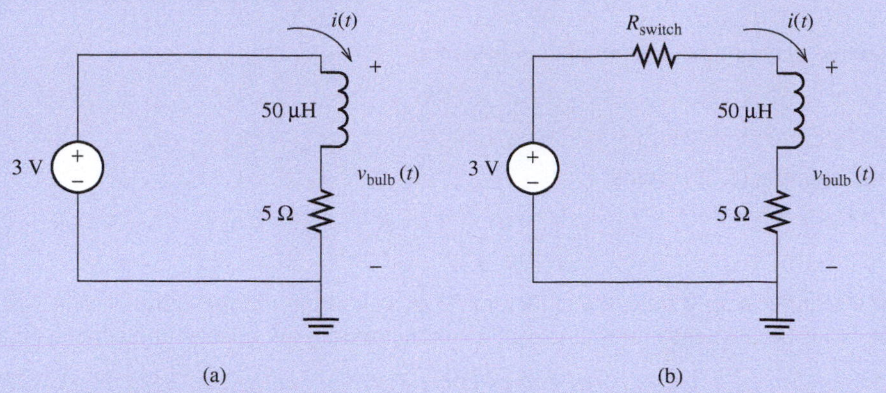

FIGURE 5.33 Equivalent circuit of Figure 5.32: (a) before opening the switch ($t < 0$); (b) when it opens.

The resistance of an open circuit is infinity. However, at the time of opening the switch, the air between the two closely located switch contacts may be modeled as a finite resistance, R_{switch}. Therefore, the 0.6 A current through the air (R_{switch}) may create a spark. The voltage across the switch is:

$$v_{switch}(t) = R_{switch} \cdot i(t)$$

If R_{switch} tends to infinity, v_{switch} also will tend to infinity. This can produce a spark. As discussed in Chapter 15, if hazardous materials are located close to the spark it may create an explosion.

5.3 DC STEADY STATE

The previous sections outlined first-order RC and RL circuits. Expressions were obtained for the voltage and current variations with time for both charging and discharging scenarios. The capacitor charging Equation (5.12) corresponds to:

$$v_C(t) = V_s - V_s e^{-t/\tau} \tag{5.39}$$

The right-hand side of Equation (5.39) consists of two terms. The second term, $V_s e^{-t/\tau}$, is the transient term and it decays to zero. The first term, V_S, is called the **steady state** or the final value and it is defined as the voltage across the capacitor as t tends to infinity. The same concept is applied to the inductor charging Equation (5.40).

$$i_L(t) = i_s - i_s e^{-t/\tau} \tag{5.40}$$

Here, i_s is the steady-state current.

In general, as stated in the previous sections, in the steady state (as $t \rightarrow \infty$):

    a. A capacitor that is charging through an independent DC source can be replaced by an open circuit.
    b. An inductor that is charging through an independent DC source can be replaced by a short circuit.

In addition:

    a. If a charged capacitor is disconnected from its DC source, it will be equivalent to a DC voltage source at $t = 0$.
    b. If a charged inductor is disconnected from its DC source, it will be equivalent to a DC current source at $t = 0$.

Finally, considering DC independent sources:

a. A discharged capacitor functions as a short circuit at $t = 0$, when the source is connected to the capacitor.
b. A discharged inductor functions as an open circuit at $t = 0$, when the source is connected to the inductor.

Four important points should be memorized:

1. Voltage across a capacitor cannot be changed suddenly.
2. Current through an inductor cannot be changed suddenly.
3. In the steady state with independent DC sources, capacitors behave as open circuits.
4. In the steady state with independent DC sources, inductors behave a short circuits.

EXAMPLE 5.13   **Inductor Current and Voltage**

Consider the circuit shown in Figure 5.34. The switch closes at $t = 0$ after being opened for a long time period. Find $i_L(0^+)$, $v_L(0^+)$, and $i_L(\infty)$.

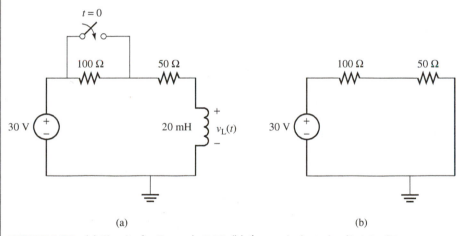

(a)   (b)

FIGURE 5.34 (a) Circuits for Example 5.13; (b) the equivalent circuit at $t = 0^-$.

SOLUTION

For $t < 0$, the switch is open for a long time period. Therefore, the inductor can be represented by a short circuit and the equivalent circuit is shown in Figure 5.34(b). The inductor current can be calculated as follows:

$$i_L(0^+) = i_L(0^-) = \frac{30}{100 + 50}$$
$$= 0.2 \text{ A}$$

Closing the switch short-circuits the 100-Ω resistor. As a result:

$$v_L(0^+) = 30 - 50 \times i_L(0^+)$$
$$= 30 - 50 \times 0.2$$
$$= 20 \text{ V}$$

At $t = \infty$, or steady state, the inductor becomes a short circuit. Thus:

$$i_L(\infty) = \frac{30}{50} = 0.6 \text{ A}$$

5.4 DC STEADY STATE FOR CAPACITIVE–INDUCTIVE CIRCUITS

Section 5.3 examined the DC steady state for circuits that contain a capacitor (RC) or inductor (RL). This section outlines the DC steady state for circuits that contain both an inductor and a capacitor (RLC circuits). As stated in Section 5.3, a capacitor in the DC steady state is equivalent to an open circuit and an inductor in the DC steady state is equivalent to a short circuit. The same rule is applied to RLC circuits. In other words, in Figure 5.35(a), if the key is closed for a long time period, the equivalent circuit will correspond to Figure 5.35(b). Equivalently, if the initial charge of all capacitors and inductors is zero in a circuit and a DC source is connected to the circuit at time $t = 0$, the capacitor reacts as a short circuit and the inductor reacts as an open circuit. In other words, at $t = 0$, the equivalent circuit of Figure 5.35(a) corresponds to Figure 5.35(c).

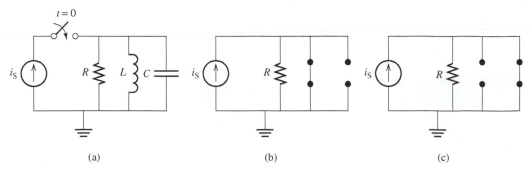

(a) (b) (c)

FIGURE 5.35 (a) RLC circuit; (b) DC steady state; (c) its equivalent circuit at $t = 0$.

| EXAMPLE 5.14 | **Initial and Final Voltage and Current** |

Consider the circuit shown in Figure 5.36. Find $v_C(0^+)$, $i_C(0^+)$, and $v_C(\infty)$.

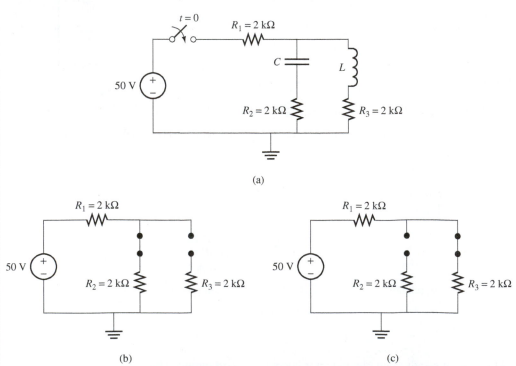

(a)

(b) (c)

FIGURE 5.36 (a) Circuit for Example 5.14; (b) its equivalent circuit at $t = 0$; (c) its equivalent at DC steady state.

SOLUTION

The equivalent circuit at $t = 0^+$ (after the switch is closed) corresponds to Figure 5.36(b). Here:

$$v_C(0^+) = 0$$

$$i_C(0^+) = \frac{50}{R_1 + R_2} = \frac{50}{4} = 12.5\,\text{mA}$$

The equivalent circuit at $t = \infty$ corresponds to Figure 5.36(c). In this case, the voltage across the capacitor is equal to the voltage across R_3. Using the voltage divider rule:

$$v_C(\infty) = 50\frac{R_3}{R_1 + R_3}$$

$$= 50\frac{2}{4} = 25 \text{ V}$$

5.5 SECOND-ORDER CIRCUITS

The previous sections discussed first-order circuits. In these circuits, KVL or KCL is written for the given circuit to develop a differential equation. This differential equation is then solved to find the required voltage or current. This section considers circuits that contain two reactive elements (a capacitor and an inductor) together with DC sources and resistors. Second-order differential equations must be solved to analyze second-order circuits. Solving second-order differential equations is not easy. The **Laplace transform** is an important tool for analysis of second-order circuits, which leads to the solution of second-order differential equations.

To analyze such circuits, first write the differential equation, find its Laplace transform, and then find the final solution in the time domain by calculating the inverse Laplace transform. A brief introduction to the Laplace transform and its inverse is presented in Appendix B. Table B.1 is used to find the Laplace transform as well as the inverse Laplace transform.

5.5.1 Series RLC Circuits with a DC Voltage Source

Consider the circuit shown in Figure 5.37. The switch is closed at $t = 0$ connecting the RLC to the DC source. The initial conditions are $i_L(0) = 0$ and $v_C(0) = 0$.

To find the voltage across the capacitor for $t > 0$, write KVL:

$$V_s - Ri(t) - v_L(t) - v_C(t) = 0 \tag{5.41}$$

Considering:

$$i(t) = i_L(t) = i_C(t) = C\frac{dv_C(t)}{dt} \tag{5.42}$$

and:

$$v_L(t) = L\frac{di_L(t)}{dt} = LC\frac{d^2v_C(t)}{dt^2} \tag{5.43}$$

Replacing $v_C(t)$ and $i_L(t)$ in Equation (5.41) with those in Equations (5.42) and (5.43) results in:

$$V_s - RC\frac{dv_C(t)}{dt} - LC\frac{d^2v_C(t)}{dt^2} - v_C(t) = 0 \tag{5.44}$$

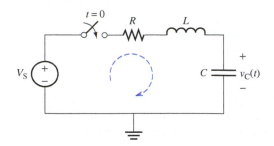

FIGURE 5.37   A series RLC circuit.

Rearranging Equation (5.44):

$$RC\frac{dv_C(t)}{dt} + LC\frac{d^2v_C(t)}{dt^2} + v_C(t) = V_s \tag{5.45}$$

Dividing both sides of Equation (5.45) by LC and reordering Equation (5.45):

$$\frac{d^2v_C(t)}{dt^2} + \frac{R}{L}\frac{dv_C(t)}{dt} + \frac{1}{LC}v_C(t) = \frac{V_s}{LC} \tag{5.46}$$

Applying Laplace transform to Equation (5.46) yields:

$$(s^2V_C(s) - sv_C(0^+) - v_C'(0^+)) + \frac{R}{L}(sV_C(s) + v_C(0^+)) + \frac{1}{LC}V_C(s) = \frac{V_s}{LCs} \tag{5.47}$$

Next, note that $v_C(0^+) = i_L(0^+)/C = 0$, thus:

$$\left(s^2 + \frac{R}{L}s + \frac{1}{LC}\right)V_C(s) = \frac{V_s}{LCs} \tag{5.48}$$

or:

$$V_C(s) = \frac{V_s/LC}{s\left(s^2 + \frac{R}{L}s + \frac{1}{LC}\right)} \tag{5.49}$$

Next, some important parameters are defined:

The **damping coefficient** is:

$$\alpha = \frac{R}{2L} \tag{5.50}$$

The **undamped resonant frequency** is:

$$\omega_0 = \frac{1}{\sqrt{LC}} \tag{5.51}$$

The **damping ratio** is:

$$\zeta = \frac{\alpha}{\omega_0} \tag{5.52}$$

Substituting Equations (5.50) and (5.51) into Equation (5.49) yields:

$$V_C(s) = \frac{V_s \times \omega_0^2}{s(s^2 + 2\alpha s + \omega_0^2)} \tag{5.53}$$

The term in the denominator of Equation (5.53), $s^2 + 2\alpha s + \omega_0^2$, is called the **characteristic equation**. Next, find $v_C(t)$ by obtaining the inverse Laplace transform of Equation (5.53). To do so, first factorize the characteristic equation. The roots of the characteristic equation are:

$$s_1 = -\alpha + \sqrt{\alpha^2 - \omega_0^2} = -\alpha\left(1 - \sqrt{1 - \frac{\omega_0^2}{\alpha^2}}\right) = -\alpha\left(1 - \sqrt{1 - \frac{1}{\zeta^2}}\right) \tag{5.54}$$

$$s_2 = -\alpha - \sqrt{\alpha^2 - \omega_0^2} = -\alpha\left(1 + \sqrt{1 - \frac{\omega_0^2}{\alpha^2}}\right) = -\alpha\left(1 + \sqrt{1 - \frac{1}{\zeta^2}}\right) \tag{5.55}$$

According to the value of the damping ratio, there are three cases:

1. **$\zeta > 1$** or equivalently $\alpha > \omega_0$: In this case, both roots are real, negative, and different. The partial fraction corresponding to Equation (5.53) is:

$$V_C(s) = \frac{V_s}{s} + \frac{k_1}{s - s_1} + \frac{k_2}{s - s_2} \tag{5.56}$$

The inverse Laplace transform of Equation (5.56) corresponds to:

$$v_C(t) = V_s + k_1 e^{s_1 t} + k_2 e^{s_2 t} \tag{5.57}$$

Because s_1 and s_2 are negative real numbers, the last two terms in Equation (5.57) decay to zero as t goes to infinity. The first term in Equation (5.57), V_s is the steady-state value of the capacitor voltage. The circuit in this case is said to be **overdamped**.

2. **$\zeta = 1$** or equivalently $\alpha = \omega_0$: In this case, both roots are real, negative, and equal to α. Accordingly, Equation (5.53) can be written as:

$$V_C(s) = \frac{V_s \times \omega_0^2}{s(s + \alpha)^2} = \frac{V_s}{s} + \frac{k_1}{(s + \alpha)} + \frac{k_2}{(s + \alpha)^2} \tag{5.58}$$

and the corresponding inverse Laplace transform is:

$$v_C(t) = V_s + k_1 e^{-\alpha t} + k_2 t e^{-\alpha t} \tag{5.59}$$

Because $\alpha = R/(2L)$ is always positive, the last two terms in Equation (5.59) decay to zero as t goes to infinity. In addition, the first term in Equation (5.59), V_s is the steady-state value of the capacitor voltage. The circuit in this case is said to be **critically damped**.

3. **$\zeta < 1$** or equivalently $\alpha < \omega_0$: In this case, the roots are complex and conjugates with a negative real part, and have the forms:

$$s_1 = -\alpha + j\sqrt{\omega_0^2 - \alpha^2} = -\alpha + j\omega_n \tag{5.60}$$

$$s_2 = -\alpha - j\sqrt{\omega_0^2 - \alpha^2} = -\alpha - j\omega_n \tag{5.61}$$

where $j = \sqrt{-1}$, and ω_n is the **natural frequency** and is given by:

$$\omega_n = \sqrt{\omega_0^2 - \alpha^2} \tag{5.62}$$

In this scenario, write Equation (5.53) as:

$$V_C(s) = \frac{\omega_0^2 V_s}{s(s + \alpha - j\omega_n)(s + \alpha + j\omega_n)}$$

$$= \frac{V_s}{s} + \frac{\omega_0^2 V_s/(j2\omega_n(-\alpha + j\omega_n))}{s + \alpha - j\omega_n} - \frac{\omega_0^2 V_s/(j2\omega_n(-\alpha - j\omega_n))}{s + \alpha + j\omega_n}$$

$$= \frac{V_s}{s} - \frac{\omega_0^2 V_s(s + 2\alpha)/(\alpha^2 + \omega_n^2)}{s^2 + 2\alpha s + \alpha^2 + \omega_n^2}, \quad \omega_0^2 = \alpha^2 + \omega_n^2$$

$$= \frac{V_s}{s} - \frac{V_s(s + 2\alpha)}{(s + \alpha)^2 + \omega_n^2}$$

$$= V_s\left(\frac{1}{s} - \frac{s + \alpha}{(s + \alpha)^2 + \omega_n^2} - \frac{\alpha}{(s + \alpha)^2 + \omega_n^2}\right)$$

$$= V_s\left(\frac{1}{s} - \frac{s + \alpha}{(s + \alpha)^2 + \omega_n^2} - \frac{\alpha}{\omega_n}\frac{\omega_n}{(s + \alpha)^2 + \omega_n^2}\right) \tag{5.63}$$

Now, the inverse Laplace transform of Equation (5.63) can be easily obtained using Table B.1 (see Appendix B) as follows:

$$v_C(t) = V_s - V_s\cos(\omega_n t)e^{-\alpha t} - \frac{\alpha V_s}{\omega_n}\sin(\omega_n t)e^{-\alpha t} \tag{5.64}$$

Similar to the previous two cases, the last two terms in Equation (5.64) decay to zero as t tends to infinity because α is always positive. In addition, the first term is the steady-state capacitor voltage. The circuit in this case is said to be **underdamped**. The current through the inductor can be calculated by differentiating the capacitor voltage and multiplying by the capacitance:

$$i_L(t) = i_C(t) = C\frac{dv_C(t)}{dt} \tag{5.65}$$

EXAMPLE 5.15   Series RLC Circuit

The DC voltage source of the circuit shown in Figure 5.38 is connected to the series RLC circuit by closing the switch at $t = 0$. The initial conditions are: $i_L(0) = 0$ and $v_C(0) = 0$. Find the voltage and the current across the capacitor if:

a. $R = 60\ \Omega$
b. $R = 40\ \Omega$
c. $R = 30\ \Omega$

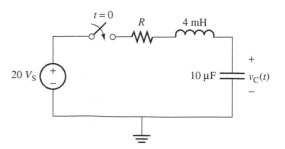

FIGURE 5.38   Circuit for Example 5.15.

SOLUTION

To find the circuit's conditions, calculate parameters α, ω_0, and ζ:

a. In this case:

$$\alpha = \frac{R}{2L} = \frac{60}{2 \times 4 \times 10^{-3}} = 7500$$

$$\omega_0 = \frac{1}{\sqrt{LC}} = \frac{1}{\sqrt{4 \times 10^{-3} \times 10 \times 10^{-6}}} = 5000$$

$$\zeta = \frac{\alpha}{\omega_0} = \frac{7500}{5000} = 1.5$$

Because $\zeta > 1$, the circuit is overdamped and the roots of the characteristic function can be calculated using Equations (5.54) and (5.55) as follows:

$$s_1 = -\alpha + \sqrt{\alpha^2 - \omega_0^2} = -1.91 \times 10^3$$

$$s_2 = -\alpha - \sqrt{\alpha^2 - \omega_0^2} = -1.31 \times 10^4$$

Substituting these values in Equation (5.53):

$$V_C(s) = \frac{20 \times 25 \times 10^6}{s(s^2 + 15 \times 10^3 s + 25 \times 10^6)} = \frac{5 \times 10^8}{s(s + 1.91 \times 10^3)(s + 1.31 \times 10^4)}$$

$$V_C(s) = \frac{20}{s} - \frac{23.42}{s + 1.91 \times 10^3} + \frac{3.42}{s + 1.31 \times 10^4}$$

The corresponding inverse Laplace transform is:

$$v_C(t) = 20 - 23.42e^{-1.91 \times 10^3 t} + 3.42e^{-1.31 \times 10^4 t}$$

The current flowing through the circuit is given by:

$$i_C(t) = C\frac{dv_C(t)}{dt} = 0 - 10 \times 10^{-6} \times 23.42 \times -1.91 \times 10^3 e^{-1.91 \times 10^3 t} + 10 \times 10^{-6}$$

$$\times 3.42 \times -1.31 \times 10^4 e^{-1.31 \times 10^4 t}$$

$$i_C(t) = 0.447e^{-1.91 \times 10^3 t} - 0.447e^{-1.31 \times 10^4 t}$$

Plots of capacitor voltage and current are shown in Figure 5.39.

b. When the resistor changes to 40 Ω:

$$\alpha = \frac{R}{2L} = \frac{40}{2 \times 4 \times 10^{-3}} = 5000$$

$$\omega_0 = \frac{1}{\sqrt{LC}} = \frac{1}{\sqrt{4 \times 10^{-3} \times 10 \times 10^{-6}}} = 5000$$

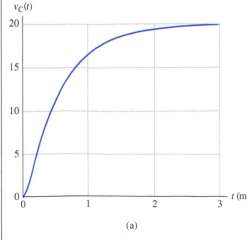
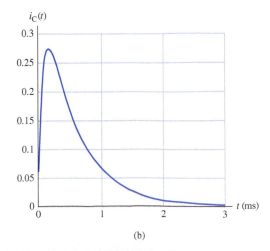

(a)   (b)

FIGURE 5.39 $R = 60\ \Omega$: (a) the capacitor voltage; (b) the current through the circuit.

(continued)

EXAMPLE 5.15 **Continued**

$$\zeta = \frac{\alpha}{\omega_0} = \frac{5000}{5000} = 1$$

Because $\zeta = 1$, the circuit is critically damped and the roots of the characteristic function are equal to $-\alpha$. Therefore:

$$s_1 = s_2 = -5 \times 10^3$$

Substituting these values in Equation (5.58):

$$V_C(s) = \frac{20 \times 25 \times 10^6}{s(s + 5 \times 10^3)^2} = \frac{V_s}{s} + \frac{k_1}{(s + 5 \times 10^3)} + \frac{k_2}{(s + 5 \times 10^3)^2}$$

The residues k_1 and k_2 are computed as follows (see Appendix B):

$$k_2 = (s + 5 \times 10^3)^2\, Y(s)\big|_{s=-5\times10^3} = \frac{20 \times 25 \times 10^6}{-5 \times 10^3} = \frac{5 \times 10^8}{-5 \times 10^3} = -10^5$$

$$k_1 = \frac{d}{ds}\big[(s + 5 \times 10^3)^2 Y(s)\big]\big|_{s=-5\times10^3} = -\frac{20 \times 25 \times 10^6}{(-5 \times 10^3)^2} = -20$$

As a result:

$$V_C(s) = \frac{20}{s} - \frac{20}{(s + 5 \times 10^3)} - \frac{10^5}{(s + 5 \times 10^3)^2}$$

The inverse Laplace transform is:

$$v_C(t) = 20 - 20e^{-5\times10^3 t} - 10^5 t e^{-5\times10^3 t}$$

$$i(t) = C\frac{dv_C(t)}{dt} = 0 - 10 \times 10^{-6} \times 20 \times -5 \times 10^3 e^{-5\times10^3 t} - 10 \times 10^{-6} \times 10^5 e^{-5\times10^3 t}$$

$$-10 \times 10^{-6} \times 10^5 \times -5 \times 10^3 t e^{-5\times10^3 t}$$

$$i(t) = e^{-5\times10^3 t} - e^{-5\times10^3 t} + 5 \times 10^3 t e^{-5\times10^3 t} = 5 \times 10^3 t e^{-5\times10^3 t}$$

Plots of capacitor voltage and current are shown in Figure 5.40.

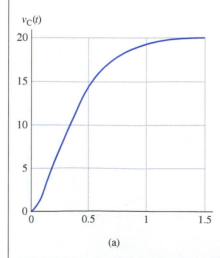

(a)

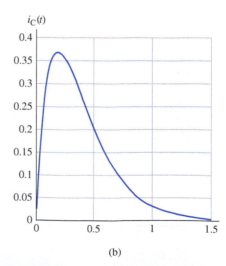

(b)

FIGURE 5.40 $R = 40\ \Omega$: (a) the capacitor voltage; (b) the current through the circuit.

c. Finally, for $R = 30\ \Omega$:

$$\alpha = \frac{R}{2L} = \frac{30}{2 \times 4 \times 10^{-3}} = 3750$$

$$\omega_0 = \frac{1}{\sqrt{LC}} = \frac{1}{\sqrt{4 \times 10^{-3} \times 10 \times 10^{-6}}} = 5000$$

$$\zeta = \frac{\alpha}{\omega_0} = \frac{3750}{5000} = 0.75$$

Because $\zeta < 1$, the circuit is underdamped. The natural frequency is calculated using Equation (5.62), which is:

$$\omega_n = \sqrt{\omega_0^2 - \alpha^2} = 3307$$

The roots of the characteristic function can be calculated using Equations (5.60) and (5.61).

$$s_1 = -\alpha + j\omega_n = -3750 + j3307$$
$$s_2 = -\alpha - j\omega_n = -3750 - j3307$$

This is an underdamped case, using Equation (5.63) results in:

$$V_C(s) = \frac{5 \times 10^8}{s(s + 3750 + j3307)(s + 3750 - j3307)}$$

and

$$V_C(s) = \frac{20}{s} - \frac{20(s + 3750)}{(s + 3750)^2 + 3307^2} - \frac{20 \times 3750}{3307}\frac{3307}{(s + 3750)^2 + 3307^2}$$

The corresponding inverse Laplace transform is:

$$v_C(t) = 20 - 20e^{-3750t}\cos(3307t) - 22.68e^{-3750t}\sin(3307t)$$

The corresponding current that flows through the circuit is given by:

$$i_C(t) = 0.75e^{-3750t}\cos(3307t) + 0.66e^{-3750t}\sin(3307t) + 0.85e^{-3750t}\sin(3307t)$$
$$- 0.75e^{-3750t}\cos(3307t)$$
$$i_C(t) = 1.51e^{-3750t}\sin(3307t)$$

Plots of capacitor voltage and current are shown in Figure 5.41.

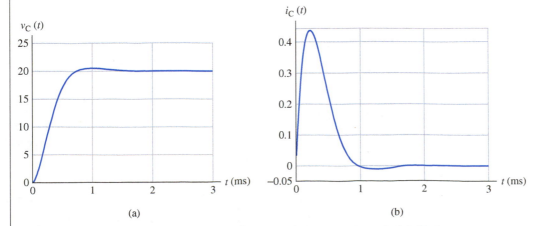

(a)   (b)

FIGURE 5.41 $R = 30\ \Omega$: (a) The capacitor voltage; (b) the current through the circuit.

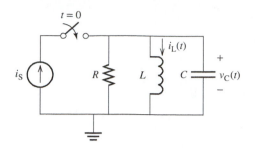

FIGURE 5.42 A parallel RLC circuit with a DC source.

5.5.2 Parallel RLC Circuits with a DC Voltage Source

This section examines circuits that contain parallel resistors, inductors, and capacitors. The previous section outlined how to compute $v_C(t)$, and how to easily find the inductor current by differentiating the voltage across the capacitor. This section describes how to compute the inductor current and then calculate capacitor voltage using differentiation.

Consider the circuit shown in Figure 5.42, in which the DC source is connected to the RLC circuit by closing the switch at time $t = 0$. To find the inductor current, write KCL to find the differential equation as follows:

$$i_R(t) + i_L(t) + i_C(t) = I_s \tag{5.66}$$

In this example:

$$i_R(t) = \frac{v_R(t)}{R} = \frac{v_L(t)}{R} = \frac{L}{R}\frac{di_L(t)}{dt} \tag{5.67}$$

$$i_C(t) = C\frac{dv_C(t)}{dt} = C\frac{dv_L(t)}{dt} = LC\frac{d^2i_L(t)}{dt^2} \tag{5.68}$$

Plugging Equations (5.67) and (5.68) into Equation (5.66) leads to:

$$\frac{L}{R}\frac{di_L(t)}{dt} + i_L(t) + LC\frac{d^2i_L(t)}{dt^2} = i_s \tag{5.69}$$

Dividing both sides of Equation (5.69) by LC and rearranging terms results in:

$$\frac{d^2i_L(t)}{dt^2} + \frac{1}{RC}\frac{di_L(t)}{dt} + \frac{1}{LC}i_L(t) = \frac{i_s}{LC} \tag{5.70}$$

Laplace transform of Equation (5.70) yields:

$$s^2I_L(s) - si_L(0) - i'_L(0) + \frac{1}{RC}(sI_L(s) - i_L(0)) + \frac{1}{LC}I_L(s) = \frac{I_s/LC}{s} \tag{5.71}$$

Setting all initial conditions to zero, the inductor current will correspond to:

$$I_L(s) = \frac{I_s/LC}{s\left(s^2 + \frac{1}{RC}s + \frac{1}{LC}\right)} \tag{5.72}$$

Next, consider:

The damping coefficient is:

$$\alpha = \frac{1}{2RC} \tag{5.73}$$

The undamped resonant frequency is:

$$\omega_0 = \frac{1}{\sqrt{LC}} \tag{5.74}$$

The damping ratio is:

$$\zeta = \frac{\alpha}{\omega_0} \tag{5.75}$$

Plugging these values into Equation (5.72) results in:

$$I_L(s) = \frac{I_s\omega_0^2}{s(s^2 + 2\alpha s + \omega_0^2)} \tag{5.76}$$

Comparing Equation (5.76) with (5.53), it is clear that the solution will be the same. The only difference between series RLC and parallel RLC circuits is the value of the damping coefficient, α. Considering the inductor's current instead of the capacitor's voltage results in the same three cases as in the previous section, and thus $i_L(t)$ can be found.

EXAMPLE 5.16   **Parallel RLC Circuit**

The DC source of the circuit shown in Figure 5.43 is connected to the parallel RLC circuit by closing the switch at $t = 0$. The initial conditions are: $i_L(0) = 0$ and $v_C(0) = 0$. Find the inductor current and the voltage across the capacitor.

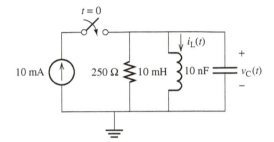

FIGURE 5.43   Circuit for Example 5.16.

SOLUTION

First, compute the key parameters using Equations (5.73) through (5.75):

$$\alpha = \frac{1}{2RC} = \frac{1}{2 \times 250 \times 10 \times 10^{-9}} = 2 \times 10^5$$

$$\omega_0 = \frac{1}{\sqrt{LC}} = \frac{1}{\sqrt{10 \times 10^{-3} \times 10 \times 10^{-9}}} = 10^5$$

$$\zeta = \frac{\alpha}{\omega_0} = \frac{2 \times 10^5}{10^5} = 2$$

Because $\zeta > 1$, the circuit is overdamped and the roots of the characteristic function can be calculated using Equations (5.57) and (5.55) as follows:

$$s_1 = -\alpha + \sqrt{\alpha^2 - \omega_0^2} = -2.6795 \times 10^4$$

$$s_2 = -\alpha - \sqrt{\alpha^2 - \omega_0^2} = -3.7321 \times 10^5$$

Substituting these values into Equation (5.76) leads to:

$$I_L(s) = \frac{10 \times 10^{-3} \times 10^{10}}{s(s^2 + 4 \times 10^5 s + 10^{10})} = \frac{10^8}{s(s + 2.6795 \times 10^4)(s + 3.7321 \times 10^5)}$$

$$I_L(s) = \frac{10 \times 10^{-3}}{s} - \frac{10.7735 \times 10^{-3}}{s + 2.6795 \times 10^4} + \frac{0.7735 \times 10^{-3}}{s + 3.7321 \times 10^5}$$

The inverse Laplace transform corresponds to:

$$i_L(t) = 10 \times 10^{-3} - 10.7735 \times 10^{-3}e^{-2.6795 \times 10^4 t} + 0.7735 \times 10^{-3}e^{-3.7321 \times 10^5 t}$$

$$v_C(t) = 2.8868e^{-2.6795 \times 10^4 t} - 2.8868e^{-3.7321 \times 10^5 t}$$

5.6 TRANSIENT ANALYSIS WITH SINUSOID FORCING FUNCTIONS

The previous sections outlined study of the transient steady state of circuits containing one or two energy storage elements and an independent DC source applied to the system at a specific time, for example, $t = 0$. This section examines first-order circuits with sinusoidal forcing functions in which the input is assumed to be a sinusoidal, rather than a constant, waveform (i.e., DC sources). For simplicity (and to avoid assumptions in the solution of differential equations), this section uses the same procedure used in Section 5.5. First, the differential equation for the circuit is found, and then Laplace transform and the inverse Laplace transform are used to find the solution.

Consider the circuit shown in Figure 5.44. The input is connected to the RC circuit by closing the switch at $t = 0$. To calculate the current through the circuit, write the KVL:

$$v_i(t) - Ri(t) - v_L(t) = 0 \tag{5.77}$$

Next, replace for $v_L(t)$ and use the fact that the current through all elements is equal to $i(t)$, because the elements are connected in series.

$$v_i(t) - Ri(t) - L\frac{di_L(t)}{dt} = 0 \tag{5.78}$$

After some mathematical manipulation:

$$\frac{di_L(t)}{dt} + \frac{R}{L}i(t) = \frac{v_i(t)}{L} \tag{5.79}$$

Applying Laplace transform to Equation (5.79) results in:

$$sI(s) - i_L(0) + \frac{R}{L}i(s) = \frac{V_i(s)}{L} \tag{5.80}$$

Rearranging and factorizing $I(s)$:

$$\left(s + \frac{R}{L}\right)I(s) = \frac{V_i(s)}{L} + i_L(0) \tag{5.81}$$

Solving for $I(s)$:

$$I(s) = \frac{V_i(s)/L}{(s + R/L)} + \frac{i_L(0)}{(s + R/L)} \tag{5.82}$$

To calculate $i(t)$, find the inverse Laplace transform of Equation (5.82).

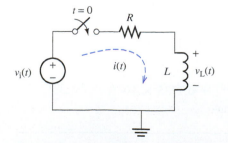

FIGURE 5.44 An RL circuit with $V_i(t)$ sinusoidal waveform.

EXAMPLE 5.17   **Capacitive Circuit with Sinusoid Input**

Find the current, which passes through the circuit shown in Figure 5.45 assuming $v_C(0) = 2$ V
and the input voltage $v_i(t) = 10 \sin(400t)$ V.

SOLUTION

First, write the voltage equation for $i > 0$ as:

$$10 \sin(400t) - 2 \times 10^3 i(t) - v_C(t) = 0$$

Next, replace for $i(t) = C\, dv_C(t)/dt$:

$$10 \sin(400t) - 2 \times 10^3 \times 10 \times 10^{-6} \frac{dv_C(t)}{dt} - v_C(t) = 0$$

After some mathematical manipulation:

$$\frac{dv_C(t)}{dt} + 50\, v_C(t) = 500 \sin(400t)$$

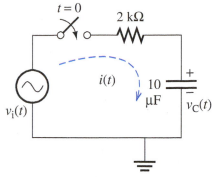

FIGURE 5.45   Circuit for Example 5.17.

Now, applying Laplace transform:

$$sV_C(s) - 2 + 50V_C(s) = \frac{400 \times 500}{s^2 + 16 \times 10^4}$$

Thus:

$$(s + 50)V_C(s) = \frac{2 \times 10^5}{s^2 + 16 \times 10^4} + 2 = \frac{2s^2 + 5.2 \times 10^5}{s^2 + 16 \times 10^4}$$

Therefore, $V_C(s)$ corresponds to:

$$V_C(s) = \frac{2s^2 + 5.2 \times 10^6}{(s^2 + 16 \times 10^4)(s + 50)}$$

The denominator can be decomposed into multiple components. This solution skips over
the details of conversion of the denominator; instead, refer to the detailed review of the process of Laplace transform prior to solving this, or other relevant problems. The procedure is
explained in detail in Example B.5 (see Appendix B). Applying this decomposition:

$$V_C(s) = \frac{2s^2 + 5.2 \times 10^6}{(s^2 + 16 \times 10^4)(s + 50)}$$

$$= \frac{3.2308}{s + 50} - \frac{0.6154 + j0.0769}{s - j400} - \frac{0.6154 - j0.0769}{s + j400}$$

$$= \frac{3.2308}{s + 50} - \frac{1.2308s - 61.5385}{s^2 + 16 \times 10^4}$$

$$= \frac{3.2308}{s + 500} - \frac{1.2308s}{s^2 + 16 \times 10^4} + \frac{61.5385}{s^2 + 16 \times 10^4}$$

Now, applying the inverse Laplace transform:

$$v_C(t) = 3.2308e^{-50t} - 1.2308\cos(400t) + \frac{61.5385}{400}\sin(400t)$$

$$= 3.2308e^{-50t} - (1.2308\cos(400t) - 0.1538 \sin(400t))$$

$$= 3.2308e^{-50t} - 1.2403\cos(400t + 7.1250)$$

In addition:

$$i(t) = 10^{-5}\frac{dv_C(t)}{dt}$$

$$= -3.2308 \times 50 \times 10^{-5}e^{-50t} + 1.2403 \times 400 \times 10^{-5} \sin(400t + 7.1250)$$

$$= -1.6154 \times 10^{-3}e^{-50t} + 4.9614 \times 10^{-3} \sin(400t + 7.1250)$$

(continued)

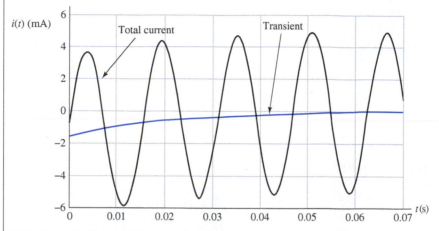

EXAMPLE 5.17   **Continued**

The first term in $i(t)$ is the **transient response**, and the second term is called the **AC steady-state response**. Chapter 6 explains how to calculate the AC steady-state response for circuits with sinusoidal inputs. Plots of the total current and the transient solution are shown in Figure 5.46.

FIGURE 5.46   The transient and the total current for Example 5.17.

APPLICATION EXAMPLE 5.18   **Disposable Camera**

Figure 5.47 shows the flash bulb circuit of a single-use disposable camera. The batteries of the camera charge the inductor continuously. After the inductor current reaches a maximum of 0.94864 the transistor allows the current to pass through the capacitor. Here, for simplicity,

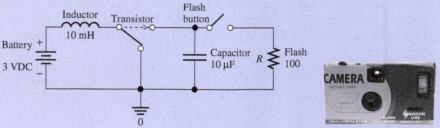

FIGURE 5.47   Circuit for a single-use disposable camera. Photo of camera (used with permission from Jeff Gynane/Shutterstock.com).

the transistor is represented as a switch. As discussed in Chapter 8, a transistor has many applications, one of which is to act as a controllable switch.

The current charges the capacitor. In practice, a fully charged capacitor may store up to 450 V. When the photographer enables the flash function on the camera and presses the shutter, the voltage stored in the capacitor goes through the flash bulb, and provides an instant high voltage. This generates a flash light at a particular instance.

Express the voltage across the capacitor during the charging state when the flash function is disabled.

SOLUTION

When the capacitor charges, there is no current flow through the flash bulb, therefore:

$$v_{flash} = 0$$

Prior to the activation of transistor, the voltage source directly charges the inductor. In this case,

$$V_S = V_L = L\frac{di_L}{dt} \longrightarrow \frac{di_L}{dt} = \frac{V_S}{L} = 300$$

Thus,

$$i_L(t) = \frac{V_S}{L}t = 300t$$

In other words, the inductor current linearly increases with time. The transistor operates as soon as $i_L(t)$ reaches 0.94864 i.e., within

$$t = \frac{0.9486}{300} = 3.16\,\text{ms}.$$

Now, the capacitor is charging through the inductor. The initial voltage of capacitor is $V_C(0^+) = 0$. However, since $i_L(0^+) = 0.94864$, and the currents of inductor and capacitor are equivalent, i.e.,

$$i_C(t) = i_L(t) = C\frac{dv_C(t)}{dt}$$

Thus,

$$V_C'(0^+) = \frac{0.9486}{10^{-5}} = 94.86 \times 10^3$$

In this case, the KVL equation is: $V_S - v_L(t) - v_C(t) = 0$

Because $v_L(t) = L\dfrac{i_L(t)}{dt}$, and $i_L(t) = i_C(t)$ replacing the current $i_L(t)$ with $i_C(t) = C\dfrac{dv_C(t)}{dt}$ results in:

$$v_L(t) = LC\frac{d^2v_C(t)}{dt^2}$$

Replacing $v_L(t)$ in the KVL equation with $v_L(t) = LCd^2v_C(t)/dt^2$ results in:

$$LC\frac{d^2v_C(t)}{dt^2} + v_C(t) = V_S$$

By taking the Laplace transform, with the initial condition of $V_C(0^+) = 0$, and $V_C'(0^+) = 94.86 \times 10^3$ the equation corresponds to:

$$s^2V_C(s) - V_C'(0^+) + \frac{1}{LC}V_C(s) = \frac{V_S}{LCs}$$

Thus, $V_C(s)$ corresponds to:

$$V_C(s) = \frac{V_S}{LCs(s^2 + 1/LC)} + \frac{V_C'(0^+)}{s^2 + 1/LC}$$

$$V_C(s) = V_S\left(\frac{1}{s} + \frac{(-1/2)}{s - j\sqrt{1/LC}} + \frac{(-1/2)}{s - j\sqrt{1/LC}}\right) + \frac{V_C'(0^+)}{s^2 + 1/LC}$$

Substituting the value of V_S, L, and C into this equation results in:

$$V_C(s) = 3\left(\frac{1}{s} - \frac{s}{s^2 + 10^7}\right) + \frac{94.86 \times 10^3}{s^2 \times 10^7}$$

Using the inverse Laplace transform, v_C expressed in the time domain is: $v_C(t) = 3(1-\cos(3162.28t)) + 30\sin(3162.28t)$. Thus, upto 30 volts is available at the capacitor.

EXERCISE 5.7

In Example 8.15, write the equation of capacitor voltage after the flash button is closed.

5.7 USING PSPICE TO INVESTIGATE THE TRANSIENT BEHAVIOR OF RL AND RC CIRCUITS

As in previous chapters, this chapter outlines how to use PSpice as a tool to sketch the voltage or current of RL and RC circuits and to study the behavior of these circuit variables. The following examples clarify the approach.

EXAMPLE 5.19	**PSpice Analysis: Initial Value of Capacitor Voltage**

In the circuit figure for Example 5.1, the switch is closed at $t = 0$, and the initial voltage across the capacitor $v_C(0) = 0$.

   a. Find the voltage across the capacitor at $t = 5$ s

   b. Find the capacitor voltage at $t = \infty$

SOLUTION

1. A PSpice schematic solution can be set up as shown in Figure 5.48. Set the initial condition for the capacitor to 0 V (Example 16 shows how to set up initial conditions).

2. Next, to choose a switch, go to "Part" and add "eval.olb" into the library (see Figure 2.63). Type "sw_topen" and choose it. SW_TOPEN will open the circuit at a given time, while SW_TCLOSE will close the circuit at a given time.

3. Double-click "TOPEN" and type 0, so that the switch will open the circuit at $t = 0$ s.

4. To plot the voltage across the capacitor at $t = 5$ s, run the simulation for 5 s (see Figure 2.71).

5. To plot the voltage across the capacitor at $t = \infty$, set the simulation time to a large value, that is, 1000 s.

6. Figure 5.49 shows the voltage across the capacitor at $t = 5$, and Figure 5.50 shows the voltage across the capacitor at $t = \infty$.

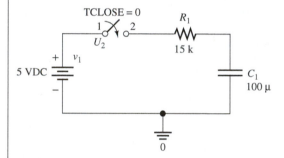

FIGURE 5.48 PSpice solution for Example 5.19.

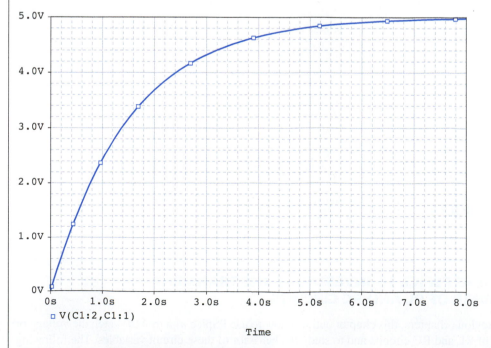

FIGURE 5.49 Voltage across the capacitor at $t = 5$ s.

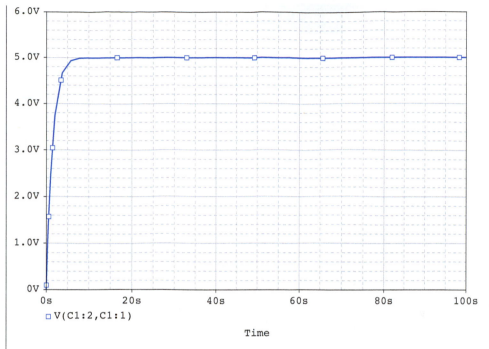

FIGURE 5.50 Voltage across the capacitor at $t = \infty$.

EXAMPLE 5.20	**PSpice Analysis**

Assume that the switch has been closed for a long period. The switch is open at $t = 0$. Use PSpice to plot the current flow through the inductor in the figure for Example 5.9.

SOLUTION

To simulate a circuit that includes a switch, which is closed for a long period, use the "SW_TOPEN" part (refer to Example 5.19) and set the open time, TOPEN to be a longer time, that is 2 s. The PSpice schematic solution is shown in Figure 5.51. Run the simulation with the run time (see Figure 2.71 for instructions on how to set up the simulation run time) that is longer than TOPEN so that an observation of change can be noticed (see Figure 5.52). Note: Remember to set the inductor's initial condition to zero (see Example 16).

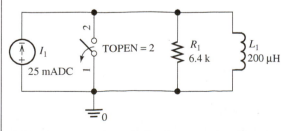

FIGURE 5.51 PSpice solution for Example 5.20.

(continued)

EXAMPLE 5.20	Continued

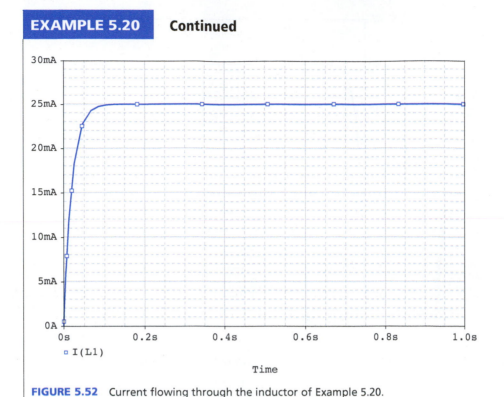

FIGURE 5.52 Current flowing through the inductor of Example 5.20.

EXAMPLE 5.21	PSpice Analysis

Consider the circuit in the figure shown for Example 5.14. Use PSpice to plot $v_C(0^+)$, $i_C(0^+)$, and $v_C(\infty)$.

SOLUTION

The PSpice schematic solution is shown in Figure 5.53. Refer to Example 5.19 to set up a switch that closes at a given time. Set the closing time TCLOSE at 0 s. Then, set the initial condition for both the capacitor and the inductor to 0 (see Example 4.16).

To observe the $i_C(0^+)$ and $v_C(0^+)$, set the simulation time to 1 μs (see Figure 2.75). Now, assume an "impulse-like" current flow through the circuit. The simulation plot can be obtained

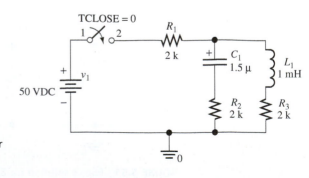

FIGURE 5.53 PSpice solution for Example 5.21.

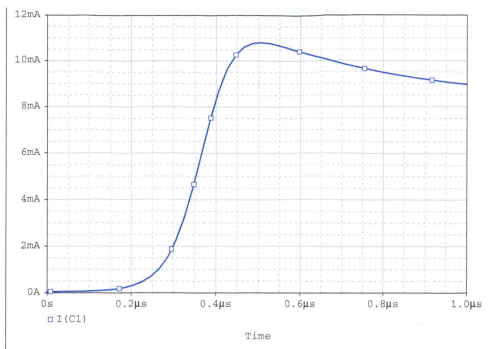

FIGURE 5.54 Current flow through the capacitor within 1 μs.

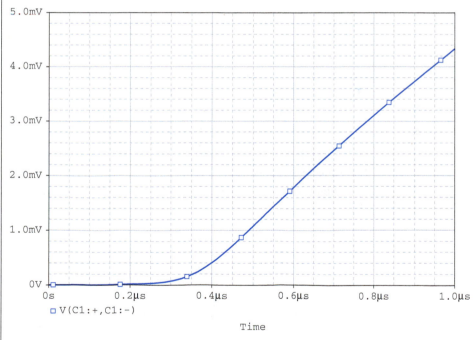

FIGURE 5.55 Voltage across the capacitor within 1 μs.

as sketched in Figures 5.54 and 5.55. Note that the input and output voltages for the capacitor are the same, so there is no voltage across the capacitor.

Next, set the simulation time to be long enough, for example, 5 s, such that a steady-state voltage, $v_C(\infty)$ is obtained. The voltage difference across the capacitor is sketched in Figure 5.56.

(*continued*)

EXAMPLE 5.21 **Continued**

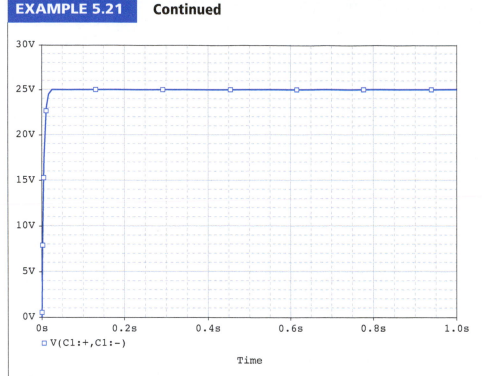

FIGURE 5.56 Voltage flow through the capacitor in a 5-s time period.

EXAMPLE 5.22 **PSpice Analysis**

Given the circuit shown in Figure 5.45, assume that $v_C(0) = 2$ V and the input voltage $v_i(t) = 10 \sin(200\pi t)$ V. Use PSpice to plot the current flowing through the resistor in the circuit for 100 ms.

SOLUTION

First, calculate the frequency of the sinusoidal voltage source. Then refer to Example 2.25 to set up the sinusoidal voltage source with its parameters.

$$f = \frac{\omega}{2\pi} \rightarrow f = \frac{200\pi}{2\pi} \rightarrow f = 100 \text{ Hz}$$

Set up the circuit as shown in Figure 5.57.

Refer to Example 5.19, set up a switch in the circuit. Because $v_C(0) = 2$ V, set the initial conditions of the capacitor to 2 V (see Example 4.16). Run the simulation for 100 ms as shown in Figure 5.58 (see Figure 2.75).

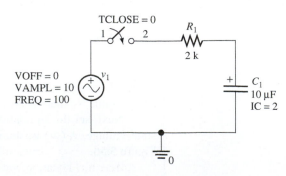

FIGURE 5.57 PSpice solution for Example 5.22.

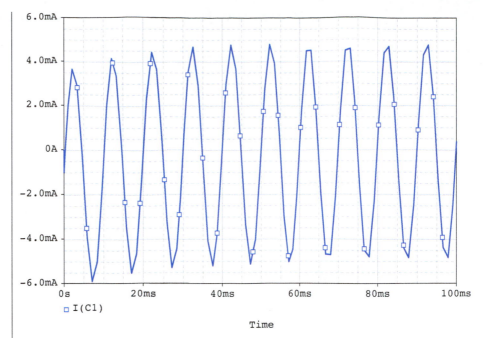

FIGURE 5.58 Current flow in the circuit for a time of 100 ms.

5.8 WHAT DID YOU LEARN?

First-Order Circuits

- An RC or RL transient circuit is said to be of the *first order*, if it contains only a single capacitor or a single inductor.
- The voltage or current of first-order circuits can be obtained by solving a first-order differential equation.
- After a period of 5τ, the capacitor or inductor is said to be fully charged or discharged. Therefore, the time constant, τ, determines the time required for the circuit to reach steady state.
- The circuit time constant of RC circuits is as shown in Equation (5.11):

$$\tau = RC$$

- The circuit time constant of RL circuits is as shown in Equation (5.32):

$$\tau = \frac{L}{R}$$

- Voltage across a capacitor cannot be changed suddenly. Likewise, current through an inductor cannot be changed suddenly.

DC Steady State

- In the steady state with independent DC sources, capacitors behave as open circuits.
- In the steady state with independent DC sources, inductors behave as short circuits.

Second-Order Circuits

- Second-order circuits contain a capacitor and an inductor.
- The voltage or current of a second-order circuit can be obtained by solving a second-order differential equation, where Laplace transform is used.

- Some important parameters for series RLC and parallel RLC circuits include:

	Series RLC Circuits	**Parallel RLC Circuits**
Damping coefficient, α	$\dfrac{R}{2L}$	$\dfrac{1}{2RC}$
Undamped resonant frequency, ω_0	$\dfrac{1}{\sqrt{LC}}$	$\dfrac{1}{\sqrt{LC}}$
Damping ratio, ζ	$\dfrac{\alpha}{\omega_0}$	$\dfrac{\alpha}{\omega_0}$

- The three types of damping are as follows. Each is shown with its corresponding second-order differential equation solution:

Overdamped [Equation (5.57)]:

$$\zeta > 1$$

$$v_C(t) = V_s + k_1 e^{-\alpha\left(1 - \sqrt{1 - \frac{1}{\zeta^2}}\right)t} + k_2 e^{-\alpha\left(1 + \sqrt{1 - \frac{1}{\zeta^2}}\right)t}$$

Critically damped [Equation (5.59)]:

$$\zeta = 1$$

$$v_C(t) = V_s + k_1 e^{-\alpha t} + k_2 t e^{-\alpha t}$$

Underdamped [Equation (5.64)]:

$$\zeta < 1$$

$$v_C(t) = V_s - V_s \cos(\omega_n t)e^{-\alpha t} - \frac{\alpha V_s}{\omega_n}\sin(\omega_n t)e^{-\alpha t}, \quad \omega_n = \sqrt{\omega_0^2 - \alpha^2}$$

Transient Analysis with Sinusoid Forcing Function

The transient analysis solution consists of two parts: the transient response and the AC steady-state response. The transient response solution is the same as the solution for first-order and second-order circuits.

Problems

*B refers to Basic, A refers to Average, H refers to Hard, and * refers to problems with answers.*

SECTION 5.2 FIRST-ORDER CIRCUITS

5.1 (B) Determine R or C from the following exponential functions:
  a. $v_C(t) = V_S(1 - e^{-6.25t})$, $R = 2000 \ \Omega$, find C
  b. $v_C(t) = V_S(1 - e^{-t/40})$, $C = 200 \ \mu F$, find R
  c. $v_C(t) = V_S(1 - e^{-5t})$, $R = 1000 \ \Omega$, find C

5.2 (B) Determine the voltage of a fully charged capacitor for the following functions:
  a. $v_C(t) = 5\left(1 - e^{-\frac{t}{20}}\right)$
  b. $v_C(t) = 10\left(1 - e^{-\frac{t}{10}}\right)$
  c. $v_C(t) = 7(1 - e^{-2t})$

5.3 (B)* Find an expression for $v_C(t)$ in Figure P5.3 given that $v_C(0) = 0$. (The switch closes a $t = 0$.)

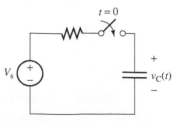

FIGURE P5.3   Circuit for Problem 5.3.

5.4 (A) Find the value of $v_C(t)$ in Figure P5.4 at $t = 42$ ms. $v_C(0) = 0$. (The switch closes at $t = 0$.)

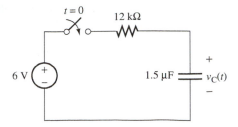

FIGURE P5.4   Circuit for Problem 5.4.

5.5 (A)* Find an expression for v_R in Figure P5.5, given that $v_C(0) = 5$ V. (The switch closes at $t = 0$.)

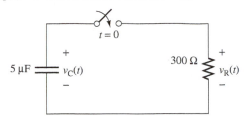

FIGURE P5.5   Circuit for Problem 5.5.

5.6 (A) Find an expression for $i(t)$ in Figure P5.6. $v_C(0) = 5$ V. (The switch closes at $t = 0$.)

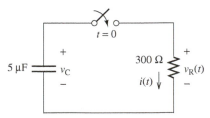

FIGURE P5.6   Circuit for Problem 5.6.

5.7 (A) Find an expression for the power supplied by the capacitor in Figure P5.7, as a function of time. $v_C(0) = 5$ V. (The switch closes at $t = 0$.)

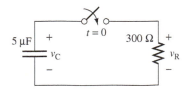

FIGURE P5.7   Circuit for Problem 5.7.

5.8 (H) At $t = 0$, the switch in Figure P5.8 is opened after being closed a long time. At what time t is $v_C(t)$ equal to ½ V_S?

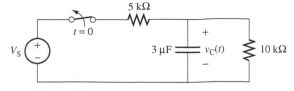

FIGURE P5.8   Circuit for Problem 5.8.

5.9 (H) Given the circuit shown in Figure P5.9, find V_R in term of R, C, and V_S.

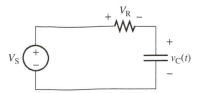

FIGURE P5.9   Circuit for Problem 5.9.

5.10 (A) Find an expression for $v_C(t)$ in Figure P5.10. The switch is opened at $t = 0$.

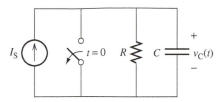

FIGURE P5.10   Circuit for Problem 5.10.

5.11 (A) Find an expression for power absorbed/consumed by the capacitor in Figure P5.11.

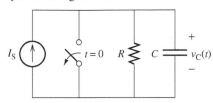

FIGURE P5.11   Circuit for Problem 5.11.

5.12 (A) Using your answers from Problems 5.10 and 5.11, find $v_C(t)$ and $p(t)$ in Figure P5.12 for $t = 10$ ms and $t = 20$ ms. $I_s = 3$ mA, $R = 5$ kΩ, $C = 2$ µF.

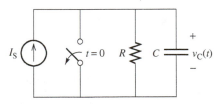

FIGURE P5.12   Circuit for Problem 5.12.

5.13 (H) Application: a sawtooth signal generator (SG)
The voltage, $v_0(t)$, across the capacitor, C, in the circuit shown in Figure P5.13 is a sawtooth waveform. The resistor, R, is adjustable. An SG represents a spark gap, which consists of two electrodes with adjustable distance. When the voltage across the SG is small, the SG is open. When that voltage exceeds the breakdown voltage, V_f, a spark will jump across the electrodes so that SG gets short-circuited. When the voltage across SG drops to zero, the SG is an open circuit. Close the switch at $t = 0$ and draw the waveform of $v_0(t)$.

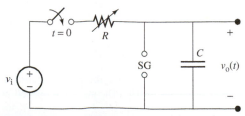

FIGURE P5.13   A sawtooth signal generator (SG).

5.14 (H)* In Problem 5.13, what conditions can make the sawtooth wave of v_o approximately linear? (*Hint:* Find the relationship of v_f and v_i and use the approximation: $e^{-\frac{T}{RC}} \approx 1 - (T/RC)$ when $e^{-\frac{T}{RC}}$ is close to 1.) If such conditions are met, find the period T of v_o.

5.15 (H)* In Problem 5.13, only R and the SG breakdown voltage V_f are adjustable.

What should you adjust in order to vary the period of v_o while keeping its amplitude? R or v_f or both? (*Hint:* Use Problem 5.14 results.)

5.16 (H)* In Problem 5.13, only R and the SG breakdown voltage V_f are adjustable.

What should you adjust in order to vary the amplitude of v_o while keeping its period? R or V_f or both? (*Hint:* Problem 5.14.)

5.17 (H) Given the circuit shown in Figure P5.13, assume that you are provided with an extra voltage source supplying v_f. Use these components to design and sketch a new circuit that can generate a voltage waveform v_o as shown in Figure P5.17.

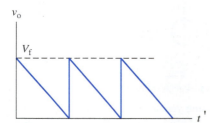

FIGURE P5.17 The sawtooth wave for Problem 5.17.

5.18 (A) Find the expression for $v_L(t)$ in Figure P5.18. (The switch is opened at $t = 0$.) $i_L(0) = 0$.

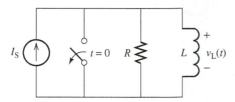

FIGURE P5.18 Circuit for Problem 5.18.

5.19 (A) Find the expression for the inductor current, $i_L(t)$, in Figure P5.19. (The switch is opened at $t = 0$.) $i_L(0) = 0$.

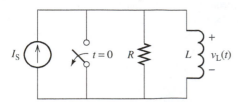

FIGURE P5.19 Circuit for Problems 5.19 to 5.21.

5.20 (A) Find the expression for the power absorbed by the inductor in Figure P5.19. (The switch is opened at $t = 0$.) $i_L(0) = 0$.

5.21 (A) Find the expression for the energy stored by the inductor as a function of t in Figure P5.19. (The switch is opened at $t = 0$.) $i_L(0) = 0$.

5.22 (H)* Determine the voltage across the inductor. Assume $v_C(0) = 0$.

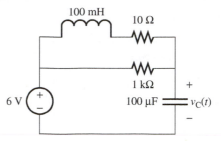

FIGURE P5.22 Circuit for problem 5.22.

SECTION 5.3 DC STEADY STATE

5.23 (B) Find the current flow through both resistor at $t(0)$ and $t(\infty)$ in the circuit shown in Figure P5.23. Assume that the switch closed at $t = 0$.

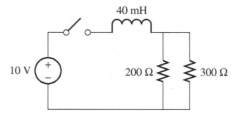

FIGURE P5.23 Circuit for Problem 5.23.

5.24 (B) Consider the circuit shown in Figure P5.24. The switch is opened at $t = 0$; find $v(0^+)$ and $v(\infty)$.

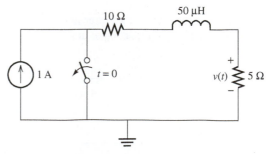

FIGURE P5.24 Circuit for Problem 5.24.

5.25 (B)* Find the current flow through resistor at $t(0)$ and $t(\infty)$ for the circuit shown in Figure P5.25. Assume that the switch is closed at $t = 0$.

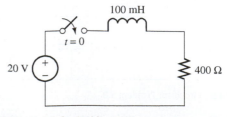

FIGURE P5.25 Circuit for Problem 5.25.

5.26 (B) Consider the circuit shown in Figure P5.26. The switch closes at $t = 0$; find $v(0^+)$, $i(0^+)$, and $v(\infty)$.

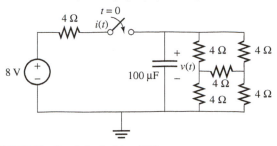

FIGURE P5.26 Circuit for Problem 5.26.

5.27 (A) Find the current flow through the resistor, $i_R(t)$, for the circuit shown in Figure P5.27. Assume that the switch is closed at $t = 0$.

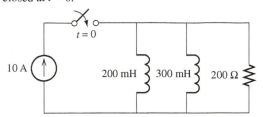

FIGURE P5.27 Circuit for Problem 5.27.

5.28 (A)* Find the voltage, $v_{out}(t)$, shown in Figure P5.28. Assume that the switch is closed at $t = 0$.

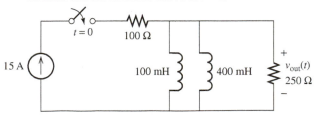

FIGURE P5.28 Circuit for Problem 5.28.

5.29 (A) Consider the circuit shown in Figure P5.29. The switch goes from 1 to 2 at $t = 0$ after connecting to 1 for a long time period. Find $i(0^+)$, $v(0^+)$, and $v(\infty)$.

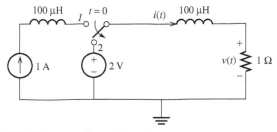

FIGURE P5.29 Circuit for Problem 5.29.

5.30 (A) Find the expression of $v_{out}(t)$ in Figure P5.30.

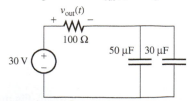

FIGURE P5.30 Circuit for Problem 5.30.

5.31 (A) Find the current flow through the resistor R_1 at $t = 0$ and $t = \infty$ in Figure P5.31. Assume that $v_C(t = 0) = 0$ V.

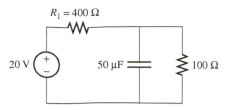

FIGURE P5.31 Circuit for Problem 5.31.

5.32 (H)* Find the expression of $v_{out}(t)$ in Figure P5.32.

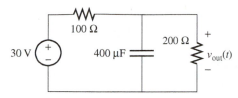

FIGURE P5.32 Circuit for Problem 5.32.

SECTION 5.4 DC STEADY STATE FOR CAPACITIVE–INDUCTIVE CIRCUITS

5.33 (B) Consider the circuit in Figure P5.33. The switch closes at $t = 0$. Determine $i_C(0^+)$, and the current flow across the resistor R at both $t = 0$ and $t = \infty$.

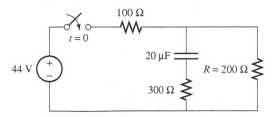

FIGURE P5.33 Circuit for Problem P5.33.

5.34 (A) Consider the circuit shown in Figure P5.34. The switch is closed at $t = 0$, draw the equivalent circuit diagram when $t = \infty$.

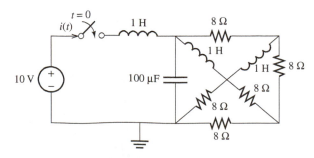

FIGURE P5.34 Circuit for Problem 5.34.

5.35 (A)* The switch closes at $t = 0$. Find the current flow through the resistor R at both $t = 0$ and $t = \infty$, and $i_L(\infty)$ for the circuit shown in Figure P5.35.

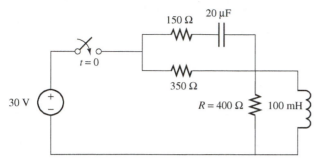

FIGURE P5.35 Circuit for Problem 5.35.

5.36 (A) Consider the circuit in Figure P5.36. The switch closes at $t = 0$ s. Sketch the current flow through the capacitor.

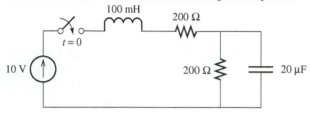

FIGURE P5.36 Circuit for Problem 5.36.

5.37 (A) Consider the circuit shown in Figure P5.37. The switch is opened at $t = 0$, find $v(0^+)$ and $v(\infty)$.

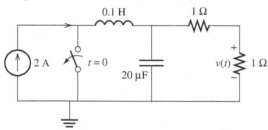

FIGURE P5.37 Circuit for Problem 5.37.

5.38 (A)* Consider the circuit shown in Figure P5.38. The switch closes at $t = 0$, find $i(\infty)$ and $v(\infty)$.

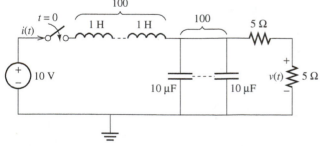

FIGURE P5.38 Circuit for Problem 5.38.

5.39 (H) Write the function of the total current flow in the circuit shown in Figure P5.39 in terms of L, C, R, and V_S.

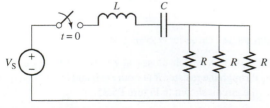

FIGURE P5.39 Circuit for Problem 5.39.

SECTION 5.5 SECOND-ORDER CIRCUITS

5.40 (B) There are three types of damping: underdamped, critically damped, or overdamped. Determine the type of damping for the following second-order differential equations:

a. $s^2 + 6s + 16$
b. $s^2 + 12s + 9$
c. $s^2 + 13s + 49$
d. $s^2 + 12s + 36$
e. $s^2 + 12s + 16$

5.41 (B) The switch closes at $t = 0$. Find the expression of $i_L(t)$ for the circuit shown in Figure P5.41.

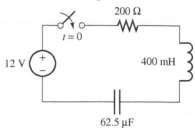

FIGURE P5.41 Circuit for Problem 5.41.

5.42 (A) Determine the range of the resistance R for the case where $v_C(t)$ is:

a. Underdamped
b. Critically damped
c. Overdamped

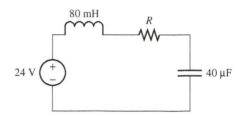

FIGURE P5.42 Circuit for Problem 5.42.

5.43 (A)* For a basic RLC circuit in Figure P5.43, what is the effect on damping ratio if we increase

a. R by twice its original value
b. C by nine multiples of its original value
c. L by four multiples of its original value

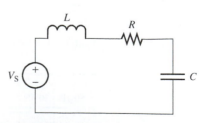

FIGURE P5.43 Circuit for Problem 5.43.

5.44 (A) Consider the circuit shown in Figure P5.44. The switch closes at $t = 0$, after the capacitance is fully charged $v_C(0-) = 6$ V:

a. Given $R = 2.5\ \Omega$, find $v_C(t)$ and $i(t)$.

b. Find the value of R, if the circuit is critically damped.

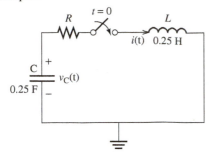

FIGURE P5.44   Circuit for Problem 5.44.

5.45 (H) Given the circuit in Figure P5.45. The switch closes at $t = 0$, find $v_C(t)$.

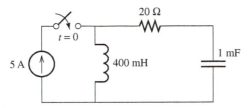

FIGURE P5.45   Circuit for Problem 5.45.

5.46 (H)* The switch in Figure P5.46 closes at $t = 0$. Determine whether $v_C(t)$ is underdamped, critically damped, or overdamped if R is 80 Ω.

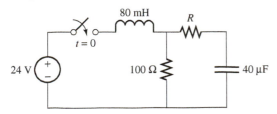

FIGURE P5.46   Circuit for Problem 5.46.

5.47 (H) Consider the circuit shown in Figure P5.47. The switch closes at $t = 0$, $v_C(0-) = 0$ V:

a. Decide whether the circuit is first order or the second order.

b. Find $v_2(t)$.

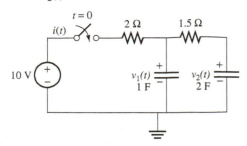

FIGURE P5.47   Circuit for Problem 5.47.

5.48 (H) The switch in Figure P5.48 closes at $t = 0$. Write the differential equation for $v_C(t)$.

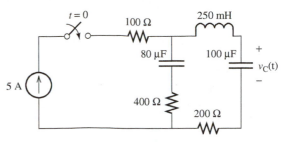

FIGURE P5.48   Circuit for Problem 5.48.

SECTION 5.6 TRANSIENT ANALYSIS WITH SINUSOID FORCING FUNCTIONS

5.49 (B)* The switch closes at $t = 0$. Given the current source $i_S(t) = 4 \sin(100t)$, find the expression of $i_L(t)$.

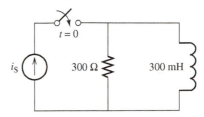

FIGURE P5.49   Circuit for Problem 5.49.

5.50 (B) The switch closes at $t = 0$. Given the voltage source $v_S(t) = 5 \sin(100t)$, find the expression of the current flow for the circuit shown in Figure P5.50.

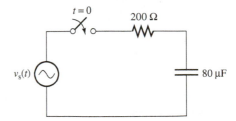

FIGURE P5.50   Circuit for Problem 5.50.

5.51 (A) Find $i(t)$ and $v(t)$ in the circuit shown in Figure P5.51 assuming $i_L(0-) = 1$ A, and the input voltage $v_i(t) = 30\cos(15t)$ V.

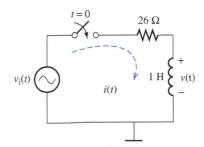

FIGURE P5.51   Circuit for Problem 5.51.

5.52 (A) The switch closes at $t = 0$ s. Find the expression of current flow for the circuit shown in Figure P5.52, if the voltage source is $v_S(t) = 10 \cos(200t)$.

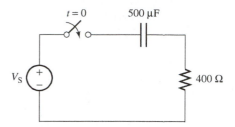

FIGURE P5.52 Circuit for Problem 5.52.

5.53 (A) The switch closes at $t = 0$ s. Given that the current source is $i_S(t) = 4\cos(100t)$. Find the voltage across the resistor R_2 for the circuit shown in Figure P5.53.

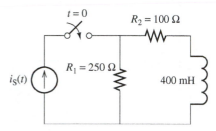

FIGURE P5.53 Circuit for Problem 5.53.

5.54 (H) The switch closes at $t = 0$ s. Find the expression of $v_C(t)$ for Figure P5.54, if the voltage source is $V_S(t) = 20 \sin(50t)$.

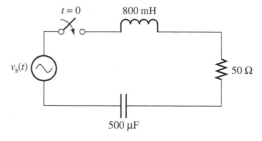

FIGURE P5.54 Circuit for Problem 5.54.

5.55 (H) The switch closes at $t = 0$ s. Given $v_S(t) = 20 \sin(200t)$, find the current flow across the resistor $i_R(t)$ for the circuit shown in Figure P5.55.

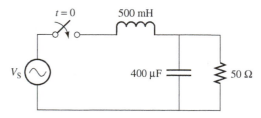

FIGURE P5.55 Circuit for Problem 5.55.

5.56 (H)* Find $i(t)$ and $v(t)$ in the circuit shown in Figure P5.56 assuming the input voltage

$$v_i(t) = 10\cos(50t) \text{ V} \qquad t < 0,$$
$$v_i(t) = 20\cos(50t) \text{ V} \qquad t > 0,$$

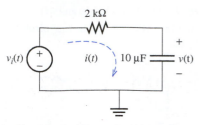

FIGURE P5.56 Circuit for Problem 5.56.

SECTION 5.7 USING PSPICE TO INVESTIGATE THE TRANSIENT BEHAVIOR OF RL AND RC CIRCUITS

5.57 (A) Given the circuit shown in Figure P5.4, with $v_o(0) = 0$. Use PSpice to find the value of $v_C(t)$ at $t = 42$ ms (the switch closes at $t = 0$).

5.58 (A) Consider the circuit shown in Figure P5.58. The switch closes at $t = 0$, find $v(0^+)$, $i(0^+)$, and $v(\infty)$ using PSpice.

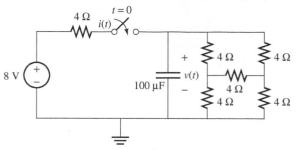

FIGURE P5.58 Circuit for Problem 5.58.

5.59 (A) Consider the circuit shown in Figure P5.59. The switch is opened at $t = 0$, find $v(0^+)$ and $v(\infty)$ using PSpice.

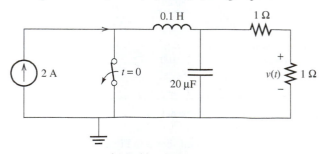

FIGURE P5.59 Circuit for Problem 5.59.

5.60 (H) Use PSpice to find $i(t)$ and $v_C(t)$ in the circuit shown in Figure P5.60 assuming the input voltage.

$$v_i(t) = 10\cos(50\pi t) \text{ V} \qquad t < 0,$$
$$v_i(t) = 20\cos(100\pi t) \text{ V} \qquad t > 0,$$

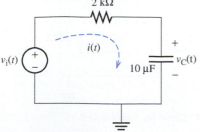

FIGURE P5.60 Circuit for Problem 5.60.

CHAPTER 6

Steady-State AC Analysis

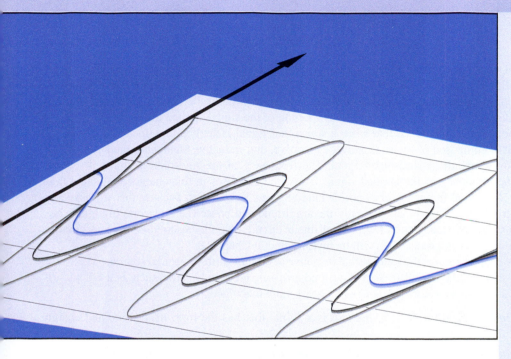

6.1 INTRODUCTION: SINUSOIDAL VOLTAGES AND CURRENTS

Sinusoidal signals (also called sinusoids) play an important role in science and engineering. Figure 6.1 shows an example of a sinusoid. Sinusoidal signals describe the characteristics of many physical processes; examples include (1) ocean waves; (2) radio or television carrier signals; (3) acoustic pressure variations of a single musical tone; and (4) voltages and currents generated by alternating-current (AC) sources. Figure 6.2 shows examples of signals formed by sinusoids.

Chapter 5 outlined the transient behavior of circuits when time-varying inputs are applied. An example of a time-varying input is when a source is applied to a circuit by turning a switch ON or OFF. Other examples are sinusoidal signals, square wave signals, and triangular signals. Chapter 5 explained that when a time-varying source is applied to a circuit that includes capacitors and/or inductors in addition to resistors, the response to the circuit will include two main components: the transient response and the steady-state response. Recall that a response can be a current that flows through any circuit component or the voltage at

FIGURE 6.1 Sinusoidal signals.

FIGURE 6.2 Examples of sinusoidal signals.

any circuit node. In addition, steady-state response represents the response to a circuit as time goes to infinity. It should be noted that circuits composed of resistors, capacitors, and inductors impact both the amplitude and the phase of the signal applied to the circuit. However, circuits that include only resistors change only the amplitude of the signal applied to the circuit.

Chapter 5 also explained that when a sinusoidal signal is applied to a linear circuit (i.e., a circuit that is composed of linear resistive, capacitive, and inductive elements), the steady-state response only includes a sinusoidal signal. The frequency of the sinusoidal signal will be the same as the frequency of the applied signal. Thus, the exercises in this chapter will show that the sinusoidal signal frequency does not include any new information because the response has the same frequency. This means that only the amplitude and phase (i.e., phasors) of signals are needed to study the steady-state response of linear circuits. Nonlinear circuits, such as circuits that include diodes and transistors, will be investigated in Chapter 8.

This chapter studies the steady-state response to circuits with sinusoidal sources. The study of steady-state AC is helpful in analyzing loudspeakers, electric heaters (such as a hot plate), motors that operate elevators, or pumps in a heart–lung machine.

Definition 1: Sinusoid   A **sinusoid** is a signal that has the form of the sine and cosine function.

Definition 2: Alternating current (AC) circuits   Circuits driven by sinusoidal current or voltage sources are called **AC circuits**.

A sinusoidal voltage is shown in Figure 6.3 and the following equation:

$$v(t) = V_m \cos(\omega t + \theta) \tag{6.1}$$

where

V_m = peak value of the voltage, also called amplitude
ω = angular frequency in rad/s.

Note that $\omega = 2\pi f = 2\pi/T$, and f is the frequency measured in hertz and T is the period measured in seconds, and θ is the phase angle.

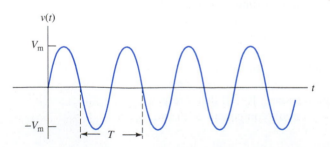

FIGURE 6.3 A sinusoidal voltage.

A sinusoid can be expressed in either sine or cosine form. Using trigonometric identities:

$$\sin\left(\alpha - \frac{\pi}{2}\right) = -\cos\alpha \qquad \text{(6.2a)}$$

$$\cos\left(\alpha - \frac{\pi}{2}\right) = \sin\alpha \qquad \text{(6.2b)}$$

$$\sin\left(\alpha + \frac{\pi}{2}\right) = \cos\alpha \qquad \text{(6.2c)}$$

$$\cos\left(\alpha + \frac{\pi}{2}\right) = -\sin\alpha \qquad \text{(6.2d)}$$

Therefore, Equation (6.1) can be expressed as:

$$v(t) = V_{\mathrm{m}}\sin\left(\frac{\pi}{2} + \omega t + \theta\right) = V_{\mathrm{m}}\sin(\omega t + \theta') \qquad \text{(6.3)}$$

where $\theta' = \left(\frac{\pi}{2}\right) + \theta$.

EXAMPLE 6.1 **Cosine-Sine Conversion**

$$\cos 35° = \sin(90° - 35°) = \sin 55°$$
$$-\cos 35° = \sin(-90° + 35°) = \sin -55°$$
$$\sin 10° = \cos(90° - 10°) = \cos 80°$$
$$-\sin 10° = \cos(90° + 10°) = \cos 100°$$

EXERCISE 6.1

Find θ in the following:

$$\cos 5° = \sin\theta$$
$$-\sin 15° = \cos\theta$$

A **periodic signal** is one that repeats itself every T seconds, that is, $f(t) = f(t + nT)$, where T is the fundamental period and n is any integer. T is called the fundamental period or simply the period of the signal. Sinusoidal signals are periodic. In Figure 6.3, T represents the signal period. The square wave shown in Figure 6.4 is an example of a periodic signal.

Note:

In general, based on the principles of Fourier series, any periodic signal such as the square wave shown in Figure 6.4 can be represented via a combination of multiple sinusoids. Each of these sinusoids have different amplitudes and angular frequencies. That is, if $v(t)$ is periodic, then we can represent $v(t)$ by:

$$v(t) = \sum_{i=-\infty}^{+\infty} V_i \cos(\omega_i t + \theta_i)$$

In this chapter, we consider the impact of a linear time invariant (LTI) circuit on a sinusoidal signal that has certain frequency component. Thus, the response of a circuit would be explicitly calculated for a certain frequency component. When, the frequency component changes, the response would also change.

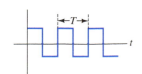

FIGURE 6.4 An example of a periodic signal: square waves.

Now, if we intend to consider the effect of all frequency components available in a periodic signal, superposition principles can be applied. For example, if $i_j(t)$ is the response of the circuit to one of the components of $v(t)$, such as $V_j \cos(\omega_j t + \theta_j)$, then the response of the circuit to all components of $v(t)$ will be $\sum_{i=-\infty}^{+\infty} i_i(t)$.

Recall that an LTI circuit is composed of components whose input–output relationships are expressed by a linear relationship as explained in Chapter 4 (Section 4.2.1). In addition, the values of the elements of an LTI circuit do not change with time.

EXAMPLE 6.2   The Period of a Sinusoid

Find the period of $x(t) = \cos(\pi t + 60)$.

SOLUTION

We know:

$$\cos(\pi t + 60) = \cos(\pi t + 60 + 2k\pi), \qquad k = 1, 2, 3, \ldots$$

Thus, it is a periodic signal. In addition:

$$\cos(\pi t + 60 + 2k\pi) = \cos(\pi(t + 2k) + 60)$$

Thus:

$$\cos(\pi t + 60) = \cos(\pi(t + 2k) + 60)$$

or

$$x(t) = x(t + 2k)$$

Based on the definition, the fundamental period (or, simply, the period) is 2.

EXERCISE 6.2

Determine if $x(t) = \cos(10\pi t + 60)$ and $y(t) = \cos(10\pi t + 60) + \sin(20\pi t + 90)$ are periodic. What is the period of these two signals?

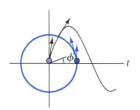

FIGURE 6.5 Creation of a sine curve.

Figure 6.5 represents the creation of sine waves. As time passes, the colored dot moves around the unit circle. Accordingly, the black dot tracks the y-coordinate of the colored dot as a function of time. Because $\sin(\phi)$ is the y-coordinate of the point on the unit circle corresponding to the angle ϕ, the black curve will be the graph of $y = \sin(\phi)$. In general, ϕ is a function of time: as time passes, ϕ increases from $0°$ to $180° = 2\pi$ rad. Thus $\phi = \omega t$, and:

$$y = \sin \omega t \tag{6.4}$$

where ω represents the angular speed of the colored dot. Considering T as the time over that the colored dot completes a turn around the circle of Figure 6.5:

$$\omega T = 2\pi \tag{6.5}$$

$$T = \frac{2\pi}{\omega} \tag{6.6}$$

Defining f as the frequency of rotation of the colored dot:

$$T = \frac{1}{f} \tag{6.7}$$

Substituting Equation (6.7) into (6.6), results in:

$$\omega = 2\pi f \tag{6.8}$$

Although a sinusoidal function can be expressed in either sine or cosine form, for uniformity, sinusoidal functions are expressed here only by the cosine function.

In general, an angle can be expressed in degrees or radians. The relationship between radians and degrees is:

$$\frac{\theta_{\text{rad}}}{\theta_{\text{deg}}} = \frac{\pi}{180} \tag{6.9}$$

EXAMPLE 6.3   Components of a Sinusoid

The purpose of this example is to understand different components of a sinusoidal signal, that is, amplitude, phase, and frequency. It should be noted that sinusoids can be fully represented by these components.

  a. Find the amplitude, phase, period, and frequency of the sinusoidal signal's voltage if:

$$v(t) = 5\cos(120\pi t + 30°)$$

  b. Compute the voltage at $t = 5$.
  c. Sketch $v(t)$.

SOLUTION

  a. The amplitude is the component that is multiplied in the cosine term:

$$V_{\text{m}} = 5 \text{ V}$$

The phase is the component that is added to the ωt term in the cosine term:

$$\theta = 30°, \quad 30° = 30 \times \frac{\pi}{180} \text{ rad} = \frac{\pi}{6} \text{ rad}$$

The angular frequency is:

$$\omega = 120\pi \text{ rad/s}$$

The period is:

$$T = \frac{2\pi}{\omega} = \frac{2\pi}{120\pi} = \frac{1}{60} = 0.01667 \text{ s}$$

The frequency is:

$$f = \frac{1}{T} = 60 \text{ Hz}$$

  b. The voltage at $t = 5$ corresponds to:

$$v(5) = 5\cos(120\pi \times 5 + 30°) = 5\cos\left(120\pi \times 5 + \frac{\pi}{6}\right) = 5\cos\left(\frac{\pi}{6}\right) = 4.3301$$

Note that the units of all angles must be the same before adding; i.e., we cannot add an angle in radians and another one in degrees.

  c. $v(t)$ is sketched in Figure 6.6. Note that the point that the signal crosses the x-axis is the point at which the voltage is zero. Finding these points helps to sketch the sinusoid. Equating the voltage to zero:

$$v(t) = 5\cos(120\pi t + 30°) = 0$$

(continued)

EXAMPLE 6.3   **Continued**

Now, recall that $\cos\phi = 0$ if $\phi = \dfrac{(2k+1)\pi}{2}$, $k = \ldots, -2, -1, 0, 1, 2, \ldots$

For this example $\phi = 120\pi t + 30°$. Thus:

$$(120\pi t + 30°) = \frac{(2k+1)\pi}{2}$$

This equation can be used to find the zero (x-axis) crossing points. For example, for $k = -1$, $t = -1/180$ as shown in the figure. Note that $\pi = 180°$, which can be used to simplify the calculations. In addition, for $k = 0$, $t = 1/360$. The cross section with the y-axis also helps to sketch Figure 6.6. This point occurs when time is set $t = 0$. Note that $\cos 30° = \sqrt{3}/2$; thus, at this point, the amplitude is $5\sqrt{3}/2$, as shown in Figure 6.6 using a dashed line.

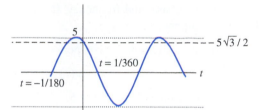

FIGURE 6.6   Figure for Example 6.3, part (c).

Sinusoids are time-varying signals and their value changes with time. Thus, sinusoid voltage, current, and power all vary with time. A time-varying power is called instantaneous power. However, instantaneous power might not be the best measure to represent the signal power. For example, to represent the voltage or power of an instrument, it is useful to have a number and not a time-varying function. This section investigates different measures of expressing a sinusoid's power, such as *root-mean-square* and *average power*.

6.1.1 Root-Mean-Square (rms) Values (Effective Values)

The **root-mean-square (rms)** for the periodic voltage, $v(t)$, is defined as:

$$V_{\text{rms}} = \sqrt{\frac{1}{T}\int_0^T v^2(t)\,\mathrm{d}t} \tag{6.10}$$

In the United States, AC power in residential wiring is distributed as a 60-Hz, 115-V sinusoid. Note that, here, 115 V is the rms value, and not the peak value. The rms value is also called the **effective value**. Substituting $v(t) = V_{\text{m}}\cos(\omega t + \theta)$ in (6.10):

$$V_{\text{rms}} = \sqrt{\frac{1}{T}\int_0^T V_{\text{m}}^2\cos^2(\omega t + \theta)\,\mathrm{d}t} \tag{6.11}$$

Note that $\cos^2\phi = (1/2)(1 + \cos 2\phi)$; therefore:

$$V_{\text{rms}} = \sqrt{\frac{V_{\text{m}}^2}{2T}\int_0^T (1 + \cos(2\omega t + 2\theta))\,\mathrm{d}t} \tag{6.12}$$

The integral of $\cos(2\omega t + 2\theta)$ over one period is zero; thus, what is left is the integral over $\mathrm{d}t$ from zero to T, which is T. This leads to the final result of:

$$V_{\text{rms}} = \frac{V_{\text{m}}}{\sqrt{2}} \tag{6.13}$$

Similarly, the rms value of the periodic current, $i(t)$, is defined as:

$$I_{\text{rms}} = \sqrt{\frac{1}{T}\int_0^T i^2(t)\mathrm{d}t} \qquad (6.14)$$

Defining $i(t) = I_{\text{m}}\cos(\omega t + \theta)$:

$$I_{\text{rms}} = \frac{I_{\text{m}}}{\sqrt{2}} \qquad (6.15)$$

Thus, given 115 V as the rms value, the peak value of the voltage of AC power in the United States is $V_{\text{m}} = V_{\text{rms}} \times \sqrt{2} = 115 \times \sqrt{2} \approx 163$ V. Again, it must be noted that usually in electric product descriptions, the voltage and current range refer to the rms value, and not the peak value. For example, the voltage for a laptop is from 100 to ~120 V, which is the rms value variation range.

EXAMPLE 6.4   **Peak and rms Value of a Sinusoid**

A sinusoidal voltage is given by:

$$v(t) = 311\cos(120\pi t + 30°)$$

Find the peak value and rms value of the sinusoidal voltage.

SOLUTION

The peak value that the amplitude corresponds to is:

$$V_{\text{m}} = 311 \text{ V}$$

Using Equation (6.13), the rms value is:

$$V_{\text{rms}} = \frac{311}{\sqrt{2}} = 220.6 \text{ V}$$

Note: In many countries such as England and France, the rms voltage of electric devices is 220 V.

6.1.2 Instantaneous and Average Power

The instantaneous power delivered by the voltage source in Figure 6.7 to the resistor, R, corresponds to:

$$p(t) = v(t) \cdot i(t) \qquad (6.16)$$

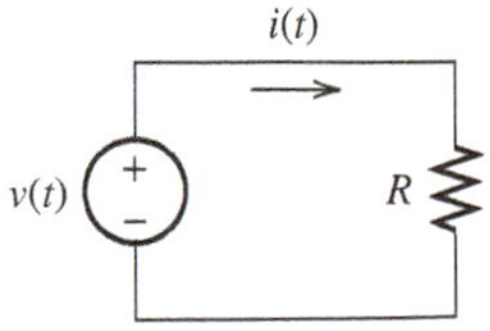

where

$$i(t) = \frac{v(t)}{R} \qquad (6.17)$$

Substituting Equation (6.17) into (6.16) results in:

$$p(t) = \frac{v^2(t)}{R} \qquad (6.18)$$

Average power corresponds to:

$$P_{\text{av}} = \frac{1}{T}\int_0^T p(t)\mathrm{d}t \qquad (6.19)$$

FIGURE 6.7   General circuit for a resistive load.

Using the same integration approach as in Equation (6.11):

$$P_{av} = \frac{V_m^2}{2R} \tag{6.20}$$

EXERCISE 6.3

Show the mathematical details of the derivation of Equation (6.20).

Note that $V_{rms}^2 = V_m^2/2$, thus:

$$P_{av} = \frac{V_{rms}^2}{R} \tag{6.21}$$

Equivalently, $v(t)$ in Equation (6.16) can be replaced by $v(t) = R \cdot i(t)$. In this case:

$$p(t) = R \times i^2(t) \tag{6.22}$$

Similarly, it can be depicted that:

$$P_{av} = R \times I_{rms}^2 \tag{6.23}$$

EXERCISE 6.4

Use the definition of Equation (6.19) to prove Equation (6.23).

EXAMPLE 6.5   **Average Delivered Power**

A sinusoidal voltage is given by:

$$v(t) = 311 \cos(120\pi t + 30°)$$

Find the average power delivered to a 100-Ω resistance.

SOLUTION

The peak value is:

$$V_m = 311 \text{ V}$$

The rms value is:

$$V_{rms} = \frac{311}{\sqrt{2}} = 220.6 \text{ V}$$

Using Equation (6.21):

$$P_{av} = \frac{V_{rms}^2}{R} = \frac{(220.6)^2}{100} = 486.6 \text{ W}$$

6.2 PHASORS

Phasors are complex numbers represented in the polar domain. Phasors are used to characterize sinusoidal voltages or currents. Considering a sinusoidal voltage:

$$v(t) = V_m \cos(\omega t + \theta) \tag{6.24}$$

its phasor is defined as:

$$V = V_m \angle \theta = V_m e^{j\theta} \tag{6.25}$$

Based on Euler's formula:

$$e^{j\theta} = \cos\theta + j\sin\theta \tag{6.26}$$

Thus (see Appendix C for a review of complex numbers):

$$\cos\theta = Re\{e^{j\theta}\}, \quad \sin\theta = Im\{e^{j\theta}\} \tag{6.27}$$

where $Re\{\cdot\}$ and $Im\{\cdot\}$ stand for the real and imaginary components, respectively. Equivalently:

$$\begin{aligned} v(t) &= V_m \cos(\omega t + \theta) \\ &= V_m \, Re\{e^{j(\omega t + \theta)}\} \\ &= Re\{V_m e^{j(\omega t + \theta)}\} \\ &= Re\{(V_m e^{j\theta}) e^{j\omega t}\} \end{aligned} \tag{6.28}$$

Thus:

$$v(t) = Re\{V e^{j\omega t}\} \tag{6.29}$$

where V is the phasor of $v(t)$. Note that sometimes phasors are represented by $\bar{V}$, i.e., a bar added on the top of the parameter. The magnitude of a phasor equals the peak value and its angle equals the phase of the sinusoid in cosine form. If a sinusoidal signal is represented in sine form, it can be converted into cosine form using the method explained in Equation (6.2).

It is important to understand why a sinusoidal voltage or current is converted to a phasor. This chapter focuses on the study of linear circuits. In the steady-state analysis of linear circuits, the frequency of the voltage across and current flow through any given element does not change. As a result, the information of amplitude and phase of a voltage or current, that is, its phasor, are sufficient to characterize that voltage or current. Note that linear circuits consist of linear elements, which maintain a linear relationship between current and voltage. Recall that a linear function is that which fulfills the properties of homogeneity $[f(\alpha x) = \alpha \cdot f(x)]$ and additivity $[f(x_1 + x_2) = f(x_1) + f(x_2)]$.

EXAMPLE 6.6   Specifying a Sinusoid from its Phasor

A voltage $v(t) = V_m \cos(2\pi \times 10^3 t)$ is applied to a linear circuit. The current phasor flowing through one of the elements is determined to be $2e^{j\pi/6}$ A. Specify the sinusoidal current flowing through the element. Can the circuit be a resistive circuit?

SOLUTION

The frequency of the current flowing through the element is the same as the applied voltage and corresponds to $f = 10^3$ Hz = 1 kHz, therefore:

$$i(t) = 2 \cos\left(2\pi \times 10^3 t + \frac{\pi}{6}\right)$$

The circuit cannot be resistive as the current and voltage phases are not the same. Later, it will be shown that the voltage and current phase in resistive circuits are equivalent.

EXAMPLE 6.7   Phasor or a Sinusoid

a. The phasor for the sinusoidal voltage, $v_1(t) = 150 \cos(120\pi t + 30^o)$, is $V_1 = 150\angle 30^\circ$.

b. The phasor for the sinusoidal voltage, $v_2(t) = 312 \sin(120\pi t + 45^o)$, is:

$$V_2 = 312\angle(45^\circ - 90^\circ) = 312\angle{-45^\circ}.$$

(continued)

EXAMPLE 6.7   **Continued**

Note that in this case, the sine function must first be converted to cosine to find its phase. The techniques of converting a sine function to cosine and vice versa are discussed in Equation (6.2).

The phasors for sinusoidal currents have the same form. For the current:

$$i(t) = I_m \cos(\omega t + \theta) \tag{6.30}$$

and the phasor is:

$$\boldsymbol{I} = I_m\angle\theta = I_m e^{j\theta} \tag{6.31}$$

Correspondingly:

$$i(t) = \text{Re}\left\{ \boldsymbol{I} e^{j\omega t} \right\} \tag{6.32}$$

6.2.1 Phasors in Additive or (Subtractive) Sinusoids

This section discusses how sinusoids can be added by adding their phasors. As an example, assume the following:

$$v_1(t) = V_{m1} \cos(\omega t + \theta_1)$$
$$v_2(t) = V_{m2} \cos(\omega t + \theta_2)$$
$$v_3(t) = V_{m3} \cos(\omega t + \theta_3)$$

Next, calculate $v(t) = v_1(t) + v_2(t) + v_3(t)$. The phasors for $v_1(t)$, $v_2(t)$, and $v_3(t)$, respectively, correspond to:

$$\boldsymbol{V}_1 = V_{m1}\angle\theta_1, \quad \boldsymbol{V}_2 = V_{m2}\angle\theta_2, \quad \boldsymbol{V}_3 = V_{m3}\angle\theta_3$$

According to Equation (6.29):

$$v_1(t) = \text{Re}\left\{ \boldsymbol{V}_1 e^{j\omega t} \right\}$$
$$v_2(t) = \text{Re}\left\{ \boldsymbol{V}_2 e^{j\omega t} \right\}$$
$$v_3(t) = \text{Re}\left\{ \boldsymbol{V}_3 e^{j\omega t} \right\}$$

Therefore:

$$v(t) = v_1(t) + v_2(t) + v_3(t) = \text{Re}(\boldsymbol{V}_1 e^{j\omega t}) + \text{Re}(\boldsymbol{V}_2 e^{j\omega t}) + \text{Re}(\boldsymbol{V}_3 e^{j\omega t})$$
$$= \text{Re}(\boldsymbol{V}_1 e^{j\omega t} + \boldsymbol{V}_2 e^{j\omega t} + \boldsymbol{V}_3 e^{j\omega t})$$
$$= \text{Re}\left[(\boldsymbol{V}_1 + \boldsymbol{V}_2 + \boldsymbol{V}_3) e^{j\omega t} \right]$$

Accordingly, the addition of sinusoids is converted into the addition of complex numbers.

EXAMPLE 6.8   **Addition of Two Sinusoids Using Their Phasors**

Assume:

$$v_1(t) = 5 \cos(\omega t + 45°)$$
$$v_2(t) = 10 \cos(\omega t + 30°)$$

Calculate:

$$v(t) = v_1(t) + v_2(t)$$

SOLUTION

The phasors of $v_1(t)$ and $v_2(t)$, respectively, are:

$$V_1 = 5\angle 45°$$
$$V_2 = 10\angle 30°$$

Thus, their addition corresponds to:

$$
\begin{aligned}
V_1 + V_2 &= 5\angle 45° + 10\angle 30° \\
&= 5 \times \cos(45°) + j5 \times \sin(45°) + 10 \times \cos(30°) + j10 \times \sin(30°) \\
&= 3.54 + j3.54 + 8.66 + j5 \\
&= 12.2 + j8.54 \\
&= \sqrt{(12.2)^2 + (8.54)^2} \angle \tan^{-1}\left(\frac{8.54}{12.2}\right) \\
&= 14.89\angle 34.99°
\end{aligned}
$$

Then, the amplitude of the signal is 14.89 and its phase is 34.99°, or:

$$v(t) = 14.89 \cos(\omega t + 34.99°)$$

6.3 COMPLEX IMPEDANCES

In the context of steady-state AC circuits, resistors, capacitors, and inductors are described using a parameter called **impedance** that is viewed as a complex resistance that changes with frequency. Note that phasors and complex impedances are applicable to sinusoidal steady-state conditions.

6.3.1 The Impedance of a Resistor

Consider that the source $v_R(t) = A \cos\omega t$ is applied to a resistor. According to Ohm's law, the current flowing through resistor R is:

$$i_R(t) = \frac{v_R(t)}{R} = \frac{A}{R}\cos\omega t \tag{6.33}$$

Using phasors to express voltage and current:

$$V_R = A\angle 0° \tag{6.34}$$

$$I_R = \frac{A}{R}\angle 0° \tag{6.35}$$

The impedance of the resistor (Z_R) is defined as the ratio of the phasor voltage across the resistor to the phasor current:

$$Z_R = \frac{V_R}{I_R} = \frac{A\angle 0°}{\dfrac{A}{R}\angle 0°} = R \tag{6.36}$$

As a result, for a resistor, $Z_R = R$. In addition, according to Equations (6.34) and (6.35), the current flowing through and the voltage across resistors are in phase.

6.3.2 The Impedance of an Inductor

As discussed in Chapter 4, the relationship of voltage and current for an ideal inductor corresponds to:

$$v_L(t) = L\frac{di_L(t)}{dt} \tag{6.37}$$

Let $i_L(t) = A\cos\omega t$, then:

$$v_L(t) = L\frac{di_L(t)}{dt}$$

$$= -A\omega L \sin\omega t$$

$$= A\omega L \cos\left(\omega t + \frac{\pi}{2}\right) \tag{6.38}$$

In other words, the voltage has a 90° ($\pi/2$ radian) lead with respect to the current flowing through it. Accordingly, voltage and current phasors are:

$$I_L = A\angle 0 \tag{6.39}$$

$$V_L = A\omega L\angle\frac{\pi}{2} \tag{6.40}$$

Equivalently, it can be shown that:

$$V_L = A\angle 0 \tag{6.41}$$

$$I_L = A\omega L\angle -\frac{\pi}{2} \tag{6.42}$$

In other words, if the voltage across an inductor is $v_L(t) = A\cos\omega t$, then the current will be $i_L(t) = (A/\omega L)\cos(\omega t - \pi/2)$. This can be verified through the following exercise.

EXERCISE 6.5

Start with the following voltage–current relationship in an inductor, that is:

$$i_L(t) = \frac{1}{L}\int v_L(t)dt + I_0$$

Take $I_0 = 0$, $v_L(t) = A\cos\omega t$ and validate the voltage and current relationship presented in Equations (6.41) and (6.42).

Based on Equations (6.39) and (6.40), or (6.41) and (6.42), the impedance of the inductor (Z_L) is defined as:

$$Z_L = \frac{V_L}{I_L} = \frac{A\angle 0}{\dfrac{A}{\omega L}\angle -\dfrac{\pi}{2}} = \omega L\angle\frac{\pi}{2} = \omega L\left(\cos\frac{\pi}{2} + j\sin\frac{\pi}{2}\right) = j\omega L \tag{6.43}$$

Equation (6.43) uses the fact that $\cos(\pi/2) = 0$, and $\sin(\pi/2) = 1$.

Note:

In Equation (6.43), $j\omega L$ is also called the **reactance** of an inductor.

6.3.3 The Impedance of a Capacitor

As discussed in Chapter 4, the relationship between voltage and current in an ideal capacitor is:

$$i_C(t) = C\frac{dv_C(t)}{dt} \tag{6.44}$$

Let $v_C(t) = A\cos\omega t$, then:

$$i_C(t) = C\frac{dv_C(t)}{dt}$$

$$= C\frac{d(A\cos\omega t)}{dt}$$

$$= -C(A\omega \sin\omega t)$$

$$= \omega CA \cos\left(\omega t + \frac{\pi}{2}\right) \tag{6.45}$$

The phasors of $v_C(t)$ and $i_C(t)$ are:

$$V_C = A\angle 0 \tag{6.46}$$

$$I_C = \omega CA\angle\frac{\pi}{2} \tag{6.47}$$

In other words, in capacitors, the voltage across the capacitor lags the current through it. The same voltage–current relationship can be established using:

$$v_C(t) = \frac{1}{C}\int i_C(t)dt + V_0 \tag{6.48}$$

EXERCISE 6.6

Use Equation (6.48), set $V_0 = 0$, and take $i_C(t) = A\cos\omega t$. Show that:

$$I_C = A\angle 0 \tag{6.49}$$

$$V_C = \frac{A}{\omega C}\angle -\frac{\pi}{2} \tag{6.50}$$

The impedance of the capacitor (Z_C) is defined as:

$$Z_C = \frac{V_C}{I_C}$$

Using Equations (6.46) and (6.47) to replace for the current and voltage phasors results in:

$$Z_C = \frac{A\angle 0}{\omega CA\angle\dfrac{\pi}{2}} = \frac{1}{\omega C}\angle -\frac{\pi}{2}$$

Now, using Euler's formula in Equation (6.26):

$$Z_C = \frac{1}{\omega C}\left[\cos\left(-\frac{\pi}{2}\right) - j\sin\left(\frac{\pi}{2}\right)\right]$$

Using the fact that $\cos(-\pi/2) = 0$, and $\sin(\pi/2) = 1$:

$$Z_C = -j\frac{1}{\omega C} = \frac{1}{j\omega C} \tag{6.51}$$

Note:

In Equation (6.51), $1/j\omega C$ is also called the *reactance* of a capacitor.

Based on Equations (6.36), (6.43), and (6.51), for the impedance of resistors, inductors, and capacitors, it can be seen that the relationship between the voltage and current phasors for each of these impedances corresponds to:

$$V = Z \times I \tag{6.52}$$

In Equation (6.52), $Z = Z_R$ [Equation (6.36)] if the element is a resistor, $Z = Z_L$ [Equation (6.43)] if the element is an inductor, and $Z = Z_C$ [Equation (6.51)], if the element is a capacitor. However, in general, Z can be a combination of resistors, capacitors, and/or inductors.

Equation (6.52) shows that the voltage–current phasor relationship is similar to the instantaneous voltage–current relationship in the resistors $\left[v(t) = R \times i(t) \right]$. As a result, the impedance of series and parallel impedances are computed similar to series and parallel resistances. Note that according to Equation (6.52), the unit of impedance is ohm (Ω), which is the same as in resistors. However, impedance is not only a real resistor, but also a complex resistor, which consists of a real part and an imaginary part.

EXAMPLE 6.9   Impedance of a Capacitance

A voltage $v_L(t) = 110 \cos(120t + 45°)$ is applied to a 47-µF capacitance. Find the impedance of the capacitance.

SOLUTION

From the equation of the voltage, it is clear that $\omega = 120$, thus:

$$Z_C = \frac{1}{j\omega C} = \frac{1}{j120 \times 47 \times 10^{-6}} = j177.305 \ \Omega$$

Note:

In general, an impedance can be represented by a complex number. The real part of this complex number represents the resistance and the imaginary part of that represents the reactance.

6.3.4 Series Connection of Impedances

The total impedance of series-connected impedances (independent of its nature, whether it comes from a resistor, capacitor, inductor or any combination of these components) shown in Figure 6.8 can be calculated using a method similar to that used for series-connected resistors.

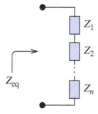

FIGURE 6.8 Series connection of impedance.

$$Z_{eq} = Z_1 + Z_2 + \cdots + Z_n = \sum_{k=1}^{n} Z_k \tag{6.53}$$

where $Z_i, i \in \{1, 2, \ldots, n\}$, can be Z_R, Z_L, Z_C, or a combination of them.

The proof of Equation (6.53) is very simple. Because the current flowing through all impedances is the same, for any impedance $Z_i, i \in \{1, 2, \ldots, n\}$, $V_i = Z_i I$. Based on Kirchhoff's voltage law:

$$V = V_1 + V_2 + \cdots + V_n = \sum_{k=1}^{n} V_k = \sum_{k=1}^{n} Z_k I$$

As a result:

$$Z_{eq} = \frac{V}{I} = \sum_{k=1}^{n} Z_k$$

6.3.5 Parallel Connection of Impedances

Simultaneously, writing the phasor voltage–current relationship in Figure 6.9 for each parallel impedance, Z_i, results in:

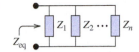

$$I_i = \frac{V}{Z_i}$$

Because, the phasor voltage, V, is constant across all impedances and the total current $\vec{I} = \sum \vec{I_i}$, it can be shown that:

FIGURE 6.9 Parallel connection of impedance.

$$Z_{eq} = \frac{1}{\displaystyle\sum_{i=1}^{n}\frac{1}{Z_i}}$$

In other words:

$$\frac{1}{Z_{eq}} = \frac{1}{Z_1} + \frac{1}{Z_2} + \cdots \frac{1}{Z_n} \tag{6.54}$$

where Z_i, $i \in \{1, 2, \ldots, n\}$, can be a resistor, a capacitor, or an inductor.

EXERCISE 6.7

Write the details of the proof of Equation (6.54).

EXAMPLE 6.10   Total Impedance

A voltage, $v(t) = 220 \cos(360t + 30°)$, is applied to the circuit shown in Figure 6.10. Find the total impedance of the series inductance and the resistor.

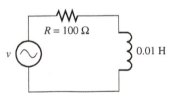

$R = 100\ \Omega$

0.01 H

v

FIGURE 6.10 Circuit for Example 6.10.

SOLUTION

Here, $\omega = 360$, thus:

$$Z = Z_R + Z_L = 100 + j360 \times 0.01 = 100 + j3.6\ \Omega$$

EXAMPLE 6.11   Total Impedance

The voltage, $v(t) = 10 \cos(500t + 60°)$, is applied to the circuit shown in Figure 6.11. Find the total impedance seen through the terminals of the voltage source.

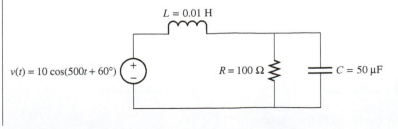

$L = 0.01$ H

$v(t) = 10 \cos(500t + 60°)$

$R = 100\ \Omega$ $C = 50\ \mu F$

FIGURE 6.11 Circuit for Example 6.11.

(*continued*)

EXAMPLE 6.11   Continued

SOLUTION

The equivalent phasor voltage is $V = 10\angle 60°$. Thus, the phasor format is shown in Figure 6.12. For the impedance of Z_L and Z_C:

$$Z_L = j\omega L = j0.01 \times 500 = j5 \ \Omega$$

$$Z_C = \frac{1}{j\omega C} = \frac{1}{j500 \times 50 \times 10^{-6}} = -j40 \ \Omega$$

The total parallel impedance, $Z_{RC} = Z_R \| Z_C$, in polar form is:

$$Z_{RC} = \frac{1}{\dfrac{1}{Z_R} + \dfrac{1}{Z_C}} = \frac{1}{0.01 + j0.025}$$

$$= \frac{1\angle 0°}{0.0269\angle 68.2°}$$

$$= 37.17\angle -68.2 \ \Omega$$

The rectangular form for Z_{RC} corresponds to:

$$Z_{RC} = 13.80 - j34.51 \ \Omega$$

Therefore, the total impedance corresponds to:

$$Z_T = Z_L + Z_{RC}$$
$$= j5 + 13.80 - j34.51$$
$$= 13.80 - j29.51$$
$$= 32.58\angle -64.94 \ \Omega$$

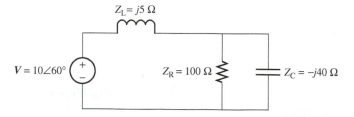

FIGURE 6.12 Transformed circuit of Figure 6.11.

APPLICATION EXAMPLE 6.12   A Loudspeaker

An audiophile wants to ensure her new speakers can shake the house, but do not harm the windows. She places a pressure sensor 0.5 m from the center of one of her new loudspeakers (see Figure 6.13). She connects a test signal of $v_{sig}(t) = 3.5 \cos(400t - 45°)$ to the amplifier. The gains of the amplifier and pressure shown in Figure 6.13 are specific for this frequency. The pressure gain (pascal per volt) of the speaker (as seen by the sensor) is available. Find the pressure seen by the pressure sensor as a function of time.

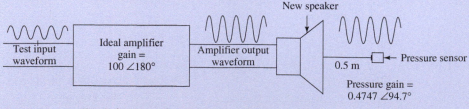

FIGURE 6.13 Circuit for Example 6.12 (loudspeaker).

SOLUTION

The test signal, $v_{sig}(t) = 3.5 \cos(400t - 45°)$, can be expressed in phasor form as $3.5\angle -45°$. Applying the amplifier and pressure gain shown in Figure 6.13, the pressure, P, corresponds to:

$$P = 3.5\angle -45° \times 100\angle 180° \times 0.4747\angle 94.7°$$
$$= 166.145\angle 229.7°$$

In the time domain, the pressure is:

$$P(t) = 166.145 \cos(400t + 229.7°) \text{ Pa (pascals or n/m}^2)$$

6.4 STEADY-STATE CIRCUIT ANALYSIS USING PHASORS

This section examines the steady-state circuit response to a sinusoidal source using phasors and impedance. If the circuit is in the steady state, the output will be sinusoidal with the same frequency as the input, as illustrated in Figure 6.14. In this figure, $v_i(t)$ is the input and $v_o(t)$ is the output. Chapter 5 explained that for a sinusoidal input, the output voltage includes two components: the transient and the steady state. As stated before, the steady-state response to a circuit is explicitly computed for a given frequency. As the frequency changes, the response would also change.

In the steady state, all voltages and currents are sinusoids. Assuming a linear circuit, the frequency of these sinusoids will be the same as the input signal frequency. Thus, they can be represented by their corresponding phasors. In addition, circuit elements, such as resistors, inductors, and capacitors, are represented by impedances, as illustrated in Figure 6.15. KVL and KCL can be applied to the voltage and current in phasor form, respectively, that is, the summation of the phasor voltages for any closed path in an electrical network is equal to zero, that is $\sum V_i = 0$.

In addition, the summation of the phasor currents $I_k^{(enter)}$ entering a node is equal to the summation of the phasor currents leaving $I_m^{(leave)}$:

$$\sum_k I_k^{(enter)} = \sum_m I_m^{(leave)}$$

Therefore, the same principles discussed in Chapters 2 and 3 to conduct steady-state analysis for sinusoidal signals can be used here. Here, all elements are in phasor form. Analyzing

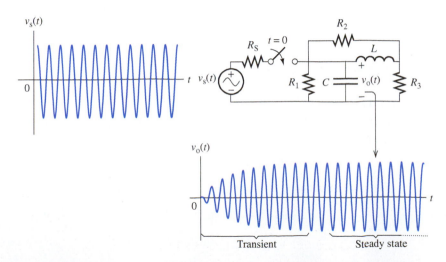

FIGURE 6.14 A steady-state circuit.

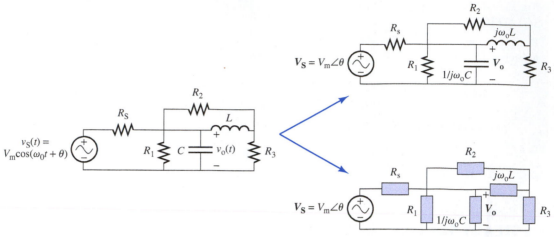

FIGURE 6.15 Steady-state circuit analysis.

circuits with impedances is similar to analyzing resistive circuits. In this analysis, real currents and voltages are replaced with complex (phasor) currents and voltages.

The procedure for steady-state analysis of circuits with sinusoidal sources is as follows:

1. Replace the instantaneous voltage and current source (s) (i.e., $v(t)$ and $i(t)$) with the corresponding phasors (i.e., V and I).
2. Replace inductance, L, with the complex impedance $Z_L = j\omega L = \omega L \angle 90°$, and capacitance, C, with the complex impedance $Z_C = 1/j\omega C = -j(1/\omega C) = (1/\omega C)\angle -90°$. Resistances have impedances equal to their resistances.
3. Analyze the circuit using techniques similar to those used for resistive circuits, and perform the calculation using complex arithmetic.

EXAMPLE 6.13   Steady-State Current

Find the steady-state current $i(t)$ and $v_L(t)$ shown in Figure 6.16.

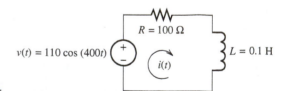

FIGURE 6.16 Circuit for Example 6.13.

SOLUTION

The voltage phasor of the source is $V = 110\angle 0°$ and the impedance of the resistor is $Z_R = 100 \ \Omega$. In addition:

$$Z_L = j\omega L = j400 \times 0.1 = j40 = 40\angle 90°$$

The total impedance of the resistor and inductor is:

$$Z_{total} = Z_R + Z_L = 100 + j40 = 107.7\angle 21.8°$$

As a result, the current phasor corresponds to:

$$I = \frac{V}{Z_{total}} = \frac{110\angle 0°}{107.7\angle 21.8°} = 1.02\angle -21.8°$$

The voltage phasor of the inductor is:

$$V_L = j\omega L \times I = 40\angle 90° \times 1.02\angle -21.8° = 40.8\angle 68.2°$$

The corresponding instantaneous current and voltage are:

$$i(t) = 1.02\cos(400t - 21.8°), \quad \text{and} \quad v_L(t) = 40.8\cos(400t + 68.2°)$$

EXAMPLE 6.14 **Steady-State Voltage**

Find the steady-state voltage, $v_A(t)$, in Figure 6.17.

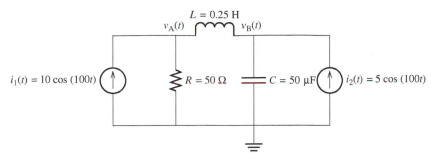

FIGURE 6.17 Circuit for Example 6.14.

SOLUTION

The equivalent phasor domain circuit of Figure 6.17 is presented in Figure 6.18. All impedances are shown in Figure 6.18.

In Figure 6.18, $Z_R = 50\ \Omega$, $Z_L = j25\ \Omega$, and $Z_C = -j200\ \Omega$. Writing KCL for nodes A and B:

$$I_A + I_{AB} = 10\angle 0°$$
$$I_B + (-I_{AB}) = 5\angle 0°$$

Replacing the currents with voltage divided by the impedance results in:

$$\frac{V_A}{50} + \frac{V_A - V_B}{j25} = 10\angle 0°$$

$$\frac{V_B}{-j200} + \frac{V_B - V_A}{j25} = 5\angle 0°$$

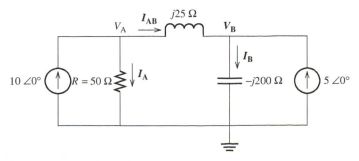

FIGURE 6.18 Equivalent circuit for Example 6.14.

(continued)

EXAMPLE 6.14   **Continued**

Converting the polar phasor to rectangular form and factorizing:

$$(0.02 - j0.04)V_A + j0.04V_B = 10$$
$$j0.04V_A + j0.035V_B = 5$$

Solving this system of equations using Cramer's rule (see Appendix A):

$$V_A = \frac{\begin{vmatrix} 10 & j0.04 \\ 5 & j0.035 \end{vmatrix}}{\begin{vmatrix} 0.02 - j0.04 & j0.04 \\ j0.04 & j0.035 \end{vmatrix}} = 48.69 \angle 76.86°$$

In the time domain:

$$v_A(t) = 48.69 \cos(100t + 76.86°)$$

EXAMPLE 6.15   **Steady-State Voltage**

Considering the circuit shown in Figure 6.19, find the steady-state voltage across the capacitor, $v_C(t)$, and the phasor current through both the inductor and the capacitor.

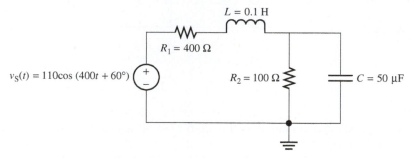

FIGURE 6.19   Circuit for Example 6.15.

SOLUTION

The equivalent circuit of Figure 6.19 in the phasor domain is shown in Figure 6.20.
In Figure 6.20:

$$Z_L = j\omega L = j0.1 \times 400 = j40$$

The equivalent series impedance of R and L is:

$$Z_{RL} = Z_{R_1} + Z_L = 400 + j40$$
$$Z_C = \frac{1}{j\omega C} = \frac{1}{j400 \times 50 \times 10^{-6}} = -j50 \ \Omega$$

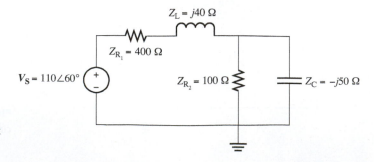

FIGURE 6.20   Equivalent circuit for Example 6.15.

The equivalent impedance of the parallel RC corresponds to:

$$Z_{R_2C} = \frac{1}{\dfrac{1}{Z_{R_2}} + \dfrac{1}{Z_C}} = \frac{1}{0.01 + j0.02} = \frac{1\angle 0°}{0.02236\angle 63.43°} = 44.6429\angle -63.43°$$

The rectangular form for Z_{RC} can be given by:

$$Z_{R_2C} = 19.968 - j39.928$$

Therefore, the circuit in Figure 6.21 will be the equivalent circuit for Figure 6.20. According to the voltage division principle:

$$
\begin{aligned}
V_C &= \frac{Z_{R_2C}}{Z_{R_1L} + Z_{R_2C}} V_S \\
&= \frac{44.6429\angle -63.43°}{419.968 + j0.072} \cdot 110\angle 60° \\
&= \frac{4910.719\angle -3.43°}{419.968\angle 0.01} \\
&= 11.693\angle -3.44
\end{aligned}
$$

Therefore:

$$v_C(t) = 11.693 \cos(400t - 3.44°)$$

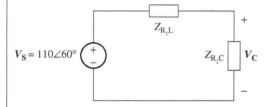

FIGURE 6.21   Equivalent circuit for Example 6.13.

APPLICATION EXAMPLE 6.16   **Elevator**

Figure 6.22 shows a simple block diagram of an elevator. The passenger selects a destination floor by pressing the floor button on the Input/Control. The Input/Control processes the message and "orders" the motor to "move" the elevator to the particular floor. The sinusoidal voltage source, $v_S(t) = 160 \cos(120t)$, the input control has a resistance of 100 Ω, and the motor has an impedance of $20 + j10$ Ω. The equivalent circuit is shown in Figure 6.23. Determine the expression for the current in the circuit.

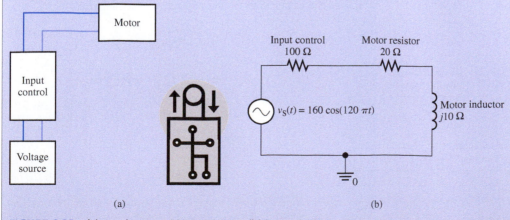

FIGURE 6.22   (a) An elevator's block diagram; (b) its equivalent circuit.

(*continued*)

APPLICATION EXAMPLE 6.16   Continued

SOLUTION

Based on the circuit shown in Figure 6.22(b), the total impedance seen through the voltage source terminals is:

$$Z_{total} = Z_{control} + Z_{resistor} + Z_{inductor}$$
$$Z_{total} = 100 \ + \ 20 \ + \ j10$$

$$Z_{total} = 120 + j10 \ \Omega$$
$$Z_{total} = 120.416\angle 4.76° \ \Omega$$

The sinusoidal voltage source is, $V_S(t) = 160\cos(120\pi t)$, which has the phasor of:

$$V_S = 160\angle 0°$$

Therefore, the current phasor in the circuit is:

$$I = \frac{V_S}{Z_{total}}$$

$$= \frac{160\angle 0°}{120.416\angle 4.76°}$$

$$= 1.33\angle -4.76°$$

Accordingly, the expression of the current in the circuit corresponds to:

$$i(t) = 1.33\cos(120\pi t \ - \ 4.76°)$$

APPLICATION EXAMPLE 6.17   Automated Traffic Light

Figure 6.23(a) represents the block diagram of an automated traffic light sensor and controller. In an automated traffic light system, the red traffic light stays ON longer if there is not any vehicle stopped at the light. However, it turns to green earlier when there is at least one vehicle waiting at the light. This system operates as follows: There is an inductor that is underneath the ground. When there is a vehicle at the stop light, the vehicle size changes the inductance. A change in the inductance is detected by a specific device, which makes the traffic light turn green.

Assume that the sinusoidal voltage source, $v_S(t) = 110\cos(120\pi t)$, the traffic light has the resistance of 300 Ω, the resistance of the detector is 20 Ω, and the inductor has an initial inductance of 100 mH. Find the expression for the voltage across the inductor.

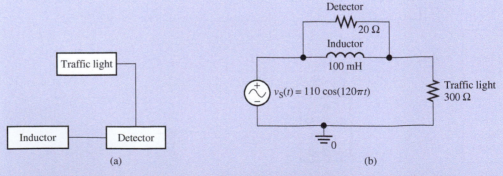

FIGURE 6.23   (a) Block diagram of an automated traffic light sensor and controller; (b) its equivalent circuit.

SOLUTION

The circuit of the automated traffic light and detector is shown in Figure 6.23 (b). The impedance of the traffic light is:

$$Z_{light} = 300 \ \Omega$$

The inductor is 100 mH or 0.1 H; thus, impedance of the inductor is:

$$Z_l = j0.1 \times 120\pi \rightarrow Z_l = j12\pi \ \Omega$$

The impedance of the parallel circuit of the detector and inductor is calculated using:

$$\frac{1}{Z_{par}} = \frac{1}{20} + \frac{1}{(j12\pi)}$$

Thus:

$$Z_{par} = \frac{j240\pi}{20 \ + \ j12\pi} \rightarrow Z_{par} = 17.667\angle 27.946° \ \Omega$$

To calculate the total impedance of the circuit, both impedances must be expressed in rectangular form. Now:

$$Z_{par} = 15.61 + j8.28 \ \Omega$$

Thus, the total impedance is:

$$Z_{total} = 300 + 15.61 + j8.28 \rightarrow Z_{total} = 315.61 + j8.28 \rightarrow Z_{total} = 315.72\angle 1.5° \ \Omega$$

Because the detector and the inductor are in parallel, the voltages across them are the same. Therefore, using voltage division principle results in:

$$V_L = \frac{Z_{par}}{Z_{total}} \times V_s \rightarrow V_L = \frac{17.667\angle 27.946°}{315.72\angle 1.5°} \times 110\angle 0° \rightarrow V_L = 6.155\angle 26.446°$$

In the time domain:

$$v_L(t) = 6.155 \cos(120\pi t + 26.446°)$$

In a circuit network, if there is more than one source, and those sources have different frequencies, the circuit can be analyzed using the superposition principle.

EXAMPLE 6.18   Output Voltage Computation

For the circuit shown in Figure 6.24, find $v_o(t)$.

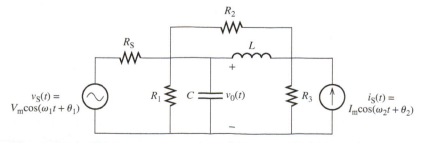

FIGURE 6.24 Circuit for Example 6.18.

(continued)

EXAMPLE 6.18 **Continued**

SOLUTION

Here, apply the superposition principle. Note that the frequencies of the two sources are different (one is ω_1 and the other is ω_2). Therefore, the superposition cannot be applied to the phasors (see Exercise 6.9 for the case in which the sources have the same frequency. Thus, two sets of analysis (one for each frequency) are needed.

1. Set $i_S(t)$ to zero and obtain the response $v_1(t)$ to $v_S(t)$. The corresponding circuit is illustrated in Figure 6.25.
2. Set $v_S(t)$ to zero to obtain the response $v_2(t)$ to $v_S(t)$. The corresponding circuit is illustrated in Figure 6.26.
   Finally, $v_o(t) = v_1(t) + v_2(t) = \mathrm{Re}\left\{V_1 e^{j\omega_1 t}\right\} + \mathrm{Re}\left\{V_2 e^{j\omega_2 t}\right\}$.

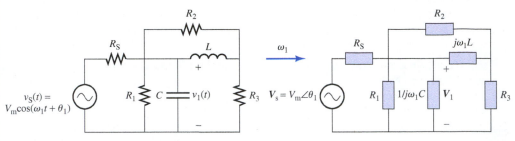

FIGURE 6.25 Circuit for Example 6.18.

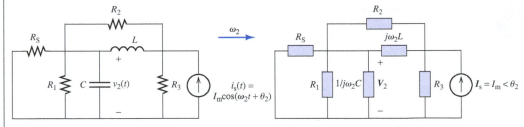

FIGURE 6.26 Circuit for Example 6.18.

EXERCISE 6.8

In Example 6.18, calculate the voltage phasors, V_1 and V_2, and the total voltage, $v_o(t)$, if $R_1 = R_2 = R_3 = 1\ \text{k}\Omega$, $L = 3$ H, and $C = 2$ F.

EXERCISE 6.9

In Example 6.18, calculate the output voltage if both sources have the same frequency (e.g., ω_1). Note that in this case the sources can be replaced with their phasor equivalents and the circuit can be analyzed considering both sources are available simultaneously. This analysis is possible using nodal analysis as discussed in Chapter 3. However, in Example 6.18, this process cannot be used because the frequencies of the two sources are different. Thus, in Example 6.18, the response to each source must be calculated separately, and the voltages in time domain are added.

6.5 THÉVENIN AND NORTON EQUIVALENT CIRCUITS WITH PHASORS

The Thévenin and Norton theorems are applied for AC steady-state analysis of circuits that include resistors, capacitors, and inductors (i.e., impedances) in the same way as they are applied to the analysis of DC circuits that include resistors as discussed in Chapter 3. A Thévenin equivalent circuit consists of a voltage source in series with a resistance. This chapter examines steady-state conditions for circuits composed of sinusoidal sources, resistances, inductances, and capacitances. In this case, the Thévenin equivalent circuit consists of a phasor voltage source in series with complex impedances.

6.5.1 Thévenin Equivalent Circuits with Phasors

6.5.1.1 THÉVENIN VOLTAGE

From the perspective of any load, the Thévenin voltage, V_{th}, equals the open-circuit voltage, V_{AB}, of the circuit, as shown in Figure 6.27.

> **Calculating the Thévenin voltage (see Figure 6.28):**
>
> 1. Remove the load, leaving the load terminals open-circuited.
> 2. Define the open-circuit voltage V_{oc} ($V_{oc} = V_{AB}$, in Figure 6.28).
> 3. Use the corresponding phasors to describe the voltage and current source.
> 4. Replace inductance, capacitance, and resistance with the corresponding impedance (Z_L, Z_C, and Z_R).
> 5. Use any method (e.g., nodal analysis, voltage division, etc.) to calculate V_{oc}.
> 6. The resulting voltage is the Thévenin equivalent voltage, that is, $V_{th} = V_{oc}$.

6.5.1.2 THÉVENIN IMPEDANCE

The Thévenin impedance, Z_{th}, seen through two terminals is found by equating the independent sources to zero. Recall that a zero-independent voltage source is equivalent to a short circuit. In addition, a zero-independent current source is equivalent to an open circuit. As discussed in Chapter 3, dependent sources cannot be set to zero.

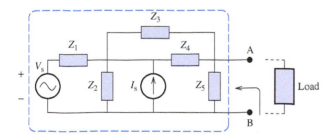

FIGURE 6.27 A Thévenin equivalent circuit.

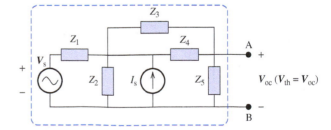

FIGURE 6.28 Thévenin equivalent voltage.

> **Calculating the Thévenin impedance (see Figure 6.28):**
>
> 1. Remove the load
> 2. Equate all independent voltage and current sources to zero
> 3. Replace all inductance, capacitance, and resistance with the corresponding impedance (Z_L, Z_C, and Z_R)

> **4.** Compute the total impedance, Z_{AB}, seen across load terminals (*with the load removed*)
> **5.** The Thévenin impedance $Z_{th} = Z_{AB}$

The Thévenin equivalent circuit is shown in Figure 6.29.

6.5.2 Norton Equivalent Circuits with Phasors

For steady-state AC circuits, another equivalent circuit is the Norton equivalent circuit, which consists of a phasor current source, I_n, in parallel with the Thévenin impedance.

6.5.2.1 NORTON CURRENT

The Norton current can be found by equating the load to zero. Thus, the load is replaced by a short circuit, and the current flowing through the short circuit can be computed.

> **Calculating the Norton current (see Figure 6.30):**
>
> **1.** Replace the load with a short circuit
> **2.** Define the short-circuit current as I_{sc}
> **3.** Use any method (e.g., nodal or mesh analysis) to calculate I_{sc}
> **4.** The Norton equivalent current $I_n = I_{sc}$

6.5.2.2 NORTON IMPEDANCE

The Norton impedance, Z_n, is the same as the Thévenin impedance, Z_{th}, and can be computed similarly. Another approach is to use the following equation:

$$Z_{th} = Z_n = \frac{V_{oc}}{I_{sc}} \tag{6.55}$$

Note that in Figure 6.29, $I_{sc} = V_{th}/Z_{th}$, $V_{th} = V_{oc}$. Figure 6.31 can be used to prove Equation (6.55). The Norton equivalent circuit is shown in Figure 6.30.

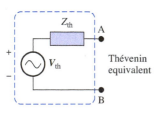

FIGURE 6.29 Thévenin equivalent impedance.

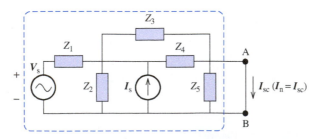

FIGURE 6.30 Norton equivalent current.

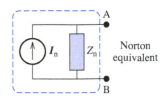

FIGURE 6.31 Norton equivalent impedance.

EXERCISE 6.10

Use Figure 6.31 to prove Equation (6.55).

EXAMPLE 6.19   **Thévenin and Norton Equivalent Circuit**

Find the Thévenin and Norton equivalent circuits for the circuit shown in Figure 6.32, as seen by the load resistor, R_L.

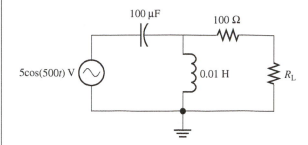

FIGURE 6.32   Circuit for Example 6.19.

SOLUTION

First, find the Thévenin impedance. The equivalent phasor circuit is shown in Figure 6.33.

Next, remove R_L and set the source equal to zero. The corresponding figure is shown in Figure 6.34.

The Thévenin impedance seen through A–B terminals is:

$$Z_{th} = 100 + \cfrac{1}{\cfrac{1}{-j20} + \cfrac{1}{j5}}$$

$$= 100 + j\frac{20}{3}$$

$$= 100.222\angle 3.8141°$$

The Norton impedance is then:

$$Z_n = Z_{th} = 100 + j\frac{20}{3} = 100.222\angle 3.8141°$$

Because the current through the 100-Ω resistor is zero, the voltage across the terminals of A and B is equal to the voltage across the inductor. Using a simple voltage dividing technique, the Thévenin voltage can be obtained:

$$V_{th} = 5\angle 0° \frac{j5}{-j20 + j5}$$

$$= \frac{5\angle 0°}{-3}$$

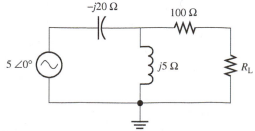

FIGURE 6.33   Equivalent circuit for Example 6.19.

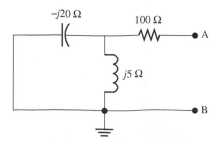

FIGURE 6.34   Equivalent circuit for Example 6.19.

(continued)

EXAMPLE 6.19   **Continued**

$$= \frac{5\angle 0°}{3\angle 180°}$$

$$= \frac{5}{3}\angle -180°$$

$$= 1.667\angle -180°$$

Finally, the Norton current is:

$$I_n = \frac{V_{th}}{Z_{th}} = \frac{1.667\angle -180°}{100.222\angle 3.8141°}$$

$$= 0.0166\angle -183.814°$$

$$= 0.0166\angle +176.186$$

The Thévenin and Norton equivalent circuits are shown in Figure 6.35.

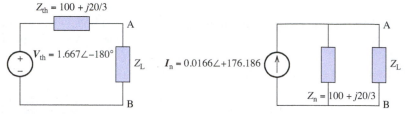

FIGURE 6.35   Thévenin and Norton equivalent circuits for Example 6.19.

APPLICATION EXAMPLE 6.20   **Controlling Building Sway**

An accelerometer is connected to a skyscraper to monitor building sway. It is wired to a computer, as shown in Figure 6.36. The parasitic capacitance and resistance of the cable connecting the sensor to the computer are also included. Find the Thévenin voltage as seen by the load.

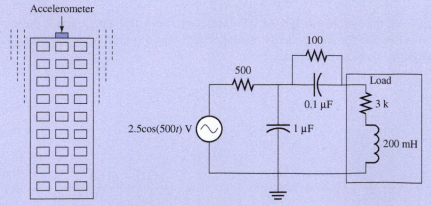

FIGURE 6.36   Controlling building sway.

SOLUTION

The equivalent circuit is shown in Figure 6.37. First, find V_{oc}.

Based on the circuit shown in Figure 6.38, V_{oc} corresponds to the voltage across the $-j2000$ impedance. Now, using the voltage division rule:

$$V_{oc} = 2.5\angle 0° \times \frac{-j2000}{500 - j2000}$$

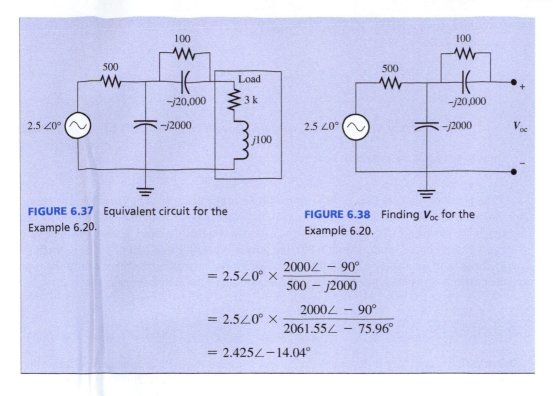

FIGURE 6.37 Equivalent circuit for the Example 6.20.

FIGURE 6.38 Finding V_{oc} for the Example 6.20.

$$= 2.5\angle 0° \times \frac{2000\angle - 90°}{500 - j2000}$$

$$= 2.5\angle 0° \times \frac{2000\angle - 90°}{2061.55\angle - 75.96°}$$

$$= 2.425\angle -14.04°$$

6.6 AC STEADY-STATE POWER

This section discusses the power delivered by the source to a general load that can be any RLC network as shown in Figure 6.39.

The first section examines the power delivered to a pure resistive load, a pure inductive load, and a pure capacitive load.

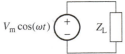

FIGURE 6.39 An RLC network.

1. *A resistive load:* **voltage, current, and power**

   Figure 6.40 is an example of a resistive load. In this case, the voltage is:

   $$v(t) = V_m \cos(\omega t) \qquad (6.56)$$

   and the current is:

   $$i(t) = \frac{V_m}{R} \cos(\omega t) = I_m \cos(\omega t), \quad \text{where} \quad I_m = \frac{V_m}{R} \qquad (6.57)$$

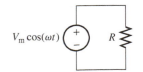

FIGURE 6.40 A resistive load.

   Thus, the instantaneous power is:

   $$p(t) = v(t)i(t) = V_m I_m \cos^2(\omega t) \qquad (6.58)$$

   For resistive loads, $p(t)$ is always positive. Therefore, the energy flows continually from the source to the resistance and is converted into heat.

2. *An inductive load:* **voltage, current, and power**

   Figure 6.41 is an example of a circuit with an inductive load. In this case, if the voltage is $v(t) = V_m \cos(\omega t)$, then the voltage phasor is:

   $$V = V_m \angle 0° \qquad (6.59)$$

   and the current phasor is:

   $$I = \frac{V}{Z} = \frac{V_m \angle 0°}{j\omega L} = \frac{V_m \angle 0°}{\omega L \angle 90°} = I_m \angle -90° \qquad (6.60)$$

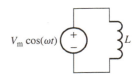

FIGURE 6.41 An inductive load.

where $I_m = V_m/\omega L = V_m/|Z|$, $Z = j\omega L = \omega L \angle 90°$. Accordingly, the time domain current is:

$$i(t) = I_m \cos(\omega t - 90°) \tag{6.61}$$

Note that using Equation (6.2), $\cos(\omega t - 90°) = \sin(\omega t)$. Thus, the instantaneous power is:

$$p(t) = v(t)i(t) = V_m I_m \cos(\omega t)\sin(\omega t) = \frac{V_m I_m}{2} \sin(2\omega t) \tag{6.62}$$

Equation (6.62) uses the equation:

$$\sin(2\omega t) = 2 \cos(\omega t)\sin(\omega t)$$

The phase of the current is $-90°$, and the phase of the voltage is $0°$. Therefore, the current lags the voltage by 90°. For the sinusoidal signal, the value is positive in one-half of a period and negative in the other half. Therefore, intuitively, it can be said that the average current will be zero.

In addition, based on Equation (6.62), the power is positive one-half of the time when the energy flows from the source to the inductance. The power is negative the other half of the time, and the energy returns back to source from the inductance. Thus, the average power is zero.

EXERCISE 6.11

Sketch $v(t)$, $i(t)$, and $p(t)$ in Equation (6.62) and compare these three functions. What is the period of $p(t)$ compared to $v(t)$ and $i(t)$? Theoretically calculate and compare the period of these three functions.

3. *A capacitance load:* **voltage, current, and power**
   Next, consider a circuit with the capacitive load shown in Figure 6.42.

   In this case, if the voltage is $v(t) = V_m \cos(\omega t)$, then the current phasor will be:

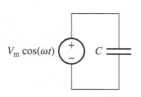

$V_m \cos(\omega t)$ C

FIGURE 6.42   A capacitance load.

$$I = \frac{V}{Z} = \frac{V_m \angle 0°}{\dfrac{1}{j\omega C}} = \frac{V_m \angle 0°}{\dfrac{1}{\omega C}\angle -90°} = I_m \angle 90° \tag{6.63}$$

where

$$I_m = \frac{V_m}{\dfrac{1}{\omega C}} = \frac{V_m}{|Z|}, \quad Z = \frac{1}{j\omega C} = \frac{1}{\omega C}\angle -90°$$

Accordingly:

$$i(t) = I_m \cos(\omega t + 90°) = -I_m \sin(\omega t) \tag{6.64}$$

Thus, the instantaneous power corresponds to:

$$p(t) = v(t)i(t) = -V_m I_m \cos(\omega t)\sin(\omega t) = -\frac{V_m I_m}{2} \sin(2\omega t) \tag{6.65}$$

In this case, the phase of the current is 90°, and the phase of the voltage is 0°. Therefore, the current leads the voltage by 90°. The instantaneous power corresponds to Equation (6.65). Therefore, the power is negative in one-half of the cycle—when the energy flows from the source to the capacitance. It is then positive in the next half cycle—when the energy returns back to the source from the capacitance. Accordingly, the average power will be zero.

4. *RLC load:* voltage, current, and power

A load that is a combination of resistance, inductance, and capacitance is called a RLC load as shown in Figure 6.43. Here, the RLC load impedance can be represented by two components, the real (resistive) and the imaginary (inductive/capacitive) components. In this case, if the voltage is $v(t) = V_m cos(\omega t + \theta_v)$, then voltage phasor will be:

$$V = V_m \angle \theta_v \tag{6.66}$$

In addition, the current phasor is:

$$I = \frac{V}{Z} = \frac{V_m \angle \theta_v}{|Z| \angle \theta_z} = I_m \angle \theta_i \tag{6.67}$$

where $I_m = V_m / |Z|$, $\theta_i = \theta_v - \theta_z$, and $Z = R + jX$. Accordingly:

$$i(t) = I_m \cos(\omega t + \theta_i) \tag{6.68}$$

Then, the instantaneous power is:

$$p(t) = v(t)i(t) = V_m I_m \cos(\omega t + \theta_v) \cos(\omega t + \theta_i) \tag{6.69}$$

Using the trigonometric equation:

$$\cos(\omega t + \theta_v)\cos(\omega t + \theta_i) = \frac{1}{2}\big[\cos(2\omega t + \theta_v + \theta_i) + \cos(\theta_v - \theta_i)\big] \tag{6.70}$$

The power is calculated to be:

$$p(t) = \frac{V_m I_m}{2} \cos(2\omega t + \theta_v + \theta_i) + \frac{V_m I_m}{2} \cos(\theta_v - \theta_i) \tag{6.71}$$

6.6.1 Average Power

The average power is calculated by averaging the instantaneous power over one period of the input signal. In Equation (6.71), the average value of the first term is zero. Considering the period of sinusoid voltage is T, $\omega = 2\pi/T$.

$$\frac{1}{T}\int_0^T \cos(2\omega t + \theta_v + \theta_i)dt = \frac{1}{T}\left[\frac{1}{2\omega}\sin(2\omega t + \theta_v + \theta_i)\right]_0^T$$

$$= \frac{1}{2\omega T}(\sin(2\omega T + \theta_v + \theta_i) - \sin(\theta_v + \theta_i))$$

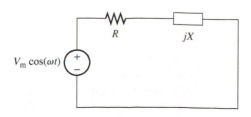

FIGURE 6.43 A general (RLC) load.

$$\left(\text{Note: } \omega = \tfrac{2n}{T}\right) \quad = \frac{1}{2\omega T}\left(\sin\left(2T\frac{2\pi}{T} + \theta_v + \theta_i\right) - \sin(\theta_v + \theta_i)\right)$$

$$= \frac{1}{2\omega T}\left[\sin(2\pi + \theta_v + \theta_i) - \sin(\theta_v + \theta_i)\right]$$

$$= \frac{1}{2\omega T}\left(\sin(\theta_v + \theta_i) - \sin(\theta_v + \theta_i)\right)$$

$$= 0$$

This calculation uses the fact that $\sin(2\pi + \theta) = \sin(\theta)$. As a result, the average power, P, [the unit is watt (W)] is calculated based on the second term of Equation (6.71) and corresponds to:

$$P = \frac{1}{T}\int_0^T p(t)\mathrm{d}t = \frac{V_m I_m}{2}\cos(\theta_v - \theta_i) \tag{6.72}$$

According to Equations (6.13) and (6.15):

$$V_{rms} = \frac{V_m}{\sqrt{2}}, \quad I_{rms} = \frac{I_m}{\sqrt{2}}$$

Using $\theta_z = \theta_v - \theta_i$, defined in Equation (6.67), results in:

$$P = V_{rms}I_{rms}\cos(\theta_v - \theta_i) = V_{rms}I_{rms}\cos(\theta_z) \tag{6.73}$$

6.6.2 Power Factor

Power factor (PF) is defined as:

$$PF = \cos\theta_z = \cos(\theta_v - \theta_i) \tag{6.74}$$

where θ_z is called the **power angle**. Because $\cos(\theta_v - \theta_i) \leq 1$, power factor is expressed as a percentage. This power is indeed the ratio of the power delivered to the load (customer) and the apparent power that is the multiplication of rms voltage and current in the circuit (see Section 6.6.5). The power factor will be **inductive** or **lagging** if current lags voltage (when θ_z is positive because θ_i is less than θ_v); the power factor will be **capacitive** or **leading** if current leads voltage (when θ_z is negative as θ_i is greater than θ_v).

6.6.3 Reactive Power

In an AC circuit, the **peak instantaneous power** (its unit is VAR, volt–ampere–reactive) associated with the energy storage elements (inductance and capacitance) for an RLC load is called **reactive power**, and corresponds to:

$$Q = V_{rms}I_{rms}\sin(\theta_z) \tag{6.75}$$

where θ_z is the power angle introduced in Equation (6.73), and V_{rms} and I_{rms} are the effective voltage and current across the load. Note that:

- For a pure resistive load, $\theta_z = 0$, and $Q = 0$
- For a pure inductive load, $\theta_z = 90°$, and $Q = V_{rms}I_{rms}$
- For a pure capacitive load, $\theta_z = -90°$, and $Q = -V_{rms}I_{rms}$

Accordingly, for a resistive load, the reactive power is zero, while it is maximum (minimum) for an inductive (capacitive) load. Therefore, it is used to express the peak instantaneous power that is associated with the inductance or capacitance. Note that the instantaneous power across a load that contains energy storage elements (inductance and capacitances) varies as voltage and current change.

6.6.4 Complex Power

Complex power is represented by:

$$S = \frac{1}{2}VI^* \tag{6.76}$$

In Equation (6.76), the * refers to the complex conjugate, that is, if $I = I_m e^{j\theta_i}$, then $I^* = I_m e^{-j\theta_i}$. Replacing V with $V_m\angle\theta_v$, and I with $I_m\angle\theta_i$ results in:

$$S = \frac{1}{2}V_m\angle\theta_v \cdot (I_m\angle\theta_i)^*$$

$$= \frac{1}{2}V_m I_m\angle\theta_v - \theta_i$$

Converting the terms from polar to rectangular coordinates:

$$S = \frac{1}{2}V_m I_m \cos(\theta_v - \theta_i) + j\frac{1}{2}V_m I_m \sin(\theta_v - \theta_i)$$

Replacing the maximum voltage and current (V_m and I_m) with their rms values:

$$S = V_{rms}I_{rms} \cos(\theta_v - \theta_i) + jV_{rms}I_{rms} \sin(\theta_v - \theta_i)$$

The real part of the S corresponds to the average power introduced in Equation (6.73), that is:

$$P = \mathrm{Re}\{S\} = \frac{1}{2}V_m I_m \cos(\theta_v - \theta_i)$$

$$= V_{rms}I_{rms} \cos(\theta_v - \theta_i)$$

$$= V_{rms}I_{rms} \cos(\theta_z) \tag{6.77}$$

where $\mathrm{Re}\{\cdot\}$ denotes the real part. The imaginary part of the SP_{comp} corresponds to the reactive power as introduced in Equation (6.75), that is,

$$Q = \mathrm{Im}\{S\} = \frac{1}{2}V_m I_m \sin(\theta_v - \theta_i)$$

$$= V_{rms}I_{rms} \sin(\theta_v - \theta_i)$$

$$= V_{rms}I_{rms} \sin(\theta_z) \tag{6.78}$$

where $\mathrm{Im}\{\cdot\}$ denotes the imaginary part. As a result:

$$S = P + jQ \tag{6.79}$$

Based on Equations (6.76) and (6.79), for complex power:

$$S = \frac{1}{2}V \cdot I^* = V_{ms}I_{ms}(\cos\theta_z + j\sin\theta_z) = P + jQ \tag{6.80}$$

Now, for an RLC circuit, voltage phasor, V, and current phasor, I, have the following relationship:

$$V = Z \times I \tag{6.81}$$

In Equation (6.81):

$$Z = |Z|\angle\theta_z = |Z|e^{j\theta_z} = R + jX \tag{6.82}$$

Substituting Equation (6.82) into Equation (6.81):

$$V = |Z| \times I \times e^{j\theta_z} \tag{6.83}$$

Substituting Equation (6.83) into Equation (6.80):

$$S = \frac{|Z|}{2} \times e^{j\theta_z} \times I \times I^* \tag{6.84}$$

Now, $I \times I^* = |I|^2 = I_m^2$, thus:

$$S = \frac{|Z|}{2} \times I_m^2 \angle \theta_z \tag{6.85}$$

Replacing $|Z| \angle \theta_z$ with its rectangular equivalent $R + jX$:

$$S = \frac{1}{2}RI_m^2 + j\frac{1}{2}XI_m^2 \tag{6.86}$$

In other words, the average (active), P, and reactive power, Q, respectively, correspond to:

$$P = \frac{1}{2}RI_m^2 \tag{6.87}$$

and:

$$Q = \frac{1}{2}XI_m^2 \tag{6.88}$$

From Equation (6.87) it is clear that the active power, P, is a function of resistive loads in the circuit while the reactive power, Q, is a function of capacitive and inductive loads in the circuit.

EXERCISE 6.12

If in Equation (6.80), V is replaced with:

$$V = I|Z| \times e^{j\theta_z} = I|Z| \times e^{j\theta_z} = I(R + jX) \tag{6.89}$$

and prove Equation (6.86).

EXERCISE 6.13

Show that Equation (6.86) can be rewritten as

$$S = RI_{rms}^2 + jXI_{rms}^2 \tag{6.90}$$

and specify the active and reactive components of the complex power.

6.6.5 Apparent Power

Apparent power (its unit is VA, volt–ampere) is defined as the product of the effective voltage and the effective current:

$$\text{Apparent power} = V_{rms}I_{rms} \qquad (6.91)$$

Actually, apparent power is the **maximal average power** that is delivered to the load. This maximum power is delivered when the load is purely resistive.

Based on the definitions in Equations (6.73) and (6.75):

$$P^2 + Q^2 = (V_{rms}I_{rms})^2 \cos^2(\theta_z) + (V_{rms}I_{rms})^2 \sin^2(\theta_z) = (V_{rms}I_{rms})^2 \qquad (6.92)$$

In addition, apparent power represents the amplitude of the complex power. Thus, according to Equation (6.80), a triangle can be used to express the relationship between average power, P, reactive power, Q, apparent power, $V_{rms}I_{rms}$, and the power angle, θ_z, as shown in Figure 6.44. Table 6.1 summarizes all discussed equations.

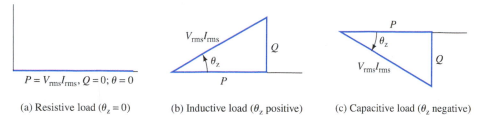

(a) Resistive load ($\theta_z = 0$)   (b) Inductive load (θ_z positive)   (c) Capacitive load (θ_z negative)

FIGURE 6.44 Power triangle.

TABLE 6.1 Summary of Power Definitions

	Formula	Unit
Instantaneous power	$p(t) = v(t)i(t) = V_m I_m \cos(\omega t + \theta_v)\cos(\omega t + \theta_i)$	watt (W)
Average power	$P = V_{rms}I_{rms} \cos(\theta_v - \theta_i) = V_{rms}I_{rms} \cos(\theta_z)$	watt (W)
Reactive power	$Q = V_{rms}I_{rms} \sin(\theta_z)$	VAR
Apparent power	$\text{Apparent power} = V_{rms}I_{rms}$	VA
Complex power	$S = \frac{1}{2}VI^* = V_{rms}I_{rms}(\cos\theta_z + j\sin\theta_z) = P + jQ$	VA

EXAMPLE 6.21 **Average, and Reactive Power and Power Factor**

Consider the circuit shown in Figure 6.45. Compute the average power and reactive power taken from the source and the power factor.

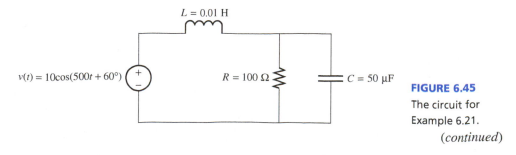

$v(t) = 10\cos(500t + 60°)$

$L = 0.01$ H

$R = 100\ \Omega$

$C = 50\ \mu$F

FIGURE 6.45
The circuit for
Example 6.21.

(*continued*)

EXAMPLE 6.21 **Continued**

SOLUTION

The equivalent circuit for the phasor domain is shown in Figure 6.46.

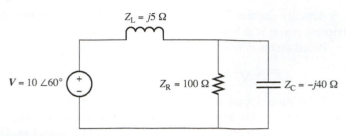

FIGURE 6.46 Equivalent circuit for Example 6.21.

where

$$Z_L = j\omega L = j0.01 \times 500 = j5 \ \Omega$$

$$Z_C = \frac{1}{j\omega C} = \frac{1}{j500 \times 50 \times 10^{-6}} = -j40 \ \Omega$$

The equivalent impedance of the parallel resistor and capacitor is:

$$Z_{RC} = \frac{1}{\dfrac{1}{Z_R} + \dfrac{1}{Z_C}} = \frac{1}{0.01 + j0.025} = \frac{1\angle 0°}{0.0269\angle 68.2°} = 37.17\angle -68.2° \ \Omega$$

In rectangular form:

$$Z_{RC} = 13.80 - j34.51 \ \Omega$$

The total impedance seen by the voltage source is $Z_L + Z_{RC}$. In addition, the current through the circuit is:

$$I = \frac{V}{Z_L + Z_{RC}} = \frac{10\angle 60°}{j5 + 13.80 - j34.51}$$

By simplifying:

$$I = \frac{10\angle 60°}{13.80 - j29.51}$$

$$= \frac{10\angle 60°}{32.58\angle -64.94°}$$

$$= 0.3069\angle 124.94°$$

Therefore, the power angle is:

$$\theta_z = \theta_v - \theta_i = 60° - 124.94° = -64.94°$$

and the power factor is:

$$PF = \cos\theta_z = 0.4236$$

The effective value of voltage and current are:

$$V_{rms} = \frac{V_m}{\sqrt{2}} = \frac{10}{\sqrt{2}} = 7.071 \ V$$

$$I_{rms} = \frac{I_m}{\sqrt{2}} = \frac{0.3069}{\sqrt{2}} = 0.217 \ A$$

Using Equations (6.73) and (6.75), the average power and reactive power correspond to:

$$P = V_{rms}I_{rms}\cos\theta_z = 7.071 \times 0.217 \times \cos(-64.94°) = 0.6499 \text{ W}$$

$$Q = V_{rms}I_{rms}\sin\theta_z = 7.071 \times 0.217 \times \sin(-64.94°) = -1.39 \text{ VAR}$$

APPLICATION EXAMPLE 6.22 Chemical Heater

An electric heater is used to control the temperature of a chemical solution. The heater is required to heat evenly with precise temperature control. The heater element is a coil of resistive material, similar to that in an electric toaster. If the circuit in Figure 6.47 is the equivalent of the heater element, find the average power that is consumed by the heater.

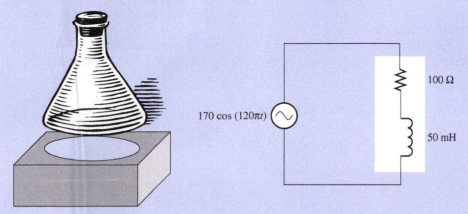

FIGURE 6.47 Chemical heater.

SOLUTION

The equivalent circuit of Figure 6.47 is shown in Figure 6.48.

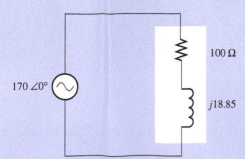

FIGURE 6.48 Equivalent circuit for Example 6.22.

The complex power corresponds to:

$$P_{comp} = \frac{1}{2}VI^* = \frac{1}{2}V\frac{V^*}{Z^*} = \frac{1}{2}\frac{V_m^2}{Z^*}$$

$$= \frac{1}{2}\frac{170^2}{100 - j18.85}$$

(continued)

$$= \frac{14{,}450}{101.76\angle -10.68°}$$

$$= 142.0\angle 10.68° = 139.54 + j26.3$$

Therefore, the average power consumed by the heater is 139.54 W, which is the real part of the complex power.

6.6.6 Maximum Average Power Transfer

Given a circuit network, the maximal average power delivered to the load by the source is a function of the load impedance. The circuit network can be represented by its Thévenin equivalent, as shown in Figure 6.49. Observe that in general, the Thévenin equivalent circuit includes a voltage source and an impedance.

The Thévenin impedance is $Z_{th} = R_{th} + jX_{th}$, and the impedance of the load is $Z_L = R_L + jX_L$. Therefore, the total impedance is $Z_{tot} = R_{th} + R_L + j(X_{th} + X_L)$ and its magnitude is:

$$|Z_{tot}| = \sqrt{(R_{th} + R_L)^2 + (X_{th} + X_L)^2}$$

The current flowing through the load is:

$$I_{out} = \frac{V_{out}}{Z_L} \tag{6.93}$$

Using the voltage division formula:

$$V_{out} = \frac{Z_L}{Z_{th} + Z_L} \cdot V_s = \frac{Z_L}{Z_{tot}} \cdot V_s \tag{6.94}$$

The complex power delivered to the load is (see Appendix C for additional information about the calculations in this section):

$$S = \frac{1}{2} V_{out} I_{out}^*$$

Replacing I_{out} using Equation (6.93) results in:

$$S = \frac{1}{2} \frac{|V_{out}|^2}{Z_L^*} \tag{6.95}$$

Replacing V_{out} using Equation (6.94):

$$S = \frac{1}{2} \left| \frac{Z_L}{Z_{tot}} \right|^2 \cdot \frac{|V_s|^2}{Z_L^*} \tag{6.96}$$

Note that $|V_s| = V_m$. Multiplying by Z_L in both the numerator and denominator of Equation (6.96) yields:

$$S = \frac{1}{2} Z_L \frac{V_m^2}{|Z_{tot}|^2}$$

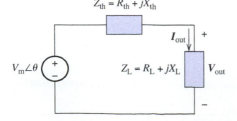

FIGURE 6.49
Computation of the maximal average power transfer to a load.

The average delivered power is the real part of P_{comp}, which is:

$$P_{av} = \frac{R_L}{2((R_{th} + R_L)^2 + (X_{th} + X_L)^2)} V_m^2 \tag{6.97}$$

Now, to find the Z_L that maximizes the transferred power, find the derivative of P_{av} with respect to R_L and X_L, and equate it to zero. The Z_L that maximizes the average power is:

$$Z_L = Z_{th}^* = R_{th} - jX_{th} \tag{6.98}$$

In other words, $R_L = R_{th}$, and $X_L = -X_{th}$. Accordingly, the maximal average power is:

$$P_{max} = \frac{V_m^2}{8R_{th}} \tag{6.99}$$

EXERCISE 6.14

Take the derivative of Equation (6.97) with respect to R_L and X_L and verify Equation (6.98). Then, use Equation (6.98) to verify Equation (6.99).

If the load is purely resistive, that is, $X_L = 0$, using Equation (6.99):

$$P = \frac{V_m^2 R_L}{2(R_{th}^2 + 2R_{th}R_L + R_L^2 + X_{th}^2)} \tag{6.100}$$

Dividing the numerator and denominator of Equation (6.100) by R_L:

$$P = \frac{V_m^2}{2\left(\dfrac{R_{th}^2 + X_{th}^2}{R_L} + R_L + 2R_{th}\right)}$$

Now, to find Z_L that maximizes the transferred power, find R_L that maximizes the power. Alternatively, find R_L that minimizes the denominator. Taking the derivative of the denominator with respect to R_L:

$$R_L^2 = R_{th}^2 + X_{th}^2$$

Therefore, for the pure resistance load, taking:

$$R_L = \sqrt{R_{th}^2 + X_{th}^2} = |Z_{th}| \tag{6.101}$$

the load can attain the maximal average power:

$$P_{max} = \frac{V_m^2}{4(\sqrt{R_{th}^2 + X_{th}^2} + R_{th})} \tag{6.102}$$

Comparing Equation (6.99) to Equation (6.102), it is clear that the power delivered to a pure resistance load is less than that delivered to a complex one because:

$$\frac{V_m^2}{8R_{th}} \geq \frac{V_m^2}{4(\sqrt{R_{th}^2 + X_{th}^2} + R_{th})} \tag{6.103}$$

In Equation (6.103), equality holds when $X_{th} = 0$, that is, the Thévenin impedance is also a pure resistance.

Note: If the load impedance is given and the equivalent R_{th} needs to be adjusted, taking R_{th} as small as possible, maximizes the maximum power transfer (see Example 6.25).

EXERCISE 6.15

Verify the inequality of Equation (6.103).

EXAMPLE 6.23   **Maximum Delivered Power**

For the circuit shown in Figure 6.50, determine the maximum average power that can be delivered to the load. Also, determine the maximum average power that can be delivered to a purely resistive load.

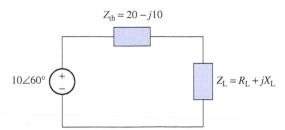

FIGURE 6.50 Circuit for Example 6.23.

SOLUTION

If the load is not purely resistive, using Equations (6.98) and (6.99):

$$Z_L = Z_{th}^* = 20 + j10$$

$$P_{max} = \frac{V_m^2}{8R_{th}} = \frac{100}{8 \times 20} = 0.625 \text{ W}$$

If the load is purely resistive, using Equations (6.101) and (6.102)

$$R_L = |Z_{th}| = \sqrt{400 + 100} = 22.3607, \quad X_L = 0$$

$$P_{max} = \frac{V_m^2}{4(\sqrt{R_{th}^2 + X_{th}^2} + R_{th})} = \frac{100}{4(\sqrt{400 + 100} + 20)} = 0.5902 \text{ W}$$

6.6.7 Power Factor Correction

Equation (6.73) represents the total power consumed by the consumer and apparent power delivered to the consumer. The consumer is indeed charged for the power that it consumes. However, the power plant generates the power that is consistent with the apparent power in Equation (6.91). Power factor ($\cos\theta_z$) maintains the relationship of Equations (6.91) and (6.73).

Now, if the power factor is small (i.e., the case for inductive loads such as motors), the power company creates more power than consumed by the customer. To avoid this problem, a power factor correction is needed. That is we should increase the power factor to 1. For example, let's say a customer has a load of 740 W, but the apparent power is 1000 VA. The power company can only bill the customer for the 740 W used even though they have to have an infrastructure to supply 1000 VA to the customer. Applying power factor correction, then the power company only has to supply 740 VA to the customer which they can fully bill since the customer is consuming 740 W.

In industrial plants, usually, a heavy inductive load (e.g., motors) increases the reactive power. When reactive power, Q, increases, the current flowing from the source to the load increases according to Equation (6.88) (see the circuit example shown in Figure 6.51). Consequently, the power factor ($\cos\theta_z$) decreases because θ_z increases when reactive power

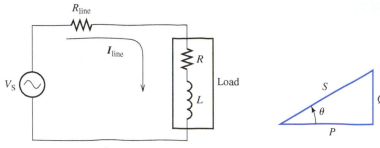

FIGURE 6.51 Circuit example.　　　　**FIGURE 6.52** Power triangle.

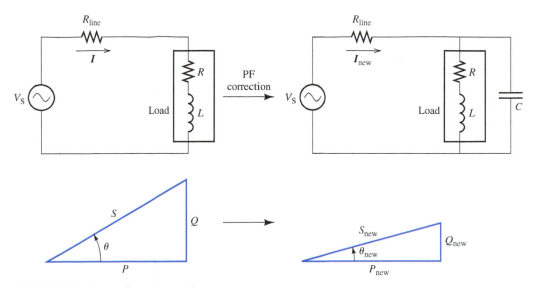

FIGURE 6.53 Power factor correction.

($\sin\theta_z$) increases. According to the power triangle shown in Figure 6.52, to increase the power factor, the reactive power should be reduced.

This can be easily achieved using the fact that this problem is mainly due to the inductance of motor windings. In addition, it is known that the effect of the positive impedance of an inductive load ($X_L = j\omega L$) can be compensated using the negative impedance of a capacitive load ($X_C = 1/j\omega C = -j/\omega C$). In other words, if a load contains inductance and capacitance with reactive powers of equal magnitude, their reactive powers will cancel. Therefore, a general approach to reduce the reactive power is to add capacitors in parallel with an inductance. This increases the power factor. This fact is illustrated in Figure 6.53. In Figure 6.53, $P = P_{new}$, thus, when Q is reduced to Q_{new}, the θ_{new} decreases. Accordingly, the power factor increases.

EXAMPLE 6.24　　Power Factor Correction Via Capacitor

A 60-W load powered by a 60-Hz, 240-V line with a power factor of 0.7071 lag is shown in Figure 6.54. Calculate the parallel capacitance required to correct the power factor to 0.9 lag.

(continued)

EXAMPLE 6.24   **Continued**

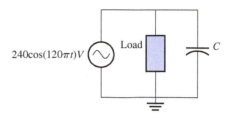

FIGURE 6.54   Circuit for Example 6.24.

SOLUTION

In order to solve this problem, find the load power angle:

$$\theta_L = \arccos(0.7071) = 45°$$

Use the power triangle in Figure 6.52 to find the reactive power Q_L.

$$Q_L = P_L \tan(\theta_L) = 60 \times \tan(45°) = 60 \text{ VAR}$$

After adding the parallel capacitor, the new power angle is:

$$\theta_{new} = \arccos(0.9) = 25.84°$$

The new value of the reactive power is:

$$Q_{new} = P_L \tan(\theta_{new}) = 29.06 \text{ VAR}$$

Note that $P_L = 60$ W is a function of the resistive elements in the circuit and does not change after adding the capacitor.

The reactive power due to the addition of the capacitance is:

$$Q_C = Q_{new} - Q_L = -30.94 \text{ VAR}$$

The reactance of the capacitor is:

$$X_C = \frac{V_{rms}^2}{Q_C} = \frac{(240/\sqrt{2})^2}{-30.94} = -930.83 \; \Omega$$

The angular frequency is:

$$\omega = 2\pi f = 2\pi \times 60 = 376.99$$

The required capacitance is then found to be:

$$C = \frac{1}{\omega |X_C|} = \frac{1}{376.99 \times 930.83} = 2.85 \; \mu\text{F}$$

APPLICATION EXAMPLE 6.25   **Heart–Lung Machine**

A heart–lung machine has an electric pump that takes over for a biological heart. Suppose a pump for a heart–lung machine is connected as shown in Figure 6.55. Determine:

a. The impedance Z needed for maximum power transfer to the pump
b. The circuit component values needed for maximum power transfer

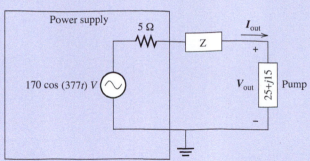

SOLUTION

a. The complex power consumed by the pump can be written as:

$$P_{comp} = \frac{1}{2} V_{out} I_{out}^*$$

By Ohm's law:

$$I_{out} = \frac{V_{out}}{25 + j15}$$

Therefore, using Equation (6.95) P_{comp} becomes:

$$P_{comp} = \frac{1}{2} \frac{|V_{out}|^2}{25 - j15} = \frac{1}{2} \frac{|V_{out}|^2 (25 + j15)}{(25 - j15)(25 + j15)} = \frac{1}{2} \frac{|V_{out}|^2 (25 + j15)}{|25 + j15|^2}$$

The average delivered power to the pump is the real part of P_{comp}, which is:

$$P_{av} = \frac{1}{2} \frac{|V_{out}|^2\, 25}{|25 + j15|^2}$$

Observing the expression of P_{av}, the power delivered to the pump is maximized when $|V_{out}|^2$ is maximized. By using the voltage division rule, we find:

$$V_{out} = \frac{25 + j15}{5 + Z + 25 + j15} 170\angle 0°$$

Therefore, $|V_{out}|^2$ is maximized when $|5 + Z + 25 + j15|^2$ is minimized. Note that the impedance, Z, always has a positive (or zero) real part. Any positive value for Z makes $|V_{out}|$ smaller. In order to maximize $|V_{out}|^2$ and P_{av}, we select $Z = -j15$.

b. In order to determine the components that make up Z, look at the value of Z. Because the imaginary part of Z is negative, it must be capacitive. Therefore, Z is made by a capacitor. The impedance of the capacitor is $-j15$ and $Z = 1/(j\omega C_z)$, and $\omega = 377$. Thus:

$$C_z = \frac{1}{j\omega Z} = \frac{1}{j377 \times (-j15)} = 176.83\ \mu F$$

The process of conversion and the final circuit are shown in Figure 6.56.

(continued)

APPLICATION EXAMPLE 6.25 Continued

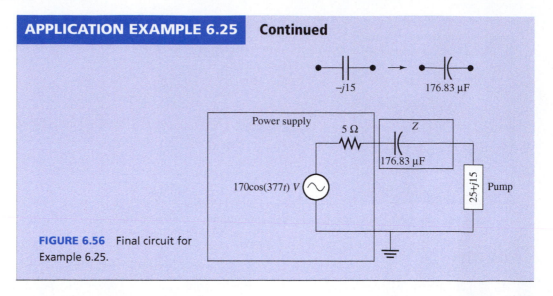

FIGURE 6.56 Final circuit for Example 6.25.

EXERCISE 6.16

Find an equivalent circuit for the following impedances:

   a. $12 - j15$, $\omega = 10$
   b. $20 + j\,25$, $\omega = 25$
   c. $+j110$, $\omega = 220\pi$

APPLICATION EXAMPLE 6.26 Elevator

A large AC motor is used to operate an elevator as shown in Figure 6.57. It is desired that the motor use the least amount of current possible. A capacitor C_{PF} is included for power factor correction due to the motor inductance. If $\omega = 377$ rad/s, determine the current I, if $C_{PF} = 150$ μF.

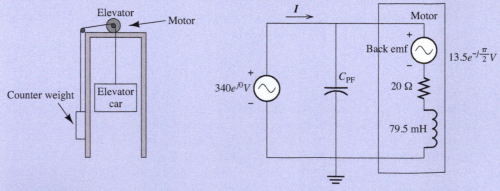

FIGURE 6.57 An elevator and its circuit.

SOLUTION

The equivalent circuit elements are shown in Figure 6.58. The impedance of the capacitor is calculated using $Z = 1/(j\omega C_{PF}) = -j17.68$. Note that the current, I, is the sum of the

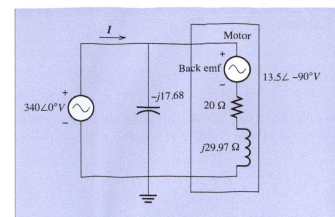

FIGURE 6.58 Equivalent circuit elements for Example 6.26.

two currents flowing through the two parallel branches. The current through the capacitor is $\dfrac{340}{-j17.68}$, and the current through the motor is:

$$I_{\text{motor}} = \frac{340 - 13.5\angle -90°}{20 + j29.97}$$

Therefore, the total current is:

$$I = \frac{340 - 13.5\angle -90°}{20 + j29.97} + \frac{340}{-j17.68}$$

Some simplifications (detailed below) arrive at the final answer:

$$
\begin{aligned}
I &= \frac{340 - 13.5\angle -90°}{20 + j29.97} + \frac{340}{-j17.68} \\[2mm]
&= \frac{340 + j13.5}{20 + j29.97} + j19.23 \\[2mm]
&= \frac{340.27\angle 2.27°}{36.03\angle 56.28°} + j19.23 \\[2mm]
&= 9.44\angle -54.01° + j19.23 \\[2mm]
&= 5.55 - j7.64 + j19.23 \\[2mm]
&= 5.55 + j11.59 \\[2mm]
&= 12.85\angle 64.4°
\end{aligned}
$$

In the time domain:

$$i(t) = 12.85 \cos(377t + 64.41°)$$

6.7 STEADY-STATE CIRCUIT ANALYSIS USING PSPICE

This section outlines the use of PSpice for steady-state circuit analysis. The tutorial and examples in Chapter 2 as well as examples in other chapters also help to develop this analysis.

EXAMPLE 6.27   **PSpice Analysis**

Use the values obtained in part (a) of Example 6.3 to set up a PSpice schematic and plot the voltage across a resistor, with the resistance of 100 Ω.

SOLUTION

To set up a sinusoidal voltage source (see Example 2.24), first determine the frequency of the source. Given the voltage source, $v(t) = 5\cos(120\pi + 30°)$, the frequency is:

$$f = \frac{120\pi}{2\pi}$$

$$= 60 \text{ Hz}$$

Next, determine the phase angle of the sinusoidal voltage source. The sinusoidal voltage source function in PSpice is "sinus." Therefore, a default phase angle of $+90°$ is needed if the given voltage source is a "cosine" function. Then, the phase angle is:

$$\theta = 90° + 30°$$

$$= 120°$$

Next, the PSpice schematic can be set up using the following steps:

1. Example 2.24 shows the process of adding a sinusoidal voltage source, "VSIN" into a circuit. Set VOFF = 0, VAMPL = 5, FREQ = 60. VOFF is the offset voltage, VAMPL is the amplitude or the peak of the voltage, and FREQ is the frequency of the voltage.
2. To set the phase angle of the voltage source, go to "edit properties" of voltage source by right-clicking on it, or double-click on the voltage source. A window similar to Figure 6.59 will appear.
3. Set the value under "PHASE" (see the black arrow) to 120. Also, go to "Display..." (see the colored arrow) and choose "Display both Name and Value."
4. Add the resistor and ground (GND) into the circuit.
5. Create a simulation profile, with the analysis type set as "Time Domain (Transient)." Set the simulation in between 50 and 100 ms and run the simulation (see Figure 2.75).
6. The PSpice schematic is shown in Figure 6.60.
7. The voltage across the resistor plot is shown in Figure 6.61.

FIGURE 6.59 Set the phase angle.

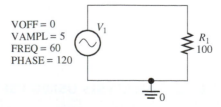

FIGURE 6.60 PSpice schematic solution for Example 6.3.

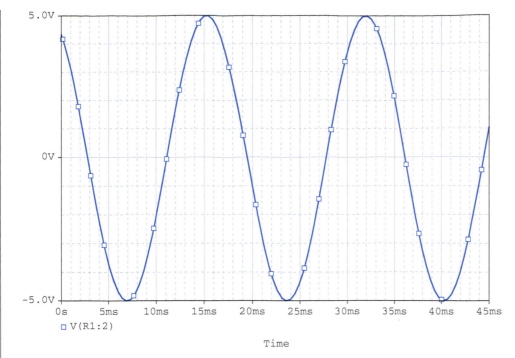

FIGURE 6.61 Sinusoidal voltage source across the resistor plot.

EXAMPLE 6.28   **PSpice Analysis**

Use PSpice to find the steady-state current $i(t)$ and $v_L(t)$ shown in the figure for Example 6.13.

SOLUTION

First, determine the frequency and the phase angle of the voltage source. The voltage source is: $v(t) = 110 \cos(400t)$. The frequency is then:

$$f = \frac{400\pi}{2\pi}$$
$$= 63.662 \text{ Hz}$$

The sinusoidal voltage source is a cosine function; therefore, the phase angle is not zero for the PSpice sinusoidal voltage source, which is:

$$\theta = 90°$$

Add the sinusoidal voltage source, VSIN, into the PSpice schematic, with VOFF = 0, VAMPL set to 110, and FREQ set to 63.662. The phase angle can be set by going to the "edit properties" of VSIN (see Figure 6.59).

Create a new simulation profile and set the analysis type to be "Time Domain (Transient)." Set the simulation time between 50 and 100 ms and run the simulation (see Figure 2.71). The PSpice schematic solution is shown in Figure 6.62, and the plots of $i(t)$ and $v_L(t)$ are shown in Figures 6.63 and 6.64, respectively.

(continued)

| EXAMPLE 6.28 | Continued |

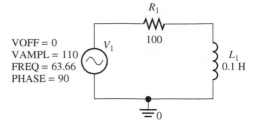

FIGURE 6.62 PSpice schematic solution for Example 6.13.

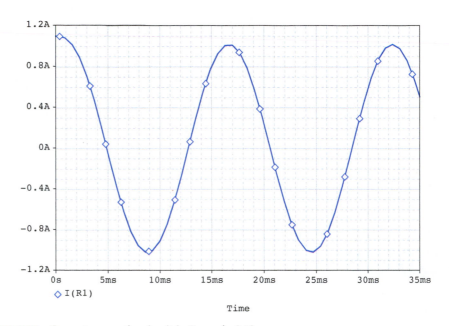

FIGURE 6.63 Current across the circuit in Example 6.13.

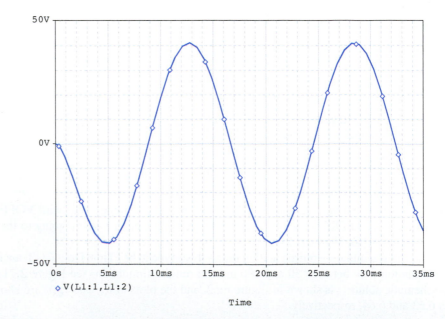

FIGURE 6.64 Voltage across the inductor in Example 6.13.

EXAMPLE 6.29	**PSpice Analysis**

Considering the circuit shown in Figure 6.64 for Example 6.15, set up a PSpice schematic plot for the $v_C(t)$ in steady state.

SOLUTION

First, determine the frequency and phase angle of the voltage source. Given that the voltage source is $v(t) = 110 \cos(400t + 60°)$, the frequency is:

$$f = \frac{400\pi}{2\pi}$$
$$= 63.662 \text{ Hz}$$

Because the sinusoidal voltage source is a cosine function, the phase angle is:

$$\theta = 60° + 90°$$
$$= 150°$$

Add the sinusoidal voltage source, VSIN, into the PSpice schematic, with VOFF = 0, VAMPL set to 110 and FREQ set to 63.662. The phase angle can be set by going to the "edit properties" of VSIN (see Figure 6.59).

Add the resistor, inductor, and capacitor into the circuit as shown in Figure 6.65. Set the initial condition for both the inductor and the capacitor to zero (see Example 4.16).

Set the simulation time to run for 100 ms (see Figure 2.75). The voltage across capacitor, C1, is shown in Figure 6.66.

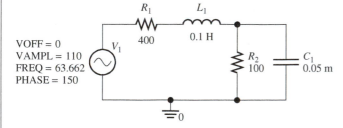

FIGURE 6.65 PSpice schematic solution for Example 6.15.

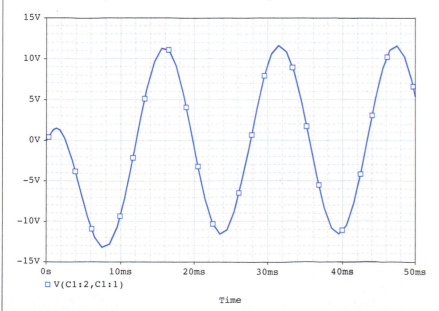

FIGURE 6.66 Voltage across the capacitor, C1, for Example 6.15.

APPLICATION EXAMPLE 6.30   Controlling Building Sway

In Example 6.20, find the Thévenin voltage as seen by the load using PSpice.

SOLUTION

First, determine the frequency and phase angle of the voltage source. Given that the voltage source, $v(t) = 2.5 \cos(500t)$, the frequency is:

$$f = \frac{500\pi}{2\pi}$$
$$= 79.577 \text{ Hz}$$

Because the sinusoidal voltage source is a cosine function, the phase angle is:

$$\theta = 90°$$

Add the sinusoidal voltage source, VSIN, into the PSpice schematic, with VOFF set to 0, VAMPL set to 2.5 and FREQ set to 79.577. The phase angle can be set by going to the "edit properties" of VSIN (see Example 6.29).

The PSpice schematic solution is shown in Figure 6.67. Notice that there is no inductor in Figure 6.67. Because the goal is to observe the Norton equivalent voltage across the series resistor and inductor, both the resistor and the inductor are replaced by a large value of resistance (100 GΩ). Also, a "Voltage Differential Marker(s)" (see Figure 6.68, noted by the arrow) is placed around the Norton resistance, so that PSpice will plot the voltage across that resistor automatically.

Set the initial condition of the capacitors to 0 V (see Example 4.16). Set the simulation time to 100 ms and press the run button (see Figure 2.71). The Norton equivalent voltage is plotted in Figure 6.69.

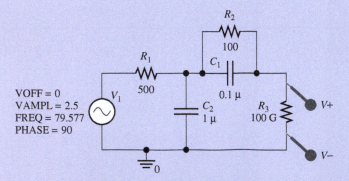

FIGURE 6.67 PSpice schematic solution for Example 6.20.

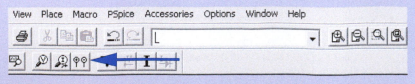

FIGURE 6.68 Voltage differential marker(s).

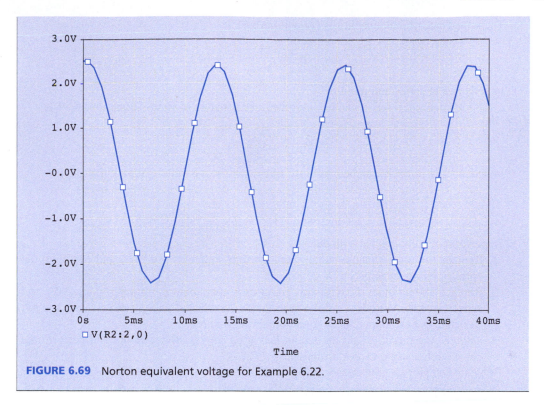

FIGURE 6.69 Norton equivalent voltage for Example 6.22.

6.8 WHAT DID YOU LEARN?

- The sinusoidal function $v(t) = V_{\mathrm{m}} \cos(\omega t + \theta)$ has an amplitude of V_{m}, a radian frequency of ω, a period of $2\pi/\omega$, and a phase angle of θ.
- The *effective value* of the sinusoidal voltage $v(t) = V_{\mathrm{m}} \cos(\omega t + \theta)$ as shown in Equation (6.13):

$$V_{\mathrm{rms}} = \frac{V_{\mathrm{m}}}{\sqrt{2}}$$

- The *instantaneous power* delivered by the voltage source to the resistor, R, corresponds to Equation (6.16):

$$p(t) = v(t) \cdot i(t)$$

The *average power* corresponds to Equation (6.11):

$$P_{\mathrm{av}} = \frac{V_{\mathrm{m}}^2}{2R}$$

- The *sinusoidal voltage*, $v(t) = V_{\mathrm{m}} \cos(\omega t + \theta)$, can be written in phasor form as shown in Equation (6.25):

$$\mathbf{V} = V_{\mathrm{m}} \angle \theta = V_{\mathrm{m}} e^{j\theta}$$

- The impedance of R, L, and C are explained as follows: The impedance of the resistor is [Equation (6.36)]:

$$Z_{\mathrm{R}} = R$$

The impedance of the inductor is [Equation (6.43)]:

$$Z_{\mathrm{L}} = j\omega L$$

The impedance of the capacitor is [Equation (6.51)]:

$$Z_C = \frac{1}{j\omega C}$$

- Impedances in series and parallel:

  Impedances in *series* can be combined like resistors in series [Equation (6.53)]:

$$Z_{eq} = \sum_{k=1}^{n} Z_k$$

Impedances in *parallel* can be combined like resistors in parallel [Equation (6.54)]:

$$\frac{1}{Z_{eq}} = \frac{1}{Z_1} + \frac{1}{Z_2} + \cdots \frac{1}{Z_n}$$

- To perform circuit analysis with phasors:
  1. Replace the instantaneous voltage and current source(s) [i.e., $v(t)$ and $i(t)$] with the corresponding phasors (i.e., V and I)
  2. Replace inductance, L, with the complex impedance $Z_L = j\omega L = \omega L\angle 90°$, and capacitance, C, with the complex impedance $Z_C = 1/j\omega C = -j1/\omega C = 1/\omega C\angle -90°$. Resistances have impedances equal to their resistances
  3. Analyze the circuit using the techniques used for resistive circuits and perform the calculation using complex arithmetic
- Thévenin and Norton equivalent circuits with phasors:

  Thévenin voltage and impedance

From the perspective of any load, the Thévenin voltage, V_{th}, equals the open-circuit voltage, V_{AB}, of the circuit. The Thévenin impedance, Z_{th}, seen through two terminals is found by setting the independent sources to zero.

  Norton current and impedance

The Norton current can be found by equating the load to zero. The Norton impedance, Z_n, is the same as the Thévenin impedance, Z_{th}, and can be computed in the same fashion.
- Power definitions (see Table 6.1).
- Maximum average power transfer:

  Given a circuit network, if the Thévenin impedance is $Z_{th} = R_{th} + jX_{th}$, and the impedance of the load is $Z_L = R_L + jX_L$, then the Z_L that maximizes the average power is [Equation (6.97)]:

$$Z_L = Z_{th}^* = R_{th} - jX_{th}$$

Accordingly, the maximal average power is [Equation (6.98)]:

$$P_{max} = \frac{V_m^2}{8R_{th}}$$

If the load is pure resistance, $X_L = 0$, then the R_L that maximizes the average power is:

$$R_L = \sqrt{R_{th}^2 + X_{th}^2} = |Z_{th}|$$

The maximal corresponding average power is then:

$$P_{max} = \frac{V_m^2}{4(\sqrt{R_{th}^2 + X_{th}^2} + R_{th})}$$

- The power factor (PF) is defined as [Equation (6.74)]:

$$PF = \cos\theta_z = \cos(\theta_v - \theta_i)$$

The power factor will be *inductive* or *lagging* if current lags voltage (when θ_z is positive); the power factor will be *capacitive* or *leading* if current leads voltage.

- *Power factor correction*

In order to increase the power factor and reduce the reactive power, a capacitor can be added in parallel with the inductor.

Problems

*B refers to Basic, A refers to Average, H refers to Hard, and * refers to problems with answers.*

SECTION 6.1 SINUSOIDAL VOLTAGES AND CURRENTS

6.1 (B)* Determine e^{2+3j}. Provide the answer to two decimal places.

6.2 (B) Determine $e^{3j} \cdot e^{1-j}$. Provide the answer to two decimal places.

6.3 (B)* A sinusoidal voltage is given by:

$$v(t) = 100 \sin(100\pi t + 30°)$$

Find the peak value of it and the voltage at $t = 10$.

6.4 (B) Find the amplitude, phase, period, and frequency of the sinusoid voltage:

$$v(t) = 10 \cos(110\pi t + 60°)$$

6.5 (B) Find the period of the following sinusoids:

  a. $v(t) = \cos(2\omega t)$
  b. $v(t) = \sin(3t)$
  c. $v(t) = \sin(t + \theta)$
  d. $v(t) = \cos(2\omega t + \theta) + \sin(2\omega t + \phi)$
  e. $v(t) = \cos(t/2 + \theta) + \sin(t/3 + \phi)$

6.6 (B) Write the following sine functions in terms of cosine:

  a. $v(t) = \sin(100t + 30°)$
  b. $v(t) = \sin(50°)$
  c. $v(t) = \sin(115°)$
  d. $v(t) = \sin(\omega t + \theta)$
  e. $v(t) = \sin(\omega t + \pi - \theta)$

6.7 (B) Write the following cosine functions in terms of sine:

  a. $v(t) = \cos(100\pi t + 10°)$
  b. $v(t) = \cos(t + 150°)$
  c. $v(t) = \cos(195°)$
  d. $v(t) = \cos(\omega t + \theta)$
  e. $v(t) = \cos(\omega t - \theta)$

6.8 (B)* A sinusoidal voltage is given by:

$$v(t) = 156 \cos(120\pi t + 45°)$$

Find the peak value and rms value of the sinusoid voltage.

6.9 (B) A sinusoidal current $i(t) = 2 \cos(120\pi t + 60°)$ passes a 100-Ω resistance; find the rms value of the sinusoid current.

6.10 (B) A sinusoidal voltage is given by:

$$v(t) = 200 \cos(120\pi t + 45°)$$

Find the average power delivered to a 1.5-kΩ resistance.

6.11 (A)* Given a sinusoidal voltage source that is $v_s(t) = 6\cos(20t) + 2$ V, what is the rms voltage of this voltage source in a period cycle?

6.12 (A) Sketch the following sinusoidal voltage source in a period cycle with appropriate label on the axes and their cross sections, $v_s(t) = 4 \cos((\pi/5)t + 30)$ V.

6.13 (H)* A voltage source plot is shown in Figure P6.13. Determine the period and function of this sinusoidal voltage.

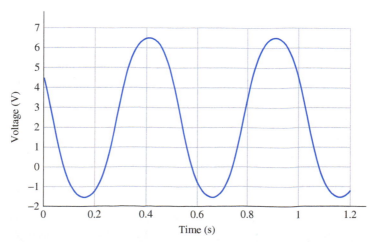

FIGURE P6.13   Voltage plot for Problem 6.13.

SECTION 6.2 PHASORS

6.14 (B) Find the phasor for the sinusoid voltages $v_1(t) = 156 \cos(110\pi t + 60°)$ and $v_2(t) = 220 \sin(120\pi t - 45°)$.

6.15 (B) Find the phasor for the sinusoid current $i(t) = 3 \cos(120\pi t + 20°)$.

6.16 (A) Comment on the phasor between these two voltage sources, $v_1(t) = 110 \cos(120\pi t + 60°)$ and $v_2(t) = 110 \cos(100\pi t + 60°)$.

6.17 (A)* Two sinusoid voltage signals are given by $v_1(t) = 100 \cos(120\pi t + 30°)$, and $v_2(t) = 250 \cos(120\pi t + 60°)$. Use phasors to compute $v_a = v_1(t) + v_2(t)$.

6.18 (A) Suppose that $v_1(t) = 100 \cos(120\pi t + 30°)$ and $v_2(t) = 250 \cos(120\pi t + 60°)$, use phasors to find $v_d = v_1(t) - v_2(t)$.

6.19 (A) Given that two phasor voltage source are $V_1 = 20\angle 15°$ V and $V_2 = 10\angle 45°$ V, find the sinusoidal voltage expression of $V_1 + V_2$. Express the solution to two decimal places.

SECTION 6.3 COMPLEX IMPEDANCES

6.20 (B)* What is the total impedance of the circuit in the Figure P6.20, if the frequency of the circuit is $200/\pi$ Hz? Provide the answer to two decimal places.

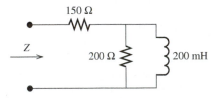

FIGURE P6.20 Circuit for Problem 6.20.

6.21 (B) Find Z in Figure P6.21 for $\omega = 200$, $\omega = 500$, and $\omega = 1000$.

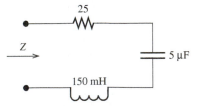

FIGURE P6.21 Circuit for Problem 6.21.

6.22 (B) Find Z in Figure P6.22 for $\omega = 200$, $\omega = 800$, and $\omega = 1500$.

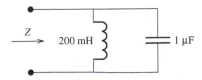

FIGURE P6.22 Circuit for Problem 6.22.

6.23 (A) If $\omega = 600$, find Z in Figure P6.23.

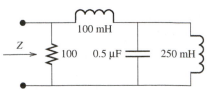

FIGURE P6.23 Circuit for Problem 6.23.

6.24 (A)* If $\omega = 250$, find Z in Figure P6.24.

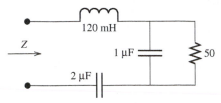

FIGURE P6.24 Circuit for Problem 6.24.

6.25 (A) Find Z in Figure P6.25.

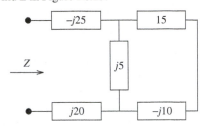

FIGURE P6.25 Circuit for Problem 6.25.

6.26 (A) Find Z in Figure P6.26.

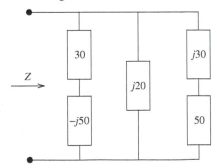

FIGURE P6.26 Circuit for Problem 6.26.

6.27 (H) Determine the power consumed by the inductor for the circuit in Figure P6.27 if the voltage source is $15 \cos(200t)$. Provide the solution up to two decimal places.

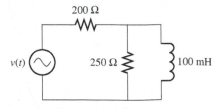

FIGURE P6.27 Circuit for Problem 6.27.

6.28 (A)* Sketch a simplified circuit for the circuit in Figure 6.28. Then, calculate the total impedance for the circuit, if the voltage source is $v_S(t) = V \cos(200t + 30°)$ V.

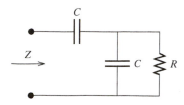

FIGURE P6.28   Circuit for Problem 6.28.

6.29   (H) What is the angular rate, ω, for the circuit in Figure P6.29, if the total impedance is $48.4125 - j137.5 \ \Omega$, the resistance R is $77.46 \ \Omega$, and capacitance C is $50 \ \mu F$?

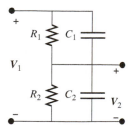

FIGURE P6.29   Circuit for Problem 6.29.

6.30   (H)* Application: voltage divider
In the circuit shown in Figure 6.30, the output and input voltage ratio is always

$$\frac{V_2}{V_1} = \frac{R_2}{R_1 + R_2}.$$

Find the relationship of R_1, R_2, C_1, and C_2.

FIGURE P6.30   Circuit for Problem 6.30.

6.31   (H) Find the impedance, Z_{AB}, as seen from terminals A and B in the circuit shown in Figure P6.31.

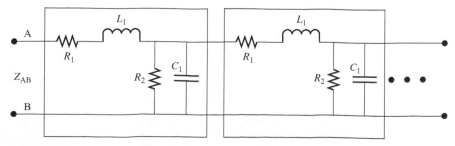

FIGURE P6.31   Circuit for Problem 6.31.

SECTION 6.4 STEADY-STATE CIRCUIT ANALYSIS USING PHASORS

6.32   (B) Find the steady-state expression of the current for the circuit in Figure P6.32 if the voltage source is $v(t) = 12 \cos(120t) \ V$.

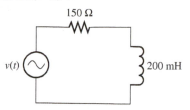

FIGURE P6.32   Circuit for Problem 6.32.

6.33   (B)* Find the steady-state expression of current for the circuit in Figure P6.33 if the voltage source is $v(t) = 10 \cos(200t + 30°) \ V$.

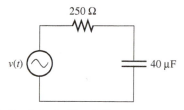

FIGURE P6.33   Circuit for Problem 6.33.

6.34   (A) Find V_1 and V_2 in Figure P6.34 using nodal analysis.

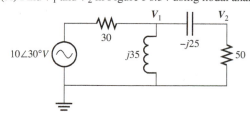

FIGURE P6.34   Circuit for Problem 6.34.

6.35   (A) Find node voltage V_1 and V_2 in Figure P6.35.

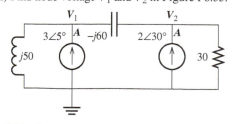

FIGURE P6.35   Circuit for Problem 6.35.

6.36 (A)* Find V_2 in Figure P6.36 using nodal analysis.

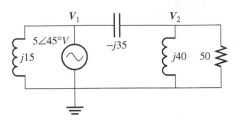

FIGURE P6.36 Circuit for Problem 6.36.

6.37 (A) Determine the rms current that flows across the resistor in Figure P6.37 if the voltage source is $v(t) = 24\cos(500t + 60°)$ V.

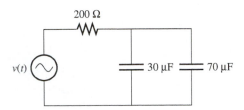

FIGURE P6.37 Circuit for Problem 6.37.

6.38 (A) Find the node voltages V_1 and V_2 in Figure P6.38.

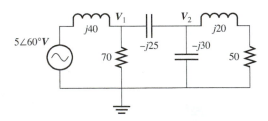

FIGURE P6.38 Circuit for Problem 6.38.

6.39 (A)* If $i_s(t) = 2\cos(200t)\,A$ in Figure P6.39, find the steady-state expressions for I_R and I_C.

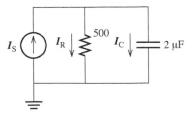

FIGURE P6.39 Circuit for Problem 6.39.

6.40 (A) An op amp is used as an amplifier to increase the magnitude of voltage output given the input voltage source. An **inverting amplifier** (see Figure P6.40) is one popular type of op amp. Given that the inverting amplifier has a gain of $G = -(Z_{out}/Z_{in})$, what is the output voltage if the voltage source is $v_S(t) = 50\cos(120\pi t + 30°)$ V and $Z_{in} = 35\ \Omega$, $Z_{out} = 105\ \Omega$?

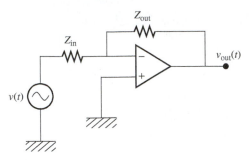

FIGURE P6.40 An inverting amplifier.

6.41 (A) Find the steady-state expressions for V_1 and V_2 in Figure P6.41.

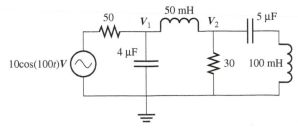

FIGURE P6.41 Circuit for Problem 6.41.

6.42 (A) Find the steady-state expressions for V_1 and V_2 in Figure P6.42.

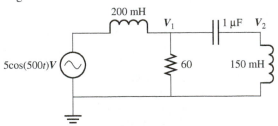

FIGURE P6.42 Circuit for Problem 6.42.

6.43 (H)* The electric fan circuit shown in Figure P6.43 is connected to US standard AC power (assume no phase angle). What is the cost of electric utility if the fan rotates at a constant speed for 12 h? Assume that the cost of electricity is 23 cents per kilowatt hour. (*Hint*: Assume the average power is $P_{av} = V_{rms}\,I_{rms}$.)

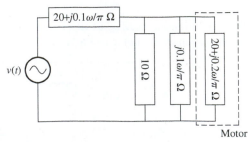

FIGURE P6.43 Circuit for Problem 6.43.

6.44 (H) The voltage source has a voltage of $v(t) = 4\cos^3 \omega t$ V. The radius frequency, ω, and the inductance, L, satisfies $\omega L = R$.

Find the voltage across the inductor at $t = 0$. (*Hint:* $4 \cos^3 \omega t = 3 \cos \omega t + \cos 3\omega t$.)

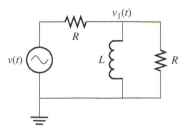

FIGURE P6.44   Circuit for Problem 6.44.

SECTION 6.5 THÉVENIN AND NORTON EQUIVALENT CIRCUITS WITH PHASORS

6.45 (A) Find the Thévenin and Norton equivalent circuits for the circuit shown in Figure P6.45, as seen by the load resistor, R_L.

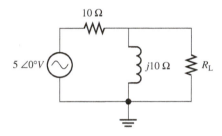

FIGURE P6.45   Circuit for Problem 6.45.

6.46 (B)* Determine the Norton equivalent current and equivalent impedance observed by R_L for the circuit in Figure P6.46 with the current source, $4 \cos(200t + 30°) \, A$.

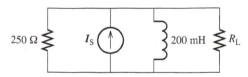

FIGURE P6.46   Circuit for Problem 6.46.

6.47 (A) Find the Norton equivalent circuits for the circuit shown in Figure P6.47.

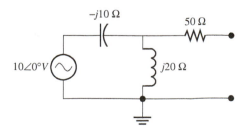

FIGURE P6.47   Circuit for Problem 6.47.

6.48 (A) Sketch the Thévenin and Norton equivalent circuit for the circuit shown in Figure P6.48, where $V_S = 20\angle 30° V$.

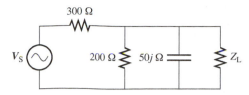

FIGURE P6.48   Circuit for Problem 6.48.

6.49 (H) Given that the Thévenin equivalent voltage, $V_{th} = 12 + j4\sqrt{6} \, V$, what are the values of Z_R and Z_C if the voltage source, $V_S = 20\angle 0° \, V$ and $Z_C/R = -\sqrt{6}/5$?

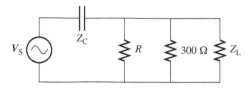

FIGURE P6.49   Circuit for Problem 6.49.

6.50 (B)* What are the Thévenin/Norton equivalent resistance observed by R_L for the circuit shown in Figure P6.50? Provide the answer to two decimal places.

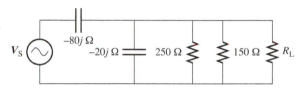

FIGURE P6.50   Circuit for Problem 6.50.

6.51 (A) Find the Thévenin equivalent circuits for the circuit shown in Figure P6.51.

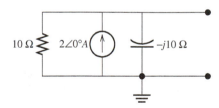

FIGURE P6.51   Circuit for Problem 6.51.

6.52 (H) Find the Thévenin and Norton equivalent circuits for the circuit shown in Figure P6.52, as seen by the load resistor, R_L.

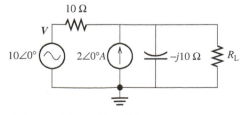

FIGURE P6.52   Circuit for Problem 6.52.

6.53 (B)* Express the Thévenin/Norton equivalent resistance, R_L, in terms of the resistance, R, and impedance of inductor Z_L for the circuit in Figure P6.53.

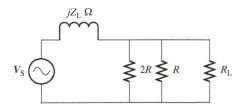

FIGURE P6.53 Circuit for Problems 6.53 and 6.54.

6.54 (H) In Figure P6.53, the voltage source is $V_{th} = 50\angle 0°$ the Thévenin voltage and Norton current are $V_{th} = 20\angle -60°$ and $I_{th} = 2\angle -90°$. Determine the value of R and Z_L.

6.55 (H) Use Thévenin equivalent circuit to solve Problem 6.44.

SECTION 6.6 AC STEADY-STATE POWER

6.56 (B) Find the equivalent circuit of the following impedances:
  a. $-j25, \omega = 110\pi$
  b. $12 + j110, \omega = 220\pi$
  c. $20 - j220, \omega = 110\pi$
  d. $+j110, \omega = 220\pi$

6.57 (B) Find the equivalent circuit of the following admittances:
  a. $j250, \omega = 110\pi$
  b. $1 + j10, \omega = 220\pi$
  c. $2 - j20, \omega = 110\pi$
  d. $1 - j, \omega = 1$

6.58 (B)* Determine the average power, reactive power, and apparent power generated by the resistor for the circuit in Figure P6.58 with the voltage source $160\angle 30°$ V.

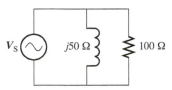

FIGURE P6.58 Circuit for Problem 6.58.

6.59 (A) Find the real and reactive power consumed by the load in Figure P6.59.

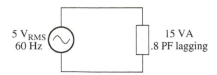

FIGURE P6.59 Circuit for Problem 6.59.

6.60 (A) Find the current, I, in Figure P6.60:

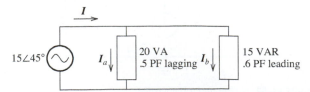

FIGURE P6.60 Circuit for Problem 6.60.

6.61 (A) Find the real, reactive, and apparent power drawn by the circuit, and the power factor for Figure P6.61.

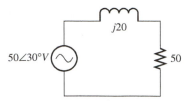

FIGURE P6.61 Circuit for Problem 6.61.

6.62 (A)* Find I, and the apparent power consumed by the circuit for Figure P6.62.

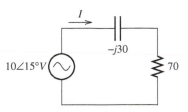

FIGURE P6.62 Circuit for Problem 6.62.

6.63 (H) Find the complex power taken from the source for the circuit shown in Figure P6.63.

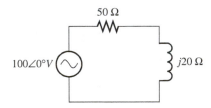

FIGURE P6.63 Circuit for Problem 6.63.

6.64 (A)* Consider the circuit shown in Figure P6.64. Determine the maximum average power that can be delivered to the load if (1) the load is pure resistance and (2) the load is the combination of resistance, inductance, and capacitance.

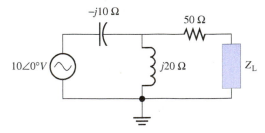

FIGURE P6.64 Circuit for Problem 6.64.

6.65 (B) What kind of load is shown in Figure P6.65?

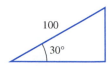

FIGURE P6.65 Triangle relationship for Problems 6.65 and 6.66.

6.66 (H) Determine θ_v and θ_i using the value shown in Figures P6.65 and P6.66.

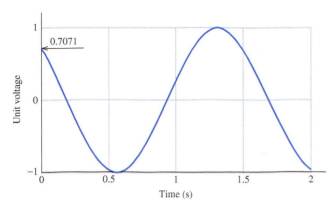

FIGURE P6.66   Unit voltage output plot for Problem 6.66.

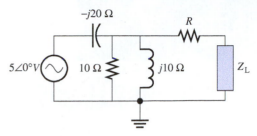

FIGURE P6.70   Circuit for Problem 6.70.

6.67 (A)* A 30-kW load is powered by a 60-Hz, 480-V_{RMS} line with a power factor of 0.8 lag. Calculate the parallel capacitance required to correct the power factor to 0.9 lag.

6.68 (A) A 100-kW load is powered by a 60-Hz, 1-kV_{rms} line, power factor of 0.5 lag. To correct the power factor, a 100-µF capacitor bank is placed in parallel with the load. What is the new power factor?

6.69 (H)* An 8-kW load is powered by a 240-V_{peak}, PF = 1.00 source.
   a. Calculate the current drawn by the load.
   b. Calculate the current if power factor is 0.4

6.70 (H) Consider the circuit shown in Figure P6.70, if the load, Z_L, is the combination of resistance, inductance, and capacitance, how can R be adjusted to make the maximum average power delivered to the load less than 0.01 W?

6.71 (H)* A lamp is rated "110 V, 15 W" and its working current is 0.7 A.
   a. Find the power factor of the lamp.
   b. What can be done to increase the lamp's power factor to 1? The voltage source has a frequency of w = 60 Hz. (*Hint:* Connect a capacitor in parallel with the lamp.)

SECTION 6.7 CIRCUIT STEADY-STATE ANALYSIS USING PSPICE

6.72 (A) A sinusoidal current $i(t) = 2\cos(120\pi t + 60°)$ A flows through a 100-Ω resistance. Set up a PSpice schematic, plot the voltage across the resistance and determine the *rms* value of sinusoid voltage from the plot.

6.73 (B) If $i_S(t) = 2\cos(200t)$ A in Figure P6.39, use PSpice to plot the current, $i_R(t)$ and $i_C(t)$.

6.74 (H) In Figure P6.41, use PSpice to plot $v_1(t)$ and $v_2(t)$.

6.75 (A) Replace the sinusoidal voltage source to the sinusoidal current that is $i_S(t) = 5\cos(100t)$ A in Figure P6.45, use PSpice to plot the Norton equivalent current flows through R_L.

Frequency Analysis

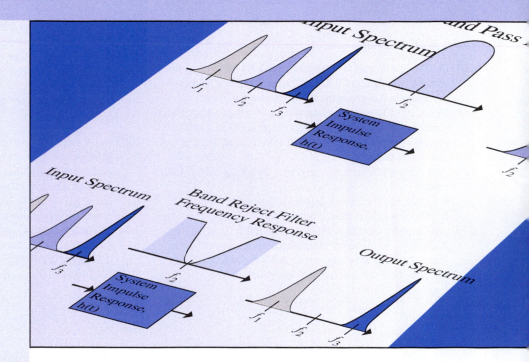

7.1 INTRODUCTION

Chapter 5 outlined the process of analysis of circuits that consist of inductors and capacitors. It focused on analyzing the transient and steady-current properties of these circuits when a DC was applied. Chapter 6 investigated steady-state response when a single-frequency sinusoid was applied to a circuit. This chapter focuses on study of the characteristics of capacitors and inductors when an AC with different (varying) frequencies is applied to circuits. These circuits are used as filters.

As it is clear from their name, in general, **filters** are devices that allow something to pass while not allowing other things to pass. Many different filters can be observed in daily life, for example, many coffee makers use simple paper filters. In the arena of electrical engineering, filters are devices that allow signals with particular frequencies to pass, while rejecting signals with other frequencies. Based on the behavior of these filters in different frequencies, they may even change the structure of a signal at the output.

In Chapter 6, it was noted that the reactance of inductors and capacitors varies with frequency. Now, assume that current and voltage signals with different frequencies are applied to circuits made by inductors and capacitors. In addition, assume that the current and the voltage of an element in the circuit are measured. Typically, the voltage or current of that branch is referred to as the circuit output. In this case, the amplitude of the output current and voltage signals will not be the same for different frequencies. In some cases, signals may be observed with certain frequencies that do not appear at the output, that is, zero amplitude.

In other words, the system filters out those frequencies and permits other frequencies. Thus, a **frequency response** can be defined for the system. The frequency response represents what frequencies appear at the output with what amplitudes. Some frequencies might be attenuated or completely reduced to zero. The parameters of a filter can be adjusted to allow desired frequencies to pass and to block any undesired frequencies.

Filters made by resistor (R), inductor (L), and capacitor (C) are called RLC filters. They are also called passive filters, because all elements consume energy. There are many applications for RLC filters, which include the tuning circuit in a radio or TV that selects and tunes on a specific frequency band. This circuit allows the desired frequency to pass through and blocks all other frequencies. Thus, only the desired radio station or TV channel is delivered to the user. Further applications include removing the low-frequency "buzz" that can be created by an AC power supply and filtering out interfering frequencies.

In general, filters are divided into two categories: (1) passive and (2) active. Active filters are made of active elements, such as diodes, transistors, and operational amplifiers as well as resistors, inductors, and capacitors. These filters are discussed in Chapters 8 and 11. Passive filters, on the other hand, do not include any active elements. Investigation of active filters is straightforward and is based on the knowledge of the passive filters as well as the principles of operational amplifiers.

This chapter focuses on passive filters. Four general types of filters: low-pass, high-pass, band-pass, and band-stop filters will be studied. These filters are summarized in Figure 7.1. This figure shows that these filters do allow some signals to pass, while stopping others. For instance, the spectrum with center frequency f_2 is the only one that passes through the band-pass filter. However, the band-reject filter allows all signals to pass except the one with center frequency f_2. Similarly, high-pass filters allow higher frequency components to pass, while low-pass filters allow lower frequency components to pass. This chapter introduces each of these filters in detail and shows how each is constructed.

7.2 FIRST-ORDER FILTERS

First-order filters are so named because they can be represented by a first-order differential equation. A first-order filter has only one capacitor or inductor. Within the four general types of filters, only low-pass and high-pass filters can be constructed as first-order filters. Band-pass and band-stop filters require at least a second-order filter and these filters are covered in Section 7.5.

7.2.1 Transfer Functions

A filter transfer function expresses the output of a filter with respect to its input. In other words, it is the ratio of the filter output (voltage or current measured at one port) to the input (voltage or current measured at the second port), and it is a function of frequency, f. Thus, a filter can be represented by a two-port system like the one shown in Figure 7.2. The transfer function is useful to determine (and plot) the frequency response of a filter. Frequency response represents the gain and the phase shift of a filter or network as a function of frequency (an example is shown in Figure 7.7, which is discussed later in this section).

Consider the two-port filter shown in Figure 7.2, which consists of input and output ports. The transfer function, $H(f)$, of a two-port filter is defined as:

$$H(f) = \frac{V_{\text{out}}}{V_{\text{in}}} \tag{7.1}$$

The inverse Fourier transform of $H(f)$, specifically $h(t)$, is called the filter impulse response. Because V_{out} and V_{in} are phasors (complex numbers), $H(f)$ is also complex. Magnitude and phase are allocated to $H(f)$.

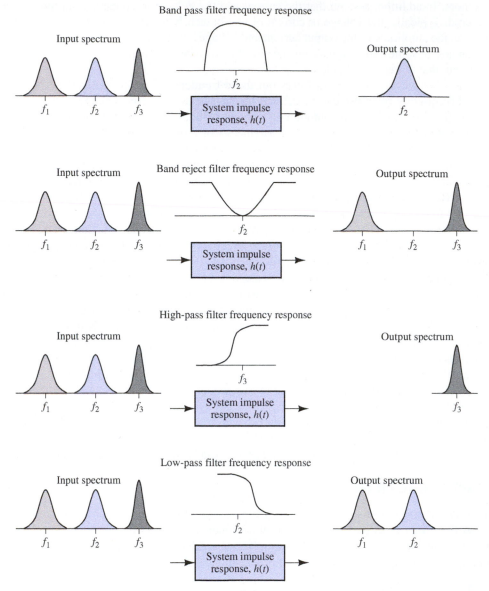

FIGURE 7.1 Different types of filters and their effects.

7.3 LOW-PASS FILTERS

Low-pass filters allow low-frequency components to pass and block high-frequency components. An RC (resistor–capacitor) first-order, low-pass filter is presented in Figure 7.3(a), and an RL (resistor–inductor) first-order, low-pass filter is shown in Figure 7.3(b).

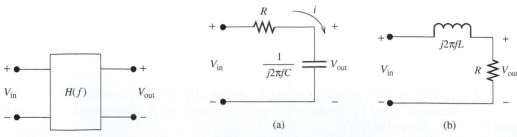

FIGURE 7.2 Two-port system.

FIGURE 7.3 Low-pass filters: (a) RC; (b) RL.

Consider the RC low-pass filter shown in Figure 7.3(a). Using the voltage divider technique, the voltage at the output corresponds to:

$$V_{\text{out}} = \frac{\frac{1}{j2\pi fC}}{R + 1/j2\pi fC} \times V_{\text{in}} \tag{7.2}$$

Given that $H(f) = V_{\text{out}}/V_{\text{in}}$, and multiplying both denominator and nominator of the right-hand side of Equation (7.2) in $j2\pi fC$;

$$H(f) = \frac{1}{1 + j2\pi fRC} \tag{7.3}$$

Similarly, for the RL low-pass filter shown in Figure 7.3(b):

$$H(f) = \frac{V_{\text{out}}}{V_{\text{in}}}$$

$$= \frac{R}{R + j2\pi fL} \tag{7.4}$$

$$= \frac{1}{1 + (j2\pi fL/R)}$$

If in Equation (7.3) the following is defined:

$$f_{\text{B}} = \frac{1}{2\pi RC} \tag{7.5a}$$

Or in Equation (7.4), the following is defined:

$$f_{\text{B}} = \frac{R}{2\pi L} \tag{7.5b}$$

Then, these two equations can be replaced by:

$$H(f) = \frac{1}{1 + j(f/f_{\text{B}})} \tag{7.6}$$

In Equation (7.6), f_{B} is called the **break frequency**. This is also known as the **half-power frequency** or the corner frequency. These two names are derived from the Bode plot. It is called the **corner frequency** because it is located at the intersection of the two asymptotes. It is called the half-power frequency because at this frequency, only half of the incoming power is available at the output. Note that $10 \log(1/2) = -3$ dB. Thus, the gain of power at this frequency is -3 dB.

EXAMPLE 7.1   Filter Transfer Function

Find the break frequency and the transfer function for the filter shown in Figure 7.4.

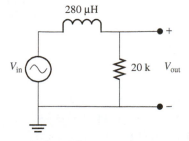

FIGURE 7.4 Circuit for Example 7.1.

(continued)

EXAMPLE 7.1 **Continued**

SOLUTION

Using Equation (7.5b):

$$f_B = \frac{R}{2\pi L} = \frac{20\text{ k}}{2\pi \cdot 280 \times 10^{-6}} = 11.4\text{ MHz}$$

Thus, Equation (7.6) corresponds to:

$$H(f) = \frac{1}{1 + j(f/11.4 \times 10^6)}$$

EXERCISE 7.1

What is the transfer function $H(f)$ for DC input, i.e., when $f = 0$?

EXAMPLE 7.2 **Filter Output Voltage**

Find the output, $v_{out}(t)$, of the filter shown in Figure 7.5 given the input

$$v_{in}(t) = 15 + 2\cos(100\pi t - 30°) + 6\cos(300\pi t + 50°) + 2.5\sin(400\pi t)$$

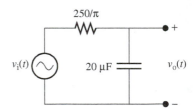

FIGURE 7.5 Filter for Example 7.2.

SOLUTION

This is a low-pass filter. Using Equation (7.5a):

$$f_B = \frac{1}{2\pi RC} = 100\text{ Hz}$$

Its transfer function is:

$$H(f) = \frac{1}{1 + j(f/100)}$$

Now, it is clear that the input is formed by four functions, each with a different frequency. The first is a DC (constant) signal with zero frequency. Thus, the frequency response for this component is $H(0) = 1$. The other three terms are cosine functions with nonzero frequency components. The second has a frequency component at $f = 50$ Hz (thus, $H(f)$ is calculated at $f = 50$ Hz); the third one at $f = 150$ Hz (thus, $H(f)$ is calculated at $f = 150$ Hz); and the fourth one at $f = 200$ Hz (thus, $H(f)$ is calculated at $f = 200$ Hz). The output can be calculated using Equation (7.1) for each frequency component, as summarized below.

- For $f = 0$ Hz: $15 \times H(0) = 15 \cdot 1 = 15$
- For $f = 50$ Hz: $2\angle -30 \times H(50) = 2\angle -30 \times 0.894\angle -26.6° = 1.79\angle -56.6°$
- For $f = 150$ Hz: $6\angle 50 \times H(150) = 6\angle 50 \times 0.5547\angle -56.3° = 3.33\angle -6.3°$
- For $f = 200$ Hz: $2.5\angle -90 \times H(200) = 2.5\angle -90 \times 0.447\angle -63.43° = 1.12\angle -153°$
  (Note that $2.5\sin(400\pi t) = 2.5\cos(400\pi t - 90°)$.)

For the final output, the results above must be converted back to cosines and all individual frequency components must be added together:

$$v_{out}(t) = 15 + 1.79 \cos(100\pi t - 56.6°) + 3.33 \cos(300\pi t - 6.3°) + 1.12 \cos(400\pi t - 153°)$$

EXERCISE 7.2

Repeat Example 7.2, assuming the input signal is:

$$v_{in}(t) = \sin(200\pi t + 60°) + \cos(300\pi t - 10°)$$

APPLICATION EXAMPLE 7.3   Engine Noise Removal

An inventor is wondering why her recently built engine is making noise. She can detect three distinct audible frequencies. She wants to hear them one at a time, so that she can fix each problem separately. She connects a microphone to the engine, which sends the noise through an electric filter and an amplifier before reaching her headphones. The equation that defines the noise is:

$$v_{mic}(t) = 20 \cos\left(500\pi t - \frac{\pi}{4}\right) + 5 \cos(4000\pi t) + 2 \cos\left(6000\pi t - \frac{3\pi}{4}\right) \text{ mV}$$

Design a filter that allows the lowest frequency to pass, but not the others.

SOLUTION

For this problem, a low-pass filter is needed with a break frequency at, or slightly above, the lowest frequency described above. The lowest frequency is 250 Hz ($500\pi = 2\pi \times 250$). Because the signal should be attenuated 3 dB at the break frequency, set the break frequency to 300 Hz. Therefore, the transfer function is:

$$H(f) = \frac{1}{1 + j(f/300)}$$

The filter is shown in Figure 7.6.

$$f_B = \frac{1}{2\pi RC} = 300 \text{ Hz}$$

Using the break frequency equation, values for R and C can be determined. There are an infinite number of possibilities. One possibility is to use $C = 10 \text{ µF}$.

$$\frac{1}{2\pi R \times 10 \times 10^{-6}} = 300 \text{ Hz}$$

$$R = \frac{1}{2\pi \times 300 \times 10 \times 10^{-6}}$$

$$= 53.05 \text{ } \Omega$$

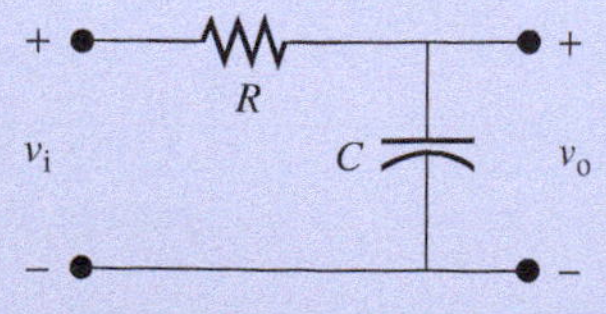

FIGURE 7.6 Filter for Example 7.3.

Using the information from Example 7.3, design a filter that blocks only the highest frequency from passing, while avoiding the other frequencies to pass.

As explained in Chapter 6, based on Fourier series principles, a periodic signal such as a square wave includes multiple frequency components. If this periodic signal is applied to an LPF that allows some frequency components to pass but avoids others, the output signal would not be a square wave any more. Apply a 60-Hz square wave to an LPF with the break frequency of 240 Hz and investigate its output signal. Study the results theoretically by writing the Fourier series of the 60-Hz periodic signal.

7.3.1 Magnitude and Phase Plots

As shown in Equation (7.6), the frequency response has a complex nature. Any complex variable has magnitude and phase. When plotting the frequency response of a filter, the magnitude and phase are plotted separately for clarity.

For the RC low-pass filter, the magnitude is [see Figure 7.7(a)]:

$$|H(f)| = \frac{1}{\sqrt{1 + (f/f_B)^2}} \tag{7.7}$$

Likewise, the phase is given by [see Figure 7.7(b)]:

$$\angle H(f) = -\tan^{-1}\left(\frac{f}{f_B}\right) \tag{7.8}$$

7.3.2 Decibels

To compare the performance of different filters, the magnitude $|H(f)|$ is often expressed in **decibels**. The decibel (dB) is a convenient way of displaying very large or small values on the same plot. Because a filter magnitude can achieve large and small values, depending on the frequency, the decibel is the method of choice for expressing filter magnitudes. The decibel corresponds to:

$$|H(f)|_{dB} = 20 \log_{10}|H(f)| \tag{7.9}$$

Likewise, the magnitude |H(f)| can be expressed in terms of its dB as:

$$|H(f)| = 10^{|H(f)|_{dB}/20} \tag{7.10}$$

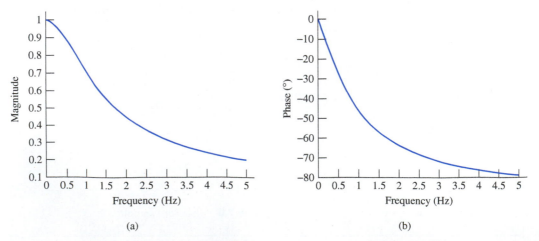

(a)   (b)

FIGURE 7.7 (a) Magnitude and (b) phase of the frequency response of a low-pass filter.

TABLE 7.1 Sample Decibel Values

$|H(f)|$	$|H(f)|_{dB}$
0.1	−20
1	0
3	9.54
8	18.06
10	20
100	40

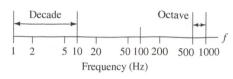

FIGURE 7.8 Logarithmic frequency scale.

Note:

Because the transfer function $|H(f)|$ is the ratio of two voltages (or currents), the logarithm is multiplied by 20. For the ratio of two powers, multiply the logarithm by 10. In other words,

$$\text{Power ratios}_{dB} = 10 \log_{10} \frac{P_1}{P_2} = 10 \log_{10}|H(f)|^2$$

$$\text{Voltage ratios}_{dB} = 20 \log_{10} \frac{V_1}{V_2} = 10 \log_{10}|H(f)|$$

Table 7.1 represents the decibel values for some $|H(f)|$ values. Note that decibel equivalents are positive for magnitudes greater than one and negative for magnitudes less than one.

Logarithmic scales are often used for plotting transfer functions. On a logarithmic scale, the variable is *multiplied* by a given factor for equal increments along the scale. However, on a linear scale, a given amount is *added* to the variable for equal increments along the scale. A sample logarithmic scale is presented in Figure 7.8.

A *decade* is defined as a range of frequencies where the ratio of the highest to the lowest frequency is 10. For example, 5 to 50 Hz is one decade, and 200 to 20,000 Hz is two decades. An *octave* is defined as a range of frequencies where the ratio of highest to lowest is 2. For example, 40 to 80 kHz is one octave and 20 to 160 Hz is three octaves. Figure 7.8 shows a decade and an octave on a logarithmic scale.

EXAMPLE 7.4   **Understanding Decibel**

a. Find the decibel equivalents of the following values:
    i. $|H(f)| = 1/3$
    ii. $|H(f)| = 1.556$
    iii. $|H(f)| = 14$
b. Given the following values for $|H(f)|_{dB}$, find $|H(f)|$.
    i. −5.72 dB
    ii. 40 dB
    iii. 25 dB

SOLUTION

Part (a)
    i. $20 \log(1/3) = -9.54$ dB
    ii. $20 \log(1.556) = 3.84$ dB
    iii. $20 \log(14) = 22.9$ dB

    Recall that:

$$\log x = \log_{10} x = \frac{\ln x}{\ln 10}$$

(continued)

EXAMPLE 7.4   **Continued**

Part (b)

i. $-5.72 = 20 \log|H(f)| \rightarrow |H(f)| = 10^{-5.72/20} = 0.518$
ii. $|H(f)| = 10^{40/20} = 100.0$
iii. $|H(f)| = 10^{25/20} = 17.78$

EXERCISE 7.5

Find $|H(f)|_{dB}$ if $|H(f)| = 100$. What will the value of $|H(f)|$ be when $|H(f)|_{dB}$ is decreased by 3 dB?

EXAMPLE 7.5   **Understanding Octave and Decode**

Given a frequency of 800 Hz, find the frequency which is:

a. Seven octaves higher
b. One decade lower
c. Four decades higher

SOLUTION

a. $800 \times 2^7 = 102.4$ kHz
b. $800 \times 10^{-1} = 80$ Hz
c. $800 \times 10^4 = 8$ MHz

7.3.3 Bode Plot

A **Bode plot** is a commonly used technique for plotting a frequency response across a large range of frequency values. This section first examines the Bode plot for amplitude. Applying Equation (7.9) to Equation (7.7) for a low-pass filter results in:

$$|H(f)|_{dB} = 20 \log_{10}|H(f)| = -10 \log\left(1 + \left(\frac{f}{f_B}\right)^2\right)$$

Now, if $f \gg f_B$, then $|H(f)|_{dB}$ can be approximated by:

$$|H(f)|_{dB} \cong -20 \log\left(\frac{f}{f_B}\right)$$

and, if $f \ll f_B$, it can be approximated by:

$$|H(f)|_{dB} \cong 0 \text{ dB}$$

The frequency scale uses a logarithmic scale, that is, it is sketched in terms of $\log(f/f_B)$ and the magnitude is expressed in decibels. Thus, there is a linear relationship between $|H(f)|_{dB}$ and $\log(f/f_B)$ for $f \gg f_B$. The coefficient of this linear equation is -20. In other words, $|H(f)|_{dB}$ decays linearly with a slope of -20 plotting $\log(f/f_B)$ in the x-axis. Accordingly, as shown in Figure 7.9, the actual Bode magnitude plot approaches two asymptotes, one of which is horizontal at 0 dB; the other starts at $f = f_B$ and has a slope of -20 dB/decade.

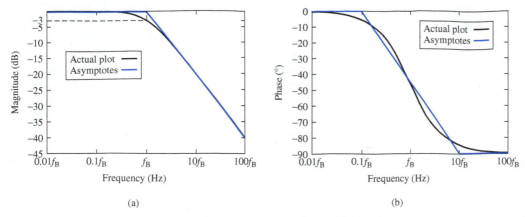

FIGURE 7.9   Bode plot for a low-pass filter (a) its magnitude and (b) its phase.

EXERCISE 7.6

Verify that the slope of the Bode plot for $f \gg f_B$ is −20 dB/decade. In other words, for example, why is it not −20 dB/octave?

Now, for the Bode plot of phase, as observed in Equation (7.8), when $f \ll f_B$, the arctangent of a small value (asymptotically zero) tends to zero. However, when $f \gg f_B$, the arctangent of a large value (asymptotically infinity) tends to −90°. Thus, the phase will change from 0 toward −90°.

Accordingly, there are three asymptotes for Bode plot of phase: (1) for very small frequencies, (2) for medium frequencies, and (3) for high frequencies. In other words, the Bode phase plot approaches an asymptote at 0° for $f \le f_B/10$, a second asymptote starting at $(f_B/10, 0°)$ with a slope of $-45°$/decade for $f_B/10 < f < 10 f_B$, and a third asymptote at −90° for $f \ge 10 f_B$.

As discussed earlier, break frequency (f_B) is also known as the half-power frequency or the corner frequency. These two names are derived from the Bode plot. It is called the corner frequency because it is located at the intersection of the two asymptotes. It is called the half-power frequency, because at this frequency only half of the incoming power is available at the output. Note that $10 \log(1/2) = -3$ dB. Thus, the gain of power at this frequency is −3 dB.

Using the same justification as discussed above, it is possible to explain how the Bode plot can be sketched for a high-pass filter or any other frequency response. High-pass filters are discussed in Section 7.4.

Figure 7.9 shows the actual gain at f_B is −3 dB. Note that if a voltage is multiplied by a factor of $1/\sqrt{2}$, the power that the voltage can deliver to a load is multiplied by a factor of 1/2. At $f = f_B$, $|H(f)| = 1/\sqrt{2} \approx 0.707$. Recall that $|H(f)|$ represents the ratio of two voltages. Thus, to calculate it in terms of decibels, use $20 \log 0.707 \approx -3$ dB. This also confirms that the break frequency can be called the half-power frequency, as only half of the input power is available. Because $|H(f)| = -3$ dB, this frequency is also called the 3-dB cut-off frequency. Based on the above explanations, the process of sketching the Bode plot can be explained as follows.

Drawing a Bode plot for magnitude:

1. Identify f_B and label it on the frequency axis.
2. Draw the asymptotes with appropriate slopes:
   a. −20 dB/decade for a first-order, low-pass filter
   b. +20 dB/decade for a first-order, high-pass filter
3. Mark the −3 dB point at f_B.
4. Starting with the low frequencies, draw a smooth curve approaching the asymptotes and passing through the point $(f_B, -3$ dB$)$.

Drawing a Bode plot for phase:

1. For the phase Bode plot, mark the frequencies $f = f_B/10$ and $f = 10f_B$.
2. Draw the horizontal asymptotes:
   a. For a first-order, low-pass filter, draw an asymptote at $0°$ for $f \leq f_B/10$, and an asymptote at $-90°$ for $f \geq 10f_B$.
   b. For a first-order, high-pass filter, draw an asymptote at $90°$ for $f \leq f_B/10$, and an asymptote at $0°$ for $f \geq 10f_B$.
3. Draw the diagonal asymptote, connecting the horizontal asymptotes.
4. Draw a smooth curve that approaches the horizontal asymptotes and crosses the diagonal asymptotes three times, including at the center of the line.

EXAMPLE 7.6 Finding the Transfer Function

A filter has the transfer function as given by the Bode plots in Figure 7.10. *Note:* This is an active filter because of the amplification (see Chapters 8 and 11).

a. Find the transfer function for this filter.
b. If the input to this filter is $x(t) = 3\cos(200\pi t - 15°) + \cos(20{,}000\pi t + 45°)$ V, find the output of the filter, $y(t)$.

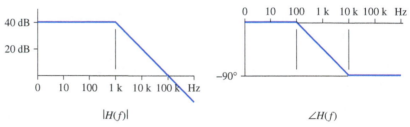

$|H(f)|$ $\angle H(f)$

FIGURE 7.10 Bode plot for Example 7.6.

SOLUTION

a. The break frequency is 1 kHz. Because the magnitude drops off at a slope of -20 dB/decade, this is a first-order filter. A first-order, low-pass transfer function is given by:

$$H(f) = \frac{H_0}{1 + j(f/f_B)}$$

Therefore, the transfer function for this filter is of the form:

$$H(f) = \frac{H_0}{1 + j(f/1000)}$$

As can be seen in the magnitude Bode plot, shown in Figure 7.10, the gain at low frequencies is not unity (0 dB). It is 40 dB or 100 dB. The transfer function needs to express this amplification. Thus, $H_0 = 100$, and

$$H(f) = \frac{100}{1 + j(f/1000)}$$

b. $x(t)$ includes two signals with two frequencies 100 Hz and 10 kHz. The gain at 100 Hz is $100\angle 0°$ (see Figure 7.10); now, because the corresponding phasor is $3\angle -15°$, the output phasor is $3\angle -15° \times 100\angle 0° = 300\angle -15°$. The gain at 10 kHz is $10\angle -90°$; thus, $1\angle 45° \times 10\angle -90° = 10\angle -45°$. Accordingly, the total output signal is:

$$y(t) = 300\cos(200\pi t - 15°) + 10\cos(20{,}000\pi t - 45°)$$

7.4 HIGH-PASS FILTERS

As the name implies, a high-pass filter passes high-frequency components and blocks low frequencies. Two models for first-order, high-pass filters are shown in Figure 7.11. The phasor of the current, i, in Figure 7.11(a) corresponds to:

$$I = \frac{V_{in}}{\frac{1}{j2\pi fC} + R} \tag{7.11}$$

The phasor of the output voltage, V_{out}, is then given by:

$$V_{out} = I \times R = \frac{V_{in} \times R}{\frac{1}{j2\pi fC} + R} = \frac{V_{in} \times j2\pi fRC}{1 + j2\pi fRC} \tag{7.12}$$

Therefore, the transfer function corresponds to:

$$H(f) = \frac{V_{out}}{V_{in}} = \frac{j2\pi fRC}{1 + j2\pi fRC} \tag{7.13}$$

Note that $H(f)$ in Equation (7.13) can also be calculated using a simple voltage division based on the impedances in Figure 7.11(a). Given $f_B = 1/2\pi RC$, for a first-order, RC high-pass filter, the transfer function corresponds to:

$$H(f) = \frac{V_{out}}{V_{in}} = \frac{j(f/f_B)}{1 + j(f/f_B)} \tag{7.14}$$

EXERCISE 7.7

Consider the RL circuit shown in Figure 7.11(b), use the same approach shown in Equations (7.11) to (7.13), and show that, assuming $f_B = 1/2\pi R$, the same transfer function as in Equation (7.14) will result.

Here, the magnitude of the high-pass filter transfer function corresponds to:

$$|H(f)| = \frac{f/f_B}{\sqrt{1 + (f/f_B)^2}} \tag{7.15}$$

Likewise, the high-pass filter phase corresponds to:

$$\angle H(f) = 90° - \arctan(f/f_B) \tag{7.16}$$

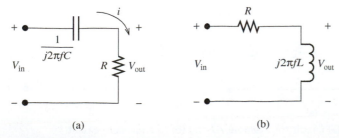

(a)   (b)

FIGURE 7.11 High-pass filters (a) RC and (b) RL.

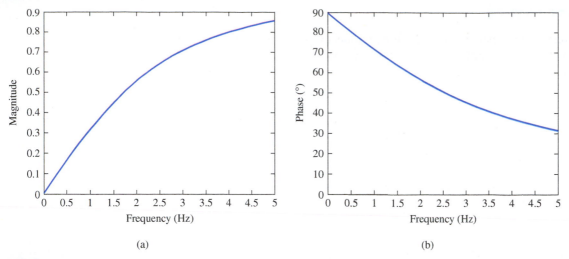

FIGURE 7.12 High-pass frequency response: (a) magnitude and (b) phase.

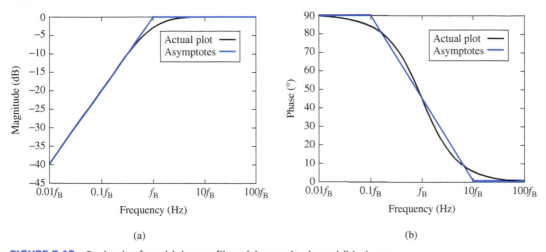

FIGURE 7.13 Bode plot for a high-pass filter: (a) magnitude and (b) phase.

The (non-Bode) magnitude and phase frequency responses are given in Figure 7.12. The Bode plot of the filter shown in Figure 7.11 is shown in Figure 7.13. Notice that the high-pass Bode plot also approaches two asymptotes. One is horizontal and starts at $f = f_B$. The other terminates at $f = f_B$ and has a slope of $+20$ dB/decade.

EXAMPLE 7.7   **High-Pass Filter**

The high-pass filter shown in Figure 7.14(a) is used to drive a load, R_L. What is its break frequency, f_B, and its transfer function?

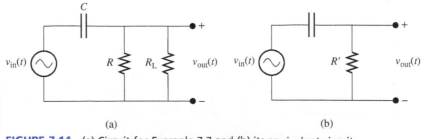

FIGURE 7.14 (a) Circuit for Example 7.7 and (b) its equivalent circuit.

SOLUTION

The equivalent resistance of parallel resistors, R and R_L, $R' = (RR_L)/(R + R_L)$. Accordingly, the equivalent circuit for the circuit of Figure 7.14(a) is shown in Figure 7.14(b). For the high-pass filter in Figure 7.14(b):

$$f_B = \frac{1}{2\pi \dfrac{RR_L}{R + R_L}C} = \frac{R + R_L}{2\pi RR_L C}$$

Thus, using Equation (7.13), its transfer function corresponds to:

$$H(f) = \frac{j(f 2\pi CRR_L/(R + R_L))}{1 + j(f 2\pi CRR_L/(R + R_L))}$$

APPLICATION EXAMPLE 7.8   **Filtering Out Power Supply Interference**

An engineer is measuring a weak signal with a frequency of 5 kHz. However, a strong signal with a frequency of 60 Hz generates interference. (*Note:* A 60-Hz power supply is often a source of such interference.)

   a. What kind of filter is needed to block the 60-Hz signal?
   b. What will the break frequency be?
   c. If the capacitor has a value of $C = 10\ \mu F$, what will the value of the resistor, R, be?
   d. Find the transfer function of the filter.

SOLUTION

   a. To avoid the low-frequency component at 60 Hz that is considered as interference in this problem and to allow reliable measurement of the 5-kHz signal, a high-pass filter is needed.
   b. The break frequency should be selected to minimize the attenuation of the desired 5-kHz signal, while effectively eliminating the 60-Hz noise. Because f_B represents the point of 3-dB drop, it should be selected between 60 and 5 kHz. For this example, choose $f_B = 4.5$ kHz.
   c. The transfer function for a first-order, high-pass RC filter is given by (see Equation 7.14)

$$H(f) = \frac{j(f/f_B)}{1 + j(f/f_B)} \quad \text{with } f_B = \frac{1}{2\pi RC}$$

Selecting $C = 10\ \mu F$:

$$4.5 \times 10^3 = \frac{1}{2\pi R \times 10\ \mu F}$$
$$R = 3.53\ \Omega$$

   d. Therefore, the transfer function is:

$$H(f) = \frac{j2.22 \times 10^{-4} f}{1 + (j2.22 \times 10^{-4} f)}$$

7.4.1 Cascaded Networks

A **cascaded network** is created by connecting the output of one two-port network to the input of another two-port network (see Figure 7.15). The transfer function of the cascaded filter network corresponds to:

$$H(f) = \frac{V_{out2}}{V_{in1}} \tag{7.17}$$

Multiplying the numerator and denominator of Equation (7.17) by V_{out1} results in:

$$H(f) = \frac{V_{out1}}{V_{in1}} \times \frac{V_{out2}}{V_{out1}} \tag{7.18}$$

Because $V_{out1} = V_{in2}$,

$$H(f) = \frac{V_{out1}}{V_{in1}} \times \frac{V_{out2}}{V_{in2}} \tag{7.19}$$

Thus, it can be seen that the transfer function of the cascaded filters is the product of the individual transfer functions, specifically:

$$H(f) = H_1(f) \times H_2(f) \tag{7.20}$$

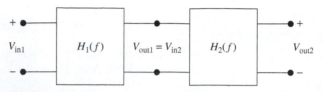

FIGURE 7.15   Cascaded filter network.

APPLICATION EXAMPLE 7.9   Temperature Sensor

A temperature sensor (thermistor) is connected to monitor the temperature of a chemical process that can vary greatly even with a small change in the process [see Figure 7.16(a)]. The goal is to monitor the temperature closely for any small change, to prevent a disaster. The temperature can be modeled by a constant and time-varying component. Any small change in this time-varying component must be closely monitored. Thus, it is important to monitor small changes in temperature. Accordingly, the thermistor-generated signal is sent through two cascaded filters before reaching the control system's computer. The equivalent circuit is shown in Figure 7.16(b). Find the expression for $v_{out}(R_T)$.

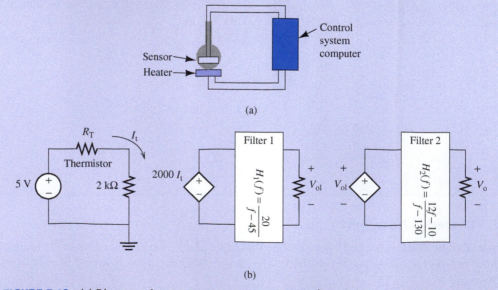

FIGURE 7.16   (a) Diagram of a temperature sensor. (b) Equivalent circuit of a temperature sensor.

SOLUTION

For the circuit in Figure 7.16(b), using a simple KVL:

$$I_t = \frac{5}{R_T + 2000}$$

Therefore, the voltage source for the middle component corresponds to:

$$2000 I_t = 2000 \frac{5}{R_T + 2000}$$

Now, based on the structure of the equivalent circuit:

$$V_{out} = \frac{V_{out}}{V_{out1}} \times \frac{V_{out1}}{2000 I_t} \times 2000 I_t$$

Replacing V_{out}/V_{out1} with $H_2(f)$ and $V_{out1}/2000 I_t$ with $H_1(f)$, and replacing for $2000 I_t$:

$$V_{out} = 2000 \frac{5}{R_T + 2000} \times \frac{20}{f - 45} \times \frac{12f - 10}{f - 130}$$

$$= \frac{2.4 \times 10^6 f - 2.0 \times 10^6}{(R_T + 2000)(f^2 - 175f + 5850)}$$

7.5 SECOND-ORDER FILTERS

An RLC combination can be used to create **band-pass** and **band-stop** filters. These filters are known as **second-order** filters, because the voltage and current through these combinations is represented by a second-order differential equation. Chapter 5 briefly introduced second-order (resonant) circuits that consist of resistance, capacitance, and inductance. Two popular resonant circuits are parallel and series RLC circuits.

7.5.1 Band-Pass Filters

RLC filters can be formed by arranging the components in two methods: parallel and series.

7.5.1.1 PARALLEL RESONANCE

A parallel resonance RLC band-pass filter is shown in Figure 7.17. The **resonance frequency** is the frequency at which the peak (minimum) magnitude occurs for band-pass (band-stop) filters. For an RLC second-order filter, the resonance frequency is given by:

$$\omega_0 = \frac{1}{\sqrt{LC}} \tag{7.21}$$

Accordingly, as discussed earlier, it is observed that the unit of hertz is equivalent to the square root of henry times farad.

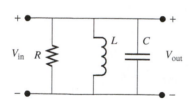

FIGURE 7.17 Parallel resonance band-pass filter.

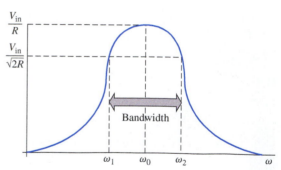

FIGURE 7.18 Amplitude frequency response of a second-order band-pass filter.

As shown in Figure 7.18, the frequency response is centered around the resonance frequency. The *3-dB bandwidth* is defined as the range of frequencies for which the output magnitude is equal to or greater than the maximum output magnitude divided by $\sqrt{2}$. In other words, 3-dB bandwidth is defined as the frequency range, where the magnitude loss is ≤ 3 dB. Equivalently, the *3-dB bandwidth* is also defined as the range of frequencies for which the output power is equal to or greater than the maximum output magnitude divided by 2.

In Figure 7.18, the bandwidth is equal to $\omega_2 - \omega_1$. This bandwidth can be calculated by equating the magnitude of the normalized frequency response $|H(f)|$ to $1/\sqrt{2}$. While in resonant circuits, $|H(f)|$ consists of polynomials of orders two, it has two roots, ω_2 and ω_1. The bandwidth is the difference between these two frequencies. The bandwidth for a parallel resonance filter corresponds to:

$$\text{BW} = \frac{1}{RC} \tag{7.22}$$

Recall that ohm times farad is equal to the time unit of seconds.

Quality factor is a measurement of the sharpness of the frequency selectivity of a filter. It is a measurement of how fast the frequency response drops off outside of the pass band. It should be noted that the sharper the frequency response drops to zero, the better. This allows signals in the desired frequency range to be extracted while avoiding other frequency components, which are interfering with the desired frequency range. The quality factor is the ratio of the resonant frequency, ω_0, to the bandwidth:

$$Q = \frac{\omega_0}{\text{BW}} \tag{7.23}$$

As bandwidth decreases, quality factor increases. This fact is clear from Figure 7.18. Replacing ω_0 and BW using Equations (7.21) and (7.22), the quality factor for a parallel resistance filter is:

$$Q = R\sqrt{\frac{C}{L}} \tag{7.24}$$

7.5.1.2 SERIES RESONANCE

RLC filters can also be arranged in series, as shown in Figure 7.19. A series band-pass filter has the same resonance frequency as its parallel equivalent [see Equation (7.21)]. The bandwidth is given by:

$$\text{BW} = \frac{R}{L} \tag{7.25}$$

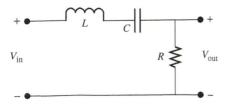

FIGURE 7.19 Series resonance band-pass filter.

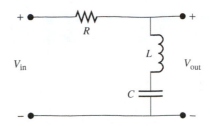

FIGURE 7.20 Second-order band-stop filter.

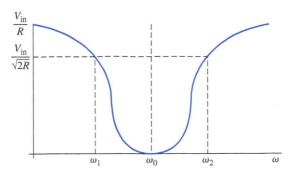

FIGURE 7.21 Amplitude frequency response of a band-stop filter.

Its quality factor corresponds to:

$$Q = \frac{1}{R}\sqrt{\frac{L}{C}} \tag{7.26}$$

7.5.2 Band-Stop Filters

A band-stop filter, also known as a **band-block** or **notch** filter, is created by rearranging a series band-pass filter, as shown in Figure 7.20. The frequency response of a band-stop filter is sketched in Figure 7.21. The resonance frequency, bandwidth, and quality factor are the same as those for series resonance band-pass filters.

APPLICATION EXAMPLE 7.10	**Filtering Out Undesired Frequencies**

An engineer intends to study an input signal with frequency components of 2, 6, and 12 kHz. The engineer desires that all other frequencies be filtered out.

  a. What kind of filter does this require?
  b. Design an appropriate filter.

SOLUTION

  a. This situation requires a band-pass filter to extract the desired frequencies. If the engineer needs to study them one-by-one, a separate filter will be required for each one. But if the engineer desires to extract a frequency range that includes all these frequencies and removes others, a band-pass filter is needed that passes all of these frequency ranges at once. For this example, assume all frequencies are needed at the same time.
  b. The band-pass filter will have a resonance frequency of 7 kHz (the average of the minimum frequency, 2 kHz, and the maximum frequency, 12 kHz). The bandwidth should be at least 12 − 2 kHz = 10 kHz. For this example, select 11 kHz to ensure that all frequencies

(continued)

APPLICATION EXAMPLE 7.10 **(Continued)**

will pass without any distortion. Based on Figure 7.18, recall that the pass band is not sharp. Thus, a bandwidth should be selected that is higher than the actual as a guard band to ensure passage of all desired frequencies.

Now, find R, L, and C for the parallel resonance filter. Using Equation (7.21), the resonance frequency corresponds to:

$$f_0 = 7 \times 10^3 = \frac{1}{\sqrt{LC}2\pi}$$

In addition, using Equation (7.22), the bandwidth is:

$$\text{BW} = 11 \times 10^3 = \frac{1}{RC}$$

Note that because there are two equations and three unknowns, there are an infinite number of solutions. If, for example, $L = 50$ mH, using the resonant equation:

$$7 \times 10^3 = \frac{1}{\sqrt{50 \times 10^{-3} \times C}\,2\pi}$$

$$\sqrt{50 \times 10^{-3} \times C} = \frac{1}{7 \times 10^3 \times 2\pi}$$

$$C = \frac{1}{(7 \times 10^3 \times 2\pi)^2 \times 50 \times 10^{-3}} = 10.34 \times 10^{-9} \text{ F}$$

To find R, use the bandwidth equation:

$$11 \times 10^3 = \frac{1}{R \times 10.34 \times 10^{-9}}$$

$$R = 8.79 \text{ k}\Omega$$

EXERCISE 7.8

In Example 7.10, the desired bandwidth was increased beyond the actual value in order to ensure the passage of all frequencies. Calculate and compare the quality factor for the two bandwidths of 10 and 11 kHz. What effect does the selection of bandwidth have on the quality factor?

Exercise 7.8 shows that the quality factor always reflects the trade-off between the passage of all desired frequencies and the effective usage of bandwidth. As the selected bandwidth increases, the quality factor decreases. In other words, sharp-edge filters are desired to ensure that the minimum bandwidth is utilized.

In the context of wireless communication systems, bandwidth plays an important role. Bandwidth in general is the most valuable resource in wireless communication systems. Wireless service providers pay a high cost to agencies such as the Federal Communication Commission (FCC) in the United States to use a specific bandwidth. Thus, it is desired to utilize the available frequencies as effectively as possible. High-quality filters that pass only the desired band and avoid all undesired interference play an important role in achieving this goal. These filters are usually designed using active elements and are called active filters (see Chapters 8 and 11).

7.6 MATLAB APPLICATIONS

This section outlines how MATLAB can be applied for analysis of filters. MATLAB can be used to plot the frequency response of various filters. To do so, first generate the output $H(f)$ for a given list of frequencies, f, and then plot the actual $H(f)$ versus f. A convenient MATLAB function can also be used to make Bode plots of filters.

EXAMPLE 7.11   **Frequency Response Plot Using MATLAB**

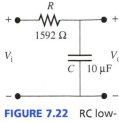

FIGURE 7.22   RC low-pass filter.

The goal of this example is to plot the frequency response of the filter shown in Figure 7.22 using MATLAB.

SOLUTION

Enter the following code into MATLAB:

```
f=[0:0.001:100];          % this creates an array of
                          % frequencies from 0
                          % to 100, incremented by 0.001
                          % The ";" is used after the expression to
                          % suppress printing
RC=1592*10e-6;            % R*C = 1592×10⁻⁶ Ω.F
lp=1./(1+j*2*pi*f*RC);    % Generates an array of
                          % transfer function outputs,
                          % given the frequencies
                          % represented in f.
                          % operator "./" means element by element
                          % division
plot(f,abs(lp));          % plot the magnitude of lp versus f
xlabel('Frequency (Hz)');
ylabel('Magnitude');
```

The plot of Figure 7.23 is the frequency response of the filter shown in Figure 7.22. Can the break frequency be distinguished in the plot shown in Figure 7.23?

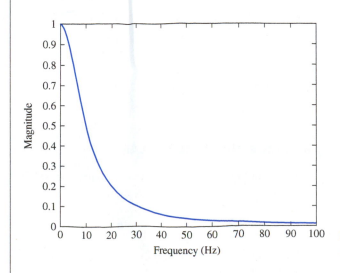

FIGURE 7.23   Frequency response of the filter in Figure 7.22.

(continued)

EXAMPLE 7.11   **Continued**

Note that the plot shown in Figure 7.23 is not a Bode plot. Next, plot the frequency response using a Bode plot. Remember that a Bode plot expresses the magnitude in decibels and uses a logarithmic scale. To do so, enter this code into MATLAB:

```
f=[0.1:0.001:1000];          % this frequency range will
                             % look better on a log scale
                             % than the previous one

lp=1./(1+j*2*pi*f*RC);
lpmag=20*log10(abs(lp));     % obtain magnitude, convert
                             % to dB
semilogx(f,lpmag);           % this plots with a log scale
                             % for the frequency

xlabel('Frequency (Hz)');
ylabel('Magnitude (dB)');
```

The Bode plot for magnitude is shown in Figure 7.24(a). Can the break frequency be distinguished in this figure?

Now, plot the Bode phase plot. Enter the following MATLAB code:

```
lpphase=180/pi*angle(lp);    % get the phase, and convert
                             % it to degrees
semilogx(f,lpphase);
xlabel('Frequency (Hz)');
ylabel('Phase (\circ)');
```

The phase Bode plot is shown in Figure 7.24(b).

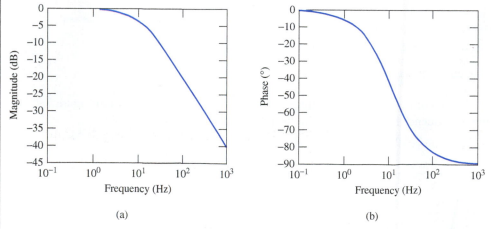

(a)   (b)

FIGURE 7.24   Bode (a) magnitude and (b) phase plot for the filter in Figure 7.22.

EXAMPLE 7.12   **Frequency Response Plot Using MATLAB**

Create the Bode plot for the high-pass filter shown in Figure 7.25 using the same component values.

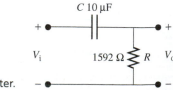

FIGURE 7.25   RC high-pass filter.

SOLUTION

To sketch the frequency response of an RC high-pass filter, use Equations (7.15) and (7.16). Similar to the approach of Example 7.11, run the MATLAB code below to sketch the plots.

```
f=[0.1:0.001:1000];

hp=(j*2*pi*f*RC)./(1+j*2*pi*f*RC);
hpmag=20*log10(abs(hp));
semilogx(f,hpmag);
xlabel('Frequency (Hz)');
ylabel('Magnitude (dB)');

hpphase=180/pi*angle(hp);
semilogx(f,hpphase);
xlabel('Frequency (Hz)');
ylabel('Phase (\circ)');
```

The high-pass Bode magnitude and phase plots are sketched in Figure 7.26.

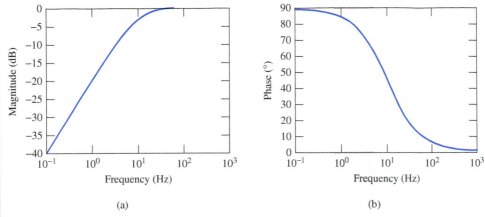

(a)   (b)

FIGURE 7.26 Bode (a) magnitude and (b) phase plot for the filter in Figure 7.25.

EXAMPLE 7.13   **Transfer Function of Cascade Filter Using MATLAB**

Cascade the low-pass and the high-pass filters from Examples 7.11 and 7.12. Sketch the transfer function using MATLAB.

SOLUTION

Remember, the transfer function of cascaded filters is the product of the transfer function of the individual filters. Cascading the filters produces a second-order filter. Thus, based on the results of Examples 7.11 and 7.12, apply the following MATLAB code.

```
cf=lp.*hp;                  % cascade the two filters
cfmag=20*log10(abs(cf));
cfphase=180/pi*angle(hp);

semilogx(f,cfmag);
xlabel('Frequency (Hz)');
ylabel('Magnitude (dB)');

semilogx(f,cfphase);
xlabel('Frequency (Hz)');
ylabel('Phase (\circ)');
```

The magnitude and phase are plotted in Figure 7.27.

(continued)

EXAMPLE 7.13 **Continued**

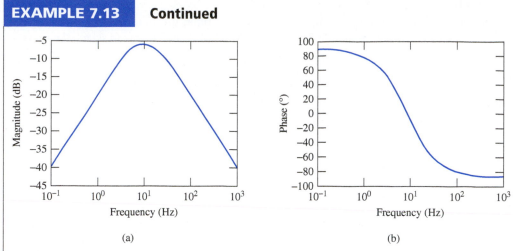

(a)   (b)

FIGURE 7.27 Bode (a) magnitude and (b) phase plot for the cascaded filter.

EXAMPLE 7.14 **Using the MATLAB Bode Plot Function**

MATLAB has its own convenient Bode plot function that will produce a Bode plot from any given transfer function. The Bode function in MATLAB, however, requires the transfer function to be created using the "tf" function; "tf" requires the use of s as the transfer function variable instead of f. *Note:* the "tf" function and the "Bode" function are set up using the control toolbox. The transfer function must be a function of s, and not a function of f. In other words, $H(f)$ must be converted to $H(s)$ (i.e., Laplace transform). Thus, using:

$$s = j2\pi f \tag{7.27}$$

the low-pass transfer function becomes

$$H_{\text{low-pass}}(s) = \frac{1}{1 + RCs} \tag{7.28}$$

In addition, the high-pass transfer function corresponds to:

$$H_{\text{high-pass}}(s) = \frac{RCs}{1 + RCs} \tag{7.29}$$

To create the transfer functions in MATLAB, enter:

```
RC=1592*10e-6;
s=tf('s');
```

This creates a transfer function containing only the variable s. This makes it easier to create the transfer functions in MATLAB. To create the transfer functions, enter

```
lp=1/(RC*s+1);
hp=RC*s/(RC*s+1);
```

Now, create the Bode plots:

```
bode(lp);
bode(hp);
```

The resulting plots are shown in Figure 7.28.

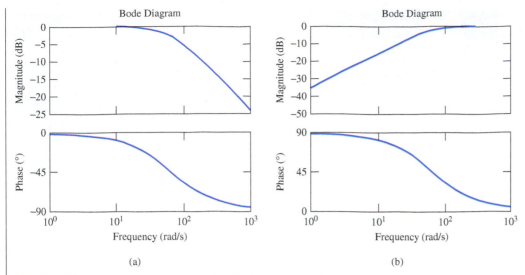

FIGURE 7.28 MATLAB (a) low-pass and (b) high-pass Bode plots.

The entire MATLAB code for this example is:

```
RC=1592*10e-6;
s=tf('s');
lp=1/(RC*s+1);
hp=RC*s/(RC*s+1);
bode(lp);
bode(hp);
```

The MATLAB Bode function plots both the frequency and phase responses of the transfer function. Note that for the MATLAB Bode function, the frequencies are given in rad/s = $2\pi f$. Thus, MATLAB is a helpful tool in analyzing the frequency response of a filter transfer function.

APPLICATION EXAMPLE 7.15   An Automatic Transmission in a Car

Figure 7.29 represents a block diagram of the automatic transmission of a car. The built-in computer detects the engine's torque and sends a signal to the servo to adjust the gear appropriately to provide the maximum transmission.

For this example, assume that:

The servo has a transfer function of $G(j\omega) = \dfrac{10}{0.1j\omega + 1}$.

The engine has a transfer function of $H(j\omega) = \dfrac{10}{20j\omega + 1}$.

1. Determine the transfer function of the system which is a combination of the servo and the engine.
2. Use MATLAB to plot the Bode diagram for the system.

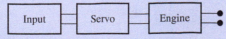

FIGURE 7.29 Block diagram of an automatic transmission of a car.

(continued)

APPLICATION EXAMPLE 7.15   Continued

SOLUTION

1. The transfer function of the system corresponds to:

$$\text{TF}(s) = G(s) \times H(s)$$

where

$$G(s) = \frac{10}{0.1s + 1}$$

$$H(s) = \frac{10}{20s + 1}$$

Therefore, the total transfer function is:

$$\text{TF}(s) = \frac{10}{0.1s + 1} \times \frac{10}{20s + 1}$$

$$\text{TF}(s) = \frac{100}{2s^2 + 20.1s + 1}$$

2. The MATLAB code to sketch the amplitude and phase diagram of this system is:

```
num = [100]
den = [2 20.1 1]
sys = tf(num,den)
bode(sys)
```

The Bode diagram from MATLAB is shown in Figure 7.30.

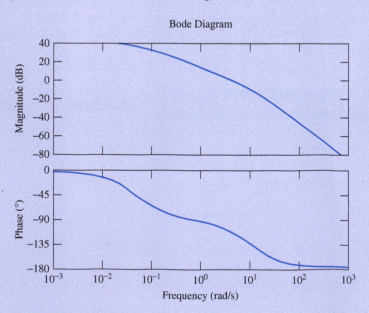

FIGURE 7.30 Bode plot for the automatic transmission system.

APPLICATION EXAMPLE 7.16   Elevator

Figure 7.31 shows a simple block diagram of an elevator. Passengers select their destination floor by choosing the floor on the input/control. The input/control processes the message and orders the motor to move the elevator to the particular floor.

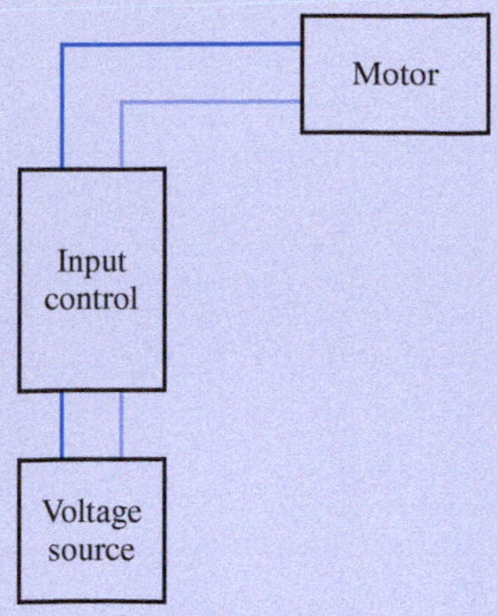

Motor

Input control

Voltage source

FIGURE 7.31   Block diagram of an elevator.

The voltage source $v_S(t) = 160\cos(120t)$, and the input and motor system has an inductance of 30 mH and a resistance of 20 Ω.

a. Determine the transfer function of the system.
b. Sketch the Bode plot of the transfer function.

SOLUTION

a. Figure 7.32 shows that due to the internal inductance of the motor, the system acts as a low-pass filter. Therefore, the break frequency for the transfer function is:

$$f_B = \frac{R}{2\pi L}$$

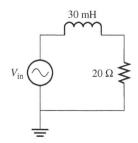

FIGURE 7.32   Elevator circuit.

R is the total resistance, where $R = 20\ \Omega$, and the inductance $L = 0.03$ H; thus, $f_B = 106.1$ Hz.

The low-pass filter transfer function corresponds to:

$$H(f) = \frac{1}{1 + j(f/f_B)}$$

For this example:

$$H(f) = \frac{1}{1 + j0.0094f}$$

b. To sketch the Bode plot, first express the transfer function in the s-domain. Given that $s = j\omega = j2\pi f$, in this example $f = s/2\pi j$. In addition, the transfer function in the s-domain is:

$$H(s) = \frac{1}{1 + 0.0015s}$$

The MATLAB code corresponds to:

```
num = [1]
den = [0.0015 1]
sys = tf(num,den)
bode(sys)
```

The Bode plot diagram is shown in Figure 7.33.

(continued)

APPLICATION EXAMPLE 7.16   **Continued**

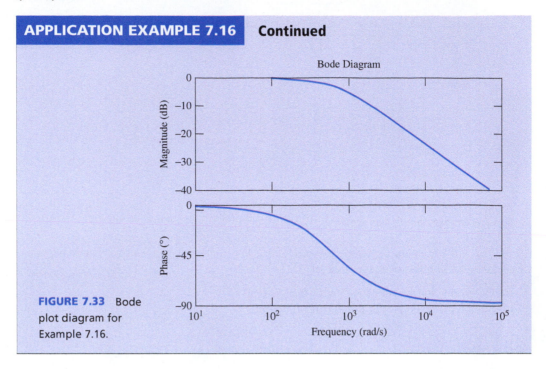

FIGURE 7.33 Bode plot diagram for Example 7.16.

7.7 FREQUENCY RESPONSE ANALYSIS USING PSPICE

This section provides a tutorial on frequency analysis using PSpice. Many examples are provided to clarify the process.

EXAMPLE 7.17   **PSpice Analysis**

For the circuit shown in Figure 7.34, with the voltage source $V(t) = 10\cos(200\pi t)$, use PSpice to plot the frequency response of the $v_{out}(t)$.

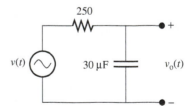

FIGURE 7.34 Circuit for Example 7.17.

SOLUTION

Given that the voltage $v(t) = 10\cos(200\pi t)$, first determine the frequency and phase angle. The frequency is:

$$2\pi f = \omega$$
$$2\pi f = 200\pi$$
$$f = 100 \text{ Hz}$$

The phase angle, θ, is 90° for the VSIN voltage source. See Examples 2.24 and 6.27 for detailed instructions about how to set up a VSIN voltage source. The PSpice schematic solution is shown in Figure 7.35.

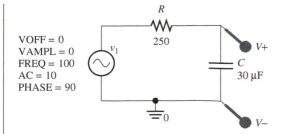

FIGURE 7.35 PSpice schematic solution for Example 7.17.

To plot the frequency response, it is important to set the magnitude of VSIN to AC. As shown in Figure 7.36, go to the "properties" of VSIN and set the appropriate value of AC (see the black arrow), which is 10 V. Go to display properties and choose "Name and Value" (see the colored arrow) as the display format.

Next, the simulation is required to be set to run in frequency analysis. Go to "New Simulation Profile," and choose the simulation analysis to be "AC Sweep/Noise." Choose "linear" as the AC sweep type, with the start and end frequency as shown in Figure 7.37.

Press "Run" to produce the simulation plot shown in Figure 7.38.

Next, go to "Edit Simulation Setting" and change the AC sweep type to "Logarithmic" as shown in Figure 7.39. The logarithmic frequency analysis is shown in Figure 7.40. The plot is also known as the Bode plot, as discussed in this chapter.

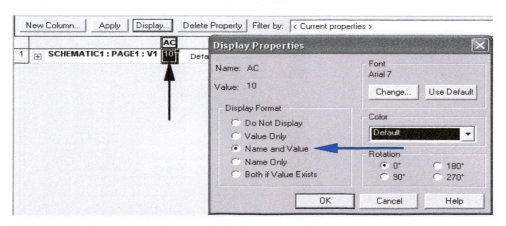

FIGURE 7.36 How to set up the magnitude in AC.

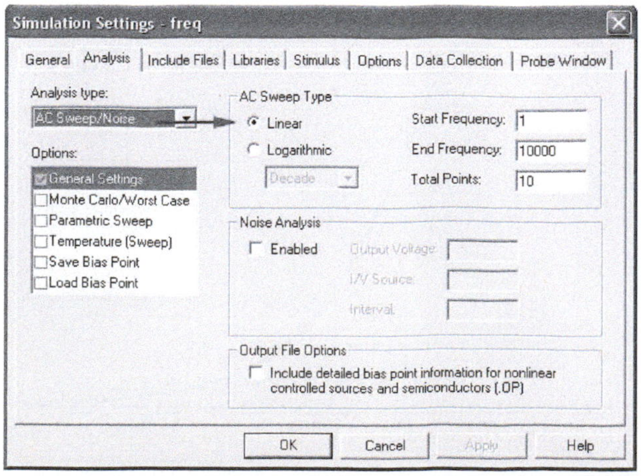

FIGURE 7.37 How to set up a linear frequency analysis.

(*continued*)

EXAMPLE 7.17   **Continued**

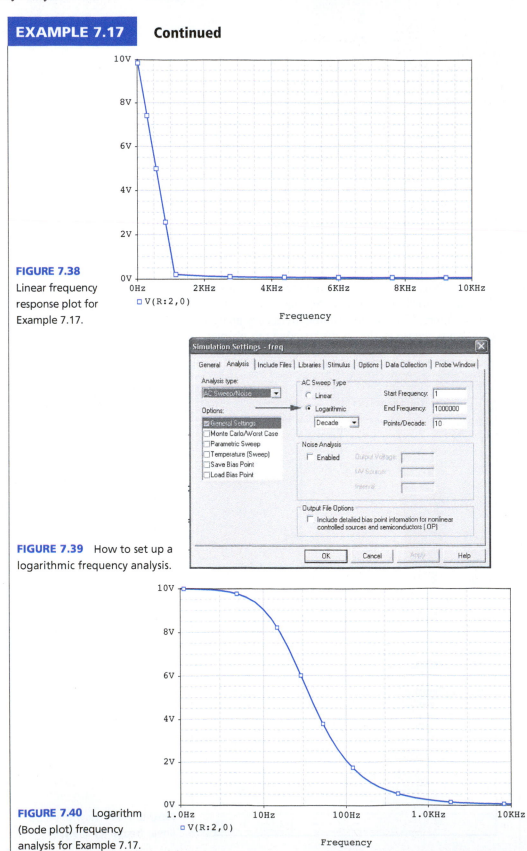

FIGURE 7.38
Linear frequency
response plot for
Example 7.17.

FIGURE 7.39 How to set up a
logarithmic frequency analysis.

FIGURE 7.40 Logarithm
(Bode plot) frequency
analysis for Example 7.17.

EXAMPLE 7.18   PSpice Analysis

Given the low-pass filter circuit shown in Figure 7.41, with a voltage source $v(t) = 3$ V $+$ $5\cos(400\pi t + 30°)$, use PSpice to sketch the output voltage Bode plot.

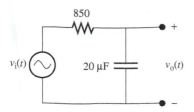

FIGURE 7.41 Circuit for Example 7.18.

SOLUTION

Given the voltage source $v(t) = 3$ V $+ 5\cos(400\pi t + 30°)$, express $v(t)$ as:

$$v(t) = V_1 + v_2(t)$$

where

$$V_1 = 3 \text{ V}$$
$$v_2(t) = 5\cos(400\pi t + 30°)$$

Then, the frequency for $V_2(t)$ is:

$$f = \frac{\omega}{2\pi}$$
$$= \frac{400\pi}{2\pi}$$
$$= 200 \text{ Hz}$$

In addition, the phase angle is:

$$\theta = 30° + 90°$$
$$= 120°$$

See Example 7.17 for instructions on how to set up the VSIN voltage source in AC. The PSpice schematic solution is shown in Figure 7.42.

Next, set up a simulation with a logarithmic frequency analysis and run the simulation (see Figure 7.37 for Example 7.17). The Bode plot consists of a magnitude and a phase angle plot. To add the phase angle plot in PSpice, go to "Plot" (see the black arrow) and choose "Add Plot to Window" (see the colored arrow) in Figure 7.43.

Next, go to "Add Trace" and click on "V2(C2)" (see Figures 2.77 and 7.44). "V2(C2)" will show up in the "Trace Expression" column. Then, change it to "V2P(C2)" (see Figure 7.44) so that the output of the plot becomes a phase angle plot, as shown in Figure 7.45.

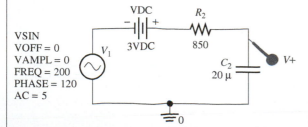

FIGURE 7.42 PSpice schematic solution for Example 7.18.

(continued)

EXAMPLE 7.18   **Continued**

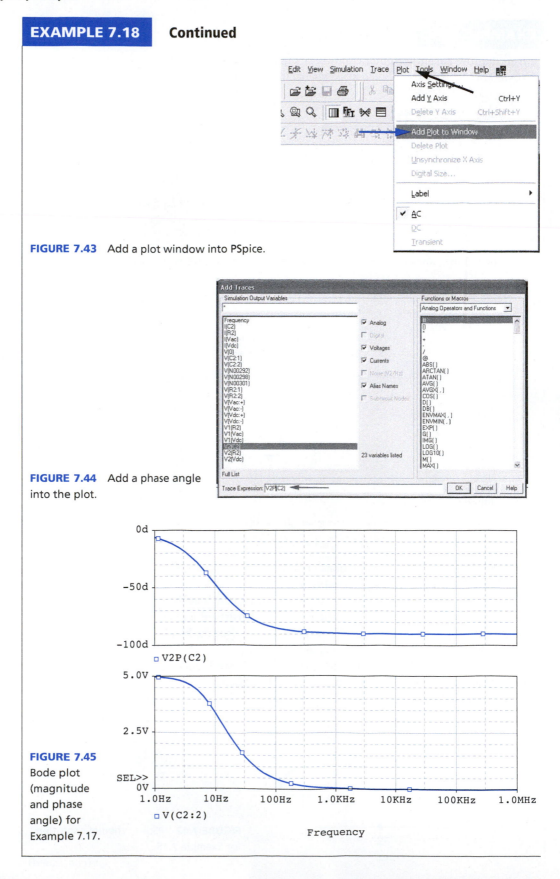

FIGURE 7.43   Add a plot window into PSpice.

FIGURE 7.44   Add a phase angle into the plot.

FIGURE 7.45
Bode plot (magnitude and phase angle) for Example 7.17.

EXAMPLE 7.19   **PSpice Analysis**

Use PSpice to sketch the Bode plot of v_{out} for the high-pass filter circuit shown in Figure 7.46, with input voltage $v_{in}(t) = 12 \cos(200\pi t + 45°)$.

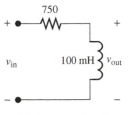

FIGURE 7.46   Circuit for Example 7.19.

SOLUTION

The voltage is given as $v_{in}(t) = 12 \cos(200\pi t + 45°)$. Thus, its frequency is:

$$f = \frac{200\pi}{2\pi}$$
$$= 100 \text{ Hz}$$

The phase angle for the VSIN in PSpice is:

$$\theta = 90° + 45°$$
$$= 135°$$

Follow the instructions for Example 7.17 to set up an AC VSIN voltage source and the instructions for Example 7.18 to set up the Bode plot. The PSpice schematic solution and the Bode plot are shown in Figures 7.47 and 7.48.

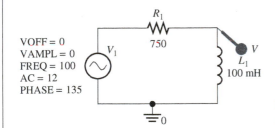

FIGURE 7.47   PSpice schematic solution for Example 7.19.

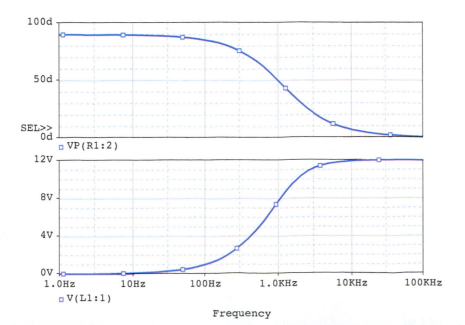

FIGURE 7.48   Bode plot (magnitude and phase angle) for Example 7.19.

EXAMPLE 7.20 PSpice Analysis

Use PSpice to plot the frequency response of voltage across the inductor and capacitor, in terms of magnitude and phase angle for a second-order, band-pass-stop filter shown in Figure 7.49. The voltage source is $v(t) = 5V + 12 \sin(200\pi t + 30°)$.

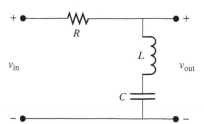

FIGURE 7.49 Circuit for Example 7.20.

SOLUTION

The voltage source $v(t) = 5 \text{ V} + 12 \sin(200\pi t + 30°)$ can be expressed as:

$$v(t) = V_1 + v_2(t)$$

where

$$V_1 = 5 \text{ V}$$
$$v_2(t) = 12 \sin(200\pi t + 30°) \text{ V}$$

Thus, the frequency of $v_2(t)$ is:

$$f = \frac{200\pi}{2\pi}$$
$$= 100 \text{ Hz}$$

In addition, the phase angle is:

$$\theta = 30°$$

Follow the steps shown in Example 7.17 to set up the AC VSIN voltage source. The PSpice schematic solution is shown in Figure 7.50.

To observe the Bode plot for the inductor and capacitor, one approach is to use the "Voltage Differential Marker(s)" and to place a V+ before the inductor and V− after the capacitor. Then, follow the steps outlined in Example 7.18 to set up the Bode plot. The Bode plot for the output is shown in Figure 7.51.

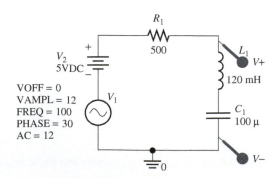

FIGURE 7.50 PSpice schematic solution for Example 7.20.

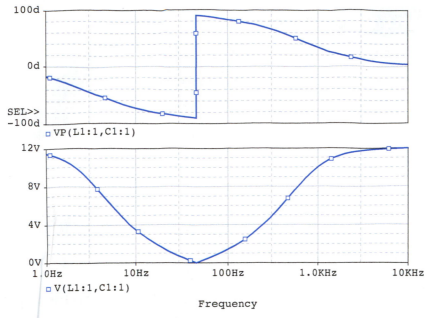

□ VP(L1:1,C1:1)

□ V(L1:1,C1:1)

Frequency

FIGURE 7.51 Bode plot (magnitude and phase angle) for Example 7.20.

EXAMPLE 7.21   **PSpice Analysis**

Figure 7.52 shows a simple block diagram of an elevator. The passenger selects a destination floor by choosing the floor on the input/control. The input/control processes the message and orders the motor to move the elevator to the particular floor.

For this example, the voltage source $v_S(t) = 160 \cos(120\pi t)$, the input control has a resistance of 100 Ω, the motor's resistance is 20 Ω, and the motor's inductance is 0.03 H.

a. Determine the transfer function of the system.

b. Use PSpice to plot the Bode diagram for the motor.

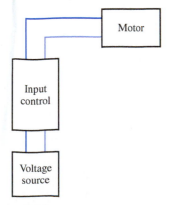

FIGURE 7.52   Block diagram of an elevator.

SOLUTION

a. Based on the equivalent circuit given in the problem for the component, Figure 7.53 represents the equivalent circuit of Figure 7.52. Thus, the system is a high-pass filter with a break frequency of:

$$f_B = \frac{L}{2\pi R}$$

(continued)

> **EXAMPLE 7.21** Continued
>
> R is the total resistance, which is:
>
> $$R = 100 + 20$$
> $$= 120 \ \Omega$$
>
> The inductance is $L = 0.03$ H. The high-pass filter transfer function is defined as:
>
> $$H(f) = \frac{j(f/f_B)}{1 + j(f/f_B)}$$
>
> with $f_B = 4 \times 10^{-4}$ Hz.
>
> $$H(f) = \frac{j2.5 \times 10^3 f}{1 + j2.5 \times 10^3 f}$$
>
> **b.** First, determine the frequency and phase angle for the VSIN in PSpice. If the voltage source, $v_S(t) = 160 \cos(120\pi t)$; the frequency is:
>
> $$f = \frac{120\pi}{2\pi}$$
> $$= 60 \text{ Hz}$$
>
> The voltage source is a cosine function; thus, the phase angle of the VSIN is $\theta = 90°$. The PSpice schematic and the Bode plot diagrams are shown in Figures 7.54 and 7.55.

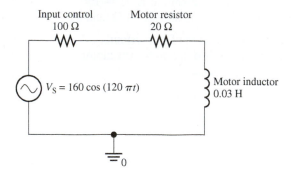

FIGURE 7.53 Circuit for the elevator circuit.

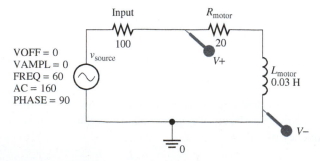

FIGURE 7.54 PSpice schematic solution for the elevator system.

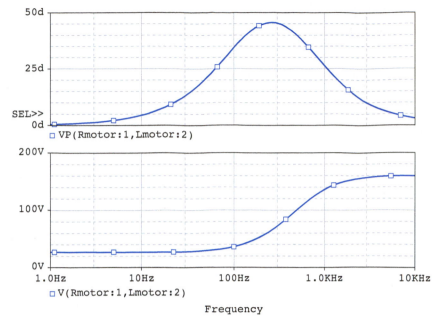

FIGURE 7.55 Bode plot diagram for the elevator system in Example 7.21.

7.8 WHAT DID YOU LEARN?

- Transfer function is defined as [Equation (7.1)]:

$$H(f) = \frac{V_{\text{out}}}{V_{\text{in}}}$$

- First-order filters have the following characteristics:

	Circuits	Break frequency, f_B	Transfer function
Low-pass filters	RC [Figure 7.2(a)]	$\dfrac{1}{2\pi RC}$	$H(f) = \dfrac{1}{1 + j(f/f_B)}$
	RL [Figure 7.2(b)]	$\dfrac{R}{2\pi L}$	
High-pass filters	RC [Figure 7.10(a)]	$\dfrac{1}{2\pi RC}$	$|H(f)| = \dfrac{f/f_B}{\sqrt{1 + (f/f_B)^2}}$
	RL [Figure 7.10(b)]	$\dfrac{L}{2\pi R}$	

- Magnitude, $|H(f)|$, is often expressed in *decibels* [Equation (7.9)]:

$$|H(f)|_{\text{dB}} = 20 \log_{10}|H(f)|$$

- A *Bode plot* is commonly used to plot a frequency response across a large range of frequency values.

    To draw a Bode plot for magnitude:

1. Identify f_B and label it on the frequency axis.
2. Draw the asymptotes with appropriate slopes:
    (a) -20 dB/decade for a first-order, low-pass filter
    (b) $+20$ dB/decade for a first-order, high-pass filter

To draw a Bode plot for phase:
1. Mark the frequencies $f = f_B/10$ and $f = 10f_B$.
2. Draw the horizontal asymptotes:
   (a) For a first-order, low-pass filter, draw an asymptote at $0°$ for $f \le f_B/10$, and an asymptote at $-90°$ for $f \ge 10f_B$.
   (b) For a first-order, high-pass filter, draw an asymptote at $90°$ for $f \le f_B/10$, and an asymptote at $0°$ for $f \ge 10f_B$.
3. Draw the diagonal asymptote, connecting the horizontal asymptotes.
• When two filters are cascaded, the total transfer function is the product of the individual transfer functions, as shown in Equation (7.20):

$$H(f) = H_1(f) \times H_2(f)$$

• Second-order filters have the following characteristics:

	Circuit	Resonance frequency, ω_0	Bandwidth, BW	Quality factor, Q
Band pass	RLC in parallel (Figure 7.16)	$\dfrac{1}{\sqrt{LC}}$	$\dfrac{1}{R \cdot C}$	$R\sqrt{\dfrac{C}{L}}$
	RLC in series (Figure 7.18)	$\dfrac{1}{\sqrt{LC}}$	$\dfrac{R}{L}$	$\dfrac{1}{R}\sqrt{\dfrac{L}{C}}$
Band stop	RLC in series (Figure 7.19)	$\dfrac{1}{\sqrt{LC}}$	$\dfrac{R}{L}$	$\dfrac{1}{R}\sqrt{\dfrac{L}{C}}$

Problems

*B refers to Basic, A refers to Average, H refers to Hard, and * refers to problems with answers.*

SECTION 7.2 FIRST-ORDER FILTERS

7.1 (B)* Find the transfer function for the filter of Figure P7.1:

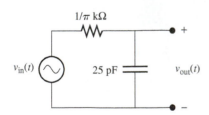

FIGURE P7.1 Circuit for Problem 7.1.

7.2 (B) Find the transfer function for the filter of Figure P7.2.

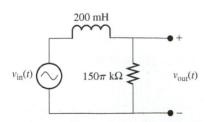

FIGURE P7.2 Circuit for Problem 7.2.

7.3 (A) Find the transfer function for the filter of Figure P7.3.

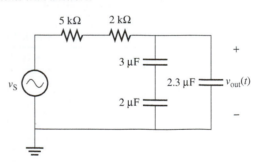

FIGURE P7.3 Circuit for Problem 7.3.

7.4 (A)* Find the transfer function for the circuit in Figure P7.4.

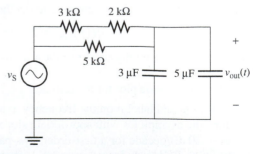

FIGURE P7.4 Circuit for Problem 7.4.

7.5 (B) What is the break frequency for the following transfer functions with given frequency?

(a) $H = \dfrac{1}{1 + j50f}$

(b) $H = \dfrac{1}{1 + j\dfrac{f}{10}}$

7.6 (H)* What is the break frequency for the following transfer functions?

(a) $H = \dfrac{1}{s + 5}$

(b) $H = \dfrac{s + 3}{s^2 + 7s + 12}$

7.7 (H) Determine the output for the following transfer function at 20 Hz, if the input voltage is $10\angle 0°$.

(a) $H = \dfrac{1}{1 + j\dfrac{f}{10}}$

(b) $H = \dfrac{1}{1 + j5f}$

SECTION 7.3 LOW-PASS FILTERS

7.8 (B) Given that the transfer function of a circuit is $1/[1 + j(f/40)]$, what is the output voltage if input voltage is:

(a) $v_{in} = 5\, v$

(b) $v_{in} = 5 + 10 \cos(100\pi t + 90°) + 6 \cos(50\pi t)$ V

7.9 (A)* If the ratio between resistance and capacitance is $R/C = 625$, what is the required capacitance if the break frequency is 800 Hz?

7.10 (A) What is the transfer function and the output voltage for the circuit in Figure P7.10 given the input voltage is $v_{in}(t) = 10 + 8 \cos(50\pi t - 90°) + 4 \cos(100\pi t + 60°)$?

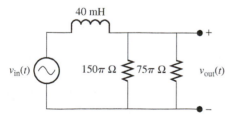

FIGURE P7.10   Circuit for Problem 7.10.

7.11 (A) Find the output, $v_{out}(t)$, of the filter in Figure P7.11, given:

$$v_{in}(t) = 15 + 2 \cos(100\pi t - 30°) + 6 \cos(300\pi t + 50°) + 2.5 \sin(400\pi t)$$

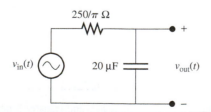

FIGURE P7.11   Circuit for Problem 7.11.

7.12 (A) Find the output, $v_{out}(t)$, of the filter in Figure P7.12, given the input:

$$v_{in}(t) = 3 + 2 \cos(40{,}000\pi t - 45°) + 5 \cos(22{,}000\pi t + 50°)$$

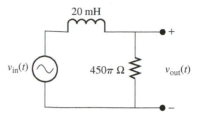

FIGURE P7.12   Circuit for Problem 7.12.

7.13 (H)* A multi-frequency signal is input to a first-order, low-pass filter. A 50-kHz component is found to be reduced in amplitude by a factor of 100.

(a) Find the break frequency of this filter.

(b) Also determine the factor of reduction for the 5-kHz component.

7.14 (A)* Find the transfer function for the circuit in Figure P7.14.

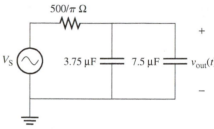

FIGURE P7.14   Circuit for Problem 7.14.

Transfer Functions

7.15 (B) What is the transfer function of the plot in Figure P7.15?

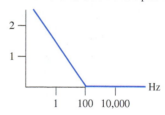

FIGURE P7.15   Magnitude plot for Problem 7.15.

7.16 (B)* Find the output of the filter represented by the frequency response in Figure P7.16, given this input:

$$v_{in}(t) = 3 \cos(2 \times 10^5 \pi t - 30°)$$

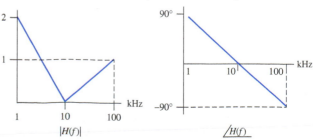

FIGURE P7.16   Frequency response for Problem 7.16.

7.17 (A) Determine type of the transfer function for Figure P7.17. Then determine the output voltage if input voltage $v_{in}(t) = 8 \cos(20\pi t - 45°) + 5 \cos(200\pi t + 30°)$.

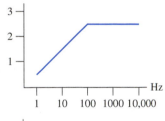

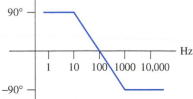

FIGURE P7.17   Frequency response for Problem 7.17.

7.18 (B) Sketch the magnitude plot of the transfer function for the circuit shown in Figure P7.18.

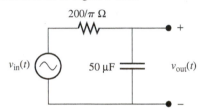

FIGURE P7.18   Circuit for Problem 7.18.

7.19 (A) Sketch the Bode plot of the transfer function for the circuit shown in Figure P7.19.

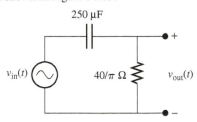

FIGURE P7.19   Circuit for Problem 7.19.

7.20 (B)* Find the output of the filter represented by the frequency response in Figure P7.20, given this input:

$$v_{in}(t) = 4 \cos(200{,}000\pi t + 60°)$$

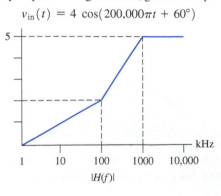

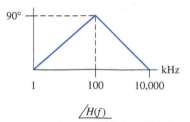

FIGURE P7.20   Frequency response for Problem 7.20.

7.21 (A) Find the output of the filter represented by the frequency response in Figure P7.21, given the input:

$$v_{in}(t) = 7 + 2 \cos(2000\pi \cdot t + 30°) + 5 \cos(20{,}000\pi t - 45°)$$
$$+ 9 \cos(2{,}000{,}000\pi t + 180°)$$

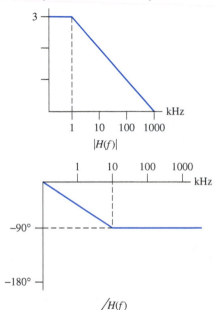

FIGURE P7.21   Frequency response for Problem 7.2.

7.22 (H) Find the output of the filter represented by the frequency response in Figure P7.22, given this input:

$$v_{in}(t) = 12 + 2 \cos(4000\pi t - 30°) + 5 \cos(6000\pi t)$$
$$+ 5 \cos(12{,}000\pi t + 45°)$$

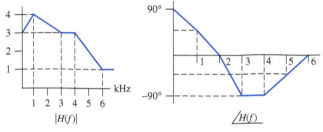

FIGURE P7.22   Frequency response for Problem 7.22.

SECTION 7.4 HIGH-PASS FILTERS

7.23 (B) Find the break frequency and transfer function for the filter of Figure P7.23.

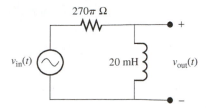

FIGURE P7.23 Filter for Problem 7.23.

7.24 (B)* Determine the break frequency and transfer function of the circuit in Figure P7.24.

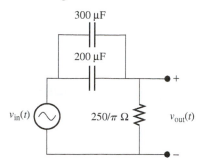

FIGURE P7.24 Filter for Problem 7.24.

7.25 (A) If the transfer function of the circuit in Figure P7.25 is $(jf/500)/[1 + (jf/500)]$, what is the capacitance, C, in the circuit?

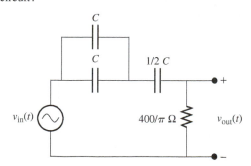

FIGURE P7.25 Filter for Problem 7.25.

7.26 (A) Find the output, $v_{out}(t)$, of the filter in Figure P7.26, given the input:

$$v_{in}(t) = 15 \cos(10{,}000\pi t - 95°) + 12 \sin(15{,}000\pi t - 15°)$$

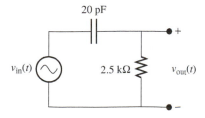

FIGURE P7.26 Filter for Problem 7.26.

7.27 (A) Find the output voltage of the circuit in Figure P7.27. Given that input voltage $v_{in} = 5 + 6 \cos(100\pi t + 30°) + 4 \sin(3000\pi t + 25°)$, what happens to the output voltage at low and high frequencies?

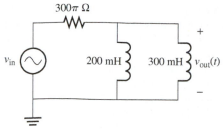

FIGURE P7.27 Filter for Problem 7.27.

7.28 (A)* What kind of filter is shown in Figure P7.28? Determine its transfer function.

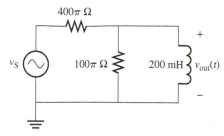

FIGURE P7.28 Circuit for Problem 7.28.

7.29 (A) A high-pass filter is used to drive a load, R_{load}, as shown in Figure P7.29. What is its break frequency and transfer function?

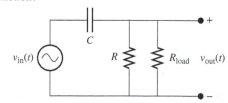

FIGURE P7.29 Circuit for Problem 7.29.

7.30 (H) As shown in Figure P7.30, a source with internal resistance is connected through a filter to drive a 15-kΩ load. Find the break frequency, f_B.

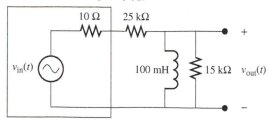

FIGURE P7.30 Circuit for Problem 7.30.

7.31 (A) What kind of combination filter is shown in Figure P7.31? Find the transfer function for the circuit between V_{bout} and V_S.

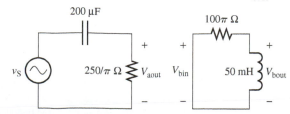

FIGURE P7.31 Circuit for Problem 7.31.

7.32 (A)* A circuit is designed such that it passes a 10-kHz frequency signal, but blocks any signal that has a frequency lower than 5 kHz. If the available resistance of the resistor in the circuit is 1 kΩ, select a suitable range of capacitance which corresponds to the appropriate range of break frequency that meets the design requirements.

7.33 (H) Two transfer functions are $H_1 = 10/[1 + j(f/10)]$ and $H_2 = 100/[1 + j(f/10)]$. Those two filters are cascaded and the new transfer function is:

$$H = H_1 \times H_2$$

(a) Find the DC gain of the cascaded filter.
(b) Draw the Bode plot of the cascaded filter and determine the type of filter: low-pass or high-pass filter.

SECTION 7.5 SECOND-ORDER FILTERS

7.34 (B) Instead of using an RLC circuit to construct a band-pass filter, is it possible to use either an RL or an RC circuit to construct a band-pass filter? If so, then how?

7.35 (A)* The quality factor of a parallel resonance band-pass filter is found to be 0.05. If the same resistor, capacitor, and inductor are used to construct a series resonance band-pass filter, what is the quality factor of the filter?

7.36 (B) A parallel resonance second-order filter has $L = 350$ mH, $R = 20$ kΩ, and $C = 1.025$ μF. Find the break frequency, f_B, and bandwidth, BW.

7.37 (A) Design a parallel resonance filter with $f_B = 1000$ kHz, and bandwidth = 10 kHz, given the inductor $L = 100$ μH. Find R and C.

7.38 (A)* From the circuit shown in Figure P7.38, determine:
(a) The type of filter.
(b) The bandwidth and break frequency.

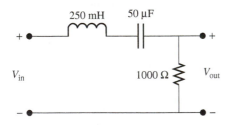

FIGURE P7.38 Circuit for Problem 7.38.

7.39 (H) (Application) An engineer intends to study an input signal with frequency components of 5, 6, and 10 kHz. It is desired that all other frequencies be filtered out.
(a) What kind of filter does this require?
(b) Design an appropriate filter, using $L = 200$ mH, and the MINIMUM bandwidth.

7.40 (H)* Application: AM radio tuner:
The circuit of an AM radio tuner is shown in Figure P7.40. The frequency range for AM broadcasting is 540 to 1600 kHz. Given that $L = 1$ μH, what must be the range of C to have the resonant frequency adjustable from one end of the AM band to another?

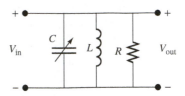

FIGURE P7.40 Circuit for Problem 7.40.

7.41 (H) AM channels are spaced in 10 kHz increments, that is, 540, 550, 1590, and 1600 kHz. The bandwidth of the AM radio tuner must not exceed 10 kHz, in order to suppress the interference of neighboring channels. The circuit of the tuner is shown in Figure P7.40. Find the conditions under which the interference from neighboring AM channels is filtered out.

SECTION 7.6 MATLAB APPLICATIONS

7.42 (B) Generate MATLAB code to generate a non-Bode frequency response plot for a filter having $R = 3$ kΩ and $C = 0.025$ μF.

7.43 (A) Generate MATLAB code to generate a Bode magnitude and phase plot for a low-pass filter having $R = 2$ kΩ and $C = 0.1$ μF. Do *not* use the MATLAB Bode function.

7.44 (A) Repeat Problem 7.39, using the MATLAB Bode function.

7.45 (A) Given the transfer function $s/(s + 1000)$, use MATLAB to plot the Bode diagram, then determine the approximate break frequency from the Bode plot diagram.

7.46 (A)* Determine the transfer function for the circuit in Figure P7.46. Then, use MATLAB to plot the Bode plot for the circuit.

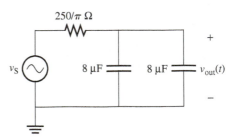

FIGURE P7.46 Circuit for Problem 7.46.

SECTION 7.7 FREQUENCY RESPONSE ANALYSIS USING PSPICE

7.47 (B) Given the circuit in Figure P7.47, with an AC voltage source (Amplitude 12 V), use PSpice to sketch the Bode plot of $v_{out}(t)$.

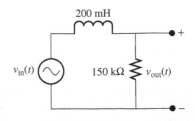

FIGURE P7.47 Circuit for Problem 7.47.

7.48 (A) Use PSpice for Bode plot analysis of the circuit shown in Figure P7.48, given the voltage source $v(t) = 15 + 5 \sin(1000\pi t) + 8 \cos(4000\pi t + 30°)$.

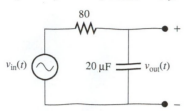

FIGURE P7.48   Circuit for Problem 7.48.

7.49 (A) Use PSpice to sketch the Bode plot of the circuit shown in Figure P7.49, given the voltage source $v(t) = 5 \sin(600\pi t + 45°) + 10 \cos(100\pi t)$.

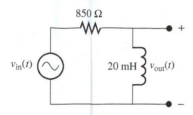

FIGURE P7.49   Circuit for Problem 7.49.

7.50 (H) Given the voltage source $v(t) = 10 + 5 \sin(2000\pi t) + 2 \sin(900\pi t + 45°)$, use PSpice to sketch the Bode plot of the circuit shown in Figure P7.50. What kind of filter is the Device A?

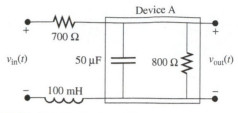

FIGURE P7.50   Circuit for Problem 7.50.

CHAPTER 8

Electronic Circuits

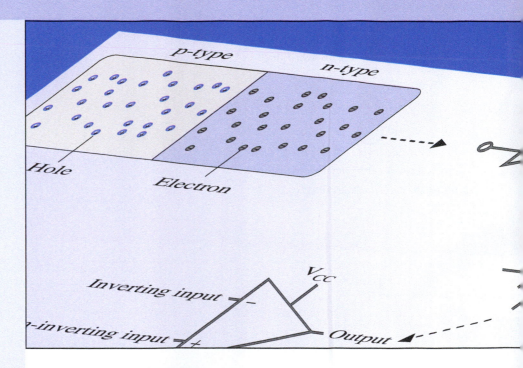

8.1 INTRODUCTION

All electronic systems and circuits, including radio, television, cell phone, and others consist of many electronic components. In general, electronic components are divided into two main categories: passive and active. The nature of the element—active or passive—is determined by the element's energy, $w(t)$, which is defined as:

$$w(t) = \int_{-\infty}^{t} p(\alpha)\, d\alpha = \int_{-\infty}^{t} v(\alpha) \cdot i(\alpha)\, d\alpha$$

Passive components are those that only absorb energy. In this case, the energy $w(t)$ [$v(t) \cdot i(t)$] is always positive. In contrast, active elements are those that may deliver energy. In this case, their energy is negative. Resistors, capacitors, and inductors are examples of passive elements. Examples of active elements include diodes and transistors.

Active elements are an important part of electronic systems. For example, they are used to make amplifiers. Amplifiers may be used to increase the power of signals; in this case, they are called power amplifiers. Amplifiers may also increase the voltage of signals; in this case, they are called voltage amplifiers. Finally, amplifiers may also be used to increase current; in this case, they are called current amplifiers. Amplifiers are integral parts of many electronic systems, such as two-way radios, cell phones, radios, and television sets. Power-assisted brakes and power steering in automobiles are mechanical examples of amplification.

Chapters 2 to 7 introduced circuits that included only passive elements. This chapter examines circuits that include active elements such as diodes and transistors. This chapter also discusses operational amplifiers (op amps), an important element of circuits. Op amps (made up of transistors) are part of many electronic circuits. For example, op amps are part of electronic amplifiers, which are used to invert signals (i.e., apply 180° phase), or to isolate one part of a circuit from another (this type of system is called a *buffer*). Chapter 7 investigated passive filters. In this chapter, op amps are discussed as an important part of active filters. Active filters are discussed in Chapter 11. Each of the elements described earlier is an essential part of electronic systems; thus, Chapter 8 is titled Electronic Circuits.

Currently, all active elements in a circuit are made of semiconductors. Semiconductors are a group of materials, which possess an electric conductivity that is intermediate between conductors (e.g., metals) and insulators (e.g., dielectrics and plastic). The variability of the electric properties of semiconductors makes these materials a natural choice for investigation of electronic devices.

The conductivity of these materials varies with temperature, illumination, and the semiconductor's impurity content. Semiconductor materials with no impurities are called *pure* or *intrinsic* semiconductors. Applying impurity atoms to semiconductors leads to the formation of two different categories of semiconductors: *p-type* and *n-type*. The category of semiconductor is specified by the nature of the impurity atoms. The process of adding impurity atoms to a pure semiconductor is called doping, which is introduced in Section 8.2.

Most semiconductor devices contain at least one junction between p-type and n-type materials. These p–n junctions are fundamental to the performance of functions, such as rectification, amplification, switching, and other operations in electronic circuits. This chapter discusses examples of these electronic devices and analyzes their operation in electric circuits.

In this chapter, direct current (DC) parameters are denoted by uppercase letters, such as "*V*" and "*I*," and AC parameters are denoted by lower case letters, such as "*v*" and "*i*."

8.2 P-TYPE AND N-TYPE SEMICONDUCTORS

This section briefly introduces the two categories of semiconductors, p-type and n-type. In addition, this section introduces two popular materials used in the construction of diodes—germanium (Ge) and silicon (Si).

Figure 8.1 shows a simplified two-dimensional representation of an intrinsic or a pure silicon crystal (i.e., does not have impurities). Specifically, Figure 8.1(a) shows the covalent bonds that form the silicon structure. Each silicon atom in the crystal is surrounded by four

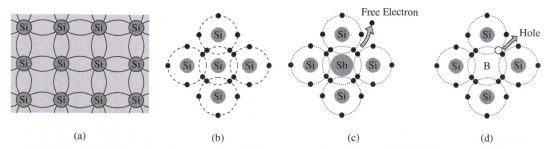

(a)   (b)   (c)   (d)

FIGURE 8.1   Crystal structure of silicon. (Courtesy of Wayne Storr, Electronics-Tutorials, www.electronics-tutorials.ws.)

other silicon atoms. Each atom includes four valence electrons (i.e., the electrons in the outermost shell of the atom), which are shared with its neighbors to form a covalent bond. At a temperature of zero degrees kelvin (0 K), Si works as a dielectric. As temperature increases, more electrons are freed and the conductivity increases. While germanium and other materials are also suitable for semiconductors, silicon is by far the most widely used, due to its cost and versatility.

Another technique used to increase the conductivity of semiconductors is *doping,* which is the addition of foreign atoms to the regular crystal lattice of silicon or germanium. The addition of pentavalent impurities (i.e., atoms with five valence electrons) such as antimony, arsenic, or phosphorous, produces n-type semiconductors by contributing extra electrons. Note that silicon can maintain a covalent bond using only four valance electrons of impurity, while the impurity has five electrons. Therefore, free electrons are produced in the n-type semiconductors, as shown in Figure 8.1(c).

Conversely, the addition of trivalent impurities (i.e., atoms with three valence electrons) such as boron, aluminum, or gallium produces p-type semiconductors by contributing extra holes. In other words, only three electrons contribute in forming the covalent bonds. Thus, one valence connection remains electron-less [see Figure 8.1(d)]. In this case, it is said that a *hole* is generated.

It should be noted that in spite of the existence of the free electrons that appear in n-type and the free holes that appear in p-type impure semiconductors, these materials are electrically neutral (i.e., the total number of electrons and protons in these materials are equivalent. However, the addition of impurity increases the conductivity).

Now, suppose that a p-type block of silicon is placed in perfect contact with an n-type block forming a p–n junction shown in Figure 8.2. In this case, free electrons from the n-type region diffuse across the junction to the p-type side where they recombine with some of the many holes in the p-type material. Similarly, holes diffuse across the junction in the opposite direction and recombine, as shown in Figure 8.2. Remember that before connecting p-type and n-type semiconductors, each was neutral. After connecting the two materials and replacing holes and electrons, the material is no longer neutral at the junction.

In addition, the recombination of free electrons and holes in the vicinity of the junction creates a narrow region on either side of the junction that contains no mobile charges. This narrow region, which has been depleted of mobile charge, is called the depletion layer. It extends into both the p-type and n-type regions as shown in Figure 8.3.

The diffusion of holes from the p-type side of the depletion layer generates fixed negative charges (the acceptor ions) in that region. Similarly, fixed positive charges (donor ions) are generated on the n-type side of the depletion layer. There is then a separation of charges: negative fixed charges on the p-type side of the depletion layer and positive fixed charges on the n-type side. Similar to a capacitor, this separation of charges creates an electric field across the depletion layer. Existence of an electric field represents the existence of a voltage difference across the depletion layer. This voltage difference is usually called *built-in* voltage and it varies according to many parameters, including temperature. It is approximately 0.7 for silicon and 0.3 for germanium.

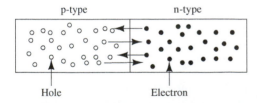

FIGURE 8.2 Electron and hole diffusion in a p–n junction.

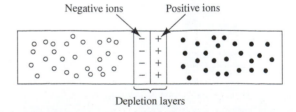

FIGURE 8.3 Depletion layer formation in a p–n junction.

8.3 DIODES

A diode is a p–n junction that allows current to pass in only one direction—from the p-type (called an *anode*) toward the n-type (called a *cathode*) semiconductor. A diode is similar to a "check valve" in a water system that allows water to pass in only one direction. The circuit diagram of a diode is shown in Figure 8.4. Equation (8.1) represents the relationship between voltage across and current through the diode.

$$I = I_o(e^{V/V_T} - 1) \qquad (8.1)$$

The current, I_o, in Equation (8.1) is called a reverse saturation current, and the voltage, V_T, is called thermal voltage and corresponds to:

$$V_T = \frac{KT}{q} \qquad (8.2)$$

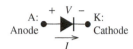

FIGURE 8.4 Circuit diagram of a diode.

where K is the Boltzmann's constant, T is the absolute temperature in kelvin, and q is the magnitude of the electron charge. The value of V_T at room temperature (i.e., $T = 300$ K) equals 0.0259 V. The reverse saturation current, I_o, is very low (typically in the microampere range). The plot of a diode's characteristics is shown in Figure 8.5.

If the voltage across the diode is negative (i.e., $V < 0$), the diode is said to be in *reverse bias*. The current that passes through the diode in reverse bias is called I_o. The reverse bias occurs when the voltage across the diode is $V < -4V_T$ or, typically, $V < -0.1$ V at room temperature. The direction of the reverse saturation current is from the cathode to the anode. The diode is said to be OFF when the voltage across it is $V < -4V_T$.

The diode is said to be in *forward bias* if its voltage is positive (i.e., $V > 0$) and it starts to pass current as the voltage increases beyond the built-in voltage or the *threshold voltage*. This threshold voltage (V_γ) is shown in Figure 8.5. The threshold voltage varies with many parameters, and it is approximately 0.7 for silicon diodes and 0.3 for germanium diodes.

Instead of using the exact relationship between voltage and current to analyze circuits that contain diodes, approximate models are used. These models simplify the process of analysis of diode circuits. One of these models is the *practical model,* which mimics practical situations. The I–V characteristics of the practical model are shown in Figure 8.6. In this figure, the effect of reverse saturation current is ignored. As discussed earlier, this current is very small (on the order of microamperes) and can be easily ignored in many practical problems. Figure 8.7 represents the diode and its voltage and current. In this figure, the voltage on the anode of the diode and cathode are V_1 and V_2, respectively. Therefore, the voltage across the diode corresponds to: $V = V_1 - V_2$.

Accordingly, the current–voltage (I–V) model of Figure 8.6 is described as follows:

1. $V < 0 \rightarrow$ Reverse bias and the diode is OFF $\rightarrow I = 0$
2. $0 < V < V_\gamma \rightarrow$ Forward bias and the diode is OFF $\rightarrow I = 0$
3. $V > V_\gamma \rightarrow$ Forward bias and the diode is ON $\rightarrow V = V_\gamma$

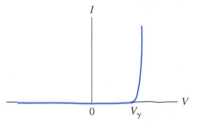

FIGURE 8.5 Diode characteristics: the relationship between voltage and current of a diode.

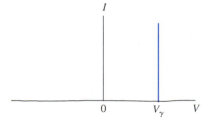

FIGURE 8.6 Diode characteristics: practical model.

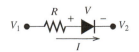

FIGURE 8.7 A simple diode circuit.

Note:

All diodes used in this chapter are assumed to be silicon diodes (unless specified).

EXAMPLE 8.1   **Current and Voltage of a Diode**

For the circuit shown in Figure 8.8, find I and V for a silicon diode when:

    a. $V_S = -10\,\text{V}$
    b. $V_S = -1\,\text{V}$
    c. $V_S = 0.5\,\text{V}$
    d. $V_S = 8\,\text{V}$

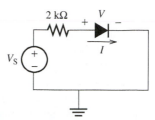

FIGURE 8.8 Circuit for Example 8.1.

SOLUTION

    a. Referring to Figure 8.7, $V_S = -10\,\text{V}$ means $V_1 - V_2 = -10\,\text{V}$. Therefore, the diode is reverse biased (OFF) and the current flowing through it $I = 0$ and the voltage across the diode $V = -10\,\text{V}$.

    b. Similar to part a, $I = 0$ and $V = -1\,\text{V}$.

    c. For the silicon diode, $V_\gamma = 0.7$

$$0 < V_1 - V_2 = V_S = 0.5 < V_\gamma = 0.7$$

Therefore, the diode is forward biased but it is OFF because there is no sufficient voltage to turn it ON. Therefore, $I = 0$ and $V = 0.5\,\text{V}$.

    d. Next, $V_1 - V_2 = V_S = 8\,\text{V} > V_\gamma = 0.7$.
    Therefore, the diode is forward biased and it is ON. Accordingly:

$$V = 0.7\,\text{V}$$

$$I = \frac{V_S - V_\gamma}{R} = \frac{8 - 0.7}{2} = 3.65\,\text{mA}$$

EXAMPLE 8.2   **The Current of Diode**

Find the current, I, flowing through the silicon diode of the circuit shown in Figure 8.9 when:

    a. $R = 1\,\text{k}\Omega$
    b. $R = 10\,\text{k}\Omega$

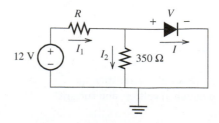

FIGURE 8.9 Circuit for Example 8.2.

SOLUTION

This problem can be solved using either of two methods.

Method 1: Because the voltage available to directly drive the diode is unknown (i.e., $V_1 - V_2$ of Figure 8.7), an assumption must be made of the status of the diode (ON or OFF). This assumption is used to solve the problem, and then the validity of the assumption can be checked by observing the direction of the diode current. For example, when assuming the diode is ON, the assumption will be valid if the current through the diode flows from the anode to the cathode, because that situation is consistent with an ON diode.

  a. $R = 1\,\text{k}\Omega$

Assume the diode is ON. Therefore, the voltage across it is:

$$V = V_\gamma = 0.7\,\text{V}$$

Accordingly, the voltage across the 350-Ω resistor is 0.7 V and the current through it corresponds to:

$$I_2 = \frac{0.7}{350} = 2\,\text{mA}$$

The current through R corresponds to:

$$I_1 = \frac{12 - 0.7}{1} = 11.3\,\text{mA}$$

Applying KCL, the current through the diode corresponds to:

$$I = I_1 - I_2 = 11.3 - 2 = 9.3\,\text{mA}$$

This current flows through the diode from anode to the cathode, which confirms the validity of the assumption.

  b. $R = 10\,\text{k}\Omega$

Again, assume the diode is ON. Therefore, the voltage across it is:

$$V = V_\gamma = 0.7\,\text{V}$$

Accordingly, the voltage across the 350-Ω resistor is 0.7 V and the current through it corresponds to:

$$I_2 = \frac{0.7}{350} = 2\,\text{mA}$$

The current through R corresponds to:

$$I_1 = \frac{12 - 0.7}{10} = 1.13\,\text{mA}$$

Applying KCL, the current through the diode corresponds to:

$$I = I_1 - I_2 = 1.13 - 2 = -0.87\,\text{mA}$$

The negative sign means that the current flows in an opposite direction to that shown in Figure 8.9 (i.e., from cathode to anode). The diode passes current only in one direction (i.e., from anode to cathode). Thus, the assumption of the diode being ON is not valid.

Now, consider another assumption: the diode is OFF. In this case: $I = 0$. Therefore, the currents I_1 and I_2 are equal and are given by:

$$I_1 = I_2 = \frac{12}{10 + 0.35} = 1.1594\,\text{mA}$$

(continued)

EXAMPLE 8.2   **Continued**

The voltage across the diode corresponds to:

$$V = 1.1594 \times 0.35 = 0.41 \text{ V}$$

It is notable that the voltage across the diode is less than V_γ. This confirms that the diode is OFF.

Method 2: For simplicity, find the Thévenin equivalent circuit between the diode terminals.
   a. $R = 1 \text{ k}\Omega$
   The Thévenin voltage corresponds to:

$$V_{th} = 12 \times \frac{0.35}{1 + 0.35} = 3.11 \text{ V}$$

In addition, the Thévenin resistance corresponds to:

$$R_{th} = \frac{1 \times 0.35}{1 + 0.35} = 0.2593 \text{ k}\Omega$$

The Thévenin equivalent circuit is shown in Figure 8.10. Because 3.11 V is greater than V_γ, the diode is ON and the current through it corresponds to:

$$I = \frac{3.11 - 0.7}{259.3} = 9.3 \text{ mA}$$

   b. $R = 10 \text{ k}\Omega$
   The Thévenin voltage corresponds to:

$$V_{th} = 12 \frac{0.35}{10 + 0.35} = 0.41 \text{ V}$$

In addition, the Thévenin resistance is:

$$R_{th} = \frac{1 \times 0.35}{10 + 0.35} = 0.3382 \text{ k}\Omega$$

The Thévenin equivalent circuit is shown in Figure 8.11.
Because 0.41 V is less than $V_\gamma = 0.7$ V, the diode is OFF and no current flows through it. Thus, $I = 0$ and $V = 0.41$ V.

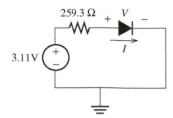

FIGURE 8.10   Thévenin equivalent of the circuit of Figure 8.9 with $R = 1 \text{ k}\Omega$.

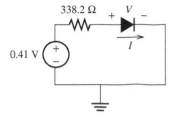

FIGURE 8.11   Thévenin equivalent of the circuit of Figure 8.9 with $R = 1 \text{ k}\Omega$.

EXERCISE 8.1

In Figure 8.9, find R to maintain 0.7 V across the diode. Find the diode current in this case.

8.3.1 Diode Applications

In some cases, diodes (or their circuits) are designed for applications such as rectification, capacitance [e.g., voltage controlled oscillators (VCOs) used in communication systems], tunneling, and light emission and detection. This section focuses only on rectifier diodes and some of their applications. As discussed in the previous section, a diode allows current flow in only one direction. Accordingly, the diode is usually called a rectifier diode or simply, a rectifier.

8.3.1.1 HALF-WAVE RECTIFIER

Consider the circuit shown in Figure 8.12. Here, the input voltage is a sinusoidal waveform with the peak voltage $V_m > 0.7$. The goal is to evaluate the voltage across the load resistance, R_L, or the output voltage, $v_o(t)$. The input voltage changes continuously between $-V_m$ and V_m. Therefore, the diode status varies between OFF and ON.

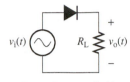

FIGURE 8.12 A half-wave rectifier.

When the value of the input voltage is less than 0.7, the diode is OFF and the current flowing through it is $i = 0$. Therefore, the output voltage corresponds to:

$$v_o(t) = i \times R = 0 \tag{8.3}$$

As the value of the input voltage increases beyond 0.7, the current starts to flow and the voltage drop across the diode is 0.7 V. Therefore, applying KVL to the loop, the output voltage corresponds to:

$$v_o(t) = v_i(t) - 0.7 \tag{8.4}$$

The transfer characteristic (i.e., the output voltage versus the input voltage) is shown in Figure 8.13(a). Plots of the input and output voltage waveforms are shown in Figures 8.13(b) and 8.13(c), respectively. Note that the maximum value of the output voltage is 0.7 V less than that of the input voltage. While the input voltage has positive and negative values, the output voltage is always positive. Therefore, the current flowing through the load resistance, $i = v_o/R$, is always positive. In other words, the current passes through the load in only one direction. Therefore, the term *rectifier* is used. The term *half wave* is used because the current through the load exists only during the positive half cycle.

8.3.1.2 BRIDGE RECTIFIER

The *bridge* rectifier circuit is shown in Figure 8.14. A bridge rectifier allows the current to pass through the load resistance in one direction during the two half cycles. The positive sign above the AC source indicates that the voltage of the upper terminal is positive with respect to the bottom terminal during the positive half cycle. Now, suppose the input voltage of the circuit shown in Figure 8.14 is a sinusoidal waveform with a peak value $V_m > 1.4$ V. During the positive half cycle, diodes D_1 and D_3 are forward biased and diodes D_2 and D_4 are reverse biased. Because diodes D_2 and D_4 are reverse biased, the current flowing through them is zero and each can be replaced with an open circuit, as shown in Figure 8.15.

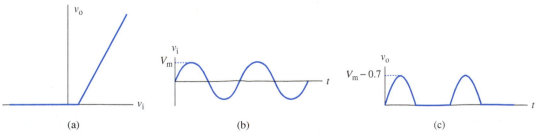

(a) (b) (c)

FIGURE 8.13 (a) Transfer characteristics for the circuit shown in Figure 8.12; (b) the input voltage; (c) the output voltage.

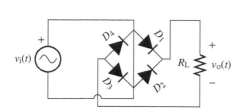

FIGURE 8.14 A bridge rectifier.

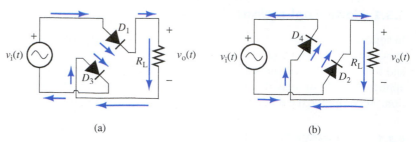

(a)   (b)

FIGURE 8.15 The equivalent circuit during (a) the positive half cycle; (b) the negative half cycle.

To turn both diodes of the circuit shown in Figure 8.15 ON, the value of the input voltage must be greater than 1.4 V (0.7 V is required for each diode). Thus, when the value of the input voltage increases beyond 1.4 V, D_1 and D_3 turn ON and the current will pass in the direction shown in Figure 8.15(a). Because the current passes through the load resistance from the top to the bottom, the output voltage has a positive value and corresponds to (consider a KVL through the loop):

$$v_o(t) = v_i(t) - 1.4 \qquad (8.5)$$

When the value of the input voltage is less than 1.4 V, both diodes will turn OFF and the current flowing through the circuit will be zero. Therefore, the output voltage $v_o(t) = 0$.

During the negative half cycle, diodes D_2 and D_4 are forward biased and diodes D_1 and D_3 are reverse biased. Similarly, because diodes D_1 and D_3 are reverse biased, each can be replaced with an open circuit as shown in Figure 8.15(b). The magnitude of the input voltage value must be greater than 1.4 V to turn both diodes ON in the circuit. In other words, the input voltage must be less than -1.4 V to enable the current to pass in the direction shown in Figure 8.15(b). In this case, the current passes through the load resistance from the top to the bottom in the same direction as in the case of positive half cycle. Referring to the circuit shown in Figure 8.15(b), and writing a simple KVL, the output voltage corresponds to:

$$v_o(t) = -v_i(t) + 1.4 \qquad (8.6)$$

When the value of the input voltage is greater than -1.4 V, both diodes turn OFF and the current flowing through the circuit will be zero. Therefore, the output voltage $v_o(t) = 0$. Plots of transfer characteristics, input voltage, and output voltage are shown in Figure 8.16.

8.3.1.3 DIODE LIMITERS

A *limiter* is a device that limits a waveform from exceeding a specified value. Limiters have applications in wave shaping and circuit protection. Many circuits cannot tolerate voltages beyond a limit; limiters protect circuits from voltages that could harm the circuit. The circuit of a limiter is similar to that of a rectifier; however, a limiter's circuit usually includes a constant (independent) voltage source as well. This section discusses examples of diode limiters, and explains how to plot their transfer characteristics as well as their input and output voltage waveforms.

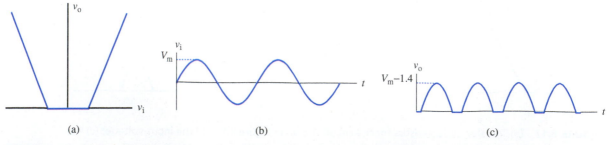

(a)   (b)   (c)

FIGURE 8.16 (a) Transfer characteristics for the circuit of Figure 8.14; (b) its input voltage; (c) its output voltage.

EXAMPLE 8.3   One-Sided Limiter

Find the output voltage of the circuit shown in Figure 8.17.

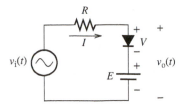

FIGURE 8.17   The limiter circuit for Example 8.3.

SOLUTION

The output voltage can be calculated by applying *KVL*. If the voltage across the diode is known, the output voltage will correspond to:

$$v_o(t) = V + E \tag{8.7}$$

On the other hand, if the current flowing through the diode is known, the output voltage will correspond to:

$$v_o(t) = v_i(t) - I \times R \tag{8.8}$$

The knowledge of v or I depends on whether the diode is ON or OFF, respectively. If the diode is ON, the voltage across the diode $v = 0.7$, and based on Equation (8.7):

$$v_o(t) = 0.7 + E \tag{8.9}$$

If the diode is OFF, the current flowing through the diode $I = 0$, and based on Equation (8.8):

$$v_o(t) = v_i(t) \tag{8.10}$$

If the voltage difference between $v_i(t)$ and E exceeds 0.7 V, the diode will turn on. In other words, to turn the diode on, the following must be true:

$$v_i(t) > E + 0.7 \tag{8.11}$$

Accordingly, the output voltage corresponds to Equation (8.9). On the contrary, the diode will be turned off if:

$$v_i(t) < E + 0.7 \tag{8.12}$$

In this case, the output voltage will correspond to Equation (8.10).

As a summary, the output voltage is given by:

$$v_o(t) = \begin{cases} E + 0.7 & v_i(t) > E + 0.7 \\ v_i & v_i(t) < E + 0.7 \end{cases} \tag{8.13}$$

Plots of the transfer characteristic, input voltage, and output voltage for the limiter circuit of Figure 8.17 are shown in Figure 8.18. As shown in the figure, it is evident that the output voltage doesn't exceed $E + 0.7$. The upper limit of the output voltage can be controlled by adjusting the value of E.

(continued)

EXAMPLE 8.3 **Continued**

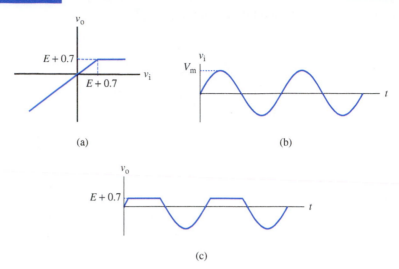

(a)

(b)

(c)

FIGURE 8.18 (a) Transfer characteristics for the circuit shown in Figure 8.17; (b) its input voltage; (c) its output voltage.

EXAMPLE 8.4 **One-Sided Limiter**

Another limiter circuit is shown in Figure 8.19. Find the output voltage.

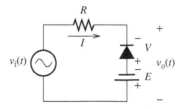

FIGURE 8.19 The limiter circuit for Example 8.4.

SOLUTION

When the voltage across the diode terminals is known (i.e., the diode is ON):

$$v_o(t) = -V - E \tag{8.14}$$

If the diode is OFF or the current flowing through it is zero:

$$v_o(t) = v_i(t) - I \times R \tag{8.15}$$

The diode is turned on as the voltage difference between $-E$ and $v_i(t)$ exceeds 0.7 V. In other words, to turn the diode on, the following must be true:

$$-E - v_i(t) > 0.7 \tag{8.16}$$

Or equivalently:

$$v_i(t) < -E - 0.7 \tag{8.17}$$

Accordingly, the output voltage corresponds to:

$$v_o(t) = -E - 0.7 \tag{8.18}$$

On the contrary, the diode is turned off if:

$$v_i(t) > -E - 0.7 \tag{8.19}$$

In this case, the output voltage will correspond to:

$$v_o(t) = v_i(t) \tag{8.20}$$

In summary, the output voltage is given by:

$$v_o(t) = \begin{cases} -E - 0.7 & v_i(t) < -E - 0.7 \\ v_i & v_i(t) > -E - 0.7 \end{cases} \tag{8.21}$$

Plots of the transfer characteristic, input voltage, and output voltage for the limiter circuit of Figure 8.19 are shown in Figure 8.20. As shown in the figure, it is notable that the output voltage doesn't drop below $-E - 0.7$. The lower limit of the output voltage can be controlled by adjusting the value of E.

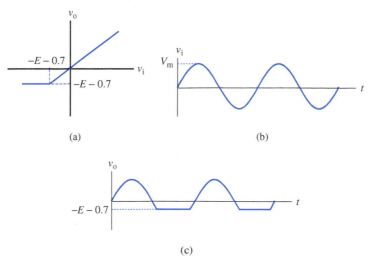

(a)

(b)

(c)

FIGURE 8.20   (a) Transfer characteristics for the circuit shown in Figure 8.19; (b) its input voltage; (c) its output voltage.

EXAMPLE 8.5   Two-Sided Limiter

For the circuit shown in Figure 8.21, sketch the transfer characteristics, input voltage, and output voltage. Assume the input voltage is a sinusoidal waveform with a peak value of 10 V and $E_1 = E_2 = 4$ V.

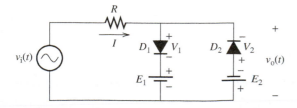

FIGURE 8.21   Circuit for Example 8.5.

(continued)

EXAMPLE 8.5   **Continued**

SOLUTION

The value of the output voltage depends on the status of diodes D_1 and D_2. Based on the distribution shown, both diodes cannot be turned ON simultaneously because:

- To turn on D_1, the value of the input voltage must be greater than E_1 by at least 0.7 V.
- To turn on D_2, the value of the input voltage must be less than $-E_2$ by at least 0.7 V.

Table 8.1 summarizes the status of diodes, the required value of the input voltage, and the corresponding output voltage.

For the given value of E_1 and E_2, the output voltage corresponds to:

$$v_o(t) = \begin{cases} 4.7 \text{ V} & v_i(t) > 4.7 \text{ V} \\ -4.7 \text{ V} & v_i(t) < -4.7 \text{ V} \\ v_i(t) & -4.7 \text{ V} < v_i(t) < 4.7 \text{ V} \end{cases}$$

Plots of the transfer characteristic, input voltage, and output voltage for the limiter circuit of Figure 8.21 are shown in Figure 8.22. As shown in the figure, it is notable that the output voltage is bounded between 4.7 V and -4.7 V.

TABLE 8.1   Analysis of the Circuit in Example 8.5

D_1	D_1	Criterion	$v_o(t)$
ON	OFF	$v_i(t) > E_1 + 0.7$	$E_1 + 0.7$
OFF	ON	$v_i(t) < -E_2 - 0.7$	$-E_2 - 0.7$
OFF	OFF	$-E_2 - 0.7 < v_i(t) < E_1 + 0.7$	$v_i(t)$

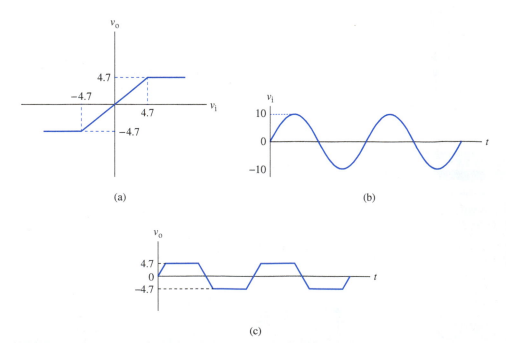

FIGURE 8.22   (a) Transfer characteristics for the circuit shown in Figure 8.21; (b) its input voltage; (c) its output voltage.

APPLICATION EXAMPLE 8.6	Protection of Electrical Devices

Electronic and electrical devices are designed to operate at a certain maximum supply voltage: They can be seriously harmed if their voltage exceeds that threshold. An excessive increase in voltage sources may be from a natural source (e.g., lightening) or a man-made source (e.g., electromagnetic induction when switching on or off inductive loads). In addition, the devices can be harmed if the polarity of the input voltage is reversed (i.e., some electronic devices cannot tolerate negative voltages).

A limiter circuit can protect devices from excess voltage. Assume a limiter circuit for the input of an electronic device needs to be designed with the following ratings: $V_{max} = 9$ V and $V_{min} = 0$.

SOLUTION

Because the input voltage to the device needs to be bounded between two values, a double limiter must be used to protect it. Considering the limiter circuit discussed in Example 8.5, and by referring to Table 8.1, the upper limit of the voltage is controlled by diode D_1 and the voltage source, E_1. The value of E_1 is calculated as follows:

$$E_1 + 0.7 = V_{max} = 9 \text{ V} \rightarrow E_1 = 8.3 \text{ V}$$

The lower limit is controlled by diode D_2 and the voltage source, E_2. The value of E_2 is calculated as follows:

$$-E_2 - 0.7 = V_{min} = 0 \rightarrow E_2 = -0.7 \text{ V}$$

The negative sign indicates that the battery should be reversed. The limiter circuit is connected between the input terminals of the device as shown in Figure 8.23.

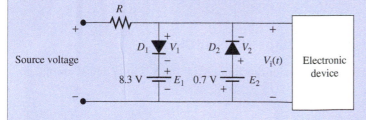

FIGURE 8.23   The limiter circuit for Example 8.6.

EXERCISE 8.2

Sketch the transfer characteristics and the output voltage for the circuit shown in Figure 8.23, assuming the input source is a sinusoid with an amplitude of 5 V.

8.3.2 Different Types of Diodes

The function of diodes is not limited to rectification. There are many types of diodes; each has a different function. Diodes are designed to exploit specific junction properties, such as capacitance or charge storage. This section provides an overview of common types of diodes and their applications.

8.3.2.1 ZENER DIODES

A *Zener diode* or a *breakdown diode* is similar to a forward-biased rectifier diode, like those discussed in Section 8.1. However, the rectifier diode does not conduct current when it is reverse biased. The Zener diode allows current to pass in the reverse direction if the reverse voltage

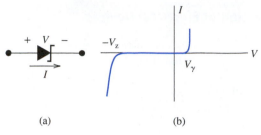

(a)   (b)

FIGURE 8.24 (a) Symbol of a Zener diode.
(b) Characteristics of a Zener diode.

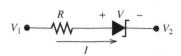

FIGURE 8.25 Circuit with a Zener diode.

TABLE 8.2 Operation of a Zener Diode

	Diode Status	V	I
$V_1 - V_2 > V_\gamma$	ON	V_γ	Positive
$-V_z < V_1 - V_2 < V_\gamma$	OFF	$V_1 - V_2$	$=0$
$V_1 - V_2 < -V_z$	Zener region	$-V_z$	Negative

exceeds a precisely defined voltage called the Zener voltage, V_z. Zener voltage ranges from 3.3 to 75 V. The circuit diagram of a Zener diode and its characteristics are shown in Figure 8.24.

To explain the operation of a Zener diode, consider the circuit shown in Figure 8.25. Assuming a practical model, Table 8.2 summarizes the operation of the Zener diode. Observe that if the reverse voltage across the diode is more than V_z, the Zener diode will be equivalent to a constant voltage source with the value of $V = -V_z$.

EXAMPLE 8.7 **Current and Voltage of Zener Diode**

For the circuit shown in Figure 8.26, the Zener diode has $V_z = 3.6$ V. Find I and V when:

a. $V_S = -1$ V
b. $V_S = 8$ V
c. $V_S = -10$ V

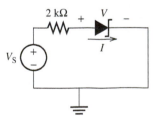

FIGURE 8.26 Circuit for Example 8.7.

SOLUTION

a. Refer to Figure 8.25, $V_S = -1$ V means -3.6 V $< V_1 - V_2 < 0$ V. Therefore, the diode is reverse biased, it is OFF, and the current flowing through it is $I = 0$. Applying KVL to Figure 8.26, the voltage across the diode corresponds to:

$$V = V_S = -1 \text{ V}$$

b. Because $V_1 - V_2 = V_S > V_\gamma = 0.7$, the diode is forward biased and it is ON. Therefore, $V = V_\gamma = 0.7$ and the current corresponds to:

$$I = \frac{8 - 0.7}{2} = 3.65 \text{ mA}$$

c. In this example, $V_1 - V_2 = V_S = -10\,\text{V} < -V_z = -3.6\,\text{V}$. Therefore, the diode is reverse biased and it is in the Zener region. Therefore, the voltage across the diode corresponds to:

$$V = -V_z = -3.6\,\text{V}$$

In addition, the current corresponds to:

$$I = \frac{V_S + V_z}{R} = \frac{-10 + 3.6}{2} = -3.2\,\text{mA}$$

EXAMPLE 8.8   **Using Zener Diodes in Limiter Circuits**

A Zener diode can be used to replace with DC sources in limiter circuits, as shown in the circuit in Figure 8.27. For this circuit, sketch the transfer characteristics, input voltage, and output voltage. Assume $V_z = 5.6\,\text{V}$ and that the input voltage is a sinusoidal waveform with a peak voltage of 12 V.

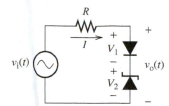

FIGURE 8.27 Limiter using a Zener diode.

SOLUTION

During the positive half cycle, the normal diode is forward biased and the Zener diode is reverse biased. To turn the rectifier diode on, and push the Zener diode to its Zener region, the input voltage must satisfy:

$$v_i(t) > V_\gamma + V_z = 0.7 + 5.6 = 6.3\,\text{V}$$

Now, because both diodes allow the flow of current: $V_1 = 0.7\,\text{V}$, and, $V_2 = 5.6\,\text{V}$. Therefore, the output voltage corresponds to:

$$v_o(t) = V_1 + V_2 = 6.3\,\text{V}$$

If the input voltage $v_i(t) < 6.3\,\text{V}$, at least one of the diodes avoids the flow of current, which leads to:

$$v_o(t) = v_i(t)$$

During the negative half cycle, the normal diode is reverse biased and the Zener diode is forward biased. However, the Zener diode can pass current and the normal diode cannot. Therefore, $v_i(t) = v_o(t)$. In general, the output voltage corresponds to:

$$v_o(t) = \begin{cases} 6.3\,\text{V} & v_i(t) > 6.3\,\text{V} \\ v_i(t) & v_i(t) < 6.3\,\text{V} \end{cases}$$

Plots of transfer characteristics, input voltage, and output voltage are shown in Figure 8.28.

The Zener diode has an important application in the design of power supplies (i.e., AC-to-DC converters) where it is used as a voltage regulator. AC-to-DC conversion is presented in Section 8.3.3.

(continued)

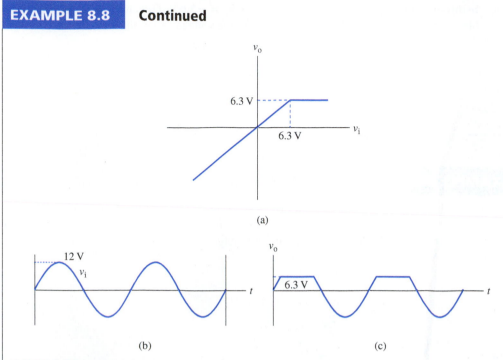

EXAMPLE 8.8 **Continued**

FIGURE 8.28 (a) Transfer characteristics for the circuit shown in Figure 8.27; (b) its input voltage; (c) its output voltage.

8.3.2.2 VARACTOR DIODE

The term *varactor* is a shortened form of *variable reactor*. Varactor diodes are typically used in a reverse bias configuration. The varactor, when biased in this fashion, acts as a variable capacitor. The capacitance of the varactor decreases with applied reverse voltage and it ranges from a few pico farads (pF) to over 100 pF. The symbol and characteristics (i.e., the capacitance versus the voltage across the diode) of a varactor are shown in Figure 8.29.

Varactors are commonly used in voltage-to-frequency converters or voltage control oscillators (VCOs). The VCO is an oscillator that changes the frequency of its output signal according to the value of the input voltage. The VCO is a widely used device in Frequency Modulation (FM) circuits. Another popular application of the varactor is in electronic tuning circuits (e.g., radio and television tuners). In these tuners, the DC control voltage varies the capacitance of the varactor, retuning the resonant circuit. A simple tuning circuit is shown in Figure 8.30.

In Figure 8.30, $v_S(t)$ is a small signal, (i.e., its amplitude is in the range of millivolts or less) that is much smaller than the DC control voltage. The small signal is a model for the received

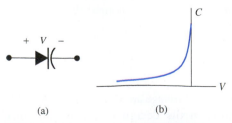

FIGURE 8.29 (a) The symbol for a varactor diode; (b) C-V characteristics of a varactor diode.

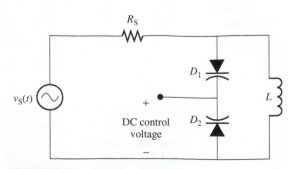

FIGURE 8.30 A tuning circuit using a varactor diode.

signal by the antenna. If the applied DC control voltage is positive, both varactors will be reverse biased. If the equivalent capacitance of both diodes is labeled C, the resonant frequency will correspond to:

$$f_r = \frac{1}{\sqrt{2\pi LC}} \qquad (8.22)$$

8.3.2.3 LIGHT-EMITTING DIODE

A *light-emitting diode* (LED) is another type of diode that emits light when current passes through it. LEDs operate in the forward bias and the emitted light intensity increases as the current flowing through it increases. The color of the light mainly depends on the material used for diode fabrication. LEDs with visible light (e.g., red, blue, yellow, and green) are commonly used as indicators and are used in many devices, such as TV power indicators, digital clocks, and seven-segment displays.

The threshold voltage, V_γ, for an LED is about 1.6 V, as it is fabricated by materials different from Si and Ge (usually gallium arsenide, or GaAs). The symbol of the LED is shown in Figure 8.31.

In many applications, the light from an LED is invisible to the human eye. Infrared emitters are well-suited to optical communication systems. For example, an LED might be used in conjunction with a photodiode (discussed in the next section) or other photosensitive devices to transfer information optically between locations.

FIGURE 8.31 The symbol of an LED.

8.3.2.4 PHOTODIODES

In contrast to LEDs, photodiodes are semiconductor devices that function as photodetectors. There are two major modes of operation for photodiodes: photovoltaic mode and photoconductive mode. In the photovoltaic mode, the diode operates under zero bias, but when light is emitted on the diode, it creates a current that leads to forward biasing. Series and parallel connection of photodiodes operating in photovoltaic mode are used in *solar cells*. A solar cell is a device that converts light or solar energy into electrical energy. Solar cells have many applications in power systems for remote areas, Earth-orbiting satellites and space probes, and consumer systems, for example, handheld calculators, wrist watches, and remote radiotelephones.

Considering the scarcity and the cost of oil resources on the earth, a significant amount of research is ongoing to develop and improve solar cells capable of efficiently converting solar energy to power. One approach that is under study by many investigators worldwide is to launch low-orbit satellites equipped with solar cells capable of efficiently collecting solar energy in outer space and transmitting it to the earth via wireless communications.

To use a photodiode in its photoconductive mode, the photodiode is reverse biased; the photodiode then allows current flow when it is illuminated. The conductivity of the photo diode varies directly with the intensity of the light. Photodiodes that operate in this mode have many applications, including smoke detectors and motor speed control.

APPLICATION EXAMPLE 8.9　Motor Speed Control

In many applications, the speed of motors needs to be kept constant. One example is the rotation of radar antennas. Radar stands for Radio Detection And Ranging. Accordingly, these devices use radio waves to detect targets and find their location. In order to locate targets in full 360° around the radar position, the radar directional antenna (usually a parabolic antenna) is rotated via a motor. The speed of this motor must be precisely monitored to avoid errors in localizing targets. In order to control the speed of a motor, the speed must be accurately measured. The speed must be corrected if it is not consistent with the desired value.

An *optical encoder* is a commonly used circuit to measure motor speed. An optical encoder produces a signal with a frequency proportional to the speed of the motor. This frequency is

(*continued*)

APPLICATION EXAMPLE 8.9   **Continued**

converted to a voltage using another circuit called a frequency-to-voltage converter. This voltage is used to control the speed of the motor. This example focuses on the optical encoder circuit.

An optical encoder is a circuit that uses light to generate a periodic signal whose frequency is proportional to the motor speed. Here, the motor speed is "encoded" in the frequency of the signal. Figure 8.32 illustrates the operation of an optical encoder. An opaque disk is mounted on a motor shaft and turns with the shaft. Slots (in this case, 500) are cut into the disk. An LED is positioned on one side of the disk. A light sensor (photo-diode) on the opposite side of the disk senses light from the LED and produces a voltage when one of the slots is positioned exactly between the LED and the sensor, allowing light to pass. When the light is blocked by the solid part of the disk between slots, the sensor's output voltage drops to zero. As the motor rotates, the sensor produces a series of pulses that take the form of a sinusoidal signal, as shown in Figure 8.32. The frequency of this signal is determined by the speed of rotation. Specifically, 500 periods of this waveform are produced during each 360° rotation of the motor. The signal is sinusoidal instead of square because the light is not interrupted instantaneously by the disk.

Assume the output signal has a frequency $f = 200$ Hz, the radius of the circle passing through the centers of slots $r = 10$ cm, and the disc has 500 slots. Find the speed of the disc at the holes.

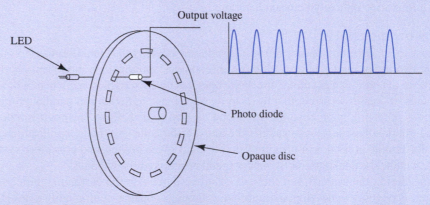

FIGURE 8.32   Optical encoder configuration.

SOLUTION

First, calculate the distance between two successive slots. The perimeter of the circle passing through the slot's centers corresponds to:

$$p = 2\pi r = 2\pi \times 0.1 = 0.6287 \text{ m}$$

Because the disc has 500 slots, the distance between two successive slots corresponds to:

$$d = \frac{p}{500} = 0.1257 \text{ cm}$$

The time between two successive peaks in the output signal is equivalent to the time to move from one slot to the next one, and it corresponds to:

$$T = \frac{1}{f} = \frac{1}{200} = 0.005 \text{ s}$$

Therefore, the speed of the encoder disc at the holes corresponds to:

$$V = \frac{d}{T} = 0.1257 \times 10^{-2} \times 200 = 0.2514 \text{ m/s}$$

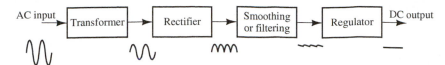

FIGURE 8.33   Block diagram of an AC-to-DC converter.

8.3.3 AC-to-DC Converter

An AC-to-DC converter is a device that converts AC signals into DC signals. Sometimes, it is called an adapter. An adapter is a DC power supply that receives 120 V/60 Hz AC signal as an input and generates a DC voltage output. Many electronic systems such as Labtop, Electronic tooth brush, radio and printer, need a DC supply (independent voltage source) for their operation. A DC supply can be offered by a battery or by converting AC source to DC. The block diagram of an AC-to-DC converter is shown in Figure 8.33. The second block, that is, the rectifier, has been discussed in the previous sections. This section examines the other three blocks.

8.3.3.1 TRANSFORMER

A transformer is a device that steps-up or steps-down an input voltage (see Section 12.4 for additional details). In DC power supplies, step-down transformers are typically used because the input voltage is 120 V AC and most power supplies have a maximum DC voltage output of 30 V. Figure 8.34 shows the circuit block diagram of the transformer. A transformer is a two-port device, with the two ports called the *primary* port and the *secondary* port. The input voltage is applied to the primary port and the output voltage is taken from the secondary port. The relationship between the input and the output voltages is given by:

$$\frac{v_o(t)}{v_i(t)} = \frac{N_2}{N_1} \tag{8.23}$$

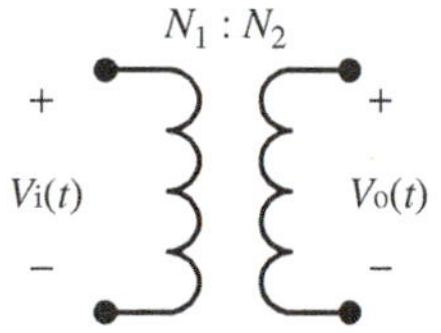

where N_2/N_1 is called the transformer's *turns ratio*. If the turns ratio is greater than unity, the transformer is called a step-up transformer and the output voltage will be greater than the input. If the turns ratio is less than unity, the transformer is called a step-down transformer and the output voltage will be less than the input.

FIGURE 8.34   Circuit diagram of a transformer.

8.3.3.2 SMOOTHING OR FILTERING

In general, the voltage output from a rectifier has an invariable polarity. It also has a nonzero average but it does not have constant magnitude (DC means size and direction are constant over time). A capacitor can be used to smooth the rectifier's output, as shown in Figure 8.35. In this case, the capacitance of the capacitor must be large enough to increase the discharge time. For an input voltage of a sinusoidal waveform, the operation of the filter shown in the circuit shown in Figure 8.35 is described as follows.

During the first half cycle, the capacitor charges through the rectifier output across zero resistance—neglecting the diode resistance—toward the peak voltage. When the rectifier output starts to decrease; the capacitor discharges through the load resistor, R_L. If the time constant, $R_L C$, is large enough compared to the input signal *period*, the capacitor will take a long time to discharge. Before losing its charge, the capacitor starts to charge again through the next half period maintaining its voltage near the peak value. The time constant of charging is relatively low because the resistance of the diodes and the transformer is very small. Plots of input voltage, secondary voltage, and output voltage are shown in Figure 8.36.

The voltage across the transformer's secondary port is related to the input by Equation (8.23). Therefore:

$$V_{sm} = \frac{N_2}{N_1} \times V_{im} \tag{8.24}$$

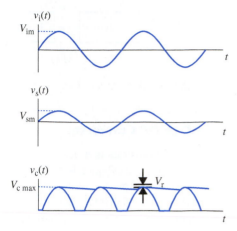

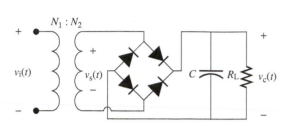

FIGURE 8.35 Using an RC filter to smooth the rectifier output.

FIGURE 8.36 Voltages at different points for the circuit shown in Figure 8.35.

As discussed in the previous section, the voltage drop across the two forward biased diodes in the bridge rectifier is 1.4 V. Therefore, the peak value of the output voltage corresponds to:

$$V_{c\,max} = V_{sm} - 1.4 \qquad (8.25)$$

The difference between the maximum and the minimum values of the output voltage is called the *ripple voltage*, V_r. As the capacitance increases, the discharge time increases, and therefore the ripple voltage decreases leading to more smooth output voltage. The minimum value of the output voltage corresponds to:

$$V_{c\,min} = V_{c\,max} - V_r \qquad (8.26)$$

As explained in Chapter 4, the voltage drop corresponds to:

$$v(t) = V_{c\,max} \cdot \left(1 - e^{-\frac{t}{R_L C}}\right)$$

Applying Taylor series to the exponential term and ignoring higher order terms:

$$v(t) \cong V_{c\,max} \cdot \left(\frac{t}{R_L C}\right)$$

Now, in a bridge rectifier, the voltage drops until about $t = T/2$, ($T = 1/f$ is the period of the sine wave). Therefore, the ripple voltage corresponds to:

$$v_r = \frac{V_{c\,max}}{2f\,R_L C} \qquad (8.27)$$

Note that in a half-wave rectifier, the ripple voltage is twice the value of Equation (8.27). Here, f is the frequency of the input signal (in North America, $f = 60$ Hz, elsewhere it is commonly 50 Hz).

EXERCISE 8.3

Start with Equation (8.26) and using the guidelines provided, prove Equation (8.27).

The ripple voltage is a function of many variables. Some are controllable, including the capacitance: as the capacitance increases, the ripple voltage decreases. In contrast, other variables are uncontrollable, including signal frequency and load resistance. Because the output voltage has a ripple voltage that cannot be completely eliminated, another circuit is required to stabilize the value of the output voltage. This circuit is called a voltage regulator.

8.3.3.3 VOLTAGE REGULATORS

A *voltage regulator* is a circuit used for voltage stabilization. A simple regulator is shown in Figure 8.37. If the Zener diode is reverse biased, it will turn on. In this case, the output voltage corresponds to:

$$v_o(t) = V_z \tag{8.28}$$

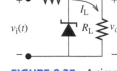

FIGURE 8.37 A simple regulator made by Zener diode.

To turn the Zener diode on in the reverse mode, the output voltage must be positive and the input voltage must satisfy:

$$v_i(t)\frac{R_L}{R_L + R_s} \geq V_z \tag{8.29}$$

or:

$$v_i(t) \geq V_z\frac{R_L + R_s}{R_L} = V_z\left(1 + \frac{R_s}{R_L}\right) \tag{8.30}$$

In other words, the minimum value voltage introduced in Equation (8.26):

$$V_{c\,min} \geq V_z\left(1 + \frac{R_s}{R_L}\right)$$

Therefore, the minimum value of the input voltage required to turn the regulator on corresponds to:

$$V_{on} = V_z\left(1 + \frac{R_s}{R_L}\right) \tag{8.31}$$

The complete AC-to-DC converter circuit is shown in Figure 8.38.

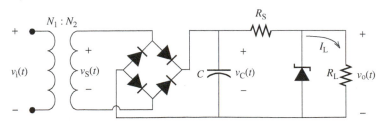

FIGURE 8.38 An AC-to-DC converter with a voltage stabilizer.

| EXAMPLE 8.10 | **DC Power Supply: Adapter** |

Design a DC power supply that supplies the DC voltage of 10 V to a load resistance of 1 kΩ. The AC input is 120 V/60 Hz. Choose suitable values for the capacitor and the transformer turns ratio. The ripple voltage should be less than 10% of the output voltage and $R_S = 200\ \Omega$.

SOLUTION

The circuit shown in Figure 8.38 is used for power supply design. Because the power supply output voltage is 10 V, the Zener diode must have $V_z = 10$ V. First, propose a value for the minimum voltage across the capacitor $V_{c\,min} > V_{on}$. Here, V_{on} is the voltage required to turn on the Zener diode and is given by Equation (8.31).

$$V_{on} = 10\left(1 + \frac{200}{1000}\right) = 12\ \text{V}$$

Let's select:

$$V_{c\,min} = 15\ \text{V}$$

(continued)

EXAMPLE 8.10 **Continued**

The ripple voltage $V_r < 0.1\ V_o = 1$ V. Select:

$$V_r = 0.5\ \text{V}$$

The maximum capacitor voltage is calculated using Equation (8.26):

$$V_{c\,\max} = V_{c\,\min} + V_r$$
$$= 15 + 0.5$$
$$= 15.5\ \text{V}$$

The ripple voltage is calculated using Equation (8.27):

$$\frac{15.5}{2 \times 60 \times 1000 \times C} = 0.5$$

Therefore:

$$C = 2600\ \mu\text{F}$$

To calculate the transformer turns ratio, first calculate the maximum voltage across the transformer's secondary port using Equation (8.25):

$$V_{s\,\max} = V_{c\,\max} + 1.4$$
$$= 15.5 + 1.4$$
$$= 16.9\ \text{V}$$

Using Equation (8.23), the transformer turns ratio corresponds to:

$$\frac{N_1}{N_2} = \frac{V_{i\,\max}}{V_{s\,\max}} = \frac{120\sqrt{2}}{16.9} = 10.42$$

8.4 TRANSISTORS

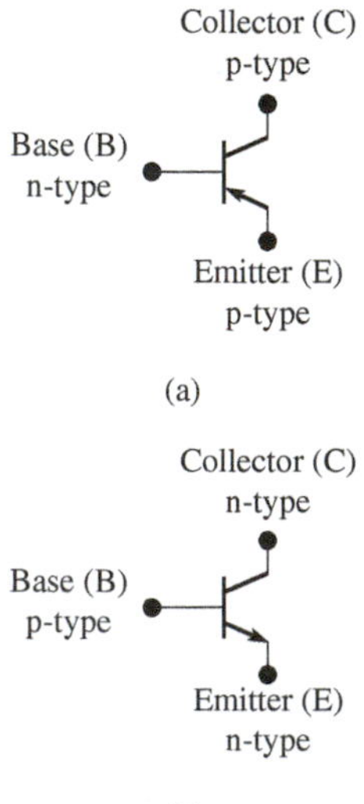

The previous section outlined diodes that consist of two terminals and are formed by connecting two types of semiconductors. A transistor is a semiconductor device that has three terminals and can be used in electrical circuits as an amplifier or as a switch. There are two major types of three-terminal semiconductor devices: the bipolar junction transistor (BJT) and the field-effect transistor (FET). This section briefly introduces the BJT and FET.

8.4.1 Bipolar Junction Transistor

A BJT has three layers of p-type and n-type semiconductors. Based on the order of these layers, there are two types of BJT: PNP and NPN transistors, named after the order of their layers. The circuit diagrams of both types are depicted in Figure 8.39.

The three terminals of a BJT are called the base, the emitter, and the collector and are abbreviated as B, E, and C, respectively. As discussed in Section 8.1, the two types of semiconductors (i.e., p-type and n-type) are connected in a junction. Accordingly, the BJT has two junctions: the first junction is between the base and the emitter (BE junction), and the second junction is between the base and the collector (BC junction).

Collector (C)
p-type

Base (B)
n-type

Emitter (E)
p-type

(a)

Collector (C)
n-type

Base (B)
p-type

Emitter (E)
n-type

(b)

FIGURE 8.39 Symbols of a BJT transistor (a) PNP type; (b) NPN type.

Biasing of BE and BC junctions (i.e., applying proper voltages to the junctions) determines the region of operation of the transistor (Table 8.3). As seen in Table 8.3, three regions are defined for the operation of the transistor: cutoff, active, and saturation.

TABLE 8.3 Regions of Operation of an NPN Transistor

Region of Operation	BE	BC	V_{BE}	I_B	I_C	V_{CE}	Application
Cutoff	Reverse	Reverse	<0.7 V	0	0	Computed by the circuit	Switch
Active	Forward	Reverse	=0.7 V	Computed by the circuit	$=\beta I_B$	Computed by the circuit >0.2 V	Amplifier
Saturation	Forward	Forward	=0.7 V	Computed by the circuit	Computed by the circuit	=0.2 V	Switch

The arrow on the emitter refers to the direction of the current. Therefore, the current flows into (toward) the emitter if the transistor is PNP (i.e., from the p-type and toward an n-type as is the case in a diode as well). However, the current flows out of the emitter in NPN transistors. The current and voltage representation for the PNP and NPN transistor is shown in Figure 8.40. Because the characteristics and operation of PNP and NPN transistors are similar, in this chapter we mainly discuss the analysis of NPN transistors.

To evaluate circuits consisting of transistors, the relationship between currents and voltages must be known. First, applying KCL to the transistor shown in Figure 8.40:

$$I_E = I_C + I_B \tag{8.32}$$

The value of I_C is determined according to the region of operation of the transistor. BJT transistors operate in three regions. Table 8.3 presents these three regions of operation of the transistor and their properties. In the active region, the value of I_C depends on the value of I_B and they are related by β, which is defined as the DC gain. Many transistor circuits are used for amplification of the input signal. In this case, the transistor is operated in the active region. In the saturation region, the collector emitter voltage has its minimum value of $V_{CESat} = 0.2$ V.

8.4.2 Transistor as an Amplifier

One of the most important functions of a transistor is *amplification* of AC signals. An amplifier is a device with an output signal equal to a factor multiplied by the input signal. This factor is called *voltage gain* or *current gain*. These gains are defined based on whether the signal to be amplified is voltage or current. Because the relationship between the output and the input is linear (e.g., $v_0(t) = A_v v_{in}(t)$; where A_v is the voltage gain), the transistor must operate in the active region when used as an amplifier. As it is observed in Table 8.3, in the saturation region the collector–emitter voltage is constant, that is, it does not vary with the input; in the cutoff region, $I_B = I_C = 0$. Thus, transistors cannot be operated as amplifiers in those regions.

Transistors must be properly biased in order to operate in the active region and function as an amplifier. Transistor bias is adjusted by applying proper DC voltage to transistor nodes. One important term, called the *operating point*, is defined as a DC voltage and current that determines the operation of transistors in the desired bias. Another well-known term for the operating point is *quiescent point* or *Q-point*. Analysis of transistor circuits is simplified by using the superposition theorem studied in Chapter 3. First, only consider DC sources to determine the operating point or the Q-point. Next, consider AC sources to determine the AC voltage or current gains. To minimize confusion, the following symbols are defined for voltage and current notations:

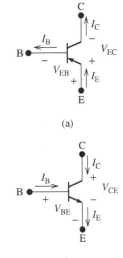

FIGURE 8.40 Current directions and voltage polarities of a BJT (a) PNP type; (b) NPN type.

DC only: Capital letters for both variable and subscript → V_{BE}, I_B, I_C, and V_{CE}
AC only: Small letters for both variable and subscript → v_{be}, i_b, i_c, and v_{ce}
DC plus AC: Small letter for variable and capital letter for subscript → v_{BE}, i_B, i_C, and v_{CE}

where:

$$v_{BE} = V_{BE} + v_{be} \tag{8.33}$$

$$i_B = I_B + i_b \tag{8.34}$$

$$v_{CE} = V_{CE} + v_{ce} \tag{8.35}$$

$$i_C = I_C + i_c \tag{8.36}$$

Based on Equations (8.33) to (8.36), the analysis of an amplifier made from a transistor circuit can be completed by:

1. *DC analysis*: Specifying its operation point and ensuring that the transistor is in its active region
2. *AC analysis*: Analyzing its voltage gain assuming it is properly adjusted in the active region

One of the most famous transistor amplifier circuits is shown in Figure 8.41. The DC voltage source, V_{CC}, together with resistors R_1, R_2, R_C, and R_E are used to determine the Q-point or the region of operation. The AC voltage signal to be amplified, v_{in}, is applied to the base and the AC amplified voltage signal, v_o, is considered across the load resistor, R_L. The capacitors C_1 and C_2 are called *coupling capacitors* while C_e is called the *bypass capacitor*.

As discussed in Chapter 4, if a DC signal is applied to a capacitive circuit, all capacitors will be equivalent to an open circuit after a short time period. Therefore, capacitors isolate the source from the transistor for DC analysis. In addition, when the signal variation is high, that is, the signal frequency is high, the capacitor will be equivalent to a short circuit. Note that the capacitor reactance corresponds to $X_c = 1/(2\pi fC)$. Therefore, for AC analysis (when f is high), the capacitors will be equivalent to a short circuit. Thus, the purpose of coupling capacitors (C_1) is to allow the AC signal to pass into and from the transistor circuit while keeping the DC circuit isolated. In order for a capacitor to function as a coupling capacitor, its capacitance must be large enough. That is, $X_c = 1/\omega C$ must be very small to maintain the capacitor as a short circuit.

A nonzero resistance of R_E reduces the amplification gain of the transistor. Thus, the purpose of the bypass capacitor, C_e, is to create an AC ground at the emitter, while the Q-point can be readjusted using R_E without affecting the AC circuit.

The DC equivalent circuit for the amplifier circuit shown in Figure 8.41 is depicted in Figure 8.42(a). Note that all capacitors are removed from the DC equivalent circuit as they act as an open circuit. To obtain the AC equivalent circuit, first replace all capacitors with short circuits. Moreover, according to the superposition theorem, the independent DC voltage source, V_{CC}, is short-circuited for AC analysis. The AC equivalent circuit is shown in Figure 8.42(b).

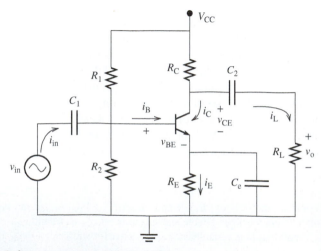

FIGURE 8.41   Circuit for a common emitter amplifier.

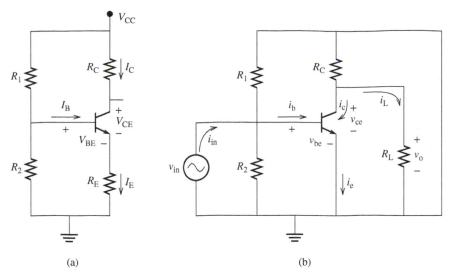

FIGURE 8.42   (a) DC; (b) AC equivalent circuits of the amplifier circuit shown in Figure 8.41.

8.4.2.1 DC ANALYSIS

In order to initiate a DC analysis, first generate a simplified version of Figure 8.42(a) and let $R_E = 0$, as shown in Figure 8.43. Create the simplified version by calculating the Thévenin equivalent circuit of Figure 8.42(a) seen through the base of the transistor. Based on Figure 8.42(a), the parameters in Figure 8.43 correspond to (see Example 8.14 for details):

$$R_B = \frac{R_1 R_2}{R_1 + R_2}$$

$$V_{BB} = \frac{R_1}{R_1 + R_2} \cdot V_{CC}$$

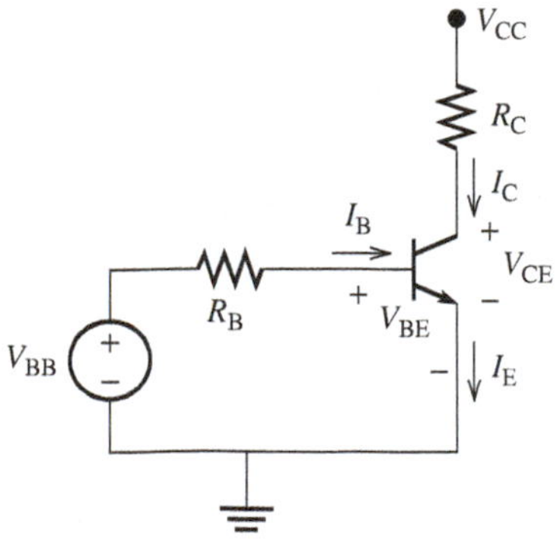

FIGURE 8.43   A basic transistor biasing circuit.

To determine the region of operation, or the Q-point (i.e., the value of voltages and currents), consider the circuit shown in Figure 8.43 and use the following procedure:

1. Replace all capacitors with open circuits.
2. Remove all independent AC sources by replacing voltage sources with short circuits and current sources with open circuits.
3. Check the biasing of the BE junction. If $V_{BB} < 0.7$ V the transistor operates at the cutoff region (OFF) and: $I_B = I_C = 0$, $V_{BE} = V_{BB}$, and $V_{CE} = V_{CC}$.
4. If $V_{BB} > 0.7$ V the transistor is ON and operates in one of two regions: the active region or the saturation region. In either region, the base emitter voltage $V_{BE} = 0.7$ V. To determine the

region of operation, make an assumption. Assume the active region and calculate I_B, I_C, and V_{CE} as follows:

Writing KVL for the input loop yields:

$$V_{BB} - I_B R_B - 0.7 = 0 \tag{8.37}$$

Thus, the current, I_B, is given by:

$$I_B = \frac{V_{BB} - 0.7}{R_B} \tag{8.38}$$

Referring to Table 8.3, with the assumption that the transistor operates in the active region:

$$I_C = \beta I_B \tag{8.39}$$

Writing KVL for the input loop yields:

$$V_{CC} - I_C R_C - V_{CE} = 0 \tag{8.40}$$

Thus, the voltage, V_{CE}, is given by:

$$V_{CE} = V_{CC} - I_C R_C \tag{8.41}$$

To check the validity of the assumption, compare V_{CE} and 0.2 V:

- If $V_{CE} > 0.2$ V, the transistor operates in the active region and the assumption is valid.
- If $V_{CE} < 0.2$ V, the transistor operates in the saturation region and the assumption is not valid.

In the saturation region, the collector–emitter voltage is constant (i.e., $V_{CE} < 0.2$ V). Thus, the value of I_C needs to be recalculated based on the properties of the saturation region. Using Equation (8.41), the collector current can be calculated as follows:

$$I_C = \frac{V_{CC} - 0.2}{R_C} \tag{8.42}$$

As shown in Table 8.3, when a transistor is used as a switch (ON–OFF), there is a transition between high voltage and low voltage. For example, using Figure 8.43, it is observed that when the transistor is in the cutoff region, $I_C = 0$; thus, $V_{CE} = V_{CC}$. When, it is in the saturation region, $V_{CE} = 0.2$, that is, the output voltage is almost zero. See Section 8.4.2 for additional details.

EXERCISE 8.4

For the circuit shown in Figure 8.43, use KVL in the output loop to validate that in the cutoff region: $V_{CE} = V_{CC}$.

EXAMPLE 8.11   Bias Point of a Transistor

For the circuit shown in Figure 8.44, find V_{BE}, I_B, I_C, and V_{CE}, if:

a. $V_{BB} = 0.3$ V
b. $V_{BB} = 2.7$ V
c. $V_{BB} = 6.7$ V

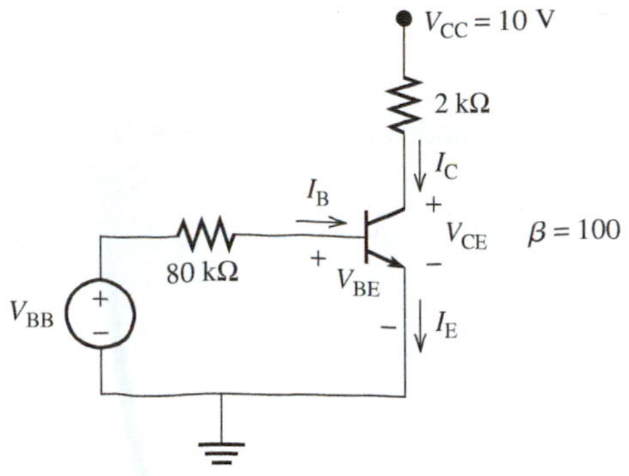

FIGURE 8.44 The circuit for Example 8.11.

SOLUTION

a. Because the value of the base–emitter voltage $V_{BB} = 0.3\,V < 0.7\,V$, the transistor operates in the cutoff region. Therefore:

$$I_B = I_C = 0$$

Applying KVL to the input loop:

$$V_{BB} - I_B \times R_B - V_{BE} = 0$$

Therefore:

$$V_{BE} = V_{BB} = 0.3\,V$$

Applying KVL to the output loop:

$$V_{CC} - I_C \times R_C - V_{CE} = 0$$

Thus:

$$V_{CE} = V_{CC} = 10\,V$$

b. Because $V_{BB} > 0.7\,V$, the transistor is ON and $V_{BE} = 0.7$. Applying KVL to the input loop:

$$2.7 - I_B \times 80 - 0.7 = 0$$

The current, I_B, is given by:

$$I_B = \frac{2.7 - 0.7}{80}$$
$$= 0.025\,mA$$

Assuming the active region:

$$I_C = \beta I_B$$
$$= 100 \times 0.025$$
$$= 2.5\,mA$$

Applying KVL to the output loop yields:

$$V_{CE} = V_{CC} - I_C R_C$$
$$= 10 - 2.5 \times 2$$
$$= 5\,V$$

Because, the value of the collector–emitter voltage $V_{CE} > 0.2\,V$, the transistor does operate in the active region and the assumption is valid.

(continued)

EXAMPLE 8.11	**Continued**

c. Again, because $V_{BB} > 0.7$ V the transistor is on and, therefore, $V_{BE} = 0.7$. Applying KVL to the input loop:

$$6.7 - I_B \times 10 - 0.7 = 0$$

The current, I_B, is given by:

$$I_B = \frac{6.7 - 0.7}{80}$$
$$= 0.075 \text{ mA}$$

Assuming the active region:

$$I_C = \beta I_B$$
$$= 100 \times 0.075$$
$$= 7.5 \text{ mA}$$

Applying KVL to the output loop yields:

$$V_{CE} = V_{CC} - I_C R_C$$
$$= 10 - 7.5 \times 2$$
$$= -5 \text{ V}$$

Because the value of the collector–emitter voltage $V_{CE} < 0.2$ V, the transistor operates in the saturation region and the assumption was not valid. In the saturation region:

$$V_{CE} = 0.2 \text{ V}$$

Therefore, the collector current corresponds to:

$$I_C = \frac{10 - 0.2}{2}$$
$$= 4.9 \text{ mA}$$

EXAMPLE 8.12	**Bias Point of a Transistor**

For the circuit shown in Figure 8.45, find V_{BE}, I_B, I_C, and V_{CE}.

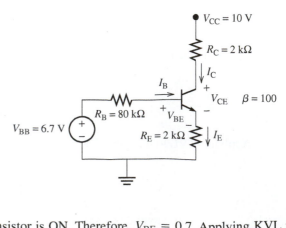

FIGURE 8.45 Circuit for Example 8.12.

SOLUTION

Because $V_{BB} = 6.7$ V > 0.7 V, the transistor is ON. Therefore, $V_{BE} = 0.7$. Applying KVL to the input loop:

$$V_{BB} - I_B R_B - V_{BE} - I_E R_E = 0$$

Assuming the active region where $I_C = \beta I_B$, and substituting Equation (8.39) into Equation (8.32):

$$I_E = I_B + \beta I_B = (1 + \beta)I_B$$

Thus, the input loop equation is:

$$V_{BB} - I_B R_B - V_{BE} - (1 + \beta)I_B R_E = 0$$

The base current corresponds to:

$$
\begin{aligned}
I_B &= \frac{V_{BB} - V_{BE}}{R_B + (1 + \beta)R_E} \\
&= \frac{6.7 - 0.7}{80 + 101 \times 2} \\
&= 0.0213 \text{ mA}
\end{aligned}
$$

In addition, the collector current corresponds to:

$$I_C = 100 \times 0.0213 = 2.13 \text{ mA}$$

Finally, the collector–emitter voltage corresponds to:

$$
\begin{aligned}
V_{CE} &= V_{CC} - I_C R_C - I_E R_E \\
&= 10 - 2.13 \times 2 - 101 \times 0.0213 \times 2 \\
&= 1.45 \text{ V}
\end{aligned}
$$

Because $V_{CE} > 0.2$ V, the transistor operates in the active region. Comparing Examples 8.11 and 8.12, it is notable that adding the resistor, R_E, prevents the transistor from saturating with higher DC voltages.

EXAMPLE 8.13 Transistor Computations

For the circuit shown in Figure 8.46, find the value of the resistor, R_B, which adjusts the collector–emitter voltage to be midway between V_{CC} and ground (i.e., $V_{CE} = V_{CC}/2$).

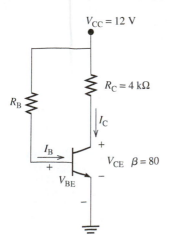

FIGURE 8.46 Circuit for Example 8.13.

SOLUTION

Applying KVL to the output loop yields:

$$12 - I_C \times 4 - V_{CE} = 0$$

(continued)

EXAMPLE 8.13   **Continued**

But, in this example, $V_{CE} = V_{CC}/2 = 6$ V. Therefore, the collector current corresponds to:

$$I_C = \frac{12 - 6}{4} = 1.5 \text{ mA}$$

Because $V_{CE} > 0.2$ V, the transistor operates in the active region. Therefore, the base current corresponds to:

$$I_B = \frac{I_C}{\beta} = \frac{1.5}{80} = 0.01875 \text{ mA}$$

Applying KVL to the input loop:

$$V_{CC} - I_B R_B - V_{BE} = 0$$

In addition, because the transistor operates in the active region, $V_{BE} = 0.7$ V and the resistor, R_B, corresponds to:

$$R_B = \frac{12 - 0.7}{0.01875} = 602.67 \text{ k}\Omega$$

EXAMPLE 8.14   **Bias Point of a Transistor**

For the circuit shown in Figure 8.47, find V_{BE}, I_B, I_C, and V_{CE}.

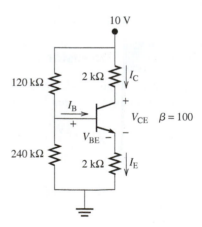

FIGURE 8.47   Circuit for Example 8.14.

SOLUTION

The circuit can be represented as shown in Figure 8.48(a). Applying the Thévenin theorem to find the equivalent circuit between A and B in the circuit at left results in:

$$V_{BB} = V_{th} = 10 \times \frac{240}{240 + 120} = 6.7 \text{ V}$$

and:

$$R_B = R_{th} = 240 \| 120 = \frac{240 \times 120}{240 + 120} = 80 \text{ k}\Omega$$

The simplified circuit is shown in Figure 8.48(b). The solution can be completed easily as discussed in Example 8.12.

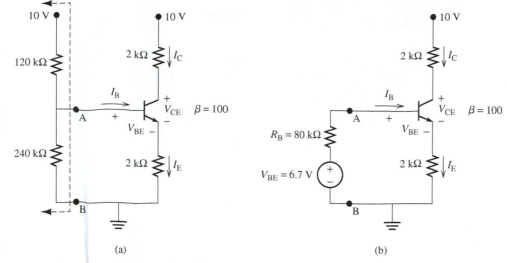

FIGURE 8.48   A simplified representation of the circuit shown in Figure 8.47.

EXERCISE 8.5

Prepare a detailed solution for Example 8.14.

APPLICATION EXAMPLE 8.15   **Gyro Sensor**

Gyno sensor and its applications was discussed in Example 5.8. Figure 8.49 shows the circuit for a single-axis gyro sensor. The gyro sensor senses a single-axis orientation of a particular moment, and sends the signal to another device that processes the information. In this problem, a gyro sensor is studied without its processing device. The circuit consists of a PNP transistor, with a current gain, β, of 200. The circuit also has two capacitors, one gyro sensor, and a resistor with a fair resistance to prevent short circuits from occurring.

Assume the voltage source is 5 V, the resistance of the resistor is 5 kΩ, the resistance of the gyro sensor is 500 Ω, the capacitance of C_1 is 33 µF, and C_2 is 0.33 µF. Determine the current and voltage across the gyro sensor.

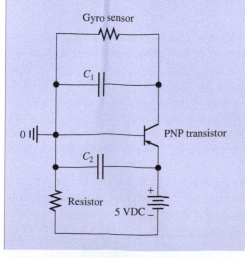

FIGURE 8.49   Circuit of a gyro sensor and photo of gyro sensor. (Photo courtesy of Sensonor Technologies AS.)

(*continued*)

APPLICATION EXAMPLE 8.15 **Continued**

SOLUTION

The circuit analysis of PNP transistors is similar to that of NPN transistors, but the directions of node currents and voltage polarities are different [see Figure 8.40(a)]. When a PNP transistor is in the active region, $V_{EB} = 0.7$ V.

Because the voltage source is 5 V > 0.7 V, the transistor is ON. In addition, the two capacitors function as open circuits. Applying KVL, the emitter current across the resistor is:

$$I_E R = 5 \text{ V} - 0.7 \text{ V}$$

$$I_E = \frac{5 - 0.7}{5000}$$

$$I_E = 0.86 \text{ mA}$$

Next, applying KCL to the transistor node, with $I_C = \beta I_B$, the base current is:

$$I_E = I_B + I_C$$

$$I_E = I_B + \beta I_B$$

$$(1 + \beta) I_B = I_E$$

$$I_B = \frac{I_E}{1 + \beta}$$

$$I_B = \frac{0.86}{1 + 200}$$

$$I_B = 0.0043 \text{ mA}$$

In addition, the collector current across the gyro sensor is:

$$I_C = 0.8557 \text{ mA}$$

Thus, the voltage across the gyro sensor is:

$$V_C = I_C \times R_{gyro}$$

$$V_C = 0.8557 \times 500$$

$$V_C = 0.4279$$

8.4.2.2 AC ANALYSIS

The previous section outlined how to determine the region of operation and how to design a circuit to adjust the Q-point (see Example 8.13). This section explains how to perform AC analysis to determine voltage and current amplification, input impedance, and output impedance of a transistor circuit. Because a transistor is used to amplify AC signals with very low amplitudes, AC signals are usually called *small signals*. One example is the amplification of radio signals in a radio system. The signal received on the radio antenna is in the order of a millivolt, while the batteries used to maintain the operating point of the transistor may have a voltage of 3 V. Comparing a 1-mV signal with 3 V, the first signal is considered to be small. Therefore, these signals are called small signals.

In AC analysis, the transistor is replaced by an equivalent circuit called the *small-signal model*. This small-signal model is appropriate for a transistor if it operates in its active region. It should be noted that the transistor is used as a voltage and/or current amplifier only in its active region. Therefore, it is important to make sure that the transistor operates in this region prior to initiating this analysis. A small-signal model for an NPN transistor is shown in Figure 8.50.

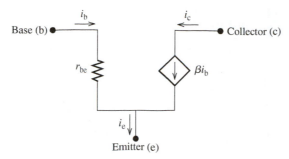

FIGURE 8.50   AC equivalent circuit (small-signal model) of an NPN transistor.

The resistance, r_{be}, is called the base–emitter resistance and it corresponds to:

$$r_{be} = \frac{0.0259}{I_{BQ}} \tag{8.43}$$

Here, I_{BQ} is the value of the base current at the Q-point. The AC collector current is represented by a dependent current source (i.e., a current-controlled current source as discussed in Chapter 3) and it corresponds to:

$$i_c = \beta i_b \tag{8.44}$$

Here, β is the AC gain. For simplicity, this text assumes that both DC and AC gains are equal and both have the same symbol, β.

The following procedure is used to perform AC analysis:

1. First, short-circuit all capacitors
2. Remove all independent DC sources by replacing voltage sources with short circuits and current sources with open circuits
3. Calculate the base-emitter resistance, r_{be}, using the DC bias current, I_{BQ}, calculated using DC analysis, as shown in Equation (8.43)
4. Replace the transistor with its AC equivalent circuit depicted in Figure 8.50
5. Evaluate the following components:

Voltage gain: Amplification of the voltage:

$$A_v = \frac{v_o}{v_{in}}$$

Current gain: Amplification of the current:

$$A_i = \frac{i_L}{i_{in}}$$

Input impedance, R_i: The equivalent impedance seen by the input source.
Output impedance, R_o: The equivalent impedance seen by the load.

Note: Impedance seen across two terminals is calculated using one of the two methods below:

1. Equate all independent sources to zero, and find the ratio of the voltage to current at the desired terminal
2. Leave all sources, and use the following equation (discussed in Chapter 4):

$$R = \frac{v_{oc}}{i_{sc}}$$

Where, v_{oc} is the voltage seen across the two terminals when it is open circuited, and i_{sc} is the current through that terminal when it is short circuited.

EXAMPLE 8.16   **Voltage and Current Gain-Input and Output Resistance**

Find A_v, A_i, R_i, and R_o for the circuit shown in Figure 8.51.

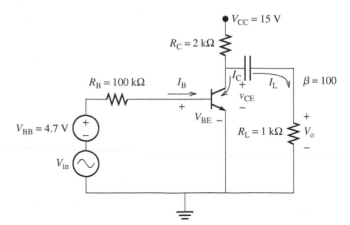

FIGURE 8.51 Circuit for
Example 8.16.

SOLUTION

First, calculate the Q-point from the DC analysis to check the region of operation. The DC equivalent circuit is shown in Figure 8.52(a). Applying the same procedure as in the previous section:

$$V_{BE} = 0.7 \text{ V}, I_B = \frac{4.7 - 0.7}{100} = 0.04 \text{ mA}, I_C = 100 \times 0.04 = 4 \text{ mA, and}$$

$$V_{CE} = 15 - 4 \times 2 = 7 \text{ V}$$

The transistor operates in the active region because $V_{CE} > 0.2$ V.

Now, replace C, V_{BB}, and V_{CC} with short circuits, which leads to the AC equivalent circuit shown in Figure 8.52(b). To complete the AC analysis, replace the transistor with the small-signal model given in Figure 8.50. The resistance, r_{be}, is calculated using Equation (8.43):

$$r_{be} = \frac{0.0259}{0.04 \times 10^{-3}} = 647.5 \text{ }\Omega$$

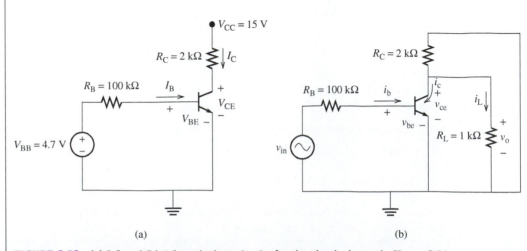

(a)   (b)

FIGURE 8.52 (a) DC and (b) AC equivalent circuits for the circuit shown in Figure 8.51.

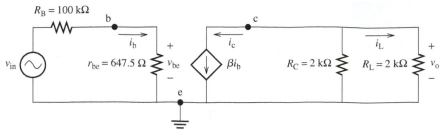

FIGURE 8.53   AC equivalent circuit for the circuit shown in Figure 8.51.

The simplified AC equivalent circuit is shown in Figure 8.53. The output voltage is given by:

$$v_o = i_L R_L$$

Applying the current division law, the load current corresponds to:

$$i_L = -i_c \times \frac{R_C}{R_C + R_L} = -\beta \frac{R_C}{R_C + R_L} \times i_b$$

Applying KVL to the input loop, the base current can be calculated:

$$i_b = \frac{v_{in}}{R_B + r_{be}}$$

Therefore, the output voltage corresponds to:

$$v_o = -\frac{\beta \cdot R_C R_L}{R_C + R_L} \times \frac{v_{in}}{R_B + r_{be}}$$

The voltage gain corresponds to:

$$A_v = \frac{v_o}{v_{in}}$$

$$= -\frac{\beta R_C R_L}{(R_C + R_L)(R_B + r_{be})}$$

$$= -\frac{100 \times 2000 \times 1000}{(2000 + 1000) \times (1 \times 10^5 + 647.5)} = -0.66$$

The negative sign indicates that the input and output voltage are out of phase. In other words, there is a 180° phase angle between the input voltage (at the base) and the output voltage (at the collector). The current gain is:

$$A_i = \frac{i_L}{i_{in}} = -\frac{\dfrac{\beta R_C}{R_C + R_L} i_b}{i_b}$$

$$= -\frac{\beta R_C}{R_C + R_L}$$

$$= -\frac{100 \times 2000}{2000 + 1000}$$

$$= -66.67$$

The negative sign indicates that the direction of the load current must be reversed.
The input impedance seen by the input source is:

$$R_i = R_B + r_{be} = 100.647 \text{ k}\Omega$$

(continued)

EXAMPLE 8.16 **Continued**

The output impedance seen by the load corresponds to:

$$R_o = R_C = 2 \, k\Omega$$

Note that the voltage gain is less than unity, which means that the transistor attenuates the input voltage signal instead of amplifying it. The reason for that is the existence of the large resistance, R_B, in the denominator. The resistance, R_B, cannot be eliminated or reduced because it is used to adjust the Q-point and limit the base current. However, its effect can be eliminated by connecting it in parallel with the input source (see the next example).

EXAMPLE 8.17 **Voltage and Current Gain-Input and Output Resistance**

Considering the circuit shown in Figure 8.54, calculate voltage gain, current gain, input impedance, and output impedance.

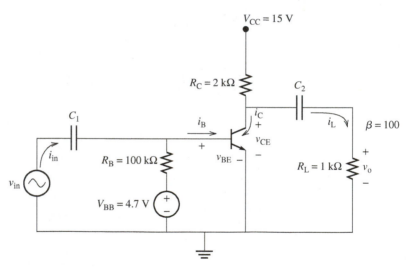

FIGURE 8.54 Circuit for Example 8.16.

SOLUTION

The DC circuit and the operating point are the same as the previous example. The AC equivalent circuits are shown in Figure 8.55. The base–emitter resistance is calculated in the previous example and is $r_{be} = 647.5 \, \Omega$. Referring to Figure 8.55(b), the output voltage corresponds to:

$$v_o = -\frac{\beta R_L R_C}{R_L + R_C} i_b$$

The base current corresponds to:

$$i_b = \frac{v_{in}}{r_{be}}$$

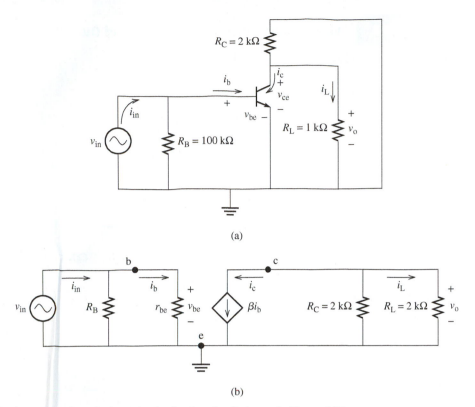

FIGURE 8.55 AC equivalent circuits for the circuit shown in Figure 8.54.

Therefore, the output voltage can be written as:

$$v_{\rm o} = -\frac{\beta R_{\rm L} R_{\rm C}}{R_{\rm L} + R_{\rm C}} \frac{v_{\rm in}}{r_{\rm be}}$$

Moreover, the voltage gain corresponds to:

$$A_{\rm v} = -\frac{\beta R_{\rm L} R_{\rm C}}{(R_{\rm L} + R_{\rm C}) r_{\rm be}}$$

$$= -\frac{100 \times 1000 \times 2000}{(1000 + 2000) \times 647.5}$$

$$= -102.96$$

Next, the current gain is:

$$A_{\rm i} = \frac{i_{\rm L}}{i_{\rm in}}$$

$$= -\frac{\dfrac{\beta R_{\rm C}}{R_c + R_{\rm L}} i_{\rm b}}{i_{\rm in}}$$

$$= -\frac{\beta R_{\rm C}}{R_c + R_{\rm L}} \frac{R_{\rm B}}{R_{\rm B} + r_{\rm be}} \quad (\text{using current division})$$

$$= -\frac{100 \times 2000}{2000 + 1000} \times \frac{10^5}{10^5 + 647.5}$$

$$= -66.24$$

$$R_{\rm o} = R_{\rm C} = 2\,{\rm k}\Omega$$

EXAMPLE 8.18 **Voltage and Current Gain-Input and Output Resistance**

For the circuit shown in Figure 8.56 find A_v, A_i, R_i, and R_o.

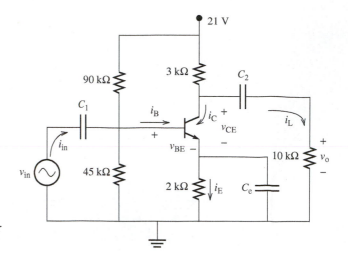

FIGURE 8.56 Circuit for Example 8.17.

SOLUTION

DC analysis

The DC equivalent circuits are shown in Figure 8.57.
The Thévenin voltage, V_{BB}, corresponds to:

$$V_{BB} = 21 \times \frac{45}{45 + 90} = 7 \text{ V}$$

Also, the Thévenin resistance corresponds to:

$$R_B = \frac{45 \times 90}{45 + 90} = 30 \text{ k}\Omega$$

Referring to Figure 8.57(b), and assuming the active region, the base current corresponds to:

$$I_B = \frac{V_{BB} - 0.7}{R_B + (1 + \beta)R_E}$$

$$= \frac{7 - 0.7}{30 + 101 \times 2}$$

$$= 0.0272 \text{ mA}$$

The collector current is given by:

$$I_C = \beta I_B = 100 \times 0.0272 = 2.72 \text{ mA}$$

The collector–emitter voltage corresponds to:

$$V_{CE} = V_{CC} - I_C R_C - I_E R_E$$

$$= 21 - 2.72 \times 3 - (2.72 + 0.0272) \times 2$$

$$= 7.37 \text{ V}$$

Because $V_{CE} > 0.2$ V, the transistor does operate in the active region; therefore, the assumption was correct.

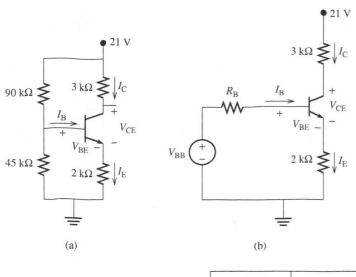

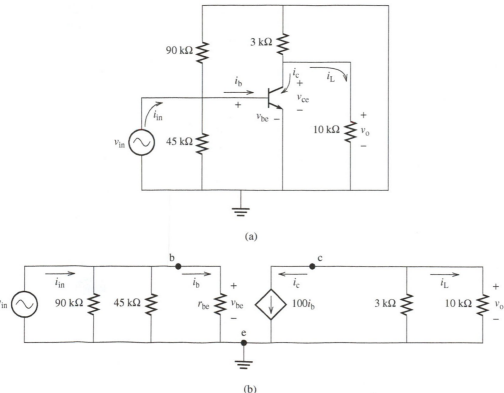

FIGURE 8.57 DC equivalent circuits for the circuit shown in Figure 8.56.

(a)   (b)

(a)

(b)

FIGURE 8.58 AC equivalent circuits for the circuit shown in Figure 8.56.

AC analysis

The AC equivalent circuits are shown in Figure 8.58. The base–emitter resistance corresponds to:

$$r_{be} = \frac{0.259}{0.0272 \times 10^{-3}} = 953.78 \ \Omega$$

Following the same procedure as in the previous examples:

$$A_v = \frac{-100 \times 3000 \times 10,000}{(3000 + 10,000) \times 953.78} = -241.95$$

(*continued*)

EXAMPLE 8.18 **Continued**

$$A_i = -241.95 \times \frac{90{,}000\|45{,}000\|953.78}{10{,}000} = -22.37$$

$$R_i = 90{,}000\|45{,}000\|953.78 = 924.39\ \Omega$$

$$R_o = R_C = 3\ \text{k}\Omega$$

8.4.3 Transistors as Switches

A transistor can function as a switch between the collector and the emitter terminals where the base terminal is used to control the status of the switch. As discussed in the previous sections, when the biasing voltage applied to the transistor base is less than 0.7 V, the transistor operates in the cutoff region, no current passes through the collector (i.e., $I_C = 0$), and the switch will be open. When the value of the biasing voltage is large enough, the transistor saturates, the voltage across the transistor, V_{CE}, is at the minimum (i.e., $V_{CE} = 0.2$ V), and the switch will be closed. The transistor that operates between cutoff and saturation is called an electronic switch. The switch is controlled by an electric voltage on the transistor base.

EXAMPLE 8.19 **Transistor Characteristic**

Consider the circuit shown in Figure 8.59. Sketch the transfer characteristics (i.e., the relationship between the input voltage, V_{BB}, and the collector–emitter voltage, V_{CE}, and calculate the minimum value of V_{BB} required such that the transistor operates in the saturation region (the switch is ON).

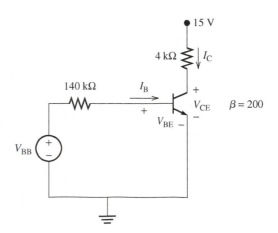

FIGURE 8.59 The circuit for Example 8.18.

SOLUTION

Three cases are possible:

1. When the input voltage $V_{BB} < 0.7$ V, the transistor operates in the cutoff region. Therefore, $I_B = I_C = 0$ and the voltage. V_{CE}, corresponds to:

$$V_{CE} = V_{CC} = 15\ \text{V}$$

2. As the value of V_{BB} increases beyond 0.7 V, both the base current and the collector current begin to flow. Therefore, the voltage, V_{CE}, will decrease. Because V_{CE} is still high (i.e., $V_{CE} > 0.2$ V), the transistor operates in the active region and the relationship between V_{BB} and V_{CE} is computed as follows:

Applying KVL to the output loop, the collector–emitter voltage corresponds to:

$$V_{CE} = V_{CC} - I_C R_C$$

Replacing I_C with βI_B yields:

$$V_{CE} = V_{CC} - \beta I_B R_C$$

Find I_B by applying KVL to the input loop. Substituting I_B by $(V_{BB} - 0.7)/R_B$:

$$V_{CE} = V_{CC} - \frac{\beta R_C (V_{BB} - 0.7)}{R_B}$$

$$= 15 - \frac{200(V_{BB} - 0.7)}{140}$$

$$= 15 - \frac{10 V_{BB}}{7} + 1$$

$$= 16 - 1.43 V_{BB}$$

This equation shows that the value of V_{CE} decreases linearly with V_{BB}. But, the transistor will operate in the saturation region as the value of V_{CE} reaches 0.2 V (see Figure 8.60). In other words, the transistor saturates when:

$$16 - 1.43 V_{BBsat} = 0.2$$

where V_{BBSat} is the value of the input voltage required to operate the transistor in the saturation region and it corresponds to:

$$V_{BBsat} = \frac{16 - 0.2}{1.43} = 11.06 \text{ V}$$

3. When $V_{BB} > V_{BBSat} = 11.06$ V, the transistor operates in the saturation region. Therefore, $V_{CE} = 0.2$ V. As a summary, the collector–emitter voltage corresponds to:

$$V_{CE} = \begin{cases} 15 \text{ V} & V_{BB} < 0.7 \text{ V} \\ 16 - 1.43 \ V_{BB} & 0.7 \text{ V} < V_{BB} < 11.06 \text{ V} \\ 0.2 \text{ V} & V_{BB} > 11.06 \text{ V} \end{cases}$$

A plot of the transfer characteristics is shown in Figure 8.60. Note that the transistor or the switch is considered OFF if $V_{BB} < 0.7$ V. Moreover, it will be ON if $V_{BB} > 0.7$ V. Therefore, the minimum voltage required to turn the switch ON is the same as the minimum voltage required to saturate the transistor and corresponds to $V_{BBSat} = 11.06$ V.

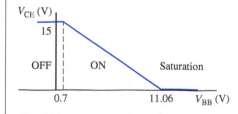

FIGURE 8.60   Transfer characteristics for the circuit shown in Figure 8.59.

8.4.4 Field-Effect Transistors

This section introduces field-effect transistors (FETs). The operational principle of the FET is based on the formation of a channel between two terminals. This channel is called the conduction channel and is formed by an electric field created between two terminals of the transistor. Therefore, it is called a *field-effect transistor*. There are many different kinds of FET (e.g., MOSFET, JFET, MESFET, MODFET, etc.), but the most commonly used FET is

(a)   (b)   (c)

FIGURE 8.61 (a) Top view of an integrated circuit (Used with permission from Stockex/Alamy.).
(b) Packaged IC microchip with large pins, ready to be mounted on a printed circuit board (PCB) (Used
with permission from Timothy Hodgkinson/Shutterstock.com.). (c) Scanning electron microscope image
of the cross-sectional view of a single MOSFET device, the most elemental unit of an IC (Image courtesy
of MuAnalysis.).

the metal–oxide–semiconductor field-effect transistor (MOSFET). Complex digital integrated
circuits (ICs) such as microprocessors and memory chips are primarily made up of MOSFETs
[see Figure 8.61(a) and 8.61(b)]. The operational principles of all FETs are approximately the
same; therefore, this section will study MOSFETs as a case study to understand FETs.

MOSFETs are the basic building block of very large-scale integration (VLSI) circuits [see
Figure 8.61(c)]. MOSFETs can operate at a much lower current than the BJTs discussed earlier
in the chapter. Therefore, they consume less power ($P = I \times V$) and generate less heat per area.
A MOSFET also occupies a smaller chip area and requires a lower number of fabrication steps
compared to a BJT. Portable handheld devices such as MP3 players, computer game players, and
laptops have limited battery lifetime. Therefore, low-power circuit design using FETs is abso-
lutely essential to enable minimum power consumption.

An analogy to understand the hierachy in an IC is to think of the MOSFET as the brick
that builds a house. If a few "bricks" are stacked together, a new digital module called a *logic
gate* (NOT, OR, AND, etc.) is built. Logic gates are discussed in detail in Chapter 10. When a
few logic gates are connected together, other more complex digital modules can be built (e.g.,
latches, flip-flops). These modules enables larger digital modules like adders and multiplexers.
The functionality of a circuit becomes more complex as its size increases. The scale of an IC is
analogous to a city. ICs are connected to each other using a PCB, which is similar to the highway
system that connects two cities. This section focuses on the "brick" (FETs) and does not discuss
more advanced digital modules. More advanced courses in electrical engineering like digital
design or very large-scale integration (VLSI) design are recommended for further understanding
of these systems. Chapter 10 offers a summary of digital circuits.

8.4.4.1 BASIC OPERATION OF MOSFETS

Similar to BJTs, FETs are used to make amplifiers and logic switches. The amplifying function
of a MOSFET is similar to that of a BJT, which was explained at length in the previous section.
This section focuses on using MOSFETs as logic switches, which are used in the world of digital
design. MOSFET analysis can be very complex. This chapter first explains MOSFET structure in
simple (but accurate) terms and then outlines the details of FET analysis.

Figure 8.62(a) shows a cross-sectional view of the physical layout of a single MOSFET
device. A MOSFET has three terminals: (a) drain, (b) source, and (c) gate. The drain and source
junctions define the direction of the electron flow, which is opposite to the direction of the current
flow. There are two kinds of MOSFET, PMOS (p-type MOSFET) and NMOS (n-type MOSFET).
These types are defined based on the nature of the device: if the source drain junctions are n-type,
the MOSFET is called NMOS; if the source and drain junctions are p-type, the MOSFET is called
PMOS. Figure 8.62(b) shows the symbols that are normally used to represent these MOSFETs.

As shown in Figure 8.62(c), an NMOS is fabricated on a p-type silicon substrate. The drain
and source junctions of the NMOS are formed by an n-type semiconductor. When a positive

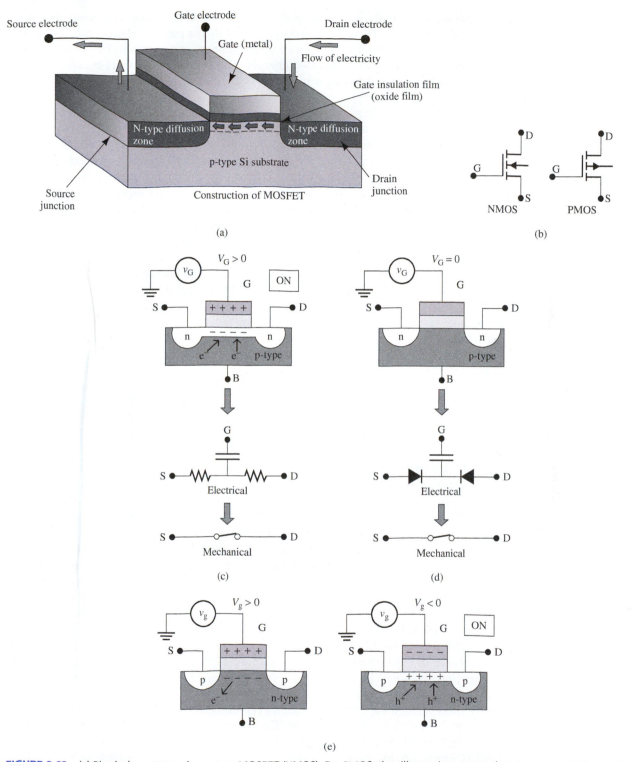

FIGURE 8.62 (a) Physical structure of an n-type MOSFET (NMOS). For PMOS, the silicon substrate needs to be n-type. (b) Symbolic representation of NMOS and PMOS. (c) When V_G is positive, electrons in the p-type substrate are attracted to the oxide–silicon interface, and form an n-type conduction channel. The electrical model is represented by resistors in series. The transistor is in its ON state. (d) When $V_G = 0$, the area underneath the oxide layer is still p-type, which forms a "back-to-back" diode with the n region, as shown in the electrical representation. The transistor is in its OFF state; (e) For PMOS, the substrate is n-type. The p-type inversion channel can be formed by applying a negative bias to the gate electrode, turning the transistor from OFF (left) to ON (right).

voltage bias is applied to the gate of an NMOS, an electric field is generated across the insulating silicon dioxide layer. The electric field generated by the positive charges attracts negative charges (electrons) from the silicon substrate to the oxide–silicon interface. The accumulation of free electrons forms an n-type conduction channel. This conduction channel is called an n-type because it has excess free electrons, similar to the concept discussed earlier in the diode section (Section 8.2).

Note that although the p-type substrate is rich with free-hole carriers, there are still some minority electron carriers available in the p-type substrate. With the conduction channel formed, current ($I = dQ/dt$) flows freely from the source to the drain junction and the NMOS is said to be in its ON state. In electrical terms, the conduction channel is formed by inverting the original p-type substrate between the two junctions to form an n-type using the accumulation of free electrons. Thus, it is also called the *inversion channel*. Conceptually, using very crude electrical terms, under positive gate bias, the drain–source channel can now be modeled by a resistor. A closed switch is a mechanical analogy of the NMOS in its ON state, as illustrated in Figure 8.62(c). A water pipe controlled by the water tap (gate) is another analogy. When the water tap is turned ON, the water (electrons) will start flowing.

When the voltage applied to the gate of the NMOS is zero, there are no electrons accumulating at the oxide–silicon interface, that is, in between the source and the drain junctions. The current flow from source to drain is now impeded and thus, the NMOS is in its OFF state. Under these voltages, hole carriers (discussed in Section 8.2) are accumulated at the oxide–silicon interface. A rough but easy-to-understand electrical model is shown in Figure 8.62(d), where the drain–source channel can now be seen as two p–n diodes connected serially in reverse direction (i.e., back-to-back diodes). This simply means that the channel is now under reverse-biased condition and there is minimal (if not zero) current flow. Again, a good mechanical analogy of this device at its OFF state is an open switch, or a water pipe with a closed tap. The electrical behavior of a PMOS is just the opposite of an NMOS. Negative gate voltage is needed to turn the PMOS to an ON state, and a positive gate voltage is needed to turn it to its OFF state, as illustrated in Figure 8.62(e).

8.4.4.2 DESIGN OF DYNAMIC RANDOM ACCESS MEMORY (DRAM) USING MOSFET

The simplest example of a practical MOSFET application is its use in the design of dynamic random access memory (DRAM). DRAM chips are used in almost every computer. An example is shown in Figure 8.63(a). A DRAM is an IC chip used to store data temporarily while the microprocessor computes the data. The beauty of DRAM is that it only requires one transistor and one capacitor to represent one bit of data (either logic 0 or logic 1) as shown in Figure 8.63(b) and 8.63(c). Because the transistor and capacitor can be made very small in size, the DRAM structure can achieve very high density compared to other kinds of memory, such as static-RAM (SRAM), which consists of six transistors.

The capacitor stores electron charges in the DRAM cell structure and discharges electrons over time; therefore, the DRAM needs to be recharged repeatedly. DRAM loses its stored data once the computer is turned off. Therefore, it is considered to be one kind of volatile memory chip. As shown in Figure 8.63(c), the voltage of the capacitor of one DRAM cell is originally zero. The data represented by zero capacitor voltage is logic 0. Once V_G is turned to high voltage ($V_G > 0$), the NMOS transistor is turned ON. Then, turning V_S to high voltage ($V_S > 0$) enables current to flow through the conduction channel of the NMOS, and then, through the capacitor. This increases the voltage of the capacitor. The high voltage of the capacitor represents one bit of data, called logic 1. The single DRAM cell in Figure 8.63(c) can be used to build huge arrays of DRAM. Figure 8.63(d) shows an example of a 4×4 array, which can store 16 bits of data.

8.4.4.3 DESIGN OF A MOSFET AS A LOGIC GATE IN A MICROPROCESSOR

The microprocessor is the heart of all electronic devices. It is capable of compiling a large number of calculations in a very short time period depending on the frequency of the processor. The frequency of the processor is related to how fast a transistor can switch from ON state to OFF state or vice versa. A 2.0-GHz processor is composed of transistors with the switching

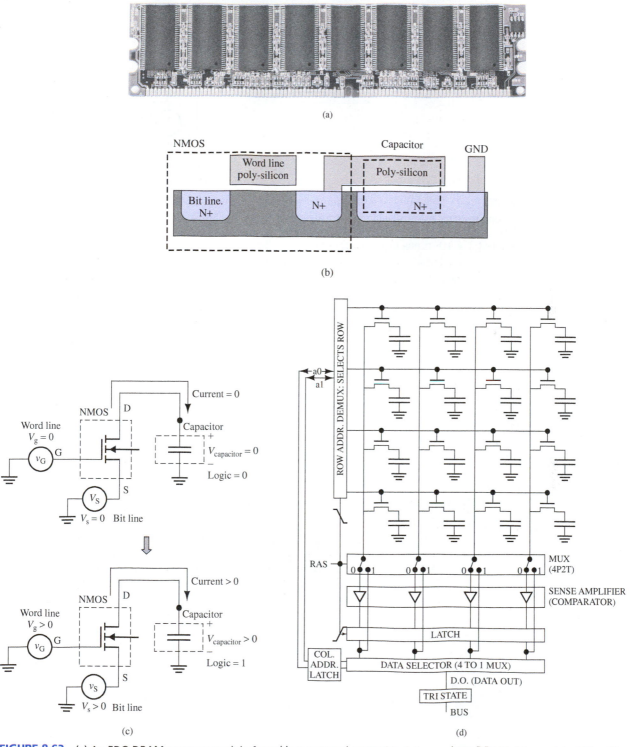

FIGURE 8.63 (a) An EDO DRAM memory module found in a personal computer. A 4-megabyte DRAM chip consist of 32 million DRAM cells. (Used with permission from Timothy Hodgkinson/Shutterstock.com.) (b) Cross-sectional view of the physical layout of a single DRAM cell. Note that the empty gap between both highly conductive poly-silicon and the n+ region of the capacitor is an insulator. (c) Electrical representation of a single DRAM cell, made of one MOSFET and one capacitor. Turning V_G and V_S to high voltage turns the data stored in the DRAM cell from logic 0 to logic 1. (d) The principles of operation of a DRAM read, for a simple 4 × 4 array. Accessing an individual DRAM cell is enabled by selecting one specific bit line and one word line. (Drawings adapted with permission from Steven Mann, University of Toronto.)

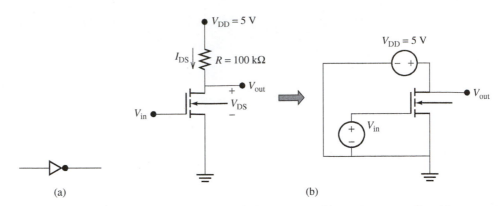

FIGURE 8.64 (a) Symbol of a NOT gate, also called an inverter. (b) A NOT gate made with one NMOS and one resistor. The equivalent electrical representation is illustrated on the right.

speed of 2×10^9 times in 1 s. As mentioned earlier, microprocessors are made of several millions of transistors. These transistors enable a very basic digital module logic gate. Examples of basic logic gates are the NOT gate, OR gate, and AND gate. These individual gates are generally interconnected in a high-density large network of IC to perform complex computations.

It is helpful to start with the simplest example of the NOT gate, also called an *inverter* because this gate produces an inverted signal. The symbolic representation of the NOT gate is shown in Figure 8.64(a) (see Chapter 10 for details). There is more than one way to design a NOT gate; the simple way is to use one NMOS and one resistor, as shown in Figure 8.64(b). The function of the NOT gate is very straightforward; it produces a voltage $V_{out} = V_{high}$ when an input voltage $V_{in} = V_{low}$ is applied to its input, and vice versa. The truth table of the NOT gate is listed in Table 8.4.

The operation of this inverter can be analyzed using the KVL rule. From Figure 8.64(b), $V_{out} = V_{DS} = V_{DD} - I_{DS}R$. When the input voltage, V_{in}, is low or zero, the NMOS is in its OFF state, and it behaves as an open circuit. The circuit shown in Figure 8.64(b) can be redrawn and represented with the crude model shown in Figure 8.65(a). In this case, $I_{DS} = 0A$ because an NMOS is an open switch. Therefore, the voltage drop across the resistor, R, is zero, and $V_{out} = V_{DD} - (0)R = V_{DD} = 5$ V. However, when V_{in} is high (assume, for example, $V_{in} = 5$ V), as shown in Figure 8.65(b); the NMOS is in its ON state, and it behaves similar to a closed circuit. In this case:

$$I_{DS} = \frac{V_{DD}}{R} = \frac{5}{100 \times 10^3} = 0.05 \text{ mA}$$

In addition, V_{out} is directly connected to the ground through the closed circuit of the NMOS, so $V_{out} = 0$ V. Note that the result of this analysis is consistent with the characteristics in the truth table of the NOT gate in Table 8.4.

TABLE 8.4 Truth Table of the NOT Gate

V_{in}	V_{out}
High voltage (logic 1)	Low voltage (logic 0)
Low voltage (logic 0)	High voltage (logic 1)

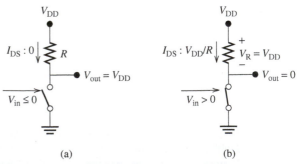

FIGURE 8.65 (a) Equivalent circuit of Figure 8.64(b) when V_{in} = Low is applied to the gate of the NMOS and it acts as an open circuit. (b) Equivalent circuit of Figure 8.64(b) when V_{in} = High is applied to the gate of the NMOS and it acts as a short circuit.

EXAMPLE 8.20   NOT Gate Analysis

The NOT gate analysis in Figure 8.65(a) and 8.65(b) assumes an ideal NMOS, that is, there is no resistance in the switch. In reality, there is a nonzero voltage drop across the NMOS channel because the resistance of the NMOS conduction channel (R_{DS}) is nonzero. Assume $V_{DD} = 5$ V, $V_{in} = 5$ V and $R = 100$ kΩ, $R_{DS} = 1$ Ω. Draw the simplified equivalent electrical circuit and calculate V_{out} and the current flowing through the transistor in its ON state.

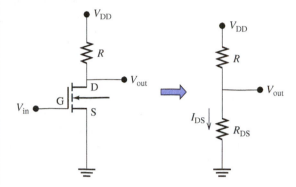

FIGURE 8.66   A simplified model of an NMOS NOT gate when $V_{in} > 0$.

SOLUTION

For an NMOS to be in its ON state, V_{in} has to be at a high voltage (5 V). In reality, as shown in Figure 8.66, the closed switch can be replaced with a resistor, representing the resistance of the conduction channel. This is a simple voltage divider problem, where:

$$V_{out} = V_{DS} = V_{DD} \times \frac{R_{DS}}{R + R_{DS}}$$

Therefore:

$$V_{out} = 5 \times \frac{1}{1 + 100 \times 10^3} = 0.05 \text{ mV}$$

which is very close to 0 V, representing logic = 0.
   In addition, writing KVL for Figure 8.66:

$$I_{DS} = \frac{V_{DD}}{R + R_{DS}} = \frac{5}{100 \times 10^3 + 1} = 0.049995 \text{ mA}$$

EXAMPLE 8.21   Power Consumption of NOT Gate

Calculate the power consumption of one NOT logic gate when $V_{in} = $ high.

SOLUTION

From the solution of Example 8.20, when V_{in} is high, NMOS is in its ON state. In this case, current flows through the resistor and also the NMOS conduction channel.
Therefore, the power consumption is:

$$P = V_{DD} \times I_{DS} = 5 \times 0.0499995 \text{ mA} = 0.249975 \text{ mW}$$

which is approximately 0.25 mW.

| **EXAMPLE 8.22** | **Logic Gate Power Consumption** |

A microprocessor consists of 1 million logic gates. Assume that this microprocessor is placed in a package that dissipates a maximum power of 4 W and all logic gates have the same power consumption. For this scenario:

    a. Compute the maximum power that is required for each logic gate. Assume that all logic gates are in the ON state.

    b. When the microprocessor functions normally, all logic gates switch between ON and OFF states. If the supply voltage is reduced to $V_{DD} = 1$ V, calculate the current that should be used by each gate. Assume that half of the logic gates are in conducting state (ON) at any given time.

SOLUTION

    a. Note that the maximum power is the largest power consumption a single logic gate can afford. If the power dissipation of each logic gates exceeds the maximum power, the total power dissipation of the IC will exceed the limit that the package can withstand and may cause heat damage to the chip. Therefore, the maximum power required for each logic gate is 4 W/1 million $= 4$ μW.

    b. Some logic gates are in their ON and some are in their OFF state. Thus, approximately 1 million/2 $= 500,000$ logic gates are active (ON state) at any given time. Power consumption for each logic gate is:

$$4W/0.5 \text{ million} = 0.008 \text{ mW} = 8 \text{ μW}$$

$$P = V_{DD} \times I_{DS}$$

Thus:

$$8 \text{ μW} = 1 \text{ V} \times I_{DS}$$

Thus, the current that should be used by each logic gate is $I_{DS} = 8$ μA.

8.4.4.4 DESIGN OF MOSFETS AS AMPLIFIERS

MOSFETs can also be used as amplifiers. Understanding a more accurate model of the MOSFET and the current–voltage characteristics of MOSFET will help to explain how they can be used as amplifiers. The electrical model in Figure 8.66 is oversimplified. An accurate model for an NMOS can be created by considering a back-to-back pair of diodes due to the n–p–n structure from the drain junction to the channel and to the source junction, as illustrated in Figure 8.67.

When the model of Figure 8.67 is considered, the current–voltage characteristic of a MOSFET corresponds to:

$$I_{DS} = 0, \text{ when } V_{GS} < V_T \text{ (cuttoff region)}$$

$$I_{DS} = K\big[2(V_{GS} - V_T)V_{DS} - V_{DS}^2\big], \text{ when } V_{DS} < V_{GS} - V_T \text{ and } V_{GS} > V_T \text{ (triode region)}$$

$$I_{DS} = K(V_{GS} - V_T)^2, \text{ when } V_{DS} \geq V_{GS} - V_T \text{ and } V_{GS} > V_T \text{ (saturation region)} \quad \textbf{(8.45)}$$

Here, V_{DS} is the drain–source voltage ($V_{DS} = V_D - V_S$) and V_{GS} is the gate–source voltage ($V_{GS} = V_G - V_S$). Here, V_S is the source voltage and normally it is connected to ground, therefore:

$$V_{DS} = V_D - 0 = V_D$$
$$V_{GS} = V_G - 0 = V_G$$

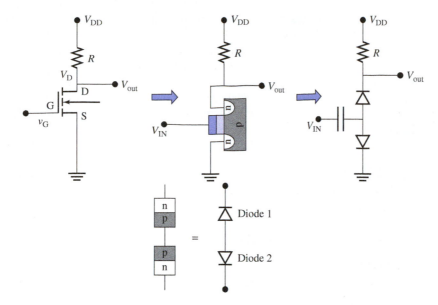

FIGURE 8.67 From left, the symbolic representation of an NMOS NOT gate, its physical layout connection, and its electrical model.

In addition, in Equation (8.45) V_T is the threshold voltage and K is a parametric constant value with a unit of A/V^2. One can see that now, the current is not only a function of V_G, but also V_D. There will be a current flow only when V_G is greater than V_T, or else $I_{DS} = 0$.

An NMOS transistor behaves somewhat similarly to, if not entirely like, a water pipe. Think of electrons as water, as water flow can be controlled by the water tap, so the electron flow (electric current) is controlled by the gate voltage. When the water tap is tightly closed, no water is flowing; this is similar to the cutoff region of a BJT transistor, where current is zero. A force that is slightly greater than the initial tightness of the water tap is needed to open up the water pipe, this scenario mirrors the fact that a gate voltage (V_G) that is greater than the threshold voltage (V_T) is required for any current to start flowing ($V_G > V_T$, I_{DS} is not zero). As the water starts dripping slowly when the water tap is slightly opened, turning the water tap a few more times will increase the water flow; this reflects the *I–V* characteristic shown in Figure 8.68, where drain current (I_{DS}) is always higher with higher V_G.

Drain voltage (V_D) is similar to the water pressure in the water pipe; when the water pressure increases slowly, so will the water flow for a given tap opening. The triode region of a transistor fits this scenario, as long as $V_D < V_G - V_T$. However, there is always a point where the transistor current comes to saturation, where current flow does not increase any more no matter how high V_D is. Naturally, this region is called the saturation region ($V_D > V_G - V_T$).

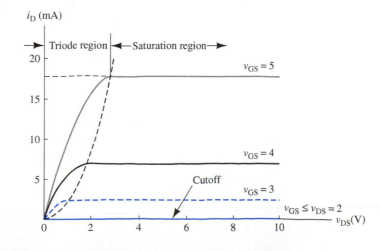

FIGURE 8.68 Current–voltage characteristics of an NMOS transistor.

Figure 8.68 represents the most important current–voltage characteristic of a MOSFET. The cutoff region actually represents the OFF state of a transistor, where I_{DS} is zero. The saturation region represents the ON state of the transistor, where I_{DS} is nonzero for a given V_G. The triode region is just a transition region between the ON and OFF state. Normally, the cutoff and saturation regions of a transistor are used to represent logic 0 and 1, respectively, as illustrated in Figure 8.63(c). Also, in the saturation region, I_{DS} is directly proportional to V_G, where different magnitude of I_{DS} can be produced by applying different V_G. Therefore, it is useful as an amplifier, a device used to increase the amplitude of the input signal.

EXAMPLE 8.23   **Current-Voltage Characteristics of NMOS**

Assume an enhancement-mode NMOS is used as an amplifier. It has $K = 2$ mA/V^2 and $V_T = 2$ V. Plot the current–voltage characteristic curve for $V_{GS} = 0, 1, 2, 3,$ and 4 V. For this example, use the NMOS represented by the model shown in Figure 8.67.

SOLUTION

Using the model in Figure 8.67, first study what happens when $V_{GS} = 0, 1,$ and 2 V.

When $V_{GS} < V_T$ that is, V_{GS} is smaller than V_T (2 V), the current $i_D = 0$.

When $V_{GS} = 3$ V, there are two choices:

$$I_{DS} = K[2(V_{GS} - V_T)V_{DS} - V_{DS}^2], \text{ when } V_{DS} < V_{GS} - V_T$$
$$I_{DS} = K(V_{GS} - V_T)^2, \text{ when } V_{DS} \geq V_{GS} - V_T$$

V_D normally starts from zero and slowly increases. Let us plot 0 V $< V_D < 10$ V.

Because $V_{GS} - V_T = 3 - 2 = 1$ V, for $0 < V_D < 1$, V_D is smaller than $V_{GS} - V_T$, and the following equation can be used to plot the colored dotted curve that forms the triode region of the $I-V$ curve:

$$I_{DS} = K[2(V_{GS} - V_T)V_{DS} - V_{DS}^2]$$

For $1 < V_D < 10$, V_D is greater than $V_{GS} - V_T$, therefore, use the equation:

$$I_{DS} = K(V_{GS} - V_T)^2 = 2\,\text{mA}$$

to plot the colored dashed line, which is the saturation region. As shown in Figure 8.69, both colored dotted and dashed lines are for $V_{GS} = 3$ V. The solid colored line is for $V_{GS} = 0, 1,$ and 2 V, representing the cutoff region.

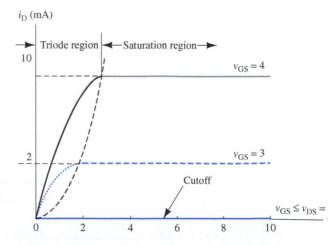

FIGURE 8.69 Current–voltage characteristic plot for Example 8.23.

When $V_{GS} = 4$ V:

$$V_{GS} - V_T = 4 - 2 = 2 \text{ V}$$

Thus, for $0 < V_D < 2$, V_D is smaller than V_{GS}. Therefore, use the equation:

$$I_{DS} = K\left[2(V_{GS} - V_T)V_{DS} - V_{DS}^2\right]$$

To plot the black curve that forms the triode region of the *I–V* curve, assuming $1 < V_D < 10$, V_D is greater than $V_G - V_T$, so use the equation:

$$I_{DS} = K(V_{GS} - V_T)^2 = 8 \text{ mA}$$

to plot the grey line, which is the saturation region. Both black and grey lines in Figure 8.69 are for $V_{GS} = 4$ V.

Using the solution in Example 8.23, and based on the setup shown in Figure 8.70(a), a simple NMOS amplifier circuit can be made.

Here, an alternating input voltage source (v_{in}) is in series with the DC source V_G. v_{in} is a sinusoid that is:

$$v_{in}(t) = 1 \times \sin(\omega t)$$

v_{in} has the amplitude of 1 V and the frequency of f. The amplifier operates similar to a BJT, except that for MOSFETs, $I_G = 0$. Therefore, a MOSFET is said to be modulated by the input voltage instead of the current.

In Figure 8.70(b), an alternating source combines with a constant DC source. This DC source produces a DC offset in the alternating sine function. When the input voltage is a combination of AC and DC components, the KVL loop produces the output voltage, V_{out}, that corresponds to:

$$v_{out}(t) = V_{DD} - i_D(t)R_D$$

Figure 8.70(c) illustrates the relationship between output and input voltages. Based on this figure, observe how the equivalent input voltage, that is:

$$v_{GS}(t) = V_{GS} + v_{in}(t)$$

creates an alternative output voltage. The straight line is called the *load line*. The load line is created by connecting two points. The first point assumes $V_{DS} = 0$, the second point assumes $I_D = 0$.

When $V_{DS} = 0$:

$$I_D = V_{DD}/R_D = 20/1 \text{ k}\Omega = 20 \text{ mA}$$

When $I_D = 0$:

$$V_{DS} = V_{DD} - 0 = V_{DD} = 20 \text{ V}$$

In general, as the current $i_D(t)$ varies, the output voltage will change. On the other hand, as explained in the previous subsections, $i_D(t)$, that is, the current flowing through the channel is controlled by the gate voltage. Thus, the load line maintains the relationship of $V_{GS}(t)$ and $V_{out}(t)$, which is V_{DS} in this case.

8.4.5 Design of NOT Gates Using NMOS Only for High-Density Integration

The design of NOT logic gates with one resistor and one NMOS transistor has been shown in Figures 8.64 to 8.67. Although, the resulting structure is very simple, there is a big disadvantage

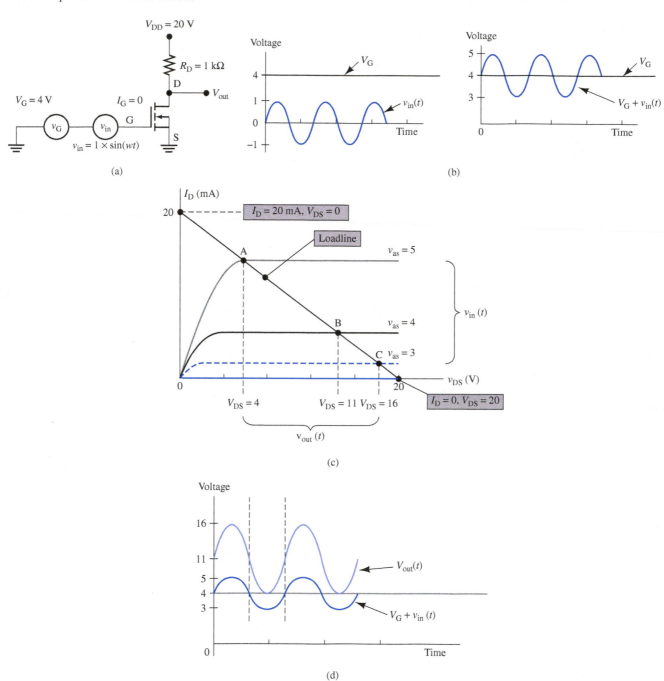

FIGURE 8.70 (a) Circuit of an NMOS amplifier. (b) Relationship between V_G and V_{in}. (c) Load-line analysis and relationship between V_{out} and V_{in}. (d) Comparison of the amplified $V_{out}(t)$ and the original signal $V_{in}(t)$.

of using resistors in logic gate design. Resistors need considerable space to be implemented on a silicon substrate using MOSFET technology. The resistance can be computed by:

$$R = \rho \times L/A$$

where ρ is the resistivity (Ω cm), L is the length of the conductor, and A is the cross-sectional area of the conductor. As an example, in Figure 8.71, a 100-kΩ resistor requires more than 10,000 μm^2 area, compared to a mere 2 μm^2 space for the NMOS. Assuming ρ is 0.001 Ω cm, for a 100-kΩ resistor, L is 10,000 μm, and A = width $\times$ depth = 1 $\mu m \times 1$ μm = 1 μm^2.

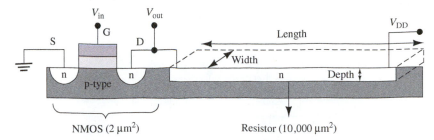

FIGURE 8.71 Physical layout of an NMOS NOT gate using one NMOS and one resistor, resulting in a low-density IC because the resistor needs a large amount of space.

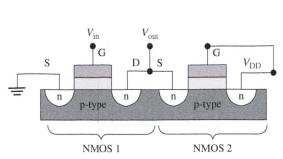

FIGURE 8.72 Physical layout of an NMOS NOT gate using NMOS only. The result is much a smaller layout area per logic gate. Therefore, a higher density IC is possible for a given chip area, for example 1 cm².

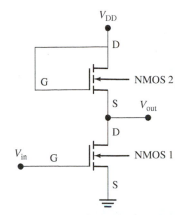

FIGURE 8.73 Symbol representation of a NOT gate using only NMOS transistors.

An alternative that allows implementation of a NOT gate without a resistor is the circuit shown in Figure 8.72. The design in this figure allows the NMOS logic gate to be implemented on a high-density IC.

Figure 8.73 represents the circuit model for the circuit shown in Figure 8.72. In this design, NMOS 2 is always ON. If V_{in} is low, then NMOS 1 will be OFF, and $V_{out} = V_{DD}$. If V_{in} is high, then V_{out} is the voltage drop across NMOS 1, which is very close to zero. NMOS 2 has to be designed such that its conduction channel resistance will be much greater than NMOS 1. In this case, the main voltage drop is across the V_{DS} of NMOS 2 instead of NMOS 1.

8.4.6 Design of a Logic Gate Using CMOS

Complementary MOSFET (CMOS) is the standard of today's ICs. It is called complementary because it has both PMOS and NMOS on the same silicon substrate instead of only NMOS or only PMOS. Its existance is due to the relatively high power consumption of NMOS-only (see Figure 8.74) and PMOS-only ICs. As the technology has progressed through the years, higher density ICs have become available. This has led to increases in the power dissipated per area. Since the surface area of a single microchip is fixed, there is a limit to how much heat can be generated due to the power consumption of each transistor before the entire circuit breaks down due to overheating.

Why does CMOS use less power than NMOS-only or PMOS-only ICs? The answer is simple. First, consider a NOT logic gate made using CMOS. In this case, there is only one transistor that is turned ON on both occasions where the input voltage, V_{in}, is low and high. There is no current path between V_{DD} and ground. In this case, the current, I_{DS}, that flows through the V_{DD} will not be too high. However, considering an NMOS-only NOT gate (see Figure 8.73), the top NMOS 1 is always turned ON. When V_{in} is high, NMOS 2 turns ON as well. In this case, a direct current path is generated between V_{DD} and the ground through these two NMOSs. Because the resistance of the two NMOSs in their ON status is low, I_{DS} flowing through them will be high.

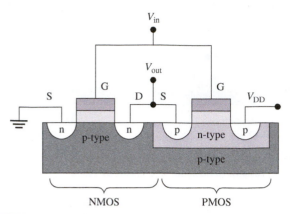

FIGURE 8.74 A cross-sectional view of the physical layout of a single CMOS. The NMOS transistor is on the left and the PMOS transistor is on the right. Note that PMOS is fabricated in a n-type region, where the p-type substrate is lightly doped with n-type dopant, thus turning it into a "minimized n-type substrate" (a so-called n-well) that is required for the PMOS to exist on a p-type substrate.

Thus, the total power consumption ($V_{DD} \times I_{DS}$) will be high and the NMOS-only logic gate leads to higher static power consumption as compared to CMOS.

As a result of overheating, a MOSFET transistor may lose its discreteness, that is, logic 1 and 0 might become indistinguishable. Digital calculation is not possible when the transistor is always displaying logic 1 when it actually should alternate between logic 1 and 0.

EXAMPLE 8.24   **CMOS Inverter**

Draw the equivalent circuit of the CMOS inverter when V_{in} is low and when it is high. Assume $V_{DD} = 5$ V. What is V_{out} for both occasions?

SOLUTION

The equivalent circuit is shown in Figure 8.75. When V_{in} is low, NMOS turns OFF, and PMOS turns ON. Applying a KVL loop leads to the conclusion that V_{out} is high. When V_{in} is high, NMOS is ON and PMOS is OFF. In this case, V_{out} is low.

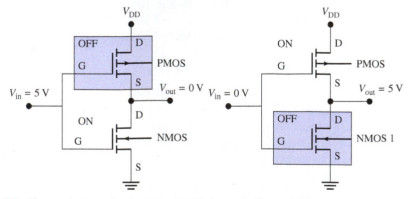

FIGURE 8.75 The equivalent circuit of the CMOS shown in Figure 8.73.

EXAMPLE 8.25   **Truth Table**

Develop the truth table of the two-input gate for the circuit shown in Figure 8.76(a). That is, determine the output voltage, V_{out}, in terms of inputs A and B when they are high (1) or low (0), for example, when both are 0, both are 1, and one is 0 and the other is 1. What is this operation?

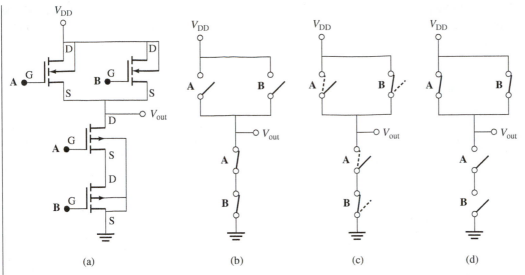

FIGURE 8.76 The logic gate for Example 8.25.

SOLUTION

It is clear that when both A and B are low (0), the NMOS will be OFF and the PMOS will be ON, thus, the output will be connected to the ground, i.e., the output will be low (0). This scenario is shown in Figure 8.76(b). When one of these inputs is high (1) and the other is low (0), the two series PMOSs block the output to be zero while one of the two parallel NMOSs allows connection to the V_{DD}. This scenario is shown in Figure 8.76(c). When both A and B are high (1), the output will be connected to V_{DD}, while the two PMOSs avoid connection to the ground. Thus, the output will be high (1) [see Figure 8.76(d)]. The solution is shown in Table 8.5. Based on this table, the output will be an OR gate (see Chapter 10).

TABLE 8.5 The Truth Table for Figure 8.76(a)

A	B	A + B
0	0	0
0	1	1
1	0	1
1	1	1

8.5 OPERATIONAL AMPLIFIERS

Operational amplifiers (or *op amps*) are high-gain voltage amplifiers with two differential inputs and a single output. Op amps have applications in almost all electronic circuits and are used to build high-power amplifiers, inverters, and buffers. Op amps are also an important part of active filters.

Compared to active filters (discussed in Chapter 11), passive filters (discussed in Chapter 7) have some disadvantages. First, the inductors needed are large and expensive for the audio frequency range in which most mechanical measurements are made. Second, passive filters restrict the sharpness of the cutoff because they do not have power amplification, which leads to loading in the measurements.

The circuit diagram of the op amp and its important terminals are shown in Figure 8.77. It should be noted that the op amp itself is made up of a combination of many transistors. Here,

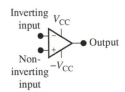

FIGURE 8.77 Circuit diagram of an operational amplifier (op amp).

for simplicity in presentation, the internal structure of an op amp is not presented. The two DC voltage sources, V_{CC} and $-V_{CC}$, are used to bias the transistors. The output voltage is limited by the values of the voltage sources where:

$$-V_{CC} \leq V_o \leq V_{CC} \qquad (8.46)$$

An op amp can be used to amplify AC signals as well as DC signals.

An ideal op amp has the following characteristics:

1. Infinite open-loop voltage gain, $A_o = \infty$
2. Infinite input impedance (i.e., the impedance between the two inputs), $R_{in} = \infty$
3. Zero output impedance (i.e., the impedance seen between the output and the ground), $R_o = 0$
4. Infinite bandwidth (i.e., its gain doesn't change with the input signal frequency)
5. Zero-input offset voltage (i.e., the output voltage is exactly zero if the input is zero)

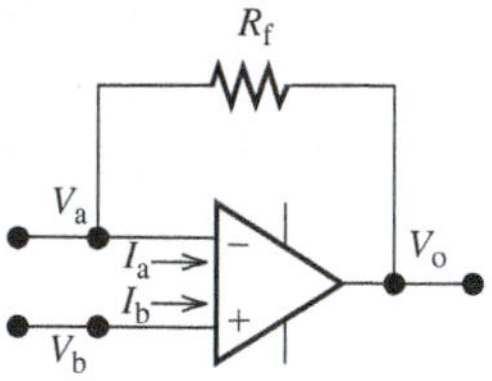

Because an op amp has a very high open-loop voltage gain, negative feedback is usually considered to control the output voltage and to limit the voltage gain. The voltage gain of the op amp with negative feedback is called closed-loop gain (or simply voltage gain) and is referred to by A_v. An op amp circuit with negative feedback is shown in Figure 8.78. To achieve negative feedback, a feedback resistance, R_f, is connected between the output and the inverting input terminals.

FIGURE 8.78   An op amp circuit with feedback.

Five properties stated above lead to two important rules called "golden rules." Referring to Figure 8.78, the golden rules are:

1. There is no input current to the op amp, that is in Figure 8.78, $I_a = I_b = 0$.
2. The voltage difference between the two inputs is zero, that is, $V_a - V_b = 0$ or $V_a = V_b$.

These two rules are very important, and are always used to analyze op amp circuits.

A simple procedure is followed to analyze any op amp circuit:

1. Calculate the voltage V_b.
2. Based on the second golden rule, $V_a = V_b$.
3. Write the *KCL* for the op amp input nodes to find the output voltage.

Note that KCL should not be written for the output node because the op amp's output current isn't known. Op amps have many applications, which are discussed in the following examples.

EXAMPLE 8.26   An Inverting Amplifier

Consider the circuit shown in Figure 8.79 and calculate the voltage gain, $A_v = V_o/V_i$.

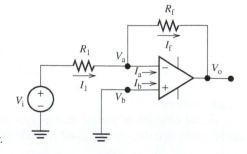

FIGURE 8.79   An inverting amplifier.

SOLUTION

Following the given procedure, first calculate V_b. Because the noninverting input is grounded:

$$V_b = 0$$

According to the second golden rule:

$$V_a = V_b = 0 \tag{8.47}$$

Applying KCL to node "a" yields:

$$I_1 = I_a + I_f \tag{8.48}$$

According to the first golden rule:

$$I_a = 0 \tag{8.49}$$

Therefore:

$$I_1 = I_f \tag{8.50}$$

Based on Ohm's law:

$$
\begin{aligned}
I_1 &= \frac{V_i - V_a}{R_1} \\
&= \frac{V_i - 0}{R_1} \\
&= \frac{V_i}{R_1}
\end{aligned}
\tag{8.51}
$$

and:

$$
\begin{aligned}
I_f &= \frac{V_a - V_o}{R_f} \\
&= \frac{0 - V_o}{R_f} \\
&= -\frac{V_o}{R_f}
\end{aligned}
\tag{8.52}
$$

Substituting Equations (8.51) and (8.52) into Equation (8.50) yields:

$$\frac{V_i}{R_1} = -\frac{V_o}{R_f} \tag{8.53}$$

Reordering Equation (8.53), the output voltage corresponds to:

$$V_o = -\frac{R_f}{R_1} V_i \tag{8.54}$$

And the voltage gain corresponds to:

$$A_v = \frac{V_o}{V_i} = -\frac{R_f}{R_1} \tag{8.55}$$

The negative sign in the voltage gain indicates that there is 180° of phase difference between the input and the output. Accordingly, this amplifier circuit is called an *inverting amplifier*.

Note that the gain and accordingly the output voltage are controlled by resistors R_1 and R_f. Although the output voltage $V_o = A_v V_i$, it is also limited by V_{CC} and $-V_{CC}$ (see Equation 8.46).

(*continued*)

EXAMPLE 8.26   **Continued**

The amplifier is said to be saturated if the absolute value of the output voltage exceeds V_{CC}. As a summary, the output voltage corresponds to:

$$V_o = \begin{cases} -\dfrac{R_f}{R_1}V_i & -V_{CC} < V_o < V_{CC} \\ V_{CC} & V_o > V_{CC} \\ -V_{CC} & V_o < -V_{CC} \end{cases}$$

This same concept can be applied for all applications.

EXAMPLE 8.27   **A Noninverting Amplifier**

For the circuit shown in Figure 8.80, follow the same procedure as the previous example to calculate the output voltage and determine the voltage gain.

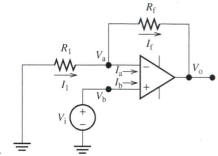

FIGURE 8.80   A noninverting amplifier.

SOLUTION
At node "b":

$$V_b = V_i \tag{8.56}$$

The second golden rule yields:

$$V_a = V_b = V_i \tag{8.57}$$

Applying KCL at node "a," and taking into account that $I_a = 0$:

$$I_1 = I_f \tag{8.58}$$

or:

$$\frac{0 - V_a}{R_1} = \frac{V_a - V_o}{R_f} \tag{8.59}$$

Substituting for V_a from Equation (8.57) into Equation (8.59):

$$-\frac{V_i}{R_1} = \frac{V_i - V_o}{R_f} \tag{8.60}$$

$$-\frac{R_f}{R_1}V_i = V_i - V_o \tag{8.61}$$

Therefore, the output voltage corresponds to:

$$V_o = V_i + \frac{R_f}{R_1}V_i \qquad (8.62)$$

or:

$$V_o = \left(1 + \frac{R_f}{R_1}\right)V_i \qquad (8.63)$$

and the voltage gain corresponds to:

$$A_v = \left(1 + \frac{R_f}{R_1}\right) \qquad (8.64)$$

Note that there is no negative sign in the voltage gain. Thus, this circuit is called a *non-inverting amplifier*. The minimum gain of the noninverting amplifier is unity, and this will happen if $R_f = 0$ and/or $R_1 = \infty$. A noninverting amplifier with unity gain is called a *buffer* or *voltage follower* and is discussed in the following example.

EXAMPLE 8.28   **Buffers or Voltage Followers**

An op amp voltage follower is shown in Figure 8.81. This circuit is similar to the circuit of Figure 8.80 after replacing R_1 with an open circuit (infinity resistance) and R_f with a short circuit (zero resistance). Using Equation (8.63), the output voltage corresponds to:

$$V_o = \left(1 + \frac{0}{\infty}\right)V_i = V_i \qquad (8.65)$$

Therefore, the voltage gain corresponds to:

$$A_v = 1 \qquad (8.66)$$

The result can be obtained following the same procedure as in the previous examples. Referring to Figure 8.81:

$$V_b = V_i \qquad (8.67)$$

The second golden rule yields:

$$V_a = V_b = V_i \qquad (8.68)$$

Because the output is directly connected to node "a" via a short circuit:

$$V_o = V_a = V_i \qquad (8.69)$$

Buffers have applications in connecting cascaded amplifiers. Because their output resistance is zero, they reduce the impact of the previous amplifier on the next one. In addition, because their input resistance is infinity, they reduce the impact of the next amplifier on the previous one.

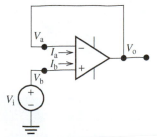

FIGURE 8.81 Buffer or voltage follower.

APPLICATION EXAMPLE 8.29 **An Audio System**

An audiophile has connected his new speakers as shown in Figure 8.82. This is only one channel; other channels have the same connections (with the exception of the subwoofer). After verifying that the speakers are up to his standards, he wants to know how much the speakers will vibrate when he blasts them.

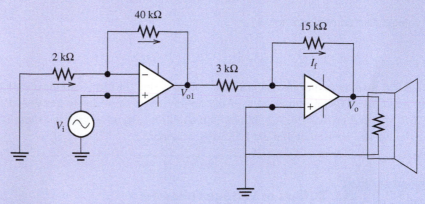

FIGURE 8.82 The circuit for Example 8.28.

SOLUTION

The input voltage is amplified using two amplifier stages. The first stage is a noninverting amplifier followed by an inverting amplifier. The output voltage from the first stage corresponds to:

$$V_{o1} = A_{v1}V_i$$

where, A_{v1} is given by Equation (8.64). In addition, the output voltage from the second stage corresponds to:

$$V_o = A_{v2}V_{o1}$$

where, A_{v2} is calculated using Equation (8.55). Therefore, the relationship between the output and the input voltage corresponds to:

$$V_o = A_{v1}A_{v2}V_i$$

Finally, the overall gain corresponds to:

$$A_v = \frac{V_o}{V_i} = A_{v1}A_{v2}$$

Using Equations (8.55) and (8.64):

$$A_v = \left(1 + \frac{40}{2}\right) \times -\frac{15}{3} = -105$$

EXERCISE 8.6

For the circuit shown in Figure 8.83, write the output phasor V_o in terms of input phasor V_i and show that the V_o/V_i represents a filter (see Chapter 7). What type of filter does it represent?

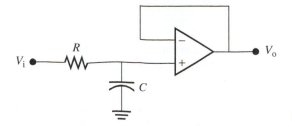

FIGURE 8.83   The circuit for Exercise 8.6.

8.6 USING PSPICE TO STUDY DIODES AND TRANSISTORS

This tutorial explains the process of editing the PSpice model to study diodes. This knowledge will also aid in any future PSpice circuit design that uses a variety of electronic devices with specific parameters. PSpice can be used to create electronic devices with the specifications of interest. For example, PSpice can be used to create a transistor with a specific DC gain, or a diode with different forward bias voltages.

EXAMPLE 8.30   **PSpice Analysis**

Use PSpice to study the circuit shown in Figure 8.9.

   a. $R = 1 \text{ k}\Omega$
   b. $R = 10 \text{ k}\Omega$

SOLUTION

The voltage source is given: $V(t) = 12V$.

   a. $R = 1 \text{ k}\Omega$

       A simple PSpice circuit that consists of a diode can be set up using the following steps:
     1. Go to "Place Part" (see Figure 2.63), under the "BREAKOUT" library, search for the part name "Dbreak."
     2. Set up the PSpice schematic shown in Figure 8.84 and run the simulation analysis as "Bias Point" (see Figure 2.71).

    Notice that there is a slight difference between the current flowing through the diode compared to its value calculated theoretically in Example 8.2. This is due to the forward bias resistor of the diode, R_f, which is nonzero. In Example 8.2, it was assumed to be zero.

   b. $R = 10 \text{ k}\Omega$

       Next, change the resistance, R1, to 10 kΩ and run the bias point analysis again. The PSpice schematic solution is shown in Figure 8.85. The small amount of current across the diode shows that the diode has been turned off.

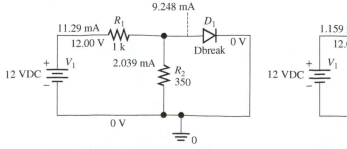

FIGURE 8.84   PSpice schematic for Example 8.30(a).

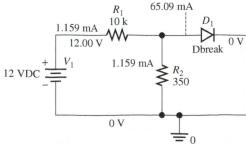

FIGURE 8.85   PSpice schematic solution for Example 8.30(b).

EXAMPLE 8.31   PSpice Analysis

Use PSpice to find the output of the voltage, $v_o(t)$, for the circuit shown in Figure 8.17 (see Example 8.3) using the following values: $v_i(t) = 5 \sin(120\pi t)$V, $R = 1$ kΩ, $E = 2$ V.

SOLUTION

The voltage source is $v_i(t) = 5 \sin(120\pi t)$V. First, determine the frequency of the voltage source.

$$f = \frac{\omega}{2\pi}$$

$$f = \frac{120\pi}{2\pi}$$

$$f = 60 \text{ Hz}$$

Follow the solution of Example 2.24 to setup the VSIN voltage source. Then, the limiter, E, can be setup by placing a direct source voltage in the direction opposite from the voltage source. The PSpice schematic solution is shown in Figure 8.86. A time analysis simulation (see Figure 2.75) has been run to observe the voltage output across the diode and limiter. This simulation is shown in Figure 8.87.

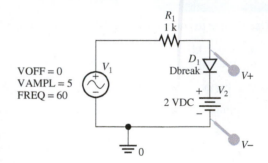

FIGURE 8.86 PSpice schematic solution for Example 8.31.

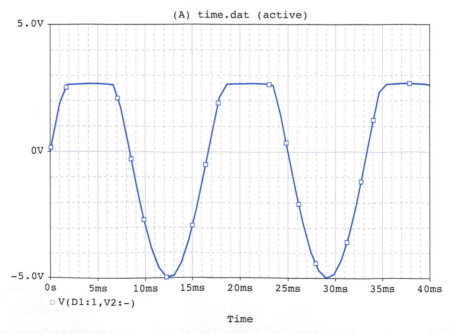

FIGURE 8.87 Voltage output across the diode and limiter for Example 8.31.

| EXAMPLE 8.32 | PSpice Analysis |

Use PSpice to study the circuit shown in Figure 8.26 in Example 8.7. The Zener diode has $V_Z = 3.6$ V. Find I and V when:

a. $V_S = -1$ V
b. $V_S = 8$ V
c. $V_S = -10$ V

SOLUTION

The PSpice model editor is required to set up a Zener diode that meets the requirements for this circuit.

1. To place a Zener diode in the circuit, go to "Place Part" (see Figure 2.65) and type "Dbreakz," which is in the "BREAKOUT" library.
2. Place the Zener diode as shown in Figure 8.88.
3. Next, set the required reverse voltage for Zener diode.
4. Click on the Zener diode or Dbreakz, and go to the menu "Edit" → "PSpice Model" (see the arrow) as shown in Figure 8.89.
5. A model window with specific parameters will appear, as shown in Figure 8.90.
6. Change the parameters in the window to the value shown in Figure 8.91 and save it. "BV" is the breakaway voltage or the reversed voltage, and "IBV" is the breakaway current.

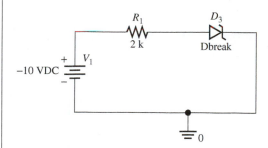

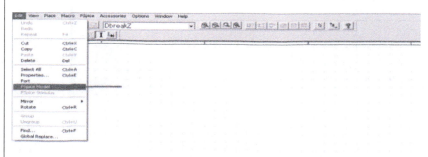

FIGURE 8.88   Place the Zener diode into the PSpice circuit.

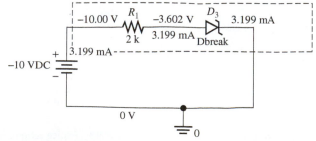

FIGURE 8.89
Enter the PSpice model editor for the Zener diode.

(continued)

7. Run the bias point analysis (see Figure 2.73). The PSpice schematic solutions for each voltage source are shown in Figures 8.92 to 8.94.
   a. $V_S = -1$ V
   b. $V_S = 8$ V
   c. $V_S = -10$ V

FIGURE 8.90 Default model parameters for the Zener diode.

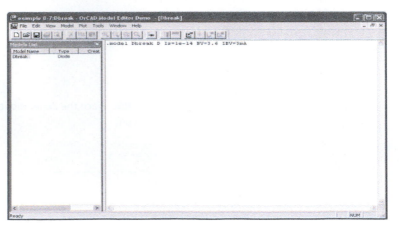

FIGURE 8.91 Desired parameters for the Zener diode in Example 8.32.

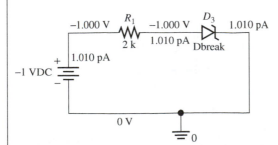

FIGURE 8.92 PSpice schematic solution for Example 8.32(a).

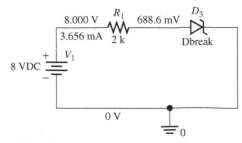

FIGURE 8.93 PSpice schematic solution for Example 8.32(b).

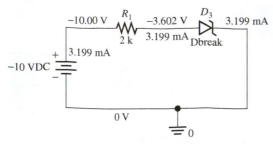

FIGURE 8.94 PSpice schematic solution for Example 8.32(c).

EXAMPLE 8.33 **PSpice Analysis**

Use PSpice to find V_{BE}, I_B, I_C, and V_{CE} in the circuit shown in Figure 8.44, if:

   a. $V_{BB} = 0.3$ V
   b. $V_{BB} = 2.7$ V
   c. $V_{BB} = 6.7$ V

   Use the PSpice model editor to set up the transistor to meet the requirements for the circuit.

SOLUTION

   1. Go to "Place Part" (see Figure 2.63) and type "Qbreakn," which is in the "BREAKOUT" library to place a transistor in the circuit.
   2. Place the transistor and set up the PSpice circuit as shown in Figure 8.95.
   3. Click on the transistor or Qbreakn, and go to the menu "Edit" → "PSpice Model" (see the arrow) as shown in Figure 8.96.
   4. Change the model parameters of the transistor to those shown in Figure 8.97. Here, Bf represents the amplifier, β.
   5. Run the bias point analysis by varying the voltage sources (see Figure 2.71). The results are shown in Figures 8.98 to 8.100.
      a. $V_{BB} = 0.3$ V
      b. $V_{BB} = 2.7$ V
      c. $V_{BB} = 6.7$ V

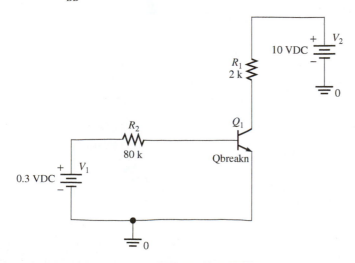

FIGURE 8.95 Placing the transistor.

(continued)

EXAMPLE 8.33 **Continued**

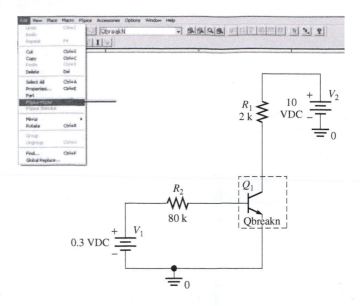

FIGURE 8.96 Enter PSpice model editor for the transistor.

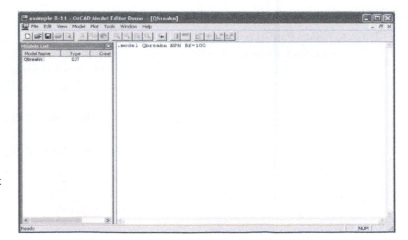

FIGURE 8.97 Set up the required parameters for transistor.

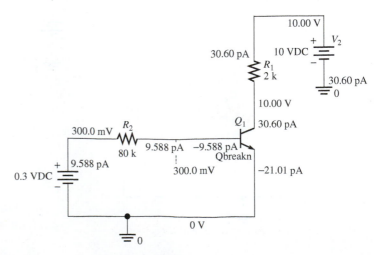

FIGURE 8.98 PSpice schematic solution for Example 8.33(a).

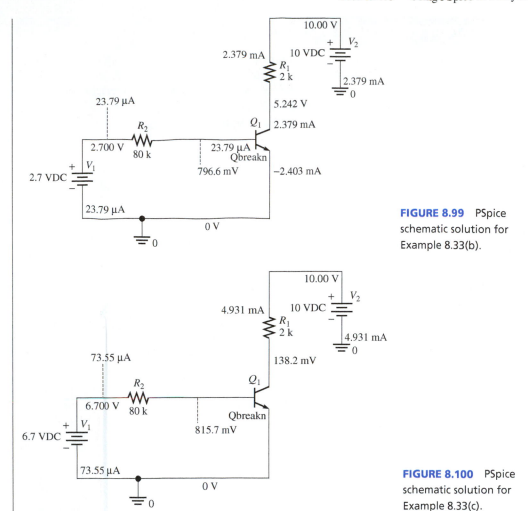

FIGURE 8.99   PSpice schematic solution for Example 8.33(b).

FIGURE 8.100   PSpice schematic solution for Example 8.33(c).

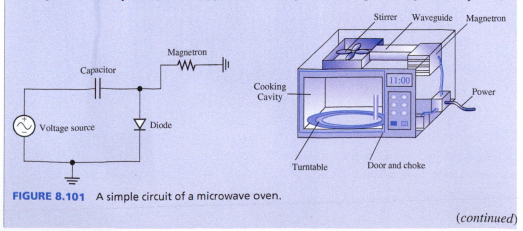

APPLICATION EXAMPLE 8.34   Microwave Oven

Figure 8.101 shows the simplified circuit of a microwave oven. A transformer at the voltage source increases the voltage input to 2800 V and charges the capacitor in the circuit. When the capacitor is fully charged, it supplies the voltage to the magnetron (power amplifier)

FIGURE 8.101   A simple circuit of a microwave oven.

(*continued*)

APPLICATION EXAMPLE 8.34 **Continued**

together with the voltage source from the transformer. A very high input voltage is supplied to the magnetron in order to generate the microwave energy to heat the food.

The voltage input after increasing by the transformer is $2800 \sin (120\pi t)$, the capacitor has the capacitance of 100 mF, and the resistance of the magnetron is 20 kΩ. Use PSpice to find the voltage across the magnetron during the charging and heating process.

SOLUTION

First, given the voltage source, $v_S(t) = 2800 \sin (120\pi t)$, the frequency is 60 Hz. Follow Example 2.24 to set up the VSIN voltage source, and follow Example 8.25 to place a diode in the circuit. The PSpice schematic solution is given in Figure 8.102.

Next, run the simulation as time transient analysis (see Figure 2.75) for 1 s, with the voltage differential markers located before and after the magnetron (see Figure 8.102). The voltage across the magnetron is shown in Figure 8.103.

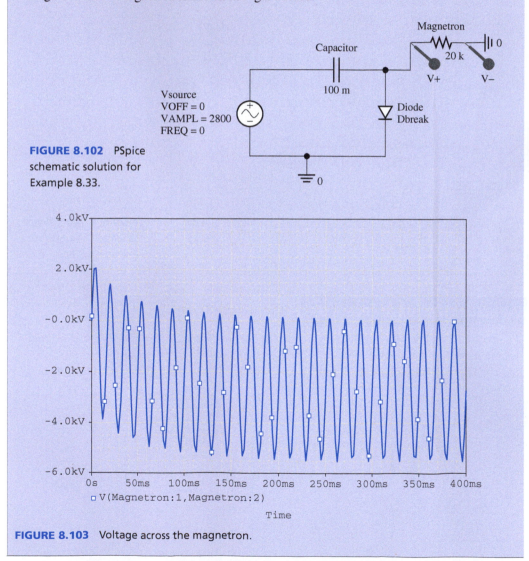

FIGURE 8.102 PSpice schematic solution for Example 8.33.

FIGURE 8.103 Voltage across the magnetron.

8.7 WHAT DID YOU LEARN?

- There are two different categories of semiconductors: p-type and n-type. The addition of pentavalent impurities (i.e., atoms with five valence electrons) produces n-type semiconductors by contributing extra electrons. The addition of trivalent impurities (i.e., atoms with three valence electrons) produces p-type semiconductors by contributing extra holes.
- A diode is a p–n junction that allows current to pass in only one direction from the p-type semiconductor toward the n-type semiconductor.
- The relationship between voltage across the diode and current through the diode is shown in Equation (8.1):

$$I = I_0(e^{V/V_T} - 1)$$

- A diode is said to be in *forward bias* if its voltage is positive and it starts to pass current as the voltage increases beyond the threshold voltage.
- Applications of diodes include rectifiers, diode limiters, and voltage regulators.
- Different types of diodes include Zener diodes, varactor diodes, light-emitting diodes, and photodiodes.
- A bipolar junction transistor (BJT) has three terminals: base, collector, and emitter.
- BJTs can operate in three regions: the active region, the saturation region, and the cutoff region. The characteristics of the three regions are summarized in Table 8.3.
- To be used as an amplifier, a BJT must be biased in the active region.
- DC analysis is used to ensure the BJT is in the active region. The DC equivalent circuit is shown in Figure 8.42a.
- AC analysis is used to find the voltage gain. The AC equivalent circuit is shown in Figure 8.50.
- The transistor that operates between cutoff and saturation is called an electronic switch.
- A MOSFET is the primary device that is used in complex digital ICs.
- The physical structure of an n-type MOSFET (NMOS) is shown in Figure 8.62(a).
- MOSFETs operate in cutoff, in triode, or in saturation (Figure 8.68).
- MOSFET applications include dynamic random access memory (DRAM), logic gates, and amplifiers.
- Complementary MOSFET (CMOS) contains both PMOS and NMOS on the same silicon substrate. It is the standard of today's ICs because it consumes less power than NMOS-only circuits.
- Op amps are characterized by having: high input resistance, low output resistance, and very high gain.
- Golden rules of op amps: (1) There is no input current to an op amp; (2) The voltage difference between the two inputs is zero.
- Voltage followers represent one application of op amps.

Further Reading

Ben Streetman and Sanjay Banerjee. 1999. *Solid state electronic devices*, 5th ed. Upper Saddle River, NJ: Prentice Hall.

Richard C. Jaeger. 1997. *Microelectronics circuit design*. New York: Mcgraw Hill.

Allan R. Hambley. 2000. *Electronics*, 2nd ed. Upper Saddle River, NJ: Prentice Hall.

Problems

B refers to basic, A refers to average, H refers to hard, and * refers to problems with answers.

SECTION 8.2 P-TYPE AND N-TYPE SEMICONDUCTORS

8.1 (B) Explain the difference between p-type and n-type semiconductors.

8.2 (B)* Which of the following is neutral?
  a. p-type semiconductor
  b. n-type semiconductor
  c. p–n junction.

8.3 (B) What is the approximate built-in voltage for silicon and germanium?

SECTION 8.3 DIODES

8.4 (B) Given that the Boltzmann's constant is 8.617343×10^{-5} eV/K, with the thermal voltage at room temperature is 0.0259V determine the thermal voltage if a diode is at temperature:
  a. 200 K
  b. 40°C
  c. 400 K

8.5 (B) Provide an application of a limiter diode and explain how it works.

8.6 (B)* What is the minimum number of standard AA batteries required to light a standard LED?

8.7 (B) What will happen to the variable reactor's resonant frequency when:
  a. The inductance in circuit increased by nine times
  b. The capacitance in the diode reduced by half

8.8 (B) Provide an application of a photovoltaic diode and explain how it works.

8.9 (B) Explain how a photoconductive diode works.

8.10 (B) For the circuit shown in Figure P8.10, find I and V when:
  a. $V_S = -5$ V
  b. $V_S = 0.3$ V
  c. $V_S = 12$ V

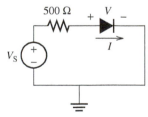

FIGURE P8.10 Circuit for Problem 8.10.

8.11 (A)* Find the current, I, flowing through the diode for the circuit shown in Figure P8.11 when:
  (a) $R = 2$ kΩ
  (b) $R = 8$ kΩ

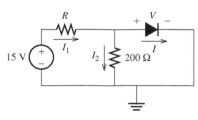

FIGURE P8.11 Circuit for Problem 8.11.

8.12 (A) Write the expression of I_B in the terms of voltage across the 10-Ω resistor, V_A, that is shown in Figure P8.12. Assume that both diodes are on and that V_S is 10 V.

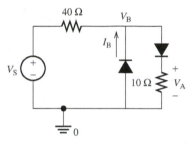

FIGURE P8.12 Circuit for Problem 8.12.

8.13 (H) Find I_D in Figure P8.13 when
  a. $V_S = 6$ V
  b. $V_S > 0$ V all the time, find the range of V_S when the diode is off.

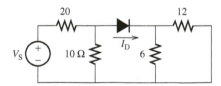

FIGURE P8.13 Diode circuit.

8.14 (H)* Determine the current through the diode for the circuit shown in Figure P8.14.

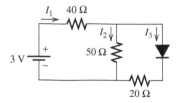

FIGURE P8.14 Circuit for Problem 8.14.

8.15 (B) For the circuit shown in Figure P8.15:
  a. Sketch the transfer characteristics when $E = 4$ V.
  b. Sketch the output voltage if the input voltage is $v_i(t) = 15 \sin(10\pi t)$V.

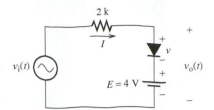

FIGURE P8.15   Circuit for Problem 8.15.

8.16 (B) For the circuit shown in Figure P8.16:
   a. Sketch the transfer characteristics when $E = 4$ V.
   b. Sketch the output voltage if the input voltage is $v_i(t) = 15 \sin(10\pi t)$V.

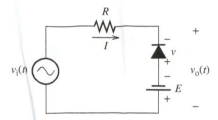

FIGURE P8.16   Circuit for Problem 8.16.

8.17 (A) For the circuit shown in Figure P8.17, sketch the transfer characteristics, input voltage, and output voltage. Assume the input voltage is a sinusoidal waveform with a peak value of 20 V and $E_1 = E_2 = 8$ V.

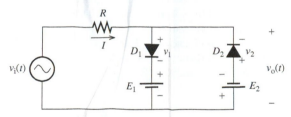

FIGURE P8.17   Circuit for Problem 8.17.

8.18 (A)* The output voltage for the circuit shown in Figure P8.18 (a) is shown in Figure P8.18(b). Given that the voltage source, v_S, is 10 V, and the resistance of the resistor is 20 Ω, determine:
   a. E_1 and E_2.
   b. The current flow through resistor, R, when the output voltage is at the upper and lower limit bounds.

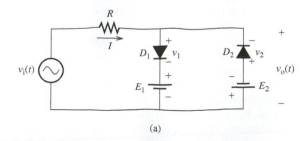

(a)

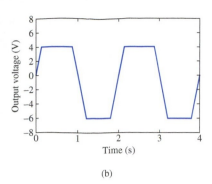

(b)

FIGURE P8.18   (a) Circuit for Problem 8.18. (b) Voltage output.

8.19 (H) Find the transfer characteristic for the circuit shown in Figure P8.19 when $E = 4.3$ V.

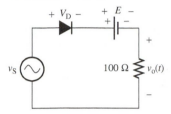

FIGURE P8.19   Voltage limiter.

8.20(H)* Find $v_o(t)$ with respect to $v_S(t)$ for Figure P8.20 if $E = 2.3$ V.

FIGURE P8.20   Voltage limiter.

8.21 (H)* Find I_1, I_2, and I_3 for the circuit shown in Figure P8.21.

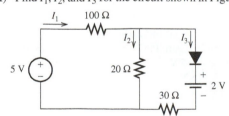

FIGURE P8.21   Circuit for Problem 8.21.

8.22 (B) Figure P8.22 shows the output of a Zener diode. From the plot, what information about the Zener diode can be obtained?

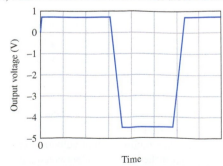

FIGURE P8.22   Zener diode output for Problem 8.22.

8.23 (A) For the circuit shown in Figure P8.23 the Zener diode has $V_z = 6.8$ V. Find I and V when:
a. $V_S = -3$ V
b. $V_S = 6$ V
c. $V_S = -20$ V

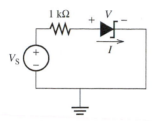

FIGURE P8.23 Circuit for Problem 8.23.

8.24 (A) Given that $V_z = 5.5$ V, find I_1 and I_2 in the circuit shown in Figure P8.24 when:
a. $V_S = -5$ V
b. $V_S = 3$ V
c. $V_S = -12$ V

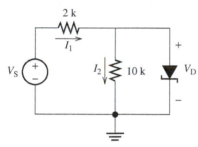

FIGURE P8.24 Circuit of the Zener diode.

8.25 (H) Given that the Zener voltage is 5.3 V for the circuit shown in Figure P8.25, find I_1 and I_2 if the input voltage is:
a. $V_S = -5$ V
b. $V_S = 8$ V
c. $V_S = 16$ V

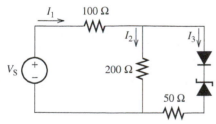

FIGURE P8.25 Circuit for Problem 8.25.

8.26 (H) Find the characteristic curve of the voltage across the diode (V_D) in Problem 8.24.
8.27 (B)* What is the minimum input voltage required for a full-wave rectifier if the required output voltage is 5 V?
8.28 (A) Determine the current through the LEDs when R is:
a. 400 Ω
b. 0 Ω

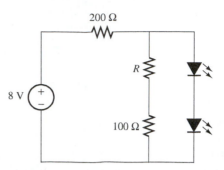

FIGURE P8.28 Circuit for Problem 8.28.

8.29 (H)* What is the maximum resistance, R, allowable to let a light-emitting diode work in the circuit shown in Figure P8.29?

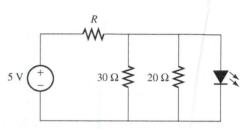

FIGURE P8.29 Circuit for Problem 8.29.

8.30 (B) Assume that the voltage input to the voltage regulator in Figure P8.30 is 10 V, with the Zener voltage of 4.5 V. Determine if the voltage regulator will turn on with the following combinations of R_L and R_S.
a. $R_L = 500$ Ω and $R_S = 20$ Ω
b. $R_L = 40$ Ω and $R_S = 90$ Ω
c. $R_L = 30$ Ω and $R_S = 33$ Ω

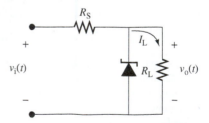

FIGURE P8.30 Circuit for Problem 8.30.

8.31 (A)* Design a DC power supply that supplies a DC voltage of 20 V to a load resistance of 500 Ω. Consider an AC input of 120 V/60 Hz. Choose suitable values for the capacitor and the transformer turns ratio. The ripple voltage should be less than 10% of the output voltage. Consider $R_S = 100$ Ω.
8.32 (H) A transformer has 200 turns of coil at its input stage and 50 turns of coil at its output stage. A voltage regulator is connected to the transformer (see Figure P8.32) to drive the load, $R_L = 300$ Ω. Given that the transformer is connected to a source, $V_S = 120 \sin(100t+60°)$ V, and the Zener voltage is 5 V, determine the capacitance required in the voltage regulator to meet the design requirements.

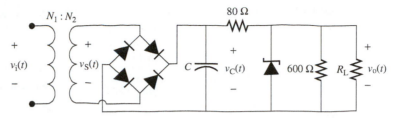

FIGURE P8.32 Circuit for Problem 8.32.

SECTION 8.4.1 BIPOLAR JUNCTION TRANSISTORS

8.33 (B) What is the direction of the current in both PNP and NPN transistors?

8.34 (H) What are the disadvantages of silicon transistors compared to electron tubes?

8.35 (A)* When a transistor, $\beta = 80$, is working in its saturation region, which one of the following statements is correct?
a. $I_C = 80I_B$
b. $I_C > 80I_B$
c. $I_C < 80I_B$
d. $V_{CE} > 0.2$ V

8.36 (B) Figure P8.36 shows the relationship between the collector current and the base–emitter voltage of a transistor. Provide the name of each region, OA, AB, BC, and CD.

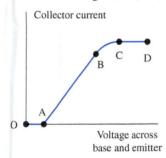

FIGURE P8.36 Collector current and the base–emitter voltage curve.

8.37 (H)* Assume that the voltage across a transistor's collector and emitter has been measured at 3 V, as shown in Figure P8.37. Other parameters are also given in Figure P8.37. Calculate the current gain, β.

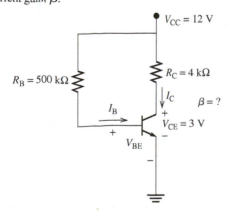

FIGURE P8.37 A simple transistor circuit.

8.38 (A) For the circuit shown in Figure P8.38, find V_{BE}, I_B, I_C, and V_{CE} when:

a. $V_{BB} = 0.5$ V
b. $V_{BB} = 1.7$ V
c. $V_{BB} = 3.7$ V

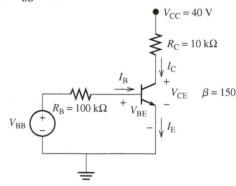

FIGURE P8.38 The circuit for Problem 8.38.

8.39 (A) For the circuit shown in Figure P8.39, find V_{BE}, I_B, I_C, and V_{CE}.

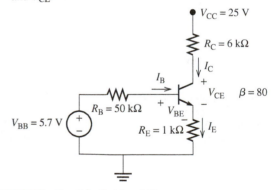

FIGURE P8.39 Circuit for Problem 8.39.

8.40 (A)* Determine the value of R_B in Figure P8.40 such that the output voltage is 3.93 V.

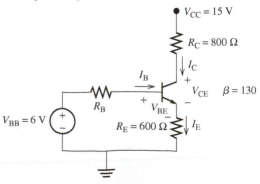

FIGURE P8.40 Transistor circuit for Problem 8.40.

8.41 (A) For the circuit shown in Figure P8.41, find the value of the resistor, R_B, which adjusts the collector–emitter voltage to be 6 V.

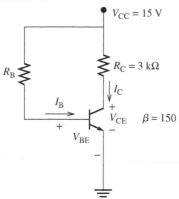

FIGURE P8.41 Circuit for Problem 8.41.

8.42 (A) For the circuit shown in Figure P8.42, find V_{BE}, I_B, I_C, and V_{CE}.

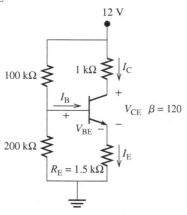

FIGURE P8.42 Circuit for Problem 8.42.

8.43 (B) Find the equivalent circuit of the one shown in Figure P8.42. (Hint: the equivalent circuit looks like the one shown in Figure P8.44.)

8.44 (A)* For the circuit shown in Figure P8.44, find V_B such that $I_C = 6$ mA. $\beta = 200$.

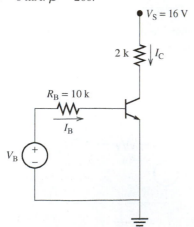

FIGURE P8.44 Transistor circuit for Problems 8.44 and 8.45.

8.45 (H) If $V_B = 4$ V for the circuit shown in Figure P8.44, find an equivalent circuit replacing V_B and R_B, having the source from V_S. (Hint: The circuit looks like Figure P8.45.)

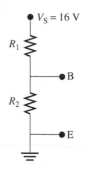

FIGURE P8.45 Hint circuit for Problem 8.45.

8.46 (B) Find the equivalent AC circuit for the one shown in Figure P8.46.

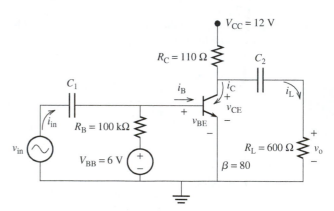

FIGURE P8.46 Biased transistor circuit.

8.47 (H) Find the relationship between R_{B1} and R_{B2}, and the range of R_C to make the circuit given in Figure P8.47 operate in the:
a. cutoff region
b. active region
c. saturate region

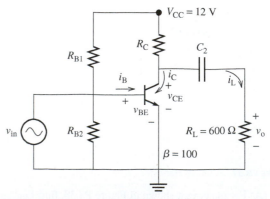

FIGURE P8.47 Circuit for Problem 8.47.

Problems **391**

8.48 (H)* Find A_v, A_i, R_i, and R_o for the circuit shown in Figure P8.48.

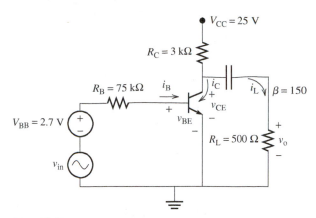

FIGURE P8.48   Circuit of Problem 8.48.

8.49 (H) For the circuit shown in P8.49, calculate voltage gain, current gain, input impedance, and output impedance.

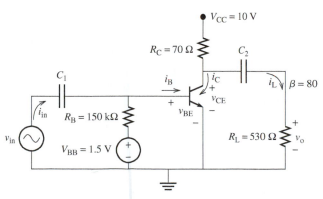

FIGURE P8.49   Circuit for Problem 8.49.

8.50 (H) For the circuit shown in Figure P8.50, find A_v, A_i, R_i, and R_o.

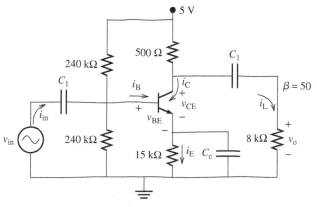

FIGURE P8.50   Circuit for Problem 8.50.

8.51 (B) Provide at least three advantages of using a transistor switch rather than a mechanical switch.

8.52 (A) A transistor switch (operating in the saturation region) is controlled by the output of a digital logic gate as shown in Figure P8.52 . The output of the logic gate is 5 V. A resistor (R_B) is needed to limit the current out from the digital logic gate to 5 mA to protect it, and a resistor (R_L) is also needed to limit the current through the photodiode to 10 mA. Calculate the value of the two protecting resistors.

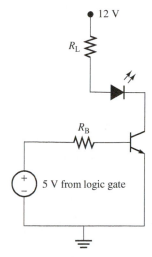

FIGURE P8.52   Logic gate-controlled photodiode.

8.53 (A)* For the circuit shown in Figure P8.51, find the minimum required value of V_{BB} to turn on the switch.

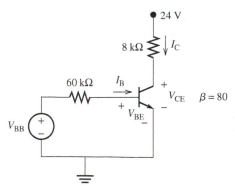

FIGURE P8.53   Circuit for Problem 8.53.

SECTION 8.4.4 FIELD-EFFECT TRANSISTORS

8.54 (B) What is the major difference between a BJT and a MOSFET?

8.55 (B) What are the three regions in which MOSFETS operate?

8.56 (A) If FETs are used as amplifiers, which region do FETs operate in? Explain the reason.

8.57 (B) Sketch the physical structure of an n-channel enhancement MOSFET. Label the channel length (L), the width (W), the terminals, and the channel region. Draw the corresponding circuit symbols.

8.58 (A) Draw a crude model of an NMOS at both its ON and OFF state, using a mechanical switch.

8.59 (A) Describe the conduction channel in an NMOS transistor and explain what causes it to form.

8.60 (A) Describe the conduction channel in a PMOS transistor and explain what causes it to form.

8.61 (H) When an NMOS transistor is in its OFF state, the drain–source channel is not conductive and it can be considered a back-to-back diode pair. Explain why this diode pair will impede the current flow.

8.62 (H) When the PMOS transistor is in its OFF state, the drain–source channel is not conductive. The channel is still a diode pair, but no longer back-to-back like NMOS. Describe the diode pair model for PMOS and explain why this diode pair will impede the current flow.

8.63 (H) When the NMOS transistor is in its ON state, the drain source channel is conductive and it can be thought of as a resistor with very low resistance. The conduction channel formation is a phenomenon that will change upon external bias. Explain (1) why the accumulation of electrons will change the p-type channel into an n-type, and (2) why an n-type channel will make the channel conductive.

8.64 (A)* Sketch the truth table of a two-input OR gate for the circuit shown in Figure P8.64. That is, determine the output voltage, V_{out}, in terms of inputs A and B when they are high (1) or low (0), for example, when both are 0, when both are 1, and when one is 0 and the other is 1. What is this operation?

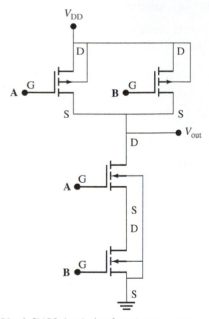

FIGURE P8.64 A CMOS circuit that functions as a gate.

8.65 (A) * Sketch the truth table of a two-input OR gate for the circuit shown in Figure P8.65. That is, determine the output voltage, V_{out}, in terms of inputs A and B when they are high (1) or low (0), for example, when both are 0, when both are 1, and when one is 0 and the other is 1. What is this operation?

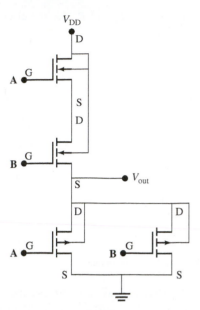

FIGURE P8.65 A CMOS circuit that functions as a gate.

8.66 (A) Repeat Problem 8.65 if PMOS is replaced by NMOS and vice versa.

8.67 (A) Design an AND gate for three inputs A, B, and C using CMOS technology. The truth table of an AND gate is shown in Table P8.67.

TABLE P8.67	The Truth Table of an AND Gate for Three Inputs		
A	**B**	**C**	**ABC**
0	0	0	0
0	0	1	0
0	1	0	0
0	1	1	0
1	0	0	0
1	0	1	0
1	1	0	0
1	1	1	1

8.68 (A) Design an OR gate for three inputs A, B, and C using CMOS technology. The truth table of an OR gate is shown in Table P8.68.

A	B	C	A + B + C
0	0	0	0
0	0	1	1
0	1	0	1
0	1	1	1
1	0	0	1
1	0	1	1
1	1	0	1
1	1	1	1

TABLE P8.68 The Truth Table of an AND Gate for Three Inputs

SECTION 8.5 OPERATIONAL AMPLIFIERS

8.69 (B) Using the circuit shown in Figure P8.69, find the following:
(a) The voltage gain in the circuit
(b) The output voltage for the following input voltages, where V_{CC} is 15 V:
  i. $V_i = 4$ V
  ii. $V_i = -3 \sin(\omega t)$V
  iii. $V_i = -7$ V

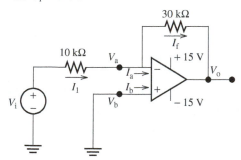

FIGURE P8.69 Circuit for Problem 8.69.

8.70 (B)* Consider the circuit shown in Figure P8.70. Find the following:
a. The voltage gain
b. The output voltage if the input voltage is 4 V

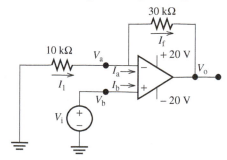

FIGURE P8.70 Circuit for Problem 8.70.

8.71 (A)* Find the output voltage for the circuit shown in Figure P8.71.

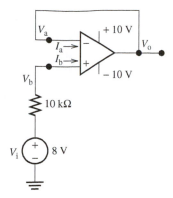

FIGURE P8.71 Circuit for Problem 8.71.

8.72 (A) Design a noninverting amplifier having a gain of 50, using a single op amp.

8.73 (H)* Find the formula for the voltage gain for the amplifier shown in Figure P8.73.

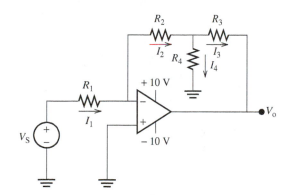

FIGURE P8.73 Advanced op amp circuit.

8.74 (H) The input voltage of the circuit shown in Figure P8.74 is 20 V. Determine the output voltage of both op amps, V_{o1} and V_{o2}.

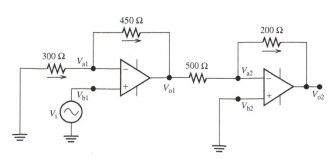

FIGURE P8.74 Circuit for Problem 8.74.

8.75 (B)* For the circuit shown in Figure P8.75, the input voltages are $V_1 = 15$ V and $V_2 = 10$ V. Determine the current, I_f, in the figure.

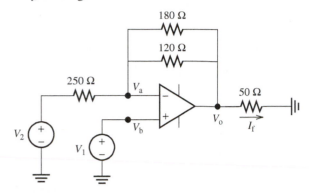

FIGURE P8.75 Circuit for Problem 8.75.

8.76 (A) Summing amplifier

For the circuit shown in Figure P8.76, verify that V_o can be represented in terms of the addition of the two input voltages, V_1 and V_2, if $R_1 = R_2$.

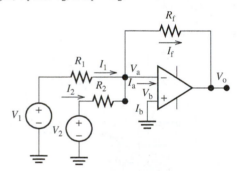

FIGURE P8.76 A summing amplifier.

8.77 (A) Difference amplifier

For the circuit shown in Figure P8.77, verify that V_{out} can be expressed in terms of the difference of V_1 and V_2 if $R_1 = R_2$, $R_f = R_g$.

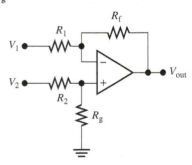

FIGURE P8.77 A difference amplifier.

8.78 (H)* Integrator

For the circuit shown in Figure P8.78, verify that the output voltage, V_{out}, is the integral of the input voltage, V_{in} (multiplied by some constants).

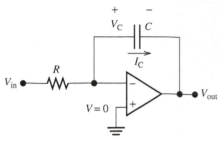

FIGURE P8.78 An integrator.

8.79 (H) Differentiator

For the circuit shown in Figure P8.79, verify that the output voltage, V_{out}, is the derivative of the input voltage V_{in} (multiplied by some constants).

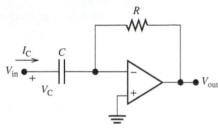

FIGURE P8.79 A differentiator.

SECTION 8.6 USING PSPICE TO STUDY DIODES AND TRANSISTORS

8.80 (B) Use PSpice to find the current, I, flowing through the diode of the circuit shown in the Figure P8.1 when:
a. $R = 2$ kΩ
b. $R = 8$ kΩ

8.81 (B) Consider the circuit shown in Figure P8.15. Use PSpice to plot the voltage output, $V_o(t)$, if the input voltage is $V_i(t) = 15 \sin(10\pi t)$ V.

8.82 (H) Use PSpice to find V_{BE}, I_B, I_C, and V_{CE} for the circuit shown in Figure P8.39.

Power Systems and Transmission Lines

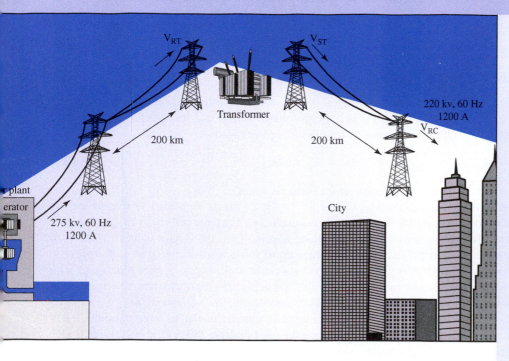

9.1 INTRODUCTION

While energy is commonly found in nature in nonelectric forms, energy is often converted to electric energy because electric energy can be transported easily with high efficiency and reasonable cost. Electricity can also be a clean source of energy; it can be produced by harnessing the potential energy of water, atomic energy, or wind. In dams, the potential energy of water is converted to electrical energy when water is allowed to rotate the core of huge electric generators.

Power can be generated from the wind in a similar fashion. Here, the wind energy is used to rotate the core of generators, which may charge a large number of batteries. Note that because wind energy is transient, some type of energy storage (e.g., batteries) is required. Next, the DC power from the batteries is converted to 50-Hz AC power and is then connected to the public transmission lines.

Generators are also important elements of vehicle engines. They generate electricity using the energy provided by fuel and charge the vehicle's battery. The vehicle's battery supplies the energy required to enable components such as the vehicle's lights, ignition system, and radio. Electric generators are discussed in Chapter 13.

Electric energy can also be produced via many other techniques such as electrochemical systems. Examples of electrochemical systems include battery and photovoltaic systems (e.g., solar cells).

The main elements of an electrical power system are generators, transformers, transmission lines, loads, and protection and control equipment. These elements are interconnected to enable the generation and transformation of electricity in sufficient quantity to satisfy customers' demand. The structure of this interconnection is critical to ensure efficient delivery of electric energy at competitive costs.

This chapter introduces three-phase systems and transmission lines and provides an overview of how to calculate inductance, capacitance, and the equivalent circuit of a transmission line. A good understanding of alternative circuits is vital for this chapter. Thus, you may wish to review Chapter 6 before reading this chapter.

9.2 THREE-PHASE SYSTEMS

9.2.1 Introduction

The voltage induced by a single coil when it is rotated in a uniform magnetic field is shown in Figure 9.1(a). This voltage is known as a single-phase voltage and its waveform is shown in Figure 9.1(b). The sinusoidal expression for the induced voltage is:

$$e(t) = E_{\mathrm{m}} \sin \omega t \qquad (9.1)$$

where E_{m} is the maximum (peak) value of the induced voltage.

Consider the voltage induced by three coils (A, B, C) which are placed 120° apart and are rotated in a uniform magnetic field as shown in Figure 9.2(a). The three coils have an equal number of turns and each coil rotates with the same angular velocity. Thus, the voltage induced across each coil has the same peak, shape, and frequency. These equal voltages are displaced by 120° from each other, as shown in Figure 9.2(b). *Three-phase systems* can be looked upon as three separate generators, each voltage out of phase by 120°. Three phase systems offer more efficient power transmission compared to single phase systems. To explain, consider the following example.

Assume, we have a 120 V source and intend to transfer power to three parallel 10 kW loads. It is clear that using single-phase systems, we need to draw 250 A ($250 \times 120 = 250$ kW) of current from the source. A transmission line capable of transferring 250 A of current needs tick copper cables and would be very heavy. In place, if a three phase system is used, each 10 kW load would be connected to one of the phases. In this case, the current through each cable would be one third of 250 A, that is, about 83.3 A. This reduces the diameter of copper cables, which in turn reduces the overall weight and cost of transmission line.

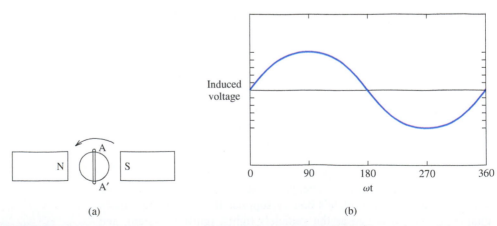

(a)   (b)

FIGURE 9.1   (a) A single coil when rotated in a uniform magnetic field; (b) waveform of the single-phase voltage.

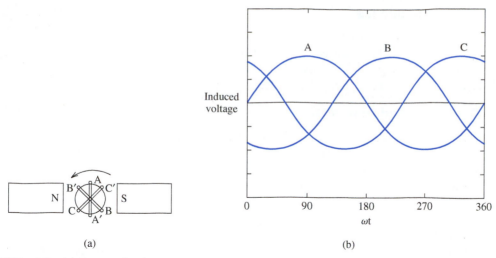

(a) (b)

FIGURE 9.2 (a) Three coils when rotated in a uniform magnetic field; (b) waveform of the three-phase voltage.

The sinusoidal expression for each of the induced voltages corresponds to:

$$e_{AN} = E_m \sin \omega t$$
$$e_{BN} = E_m \sin(\omega t - 120°)$$
$$e_{CN} = E_m \sin(\omega t - 240°) = E_m \sin(\omega t + 120°)$$

(9.2)

where E_m is the maximum (peak) value of the induced voltage as introduced in Equation (9.1). As seen in Figure 9.2(a), a generator may include three sets of coils. The rotation of these sets of coils within a magnetic field creates a time-varying voltage [Figure 9.2(b)]. These three coils supply a given load. The fundamentals of generators will be explained in Chapter 13. This section explains how the coils can be connected to a given load in either a star (Y) or a delta (Δ) shape.

Figure 9.3(a) represents a simple three-phase circuit. Usually, generator coils are connected in star (Y) to avoid power loss and lower production costs. As explained in this section, generators are rarely Δ-connected [see Figure 9.3(b)]. Note that the Δ-shaped wiring creates a loop that allows the circulation of the current. In this formation, if the phase voltages of each Δ-connected side are not perfectly balanced, for example, due to the phase-winding imbalance across these voltages, there will be a net voltage, and consequently a circulating current around the Δ-shaped wiring. This results in a power loss. Moreover, the phase voltages are lower in a Y-connected generator, and thus, less insulation is required to maintain safe transmission of power through the transmission lines. This leads to lower cost of transmission lines and cables. The next subsection explains the notion of star and Δ connection of generators and loads.

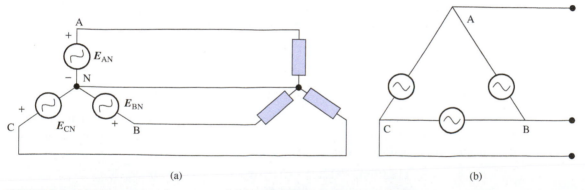

(a) (b)

FIGURE 9.3 (a) Star-connected generator feeds a star (Y-connected) load. (b) Star-connected generator.

9.2.2 Phase Sequence

The *phase sequence* of a three-phase circuit is the order in which the voltages or currents attain their maximum value. For instance, Figure 9.2(b) shows that e_{AN} peaks first, then e_{BN} then e_{CN} which results in an order of ABCABCABC.... Any three adjacent letters can be selected to designate the phase sequence, but usually the selected three are A, B, and C.

9.2.3 Y-Connected Generators

At power generation stations, three sinusoidal voltages are generated having the same amplitude but displaced in phase by $120°$. This is called a balanced voltage source, as shown in Figure 9.3. The point at which all coils are connected is called the neutral point, N.

   The voltages E_{AN}, E_{BN}, and E_{CN} created by the generator are called *phase voltages* or *line-to-neutral* voltages. Phase voltages are generally denoted by E_P while the voltages E_{AB} (defined as $E_{AN} - E_{BN}$), E_{BC} (defined as $E_{BN} - E_{CN}$), and E_{CA} (defined as $E_{CN} - E_{AN}$) are called *line voltages* and generally denoted by E_L.

   A four-wire system consists of the Y connection of three phases of a supply and a neutral conductor (located at the middle of the connection, at point N). A four-wire system allows the use of two voltages: the phase voltage and the line voltage. A four-wire system is used when the load is not balanced (see Section 9.2.4).

9.2.4 Y-Connected Loads

A Y-connected load is shown in Figure 9.4 where the three line conductors are each connected to a load and the outlets from the loads are connected at N to form what is called the neutral point. (Note that Y can also be referred to as "wye.") In this text, Y and "star" are used interchangeably.

   The voltages V_{AN}, V_{BN}, and V_{CN} are called phase voltages or line-to-neutral voltages. Phase voltages are generally denoted by V_P while the voltages V_{AB}, V_{BC}, and V_{CA} are called line voltages and are generally denoted by V_L.

   For a balanced system: $I_A = I_B = I_C$, $V_{AN} = V_{BN} = V_{CN}$, $V_{AB} = V_{BC} = V_{CA}$, $Z_A = Z_B = Z_C$ and the current in the neutral conductor $I_N = 0$. Therefore, the neutral conductor is unnecessary and it is often omitted.

   To find the relationship between line and phase voltages, first assume the typical ABC sequence and arbitrarily select the line-to-neutral voltage of the A-phase as the reference. In this case, the relationship between voltage phasors of different voltages corresponds to:

$$V_{AN} = |V_P|\angle 0°$$
$$V_{BN} = |V_P|\angle -120° \tag{9.3}$$
$$V_{CN} = |V_P|\angle 120°$$

where $|V_P|$ represents the magnitude of the phase voltage. Thus, for the phase voltage:

$$|V_P| = V_P = V_{AN}$$

In other words, the phase of V_P is zero. The line voltages at the load terminals in terms of phase voltages are obtained by applying Kirchhoff's voltage law to Figure 9.4:

$$V_{AB} = V_{AN} - V_{BN} = |V_P|(1\angle 0° - 1\angle -120°) = \sqrt{3}|V_P|\angle 30°$$
$$V_{BC} = V_{BN} - V_{CN} = |V_P|(1\angle -120° - 1\angle -240°) = \sqrt{3}|V_P|\angle -90° \tag{9.4}$$
$$V_{CA} = V_{CN} - V_{AN} = |V_P|(1\angle -240° - 1\angle 0°) = \sqrt{3}|V_P|\angle 150°$$

   The voltage phasor diagram of the Y-connected loads of Figure 9.4 is shown in Figure 9.5. The relationship between the line voltages and the phase voltages is demonstrated graphically. By using the first equation of Equation set (9.4), note that:

$$(1\angle 0° - 1\angle -120°) = 1 - (\cos 120 - j\sin 120)$$
$$= 1.5 + j0.886$$

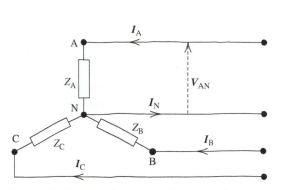

FIGURE 9.4 Y-connection load with a neutral conductor.

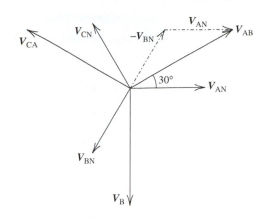

FIGURE 9.5 Phasor diagram showing phase and line voltages.

$$= \sqrt{1.5^2 + 0.886^2} \cdot e^{+j \tan^{-1} 0.886/1.5}$$
$$= \sqrt{3} \cdot e^{+j30.59}$$
$$= \sqrt{3} \angle 30.59°$$

EXERCISE 9.1

Prove the second and the third equations in Equation set (9.4), that is, show:

$$(1\angle -120° - 1\angle -240°) = \sqrt{3} \angle -90°$$
$$(1\angle -240° - 1\angle 0°) = \sqrt{3} \angle 150°$$

Using Equations (9.3) and (9.4), note that:

$$V_{AB} = \sqrt{3} \, V_{AN} \angle 30°$$

or:

$$V_L = \sqrt{3} \cdot |V_P| \angle 30°$$
$$= \sqrt{3} \cdot V_P \angle 30 \tag{9.5}$$

Equation (9.5) represents that the phase of the set of line voltages leads the set of phase voltages by 30°. In addition, considering Y-connected loads, the magnitude (and the rms) of the line voltage is $\sqrt{3}$ times higher than the magnitude (and the rms) of the phase voltage. This fact can also be easily observed in Figure 9.5. In other words:

$$V_{L,rms} = \sqrt{3} \times V_{P,rms}$$

In Figure 9.4, observe that the phase currents are equal to their respective line currents I_A, I_B, and I_C. Therefore, for a balanced Y connection, denoting the phase current by I_p, and the impedance of the load by Z:

$$I_p = V_p/Z \text{ and } I_L = I_A = I_P$$

It is clear that, assuming a balanced Y connection, the current through three elements of a star connection has a phase difference. This phase difference is due to the 120° of phase difference of the voltages in Equation (9.3) and the phase difference due to the nature of the load. For example, refer to Figure 9.4, considering a resistive load:

$$I_B = I_A \angle -120°$$

EXAMPLE 9.1 Star Connected Loads

Three 30-Ω (resistive) loads are star connected to a 415-V, three-phase supply. Determine the system's (a) phase voltage and current, and (b) line current. Note that all of these values are rms values.

SOLUTION

a. Usually, when a 415-V, three-phase supply is discussed, 415 V is the line rms voltage, $V_{L,rms}$. For a star connection, using Equation (9.5), $V_{L,rms} = \sqrt{3}V_{P,rms}$. Therefore, the phase voltage corresponds to:

$$V_{P,rms} = \frac{V_{L,rms}}{\sqrt{3}} = \frac{415}{\sqrt{3}} = 239.6 \text{ V or } 240 \text{ V}$$

For phase current, because the load is resistive, current and voltage will have the same phase, and because the phase of V_P is zero:

$$I_{P,rms} = \frac{V_{P,rms}}{R_P} = \frac{240}{30} = 8 \text{ A}$$

b. For a star connection, $I_{L,rms} = I_{P,rms}$. Thus, the line current is $I_{L,rms} = 8$ A.

EXAMPLE 9.2 Star Connected Loads

A star-connected load consists of three identical coils each of resistance 30 Ω and inductance 127.3 mH. If the line current is 5.08 A, and the supply frequency is 50 Hz, calculate the rms magnitude of the line voltage.

SOLUTION

The inductive reactance is $X_L = 2\pi \cdot f \cdot L = 2\pi (50) \cdot (127.3 \times 10^{-3}) = 40 \Omega$. Thus, the magnitude of the impedance of each phase is:

$$|Z_P| = \sqrt{R^2 + X_L^2} = \sqrt{30^2 + 40^2} = 50 \ \Omega$$

For a star connection, the line and phase currents are equal and correspond to:

$$I_{L,rms} = I_{P,rms} = \frac{V_{P,rms}}{|Z_P|}$$

Thus, the phase voltage is:

$$V_{P,rms} = I_{P,rms}|Z_P| = 5.08 \times 50 = 254 \text{ V}$$

and the line voltage is:

$$V_{L,rms} = \sqrt{3}V_{P,rms} = \sqrt{3}(254) = 440 \text{ V}$$

EXAMPLE 9.3 Star Connected Loads

A 415-V, three-phase, four-wire, star-connected system supplies three resistive loads with the average power shown in Figure 9.6. Determine the current in (a) each line and (b) the neutral conductor.

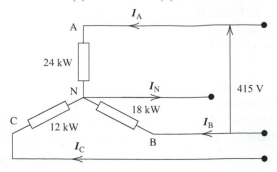

FIGURE 9.6 Y-connected resistive load.

SOLUTION

a. For a star connection:

$$V_{L,rms} = \sqrt{3}V_{P,rms}$$

Therefore, the phase voltage is:

$$V_{P,rms} = \frac{V_{L,rms}}{\sqrt{3}} = \frac{415}{\sqrt{3}} = 239.6 \text{ V or } 240 \text{ V}$$

For resistive loads [see Equation (6.73), or Table 6.1]:

$$\text{rms current}\,(I) = \frac{\text{Average power}\,(P)}{\text{rms voltage}\,(V)}$$

Again, note that all currents and voltages are considered rms values. Next, calculate:

$$I_{A,rms} = \frac{P_A}{V_{A,rms}} = \frac{24,000}{240} = 100 \text{ A}$$

$$I_{B,rms} = \frac{P_B}{V_{B,rms}} = \frac{18,000}{240} = 75 \text{ A}$$

$$I_{C,rms} = \frac{P_C}{V_{C,rms}} = \frac{12,000}{240} = 50 \text{ A}$$

b. Because each load is resistive, the currents are in phase with the phase voltages. Accordingly, their phases are mutually displaced by 120°. The three line currents as phasors (magnitude and direction) are:

$$I_A = 100 \cdot \sqrt{2}\angle 0° = 141.4\angle 0°$$
$$I_B = 75 \cdot \sqrt{2}\angle -120° = -53 - j91.85$$
$$I_C = 50 \cdot \sqrt{2}\angle 120° = -35.35 + j61.23$$

Note that a phasor (voltage or current) is represented by the amplitude and the phase (see Chapter 6). The amplitude of a current phasor and its rms value differ by a coefficient of $\sqrt{2}$. Now, using a simple KCL:

$$I_N = I_A + I_B + I_C = (141.4 - 53 - 35.35) + j(-91.85 + 61.23) = 53.05 - j30.62$$
$$\Rightarrow I_N = 61.25\angle -30°$$

9.2.5 Δ-Connected Loads

A Δ-connected load is shown in Figure 9.7 where one port of each load is connected to one port of the next load (note that loads are considered two-port elements). A balanced Δ-connected load consists of equal phase impedances. Figure 9.7 shows that line voltages are the same as phase voltages. Therefore, for a Δ connection:

$$V_L = V_P$$

To find the relationship between line currents and phase currents, assume the standard ABC sequence and arbitrarily select the phase current, I_{AB}, as the reference; thus,

$$I_{AB} = |I_P|\angle 0°$$
$$I_{BC} = |I_P|\angle -120° \qquad (9.6)$$
$$I_{CA} = |I_P|\angle 120°$$

where $|I_p|$ represents the magnitude of the phase current. To find the line currents in terms of the phase currents, apply Kirchhoff's voltage law, which corresponds to:

$$I_A = I_{AB} - I_{CA} = |I_P|(1\angle 0° - 1\angle 120°) = \sqrt{3}|I_P|\angle -30°$$
$$I_B = I_{BC} - I_{AB} = |I_P|(1\angle -120° - 1\angle 0°) = \sqrt{3}|I_P|\angle -150° \qquad (9.7)$$
$$I_C = I_{CA} - I_{BC} = |I_P|(1\angle -240° - 1\angle -120°) = \sqrt{3}|I_P|\angle 90°$$

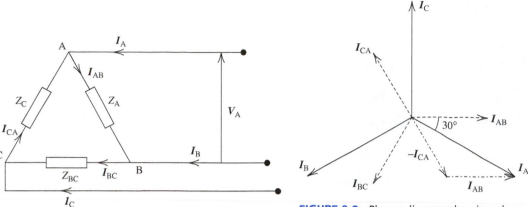

FIGURE 9.7   Δ-Connected load.

FIGURE 9.8   Phasor diagram showing phase and line currents.

The relationship between the line currents and the phase currents is demonstrated graphically in Figure 9.8. Using trigonometry, or via measurement, it can be shown that:

$$I_A = \sqrt{3} I_{AB} \angle -30°$$

Here, I_A is the line current and I_{AB} is the phase current. Therefore, for a balanced Δ connection:

$$I_L = \sqrt{3} |I_P| \angle -30° \qquad (9.8)$$

Thus, in the case of Δ-connected loads for the discussed ABC sequence, the set of line currents lags the set of phase currents by 30°. In addition, the rms magnitude of the line current is $\sqrt{3}$ times that of the phase current, that is:

$$I_{L,rms} = \sqrt{3} \cdot I_{P,rms}$$

Assuming a balanced Δ connection, the voltages across the three elements of the Δ connection have a phase difference composed of two components: the 120° phase difference of the currents in Equation (9.6) plus the phase difference due to the impedance of the load. For example, refer to Figure 9.7, considering a resistive load (i.e., zero phase difference due to the load):

$$V_{BC} = V_{AB} \angle -120°$$

EXAMPLE 9.4   Delta Connected Loads

Three identical coils, each of resistance 30 Ω and inductance 127.3 mH, are connected in Δ to a 440-V, 50-Hz, three-phase supply. Determine (a) the phase current and (b) the line current.

SOLUTION

First, calculate the inductive reactance that corresponds to:

$$X_L = 2\pi f \times L = 2\pi(50) \times (127.3 \times 10^{-3}) = 40\ \Omega$$

Next, calculate the magnitude of the impedance of each phase, which consists of a resistance and an inductance, that is:

$$|Z_P| = \sqrt{R^2 + X_L^2} = \sqrt{30^2 + 40^2} = 50\ \Omega$$

Note that for a Δ connection $V_P = V_L$.

a. Next, find the phase current using ohm's law:

$$I_{P,rms} = \frac{V_{P,rms}}{|Z_P|} = \frac{V_{L,rms}}{|Z_P|} = \frac{440}{50} = 8.8\ A$$

b. For a Δ connection:

$$I_{L,\text{rms}} = \sqrt{3}\,I_{P,\text{rms}} = \sqrt{3}(8.8) = 15.24\,\text{A}$$

Thus, when loads are connected in Δ, a larger line current is taken from the supply compared to a star-connected load (see Example 9.2).

EXAMPLE 9.5 **Delta Connected Loads**

Three identical capacitors are connected in Δ to a 415-V, 50-Hz, three-phase supply. If the line current is 15 A, determine the capacitance of each of the capacitors.

SOLUTION

In a Δ connection, $I_{L,\text{rms}} = \sqrt{3}\,I_{P,\text{rms}}$; thus, the phase current corresponds to:

$$I_{P,\text{rms}} = \frac{I_{L,\text{rms}}}{\sqrt{3}} = \frac{15}{\sqrt{3}} = 8.66\,\text{A}$$

Because in a Δ connection $V_P = V_L$, the capacitive reactance per phase is:

$$X_C = \frac{V_{P,\text{rms}}}{I_{P,\text{rms}}} = \frac{V_{L,\text{rms}}}{I_{P,\text{rms}}} \Rightarrow X_C = \frac{415}{8.66} = 47.92\,\Omega$$

Next, use $X_C = 1/2\pi f C$ to calculate the capacitance:

$$C = \frac{1}{2\pi f X_C} = \frac{1}{2\pi(50)(47.92)} = 66.43\,\mu\text{F}$$

EXAMPLE 9.6 **Star and Delta Connection Current and Voltage**

Three coils each having 3-Ω resistance and 4-Ω inductive reactance are connected, (i) in star and (ii) in Δ, to a 415-V, three-phase supply. For each connection, calculate (a) the magnitude (rms value) of line and phase voltages, and (b) the magnitude (rms value) of phase and line currents.

SOLUTION

Each phase includes a resistance and an inductance. The impedance of each phase is $R + jX_L$; thus, the impedance of each phase is:

$$|Z_P| = \sqrt{R^2 + X_L^2} = \sqrt{3^2 + 4^2} = 5\,\Omega$$

i. For a star connection, recall that $I_{L,\text{rms}} = I_{P,\text{rms}}$, $V_{L,\text{rms}} = \sqrt{3}\,V_{P,\text{rms}}$
  a. A 415-V, three-phase supply means that to calculate voltage:
  1. Line voltage: $V_{L,\text{rms}} = 415$ V
  2. Phase voltage: $V_{P,\text{rms}} = \dfrac{V_{L,\text{rms}}}{\sqrt{3}} = \dfrac{415}{\sqrt{3}} = 240$ V

(continued)

EXAMPLE 9.6   **Continued**

    b. To determine current:

       1. Phase current: $I_{P,rms} = \dfrac{V_{P,rms}}{|Z_P|} = \dfrac{240}{5} = 48\ A$

       2. Line current: $I_{L,rms} = I_{P,rms} = 48\ A$

  ii. For a Δ connection, recall that: $V_{L,rms} = V_{P,rms},\ I_{L,rms} = \sqrt{3}\,I_{P,rms}$

    a. To determine voltage:

       1. Line voltage: $V_{L,rms} = 415\ V$

       2. Phase voltage: $V_{P,rms} = V_{L,rms} = 415\ V$

    b. To find current:

       I. Phase current: $I_{P,rms} = \dfrac{V_{P,rms}}{|Z_P|} = \dfrac{415}{5} = 83\ A$

      II. Line current: $I_{L,rms} = \sqrt{3}\,I_{P,rms} = \sqrt{3}\,(83) = 144\ A$

9.2.6 Δ-Star and Star-Δ Transformations

It is possible to transform a star-connected circuit into an equivalent Δ-connected circuit and vice versa. The corresponding circuits are equivalent only for voltages and currents external to the Y and Δ circuits. Internally, the voltages and currents are different.

    Figure 9.9 shows that Y impedances are Z_A, Z_B, Z_C, while Δ impedances are Z_{AB}, Z_{BC}, Z_{CA}. To transform a Δ-connected circuit to an equivalent star-connected circuit, the following equations are used:

$$
\begin{aligned}
Z_A &= \frac{Z_{AB}Z_{CA}}{Z_{AB} + Z_{BC} + Z_{CA}} \\[4pt]
Z_B &= \frac{Z_{BC}Z_{AB}}{Z_{AB} + Z_{BC} + Z_{CA}} \\[4pt]
Z_C &= \frac{Z_{CA}Z_{BC}}{Z_{AB} + Z_{BC} + Z_{CA}}
\end{aligned}
\tag{9.9}
$$

    To transform a star-connected circuit to an equivalent Δ-connected circuit, the following equations are used:

$$
\begin{aligned}
Z_{AB} &= \frac{Z_A Z_B + Z_B Z_C + Z_C Z_A}{Z_C} \\[4pt]
Z_{BC} &= \frac{Z_A Z_B + Z_B Z_C + Z_C Z_A}{Z_A} \\[4pt]
Z_{CA} &= \frac{Z_A Z_B + Z_B Z_C + Z_C Z_A}{Z_B}
\end{aligned}
\tag{9.10}
$$

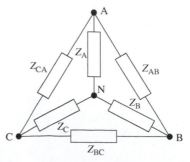

FIGURE 9.9 Y–Δ and Δ–Y transformations.

Note that all Δ to Y transformation formulas in Equation (9.9) have the same denominator, which is the sum of the Δ impedances. Each numerator is the product of the two Δ impedances (as shown in Figure 9.9) that are adjacent to the Y impedance being found. On the other hand, all Y to Δ transformation formulas in Equation (9.10) have the same numerator, which is the sum of the different products of the pairs of the Y impedances. Each denominator is the Y impedance that is opposite to the calculated impedance (see Figure 9.9). If Δ impedances are equal, all Y impedances will be equal and Equation (9.9) reduces to:

$$Z_Y = \frac{Z_\Delta}{3} \qquad\qquad \textbf{(9.11)}$$

Similarly, if Y impedances are equal, all Δ impedances will be equal and Equation (9.10) reduces to:

$$Z_\Delta = 3Z_Y \qquad\qquad \textbf{(9.12)}$$

EXAMPLE 9.7	**Delta-to-Star Conversion**

Replace a Δ-connected circuit whose impedances are $Z_{AB} = 20 \ \Omega$, $Z_{BC} = (10 + j10) \ \Omega$, and $Z_{CA} = -j20 \ \Omega$ with an equivalent star connection.

SOLUTION

Using Equation (9.9):

$$Z_A = \frac{Z_{AB}Z_{CA}}{Z_{AB} + Z_{BC} + Z_{CA}}$$

$$= \frac{20(-j20)}{20 + (10 + j10) - j20}$$

$$= \frac{-j400}{30 - j10}$$

Thus, calculate:

$$Z_A = \frac{400\angle{-90}}{31.62\angle{-18.43}} = 12.65\angle{-71.57} \ \Omega$$

In addition:

$$Z_B = \frac{Z_{BC}Z_{AB}}{Z_{AB} + Z_{BC} + Z_{CA}}$$

$$= \frac{20(10 + j10)}{20 + (10 + j10) - j20}$$

$$= \frac{200 + j200}{30 - j10}$$

Thus:

$$Z_B = \frac{282.84\angle{45}}{31.62\angle{-18.43}} = 8.945\angle{63.43} \ \Omega$$

Finally, calculate:

$$Z_C = \frac{Z_{CA}Z_{BC}}{Z_{AB} + Z_{BC} + Z_{CA}}$$

$$= \frac{-j20(10 + j10)}{20 + (10 + j10) - j20}$$

$$= \frac{200 - j200}{30 - j10}$$

(continued)

EXAMPLE 9.7 **Continued**

Therefore:

$$Z_C = \frac{282.84\angle -45}{31.62\angle -18.43} = 12.65\angle -26.57 \ \Omega$$

EXAMPLE 9.8 **Delta-to-Star Conversion**

Replace a star-connected circuit whose impedances are $Z_A = 10 \ \Omega$, $Z_B = 20 \ \Omega$, and $Z_C = j5 \ \Omega$ with an equivalent Δ connection.

SOLUTION

Using Equation (9.10):

$$Z_{AB} = \frac{Z_A Z_B + Z_B Z_C + Z_C Z_A}{Z_C}$$

$$= \frac{(10)(20) + (20)(j5) + (10)(j5)}{j5}$$

$$= \frac{200 + j150}{j5}$$

Thus:

$$Z_{AB} = \frac{250\angle 36.87}{5\angle 90} = 50\angle -53.13 \ \Omega$$

In addition:

$$Z_{BC} = \frac{Z_A Z_B + Z_B Z_C + Z_C Z_A}{Z_A}$$

$$= \frac{(10)(20) + (20)(j5) + (10)(j5)}{10}$$

$$= \frac{200 + j150}{10}$$

Accordingly:

$$Z_{BC} = \frac{250\angle 36.87}{10\angle 0} = 25\angle 36.87 \ \Omega$$

Finally, calculate:

$$Z_{CA} = \frac{Z_A Z_B + Z_B Z_C + Z_C Z_A}{Z_B}$$

$$= \frac{(10)(20) + (20)(j5) + (10)(j5)}{20}$$

$$= \frac{200 + j150}{20}$$

Thus:

$$Z_{CA} = \frac{250\angle 36.87}{20\angle 0} = 12.5\angle 36.87 \ \Omega$$

9.2.7 Power in Three-Phase Systems

The average power dissipated in a three-phase load is equal to the sum of the power dissipated in each phase. If a load is balanced, then the total power, P, will correspond to three times the power consumed by one phase. The power consumed in a one-phase system (assuming a resistive load) is (see Table 6.1):

$$P_\text{p} = V_\text{P,rms} I_\text{P,rms} \cos\varphi \tag{9.13}$$

Replacing for the power factor $\cos\varphi = R_\text{P}/|Z_\text{p}|$ [see Equations (6.73) and (6.74); R_p is the real part of Z_p, and using $V_\text{P,rms} = I_\text{P,rms} \cdot |Z_\text{p}|$:

$$P_\text{p} = I_\text{P,rms}^2 R_\text{P} \tag{9.14}$$

where ϕ is the phase angle between the voltage, $\boldsymbol{V}_\text{P}$, and current, $\boldsymbol{I}_\text{P}$.

EXERCISE 9.2

Start with $Z_\text{p} = \boldsymbol{V}/\boldsymbol{I}$, $\boldsymbol{V} = |\boldsymbol{V}|e^{j\theta_\text{v}}$, $\boldsymbol{I} = |\boldsymbol{I}|e^{j\theta_\text{i}}$, and $\phi = \theta_\text{v} - \theta_\text{i}$, and show that $\cos\varphi = R_\text{P}/|Z_\text{p}|$. Here, R_p is the real part of Z_p.

The total power is three times that of the power consumed in each phase, that is,

$$P_\text{T} = 3P_\text{P} = 3\,V_\text{P,rms} I_\text{P,rms} \cos\varphi\,\text{W} \tag{9.15}$$

Using the fact that $\cos\varphi = R_\text{P}/|Z_\text{p}|$ and $I_\text{p,rms} = V_\text{p,rms}/|Z_\text{p}|$:

$$P_\text{T} = 3\,I_\text{P,rms}^2 R_\text{P}\,\text{W} \tag{9.16}$$

Equation (9.15) represents the total power for either a star- or a Δ-balanced connection. Now, find the total consumed power for the star and Δ connection scenarios.

For a star connection: $I_\text{P,rms} = I_\text{L,rms}$, $V_\text{P,rms} = \dfrac{V_\text{L,rms}}{\sqrt{3}}$; thus:

$$P_\text{T} = 3\left(\frac{V_\text{L,rms}}{\sqrt{3}}\right) I_\text{L,rms} \cos\varphi = \sqrt{3}\,V_\text{L,rms} I_\text{L,rms} \cos\varphi$$

For a Δ connection: $V_\text{P,rms} = V_\text{L,rms}$, $I_\text{P,rms} = \dfrac{I_\text{L,rms}}{\sqrt{3}}$; thus:

$$P_\text{T} = 3\,V_\text{L,tms}\left(\frac{I_\text{L,rms}}{\sqrt{3}}\right) \cos\varphi = \sqrt{3}\,V_\text{L,rms} I_\text{L,rms} \cos\varphi$$

Therefore, for both star and Δ connections, the total power corresponds to:

$$P_\text{T} = \sqrt{3}V_\text{L,rms} I_\text{L,rms} \cos\varphi\,\text{W} \tag{9.17}$$

The apparent power (total volt-amperes) is defined as:

$$S = \sqrt{3}V_\text{L,rms} I_\text{L,rms}\,\text{Volt-amperes} \tag{9.18}$$

EXAMPLE 9.9   Dissipated Power

Three 12-Ω resistors are connected in star to a 415-V, three-phase supply. Determine the total power dissipated by the resistors.

SOLUTION

For a star connection, $V_\text{L,rms} = \sqrt{3}V_\text{P,rms}$. Thus, the phase voltage is:

$$V_\text{P,rms} = \frac{V_\text{L,rms}}{\sqrt{3}} = \frac{415}{\sqrt{3}} = 240\,\text{V}$$

(continued)

EXAMPLE 9.9 Continued

and the phase current is:

$$I_{P,rms} = \frac{V_{P,rms}}{R_P} = \frac{240}{12} = 20 \text{ A}$$

For a star connection, $I_{L,rms} = I_{P,rms}$. Thus, the line current is $I_{L,rms} = 20$ A. For a purely resistive load, the power factor $= \cos \phi = 1$. Therefore, three methods to compute the total dissipated power include:

$$P_T = 3V_{P,rms}I_{P,rms} \cos \varphi = 3 \times 240 \times 20 = 14.4 \text{ kW}$$

or

$$P_T = \sqrt{3}V_{L,rms}I_{L,rms} \cos \varphi = \sqrt{3}(415) \times 20 = 14.4 \text{ kW}$$

or

$$P_T = 3I_{P,rms}^2 R_P = 3 \times 20^2 \times 12 = 14.4 \text{ kW}$$

EXAMPLE 9.10 Power Factor

The input power to a three-phase AC motor is measured as 5 kW. If the rms voltage and current of the motor are 400 V and 8.6 A, respectively, determine the power factor of the system.

SOLUTION

Power, $P_T = 5000$ W; Line voltage, $V_{L,rms} = 400$ V; Line current, $I_{L,rms} = 8.6$ A. Power is defined as:

$$P_T = \sqrt{3}V_{L,rms}I_{L,rms} \cos \varphi$$

Therefore, the power factor corresponds to:

$$\cos \varphi = \frac{P_T}{\sqrt{3}\, V_{L,rms}I_{L,rms}} = \frac{5000}{\sqrt{3}(400)(8.6)} = 0.839$$

EXAMPLE 9.11 Dissipated Power of Δ- and Star-Connected Lines

Three identical coils, each with a resistance of 10 Ω and an inductance of 42 mH are connected (a) in star and (b) in Δ to a 415-V, 50-Hz, three-phase supply. Find the total power dissipated in each case.

SOLUTION

The reactance (impedance) of an inductance corresponds to:

$$X_L = 2\pi f \cdot L = 2\pi \cdot (50) \cdot (42 \times 10^{-3}) = 13.19 \text{ Ω}$$

The impedance of each phase is $R_P + jX_L$, and the magnitude of the impedance of each phase is:

$$|Z_P| = \sqrt{R_P^2 + X_L^2} = \sqrt{10^2 + 13.19^2} = 16.55 \text{ Ω}$$

In addition, the power factor corresponds to:

$$\cos \varphi = \frac{R_P}{|Z_P|} = \frac{10}{16.55} = 0.6042 \text{ lagging (inductive load)}$$

a. Now, considering a star connection, the line voltage is $V_{\text{L,rms}} = 415$ V. Therefore, the phase voltage is:

$$V_{\text{P,rms}} = \frac{V_{\text{L,rms}}}{\sqrt{3}} = \frac{415}{\sqrt{3}} = 240 \text{ V}$$

In addition, using ohms law, the phase current can be found to be:

$$I_{\text{P,rms}} = \frac{V_{\text{P,rms}}}{|Z_{\text{P}}|} = \frac{240}{16.55} = 14.5 \text{ A}$$

For a star connection, the line current and phase current are equal, that is,

$$I_{\text{L,rms}} = I_{\text{P,rms}} = 14.5 \text{ A}$$

Accordingly, the dissipated power is:

$$P_{\text{T}} = 3V_{\text{P,rms}}I_{\text{P,rms}} \cos \varphi = 3(240) \times 14.5 \times 0.6042 = 6.3 \text{ kW}$$

or

$$P_{\text{T}} = \sqrt{3}V_{\text{L,rms}}I_{\text{L,rms}} \cos \varphi = \sqrt{3} \times (415) \times 14.5 \times 0.6042 = 6.3 \text{ kW}$$

or

$$P_{\text{T}} = 3I_{\text{P,rms}}^2 R_{\text{P}} = 3 \times 14.5^2 \times 10 = 6.3 \text{ kW}$$

b. For a Δ connection, the line voltage is $V_{\text{L,rms}} = 415$ V. Therefore, the phase voltage corresponds to $V_{\text{P,rms}} = V_{\text{L,rms}} = 415$ V, and the phase current is:

$$I_{\text{P,rms}} = \frac{V_{\text{P,rms}}}{|Z_{\text{P}}|} = \frac{415}{16.55} = 25.08 \text{ A}$$

For a Δ connection, the line current is:

$$I_{\text{L,rms}} = \sqrt{3}I_{\text{P,rms}} = \sqrt{3}(25.08) = 43.44 \text{ A}$$

Therefore, the dissipated power corresponds to (using three approaches):

$$P_{\text{T}} = 3V_{\text{P,rms}}I_{\text{P,rms}} \cos \varphi = 3(415) \times 25.08 \times 0.6042 = 18.87 \text{ kW}$$

or

$$P_{\text{T}} = \sqrt{3}V_{\text{L,rms}}I_{\text{L,rms}} \cos \varphi = \sqrt{3}(415) \times 43.44 \times 0.6042 = 18.87 \text{ kW}$$

or

$$P_{\text{T}} = 3I_{\text{P,rms}}^2 R_{\text{P}} = 3 \times 25.08^2 \times 10 = 18.87 \text{ kW}$$

Thus, loads connected in Δ dissipate three times greater power compared to star-connected loads. In addition, they draw a line current three times higher, which, in turn, requires greater cable costs.

EXAMPLE 9.12   **Efficiency**

A 415-V, three-phase AC motor has a power output of 12.75 kW and operates at a power factor of 0.77 lagging and with an efficiency of 85%. If the motor is Δ-connected, determine (a) the power input, (b) the line current, and (c) the phase current.

(continued)

EXAMPLE 9.12 | **Continued**

SOLUTION

a. Efficiency is defined as the ratio of the output power to the input one; thus, using 85% efficiency:

$$\text{Efficiency} = \frac{\text{Output power}}{\text{Input power}} \Rightarrow \frac{85}{100} = \frac{12{,}750}{\text{Input power}}$$

Input power corresponds to:

$$\text{Input power} = P_T = \frac{12{,}750 \times 100}{85} = 15 \text{ kW}$$

b. Therefore, the input power is:

$$P_T = \sqrt{3} V_{L,rms} I_{L,rms} \cos \varphi$$

As a result, the line current is:

$$I_{L,rms} = \frac{P_T}{\sqrt{3} \, V_{L,rms} \cos \varphi} = \frac{15{,}000}{\sqrt{3}(415)(0.77)} = 27.1 \text{ A}$$

c. For a Δ connection, $I_{L,rms} = \sqrt{3} I_{P,rms}$. Therefore, the phase current is:

$$I_{P,rms} = \frac{I_{L,rms}}{\sqrt{3}} = \frac{27.1}{\sqrt{3}} = 15.65 \text{ A}$$

EXAMPLE 9.13 | **Delta Connection Computations**

Each phase of a Δ-connected load comprises a resistance of 30-Ω and an 80-μF capacitor in series. The load is connected to a 400-V, 50-Hz, three-phase supply. Calculate (a) the phase current, (b) the line current, (c) the total power dissipated, and (d) the kVA rating (apparent power) of the load.

SOLUTION

a. First, find the capacitive reactance in order to find the impedance, which is:

$$X_C = \frac{1}{2\pi f C} = \frac{1}{2\pi(50)(80 \times 10^{-6})} = 39.79 \ \Omega$$

The impedance of each phase is:

$$|Z_P| = \sqrt{R^2 + X_C^2} = \sqrt{30^2 + 39.79^2} = 49.83 \ \Omega$$

The power factor is:

$$\cos \varphi = \frac{R_P}{|Z_P|} = \frac{30}{49.83} = 0.602 \text{ leading (capacitive load)}$$

For a Δ connection, $V_{P,rms} = V_{L,rms} = 400$ V, and the phase current is:

$$I_{P,rms} = \frac{V_{P,rms}}{|Z_P|} = \frac{400}{49.83} = 8.027 \text{ A}$$

b. For a Δ connection:

$$I_{L,rms} = \sqrt{3} I_{P,rms} = \sqrt{3}(8.027) = 13.9 \text{ A}$$

c. The total power dissipated corresponds to:

$$P_T = 3V_{P,rms}I_{P,rms}\cos\varphi = 3 \times 400 \times 8.027 \times 0.602 = 5.8\text{ kW}$$

or

$$P_T = \sqrt{3}\,V_{L,rms}I_{L,rms}\cos\varphi = \sqrt{3} \times 400 \times 13.9 \times 0.602 = 5.8\text{ kW}$$

or

$$P_T = 3I_{P,rms}^2 R_P = 3 \times 8.027^2 \times 30 = 5.8\text{ kW}$$

d. The total kVA can be found to be:

$$S = \sqrt{3}\,V_{L,rms}I_{L,rms} = \sqrt{3}(400) \times 13.9 = 9.63\text{ kVA}$$

9.2.8 Comparison of Star and Δ Load Connections

As shown in Example 9.11, loads connected in Δ dissipate three times more power than those same loads connected in star to the same three-phase supply. Note that power dissipation corresponds to $P_T = 3I_{P,rms}^2 R_P$. Therefore, the line current in a Δ-connected system is $\sqrt{3}$ times that of a star-connected system.

Now based on $P_T = 3I_{P,rms}^2 R_P$, equating the power dissipation of the star-connected and Δ-connected loads, is equivalent to equating their phase currents. To do so, the line voltage in the star system must be $\sqrt{3}$ times that of the Δ system. Thus, for a given power dissipation, a Δ system is associated with larger line currents and a star system is associated with larger line voltages.

Larger line currents in Δ system result in more voltage drop across the transmission line and the voltage drop is undesired. In order to maintain a small voltage drop across the transmission line, the transmission line's resistance has to be reduced. As discussed in Section 9.3, the transmission line's resistance can be reduced by increasing its cross section. In addition, a larger cross section results in a higher cost and greater weight. Thus, a star connection is generally preferred over a Δ connection.

9.2.9 Advantages of Three-Phase Systems

Three-phase systems offer several advantages:

1. For a given amount of power transmitted through a system, three-phase systems require conductors with a smaller cross section than single-phase systems. This means saving copper (or aluminum) and thus lowers original installation costs. In addition, the weight of cables in a three-phase system is lower, also lowering the installation costs.
2. Two voltages (phase voltage/line voltage) are available.
3. When compared to single-phase motors, three-phase motors, as loads, are very robust, relatively cheap, are generally smaller, have self-starting properties, provide a steadier output, and require little maintenance. Three-phase motors are discussed in Chapter 13.

EXAMPLE 9.14   **Hydroelectric Dam Design**

A hydroelectric power dam is planned for the purpose of supplying electricity to a city area. After initial calculations, the technician finds that the city requires a minimum power supply of 2.5 GW to operate and that there is a transmission loss of 2% between the dam and the city.

Given that the flow rate of the water is $q = 8 \times 10^6$ liter/sec, and the power generator has an efficiency of $\eta_{generator} = 90\%$, what is the required height, h, of the hydroelectric dam so that it meets the power supply requirements?

(continued)

EXAMPLE 9.14   **Continued**

SOLUTION

The relationship between power generated and the water flow rate, height, gravity, and efficiency corresponds to:

$$P_{gen} = \rho h g q \eta_{generator}$$

Here, h is the required height of the dam, $g \leq 9.81$ m/s^2, q is the water flow rate and $\eta_{generator}$ is the generator efficiency. In addition, the relationship between the power required, power generated, and transmission efficiency is:

$$\frac{P_{req}}{P_{gen}} = \eta_{trans}$$

The transmission loss is 2%; therefore, the power generated should be greater than the required power at the city by 2%, or the transmission efficiency is:

$$\eta_{trans} = \frac{P_{req}}{P_{gen}} = 0.98$$

Therefore, the generated power corresponds to:

$$P_{gen} = 2.5 \times 10^9 / 0.98 \rightarrow P_{gen} = 2.55 \times 10^9$$

The following information is known:

$$\text{Water flow rate, } q = 8 \times 10^6 \text{ liter/sec } = 8 \times 10^3 \text{ m}^3\text{/sec}$$

In addition, the water density corresponds to $\rho = 1000$ Kg/m^3

$$\text{Efficiency, } \eta_{generator} = 0.90$$

Therefore, the required height for the dam is:

$$h = \frac{P_{gen}}{\rho g q \eta_{generator}}$$

$$h = \frac{2.55 \times 10^9}{10^3 (8 \times 10^3)(9.81)(0.90)}$$

$$h = 36.1 \text{ m}$$

9.3 TRANSMISSION LINES

9.3.1 Introduction

The purpose of a transmission network is to transfer electric energy from generating units such as dams or atomic power plants at various locations to distribution systems, which ultimately supply the load to end users, for example, city residents, as shown in Figure 9.10. Transmission lines also interconnect neighboring utilities, which permit not only economic dispatch of power within regions during normal conditions, but also the transfer of power between regions during emergencies.

In order to increase the power transmission efficiency, the voltage at the generator is stepped up to a high level (such as 230 kV) by a transformer for transmission. Moreover, at the transformer near the city, the voltage is stepped down to a low level for household use (such as 110 V in United States). Assuming there is no power loss in the transformer, a high transmission voltage would result in a low transmission current. For a certain transmission line resistance, R, the power consumed by the transmission line is $P = I^2 R$. Therefore, power loss in the transmission is low for a low transmission current and the use of "step up" and "step down" transformers increases power transmission efficiency.

Let us see what happens if we transfer the power through the lines without using transformer. For simplicity, assume the load per each phase is 110 kW that is a typical power needed by a small town. If the voltage is selected to be 110 V, then the current through the transmission line needs to be 1 kA (1 kA × 110 V = 110 kW). This huge current needs a very thick copper cable. It should also be mentioned that higher currents are hazardous. Chapter 15 discusses

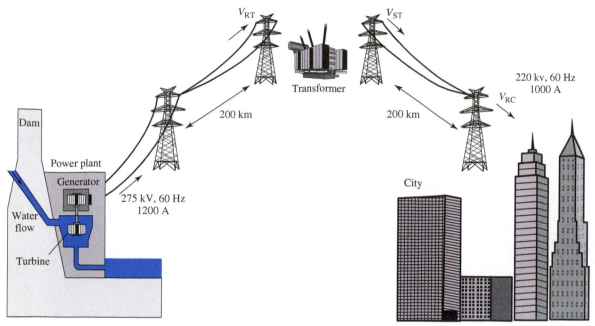

FIGURE 9.10   The transmission line between a dam and city.

environmental problems associated with high-level currents. To reduce cable surface currents and its environmental effects, the diameter of the cable should be kept larger. This highly increases the cost of the cables. In addition, it increases the weight per feet of cables, and accordingly the cost of installation. Ultimately, it highly reduces power transmission efficiency.

Therefore, indeed transformers are used to reduce the current through the cables and accordingly the cable diameter, and maintain power distribution costs in a reasonable level. This argument is similar to our earlier argument in Section 9.2.1, to use three phase in place of single-phase transmission.

The selection of a cost-effective voltage level for a transmission line depends on the power level and the distance between the source and destination. In the United States, standard transmission voltages are established by the American National Standards Institute (ANSI). Transmission voltages above 230 kV are usually referred to as extra-high voltage (EHV) and those at 765 kV and above are referred to as ultra-high voltage (UHV). Construction of homes in close proximity of EHF and UHF voltage lines may lead to health issues. Engineers advise on keeping at least a minimum recommended distance from high-voltage lines. Chapter 15 discusses standards for safe distances between residences and high-voltage lines.

All transmission lines in a power system exhibit the electrical properties of resistance, inductance, capacitance, and conductance. The series inductance and shunt capacitance are due to the effects of magnetic and electric fields around the conductor. The shunt conductance accounts for leakage currents flowing across insulators and ionized pathways in the air. The leakage currents are negligible compared to the current flowing in the transmission lines. Therefore, the shunt conductance that represents the leakage current is often ignored.

A transmission line is considered as a distributed (and not a lumped) element. An element is called lumped if its dimensions are much smaller than the wavelength of the current or voltage. If the dimension of an element is in the order of wavelength, it is called distributed. Typically, at 60 Hz frequency, the wavelength is $3 \times 10^8/60 = 5000$ km. The length of a transmission line is not considered much smaller than 5000 km. Thus, it is considered as distributed element.

A distributed element such as transmission line can be modeled as a combination of resistors, capacitors, and inductors. These components are essential for the development of transmission line models used in power system analysis. The resistance (R) and inductance (L) that are uniformly distributed along the line form the series impedance. The conductance (G) and capacitance (C) between conductors of a single-phase line or from a conductor to the neutral line of a three-phase line form the shunt admittance.

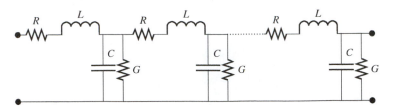

FIGURE 9.11
Equivalent circuit of a
transmission line.

Figure 9.11 represents the equivalent circuit of a transmission line. The equivalent circuit of the transmission line will be discussed in detail in Section 9.3.6. Section 9.3.3 to 9.3.5 discuss the values of the R, L, and C elements in the model of Figure 9.11.

9.3.2 Resistance (R)

The resistance of a conductor is very important for economic studies and for evaluating its transmission efficiency. The DC resistance of a solid round conductor at a specified temperature corresponds to:

$$R_{dc} = \frac{\rho l}{A} \tag{9.19}$$

where

ρ = conductor resistivity
l = conductor length
A = conductor cross-sectional area.

Conductor resistance is affected by spiraling and temperature as detailed below.

Spiraling: Cable spiraling (shown in Figure 9.12) reduces cable manufacturing costs and increases cable flexibility. The length of each strand in a spiraled conductor is longer than the produced transmission line. This slightly increases the resistance compared to the value calculated by Equation (9.19). In addition, when AC flows in a conductor, due to the time variation of the current, an electromagnetic wave is generated, which creates nonuniformity in the current distribution over the conductor's cross-sectional area: the current density is higher at the surface of the conductor compared to its core. Accordingly, the effective cross section of the wire decreases. This makes the AC resistance slightly higher than the DC resistance. This behavior is known as the *skin effect*. At 60 Hz, the AC resistance is about 2% higher than the DC resistance.

Temperature: The conductor's resistivity increases as temperature increases. Refer to Equation (9.19), the conductor's resistance increases as the temperature increases. This change can be considered linear over the range of temperature normally encountered. The resistance, R, at the temperature, T, may be calculated using the following equation:

$$R = R_0 \left[1 + \alpha (T - T_0) \right] \tag{9.20}$$

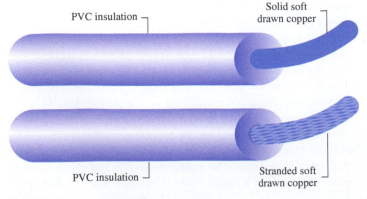

PVC insulation

Solid soft
drawn copper

PVC insulation

Stranded soft
drawn copper

FIGURE 9.12 A spiral cable.

T_0 is usually selected to be 20°C at which R_0 is measured. T is also measured in °C. The constant α depends on the conductor material. For aluminum $\alpha = 0.0043$, and for copper it is 0.004041.

EXERCISE 9.3

Investigate the impact of frequency on the resistance and impedance of a transmission line.

EXAMPLE 9.15   **Copper Wire Resistance**

Find the resistance of a copper wire at 16°C if its resistance at 20°C is 0.4 Ω.

SOLUTION

Here, $R_0 = 0.4$ Ω, and for copper $\alpha = 0.004041$. Thus, by referring to Equation (9.20):

$$R = 0.4 \times (1 + 0.004041 \times (16 - 20)) = 0.3935 \ \Omega$$

EXAMPLE 9.16   **Temperature and Resistance**

Find the temperature at which the resistance of a copper conductor increases to 1 Ω from a level of 0.8 Ω at 20°C.

SOLUTION

For copper $\alpha = 0.004041$, and for this problem: $R_0 = 0.8$ Ω and $R = 1$ Ω; applying these values to Equation (9.20) results in:

$$1 = 0.8 \times (1 + 0.004041 \times (T - 20))$$

Thus:

$$T = 81.86°C$$

9.3.3 Different Types of Conductors

In the early days of transmission of electric power, conductors were usually made from copper. Now, aluminum conductors have completely replaced copper for overhead lines because of their lower cost and lighter weight compared to copper conductors of the same resistance. For cables of the same resistance, aluminum cables have larger diameters compared to copper cables, which is an advantage. While aluminum is a good conductor, when used without chemical flux, the connection rapidly becomes oxidized causing a bad connection. This, in turn, can result in a fire hazard. For this reason, while it is used in overhead lines, aluminum wire is no longer used in houses. When aluminum is used, the connectors are relatively large and flux is added to keep the connection from oxidizing.

With a larger diameter, the electric field at the conductor's surface is reduced. This creates a lower tendency to ionize the air around the conductor. Ionized air is toxic. Ionization produces the undesirable effect called *corona*. Corona refers to the ionization caused by the electric field around power lines. It is undesirable in power lines because it leads to energy loss (see Chapter 15).

The most commonly used conductor materials for high-voltage transmission lines are ACSR (aluminum conductor steel-reinforced), AAC (all-aluminum conductor), AAAC (all-aluminum alloy conductor), and ACAR (aluminum conductor alloy-reinforced). These materials are popular because, compared to copper conductors, their cost is lower and their strength-to-weight ratio is higher. In addition, aluminum is in abundant supply, while the copper supply is limited.

The conductors are stranded to have flexibility. An ACSR conductor consists of a center core of steel strands surrounded by layers of aluminum. Each layer of strand is spiraled in the opposite direction of its adjacent layer. This spiraling holds the strands in place.

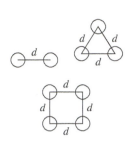

FIGURE 9.13 Bundle arrangements.

At voltages above 230 kV (Extremely High Voltage, EHV), it is preferable to use more than one conductor per phase, which is known as bundling of conductors. Typically, bundled conductors consist of two, three, or four subconductors symmetrically arranged in configurations shown in Figure 9.13. Bundling increases the effective radius of line conductors and reduces the line reactance, which improves the line performance and increases the power capability of the line.

Bundling also reduces corona loss, radio interference, and line characteristic impedance. Assuming a uniform transmission line, its characteristic impedance (also called surge impedance) is the ratio of the amplitudes of voltage and current waves propagating along the line when the line is perfectly matched (no reflection along the line). Impedance matching was discussed Section 6.6.6.

Consider two bundled conductors, X and Y, which are composed of m and n subconductors, respectively. The geometric mean radius (GMR) of a conductor (X or Y) is used as a measure to compute its equivalent radius. The GMR is represented by the parameter D_s.

The geometric mean distance (GMD) between two bundled conductors, X and Y, is used to calculate the equivalent separation between them. The GMD is represented by the parameter D_m. Thus, GMD is the geometric mean of all possible distances between each conductor carrying opposite currents. See Example 9.17 for details.

The above-mentioned parameters are summarized in Table 9.1. As an example, to introduce the equivalent radius (D_s), let r be the radius of each subconductor and d be the bundle spacing (as shown in Figure 9.13).

Based on the rule presented in Table 9.1, for a two-subconductor bundle, there are two distances, one is the radius of each subconductor, r, and the other is the distance between the two. Thus:

$$D_s = \sqrt[4]{(r \times d)^2} = \sqrt{r \times d} \tag{9.21}$$

For the three-subconductor bundle shown in Figure 9.13, there are three distances. One is the radius of each subconductor, r, and the other two are the distances between each subconductor and the other two subconductors, which is d. Thus:

$$D_s = \sqrt[9]{(r \times d \times d)^3} = \sqrt[3]{r \times d^2} \tag{9.22}$$

For the four-subconductor bundle, it can be shown that (see Exercise 9.4):

$$D_s = \sqrt[16]{(r \times d \times d \times d \times 2^{1/2})^4} = 1.09 \sqrt[4]{r \times d^3} \tag{9.23}$$

EXERCISE 9.4

Prove Equation (9.23).

TABLE 9.1 The Definition of the GMR and GMD in Transmission Lines

GMR (D_s) of the conductor X = the $m^{2\text{th}}$ root of the product of the m^2 distances.

GMR (D_s) of the conductor Y = the $n^{2\text{th}}$ root of the product of the n^2 distances.

GMD (D_m) between X and Y = the mn^{th} root of the product of the mn distances.

The parameters introduced in this subsection, that is, GMR (D_s) and GMD (D_m), are used to calculate the inductance of the line. See Examples 9.17 to 9.19 for applications.

9.3.4 Inductance (L)

Figure 9.11 discusses how a transmission line can be modeled as a combination of capacitors, inductors, and resistors. The previous section discussed how the resistance of the lines is computed. This section examines how the capacitance of the lines can be calculated.

9.3.4.1 THE INDUCTANCE OF SINGLE-PHASE TRANSMISSION LINES

Consider a 1-m length of a single-phase line consisting of two solid round conductors of radiuses, r_1 and r_2, as shown in Figure 9.14. The two conductors are separated by the distance, D. Assume that conductor 1 carries the phasor current I_1 and its direction is into the page and conductor 2 carries the return current $I_2 = -I_1$. These currents create the magnetic field lines that maintain the link between the two conductors. Now, to find the total inductance of the line, first find the inductance of each conductor and then use those results to find the total inductance. The total inductance of Conductor 1 per each meter of the line is:

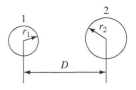

FIGURE 9.14 Single-phase, two-wire line.

$$L_1 = 2 \times 10^{-7} \ln \frac{D}{r_1{}'} \text{ H/m} \tag{9.24}$$

where $r_1{}' = r_2 e^{-1/4} = 0.7788\, r_1$. (Note that Equation (9.24) can be proven through advanced electromagnetic theory, which is beyond of the scope of this chapter.) Similarly, the inductance of conductor 2 per meter of line is:

$$L_2 = 2 \times 10^{-7} \ln \frac{D}{r_2{}'} \text{ H/m} \tag{9.25}$$

where $r_2{}' = r_2 e^{-1/4} = 0.7788\, r_2$. Therefore, the total inductance per meter of the line corresponds to the sum of the two inductances in Equations (9.24) and (9.25), that is:

$$L_T = L_1 + L_2$$

$$= 4 \times 10^{-7} \ln \frac{D}{\sqrt{r_1' r_2'}} \tag{9.26}$$

If $r_1 = r_2 = r$, which is often the case in real-world lines, then the total line inductance of the single-phase lines corresponds to:

$$L_T = 4 \times 10^{-7} \ln \frac{D}{r'} \text{ H/m} \tag{9.27}$$

9.3.4.2 THE INDUCTANCE OF THREE-PHASE TRANSMISSION LINES
WITH SYMMETRICAL SPACING

Consider a 1-m length of a three-phase line with three conductors, each with radius r, symmetrically spaced in a triangular configuration as shown in Figure 9.15. Assuming balanced three-phase currents, $I_a + I_b + I_c = 0$, the total inductance of each phase of the conductor per each meter of the line is therefore:

$$L_a = 2 \times 10^{-7} \ln \frac{D}{r'} \text{ H/m} \tag{9.28}$$

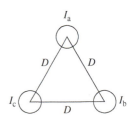

FIGURE 9.15 Three-phase line with symmetrical spacing.

where $r' = re^{-1/4} = 0.7788\, r$, because of symmetry, $L_a = L_b = L_c$, and the three inductances are identical. Equation (9.28) represents the inductance per phase of a three-phase line. Comparison of Equations (9.24) and (9.25) with Equation (9.28) shows that the inductance per phase for a three-phase circuit with equilateral spacing is the same as a one-conductor, single-phase scenario.

9.3.4.3 INDUCTANCE OF THREE-PHASE TRANSMISSION LINES WITH ASYMMETRICAL SPACING

Practical transmission lines cannot maintain symmetrical spacing of conductors due to construction considerations. With asymmetrical spacing, even with balanced currents, the voltage drop due to line inductance will be unbalanced. In this case, balance of the three phases can be restored by exchanging the positions of the conductors at regular intervals along the line so that each conductor occupies the original position of every other conductor over an equal distance. Such an exchange of conductor position is called *transposition*. A complete transposition cycle is shown in Figure 9.16. Transposition maintains the same average inductance over the whole cycle in each conductor.

Consider a 1-m length of a three-phase line with three conductors, each with radius r, asymmetrically spaced in a triangular configuration as shown in Figure 9.17. In this case, the average inductance per phase of Equation (9.28) is replaced by:

$$L_a = 2 \times 10^{-7} \cdot \ln \frac{D_{eq}}{r'} \, \text{H/m} \tag{9.29}$$

where $D_{eq} = \sqrt[3]{D_{ab}D_{bc}D_{ca}}$. Here, D_{eq} is the geometric mean of the three distances of the asymmetrical line. This is called equivalent equilateral spacing. Table 9.2 summarizes the inductance formulas for different configurations. In this table, D_s is the geometric mean radius of a bundle conductor [Equations (9.21) to (9.23)], where r is replaced with r', and D_m is the geometric mean distance (GMD) between bundled conductors carrying opposite currents as defined in Table 9.1.

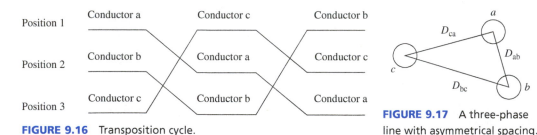

FIGURE 9.16 Transposition cycle.

FIGURE 9.17 A three-phase line with asymmetrical spacing.

TABLE 9.2 The Inductance (per Meter) Formulas for Different Configurations

	Solid Conductor	Bundled Conductor
The total inductance of single-phase transmission lines (H/m)	$L_T = 4 \times 10^{-7} \ln \frac{D}{r'}$	$L_T = 4 \times 10^{-7} \ln \frac{D_m}{D_s}$
The inductance per phase of three-phase transmission lines with symmetrical spacing (H/m)	$L_a = 2 \times 10^{-7} \ln \frac{D}{r'}$	$L_a = 2 \times 10^{-7} \ln \frac{D_m}{D_s}$
The average inductance per phase of three-phase transmission lines with asymmetrical spacing (H/m)	$L_a = 2 \times 10^{-7} \ln \frac{D_{eq}}{r'}$	$L_a = 2 \times 10^{-7} \ln \frac{D_m}{D_s}$

EXAMPLE 9.17   Line Inductance

One circuit of a single-phase transmission line (as shown in Figure 9.18) is composed of three solid $D_{aa} = D_{bb} = D_{cc} = 0.25$ cm radius wires. The return circuit is composed of two $D_{a'a'} = D_{b'b'} = D_{c'c'} = 0.5$ cm radius wires. Find the inductance in each side of the line and the inductance of the complete line.

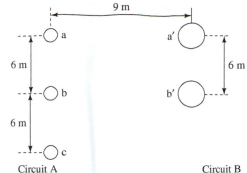

FIGURE 9.18 Arrangement of conductors for Example 9.17.

SOLUTION

This is an asymmetric, single-phase bundled line. Each side includes different bundling structures. Based on Table 9.2, the total inductance per meter of each line is:

$$L_{\mathrm{T}} = 4 \times 10^{-7} \ln \frac{D_{\mathrm{m}}}{D_{\mathrm{s}}}$$

In order to find L_{T}, first find D_{m} and D_{s}. There are six possible distances in a two-by-three conductor set. Here, D_{m} is the line's GMD defined in Table 9.1 and corresponds to:

$$D_{\mathrm{m}} = \sqrt[6]{D_{\mathrm{aa'}} D_{\mathrm{ab'}} D_{\mathrm{ba'}} D_{\mathrm{bb'}} D_{\mathrm{ca'}} D_{\mathrm{cb'}}}$$

where, based on Figure 9.18:

$$D_{\mathrm{aa'}} = D_{\mathrm{bb'}} = 9 \text{ m}$$
$$D_{\mathrm{ab'}} = D_{\mathrm{ba'}} = D_{\mathrm{cb'}} = \sqrt{6^2 + 9^2} = \sqrt{117} \text{ m}$$
$$D_{\mathrm{ca'}} = \sqrt{12^2 + 9^2} = 15 \text{ m}$$

Thus:

$$D_{\mathrm{m}} = 10.743 \text{ m}$$

The GMR is the geometric mean of all distances within a bundle as defined in Table 9.1. Here, $m = 3$; therefore, $3^2 = 9$ distances are available, which for circuit A leads to:

$$
\begin{aligned}
(D_{\mathrm{s}})_{\mathrm{A}} &= \sqrt[3^2]{D_{\mathrm{aa}} D_{\mathrm{ab}} D_{\mathrm{ac}} D_{\mathrm{ba}} D_{\mathrm{bb}} D_{\mathrm{bc}} D_{\mathrm{ca}} D_{\mathrm{cb}} D_{\mathrm{cc}}} \\
&= \sqrt[9]{(0.25 \times 10^{-2} \times e^{-1/4})^3 \times 6^4 \times 12^2} \\
&= 0.481 \text{ m}
\end{aligned}
$$

Note that in table 9.1 D_{s} is the geometric mean radius where r is replaced with $r' = re^{-1/4}$. For circuit B, $m = 2$; therefore, 2^2 distances are available. Thus:

$$
\begin{aligned}
(D_{\mathrm{s}})_{\mathrm{B}} &= \sqrt[2^2]{D_{\mathrm{a'a'}} D_{\mathrm{a'b'}} D_{\mathrm{b'b'}} D_{\mathrm{b'a'}}} \\
&= \sqrt[4]{(0.5 \times 10^{-2} \times e^{-1/4})^2 \times 6^2} \\
&= 0.153 \text{ m}
\end{aligned}
$$

(continued)

EXAMPLE 9.17 **Continued**

Next, by referring to Table 9.2, the inductance of circuit A can be calculated to be:

$$L_A = 2 \times 10^{-7} \ln \frac{D_m}{(D_s)_A}$$

$$= 2 \times 10^{-7} \ln \frac{10.743}{0.481}$$

$$= 6.212 \times 10^{-7} \text{ H/m}$$

In addition, the inductance of circuit B is:

$$L_B = 2 \times 10^{-7} \ln \frac{D_m}{(D_s)_B}$$

$$= 2 \times 10^{-7} \ln \frac{10.743}{0.153}$$

$$= 8.503 \times 10^{-7} \text{ H/m}$$

The total inductance is then:

$$L = L_A + L_B = 14.715 \times 10^{-7} \text{ H/m}$$

EXERCISE 9.5

Repeat Example 9.17, if circuit B in Figure 9.18 has the same number of lines and structure as circuit A does.

EXAMPLE 9.18 **Transmission Line Inductive Reactance**

A three-phase line operated at 60 Hz is arranged as shown in Figure 9.19. The conductors have the radius of $r' = r \cdot e^{-1/4} = 0.0373$ ft. Find the inductive reactance per phase, per mile.

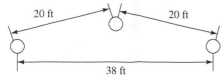

FIGURE 9.19 Arrangement of conductors for Example 9.18.

SOLUTION

Refer to Equation (9.29), and Table 9.2. The geometric mean of the phase conductor spacing is:

$$D_{eq} = \sqrt[3]{20 \times 20 \times 38} = 24.8 \text{ ft}$$

The inductance per phase, per meter of the line is:

$$L = 2 \times 10^{-7} \ln \frac{24.8}{0.0373} = 13 \times 10^{-7} \text{ H/m}$$

The inductive reactance per phase, per mile is:

$$X_L = \omega L$$

$$= 2\pi \times 60 \times 1609 \times 13 \times 10^{-7}$$

$$= 0.788 \text{ }\Omega/\text{mile}$$

Note that the multiplication by 1609 is required to convert the reactance per meter into the reactance per mile.

EXAMPLE 9.19   Transmission Line Inductive Reactance

A three-phase line operated at 60 Hz is arranged as shown in Figure 9.20. Each subconductor of the bundled conductor line has the radius of $r' = 0.0466$ ft. Find the inductive reactance in Ω/km and Ω/mile per phase for $d = 45$ cm.

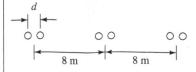

FIGURE 9.20   Arrangement of conductors for Example 9.19.

SOLUTION

The distances in feet are:

$$d = \frac{0.45}{0.3048} = 1.476 \text{ ft}$$

$$D = \frac{8}{0.3048} = 26.25 \text{ ft}$$

Refer to Equation (9.21). The GMR for a bundle of conductors is:

$$D_s = \sqrt{r' \times d} = \sqrt{0.0466 \times 1.476} = 0.2623 \text{ ft}$$

There are three distances in this configuration, which include the two distances between the two nearby lines (26.25 ft) and the distance between the two nonnearby one (52.49 ft). The geometric mean of the phase conductor spacing is therefore:

$$D_m = D_{eq} = \sqrt[3]{26.25 \times 26.25 \times 52.49} = 33.07 \text{ ft}$$

Refer to Table 9.2; the inductance per phase, per meter of the line is:

$$L = 2 \times 10^{-7} \ln\frac{D_m}{D_s}$$

$$= 2 \times 10^{-7} \ln\frac{33.07}{0.2623}$$

$$= 9.674 \times 10^{-7} \text{ H/m}$$

The inductive reactance per phase is:

$$X_L = 2\pi fL$$

$$= 2\pi \times 60 \times 9.674 \times 10^{-7}$$

$$= 3.647 \times 10^{-4} \ \Omega/\text{m}$$

$$= 0.3647 \ \Omega/\text{km}$$

$$= 0.5868 \ \Omega/\text{mile}$$

9.3.5 Capacitance

Earlier, this chapter outlined how a transmission line can be modeled as a combination of capacitors, inductors, and resistors as shown in Figure 9.11. Section 9.3.2 discussed the method used to compute the resistance of the lines. In addition, prior sections examined how resistance may vary with parameters such as temperature and how line inductance can be computed. Moreover, the previous section studied the line inductance. This section outlines how to calculate the capacitance of the lines.

9.3.5.1 CAPACITANCE OF SINGLE-PHASE TRANSMISSION LINES

Consider a 1-m length of a single-phase line consisting of two solid round conductors of radiuses r_1 and r_2 as shown in Figure 9.21. The two conductors are separated by distance D. Conductor 1 carries a charge of q_1 C/m and conductor 2 carries a charge of q_2 C/m. These currents set up magnetic field lines that maintain a link between the conductors. Assuming one solid round conductor, the capacitance per meter between the two conductors is:

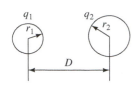

FIGURE 9.21 Single-phase, two-wire line.

$$C_{12} = \frac{\pi \varepsilon_o}{\ln(D/\sqrt{r_1 r_2})} \text{ F/m} \tag{9.30}$$

Equation (9.30) gives the line-to-line capacitance between the conductors. For the purpose of transmission line modeling, it is convenient to define a capacitance, C, between each conductor and a neutral (point n), as shown in Figure 9.22.

Because the voltage to neutral is half of the voltage difference between two conductors, V_{12}, the capacitance to neutral $C = 2C_{12}$, or:

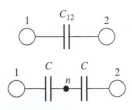

FIGURE 9.22 Illustration of capacitance to neutral (point n).

$$C = \frac{2\pi \varepsilon_o}{\ln(D/\sqrt{r_1 r_2})} \text{ F/m} \tag{9.31}$$

If $r_1 = r_2 = r$, Equations (9.30) and (9.31) will reduce to:

$$C_{12} = \frac{\pi \varepsilon_o}{\ln(D/r)} \text{ F/m} \tag{9.32}$$

and

$$C = \frac{2\pi \varepsilon_o}{\ln(D/r)} \text{ F/m} \tag{9.33}$$

9.3.5.2 CAPACITANCE OF THREE-PHASE TRANSMISSION LINES WITH SYMMETRICAL SPACING

Consider a 1-m length of a three-phase line with three conductors, each with radius r, symmetrically spaced in a triangular configuration, as shown in Figure 9.23. The capacitance per phase to neutral is:

$$C = \frac{2\pi \varepsilon_o}{\ln(D/r)} \text{ F/m} \tag{9.34}$$

Equations (9.33) and (9.34) are identical. These equations express the capacitance to neutral for single-phase and the capacitance per phase for equilaterally spaced three-phase lines, respectively.

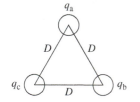

FIGURE 9.23 Three-phase line with symmetrical spacing.

9.3.5.3 CAPACITANCE OF THREE-PHASE TRANSMISSION LINES WITH ASYMMETRICAL SPACING

When the conductors of a three-phase line are not equilaterally spaced, the problem of calculating capacitance becomes more difficult. Consider a 1-m length of a three-phase line with three conductors, each with radius r, asymmetrically spaced in a triangular configuration as shown in Figure 9.24.

The average capacitance per phase, per meter to neutral is:

$$C = \frac{2\pi \varepsilon_o}{\ln(D_{eq}/r)} \text{ F/m} \tag{9.35}$$

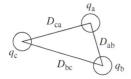

FIGURE 9.24 Three-phase line with asymmetrical spacing.

where $D_{eq} = \sqrt[3]{D_{ab} D_{bc} D_{ca}}$. Here, D_{eq} is the geometric mean of the three distances of the asymmetrical line. Compare Equations (9.35) and (9.34) and note similarities and differences.

9.3.5.4 THE EFFECT OF EARTH ON THE CAPACITANCE

The Earth's surface is an equipotential surface; therefore, electric flux lines are forced to cut the surface of the earth orthogonally. Earth increases the capacitance, but normally, the height of the conductors in a transmission line with respect to the earth is large (typically 15 to 30 ft) compared to the distance between the conductors. Accordingly, in a steady-state analysis of all *symmetrical* lines, the earth's effect on the capacitance can be neglected. However, the effects of the earth must be considered in a steady-state analysis of *asymmetrical* lines.

Table 9.3 summarizes the capacitance per meter for different line configurations. In this table, D_s is the geometric mean radius of a bundle conductor that is defined in Table 9.1 [also see Equations (9.21) to (9.23)]. Note that in Table 9.2 r is replaced with r'. However, here, the actual value of r is used. In addition, D_m is the geometric mean distance between two bundled conductors defined in Table 9.1.

TABLE 9.3   The Capacitance per Meter Formulas for Different Configurations

	Solid Conductor	Bundled Conductor
The capacitance between the conductors of single-phase transmission lines (F/m)	$C_{12} = \dfrac{\pi \varepsilon_0}{\ln D/r}$	$C_{12} = \dfrac{\pi \varepsilon_0}{\ln D_m/D_s}$
The capacitance between each conductor and a neutral of single-phase transmission lines (F/m)	$C = \dfrac{2\pi \varepsilon_0}{\ln D/r}$	$C = \dfrac{2\pi \varepsilon_0}{\ln D_m/D_s}$
The capacitance per phase to neutral of three-phase transmission lines with symmetrical spacing (F/m)	$C = \dfrac{2\pi \varepsilon_0}{\ln D/r}$	$C = \dfrac{2\pi \varepsilon_0}{\ln D_m/D_s}$
The average capacitance per phase to neutral of three-phase transmission lines with asymmetrical spacing (F/m)	$C = \dfrac{2\pi \varepsilon_0}{\ln D_{eq}/r}$	$C = \dfrac{2\pi \varepsilon_0}{\ln D_m/D_s}$

EXAMPLE 9.20   **Transmission Line Capacitive Reactance**

Find the capacitive reactance per phase to neutral of the line described in Example 9.18.

SOLUTION

Here, r' is given and r, the actual radius, must first be found. Given $r' = e^{-1/4} \cdot r = 0.7788 \cdot r$, the actual conductor radius corresponds to:

$$r = \frac{r'}{0.7788} = \frac{0.0373}{0.7788} = 0.04789 \text{ ft}$$

The geometric mean of the phase conductor spacing is:

$$D_{eq} = \sqrt[3]{20 \times 20 \times 38} = 24.8 \text{ ft}$$

Based on Table 9.3, the capacitance per phase to neutral is:

$$C = \frac{2\pi \varepsilon_0}{\ln \dfrac{D_{eq}}{r}} = \frac{2\pi \times 8.854 \times 10^{-12}}{\ln \dfrac{24.8}{0.04789}} = 8.9 \times 10^{-12} \text{ F/m}$$

Accordingly, the capacitive reactance per phase to neutral is:

$$X_C = \frac{10^{12}}{2\pi \times 60 \times 8.9 \times 1609} = 0.185 \times 10^6 \ \Omega \cdot \text{mile}$$

| EXAMPLE 9.21 | **Transmission Line Capacitive Reactance** |

Find the capacitive reactance per phase in Ω.km and Ω.mile of the line described in Example 9.19.

SOLUTION

Again, first find the conductor's radius using $r' = re^{-1/4} = 0.7788\ r$. Thus, the actual conductor radius is:

$$r = \frac{r'}{0.7788} = \frac{0.0466}{0.7788} = 0.0598\ \text{ft}$$

The distances in ft are:

$$d = \frac{0.45}{0.3048} = 1.476\ \text{ft}$$

$$D = \frac{8}{0.3048} = 26.25\ \text{ft}$$

The GMR for a bundle of conductors is:

$$D_\text{s} = \sqrt{r \times d} = \sqrt{0.0598 \times 1.476} = 0.297\ \text{ft}$$

The geometric mean of the phase conductor spacing is:

$$D_\text{m} = D_\text{eq} = \sqrt[3]{26.25 \times 26.25 \times 52.49} = 33.07\ \text{ft}$$

The capacitance per phase to neutral is then:

$$C = \frac{2\pi\varepsilon_\text{o}}{\ln \dfrac{D_\text{m}}{D_\text{s}}} = \frac{2\pi \times 8.845 \times 10^{-12}}{\ln \dfrac{33.07}{0.297}} = 11.8 \times 10^{-12}\ \text{F/m}$$

The capacitive reactance per phase to neutral is:

$$X_\text{C} = \frac{1}{2\pi f C}$$

$$= \frac{10^{12}}{2\pi \times 60 \times 11.8}$$

$$= 0.2248 \times 10^9\ \Omega \cdot \text{m}$$

$$= 0.2248 \times 10^6\ \Omega \cdot \text{km}$$

$$= 0.1397 \times 10^6\ \Omega \cdot \text{mile}$$

9.3.6 Transmission Line Equivalent Circuits

Based on their length, transmission lines are represented by an equivalent model, which is used to calculate voltages, currents, and power flows. For three-phase systems, this model is built with appropriate circuit parameters on a per-phase basis. The terminal voltages are expressed from one line to neutral and the current is defined for one phase. Thus, the three-phase system is reduced to an equivalent single-phase system.

The transmission line may be represented by a two-port network as shown in Figure 9.25. Any two-port system can be represented via equations written in terms of the generalized circuit constants commonly known as the *ABCD* constants.

These constants are, in general, complex numbers: *A* and *D* are dimensionless while the dimensions of B and C are in ohms and mhos or siemens, respectively. Based on Figure 9.25, the general equations relating the phasor of voltages and currents at the two ports (transmitting and receiving ports) of transmission lines correspond to:

$$V_\text{S} = AV_\text{R} + BI_\text{R} \tag{9.36}$$

$$I_\text{S} = CV_\text{R} + DI_\text{R} \tag{9.37}$$

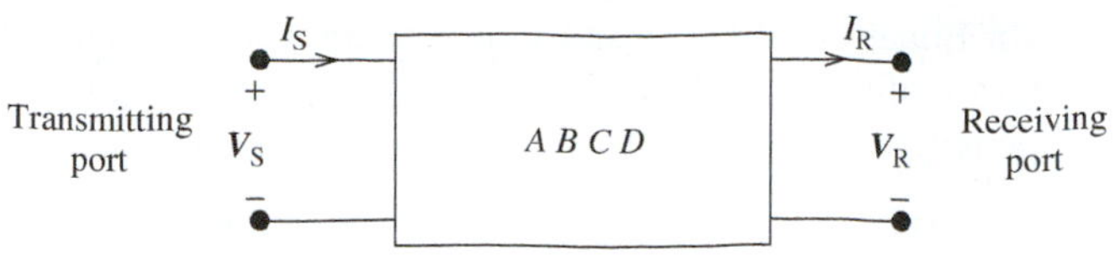

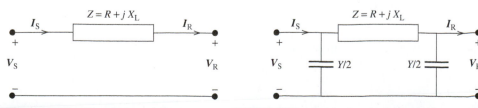

I_S

$+$

Transmitting port V_S

$A B C D$

V_R Receiving port

$+$

I_R

FIGURE 9.25 Two-port representation of transmission line.

Voltage regulation of a transmission line may be defined as the percentage of change in voltage at the receiving end of the line (expressed in terms of the percent of full-load voltage).

$$\text{Voltage regulation} = \frac{|(V_R)_{NL}| - |(V_R)_{FL}|}{|(V_R)_{FL}|} \times 100 \tag{9.38}$$

Here, $|(V_R)_{NL}|$ is the magnitude of the receiving-end voltage at no-load, and $|(V_R)_{FL}|$ is the magnitude of receiving-end voltage at full-load when V_S is constant. The transmission line efficiency corresponds to:

$$\eta = \frac{(P_R)_T}{(P_S)_T} \times 100 \tag{9.39}$$

Here, $(P_R)_T$ and $(P_S)_T$ are the total power at the receiving and sending ends, respectively.

9.3.6.1 SHORT-LENGTH LINES

Lines below 80 km (50 miles) in length are called *short-length lines*. For short-length lines, the capacitance and the conductance between lines are often ignored. The error in the results generated by the model due to omitting the capacitance is minimal. Figure 9.26 represents the equivalent circuit for these lines. Applying KVL to this figure results in $V_S = V_R + ZI_R$ and $I_S = I_R$. Thus, for short lines, the parameters in Equations (9.36) and (9.37) correspond to:

$$A = 1, \quad B = Z, \quad C = 0, \quad D = 1 \tag{9.40}$$

Here, Z is the total series impedance of the line and is obtained by multiplying the series impedance per unit length by the line length as follows:

$$Z = (r + j\omega L) \times \ell = R + jX_L$$

where R and X_L are the total series resistance and inductive reactance of the line, respectively.

9.3.6.2 MEDIUM-LENGTH LINES

As the length of line increases, the effect of shunt capacitance must be taken into account. Lines above 80 km (50 miles) and below 250 km (150 miles) in length are called *medium-length lines*. For medium-length lines, half of the shunt capacitance is considered to be lumped at each end of the line. This is called the π model and is shown in Figure 9.27. For the medium line π model of Figure 9.27, the parameters in Equations (9.36) and (9.37) correspond to:

$$A = \left(1 + \frac{ZY}{2}\right), \quad B = Z, \quad C = Y\left(1 + \frac{ZY}{4}\right), \quad D = \left(1 + \frac{ZY}{2}\right) \tag{9.41}$$

I_S $Z = R + jX_L$ I_R

$+$

V_S

$+$

V_R

$-$

$-$

I_S $Z = R + jX_L$ I_R

$+$

V_S $Y/2$ $Y/2$ V_R

$+$

$-$

$-$

FIGURE 9.26 Short line model.

FIGURE 9.27 Nominal π model for medium-length line.

In Equation (9.41), Z and Y are the total series impedance of the line and the total shunt admittance of the line. Here, Y is obtained by multiplying the shunt admittance per unit length by the line length (ℓ) as follows:

$$Y = (G + j\omega C) \times \ell$$

Under normal conditions, the shunt conductance per unit length (G), which represents the leakage current over the insulators and due to corona will be negligible and is therefore assumed to be zero. Note that the unit of Y (admittance) is siemense (S), i.e., $S = 1/\Omega$

EXERCISE 9.6

Use KVL and KCL to prove Equation (9.41).

9.3.6.3 LONG-LENGTH LINES

For short- and medium-length lines, reasonably accurate models are obtained by assuming resistivity, capacity, and inductivity of the line. Lines 250 km (150 miles) and longer are called *long-length lines*. For long-length lines, or for more accurate solutions for lines of any length, the exact effect of the distributed parameters must be considered. For a long-line model:

$$A = \cosh(\gamma l), \quad B = Z_C \sinh(\gamma l), \quad C = \frac{1}{Z_C}\sinh(\gamma l), \quad D = \cosh(\gamma l) \qquad \textbf{(9.42)}$$

and $\gamma = \sqrt{zy}$, $Z_C = \sqrt{z/y}$ where z, y, and ℓ are the series impedance per unit length, the shunt admittance per unit length, and the total length of the line, respectively.

EXERCISE 9.7

Use KVL and KCL to prove Equation (9.42).

EXAMPLE 9.22   Voltage Regulation and Efficiency

A 220-kV, 60-Hz three-phase, star-connected transmission line is 40 km long. The resistance per phase is 0.15/km and the inductance per phase is 1.3263 mH/km. The shunt capacitance is negligible. Use the short-line model to find the voltage and power at the transmitting end, and the voltage regulation and efficiency when the line is sullying a three-phase load of:

a. 381 MVA at 0.8 power factor lagging at 220 kV.
b. 381 MVA at 0.8 power factor leading at 220 kV.

SOLUTION

The equivalent impedance of the line (see Figure 9.25) is:

$$Z = (r + j\omega L) \times \ell$$
$$= (0.15 + j2 \times \pi \times 60 \times 1.3263 \times 10^{-3}) \times 40$$
$$= 6 + j20 \ \Omega$$

The line voltage is 220 kV; therefore, the receiving end voltage per phase (rms) is:

$$V_{R,rms} = \frac{220}{\sqrt{3}} = 127 \text{ kV}$$

381 MVA at the load is three times the phase rms current and voltage multiplication at the load. Accordingly:

$$I_{R,rms} = \frac{S_R}{3V_{R,rms}} = \frac{381 \times 10^3}{3 \times 127} = 1000A$$

By using the receiving-end voltage as a reference, the receiving-end peak voltage per phase is:

$$V_R = 127\sqrt{2}\angle 0° \text{ kV}$$

Recall that the peak voltage is the square root of two multiplied by the rms voltage (see Chapter 6).

a. Recall that in this example, the power factor is 0.8 lagging. This means that the current phasor has $-\cos^{-1} 0.8$ phase difference with voltage. Thus, considering $V_R = 127\sqrt{2}\angle 0°$ kV, the current phasor corresponds to:

$$I_R = 1000\sqrt{2}\angle -\cos^{-1}0.8 = 10,000\sqrt{2}\angle -36.87° \text{ A}$$

The negative sign is selected because the power factor is lagging. Now, based on Equation (9.36), and using the short-line model in Figure 9.25, and considering that the impedance $Z = 6 + j20 = 20.88 \, e^{j73.3}$:

$$
\begin{aligned}
V_S &= V_R + ZI_R \\
&= 127\sqrt{2}\angle 0° + (20.88\angle 73.3°)(1000\sqrt{2}\angle -36.87°)(10^{-3}) \\
&= 204.11\angle 4.93°
\end{aligned}
$$

V_S represents the transmitted voltage phasor (see Figure 9.25). Thus, the transmitting-end, line-to-line voltage magnitude is [see Equation (9.5)]:

$$|V_{S(L-L)}| = \sqrt{3}|V_S| = 353.53 \text{ kV}$$

Because the current is assumed constant over the line, that is, $I_{S,rms} = I_{R,rms}$, the transmission volt-amperes corresponds to:

$$
\begin{aligned}
S_s &= 3V_{s,rms}I_{s,rms} = 3 \times 144.33 \times 1000 \\
&= 433 \text{ MVA}
\end{aligned}
$$

Given that $I_S = I_R$, the power factor at the sending end is:

$$\cos(\angle V_s - \angle I_R) = \cos(4.93 - (-36.87)) = \cos(31.94) = 0.745$$

which is a lagging power factor due to the positive sign of $+31.94$ (see Section 6.6.2). Now, refer to Equation (9.38) to see that:

$$\text{Voltage regulation} = \frac{353.53 - 220\sqrt{2}}{220\sqrt{2}} \times 100 = 13.6\%$$

In addition, refer to Equation (9.39) to see that transmission line efficiency is:

$$\eta = \frac{P_R}{P_S} \times 100 = \frac{S_R \times (P.f)_R}{S_S \times (P.f)_S} \times 100 = \frac{381 \times 0.8}{433 \times 0.745} \times 100 = 94.48\%$$

b. In this part of the example, the power factor is leading. Thus:

$$I_R = 1000\sqrt{2}\angle \cos^{-1}0.8 = 1000\sqrt{2}\angle 36.87° \text{ A}$$

(continued)

EXAMPLE 9.22   **Continued**

Using Equation (9.36) for a short-line model:

$$V_S = V_R + Z \cdot I_R$$
$$= 127\sqrt{2}\angle 0° + (6 + j20)(1000\sqrt{2}\angle +36.87°)(10^{-3})$$
$$= 171.67\angle 9.29 \text{ kV}$$

The sending end line-to-line voltage magnitude is:

$$|V_{S(L-L)}| = \sqrt{3}|V_S| = 297.34 \text{ kV}$$

Because the current is assumed constant over the line, that is, $I_{S,\text{rms}} = I_{R,\text{rms}}$, the transmission power is:

$$S_s = 3V_{s,\text{rms}}I_{s,\text{rms}} = 3 \times 121.39 \times 1000$$
$$= 364.17 \text{ MVA}$$

The power factor at the sending end $= \cos(\angle V_S - \angle I_R) = \cos(9.29 - 36.87) = \cos(27.58) = 0.886$, which is a leading power factor due to the negative sign of -31.94.

$$\text{Voltage regulation} = \frac{297.34 - 220\sqrt{2}}{220\sqrt{2}} \times 100 = -4.43\%$$

Transmission line efficiency is:

$$\eta = \frac{P_R}{P_S} \times 100 = \frac{S_R \times (\text{P.f})_R}{S_S \times (\text{P.f})_S} \times 100 = \frac{381 \times 0.8}{364.17 \times 0.886} \times 100 = 94.46\%$$

EXAMPLE 9.23   **Transmission Line Impedance Matching**

A 200-kV, 60-Hz load is applied to a single-phase transmission line modeled by the short-line model of Figure 9.28. Find the load impedance that perfectly matches the line. Find the maximum power transferred to this load.

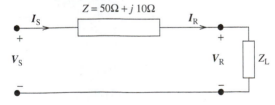

FIGURE 9.28 Short-line model of Example 9.23.

SOLUTION

Based on Section 6.6.6 (Chapter 6), to maximize the transferred power to this load:

$$Z_L = Z^* = 50 \ \Omega - j10 \ \Omega$$

In this case, the maximum transferred power (Equation (6.98)) is:

$$P_L = \frac{1}{8} \times \frac{V_L^2}{\text{Re}\{Z_L\}}$$
$$= \frac{1}{8} \times \frac{(200\sqrt{2} \times 10^3)^2}{50}$$
$$= 200 \text{ MW}$$

APPLICATION EXAMPLE 9.24	City Power Delivery

As shown in Figure 9.29, a power station delivers a source with the voltage, V_S, of 275 kV and current, I_S, of 1200 A at 60 Hz to a city that is 400 km away from the power station. A transformer is located 200 km away from the power station to maintain 220 kV as the minimum voltage delivered to the city, with the peak current, I_S, of 1000 A. The 200-km transmission line has a total series impedance of $40 + j\,50\ \Omega$ and shunt admittance of $j6 \times 10^{-6}$ S/km.

Determine:

1. The ratio of the number of secondary turns to primary turns (called the step-up coefficient) for the transformer.
2. The efficiency of the transformer needed in order to meet the delivery voltage requirement.

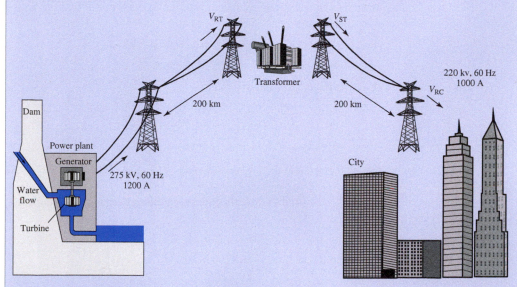

FIGURE 9.29 Electricity delivered to a city from a power station.

SOLUTION

First, find the voltage at the input and at the output of the transformer before finding the ratio of the secondary to primary turns. Two pieces of information are needed.

Here, the source voltage and current of the primary of the transformer are known and the line's equivalent circuit is known. For the secondary, the end voltage and current are known. We use this information to find the secondary voltage and current.

Thus, to solve this problem, first we find the equivalent circuit of the transmission line. The 200-km line is a medium-length line. Thus, A, B, C, and D constants can be determined using Equation (9.42) that corresponds to:

$$A = \left(1 + \frac{ZY}{2}\right), B = Z,$$

$$C = Y\left(1 + \frac{ZY}{4}\right), D = \left(1 + \frac{ZY}{2}\right)$$

Here, $Z = 40 + j50\ \Omega$ and $Y = j6 \times 10^{-6} \times 200 = j1.2 \times 10^{-3}$ S; thus, the ABCD constants correspond to:

$$A = 0.97 + j0.024, B = 40 + j50,$$

$$C = -0.0000144 + j0.001182, D = 0.97 + j0.024$$

(continued)

APPLICATION EXAMPLE 9.24 **Continued**

Using the current and voltage relationship of Equation (9.37), we express the equation of voltage and current received in terms of the voltage and current source.

$$I_S = CV_R + DI_R \rightarrow I_R = \frac{I_S - CV_R}{D}$$

Now, by replacing for I_R in Equation (9.36) and performing some mathematical manipulations:

$$V_S = AV_R + BI_R \rightarrow V_S = AV_R + B \times \frac{I_S - CV_R}{D} \rightarrow V_R = \frac{DV_S - BI_S}{DA - BC}$$

Next, we find V_R from this equation and replace it into the I_R equation to find the voltage and current delivered to the transformer:

$$V_{RT} = 218,750 - j53,400 = 225.17\angle -13.71° \text{ kV}$$
$$I_{RT} = 1168 - j296.25 = 1205\angle -14.23° \text{ A}$$

The average power delivered to the transformer is:

$$P_{ave} = V_{rms}I_{rms} \cos(\theta_v - \theta_i) = V_{RT}I_{RT} \cos(\theta_v - \theta_i)$$

Thus:

$$P_{ave} = 225.173 \times 10^3 \times 1205 \cdot \cos(-13.71° + 14.23°) \rightarrow P_{ave} = P_{in} = 0.272 \text{ MW}$$

The delivered voltage and current at the city corresponds to $V_{RC} = 220\angle 0°$ kV, $I_{RC} = 1000\angle 0°$ A.

Equations (9.36) and (9.37) are used to directly determine the required voltage and current output by the transformer:

$$V_{ST} = 253,400 + j55,280 = 259.36\angle 12.3° \text{ kV}$$
$$I_{ST} = 966.83 + j284 = 1007.69\angle 16.37° \text{ A}$$

Next, we determine the average power output from the transformer:

$$P_{ave} = V_{rms}I_{rms} \cos(\theta_v - \theta_i)$$
$$P_{ave} = 259.36 \times 10^3 \times 1007.69 \cos(12.3° - 16.37°) \rightarrow P_{ave} = P_{out} = 0.260 \text{ MW}$$

Therefore, the step-up voltage coefficient for the transformer is:

$$\eta = \frac{|V_{ST}|}{|V_{RT}|} \rightarrow \eta = \frac{259.36}{225.17} \rightarrow \eta = 1.152$$

And the efficiency of the transformer is:

$$\xi = \frac{P_{out}}{P_{in}} \times 100\% \rightarrow \xi = \frac{0.260}{0.272} \times 100\% \rightarrow \xi = 95.59\%$$

APPLICATION EXAMPLE 9.25 **City Power Delivery**

As shown in Figure 9.30, a town that is 400 km away from a power station is expected to receive the power source of 162 kV at 707 A with a 0.866 lagging factor. What is the minimum voltage and current required to be created by the power station to meet this need? Also, what is the transmission line efficiency, given the impedance and admittance per unit length of transmission line are:

$$Z = 0.00125 + j0.00005 \text{ }\Omega/\text{km}$$
$$Y = 5 \times 10^{-5} - j5 \times 10^{-5} \text{ S/km}$$

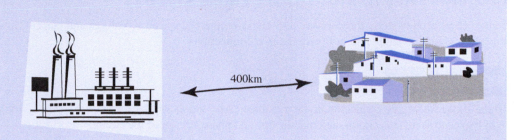

FIGURE 9.30 A long transmission line between a power plant and a city.

SOLUTION

The root mean square voltage and current delivered to the town are $V_{R,rms} = 162$ kV and $I_{R,rms} = 707$ A with lag of 0.866. The magnitude and phase angle for the voltage and current received by the town are:

$$V_R = 162 \times \sqrt{2} = 229\angle 0° \text{ kV}$$

$$I_R = 707 \times \sqrt{2}\angle -\cos^{-1}0.866 = 1000\angle -30° \text{ A}$$

The impedance and admittance per unit length of the transmission line are given. The coefficients Z_C and γ defined in Equation (9.42) are:

$$Z_C = \sqrt{Z/Y} = 3.853 + j1.687$$

$$\gamma = \sqrt{ZY} = 2.77 \times 10^{-4} - j1.083 \times 10^{-4} \ \Omega/\text{km}$$

Using these coefficients, and given that the length of transmission line, l, is 400 km, Equation (9.42) can be used to find the transmission line parameters:

$$A = \cosh(\gamma l) = 1.005 - j0.005; B = Z_C \sinh(\gamma l) = 0.5 + j0.0019$$

$$C = \frac{1}{Z_C}\sinh(\gamma l) = 0.02 - j0.02; D = \cosh(\gamma l) = 1.005 - j0.005$$

Now, given V_R and I_R, the required minimum voltage and current sources are:

$$V_S = AV_R + BI_R \rightarrow V_S = 673.6 - j234.9 = 713.38\angle -19.22° \text{ V}$$

$$I_S = CV_R + DI_R \rightarrow I_S = 871.06 - j564.15 = 1037.9\angle -32.94°$$

Using Equation (9.39), the transmission line efficiency can be determined using the equation of average power for both power sources, which is:

$$\eta = \frac{162 \times 707 \times 0.866}{\frac{1}{2}713.38 \times 1037.9 \times \cos(-19.22° + 32.94°)} \times 100\% \rightarrow$$

$$\eta = 27.58\%$$

The 1/2 in the denominator is added because the voltage and current values in the denominator are peak values while the voltage and current in the numerator are rms values. The result shows that long-line-length transmission is not feasible because the efficiency of transmission is too low.

9.4 USING PSPICE TO STUDY THREE-PHASE SYSTEMS

PSpice can be used to study three-phase star- and Δ-connected systems. Voltage and current flow plots are provided to study and compare the two connections.

EXAMPLE 9.26   **PSpice Analysis**

Three 1-kΩ resistors are connected in star to a three-phase supply of phase voltage 5V cos(120πt). Use PSpice to sketch the circuit, and plot the line and phase voltage of the circuit.

SOLUTION

The amplitude of the phase voltage source is 5 V, with a frequency of 60 Hz. Follow Example 6.27 to set up a VSIN voltage source with phase angle. The PSpice schematic for a star-connected load is shown in Figure 9.31.

Set the simulation time to at least 20 ms and run the simulation (see Figure 2.75). The line and phase voltage plot is shown in Figure 9.32. The peak voltage of the phase voltage is about 8.6 V while the peak voltage of the line voltage is 5 V. The ratio between them is close to $\sqrt{3}$, which is consistent with the star connection.

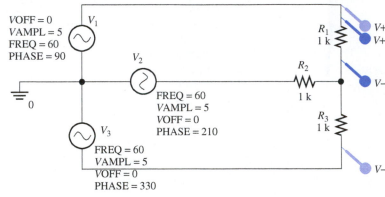

FIGURE 9.31 PSpice schematic solution for Example 9.26.

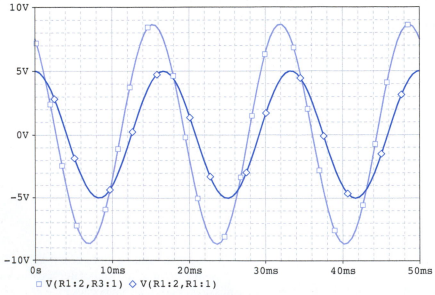

FIGURE 9.32 Line and phase voltage plot for Example 9.26.

EXAMPLE 9.27   **PSpice Analysis**

Three 1 kΩ resistors are connected in Δ connection to a three-phase supply of phase voltage 10 V. Use PSpice to sketch the circuit, and plot the line and phase current flow over time. Assume the supply has a frequency of 60 Hz.

SOLUTION

The amplitude of the phase voltage source is 10 V, with a frequency of 60. Follow Example 6.25 to set up a VSIN voltage source with phase angle. The PSpice schematic for a Δ-connected load is shown in Figure 9.33.

Set the simulation time to at least 20 ms and run the simulation (see Figure 2.75). Figure 9.34 shows the current flow through the voltage source and the resistor.

In the figure, notice that the phase current flow through the resistor is about 17 mA, while the line current flow through the voltage source is 30 mA. The ratio between line and phase current, $I_L/I_P = 1.7321$ is close to $\sqrt{3}$, which satisfies the condition for a Δ connection.

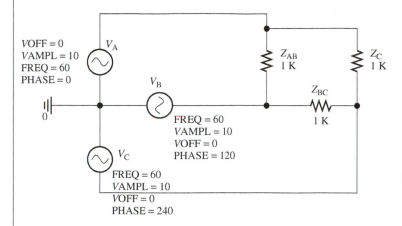

FIGURE 9.33 PSpice schematic solution for Example 9.26.

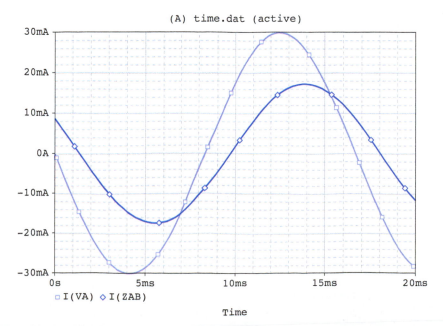

FIGURE 9.34 Line and phase current for Example 9.27.

EXAMPLE 9.28   **PSpice Analysis**

A star-Δ connection load is connected to a three-phase supply, with a phase voltage source of 10 V. Consider the $Z_{AB} = Z_{CA} = Z_{BC}$ and $Z_A = Z_B = Z_C$, with $Z_{AB} = 900 \, \Omega$. Use PSpice to compare the current and voltage flow through the resistors Z_A and Z_{AB}.

SOLUTION

First, given the resistance of $Z_{AB} = Z_{CA} = Z_{BC} = 900 \, \Omega$. The resistances of Z_A, Z_B, and Z_C correspond to:

$$Z_A = \frac{Z_{AB}}{3}$$

$$Z_A = Z_B = Z_C = 300 \, \Omega$$

Follow the Example 6.25 to set up a VSIN voltage source with phase angle. The PSpice schematic solution is shown in Figure 9.35.

Set the simulation time to at least 20 ms and run the simulation (see Figure 2.75). The current and voltage flow through resistors Z_A and Z_{AB} are shown in Figures 9.36 and 9.37, respectively.

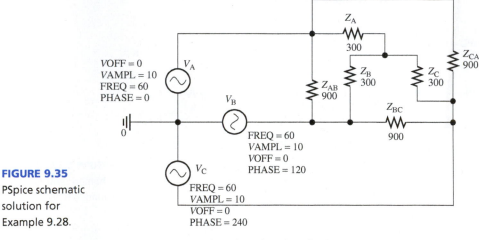

FIGURE 9.35
PSpice schematic solution for Example 9.28.

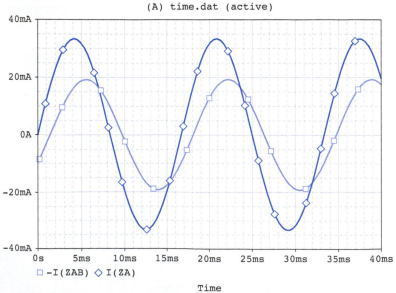

FIGURE 9.36   Current flow through resistors Z_A and Z_{AB}.

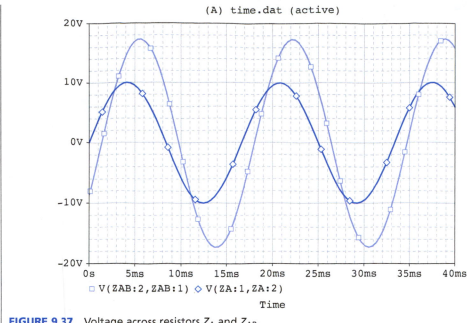

FIGURE 9.37   Voltage across resistors Z_A and Z_{AB}.

9.5 WHAT DID YOU LEARN?

- A balanced, three-phase voltage source has three sinusoidal voltages of same magnitude and frequency. Each voltage is 120° out of phase with the others.
- A generator is always Y-connected to avoid loop.
- The voltage, current, and power relationships for Y-connected and Δ-connected loads are as follows:

	Y-Connected Load	**Δ-Connected Load**
Line voltage	$V_L = \sqrt{3} \cdot |V_P| \angle 30°$	$V_L = V_P$
Line current	$I_L = I_P$	$V_L = V_P$
Power dissipation	$P_T = \sqrt{3}\, V_{L,rms}\, I_{L,rms} \cos\varphi$ W	

- Star-connected (Y-connected) loads and delta-connected loads are interchangeable (Figure 9.9).
- The advantages of three-phase systems include (1) a smaller amount of conductor is used; (2) two voltages (phase voltage and line voltage) are available; and (3) three-phase motors are more robust than the single-phase ones.
- The resistance, inductance, and capacitance of transmission lines can be determined for any given configuration [see Equation (9.19) and Tables 9.2 and 9.3].

Equivalent circuits for short length, medium length, and long length transmission line situations are discussed in Figures 9.26 and 9.27, and Equation (9.42).

Further Reading

Hadi Saadat. 2002. *Power system analysis*. McGraw Hill, NY, USA.

John J. Grainger, and William D. Stevenson. 1994. *Power system analysis*. McGraw Hill, NY, USA.

John Bird. 2003. *Electrical circuit theory and technology*. 3rd Edition. Oxford, Oxford, UK.

John O'Malley. 1992. *Basic circuit analysis*. 2nd Edition. McGraw Hill, NY, USA.

Problems

B refers to Basic, A refers to Average, H refers to Hard, and * refers to problems with answers.

SECTION 9.1 INTRODUCTION

9.1 (A) Explain two difficulties in collecting energy from lightening as a source of electricity.

SECTION 9.2 THREE-PHASE SYSTEMS

9.2 (B) Three loads, each of resistance 50 Ω are connected (a) in star and (b) in Δ to a 400-V, three-phase supply. Determine (i) the phase voltage, (ii) the phase current, and (iii) the line current.

9.3 (A)* Three identical capacitors are connected (a) in star and (b) in Δ to a 400-V, 50-Hz, three-phase supply. If the line current is 12 A, determine the capacitance of each of the capacitors for both (a) star and (b) Δ connections.

9.4 (A) Three coils each having a resistance of 6 Ω and an inductance of L H are connected (a) in star and (b) in Δ to a 415-V, 50-Hz, three-phase supply. If the line current is 30 A, find for each connection the value of L.

9.5 (B)* Determine the total power dissipated by three 20-Ω resistors when connected (a) in star and (b) in Δ to a 440-V, three-phase supply.

9.6 (A) Three inductive loads, each of resistance 4 Ω and reactance 9 Ω are connected in Δ. When connected to a three-phase supply, the loads consume 1.2 kW. Calculate (a) the phase current, (b) the line current, (c) the supply voltage, and (d) the power factor of the load.

9.7 (A) A 440-V, three-phase AC motor has an output power of 11.25 kW, operates at a power factor of 0.8 lagging, and with an efficiency of 84%. If the motor is Δ connected, determine (a) the input power, (b) the line current, and (c) the phase current.

9.8 (H) Each phase of a Δ-connected load comprises a resistance of 40 Ω and a 40-μF capacitor in series. When connected to a 415-V, 50-Hz, three-phase supply determine (a) the phase current, (b) the line current, (c) the total power dissipated, and (d) the kVA rating of the load.

9.9 (H)* Three 24 μF capacitors are connected in star across a 400-V, 50-Hz, three-phase supply. What value of capacitance must be connected in Δ in order to handle the same line current?

9.10 (H) In a balanced Y connection, why is $I_N = 0$?

9.11 (H) In a three-phase 50-Hz supply, if the cable for phase A is 300 m (ΔL) longer than the other two (Phases B and C), what will the phase mismatch be between A and B (assuming the electromagnetic wave speed in the cable is the same as the speed of light)?

9.12 (B)* Three 95.5-mH inductors in star connection are connected to a three-phase, 415-V, 50-Hz supply. Calculate the input power.

9.13 (B) A three-phase motor is in star connection, its efficiency is 80%, and its output power is 16 kW. If it is connected to a three-phase, 415-V, 50-Hz supply, calculate the current in each phase.

9.14 (A)* A star-connected load includes a 20-Ω resistor and a 63.7-mH inductor series in phase A; a 31.8-mH inductance in phase B; and, a 10-Ω resistor and a 12.7-mH inductor series in phase C. If the load is connected to a three-phase, 415-V, 50-Hz supply, calculate the impedance and the current in each phase.

9.15 (A) A three-phase motor in Δ connection with 3-Ω resistance and 10.6-mH inductance in each phase is connected to a three-phase, 415-V, 60-Hz supply. Calculate the efficiency of the motor.

9.16 (A) A coil (inductance) in motors usually has some resistance that is not desired. Explain why the resistance is not desired. What is an ideal motor coil? (*Hint:* Explain it in terms of input power and output power of a motor.)

9.17 (A) In a star connection, when the frequency of a three-phase supply is increased, how will the input power change if the load is (a) balanced pure resistance (R), (b) balanced capacitance (C), and (c) balanced inductance (L)?

9.18 (H)* A balanced, Δ-connected load includes an unknown value resistor and an 80-μF capacitor in each phase. When the load is connected to a three-phase, 415-V, 50-Hz supply, its input power is 3.46 kW. Calculate the value of the resistor.

9.19 (H) Assume a three-phase supply with 415 V voltage and 50 Hz frequency with a balanced, star-connected load with 4 Ω resistance and unknown inductance in each phase. When the input power is 34.56 kW, calculate the capacitor's capacitance.

9.20 (B)* Three 42-mH inductors are connected in star. In the equivalent Δ connection, what is the inductance of the three inductors?

9.21 (A) Assume a Δ connection with phase impedance of $Z_{AB} = (10 + j20)$ Ω, $Z_{BC} = (20 + j10)$ Ω, and $Z_{CA} = (30 + j30)$ Ω. What are the phase impedances in the equivalent star connection?

9.22 (H) Assume a star connection with phase impedance of $Z_A = (3 + j4)$ Ω, $Z_B = (10 - j10)$ Ω, and $Z_C = (20 + j15)$ Ω is connected to a 415-V, 50-Hz, three-phase supply. (1) Calculate the current in each phase and the total current in the neutral conductor. (2) Calculate the power factor in each phase and the power dissipated in each phase.

9.23 (B) Compare the differences of star and Δ connections in terms of the relationship between line voltage (current) and phase voltage (current).

9.24 (B) What is the primary advantage of a three-phase supply system compared to a single-phase supply system?

9.25 (H)* Power companies have to maintain the rms voltage level at each customer's location. Assume the acceptable deviation is ±5.8%. Thus, a nominal rms voltage of 120 V will range from 113 to 127 V. Suppose the line-to-line voltage at customer A is 13.8 kV, the load at customer A consumes complex power $(1.2 + j1.2) \times 10^6$ W. Customer A is connected to a power plant via a three-phase line and each phase line has the impedance of $1.8 + j14.4$ Ω.

(a) What line-to-line voltage at the plant is needed to generate a line-to-line voltage of 13.8 kV at the customer?

(b) Could the voltage found in (a) create problems for other customers requiring 13.8 kV (such as those who don't suffer the transmission line voltage loss)? What measures can be taken to solve this problem? (*Hint:* Add a capacitive load at customer A.)

9.26 (H) Application: three-phase power measurement
In a balanced, three-phase system as shown in Figure P9.26, the load power can be measured by using two wattmeters. The power measured by W_1 is P_1, $P_1 = Re(V_{ab}I_a{}^*)$ and the power measured by W_2 is P_2, $P_2 = Re(V_{cb}I_c{}^*)$. Show that the total real power $P_T = P_1 + P_2$.

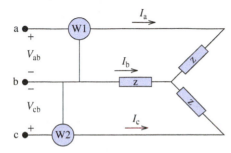

FIGURE P9.26   Two-wattmeters are used to measure three-phase power.

9.27 (H)* Find $P_2 - P_1$ in the Problem 9.26, and explain how to tell if the load is resistive, inductive, and capacitive from the result.

SECTION 9.3 TRANSMISSION LINES

9.28 (A) The conductor of a single-phase 60-Hz line is a solid round aluminum wire having a diameter of 0.412 cm. The conductor spacing is 3 m. Determine the inductance of the line in millihenry per mile.

9.29 (A) A three-phase 60-Hz line has flat horizontal spacing. The conductors have a GMR of 0.0133 m with 10 m between adjacent conductors. Determine the inductive reactance per phase in ohms per kilometer.

9.30 (H)* A three-phase 60 Hz line has flat horizontal spacing. The conductors have an outside diameter of 3.28 cm with 12 m between conductors. Determine the capacitive reactance to neutral in ohm-meters and the capacitive reactance of the line in ohms if its length is 125 miles.

9.31 (A) Calculate the inductance reactance in ohms per kilometer and the capacitive reactance in ohm kilometers of a bundled, 60-Hz, three-phase line having three conductors per bundle. The conductors have a diameter of 2.96 cm with 45 cm between conductors of the bundle. The spacing between bundle centers is 9, 9, and 18 m.

9.32 (B) Given a 60-Hz three-phase transmission line with symmetrical spacing of 20 cm and a 5 cm diameter of each solid conductor, what is the capacitance reactance?

9.33 (B) A 50-mile long, 60-Hz three-phase, star-connected transmission line has a total impedance of $30 + j150$ Ω. The delivered voltage is 200 kV and the delivered current is $100\angle-15°$ A. What is the sending voltage?

9.34 (A)* The sending voltage magnitude of a 43 miles long, 60-Hz, three-phase, star-connected transmission line is measured as 367.42 kV, and the phase angle is 5°. The total impedance is $30 + j120$ Ω, and the transmission power is 450 MVA with 0.9 power factor lagging at the receiving node. What is the delivered voltage?

9.35 (H) A 60-Hz, three-phase transmission line is 175 miles long. It has a total series impedance of $35 + j140$ Ω and a shunt admittance of $930 \times 10^{-6}\angle 90°$ S. It delivers 40 MW at 220 kV, with 90% power factor lagging. Find the voltage at the sending-end by using (a) short-line approximation; (b) the nominal π approximation; and (c) long-line equations.

9.36 (B) Explain the reason for not using a solid cable with the same cross-sectional area to take the place of the typically used spiraled cable with multiple strands. Why has aluminum cable taken the place of copper cable in overhead power transmission?

9.37 (B) Why is the diameter of a transmission line for high-voltage power (e.g., 220 kV) larger than the diameter of a transmission line for low-voltage power (e.g., 120 V)?

9.38 (B)* Calculate the resistance of a cable with a 300 m length, 1 mm cross-sectional radius, and 0.0001 Ω·m conductor resistivity.

9.39 (B) A copper cable's resistance is 0.5 Ω at 15°C. Calculate its resistance at 30°C.

9.40 (B) Explain why bundled cable is better than single cable in terms of inductive reactance even though material costs are the same (i.e., the same amount of material is needed to build every meter of these two kinds of cables).

9.41 (A) Assume a set of bundled conductors is configured as shown in Figure P9.41. Each subconductor's radius is 10 cm, and the distance between neighboring conductors is 2 m (the length of each of the hexagon's sides). Calculate the equivalent radius of the bundled conductors.

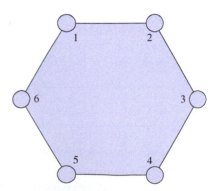

FIGURE P9.41   Symmetrical bundled conductors' configuration.

9.42 (A)* Calculate the equivalent radius of the bundled conductors shown in Figure P9.42. Each subconductor's radius is 20 cm.

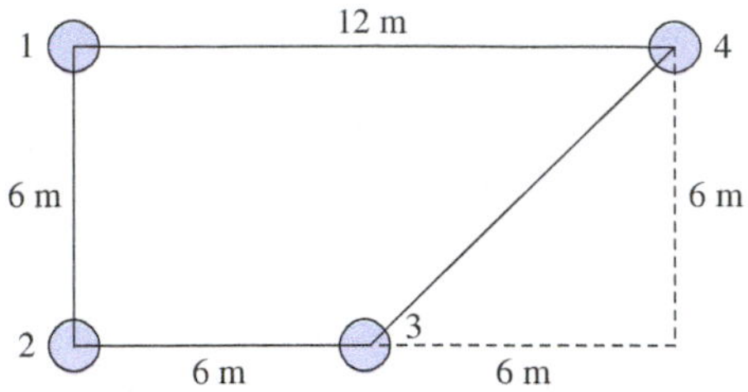

FIGURE P9.42 Asymmetrical bundled conductors' configuration.

9.43 (A) Calculate the inductive reactance per meter of a single-phase, 50-Hz, two-wire transmission line. Assume each conductor's radius is 5 cm and the distance between the two conductors is 2 m.

9.44 (H) Assume a 40-km single-phase transmission line with two 2-cm radius solid conductors ($\rho = 16.8$ nΩm) and that the distance between the two conductors is 3 m. The 60-Hz, 220-kV source internal impedance is $0.3 + j0.4\ \Omega$. Calculate the load matching with the source and the transmission line.

9.45 (A)* A single-phase, 60-Hz transmission line includes two conductors, one conductor's radius is 5 cm, and the other's radius is 8 cm. The distance between the two conductors is 5 m. Calculate the inductive resistance of the transmission line for each mile.

9.46 (H) A single-phase, 60-Hz power will be transmitted through the transmission line given in Figure P9.46. Each subconductor's radius is 10 cm. There are two configurations, the first one is setting subconductors A and B as a bundle, and C and D as another bundle; the second is setting subconductors A and C as a bundle, and B and D as another bundle. Which should be selected to achieve smaller inductance?

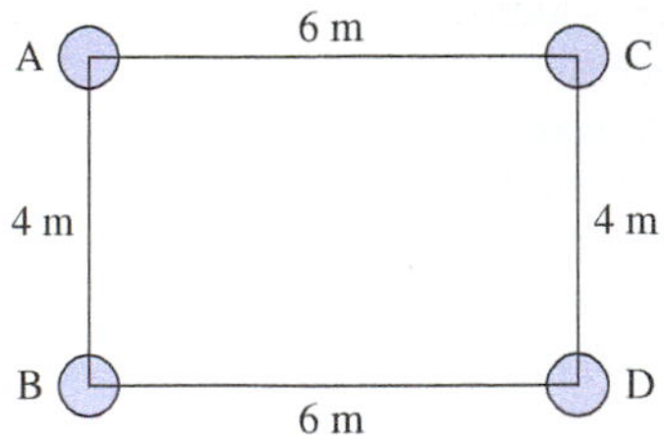

FIGURE P9.46 Single-phase transmission line configuration.

9.47 (A) Calculate the capacitance per meter for the configuration in Figure P9.47. Each subconductor's radius is 5 cm; A and B are in a bundle, and C and D are in the other bundle.

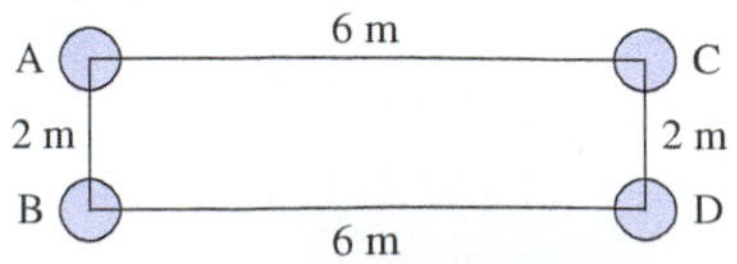

FIGURE P9.47 Single-phase, two-bundle transmission line.

9.48 (H)* A single-phase, 60-Hz transmission line is shown in Figure P9.48. Each subconductor's radius is 10 cm, and the distance between neighboring subconductors is 2 m ($D_{12} = D_{23} = D_{34} = D_{41} = D_{56} = D_{67} = D_{78} = D_{85} = 2$ m). A subconductor in bundle B (number 8) is missing for some reason, that is, only subconductors numbered 5, 6, and 7 are available in bundle B. Calculate the unit capacitance difference between the scenarios when the number 8 subconductor is available and when it is missing.

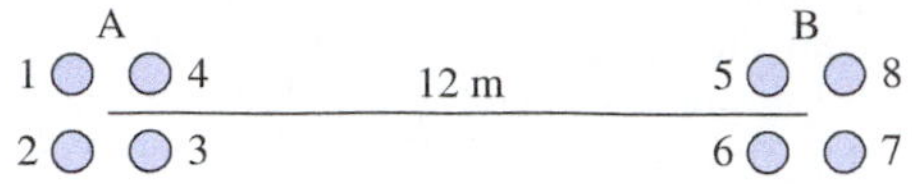

FIGURE P9.48 Single-phase, two-bundle asymmetrical transmission line.

9.49 (H) A 60-Hz, three-phase, star-connected transmission line is 45 miles long. Its resistance per-phase, per-mile is 0.2 Ω and its inductance per-phase, per-mile is 1.179 mH. Calculate the transmission line efficiency when it is supplying 346.4 kV, 600 MVA at 0.9 power factor lagging.

9.50 (H) A 60-Hz, single-phase transmission line has a length of 80 miles. Its shunt conductance (G) is 3.4×10^{-11}S/mile, resistance (R) is 0.1 Ω/mile, capacitance (C) is 16 pF/mile, and inductance (L) is 0.56 mH/mile. If 220 kV, 220 MVA with power factor 0.8 lagging is determined to be at the receiving end, calculate the voltage and current at the transmitting end.

9.51 (H) A 60-Hz, three-phase, star-connected transmission line's length is 200 miles. Its shunt conductance (G) is 3.8×10^{-13} S/mile, resistance (R) is 0.01 Ω/mile, capacitance (C) is 10 pF/mile, and inductance (L) is 0.16 mH/mile. If 381 kV, 660 MVA, with power factor 0.8 lagging is at the receiving end, calculate the voltage regulation and transmission line efficiency.

9.52 (H)* Pelamis P-750 in Figure P9.52 is a machine that converts wave energy into electric power. Three machines have been installed in the sea of Portugal and generate a total power of 2.25 MW. The installation of Pelamis P-750 has been proposed to supply the electricity at 110 V/14 A/60 Hz to a seaside town that has a household population of about 2200. In the proposal, it states that the duration of the ocean wave in the area is only 1/3 of that in Portugal. In addition, the power station will be built in an area that is 2 km away from the town. The power generated by wave converters is proportional to the duration of the ocean wave. Assume the transmission line's resistance is 0.25 Ω/km and its inductance is ignorable. What number of Pelamis P-750 units is required to supply the electricity to the town?

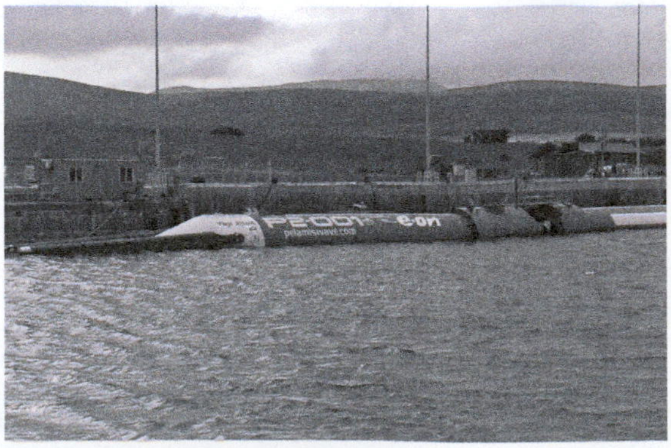

FIGURE P9.52 Pelamis P-750 wave energy converter. (Used with permission from © Doug Houghton/Alamy.)

SECTION 9.4 USING PSPICE TO STUDY THREE-PHASE SYSTEMS

9.53 (B) Three 3-kΩ resistors are connected in star to a three-phase supply of phase voltage 120 V. Use PSpice to sketch the circuit and plot the line and phase voltage of the circuit.

9.54 (B) Three 700-Ω resistors are connected in Δ connection to a three-phase supply of phase voltage 240 V. Use PSpice to sketch the circuit and to plot the line and phase current of the circuit.

9.55 (A) A star-Δ connection load is connected to a three-phase supply, with a phase voltage source of 120 V. The impedance of the load Z_{AB}, Z_{CA}, and Z_{BC} are defined as below:

$$Z_{AB} = 900 \ \Omega$$
$$Z_{BC} = 2 \ k\Omega$$
$$Z_{CA} = 1.3 \ k\Omega$$

(a) Calculate the impedance of Z_A, Z_B, and Z_C.

(b) Use PSpice to sketch the circuit and plot the current flow across the resistors Z_A, Z_B, and Z_C.

CHAPTER 10

Fundamentals of Logic Circuits

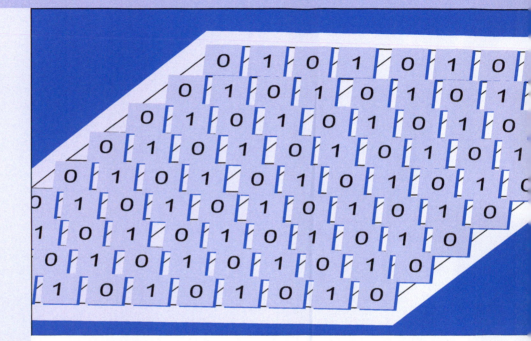

10.1 INTRODUCTION

So far, this book has examined circuits that process analog signals. As you have learned, a signal represents the variation of amplitude with time. For example, the amplitude of a sinusoid varies periodically with time. An analog signal is continuous in both the time domain and the amplitude domain. An example of such a signal is shown in Figure 10.1. In this example, no discontinuity is observed in amplitude and time. A discrete signal or a discrete-time signal (Figure 10.2) is a time series that is sampled from an analog signal. Note that a discrete-time signal may achieve any value for its amplitude. Thus, while it is discrete in time, it is not discrete in amplitude. On the other hand, a digital signal (Figure 10.3) is a discrete-time signal that attains only a discrete set of values. Thus, it is discrete in both the amplitude domain and the time domain.

The process of converting an analog signal to a digital signal is called *digitization*. Digitization includes two steps: sampling and quantization. In the *sampling* stage, the continuous signal is sampled at regular intervals. This creates a time series (discrete signals) as shown in Figure 10.2. *Quantization* refers to the adjustment of sampled levels to some predefined signal levels (Figure 10.3). As an example, consider three predefined signal levels of +5 V,

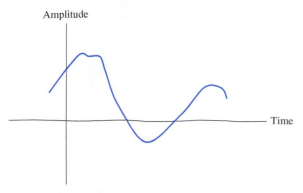

FIGURE 10.1   Analog signal.

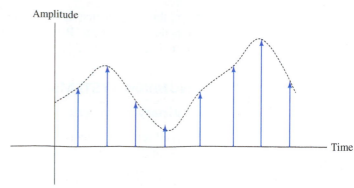

FIGURE 10.2   Discrete time signal.

0 V, and −5 V. If the sampled voltage level is 3.2 volts, it will be quantized to 5 V because that is the closest predefined level. Likewise, if the sampled signal is closest to the 0 V level (e.g., 1.2 V), it will be quantized as 0 V. Although, in general, a digital signal may have multiple voltage levels, this chapter is interested in two-level digital signals, namely '0' and '1,' as shown in Figure 10.4. These two levels refer to different voltage levels. For example, 1 may refer to +5 V and 0 may refer to 0 V.

Note that a multi-level digital signal may also be formed by a two-level one. For instance, each level of an eight-level quantized signal can be represented by three (0 or 1) digits. As an example, if the eight levels start at 1 V and go to 8 V, 1 V can be represented by 000, 2 V by 001, 3 V by 010, and 8 V can be represented by 111. The process of converting these eight predefined voltages to 0-1 (digital) is called analog-to-digital conversion. In general, the devices that convert analog signals to digital are called ADC (analog-to-digital convertor). For the remainder of this chapter, digital signals will be referred to as two-level (digital) signals.

In order to process analog signals using digital signal processors (DSP), analog signals must be converted to digital. In order to apply the signal to a DSP, the original analog (continuously varying) signal is first sampled, then quantized, and then converted to digital (0 or 1).

Today, almost all systems are digital. Examples include digital cell phones, digital TV, digital cameras, and so on. Digital representation of signals offers many advantages over analog, including: fewer components, stability, and a wide range of applications. In addition, signal processing of analog signals is not as straightforward as processing of digital signals. In digital communication systems, analog signals are first converted to digital. For example, in today's cellular systems, after reception at the radio's RF front-end, all signals are converted to digital to allow further processing. This eases the reconstruction process of transmitted signals that are altered by interference and noise effects. Thus, it improves performance and supports transmission of data (e.g., an image).

In digital logic circuits, the binary numbers 1 and 0 are used to represent the voltage or current. The logic 1 is also called high, true, or ON; logic 0 is also called low, false, or OFF. However, in daily life, decimal numbers are most commonly used to represent quantities.

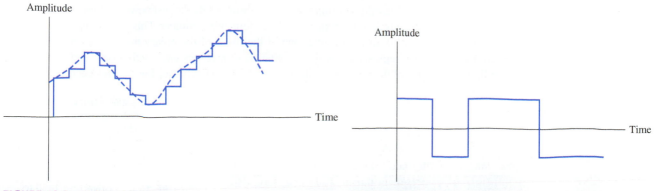

FIGURE 10.3   Multi-level digital signal.

FIGURE 10.4   Two-level digital signal.

Therefore, this chapter first discusses how to convert decimal numbers into binary. The chapter introduces number systems, Boolean algebra, basic logic gates, and sequential logic circuits that are used in digital circuits. PSpice examples are provided to help construct digital systems.

10.2 NUMBER SYSTEMS

10.2.1 Binary Numbers

Currently, all computer data is stored in binary format (using only the numbers 0 or 1). Examples of this type of data include the following:

- Data on a floppy disk, hard disk, and so on (magnetic storage)
- Data in flash memory (semiconductor storage)
- Data written to a CD, DVD, and so on (optical disc storage).

Therefore, most numerical data is represented in the binary format (base 2). However, in daily life, decimal numbers (base 10) are more commonly used to represent numerical data. Therefore, it is important to define the relationship between decimal and binary numbers.

First, let's see how decimal numbers can be converted to binary. The decimal number 1450 can be represented as:

$$1450 = 1 \times 10^3 + 4 \times 10^2 + 5 \times 10^1 + 0 \times 10^0$$

In this representation, the base is 10 (ten). The number 1450 can also be represented in a similar way in base 2:

$$1450 = 1 \times 2^{10} + 0 \times 2^9 + 1 \times 2^8 + 1 \times 2^7 + 0 \times 2^6 + 1 \times 2^5 + 0 \times 2^4 + 1 \times 2^3 + 0 \times 2^2$$
$$+ 1 \times 2^1 + 0 \times 2^0$$

The coefficients of each term form the binary format of the number 1450 (see Figure 10.5).

Therefore, the binary form of the number 1450 is 10110101010. Two methods are available for converting a decimal number to binary form that are detailed in the next subsections.

10.2.1.1 CONVERTING A DECIMAL INTEGER TO BINARY FORM

Method 1: Figure 10.6 summarizes Method 1 using 1450 as the number to be converted. Therefore, 1450 can be represented as:

$$1450 = 1 \times 2^{10} + 0 \times 2^9 + 1 \times 2^8 + 1 \times 2^7 + 0 \times 2^6 + 1 \times 2^5 + 0 \times 2^4 + 1 \times 2^3$$
$$+ 0 \times 2^2 + 1 \times 2^1 + 0 \times 2^0$$

Thus, the binary form of 1450 is 10110101010.

In summary, to use Method 1 to convert a decimal number to binary, first select the largest power of 2 (e.g., 1024, 512, 256, …) for which the result is smaller than the original decimal number. This prevents obtaining negative values during conversion. Subtract the largest power of 2 from the original decimal number and then repeat this process with the remaining number. This process is repeated until the subtraction results in 0. In each round, one of the digits of the binary number is generated. The decimal number is then represented as the summation of numbers each with a power of 2. The corresponding coefficients of the numbers to the power of 2 form the binary form of this decimal number.

10th	9th	8th	7th	6th	5th	4th	3th	2th	1th	0th
1	0	1	1	0	1	0	1	0	1	0

FIGURE 10.5 Coefficients of binary form of 1450 (decimal).

$$1450 - 1024 = 426 \quad (1024 = 2^{10})$$
$$426 - 256 = 170 \quad (\ 256 = 2^8)$$
$$170 - 128 = 42 \quad (\ 128 = 2^7)$$
$$42 - 32 = 10 \quad (\ 32 = 2^5)$$
$$10 - 8 = 2 \quad (\ 8 = 2^3)$$
$$2 - 2 = 0 \quad (\ 2 = 2^1)$$

FIGURE 10.6 Converting 1450 to binary form (Method 1).

	Quotient	Remainder
1450/2 =	725	0
725/2 =	362	1
362/2 =	181	0
181/2 =	90	1
90/2 =	45	0
45/2 =	22	1
22/2 =	11	0
11/2 =	5	1
5/2 =	2	1
2/2 =	1	0
1/2 =	0	1

FIGURE 10.7 Converting 1450 to binary form (Method 2).

Method 2:   Method 2 is summarized in Figure 10.7, again for the decimal number of 1450.

In Method 2, the quotient (the first quotient is the original number) is repeatedly divided by 2 until quotient is 0. The reverse sequence of the remainders forms the binary number (10110101010) of the decimal number.

10.2.1.2 CONVERTING A DECIMAL FRACTION TO BINARY FORM

Method 1:   To explain this method we will use 0.2111 as a sample number (see Figure 10.8).

As shown in Figure 10.8, the process outlined above can continue to the required precision to convert decimal numbers. The process is stopped when it reaches the precision requirement.

$$0.2111 \approx 0 \times 2^{-1} + 0 \times 2^{-2} + 1 \times 2^{-3} + 1 \times 2^{-4} + 0 \times 2^{-5} + 1 \times 2^{-6} + 1 \times 2^{-7}$$

Therefore, the binary form for 0.2111 is 0.0011011.

In Method 1, when converting a decimal fraction number to binary, choose the largest negative number of power of 2 (e.g., 0.5, 0.25, 0.125 . . .), that is smaller than the value you are subtracting from. This prevents obtaining negative values during conversion. The process is repeated until 0 is obtained, or the value of the binary form of the fraction number approximates the original number with the required precision. Finally, the decimal fraction number is represented as the summation of numbers of negative power of 2. The corresponding coefficients of power of 2 form the binary number consistent with the desired decimal number.

Method 2:   Method 2 is summarized in Figure 10.9. As shown in this figure, the binary form of 0.2111 is 0.0011011.

In Method 2, the fraction part of the number is repeatedly multiplied by 2 (the first fraction part to be multiplied is the original number) until the fraction part becomes 0 or until the desired precision is reached. As shown, the equivalent binary number in our example corresponds to 0.0011011.

In order to convert a decimal number consisting of both integer and fraction parts, each part is separately converted to binary.

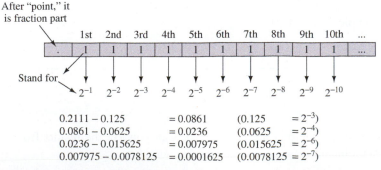

FIGURE 10.8 Converting 0.2111 to binary (Method 1).

	Integer part	Fraction part
$0.2111 \times 2 =$	0	0.4222
$0.4222 \times 2 =$	0	0.8444
$0.8444 \times 2 =$	1	0.6888
$0.6888 \times 2 =$	1	0.3776
$0.3776 \times 2 =$	0	0.7552
$0.7552 \times 2 =$	1	0.5104
$0.5104 \times 2 =$	1	0.0208

FIGURE 10.9 Converting 0.2111 to binary form (Method 2).

EXAMPLE 10.1   **Decimal to Binary Conversion**

Convert the decimal integer number 626 into the binary form.

SOLUTION

The approach has been summarized in Figure 10.10.

According to the results shown in Figure 10.10, 626 (decimal) = 1001110010 (Binary).

		Quotient	Remainder
626/2	=	313	0
313/2	=	156	1
156/2	=	78	0
78/2	=	39	0
39/2	=	19	1
19/2	=	9	1
9/2	=	4	1
4/2	=	2	0
2/2	=	1	0
1/2	=	0	1

FIGURE 10.10   Converting 626 to binary form.

EXAMPLE 10.2   **Decimal to Binary Conversion**

Convert the decimal fraction number 0.47 into binary (stop when the fraction part has 8 bits).

SOLUTION

The process has been summarized in Figure 10.11.

Accordingly: 0.47(decimal) = 0.01111000 (binary).

		Integer part	Fraction part
0.47×2	=	0	0.94
0.94×2	=	1	0.88
0.88×2	=	1	0.76
0.76×2	=	1	0.52
0.52×2	=	1	0.04
0.04×2	=	0	0.08
0.08×2	=	0	0.16
0.16×2	=	0	0.32

FIGURE 10.11   Converting 0.47 to binary form.

EXAMPLE 10.3   **Decimal to Binary Conversion**

Convert the decimal number 626.47 into binary form (stop when the fraction part has 8 bits).

SOLUTION

Based on the results of Example 10.1 and 10.2:

626.47(decimal) = 1001110010.01111000 (binary).

EXERCISE 10.1

Determine the binary form of the decimal number 627.47. Can you use the results from Example 10.3 to solve this problem?

10.2.1.3 CONVERTING A BINARY NUMBER TO A DECIMAL NUMBER

Here, we use the number 110110.011011(binary) as an example. The solution is shown in Figure 10.12.

As shown in Figure 10.12:

$$110110.011011 = 1 \times 2^5 + 1 \times 2^4 + 0 \times 2^3 + 1 \times 2^2 + 1 \times 2^1 + 0 \times 2^0 + 0 \times 2^{-1} + 1 \times 2^{-2}$$
$$+ 1 \times 2^{-3} + 0 \times 2^{-4} + 1 \times 2^{-5} + 1 \times 2^{-6}$$
$$= 54.421875 \text{ (decimal)}$$

To convert a binary number to a decimal, each bit (0 or 1) is multiplied by the corresponding power of 2. Then, the results are added together. The summation will be the desired decimal number.

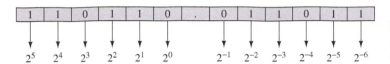

FIGURE 10.12 Converting a binary number to decimal form.

APPLICATION EXAMPLE 10.4 | Computing Device

An engineer intends to design his own computing device for calculating the daily income from his store. He wants the device to have a memory to store the input data. The maximal incoming data is 1000.00, and he wants the device to have 2-degrees of precision. If one memory cell only can store one binary bit, how many memory cells are required for his computing device? (No memory cell is required to store the point '.' in the fraction number. In addition, to minimize the error and use less memory cells, make the absolute value of the error between the decimal value and the corresponding binary form less than 0.005.)

SOLUTION

For the integer part (using Method 1), converting 1000 to binary as shown in Figure 10.13 results in: 1000 (decimal) = 1111101000 (binary). Thus, 10 memory cells are required for the integer part.

For the fraction part (using Method 2): Assuming two-degrees of precision, the maximal value for the fraction part will be 0.99, and the minimal valid fraction part is 0.01. The conversion process of 0.99 to binary is shown in Figure 10.14.

For the fraction part, based on Figure 10.14, if 6 cells are used, the absolute value of error will be $0.99 - (2^{-1} + 2^{-2} + 2^{-3} + 2^{-4} + 2^{-5} + 2^{-6}) = 0.005625$, which is greater 0.005. If 7 cells are used, the absolute value of error is $|-0.002187| < 0.005$. Considering the minimal valid fraction of 0.01, the last bit (0000001) corresponds to 2^{-7}; thus, the absolute value of error is $|0.01 - 2^{-7}| = 0.0021875 < 0.005$.

Therefore, 7 memory cells are required to store the fraction part. In total, the engineer needs 17 memory cells for his computing device.

	Quotient	Remainder	
1000/2 =	500	0	
500/2 =	250	0	
250/2 =	125	0	
125/2 =	62	1	
62/2 =	31	0	
31/2 =	15	1	
15/2 =	7	1	
7/2 =	3	1	
3/2 =	1	1	
1/2 =	0	1	

0.99–0.5	= 0.49	$(0.5 = 2^{-1})$
0.49–0.25	= 0.24	$(0.25 = 2^{-2})$
0.24–0.125	= 0.115	$(0.125 = 2^{-3})$
0.115–0.0625	= 0.0525	$(0.0625 = 2^{-4})$
0.0525–0.03125	= 0.02125	$(0.03125 = 2^{-5})$
0.02125–0.015625	= 0.005625	$(0.015625 = 2^{-6})$
0.005625–0.0078125	= – 0.0021875	$(0.0078125 = 2^{-7})$

FIGURE 10.13 Converting 1000 to binary. **FIGURE 10.14** Converting 0.99 to binary form.

10.2.1.4 BINARY ARITHMETIC

Addition. The addition of binary numbers is similar to that of decimal numbers. The rules for the addition of binary bits are summarized in Table 10.1.

TABLE 10.1 Addition Rules for Binary Bits

	Carry	Sum
$0 + 0 =$	0	0
$0 + 1 =$	0	1
$1 + 1 =$	1	0
$1+1+1=$	1	1

EXAMPLE 10.5   **Binary Addition**

Use binary addition to calculate 12 (decimal) + 14 (decimal).

SOLUTION

The solution has been summarized in Figure 10.15.

To conduct simple addition using binary numbers, first convert the decimal numbers to binary and then use the binary addition rules. In this example, the decimal answer should be 26, which will serve as a check by converting the binary answer back to decimal form.

Figure 10.16 shows how the answer is converted to decimal form to check the accuracy of the binary addition.

Thus, the final answer is correct.

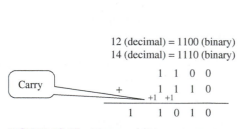

12 (decimal) = 1100 (binary)
14 (decimal) = 1110 (binary)

FIGURE 10.15   Binary addition.

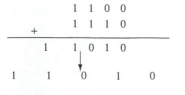

$1 \times 2^4 + 1 \times 2^3 + 0 \times 2^2 + 1 \times 2^1 + 0 \times 2^0 = 26$

FIGURE 10.16   Converting the answer to decimal form.

Subtraction.   To understand binary subtraction, two important terms must be understood: *one's complement* and *two's complement*.

One's complement   For a given binary number, if all "0" bits are replaced with "1" and all "1" bits are replaced with "0", the transformed binary number is called the *one's complement* of the original binary number. For example:

Binary number:	01100101
Its one's complement:	10011010

Two's complement   The *two's complement* of a given binary number is obtained by adding 1 to the one's complement of the binary number and neglecting the carry out (if any) of the most significant bit.

The most significant bit is the first bit from the left. In general, the most significant integer in any number is the integer on the left. For example, in the number 245, 2 is the most significant integer. It is clear that we can write 245 as $200 + 40 + 5$. Thus, integer 2 represents the number 200. Accordingly, it is called the most significant integer. Equivalently, integer 5 in 245 is the least significant integer.

Now, consider the binary number 01100101:

Its one's complement is	10011010
Its two's complement is	10011011

EXAMPLE 10.6   **Two's Complement**

Find the two's complement of the binary number 0110101101.

SOLUTION

The solution has been summarized in Figure 10.17.
Therefore, the two's complement of the given binary number is 1001010100.

```
 0 1 1 0 1 0 1 1 0 0
         ↓
 1 0 0 1 0 1 0 0 1 1 (One's complement)
+                  1
―――――――――――――――――――
 1 0 0 1 0 1 0 1 0 0 (Two's complement)
```

FIGURE 10.17   Finding the two's complement of 0110101101.

Subtraction of binary numbers utilizes the two's complement.

Subtraction using the two's complement   The two's complement can be used to represent negative numbers and perform the subtraction. The first bit (i.e., the most significant bit) is taken as the sign bit (i.e., if the number is positive or negative) of the two's complement. The first bit is 0 for positive numbers and 1 for negative numbers. Negative numbers are represented as the two's complement of the corresponding positive number. For example, the decimal number +2 is equivalent to 00000010 in binary form (using 8 bits), and the decimal number −2 is 11111110 in binary form which is exactly the two's complement of number 2 in binary form. Therefore, subtraction can be performed by finding the two's complement of the subtrahend and then adding. For example:

$$33 - 31 = 33 + (-31) = 33 \text{ (binary)} + \text{Two's complement of 31 (binary)}$$
$$= 100001 + 100001 \text{ (Two's complement of 31 is 100001)}$$
$$= 000010$$
$$= 2 \text{ (decimal)}$$

Note that again the binary answer can be checked by converting the answer to decimal form.

EXAMPLE 10.7   **Binary Subtraction**

a. Find $2110 - 1250$ using the binary form.
b. Find $1250 - 2110$ using the binary form.

SOLUTION

a. First, convert 2110 and 1250 to binary form.

$$2110 \text{ (decimal)} = 100000111110 \text{ (binary)}$$

$$1250 \text{ (decimal)} = 010011100010 \text{ (binary)}$$

The two's complement of 1250 (decimal) is 101100011110. The operation is shown in Figure 10.18.

The operation $2110 - 1250$ is equivalent to 2110 (binary) + the two's complement of 1250 (binary). This operation has been shown in Figure 10.19.

The answer in binary is positive because the leading binary is 0. Here, we ignore the leading 1 (from the last carry) in Figure 10.19.

Next, convert the result back to decimal to check the validity of our operation. This process has been shown in Figure 10.20.

b. The two's complement of 2110 (decimal) is 101100011110. The process of generating the two's complement is shown in Figure 10.21.

The operation $1250 - 2110$ is equivalent to 1250 (binary) + the two's complement of 2110 (binary). This operation is shown in Figure 10.22.

The result is the signed two's complement representation for -860.

```
    010011100010                  Complement means to flip all
         ↓                        the bits (0 to 1, 1 to 0)
    101100011101 ──────→ 1's complement
  +   000000000001              1's complement + 1 = 2's complement
    ─────────────
    101100011110 ──────→ 2's complement
```

FIGURE 10.18   Finding the two's complement of 1250.

```
         2110 − 1250 =

         100000111110  ◄──── 2110 in binary
    +    101100011110  ◄──── 2's complement of
         ────────────        1250 in binary
        1001101011100
```

FIGURE 10.19   Binary subtraction using the two's complement.

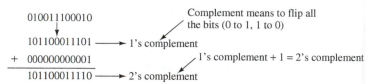

$$0 \times 2^{11} + 0 \times 2^{10} + 1 \times 2^9 + 1 \times 2^8 + 0 \times 2^7 + 1 \times 2^6 + 0 \times 2^5 + 1 \times 2^4 + 1 \times 2^3 + 1 \times 2^2 + 0 \times 2^1 + 0 \times 2^0 = 860$$

FIGURE 10.20   Binary subtraction result verification.

```
    100000111110 ──────→ 2110 in binary form
         ↓
    011111000001 ──────→ 1's complement
  +   000000000001
    ────────────
    011111000010 ──────→ 2's complement
```

FIGURE 10.21   Finding the two's complement of 2110.

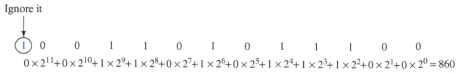

```
         1250 − 2110 =

         010011100010  ◄──── 1250 in binary form
    +    011111000010  ◄──── 2's complement of 1210
         ────────────
         110010100100  ◄──── 2's complement of 860
```

FIGURE 10.22   Binary subtraction.

EXERCISE 10.2

Check the validity of the answer in Example 10.7, part b.

Note that if the result is beyond the maximal value that the limited number of bits can represent, *overflow* occurs. For example, 55 (decimal) + 96 (decimal) = 151 (decimal); the result 151 (decimal) is beyond the maximal value that can be represented by eight bits [+127 (decimal)].

$$
\begin{array}{rll}
55 & 00110111 & \text{(Positive Sign)} \\
+96 & +01100000 & \text{(Positive Sign)} \\
\hline
151 & 10010111 & \text{(Negative Sign)}
\end{array}
\qquad \text{(Overflow occurs)}
$$

Similarly, if the result is less than the minimal value that the limited number of bits can represent, *underflow* occurs. For example, −100 (decimal) −32 (decimal) = −132 (decimal); the result −132 (decimal) is less than the minimal value that the limited number of bits can represent (−128 (decimal)).

$$
\begin{array}{rll}
-100 & 10011100 & \text{(two's complement, negative sign)} \\
-32 & +\ 11100000 & \text{(two's complement, negative sign)} \\
\hline
-132 & 01111100 & \text{(Positive sign)}
\end{array}
\qquad \text{(Underflow occurs)}
$$

In general, if the two numbers that are added have the same sign and the result has the opposite sign, *underflow or overflow* has occurred.

10.2.2 Hexadecimal Numbers

Hexadecimal numbers are converted from binary numbers, and can represent information more efficiently. The base is 16 for hexadecimal numbers; therefore, 16 symbols are required for a hexadecimal number (A through *F* are used to represent the digits for symbols 10 through 15). Table 10.2 shows the symbols used for hexadecimal numbers.

10.2.2.1 CONVERTING HEXADECIMAL NUMBERS TO BINARY

Based on Table 10.2, for a given hexadecimal number, replace each digit with its binary equivalent (4 bits). For example:

$$
A \quad 1 \quad F \ . \ E
$$
$$
\text{A1F} \cdot E \text{ (hex)} = \underline{1010}\ \underline{0001}\ \underline{1111}.\underline{1110} \text{ (binary)} = 101000011111.1110 \text{ (binary)}.
$$

10.2.2.2 CONVERTING BINARY NUMBERS TO HEXADECIMAL

To convert a given binary number to hexadecimal, first, group each four bits starting from the binary point and working outward. If the number of bits is less than four in the leading and/or tailing group, insert 0 bits in the leading group and append 0 bits in the tailing group. Next, replace each group of four bits with its hexadecimal equivalent symbol. For example, the process of converting the binary number 10011000101010.0010011 to hexadecimal is shown in Figure 10.23.

As shown, the hexadecimal form for the given binary number is 262A.26.

TABLE 10.2 Symbols for Hexadecimal Numbers

Hexadecimal	Binary (4 bits)	Decimal
0	0000	0
1	0001	1
2	0010	2
3	0011	3
4	0100	4
5	0101	5
6	0110	6
7	0111	7
8	1000	8
9	1001	9
A	1010	10
B	1011	11
C	1100	12
D	1101	13
E	1110	14
F	1111	15

FIGURE 10.23
Converting a binary
number to hexadecimal.

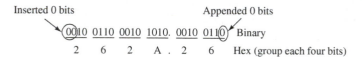

Inserted 0 bits   Appended 0 bits

0010 0110 0010 1010 . 0010 0110 Binary
2   6   2   A . 2   6   Hex (group each four bits)

EXAMPLE 10.8 **Conversion of Binary to Hexadecimal**

Convert 100001110010010 and 0011111101110001000011110 to hexadecimal.

SOLUTION

$$100001110010010 \text{ (binary)} = 4392 \text{ (hex)}.$$

$$0011111101110001000011110 \text{ (binary)} = 3F711E \text{ (hex)}.$$

EXERCISE 10.3

Convert binary hexadecimal numbers 4392 and 3F711E in Example 10.8 back to binary.

10.2.3 Octal Numbers

The base of octal numbers is 8. Therefore, 8 symbols are required for these numbers. Table 10.3 shows the symbols used for the octal numbers.

10.2.3.1 CONVERTING OCTAL NUMBERS TO BINARY

Based on Table 10.3, for a given octal number, replace each digit with its binary equivalent (3 bits). For example:

$$\qquad 2 \quad 7 \quad 1 \; . \; 5 \quad 3$$

271.53 (octal) = 010 111 001.101 011 (binary) = 010111001.101011 (binary).

TABLE 10.3 Symbols for Octal Numbers

Octal	Binary (3 bits)	Decimal
0	000	0
1	001	1
2	010	2
3	011	3
4	100	4
5	101	5
6	110	6
7	111	7

10.2.3.2 CONVERTING BINARY NUMBERS TO OCTAL

To convert a given binary number to octal, first, group each three bits starting from the binary point and working outward. If the number of bits is less than three in the leading and/or tailing groups, insert bits of 0 in the leading group and append bits of 0 in the tailing group. Next, replace each group of three bits with its octal equivalent symbol. For example, the process of converting binary number 11110010101.00101 to octal form is shown in Figure 10.24.

The octal form for the given binary number is 3625.12.

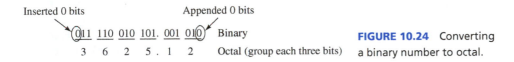

FIGURE 10.24 Converting a binary number to octal.

EXAMPLE 10.9 **Conversion of Binary to Octal**

Convert 111100101011001000 and 010010111011110111 to octal form.

SOLUTION

$$\underline{111}\ \underline{100}\ \underline{101}\ \underline{011}\ \underline{001}\ \underline{000}\ (binary) = 745310\ (octal)$$
$$\quad 7 \quad\ \ 4 \quad\ \ 5 \quad\ \ 3 \quad\ \ 1 \quad\ \ 0$$
$$\underline{010}\ \underline{010}\ \underline{111}\ \underline{011}\ \underline{110}\ \underline{111}\ (binary) = 227367\ (octal)$$
$$\quad 2 \quad\ \ 2 \quad\ \ 7 \quad\ \ 3 \quad\ \ 6 \quad\ \ 7$$

10.3 BOOLEAN ALGEBRA

Boolean algebra defines the arithmetic operations on the binary bits 0 and 1, which are sometimes referred to as true (1) and false (0).

10.3.1 Boolean Inversion

The inversion of 0 is 1 and the inversion of 1 is 0, that is,

$$\bar{1} = 0 \qquad\qquad (10.1)$$
$$\bar{0} = 1 \qquad\qquad (10.2)$$

10.3.2 Boolean AND Operation

The AND operation on two variables (A and B) is represented by AB or $A \cdot B$. It is also called logical multiplication. The AND operation can be denoted as $F = A \cdot B$. Here, F is equal to 1 only when A and B are both 1, otherwise, F is equal to 0. The truth table for AND operation is shown in Table 10.4.

TABLE 10.4 Truth Table for the AND Operation

A	B	F = AB
0	0	0
0	1	0
1	0	0
1	1	1

A truth table lists all possible combinations of the input variables and the corresponding output values. Logic circuits can be shown to be equivalent by using a truth table and/or through Boolean algebra.

10.3.3 Boolean OR Operation

The OR operation on two variables (A and B) is represented by $A + B$. It is also called logical addition. The OR operation can be denoted as $F = A + B$: F equals 0 only when A and B are both 0, otherwise, F equals 1. The truth table for the OR operation is shown in Table 10.5.

10.3.4 Boolean NAND Operation

The NAND operation on two variables (A and B) is the inversion of the result obtained by the Boolean AND operation. The NAND operation can be denoted as $F = \overline{A \cdot B}$, and F is equal to 0 only when A and B are both 1, otherwise, F is equal to 1. The truth table for the NAND operation is shown in Table 10.6.

10.3.5 Boolean NOR Operation

The NOR operation on two variables (A and B) is the inversion of the result obtained by the Boolean OR operation. The NOR operation can be denoted as $F = \overline{A + B}$, and F is equal to 1 only when A and B are both 0, otherwise, F is equal to 0. The truth table for the NOR operation is shown in Table 10.7.

10.3.6 Boolean XOR Operation

The XOR operation on two variables (A and B) is also called modulo-two addition. If A and B are equal, the result of the XOR operation is 0; if A and B are different, the result of the XOR operation is 1. The XOR operation can be denoted as $F = A \oplus B$. The truth table for the XOR operation is shown in Table 10.8. $F = A \oplus B$ corresponds to $F = A\overline{B} + \overline{A}B$ as well. In other words:

$$A \oplus B = A\overline{B} + \overline{A}B \qquad (10.3)$$

10.3.7 Summary of Boolean Operations

$$\overline{1} = 0 \qquad (10.4)$$
$$\overline{0} = 1 \qquad (10.5)$$
$$0 \cdot 0 = 0 \qquad (10.6)$$
$$0 \cdot 1 = 0 \qquad (10.7)$$
$$1 \cdot 0 = 0 \qquad (10.8)$$
$$1 \cdot 1 = 1 \qquad (10.9)$$
$$0 + 0 = 0 \qquad (10.10)$$
$$0 + 1 = 1 \qquad (10.11)$$
$$1 + 0 = 1 \qquad (10.12)$$
$$1 + 1 = 1 \qquad (10.13)$$
$$0 \oplus 0 = 0 \qquad (10.14)$$
$$0 \oplus 1 = 1 \qquad (10.15)$$
$$1 \oplus 0 = 1 \qquad (10.16)$$
$$1 \oplus 1 = 0 \qquad (10.17)$$

10.3.8 Rules Used in Boolean Algebra

$$\overline{\overline{A}} = A \qquad (10.18)$$
$$A \cdot A = A \qquad (10.19)$$

TABLE 10.5 Truth Table for the OR Operation

A	B	F = A + B
0	0	0
0	1	1
1	0	1
1	1	1

TABLE 10.6 Truth Table for the NAND Operation

A	B	$F = \overline{A \cdot B}$
0	0	1
0	1	1
1	0	1
1	1	0

TABLE 10.7 Truth Table for the NOR Operation

A	B	$F = \overline{A + B}$
0	0	1
0	1	0
1	0	0
1	1	0

TABLE 10.8 Truth Table for the XOR Operation

A	B	$F = A \oplus B$
0	0	0
0	1	1
1	0	1
1	1	0

$$A \cdot \overline{A} = 0 \tag{10.20}$$
$$A \cdot 0 = 0 \tag{10.21}$$
$$A \cdot 1 = A \tag{10.22}$$
$$A + \overline{A} = 1 \tag{10.23}$$
$$A + 0 = A \tag{10.24}$$
$$A + 1 = 1 \tag{10.25}$$
$$A + A = A \tag{10.26}$$

10.3.9 De Morgan's Theorems

$$\overline{A \cdot B} = \overline{A} + \overline{B} \tag{10.27}$$
$$\overline{A + B} = \overline{A} \cdot \overline{B} \tag{10.28}$$

De Morgan's theorems also apply to three or more variables.

$$\overline{A \cdot B \cdot C} = \overline{A} + \overline{B} + \overline{C} \tag{10.29}$$
$$\overline{A + B + C} = \overline{A} \cdot \overline{B} \cdot \overline{C} \tag{10.30}$$

De Morgan's theorems can be stated as: if the variables in a logic expression are replaced by their inverters (e.g., A is replaced by $\overline{A}$, $\overline{B}$ is replaced by B), then the AND operation is replaced by OR, and the OR operation is replaced by AND. Here, any term consisting of two or more variables must be parenthesized before changing the operator (AND, OR).

A truth table can be used to prove the theorems for two variables (Tables 10.9 and 10.10). Sometimes in truth tables "1" is represented by "T" (true) and "0" is represented by "F" (false). Indeed, this is why traditionally these tables are called "truth" tables.

EXERCISE 10.4

Use a truth table to prove De Morgan's theorems for three variables.

TABLE 10.9 Truth Table for Proving De Morgan's Theorems (10.27)

A	B	$A \cdot B$	$\overline{A \cdot B}$	$\overline{A}$	$\overline{B}$	$\overline{A} + \overline{B}$
0	0	0	1	1	1	1
0	1	0	1	1	0	1
1	0	0	1	0	1	1
1	1	1	0	0	0	0

TABLE 10.10 Truth Table for Proving De Morgan's Theorems (10.28)

A	B	$A + B$	$\overline{A + B}$	$\overline{A}$	$\overline{B}$	$\overline{A} \cdot \overline{B}$
0	0	0	1	1	1	1
0	1	1	0	1	0	0
1	0	1	0	0	1	0
1	1	1	0	0	0	0

10.3.10 Commutativity Rule

The AND, OR, and XOR operations satisfy the commutativity rule.

$$A \cdot B = B \cdot A \tag{10.31}$$

$$A + B = B + A \tag{10.32}$$

$$A \oplus B = B \oplus A \tag{10.33}$$

A truth table can be used to prove the commutativity rule (see Tables 10.11, 10.12, and 10.13).

10.3.11 Associativity Rule

The AND and OR operations satisfy the associativity rule.

$$A \cdot (B \cdot C) = (A \cdot B) \cdot C = A \cdot B \cdot C \tag{10.34}$$

$$A + (B + C) = (A + B) + C = A + B + C \tag{10.35}$$

A truth table can be used to prove the associativity rule (see Tables 10.14 and 10.15).

10.3.12 Distributivity Rule

$$A + B \cdot C = (A + B) \cdot (A + C) \tag{10.36}$$

$$A \cdot (B + C) = A \cdot B + A \cdot C \tag{10.37}$$

A truth table can be used to prove the distributivity rule (see Tables 10.16 and 10.17).

We have seen that given the expression, we can generate the truth table. Inversely, given the truth table we can generate the expression. To do so, we simply add the expressions corresponding to the outcomes 1 of the table. For example, suppose we have the outcomes of the fifth

TABLE 10.11 A truth Table Used to Prove the Commutativity Rule for the AND Operation (10.31)

A	B	$A \cdot B$	$B \cdot A$
0	0	0	0
0	1	0	0
1	0	0	0
1	1	1	1

TABLE 10.12 A Truth Table Used to Prove the Commutativity Rule for the OR Operation (10.32)

A	B	$A + B$	$B + A$
0	0	0	0
0	1	1	1
1	0	1	1
1	1	1	1

TABLE 10.13 A Truth Table Used to Prove the Commutativity Rule for the XOR Operation (10.33)

A	B	$A \oplus B$	$B \oplus A$
0	0	0	0
0	1	1	1
1	0	1	1
1	1	0	0

TABLE 10.14 A Truth Table Used to Prove the Associativity Rule for the AND Operation (10.34)

A	B	C	B · C	A · (B · C)	A · B	(A · B) · C	A · B · C
0	0	0	0	0	0	0	0
0	0	1	0	0	0	0	0
0	1	0	0	0	0	0	0
0	1	1	1	0	0	0	0
1	0	0	0	0	0	0	0
1	0	1	0	0	0	0	0
1	1	0	0	0	1	0	0
1	1	1	1	1	1	1	1

TABLE 10.15 A Truth Table Used to Prove the Associativity Rule for the OR Operation (10.35)

A	B	C	B + C	A + (B + C)	A + B	(A + B) + C	A + B + C
0	0	0	0	0	0	0	0
0	0	1	1	1	0	1	1
0	1	0	1	1	1	1	1
0	1	1	1	1	1	1	1
1	0	0	0	1	1	1	1
1	0	1	1	1	1	1	1
1	1	0	1	1	1	1	1
1	1	1	1	1	1	1	1

TABLE 10.16 A Truth Table Used to Prove the Distributivity Rule for (10.36)

A	B	C	B · C	A + B · C	A + B	A + C	(A + B) · (A + C)
0	0	0	0	0	0	0	0
0	0	1	0	0	0	1	0
0	1	0	0	0	1	0	0
0	1	1	1	1	1	1	1
1	0	0	0	1	1	1	1
1	0	1	0	1	1	1	1
1	1	0	0	1	1	1	1
1	1	1	1	1	1	1	1

TABLE 10.17 A Truth Table Used to Prove the Distributivity Rule for (10.37)

A	B	C	B + C	F = A · (B+C)	A · B	A · C	A · B + A · C
0	0	0	0	0	0	0	0
0	0	1	1	0	0	0	0
0	1	0	1	0	0	0	0
0	1	1	1	0	0	0	0
1	0	0	0	0	0	0	0
1	0	1	1	1	0	1	1
1	1	0	1	1	1	0	1
1	1	1	1	1	1	1	1

column of Table 10.17, but we don't know that it corresponds to $F = A \cdot (B + C)$. Here, only the last three rows correspond to 1; thus, F is:

$$F = A\overline{B}C + AB\overline{C} + ABC$$

Now, factorizing AB from the last two terms, we have:

$$F = A\overline{B}C + AB(\overline{C} + C)$$

Now, $(\overline{C} + C) = 1$. Factorizing A we have:

$$F = A\overline{B}C + AB = A(\overline{B}C + B)$$

Using distributivity rule,

$$F = A(\overline{B} + B)(C + B)$$

Since $(\overline{B} + B) = 1$, we have:

$$F = A(B + C)$$

In general, it is the role of a logic engineer to simplify a logic such that a lower number of gates is needed to create that.

EXERCISE 10.5

Using the outcomes of the Truth Table 10.17, show that the last column of this table corresponds to $A \cdot B + A \cdot C$.

EXAMPLE 10.10   Simplifying Logic Functions Using De Morgan's Law

Use De Morgan's laws to change the expression of the following logic function.

$$F = \overline{A} \cdot B + A + \overline{B}$$

SOLUTION

Consider $\overline{A} \cdot B$ as the first term and $A + \overline{B}$ as the second term. Replace $\overline{A} \cdot B$ with its inversions; change "·" to "+". We have:

$$F = \overline{\overline{A} + \overline{\overline{B}}} + A + \overline{B}$$

Now, take $A + \overline{B} = D$,

$$F = \overline{D} + D = 1$$

EXAMPLE 10.11   **Simplifying Logic Functions**

Simplify the logic expression:

$$F = A\overline{B} + B\overline{C} + \overline{B}C + \overline{A}B$$

SOLUTION

Because $(A + \overline{A}) = 1$, and $(C + \overline{C}) = 1$:

$$F = A \cdot \overline{B} + B \cdot \overline{C} + \overline{B} \cdot C + \overline{A} \cdot B$$
$$= A \cdot \overline{B} + B \cdot \overline{C} + \overline{B} \cdot C \cdot (A + \overline{A}) + \overline{A} \cdot B \cdot (C + \overline{C})$$

Using the distributivity rule:

$$F = A \cdot \overline{B} + B \cdot \overline{C} + A \cdot \overline{B} \cdot C + \overline{A} \cdot \overline{B} \cdot C + \overline{A} \cdot B \cdot C + \overline{A} \cdot B \cdot \overline{C}$$

Rearrange the expression using the distributivity:

$$F = A \cdot \overline{B} + A \cdot \overline{B} \cdot C + B \cdot \overline{C} + \overline{A} \cdot B \cdot \overline{C} + \overline{A} \cdot \overline{B} \cdot C + \overline{A} \cdot B \cdot C$$
$$= A \cdot \overline{B} \cdot (1 + C) + (1 + \overline{A}) \cdot B \cdot \overline{C} + \overline{A} \cdot C \cdot (\overline{B} + B)$$
$$= A \cdot \overline{B} + B \cdot \overline{C} + \overline{A} \cdot C$$

APPLICATION EXAMPLE 10.12   **Stair-Way Light System**

A civil engineer designs the light system for a building as shown in Figure 10.25. Switch A is installed upstairs and switch B is installed downstairs. The initial position of each switch is shown in Figure 10.26(a). Switch A touches node a, and switch B touches node d. When the engineer is upstairs, he turns on the light shown in Figure 10.26(b), and when he goes downstairs, he can turn off the light as shown in Figure 10.26(c). The status of each switch is shown in Table 10.18. Find the truth table for the switch and logic expression for switches A and B and the lamp. Find alternative expressions for switches A and B and the lamp using AND, OR, and NOT operations.

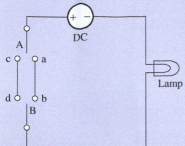

FIGURE 10.25 Light system for Example 10.12.

(continued)

APPLICATION EXAMPLE 10.12   **Continued**

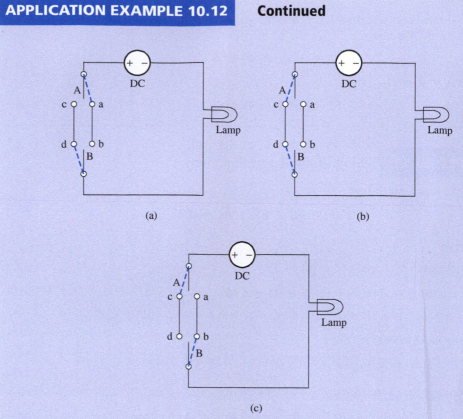

FIGURE 10.26   The status of each switch in the light system.

SOLUTION

For switch A:

Use 0 to denote the status when switch A touches node c.
Use 1 to denote the status when switch A touches node a.

For switch B:

Use 0 to denote the status when switch B touches node d.
Use 1 to denote the status when switch B touches node b.

Furthermore, use 1 to denote that the light is ON and 0 to denote that the light is OFF. Therefore, the truth table corresponds to Table 10.19.

TABLE 10.18 Switch Status		
Switch A	**Switch B**	**Lamp**
c	d	ON
c	b	OFF
a	d	OFF
a	b	ON

TABLE 10.19 Switch Status		
A	**B**	**F**
0	0	1
0	1	0
1	0	0
1	1	1

The expression to describe the relationships of A, B, and F corresponds to:

$$F = \overline{A \oplus B}$$
$$= \overline{A\overline{B} + \overline{A}B} \quad \text{(see Equation (10.3))}$$

Using De Morgan's theorem, an alternative expression is:

$$F = (\overline{A} + B)(A + \overline{B})$$
$$= \overline{A}\cdot\overline{B} + A\cdot B$$

10.4 BASIC LOGIC GATES

A logic gate is a physical device. Logic gates are the elementary building blocks of digital circuits. Any complex digital circuit can be built using basic logic gates. Most logic gates have two (or more) inputs and one output. The input or output terminal is in one of the two binary conditions *low* (0) *or high* (1), represented by different voltage levels. In most logic gates, the low state is approximately zero volts (0 V), while the high state is approximately five volts positive (+5 V).

This section discusses basic logic gates, including the AND gate, OR gate, XOR gate, NOT gate, NAND gate, and NOR gate. Table 10.20 summarizes the symbols used for these gates. These logic gates perform basic Boolean operations.

10.4.1 The NOT Gate

The NOT gate performs the NOT Boolean operation on a single input. Thus, it is also called an inverter. The NOT circuit symbol is shown in Figure 10.27, and its truth table is shown in Table 10.21.

TABLE 10.20 Circuit Symbols for Gates

Gate	Circuit symbol
NOT	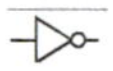
AND	
OR	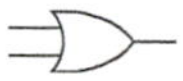
NAND	
NOR	
XOR	
XNOR	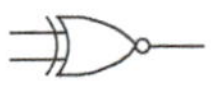

FIGURE 10.27 Circuit symbol for an inverter gate.

TABLE 10.21 Truth Table for the NOT Gate

A	$\overline{A}$
0	1
1	0

Note:

Complementing can be accomplished using a NOT gate. Here, "1" is changed to "0" and "0" is changed to "1".

10.4.2 The AND Gate

The AND gate performs the Boolean AND operation, which is applied to the input variables. Its circuit symbol is shown in Figure 10.28, and its truth table is shown in Table 10.22.

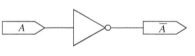

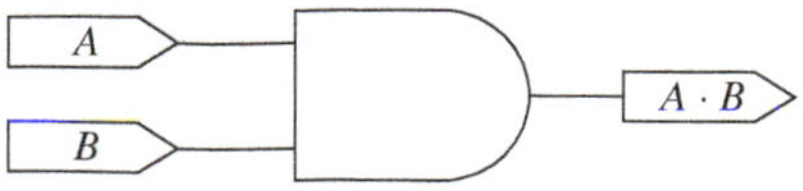

FIGURE 10.28 Circuit symbol for a two-input AND gate.

TABLE 10.22 Truth Table for an AND Gate

A	B	$A\cdot B$
0	0	0
0	1	0
1	0	0
1	1	1

Note:

The result is "1" if and only if both A and B are "1".

10.4.3 The OR Gate

The OR gate performs the Boolean OR operation, which is applied to the input variables. The OR circuit symbol is shown in Figure 10.29, and its truth table is shown in Table 10.23.

TABLE 10.23	Truth Table for an OR Gate	
A	B	A + B
0	0	0
0	1	1
1	0	1
1	1	1

FIGURE 10.29 Circuit symbol for a two-input OR gate.

Note:

The result is "0" if and only if A and B are both "0".

10.4.4 The NAND Gate

The NAND gate performs the Boolean NAND operation, which is applied to the input variables. The NAND circuit symbol is shown in Figure 10.30, and its truth table is shown in Table 10.24.

TABLE 10.24	Truth Table for a NAND Gate		
A	B	A · B	$\overline{A \cdot B}$
0	0	0	1
0	1	0	1
1	0	0	1
1	1	1	0

FIGURE 10.30 Circuit symbol for a two-input NAND gate.

Note:

The result is "0" if and only if A and B are both "1". The NAND gate is the reverse of the AND gate.

10.4.5 The NOR Gate

The NOR gate performs the Boolean NOR operation, which is applied to the input variables. The NOR circuit symbol is shown in Figure 10.31, and its truth table is shown in Table 10.25.

TABLE 10.25	Truth Table for a NOR Gate		
A	B	A + B	$\overline{A + B}$
0	0	0	1
0	1	1	0
1	0	1	0
1	1	1	0

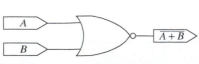

FIGURE 10.31 Circuit symbol for a two-input NOR gate.

Note:

The result is "1" if and only if A and B are both "0". The NOR gate is the reverse of the OR gate.

Important Notes:

(1) It is important to note that all Boolean operations can be represented by NAND and NOR operations. For example, we can represent $\overline{A}$ via $\overline{1 \cdot A}$ (NAND) or via $\overline{0 + A}$ (NOR). In addition, any expression can be solely represented by NAND or NOR gates. For example, using De Morgan's theorem:

$$A + B = \overline{\overline{A} \cdot \overline{B}}$$

Thus, $A + B$ can be represented by NAND operations, by creating $\overline{A}$ and $\overline{B}$ using $\overline{1 \cdot A}$ and $\overline{1 \cdot B}$ then applying another NAND to create $\overline{\overline{A} \cdot \overline{B}}$. Accordingly, the NAND and NOR operations are called *functionally complete*.

(2) As mentioned in Exercise 10.6 and Note (1), NAND and NOR are *functionally complete*. Thus, NAND and NOR gates can be used to represent any other gates. For example, we can represent a NOT gate using a NAND gate while one input of the gate is connected to a high voltage. As shown in Figure 10.32(a), we can represent XY using NAND gates. In addition, as shown in Figure 10.32(b), we can represent XY using NOR gates.

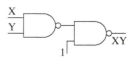

(a)

EXERCISE 10.6

Represent the expression of Example 10.12 using (a) NAND gate only and (b) using NOR gate only.

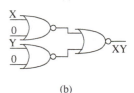

(b)

FIGURE 10.32

(a) Representing XY using (a) NAND gates and (b) NOR gates only.

EXERCISE 10.7

Show with details that in Figure 10.32, XY has been indeed produced via NAND and NOR gates. What theorems and operations are you using?

EXAMPLE 10.13   Representing Logic Functions Using NOR Gate

Use NOR gates only to represent $F = (A + \overline{B})(\overline{A} + C)$.

SOLUTION

First we represent F in terms of NOR operations only, and then we will use NOR gates to create F. To find F in terms of NOR gates, there are two approaches:

    Approach 1:   Convert F to:

$$F = (A + \overline{B})(\overline{A} + C) = AC + \overline{A}\,\overline{B} + \overline{B}C = F_1 + F_2 + F_3$$

(continued)

EXAMPLE 10.13   **Continued**

Using De Morgan's theorem,

$$F_1 = AC = \overline{(\overline{A} + \overline{C})} = \overline{(\overline{(0 + A)} + \overline{(0 + C)})}$$

Similarly, $F_2 = \overline{A}\,B = \overline{(A + \overline{B})}$, and

$$F_3 = \overline{B}C = \overline{(B + \overline{(0 + C)})}$$

Up to this point, we have created F_1, F_2, and F_3 in terms of NOR gates. Now, we should create $F = F_1 + F_2 + F_3$ in terms of NOR gates. Because $\overline{(\overline{F_1 + F_2})} = \overline{\overline{F_1}\,\overline{F_2}} = F_1 + F_2$, it is easy to show:

$$F = \overline{(\overline{(\overline{F_1 + F_2})} + F_3)}$$

Figure 10.33(a) represents the creation of F using NOR gates only. As it is observed, this approach is very complex.

Approach 2:   Directly use $F = (A + \overline{B})(\overline{A} + C)$ and write it as (use De Morgan's rule):

$$F = \overline{\overline{(A + \overline{B})} + \overline{(\overline{A} + C)}}$$

Figure 10.33(b) represents F using the second method. From this example, it is clear that if the approach is selected properly, lower number of gates can be used to create the same logic. It is the role of a good engineer to properly design the logic using lower number of gates.

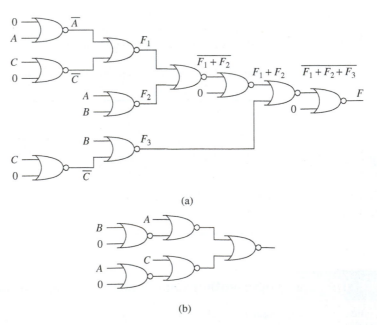

(a)

(b)

FIGURE 10.33 $F = (A + \overline{B})(\overline{A} + C)$ created using NOR gates only. (a) Approach 1 and (b) approach 2.

Conclusion: In order to implement a representation using NOR gates, it is better to write it first in terms of *multiplication of additions*. In order to implement a representation using NAND gates, it is better to first write it in terms of *addition of multiplications*.

EXERCISE 10.8

Repeat Example 10.13 using NAND only gates.

10.4.6 The XOR Gate

The XOR gate performs the Boolean XOR operation, which is applied to the input variables. The XOR circuit symbol is shown in Figure 10.34, and its truth table is shown in Table 10.26.

FIGURE 10.34 Circuit symbol for a two-input XOR gate.

TABLE 10.26	Truth Table for an XOR Gate	
A	**B**	**A ⊕ B**
0	0	0
0	1	1
1	0	1
1	1	0

Note:

The result is "1" if A is "0" and B is "1" or vice versa, that is, A and B are not equal.

10.4.7 The XNOR Gate

The XNOR gate inverts the result obtained by a Boolean XOR operation. The XNOR circuit symbol is shown in Figure 10.35, and its truth table is shown in Table 10.27.

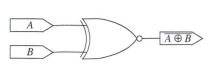

FIGURE 10.35 Circuit symbol for a two-input XNOR gate.

TABLE 10.27	Truth Table for an XNOR Gate		
A	**B**	**A ⊕ B**	**$\overline{A \oplus B}$**
0	0	0	1
0	1	1	0
1	0	1	0
1	1	0	1

Note:

The XNOR gate inverts the output of the XOR gate.

As mentioned before, any complex digital circuit can be built using these basic logic gates. From De Morgan's rule, any logic function can be implemented by AND gates and inverter gates. Similarly, any logic function can be implemented by OR gates and inverter gates. Karnaugh maps can be used to simplify a logic expression that includes multiple variables [1].

EXAMPLE 10.14   **Using AND and OR Gates to Implement a Logic**

Assume four inputs: A, B, C, and D. Use AND, OR, and inverter gates to implement the logic function:

$$F = A \cdot B + C \cdot \overline{D} + B \cdot C \cdot D$$

(*continued*)

EXAMPLE 10.14   **Continued**

SOLUTION

The process is very simple. Start with the inputs A, B, C, and D. Then, make $\overline{D}$, then, $A \cdot B$, C $\cdot$ $\overline{D}$, and, $B \cdot C \cdot D$. Finally, make F the output. The logic circuit is shown in Figure 10.36.

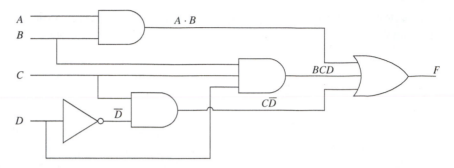

FIGURE 10.36   Circuit for Example 10.14.

EXERCISE 10.9

Use De Morgan's theorem to find an equivalent expression for F in Example 10.14. Then, sketch the equivalent logic circuit of that expression.

EXAMPLE 10.15   **Using AND and OR Gates to Implement a Function**

Assume four inputs: A, B, C, and D. Use AND, OR, and inverter gates to implement the logic function:

$$F = A \cdot B \cdot \overline{C} + C \cdot \overline{D} + (A + B + C) \cdot (\overline{C} + D)$$

SOLUTION

Sometimes, the logic function can be simplified to find an expression that is equivalent and easy to implement. In this case:

$$F = A \cdot B \cdot \overline{C} + C \cdot \overline{D} + (A + B + C) \cdot (\overline{C} + D)$$
$$= A \cdot B \cdot \overline{C} + C \cdot \overline{D} + A \cdot \overline{C} + B \cdot \overline{C} + C \cdot \overline{C} + A \cdot D + B \cdot D + C \cdot D$$

The term $C \cdot \overline{C}$ is always 0. Thus, it can be simplified to:

$$F = A \cdot B \cdot \overline{C} + C \cdot \overline{D} + A \cdot \overline{C} + B \cdot \overline{C} + A \cdot D + B \cdot D + C \cdot D$$
$$= A \cdot B \cdot \overline{C} + C \cdot (\overline{D} + D) + A \cdot \overline{C} + B \cdot \overline{C} + A \cdot D + B \cdot D$$

The term $\overline{D} + D$ is always 1. Further simplification yields:

$$F = A \cdot B \cdot \overline{C} + C + A \cdot \overline{C} + B \cdot \overline{C} + A \cdot D + B \cdot D$$
$$= (A \cdot B + A) \cdot \overline{C} + C + B \cdot \overline{C} + A \cdot D + B \cdot D$$
$$= A \cdot (B + 1) \cdot \overline{C} + C + B \cdot \overline{C} + A \cdot D + B \cdot D$$

The term $B + 1$ is always 1, thus:

$$F = A \cdot \overline{C} + C + B \cdot \overline{C} + A \cdot D + B \cdot D$$
$$= (A + B) \cdot \overline{C} + C + (A + B) \cdot D$$
$$= (A + B) \cdot (\overline{C} + D) + C$$

The corresponding logic circuit is shown in Figure 10.37.

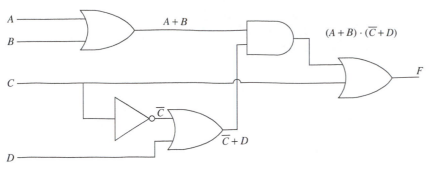

FIGURE 10.37 Circuit for Example 10.15.

APPLICATION EXAMPLE 10.16   **Polling System**

An association group has four members. They want to design a polling system to elect a leader among the four members. In the election meeting, each member holds an input terminal, and there is a button on the terminal device. If the button is pressed down, the terminal device output will be the logic 1 (e.g., a 5-V voltage). All the terminals are connected to a device equipped with a light. The light will be ON if it receives a 1. Therefore, whenever a member's name appears on the screen, if a member agrees to elect this member as their leader, each member can press the button on the terminal device. Design the circuit for the polling system. A member will be selected as leader if all the members simultaneously choose to elect him, thereby illuminating the light.

SOLUTION

Two-input AND gates can be used to design the circuit shown in Figure 10.38. When all members press down on their buttons, that is, $A = B = C = D = 1$, the output to the light is 1. In all other cases, the output to the light is 0.

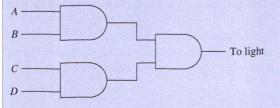

To light

FIGURE 10.38 Circuit for Example 10.16.

APPLICATION EXAMPLE 10.17   **Labtop Processor Design**

A new labtop design has two processors P_1 and P_2 (core-duo) and four slots (S_1, S_2, S_3, S_4) for memory chips. When the processor P_1 is running, it uses the memory in slot S_1. When the processor P_2 is running, it uses the memory in slot S_4. When processors P_1 and P_2 are running simultaneously, they use the memory in slots S_2 and S_3. Design a circuit to connect the processors and the memory slots. If a processor is running, it produces an active, high output to activate the memory slots.

(continued)

APPLICATION EXAMPLE 10.17   **Continued**

SOLUTION

The truth table for the circuit is shown in Table 10.28.

Based on this table, it is clear that $S_1 = P_1 \bar{P_2}$, and $S_2 = S_3 = P_1 P_2$, and $S_4 = \bar{P_1} P_2$. Therefore, the circuit is shown in Figure 10.39.

TABLE 10.28   Truth Table for Example 10.17

P_1	P_2	S_1	S_2	S_3	S_4
0	0	0	0	0	0
0	1	0	0	0	1
1	0	1	0	0	0
1	1	0	1	1	0

FIGURE 10.39   Circuit for Example 10.17.

Note:

As observed in these examples, the gates create the desired output as soon as an input is applied to them. Thus, their operation speed is very high. Today, field-programmable gated arrays (FPGA) are used for super high speed data/information processing. FPGA consists of a large number (e.g., hundreds) of gates implemented on an integrated circuit (IC). They are programmed to perform a digital task. Programming FPGA efficiently to use minimum number of gates requires specific skills. Many companies pay a good salary to the skilled FPGA programmers.

10.5 SEQUENTIAL LOGIC CIRCUITS

At any time instance, the output of combinational logic circuits, such as gates, depends only on the input values at that instant. Thus, these circuits are called *memoryless*. However, the output of sequential logic circuits depends on both past and present inputs. Sequential logic circuits do have memory. Thus, they are different from memoryless circuits such as those composed of basic gates.

A sequential circuit network is defined as a two-valued network. In these systems, the outputs at any time instance depend not only on the present inputs at that instant but also on the history of inputs. In other words, sequential circuit networks have the ability to remember past inputs. A sequential circuit is controlled by a periodic clock signal as shown in Figure 10.40.

10.5.1 Flip-flops

The core element of sequential networks is the flip-flop. The flip-flop is a device with two stable states. It remains in a stable state until its state is changed by new inputs. Different flip-flops respond differently to the clock signal or to input signals.

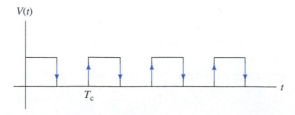

FIGURE 10.40   Clock signal used in the sequential circuit.

Complement outputs

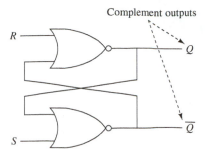

FIGURE 10.41 The SR latch.

10.5.1.1 THE LATCH

The latch is a basic flip-flop. A latch is a logic circuit with two outputs that are complements of each other. A gated latch or clocked latch is a latch which is synchronized by a control signal (usually a clock). The SR (set-reset) latch is used here as an example.

A SR latch consists of two NOR gates, and is shown in Figure 10.41. In this circuit:

1. Assume both inputs are initially 0; output Q is at 1 and $\overline{Q}$ is at 0; and that the circuit is stable with Q at 1 and $\overline{Q}$ at 0.
2. If R is set to 1, Q will be changed to 0, and $\overline{Q}$ changes to 1. If the R changes back to 0, the circuit remains in its stable state.
3. If S is set to 1, $\overline{Q}$ will change to 0 and Q will be 1.
4. S and R are not allowed to be equal to 1 simultaneously because in this case both Q and $\overline{Q}$ will be at 0 state. As a result, they will not be complemented.

   In addition, in this case, if S and R are changed to 0, the output will be ambiguous, that is, it is not clear whether they will be 0 or 1. For example, if Q is changed to 1 first (the two NORs have different delays), then $\overline{Q} = 0$; if $\overline{Q}$ is changed to 1 first, Q will be forced to 0. Thus, it is impossible to predict the output.

The logic symbol for the SR latch is shown in Figure 10.42. The truth table for the SR latch is sketched in Table 10.29.

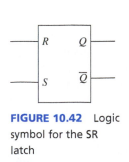

FIGURE 10.42 Logic symbol for the SR latch

TABLE 10.29 Truth Table for the SR Latch

Inputs	Outputs
S R	$Q\ \overline{Q}$
0 0	No change
0 1	0 1
1 0	1 0
1 1	Ambiguous

Stable states ← 0 1 / 1 0

APPLICATION EXAMPLE 10.18   **Switch De-Bouncing in a Circuit**

Switch contacts are usually made of springy metals that are forced into contact by an actuator. When the contacts strike together, their momentum and elasticity act together to cause bounce. Therefore, when the switch is closed or opened, it may bounce for a period of time before settling down. This bouncing effect is harmful and should be prevented.

Switch de-bouncing has been discussed in Chapter 5 (see Example 5.2). Example 5.2 presented an RC circuit to prevent the harmful effects of switch bouncing. In addition, it will be discussed in Chapter 14 (Example 14.10) and Chapter 15 where the safety features of switches are discussed. Here, an RS flip-flop is used to avoid bouncing effects.

(continued)

APPLICATION EXAMPLE 10.18 **Continued**

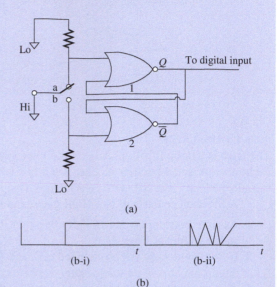

FIGURE 10.43 (a) A de-bouncing circuit constructed using an RS flip-flop. (b) (b-i) Ideal switch responses; (b-ii) switch exhibiting switch bounce.

The goal is to achieve a stable 1 at the input when a switch is turned on and to achieve a stable 0 at the input when the switch is turned off. A de-bouncing circuit constructed with an RS flip-flop for a digital system is shown in Figure 10.43(a). Figure 10.43(b) represents the bounce of switch. Explain the operation of this system.

SOLUTION

The operation of the de-bounce system is as follows. In general, when the switch is at position a, one input of the #1 NOR is 1, thus, the #1 NOR's output (Q) is 0. In this case, all inputs of the #2 NOR are 0 and its output ($\overline{Q}$) is 1. Therefore, the "digital input" will be stable on 0. When the switch is adjusted to position b, one of the #2 NOR's inputs is 1 and its output is 0. Both inputs of the #1 NOR are 0 and its output is 1. The "digital input" will be 1.

At the start, the switch is turned to position b and it bounces at position b. As a result, while the input of the #1 NOR becomes zero, the input of the #2 NOR that comes from the switch will change from 1 to 0, then 1, then 0, . . . , and finally it stabilizes on 1 after switch bouncing process is over [see Figure 10.43(b-ii)].

The first time the input b becomes 1, the output of #2 NOR gate becomes zero, and Q will be 1. When the input of the NOR gate becomes 0 due to bouncing, because the input of both gates will be zero, the flip-flop will stay in a no-change situation (see Table 10.29). Thus, Q will not bounce and instead will stay on 1. Thus, by applying an RS flip-flop, the switch bouncing effect is removed, resulting in a stable digital input.

10.5.1.2 SIMPLE JK FLIP-FLOP

A simple JK flip-flop is shown in Figure 10.44. The clock (Clk) is either "0" or "1" (see Figure 10.40).

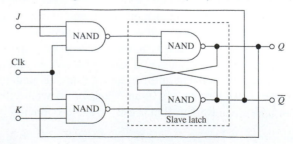

FIGURE 10.44 JK flip-flop.

J and *K* inputs control the mode in which the device operates.

1. When both *J* and *K* are 0, the flip-flop remains unchanged upon receipt of the next clock pulse;
2. When *J* is 1 and *K* is 0, upon receipt of the next clock pulse, the output of *Q* changes to 1, and $\overline{Q}$ changes to 0;
3. When *J* is 0 and *K* is 1, upon receipt of the next clock pulse, the output of *Q* changes to 0, and $\overline{Q}$ changes to 1;
4. When both *J* and *K* are 1, upon receipt of the next clock pulse, the output of *Q* and $\overline{Q}$ will flip to the opposite state.

The logic symbol for a JK flip-flop is shown in Figure 10.45. The truth table for a JK flip-flop is shown in Table 10.30.

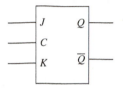

FIGURE 10.45
Logic symbol for a JK flip-flop.

TABLE 10.30   Truth Table for a JK Flip-Flop

Inputs		Outputs		
J	K	Q	$\overline{Q}$	
0	0	Q	$\overline{Q}$	←—— unchanged
0	1	0	1	
1	0	1	0	
1	1	$\overline{Q}$	Q	←—— flip

10.5.1.3 EDGE-TRIGGERED FLIP-FLOP

The inputs of the discussed JK flip-flop are enabled by the level of the clock signal. However, in some flip-flops, the changes of output are triggered by the leading or the trailing edge of the clock signal. Leading edge (rising edge) refers to the transition from low to high for the clock signal, and the trailing edge (falling edge) refers to the transition from high to low as shown in Figure 10.46. If the inputs are enabled when the clock signal is at the leading edge, the flip-flops are called *positive-edge-triggered flip-flops*. If the inputs are enabled when the clock signal is at the trailing edge, the flip-flops are called *negative-edge-triggered flip-flops*. For edge-triggered flip-flops, if the clock signal is steady (not at the leading edge or trailing edge), the inputs are disabled. Positive-edge-triggered D flip-flops and T flip-flops (see the next sections) are examples of edge-triggered flip-flops.

10.5.1.4 D FLIP-FLOP

A D flip-flop is also called a *delay flip-flop*. For the positive-edge-triggered D flip-flop, when the clock signal is at its leading edge, the D flip-flop output will be equal to the value of the input just prior to the triggering clock transition. The logic symbol for a positive-edge-triggered D flip-flop is shown in Figure 10.47.

In the logic symbol, if the clock signal is with the symbol ">", the flip-flop will be an edge-triggered flip-flop; otherwise, it will be a level-enabled flip-flop. The truth table for a positive-edge-triggered D flip-flop is shown in Table 10.31. Here, "↑" means the leading edge triggers the change of output of the flip-flop, and "↓" means the trailing edge triggers the change of output of the flip-flop. In addition, the sign X refers to 0 or 1.

10.5.1.5 T FLIP-FLOP

For the positive-edge-triggered T flip-flop, when $T = 0$, the flip-flop maintains its current state (holding), that is, *Q* is kept the same as it was before the clock edge. When $T = 1$, the output *Q* is

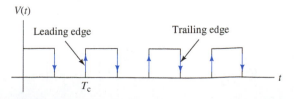

FIGURE 10.46   Diagram of a clock signal.

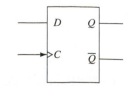

FIGURE 10.47   A positive-edge-triggered D flip-flop.

TABLE 10.31	Truth Table for a Positive-Edge-Triggered D Flip-Flop		
Inputs		Outputs	
D	C	Q	$\overline{Q}$
0	↑	0	1
1	↑	1	0
x	0	Q	$\overline{Q}$
x	1	Q	$\overline{Q}$

TABLE 10.32	Truth Table for a Positive-Edge-Triggered T Flip-Flop		
Inputs		Outputs	
T	C	Q	$\overline{Q}$
0	↑	Q	$\overline{Q}$
1	↑	$\overline{Q}$	Q ←—Toggle
x	0	Q	$\overline{Q}$
x	1	Q	$\overline{Q}$

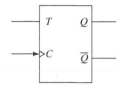

FIGURE 10.48 A positive-edge-triggered T flip-flop.

negated after the clock edge, compared to the value before the clock edge. Thus, the T flip-flop either maintains the current state's value for another cycle, or toggles the value (negates it) at the next clock edge. The logic symbol for a positive-edge-triggered T flip-flop is shown in Figure 10.48. The truth table for the positive-edge-triggered T flip-flop is shown in Table 10.32. The T and D flip-flop can be used to construct counters.

10.5.2 COUNTER

The edge of a clock signal can trigger the change of the output of a flip-flop. Therefore, a circuit consisting of flip-flops can be used to count the number of pulses of an input signal. The output of cascade flip-flops can produce a specified pattern. For example, for a binary counter, the output sequence of three flip-flops corresponds to the binary numbers (000, 001, 010, . . . , 111). A modulo-8 counter is shown in Figure 10.49.

The output sequence of this counter is shown in Figure 10.50. The original value of Q_0, Q_1, and Q_2 is 000. At the leading edge of the first pulse of the clock signal, the output of Q_0 changes from 0 to 1, and the output $\overline{Q}$ of the flip-flop *1* changes from 1 to 0. Therefore, the output of Q_1 does not change because the clock signal is not at the transition from 0 to 1.

Note that, the output $\overline{Q}$ of flip-flop *1* is the clock signal of flip-flop *2*. Consequently, the output of Q_2 does not change because the clock signal of flip-flop *2* does not change. At the leading edge of the second pulse of the clock signal, the output Q_0 changes back to 0, and the output of $\overline{Q}$ of flip-flop *1* changes from 0 to 1. In other words, the input C of the second flip-flop changes from 0 to 1. This makes the output of Q_1 change from 0 to 1. But the output of Q_2 does not change. Why?

To find the answer, look at the input C of the third flip-flop; does it change from 0 to 1? Note that, the output of the flip-flop changes if the C input changes from 0 to 1.

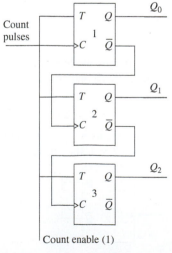

FIGURE 10.49 A modulo-8 counter constructed from T flip-flops.

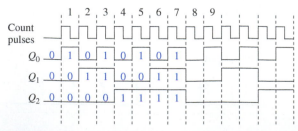

FIGURE 10.50 Output sequence of a modulo-8 counter.

At the leading edge of the third pulse, the output Q_0 changes to 1 and the output Q_1 stays at 1; therefore, the output of Q_2 remains unchanged. At the leading edge of the fourth pulse of the clock signal, the output Q_0 changes to 0, and the output of Q_1 changes from 1 to 0. The corresponding $\overline{Q}$ of the flip-flop changes from 0 to 1, and the output of Q_2 changes from 0 to 1. A thorough understanding of the details of this paragraph is important to understand the subsequent material.

Based on the analysis above, the change period for Q_0 is one period of the clock signal, the change period for Q_1 is two periods of the clock signal, and the change period for Q_2 is four periods of the clock signal. The simultaneous output of Q_2, Q_1, and Q_0 is shown in Figure 10.51.

Figure 10.51 shows that the output of the counter returns to a 000 state at the eigth clock pulse. In other words, the output of Q_2, Q_1, and Q_0 will repeat the pattern in Figure 10.51, therefore, the counter is called a modulo-8 counter.

D flip-flops can also be used to construct a 3-bit ring counter as shown in Figure 10.52. In a 3-bit ring counter, a single clock signal is applied to all three D flip-flops. Note that the Q output of each flip-flop is the D input (D_{in}) of the next flip-flop (the Q output of the flip-flop 1 is the D input of flip-flop 3).

A *reset* signal is connected to the reset inputs of flip-flops 1 and 2 to "0", and the *preset* input of flip-flop 3 to "1". An active *reset* signal can trigger the flip-flops to output logic "0" and an active *preset* signal can trigger the flip-flops to output logic "1." In Figure 10.52, the reset input of flip-flop 3 is high-active, that is, the reset signal is active when its values is logic "1"; the reset inputs of flip-flop 1, 2 and the preset input of flip-flop 3 are low-active, that is, those signals are active when their values are logic "0." Note that "reset" and "preset" can also be referred to as "clear" and "set."

When the reset signal is active, flip-flops 1 and 2 are cleared and flip-flop 3 is preset, that is, the output Q_1 and Q_2 is 0 and the output Q_3 is 1. When the reset signal changes to 1, the clock signal controls the outputs of the three flip-flops as follows:

1. Initial state: the output of Q_3 is 1, and the outputs of Q_2 and Q_1 are 0, that is:

$$Q_1\, Q_2\, Q_3 = 001$$

2. At the positive edge of the first pulse of the clock signal: the output of Q_1 in the last state is applied to flip-flop 3, and the output of Q_3 will be 0; the output of Q_3 in the last state is applied to flip-flop 2, therefore, the output of Q_2 will be 1; the output of Q_2 in the last state is applied to flip-flop 1 and the output of Q_1 is 0, that is:

$$Q_1\, Q_2\, Q_3 = 010$$

3. At the positive edge of the second pulse of the clock signal, the output of Q_1, Q_2, and Q_3 will be 100, that is:

$$Q_1\, Q_2\, Q_3 = 100$$

4. At the third pulse of the clock signal, the counter returns to the state 001, that is:

$$Q_1\, Q_2\, Q_3 = 001$$

The repeating pattern for the 3-bit ring counter is 001, 010, and 100, as shown in Figure 10.53. Because a 1 is moving periodically in a ring formed by the three flip-flops, this counter is called a *ring counter*.

Q_2	Q_1	Q_0
0	0	0
0	0	1
0	1	0
0	1	1
1	0	0
1	0	1
1	1	0
1	1	1

FIGURE 10.51 Output of Q_2, Q_1, and Q_0 in a modulo-8 counter.

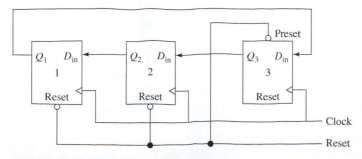

FIGURE 10.52   Circuit for a 3-bit ring counter.

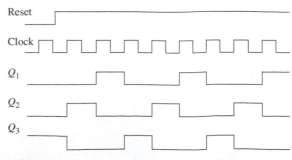

FIGURE 10.53   Repeated pattern for a 3-bit ring counter.

| EXAMPLE 10.19 | D Flip Flop |

For the negative-edge-triggered D flip-flop shown in Figure 10.54, the inputs are enabled when the clock signal is at the trailing edge (falling edge). The clock signal and the input data signal are shown in Figure 10.55. Sketch the output Q (the initial value of Q is zero).

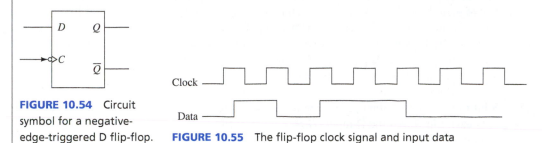

FIGURE 10.54 Circuit symbol for a negative-edge-triggered D flip-flop.

FIGURE 10.55 The flip-flop clock signal and input data

SOLUTION

The output, Q, is sketched in Figure 10.56.

FIGURE 10.56 Output, Q, for Example 10.19.

| APPLICATION EXAMPLE 10.20 | Signal Light |

A mechanical engineer intends to design a signal light for a small car model. He wants to use two bulbs for the signal light. If the bulbs blink as shown in Figure 10.57, the car will turn right. Design the circuit for the signal light system.

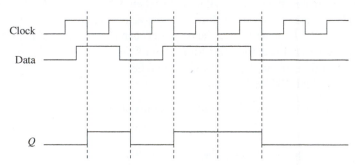

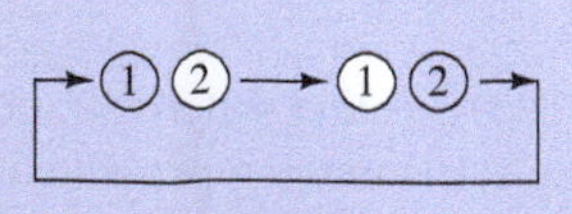

FIGURE 10.57 Signal demonstration for the car model.

SOLUTION

A 2-bit ring counter can be used to control the bulb. The circuit is shown in Figure 10.58.

Here, the output of Q_2 controls light bulb 2, and the output of Q_1 controls light bulb 1. The output of Q_1 and Q_2 will be zero if the car does not turn right (e.g., no power input to the device). When the car starts to turn right, the circuit is reset and the output of Q_2 will be 1 and the output of Q_1 will be 0, and therefore signal light bulb 2 will be ON, while light bulb 1 remains OFF. After the next clock signal, the output of Q_2 is 0 and the output of Q_1 is 1; therefore, signal light bulb 2 will be OFF and light bulb 1 will be ON. The process will repeat until the car finishes turning (power off for this device).

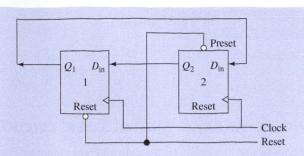

FIGURE 10.58   Circuit for a 2-bit ring counter.

APPLICATION EXAMPLE 10.21   Packing Machine

A packing factory uses a machine to fill a box with 12 bottles of soft drink. Sketch the simple logic circuit to demonstrate how this machine works.

SOLUTION

A counter flip-flop can be used for the logic circuit. The binary output for number 12 is 1100. Thus, given that the machine needs to fill the package with 12 bottles of the soft drink, the required binary output to trigger the machine is 1100. Therefore, given four flip-flops, with the outputs of Q_1, Q_2, Q_3, and Q_4, the instruction for packing is sent when Q_1 and Q_2 are 0, and Q_3 and Q_4 are 1. In other words, considering $A = \overline{Q}_1\overline{Q}_2Q_3Q_4$, A would be 1 if $Q_1 = Q_2 = 0$, and $Q_3 = Q_4 = 1$. Using De Morgan's theorem:

$$A = \overline{Q}_1\overline{Q}_2 \cdot Q_3Q_4$$
$$= \overline{(Q_1 + Q_2) + (\overline{Q}_3 + \overline{Q}_4)}$$
$$= \overline{Q_1 + Q_2} \cdot \overline{(\overline{Q}_3 + \overline{Q}_4)}$$
$$= \overline{Q_1 + Q_2} \cdot (Q_3 \cdot Q_4)$$

Therefore, the solution is shown in Figure 10.59. When Q_1 and Q_2 are 0, the output of the first NOR that is connected to Q_1 and Q_2 will be 1. In addition, when Q_3 and Q_4 are both 1, the

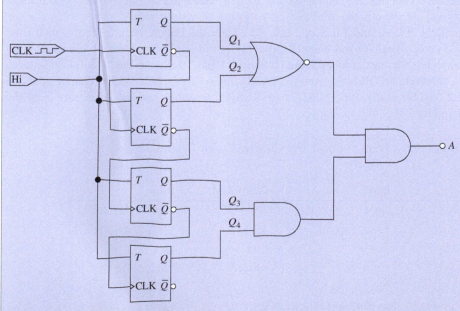

FIGURE 10.59   Logic circuit for application Example 10.21.

(continued)

APPLICATION EXAMPLE 10.21 **Continued**

output of the AND gate that is connected to Q_3 and Q_4 will be 1 as well. In this case, the inputs of the second AND logic circuit will be 1, and the resulting output of the AND will also be 1. This triggers the machine to close the box.

10.6 USING PSPICE TO ANALYZE DIGITAL LOGIC CIRCUITS

This section outlines how to use PSpice to set up a digital logic circuit. Unlike analog circuits, digital circuits do not need to be connected to a ground in order for the circuit to become a complete circuit. However, to study a bias point analysis of digital circuits, an analog circuit must be used in PSpice to force bias point to appear in the digital logic circuit. Because of this, the circuit must be connected to a ground at the end of the analog circuit (see Examples in this section). Note that use of an analog circuit is not required for time domain analysis in PSpice. This section examines two different kinds of simulation analysis in PSpice:

1. Bias point analysis: A set of fixed inputs is applied to study how the digital algorithm behaves.
2. Time domain analysis: The input varies over time to study the change of output with respect to the input.

At least one PSpice library file (*.lib and/or *.olb) is required to analyze a digital circuit using PSpice. However, it is advisable to have both of these library files installed. These files can be downloaded from the web. At least one of the following set of library files should be downloaded:

1. EVAL.lib and EVAL.olb (default in PSpice program)
2. 74LS.lib and 74LS.olb
3. DIG_PRIM.lib and DIG_PRIM.olb

To install these library files, choose "save" or "copy and paste" both the *.lib and *.olb files into the folder at file path C:\Program Files\OrCAD\Capture\Library\PSpice, or C:\Program Files\OrCAD_Demo\Capture\Library\PSpice for the demo version.

Table 10.33 summarizes some gates and their respective library that can be used for building digital logic circuits discussed in this chapter. There are other similar parts with different part names usable for digital logic circuits in the libraries. Students are encouraged to discover these unlisted parts and use them in the exercises.

EVAL, DIG_PRIM, and 74LS in Table 10.33 are the PSpice libraries that contain the required logic components for this chapter. Students can obtain the respective logic parts from PSpice by applying the part number/name shown in Table 10.33 into the "Place Part" menu. For example, to obtain an OR logic from the EVAL library, type "7432" in the Place Part menu (see Figure 2.63). Similarly, type "OR2" or "74LS32" to obtain the OR logic part from the DIG_PRIM and the 74LS library, respectively.

TABLE 10.33 PSpice Logic Part Names with Their Respective Library

	EVAL	DIG_PRIM	74LS
NOT	7404, 7405, 7406	INV	74LS04, 74LS05
AND	7408, 7409	AND2	74LS08, 74LS09
OR	7432	OR2	74LS32
NAND	7400, 7401, 7403	NAN2	74LS00, 74LS01
NOR	7402, 7428	NOR2	74LS02
XOR	74128, 74136	XOR	74LS136
XNOR	N/A	N/A	74LS266

| EXAMPLE 10.22 | **PSpice Analysis** |

Let *A* and *B* represent the inputs for three digital logic circuits: AND, OR, and XOR. Use PSpice to sketch the digital circuits considering the following cases:

a. Bias point analysis for an AND logic circuit.
b. Time domain analysis for AND, OR, and XOR digital logic circuits.

SOLUTION

Part (a)

1. First, go to "Place Part" (see Figure 2.63) and use the reference in Table 10.33; pick an AND logic gate from one of the libraries and place it on the PSpice circuit board.
2. For bias point analysis, use the source that has only the output of 1 or 0 instead of a clock generator. Go to "Place Ground" and type "$D_HI" to choose the source that has the output of 1. Place and connect it to one of the inputs of AND, as shown in Figure 10.60.
3. Go to "Place Ground" again, and type "$D_LO" to choose the source with the output of 0. Place and connect it to another input of AND, as shown in Figure 10.60.
4. To force PSpice to show the bias point analysis results, an analog circuit part must be added to the digital logic circuit.
5. A logic part, "7407" in EVAL.lib library is used so that the PSpice circuit is forced to generate a digital output.
6. Go to "Place Part" and type "7407" under the "Part Name." Place it as shown in Figure 10.60, and also place the resistor, voltage source, and ground.

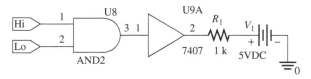

FIGURE 10.60 PSpice schematic circuit for Example 10.22 Part (a).

7. Set up a bias point simulation and press run (see Figure 2.71). The simulation results are shown in Figure 10.61.

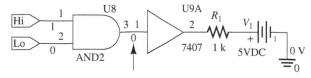

FIGURE 10.61 PSpice result for Example 10.22 Part (a).

8. The output of the AND logic is shown in the figure by the arrow; the output is 0 because one input is 1 and the other is 0.

Part (b)

1. First, go to "Place Part" (see Figure 2.63), and use Table 10.33 as reference. Place the logic parts AND, OR, and XOR onto the PSpice circuit board.
2. For time domain analysis, use the "STIM1" source. Go to "Place Part" and type "STIM1" under the "Part Name" and place it as shown in Figure 10.62.
3. Next, place the voltage level marker (shown by the arrow in Figure 10.63) as shown in Figure 10.62.

(continued)

EXAMPLE 10.22 Continued

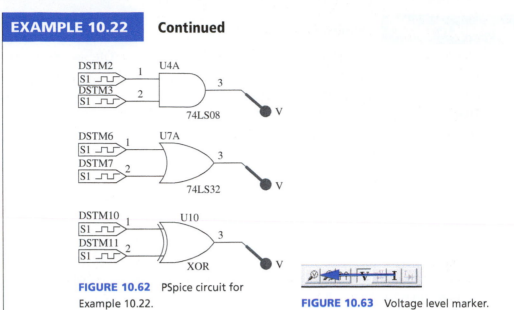

FIGURE 10.62 PSpice circuit for Example 10.22.

FIGURE 10.63 Voltage level marker.

4. Next, set up the output signal sequence of the DTSM (STIM1) source in the logic circuit. Click on the first STIM1 digital source of the AND logic part and go to its properties through "Edit Properties." Enter the values, "10 ms 0," "20 ms 1," and "30 ms 1" inside the Command 2, 3, and 4 columns (see Figure 10.64).

COMMAND1	COMMAND10	COMMAND11	COMMAND12	COMMAND13	COMMAND14	COMMAND15	COMMAND16	COMMAND2	COMMAND3	COMMAND4	COMM
0s 0	<<<<<<<<<<These are Command 10 to 16>>>>>>>>>>>>>>							10ms 0	20ms 1	30ms 1	

Start at here

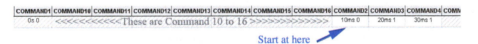

FIGURE 10.64 Command value of first digital source.

5. These value sets represent that at 10 ms, the digital source will generate the output of 0; at 20 ms the digital source will switch to output 1, etc.
6. Next, go to the properties of the second STIM1 digital source of the AND logic part. Enter "10 ms 1," "20 ms 0," and "30 ms 1" as shown in Figure 10.65. At 10 ms, the digital source switches to output 1; at 20 ms, the digital source switches the output back to 0. Finally, at 30 ms, the digital source switches to output 1 again.

COMMAND1	COMMAND10	COMMAND11	COMMAND12	COMMAND13	COMMAND14	COMMAND15	COMMAND16	COMMAND2	COMMAND3	COMMAND4
0s 0	<<<<<<<<<<These are Command 10 to 16>>>>>>>>>>>>>>							10ms 1	20ms 0	30ms 1

Start at here

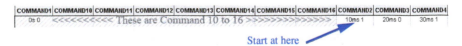

FIGURE 10.65 Command value of second digital source.

7. Note that "Command 10 to 16" is beside the Command 1's column; while the Command 2's column is after the Command 16's column. PSpice simulates in the sequence of Command 1, 2, 3, and so on; therefore, the output of the digital source will not switch if the Command 2, 3, etc. are blank while there is value inside Command 10, 11, and so on.
8. Repeat steps 4 to 6 for the digital source for the OR and XOR logic parts.
9. Set the simulation to be time domain analysis (see Figure 2.75) with the simulation time at least 40 ms.
10. The simulation plot is shown in Figure 10.66. Notice that there are only two values for each output. The line with higher level is 1, while the lower level is 0.

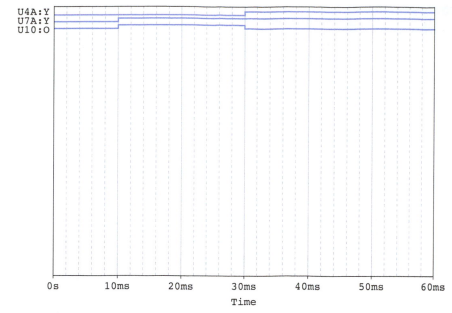

U4A:Y
U7A:Y
U10:O

FIGURE 10.66   Digital output of Example 10.22 Part (b).

| EXAMPLE 10.23 | **PSpice Analysis** |

An association group has four members, and they intend to design a polling system to elect a leader among the four members. In the election meeting, each member holds an input terminal, and there is a button on the terminal device. If the button is pressed down, the terminal device will output logic 1 (e.g., a 5-V voltage). Each terminal is connected to a device equipped with a light. The light will be ON, if it receives a 1 from all connected input terminals. Therefore, whenever a member's name appears on the screen, if all members agree to elect this member as their leader, they can press the button on the terminal device. Design the circuit for the polling system. The member whose name is listed on the screen will be the leader if all members elect him.

SOLUTION

The digital logic circuit for the question is given in Figure 10.67.

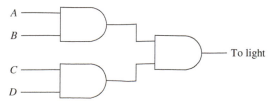

FIGURE 10.67   Circuit for Example 10.23.

1. Use the Part reference in Table 10.33 and follow the steps in Example 10.22 to set up a PSpice digital logic circuit, as shown in Figure 10.68.

(continued)

EXAMPLE 10.23 Continued

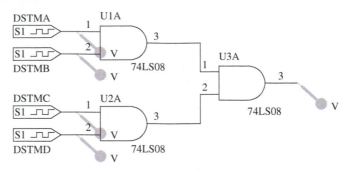

FIGURE 10.68 PSpice digital logic circuit for Example 10.23.

2. Using 10 ms as the time step, the output of each digital source is shown in Table 10.34, in order to study the output of the voting result. In each time step, a vote is allocated by the four members to each candidate. For example, at 0 s nobody votes for candidate 1 and at 10 ms two people are voting for candidate two, and so on.

TABLE 10.34 Setup of the Output Set of Each Digital Source

Time	0 s	10 ms	20 ms	30 ms	40 ms	50 ms	60 ms
DSTMA	0	0	1	1	1	0	1
DSTMB	0	1	1	0	1	1	0
DSTMC	0	1	1	1	0	0	1
DSTMD	0	0	1	1	0	0	1

3. Set the simulation analysis as time domain (see Figure 2.75), with a simulation time of at least 70 ms. The simulation output plot is shown in Figure 10.69.

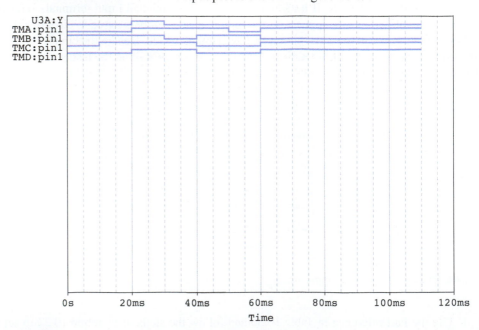

FIGURE 10.69 Digital output of Example 10.23.

4. TMA, TMB, TMC, and TMD are the input sources into the digital circuit. Use the reference in Table 10.34 and compare it to Figure 10.69. The result shows that there is only one member elected, which is the third vote.

EXAMPLE 10.24   **PSpice Analysis**

A new design of labtop has two processors P_1 and P_2 (core-duo) and four slots (S_1, S_2, S_3, and S_4) for the memory chips. When only processor P_1 is running, it uses only the memory in slot S_1; when only processor P_2 is running, it uses only the memory in slot S_4; when processors P_1 and P_2 are running simultaneously, they use the memory in slots S_2 and S_3. Design a circuit to connect the processors and the memory slots (if a processor is running, it produces an active, high output to activate the memory slots).

SOLUTION

The digital circuit of Example 10.17 is shown in Figure 10.70; the input values of P_1 and P_2 are shown in Table 10.35.

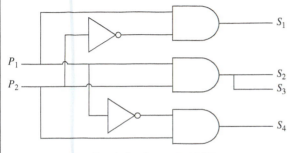

FIGURE 10.70   Circuit for Example 10.17.

TABLE 10.35	Input Value of P_1 and P_2 Sources	
P_1		P_2
0		0
0		1
1		0
1		1

Use Table 10.33 as the Part reference and follow the steps in Example 10.22 to set up the PSpice digital logic circuit as shown in Figure 10.71. Edit the properties command for the DSTM1 and DSTM2 so that the input changes according to the values in Table 10.35. Select a constant time step, such as 10 ms. The output of S_1, S_2, S_3, and S_4 are shown in Figure 10.72. Figure 10.72 shows that both S_2 and S_3 share the same output, which is consistent with Figures 10.70 and 10.71.

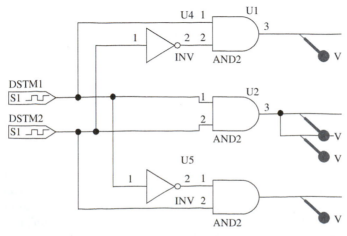

FIGURE 10.71   PSpice digital logic circuit for Example 10.24.

(continued)

EXAMPLE 10.24	Continued

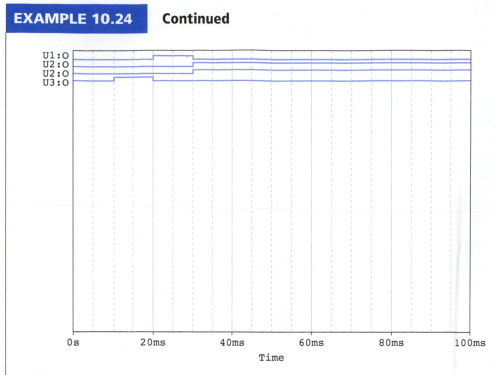

FIGURE 10.72 The output of Example 10.24.

EXAMPLE 10.25	PSpice Analysis

Use PSpice to set up an SR (NOR) latch as discussed in Section 10.5(c) and shown in Figure 10.73 with the input value given in Table 10.36.

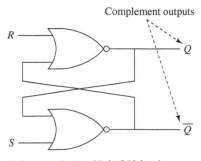

Complement outputs

FIGURE 10.73 SR (NOR) latch.

TABLE 10.36 Input Value for S and R

R	S
1	0
0	0
0	1
1	1
1	0
0	0

SOLUTION

Use Table 10.33 and follow the steps in Example 10.22 to set up the PSpice logic circuit that is shown in Figure 10.17. Set the properties command for both input sources based on the values in Table 10.36 with a reasonable time step. Next, set the simulation type as time analysis (see Figure 2.75). Then, go to "Option," and choose "Gate-level Simulation" under the Category option as shown in Figure 10.74. Set the "Initialize all flip-flops to" as 0 for the initial condition. The setup, circuit, and simulation results are shown in Figures 10.75 and 10.76.

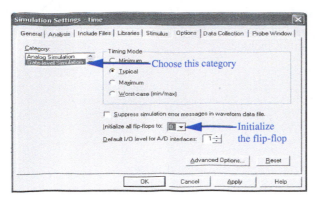

FIGURE 10.74 Setting the simulation option.

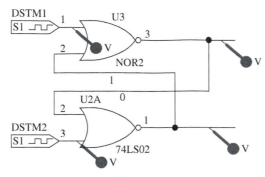

FIGURE 10.75 PSpice logic circuit for Example 10.25.

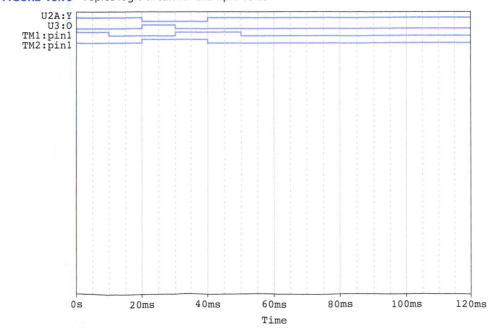

FIGURE 10.76 The output of Example 10.25.

10.7 WHAT DID YOU LEARN?

- In digital logic, the higher voltage represents logic 1 and the lower voltage represents logic 0.
- Numerical data can be represented in decimal, binary, octal, or hexadecimal forms.

- Subtraction of binary numbers are conducted by using the two's complement representation.
- A truth table lists all combinations of input variables and the corresponding output of each.
- Boolean operations include:

$$\overline{1} = 0$$
$$\overline{0} = 1$$
$$0 \cdot 0 = 0$$
$$0 \cdot 1 = 0$$
$$1 \cdot 0 = 0$$
$$1 \cdot 1 = 1$$
$$0 + 0 = 0$$
$$0 + 1 = 1$$
$$1 + 0 = 1$$
$$1 + 1 = 1$$
$$0 \oplus 0 = 0$$
$$0 \oplus 1 = 1$$
$$1 \oplus 0 = 1$$
$$1 \oplus 1 = 0$$

- Rules used in Boolean algebra are defined as:

$$\overline{\overline{A}} = A$$
$$A \cdot A = A$$
$$A \cdot \overline{A} = 0$$
$$A \cdot 0 = 0$$
$$A \cdot 1 = A$$
$$A + \overline{A} = 1$$
$$A + 0 = A$$
$$A + 1 = 1$$
$$A + A = A$$

- De Morgan's theorems are:

$$\overline{A \cdot B} = \overline{A} + \overline{B}$$
$$\overline{A + B} = \overline{A} \cdot \overline{B}$$

De Morgan's theorems also apply to three or more variables, as shown here:

$$\overline{A \cdot B \cdot C} = \overline{A} + \overline{B} + \overline{C}$$
$$\overline{A + B + C} = \overline{A} \cdot \overline{B} \cdot \overline{C}$$

- Basic logic gates include the AND gate, OR gate, XOR gate, NOT gate, NAND gate, and NOR gate. These gates perform the corresponding Boolean operations.
- Sequential logic circuits have memory because the output of sequential logic circuits depends on both past and present inputs.
- Flip-flops include SR latch, JK flip-flop, D flip-flop, and T flip-flop.
- Flip-flops can be used to construct a counter.

References

1. Parag K. Lala. 1996. *Practical digital logic design and testing*. Englewood Cliffs, NJ: Prentice Hall.

Problems

*B refers to basic, A refers to average, H refers to hard, and * refers to problems with answers.*

SECTION 10.2 NUMBER SYSTEMS

10.1 (B) State the difference between an analog signal, a discrete signal, and a digital signal.

10.2 (A)
(a) Convert 1209 (decimal) to binary form.
(b) Convert 1225 (decimal) to binary form.

10.3 (B)* Convert 10001011(binary) to decimal form.

10.4 (A) Convert the decimal fraction number 0.68 into binary form (stop when the fraction part has 8 bits).

10.5 (A) Convert the decimal fraction number 0.141 into binary form and then convert it back into decimal fraction form. Stop when the fraction part has:
(a) 8 bits,
(b) 16 bits.
Compare both answers.
Express the decimal answer to the accuracy of 5 decimal points.

10.6 (B)* Convert the binary number 1101.1101 to a decimal number.

10.7 (A) Convert the decimal number 88.65 into binary form (stop when the fraction part has 4 bits).

10.8 (B) Use binary addition to calculate 11 (decimal) + 77 (decimal).

10.9 (B) Find both the one's complement and the two's complement of 120 (decimal).

10.10 (A) Find 168 − 66 using the binary form.

10.11 (A) Find 66 − 168 using the binary form.

10.12 (B) Convert the hexadecimal number 1FA3.0C42 to a binary number.

10.13 (B)* Convert the binary number 1010001101.11010011 to hexadecimal form.

10.14 (B) Convert the octal number 734.216 to a binary number.

10.15 (B) Convert the binary number 101110101.11010101 to octal form.

10.16 (A) Convert 1424 (decimal) to hexadecimal form.

10.17 (A) Convert 1211 (decimal) to octal form.

10.18 (B) Convert the octal number 272.61 to decimal form.

10.19 (B)* Convert the hexadecimal number 1FC2 to decimal form.

10.20 (H) Implement the conversion between a decimal number and a quaternary number (base 4): convert 123 (decimal) to quaternary form.

10.21 (B) Convert the octal number 742 to hexadecimal form.

10.22 (B) Convert the hexadecimal number FE021 to octal form.

10.23 (B)* Convert the quaternary number 3221.3 to hexadecimal form.

10.24 (B)* Convert the hexadecimal number 7F.32 to quaternary form.

10.25 (A) Calculate quaternary subtraction of 32 − 12 using binary form.

10.26 (A) Calculate octal subtraction of 45 − 32 using binary form.

10.27 (A) Calculate hexadecimal subtraction of 89 − 29 using binary form.

10.28 (A) Transform the quaternary number 20 (Q), octal number 20 (O), and hexadecimal number 20 (H) to decimal form and explain why they have the same form in quaternary, octal, and hexadecimal form but different values in decimal form.

10.29 (H) Provide an example to explain the limitations of a short-length word processor, for example an 8-bit word length processor. (Hint: consider the processing overflow.)

SECTION 10.3 BOOLEAN ALGEBRA

10.30 (B) State De Morgan's laws.

10.31 (B) Develop truth tables for the following Boolean expressions:

$$D = AB + B\overline{C}$$
$$F = BC + A\overline{D}$$

10.32 (B)* If $A = 1$, $B = 0$, and $C = 1$, find $F = A \cdot B + \overline{A} \cdot \overline{B} + C \cdot \overline{B} + ABC$.

10.33 (A) If $A = 0$, $B = 0$, and $C = 1$, find $F = (A + \overline{B} + C) \cdot (\overline{A \cdot B \cdot C})$.

10.34 (H) Show that $A \cdot (\overline{\overline{B} \cdot (A + C)}) = AB$.

10.35 (H)* Simplify the logic circuit $F = \overline{A} \cdot B + A \cdot \overline{C}$.

10.36 (H) What is the output of $F = (\overline{C} \cdot B) \cdot [(A + C) \cdot \overline{(A + B)}]$?

10.37 (A) Find the value of $E = \overline{A} + \overline{C} + A \cdot (\overline{D} + B) + \overline{C} + B \cdot \overline{D}$ when $A = 1, B = 0, C = 0, D = 0$.

10.38 (A)* Find the value of $E = \overline{\overline{A} + B} + \overline{B} \cdot \overline{C} + \overline{C} + B\overline{D}$ when $A = 0, B = 1, C = 0, D = 1$.

10.39 (H) From the truth table of Table 10.39, find the Boolean function $D = f(A, B, C)$.

TABLE P10.39   Truth Table

A	B	C	D
0	0	0	0
0	0	1	0
0	1	0	0
0	1	1	1
1	0	0	0
1	0	1	1
1	1	0	0
1	1	1	1

10.40 (H) Prove that $\overline{\overline{\overline{A + B} + C \cdot A} + B \cdot C} = 1$.

10.41 (H)* Simplify $D = \overline{\overline{A + \overline{B} \cdot \overline{C} \cdot A} + \overline{A} + B}$.

10.42 (H)* Simplify $C = B + \overline{A \cdot B \cdot \overline{A}}$.

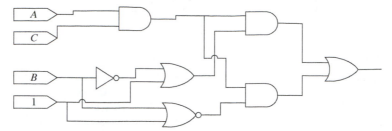

FIGURE P10.49 Logic circuit for Problem 10.49.

SECTION 10.4 BASIC LOGIC GATES

10.43 (A) Find the expression for the output of the logic circuits shown in Figure P10.43.

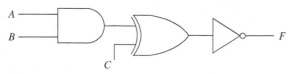

FIGURE P10.43 Circuit for Problem 10.43.

10.44 (A) Draw a circuit to realize the following expressions using AND gates, OR gates, and inverters:

$$F = A\overline{B}C + \overline{A}B$$

10.45 (A)* If $A = 0$, $B = 1$, which circuit as shown in Figure P10.45 will output 1?

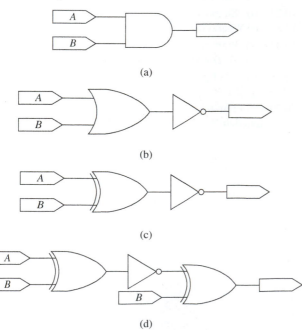

(a)

(b)

(c)

(d)

FIGURE P10.45 Circuit for Problem 10.45.

10.46 (A) Four inputs are A, B, C, and D. Use AND gates (two inputs), OR gates (two inputs), and inverter gates to implement the logic function:

$$F = A \cdot B \cdot D + B \cdot \overline{D} + C \cdot D$$

10.47 (H) Use only NAND gates to implement the logic function of Problem 10.46.

10.48 (A)* Simplify $(A + C) \cdot B + (\overline{B} + \overline{C}) \cdot C$ then draw the logic circuit for the solution.

10.49 (H) Write the logic expression for the circuit shown in Figure P10.49. Then, simplify the expression and draw the logic circuit that corresponds to the expression.

10.50 (H) A new design for an emergency demonstration system is composed of three bulbs installed on the top of a car, shown in Figure P10.50. A control console is installed inside the car. The console has two switches each with an output of 0 or 1. If the emergency level is 0 (no emergency), the green light is on. If the emergency level is 1, one red light is ON; if the emergency level is 2, both red lights are ON. Design the control system (assume that NOT gates, two-input AND gates, and OR gates are available).

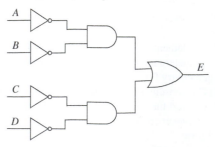

FIGURE P10.50 Emergency demonstration car.

10.51 (B)* Use the smallest number of logic circuits to express $E = A\overline{C} + A\overline{D} + B\overline{C} + B\overline{D}$.

10.52 (H) Find the expression of the output of the given logic circuit, and then use the smallest number of logic circuits (a two-input AND operator, a two-input OR operator, and inversion operator) to express it.

FIGURE P10.52 Logic circuit.

10.53 (A) Find the expression of the output of the logic circuit shown in Figure P10.53.

10.54 (H)* Simplify the result of Problem 10.53 to make it only include AND, OR, and inverse operators, and draw the corresponding logic circuit.

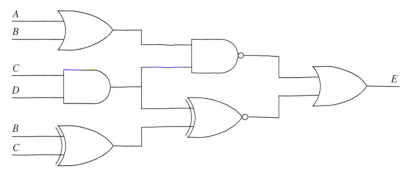

FIGURE P10.53   Logic circuit for P10.53.

10.55 (H) Using XOR, NAND, and AND operators, draw the logic circuit for the logic equation $E = A \cdot \overline{B} \cdot \overline{C} + A \cdot \overline{B} \cdot \overline{D} + \overline{A} \cdot B \cdot \overline{C} + \overline{A} \cdot B \cdot \overline{D}$.

10.56 (A) Simplify $E = A \oplus B + A \cdot B + \overline{A} \cdot \overline{B}$

10.57 (A) Find the truth table for the logic circuit shown in Figure P10.57.

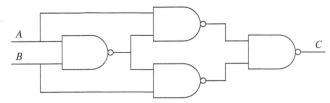

FIGURE P10.57   Logic circuit constructed with NAND and NOR.

10.58 (H)* Design a logic circuit $D = f(A, B, C)$ to satisfy the following truth table.

TABLE P10.58   Truth Table

A	B	C	D
0	0	0	0
0	0	1	1
0	1	0	0
0	1	1	1
1	0	0	0
1	0	1	1
1	1	0	1
1	1	1	1

SECTION 10.5 SEQUENTIAL LOGIC CIRCUITS

10.59 (B) For the JK flip-flop shown in Figure P10.59, the current values of J and K are $J = 0$, $K = 0$, and $Q = 0$. What is the value of Q when J changes to 1, and K changes to 0 on the arrival of another clock pulse?

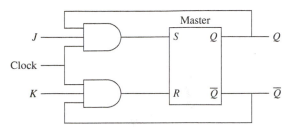

FIGURE P10.59   JK flip-flop for Problem 10.59.

10.60 (A)* How many D flip-flops are needed to represent a 4-bit ring counter?

10.61 (H) A student wants to use a simple logic circuit to design a three-second countdown for a car racing competition. The countdown is initiated by a red light. The initial light is followed by a yellow light when there is 1 s left. Finally, a green light represents the beginning of the race. Design a control system that meets the requirements.

10.62 (A) Determine the output, Q, of the JK flip-flop if the input is as shown in Table P10.62, given that the initial output, Q, of the flip-flop is 0.

TABLE P10.62   JK Flip-Flop Input for Problem 10.62

J	1	0	1	1	0	1	0	1	1	0
K	0	0	1	0	1	1	0	1	0	1

10.63 (H) Figure P10.63(a) shows the sequence of clock signal for both clock inputs, T_1 and T_2, while Figure P10.63(b) shows the JK flip-flop circuit. What is the output, Q_2 of JK flip-flop 2 immediately before the second trailing edge of the clock signal 2, T_2? Both the initial outputs Q_1 and Q_2 are zero. Assume that the input from the clock signal is 1 after the leading edge of the clock signal, and the input becomes 0 after the trailing edge of the clock signal. Also, assume a clock signal of 1, T_1 starts at the leading edge.

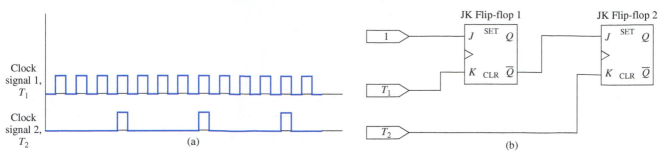

FIGURE P10.63 (a) Clock signal input for Problem 10.63; (b) logic circuit for Problem 10.63.

10.64 (A) A logic circuit and its input wave form are given in Figure P10.64. Assuming its initial output is 0, draw the output of the logic circuit with these input signals.

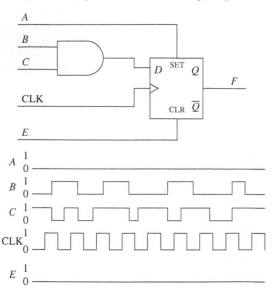

FIGURE P10.64 Logic circuit and its input.

10.65 (H) A frequency divider and the input clock are shown in Figure P10.65, and its initial outputs are all zeros. Draw the output waveform of A, B, and C.

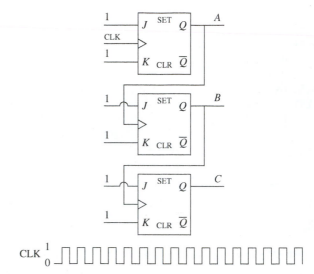

FIGURE P10.65 Frequency divider and clock wave.

10.66 (A) A 5-bit shift register constructed with D flip-flops is shown in Figure P10.66. At the first CLK rising edge, the output is as follows: $Q_1 = 0$, $Q_2 = 1$, $Q_3 = 1$, $Q_4 = 0$, $Q_5 = 0$. Provide the output at the second, third, fourth, and fifth CLK rising edges.

10.67 (H) A counter constructed with D flip-flops is shown in Figure P10.67. Assume that each D flip-flop's initial output is 0. Draw its output wave form.

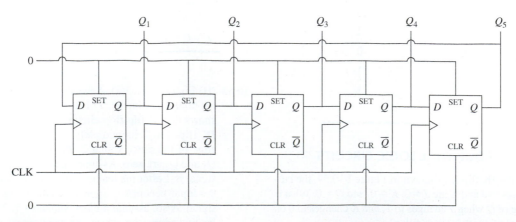

FIGURE P10.66 A 5-bit shift register.

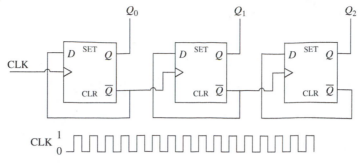

FIGURE P10.67   Counter and its input.

10.68 (H) Analyze the circuit shown in Figure P10.68 and provide the output waveform based on the input waveform.

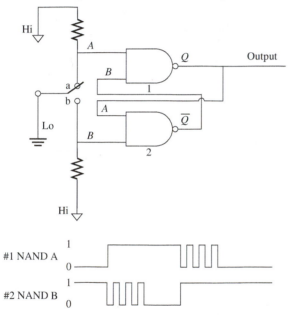

FIGURE P10.68   A logic circuit and its input.

SECTION 10.6 USING PSPICE TO ANALYZE DIGITAL LOGIC CIRCUITS

10.69 (A)* Use PSpice to solve the digital logic circuit, $F = \overline{A} \cdot \overline{B} + C \cdot \overline{B} + A \cdot B \cdot C$ if $A = 1$, $B = 0$, $C = 1$.

10.70 (H)* Use PSpice to solve the digital logic circuit, $F = (A + \overline{\overline{B} + C}) \cdot (\overline{A \cdot B} \cdot C)$ if $A = 0$, $B = 0$, $C = 1$.

10.71 (B)* If $A = 0$ and $B = 1$, use PSpice to find the logic circuit which generates an output of 1 in Figure P10.45.

10.72 (H) For the JK flip-flop shown in Figure P10.59, the current values of J and K are $J = 0$, $K = 0$, and $Q = 0$. Use PSpice to set up the digital logic circuit. The JK flip-flop circuit can be obtained from part "JKFF" in the DIG_PRIM library. Solution

Computer-Based Instrumentation Systems

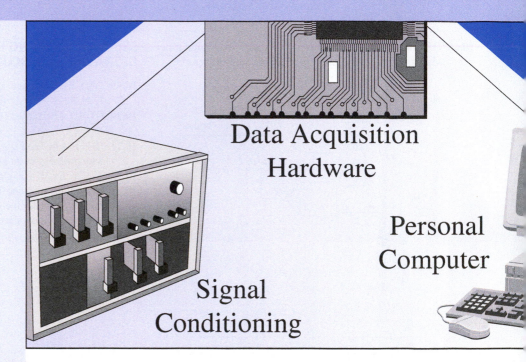

Data Acquisition Hardware

Personal Computer

Signal Conditioning

11.1 INTRODUCTION

In our daily lives, data are collected from many sources. Examples include measuring the tension of a bridge brace to gather data used to maintain the bridge (Figure 11.1) or noting the wind speed or humidity of the air to forecast the weather (Figure 11.2). The collected data are usually continuous in nature, and thus, it is represented by analog signals.

To obtain data about the tension on a bridge brace or wind speed, devices can be employed within the structure to measure the tension, or devices can be installed on poles in the air to measure the wind speed. Devices that collect data from the environment are called *sensors*. Many of these devices are capable of transmitting collected data wirelessly to a destination where the collected data are processed and deductions are made. Examples of real-world applications of sensors include temperature monitoring systems, smart buildings fire alarm systems, seismic pressure sensing in earthquake prediction, and crash-detecting systems in cars.

In general, due to the kind and amount of field data collected by sensors, advanced computer technology is incorporated to process the data (Figure 11.3). Statistical science techniques are used to summarize the collected data and make conclusions. For example, after analyzing

FIGURE 11.1 Bridge maintenance. (Used with permission from © Emily Lai/Alamy.)

FIGURE 11.2 Weather forecast. (Used with permission from © peter dazeley/Alamy.)

FIGURE 11.3 Data processing. (Courtesy Alamy Images.)

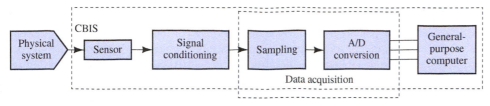

FIGURE 11.4 Computer-based instrumentation system (CBIS).

the data, the analyst may decide that a bridge brace needs replacement, a tornado is likely to form, a fire is ongoing, and so on.

Finally, assuming that the installed sensors have collected analog signals, the question is: How can digital signals be generated and processed when analog signals have been collected? The process of converting analog signals to digital (digitizing) includes sampling and coding. The coder is called an analog-to-digital converter (A/D converter; ADC). The signal generated by the ADC is applied to a computer or digital signal processor (DSP) for processing, making conclusions, and taking actions.

As summarized in Figure 11.4, the entire structure starting from the sensor through the computer (or DSP) forms a computer-based instrumentation system (CBIS).

Physical quantities such as pressure, temperature, stress, flow, angular speed, and displacement can be measured using sensors. These physical quantities produce changes in the voltages, currents, resistances, capacitances, and/or inductances of the sensors, which convert the change of the physical quantity into a corresponding change in an electrical quantity.

Signal conditioning transforms the changes in the electrical quantities into voltages, when the electrical quantity is not a voltage. Furthermore, signal conditioning alters the sensor output into an appropriate form through actions, such as amplifying, filtering, and so on. The conditioned signals are sent to the *data-acquisition system* that periodically samples the signal and converts the sampled value to digital "words" using an ADC. The digital words are then read by the computer for further processing. This chapter discusses the instrumentation system in detail and explains each of the components shown in Figure 11.4.

11.2 SENSORS

A sensor, also called a *transducer*, is defined as a device that responds to a physical stimulus (e.g., heat, light, sound, pressure, magnetism, or motion) and transmits a resulting impulse for measurement or control. Sensors/transducers detect and measure a signal or energy and convert it to a desirable output signal or energy. A microphone is one example of a transducer. Generally, sensors can be grouped according to their physical characteristics (e.g., electronic sensors or resistive sensors) or by the physical variable or quantity measured by the sensor (e.g., pressure, flow, or temperature). Sensors can also be grouped based on the domains to which they belong, such as thermal, mechanical, chemical, magnetic, radiant, or electrical. This section examines typical physical sensors.

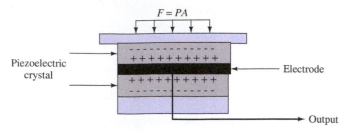

FIGURE 11.5 A piezoelectric pressure sensor.

11.2.1 Pressure Sensors

Pressure is a mechanical quantity defined as the normal force per unit area. Pressure can be sensed by elastic mechanical elements, such as plates, shells, or tubes, which are designed and constructed to deform when pressure is applied. This deformation is the basic mechanism of converting pressure to physical movement. The movement can then be transduced to obtain an electrical signal. Therefore, a pressure sensor is defined as a device that responds to the pressure applied to its sensing surface and converts the pressure to a measurable signal. The most common family of force and pressure sensors is made up of those based on strain gauges and piezoelectric sensors. This section examines how piezoelectric pressure sensors convert force or pressure to an electrical signal.

A representation of a typical piezoelectric pressure sensor is shown in Figure 11.5. Pressure and force sensors are nearly identical and rely on an external force to strain the crystals. A major difference is that pressure sensors use a diaphragm to collect pressure. Pressure is simply the force applied over a unit area, that is, force = pressure × area.

The piezoelectric pressure sensor contains a piezoelectric crystal that generates an electric charge in response to deformation. When a force or pressure is applied to the crystal, which produces a displacement, charges are generated within the crystal. If the external force generates a displacement, x_i, then the sensor will generate a charge, q, according to the expression:

$$q = K_p x_i \tag{11.1}$$

where K_p is the pressure sensitivity in coulomb/displacement unit. The magnitude of K_p depends on the piezoelectric material and the structure of the sensor. Figure 11.6 depicts a basic model of a piezoelectric sensor and the corresponding circuit. The model of a piezoelectric sensor consists of a piezoelectric crystal and two conducting electrodes, while the simple circuit contains a current source and a capacitor; V_0 is the voltage output. In the circuit model, the current source represents the rate of change of the charge due to the motion of the crystal when an external force is applied.

In addition, the capacitance represents the effect of the sensor's structure responding to the external force. The sensor's output voltage, V_0, corresponds to:

$$V_0 = \frac{1}{C} \int i \, \mathrm{d}t = \frac{1}{C} \int \frac{\mathrm{d}q}{\mathrm{d}t} \mathrm{d}t = \frac{q}{C} = \frac{K_p x_i}{C} \tag{11.2}$$

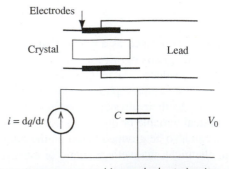

FIGURE 11.6 A piezoelectric pressure sensor and its equivalent circuit model.

In addition to piezoelectric sensors, potentiometric pressure sensors, inductive pressure sensors, capacitive pressure sensors, and strain-gauge pressure sensors are also widely used in practical applications.

Application of Pressure Sensors

Describe some applications of pressure sensors.

SOLUTION

Some applications of pressure sensors are listed below:

1. *Fuel pressure:* Pressure sensors are used to measure fuel pressure during fuel pumping.
2. *Fluid flow:* Pressure sensors are used to measure the differential pressure across an orifice. For example, intake airflow or engine coolant flow is measured in this way.
3. *Air pressure:* An aneroid barometer (a kind of pressure sensor) is used to measure the air pressure outside an aircraft.
4. *Tire pressure:* Pressure sensors are used to monitor the pressure in vehicle tires.
5. *Blood pressure:* A small sensor is penetrated into blood vessels to measure blood pressure.
6. *Altitude sensing:* The relationship between changes in pressure relative to altitude is used to measure the altitude in aircraft, rockets, satellites, and weather balloons.

EXAMPLE 11.2   **Pressure Sensors**

In piezoelectric sensors, the output voltage of the sensors can be measured to obtain the pressure from the object according to the equation:

$$V_{\mathrm{o}} = K_{\mathrm{E}} p x$$

where K_{E} is the voltage sensitivity of the sensor. For this example, assume $K_{\mathrm{E}} = 0.055$ V m/N. In this equation, p is the pressure in N/m^2 and x is the displacement in m. If an object is placed on the sensor and measures $V_{\mathrm{o}} = 3$ V, and $x = 1$ mm, determine the pressure from the object.

SOLUTION

Using the equation discussed in this section, and replacing for the voltage, displacement, and the parameter K_{E}, note that:

$$p = \frac{V_{\mathrm{o}}}{K_{\mathrm{E}} x} = \frac{3}{0.055 \times 0.001} = 5.45 \times 10^4 \text{ N/m}^2$$

11.2.2 Temperature Sensors

Temperature sensors measure the temperature, which is one of the most frequently measured physical quantities. Thermocouples, resistance temperature devices (RTDs), thermistors, and infrared thermometers represent different types of temperature sensors. The most common types of temperature sensors are thermocouples and RTDs. Thermistors are part of the RTD family.

11.2.2.1 THERMOCOUPLES
Thermocouples are the most common electrical output sensors used to that measure temperature. A thermocouple is formed by two dissimilar metals that are connected at one end (the measuring junction) and connected to a voltage-measuring instrument at the other end (the reference junction). This configuration is illustrated in Figure 11.7. Whenever the measuring junction is at a

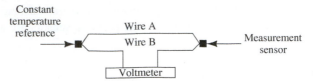

FIGURE 11.7 Thermocouple configuration.

temperature different from the reference junction, a voltage difference is produced across the two metals. Therefore, the temperature difference between the two junctions can be detected by measuring the change in the voltage across two dissimilar metals at the temperature measurement junction. This voltage varies with the temperature disparity of the junctions. If the temperature at one junction is known, the temperature at the other junction can be calculated.

Thermocouples generate an open-circuit voltage, called *Seebeck* voltage. This voltage is proportional to the temperature difference between the measuring and reference junctions.

Because thermocouple voltage is a function of the temperature difference between junctions, the voltage and reference junction temperature must be known to determine the temperature at the measuring junction. Consequently, a thermocouple measurement system must either measure the reference junction temperature or maintain it at a fixed, known temperature. Most industrial thermocouple measurement systems choose to measure, rather than to control, the reference junction temperature due to cost.

Thermocouples function based on the fact that the electromotive force (EMF) between two dissimilar metals is a function of their temperature difference. Three major effects are involved in a thermocouple circuit: the *Seebeck*, *Peltier*, and *Thomson effects*.

Seebeck effect: When two dissimilar conductors, A and B, comprise a circuit as shown in Figure 11.8, a current will flow in that circuit as long as the two junctions are at different temperatures, one junction at a temperature, T, and the other at a higher temperature $T + \Delta T$. The current will flow from A to B at the colder junction when conductor A has a higher potential with respect to B.

The Peltier, and Thomson effects, are related to the Seebeck effect.

Peltier effect: When current flows across a junction of two dissimilar conductors, heat is absorbed or created at the junction, depending upon the direction of current flow.

Thomson effect: When current flows through a conductor (generally in a long, thin bar to reduce thermal conductivity) along which a temperature gradient exists, heat is absorbed by or delivered from the wire.

The Seebeck effect describes the electromotive force (EMF) created across two dissimilar metallic materials. The change in material EMF with respect to a change in temperature is called the *Seebeck coefficient* or thermoelectric sensitivity. This coefficient is usually a nonlinear function of temperature.

A sample *thermocouple circuit* is illustrated in Figure 11.9. (Note: the ice bath shown is not a necessary component of every thermocouple, but is shown here for demonstration purposes.)

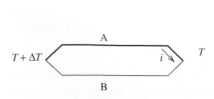

FIGURE 11.8 Seebeck effect.

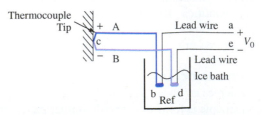

FIGURE 11.9 Example of a thermocouple circuit.

The voltage output, V_o, measured at the output of the thermocouple corresponds to:

$$V_o = \int_a^b S_{Lead}(T)\,dT + \int_b^c S_A(T)\,dT + \int_c^d S_B(T)\,dT + \int_d^e S_{Lead}(T)\,dT \qquad (11.3)$$

where b and d represent the reference points; c represents the probe tip; and a and e represent output terminals (see Figure 11.9). In addition, $S_A(t)$, $S_B(t)$, and $S_{Lead}(t)$ are the Seebeck coefficients (unit: volt/degree C) of two dissimilar metallic materials, metals A and B, and the lead wires, respectively. In general, all three Seebeck coefficients are nonlinear functions of temperature, T.

The voltage induced by the temperature of the lead wires cancels with each other, that is, $\int_a^b S_{Lead}(T)\,dT + \int_d^e S_{Lead}(T)\,dT = 0$. Thus:

$$V_o = \int_b^c S_A(T)\,dT + \int_c^d S_B(T)\,dT \quad = \int_b^c \left[S_A(T) - S_B(T) \right] dT$$

The second equality is due to the fact that the temperature b equals the temperature d. If the Seebeck coefficient functions of the two thermocouple wire materials are pre-calibrated and the reference temperature is known (usually set by a 0°C ice bath), the temperature at the probe tip becomes the only unknown and can be directly related to the voltage readout.

If the Seebeck coefficients are nearly constant across the targeted temperature range, Equation (11.3) can be simplified to:

$$V_o = (S_A - S_B)(c - b) \Rightarrow c = b + \frac{V_o}{S_A - S_B} \qquad (11.4)$$

Here, C is the temperature at the tip and b is the reference temperature; thus,

$$T_{Tip} = T_{Ref} + \frac{V_o}{S_A - S_B} \qquad (11.5)$$

Equation (11.5) represents a linear relationship between the tip temperature and the output voltage. In practice, this linear relationship is not available and vendors provide calibration functions for their products. These functions are usually high-order polynomials and are calibrated with respect to a certain reference temperature, for example, 0°C (32°F). Assuming the coefficients of the calibration polynomials are $\alpha_0, \alpha_1, \alpha_2, \ldots, \alpha_n$, the relationship of the temperature at the probe tip and the output voltage corresponds to:

$$T_{Tip} = \alpha_0 + \alpha_1 V_o + \alpha_2 V_o^2 + \cdots + \alpha_n V_o^n \qquad (11.6)$$

where V_o is the thermocouple voltage in volts, T_{Tip} is the temperature in degree Celsius, and α_0 through α_n are coefficients that are specific to each thermocouple type. Note that Equation (11.6) is effective only if the reference temperature in the experiment is kept constant.

EXAMPLE 11.3 Thermocouple

The coefficients of calibration polynomials of a K-type thermocouple (commonly used due to their low cost) are shown in Table 11.1, and the voltage reading of the thermocouple is 3.47 mV. Find the room temperature computed by a K-type thermocouple, if the thermocouple considers the boiling water temperature as a reference temperature, while the other end is measuring the room temperature. Assuming the actual room temperature is 16°C, discuss the error in the thermocouple reading. Table 11.1 represents Equation (11.6) coefficients.

(continued)

EXAMPLE 11.3   **Continued**

TABLE 11.1 α_n Coefficient for a K-Type Thermocouple

α_n	Coefficient
0	0.226584602
1	24152.10900
2	67233.4248
3	2210340.682
4	−860963914.9
5	4.83506×10^{10}
6	$−1.18452 \times 10^{12}$
7	1.38690×10^{13}
8	$−6.33708 \times 10^{13}$

SOLUTION

Using Equation (11.6):

$$T = \sum_{n=0}^{\infty} a_n V^n$$

the thermocouple reference temperature is $T = 84.83°C$. Assuming that the water is boiled at 100°C, the room temperature reading with thermocouple reference is:

$$T = 100 - 84.83$$
$$T = 15.17°C$$

For this K-type thermocouple, the reading error is in the order of 0.83°C (that is less than 1°C). Therefore, the answer obtained from the thermocouple is reasonable.

11.2.2.2 THERMISTORS

A resistance temperature detector (RTD) is a variable-resistance device whose resistance is a function of temperature (Figure 11.10). RTDs are more accurate and stable than thermocouples.

Similar to an RTD, the **thermistor** (bulk semiconductor sensor) uses resistance to detect temperature. A thermistor is a resistance thermometer and is usually manufactured in the shape of beads, discs, or rods. It is made by combining two or more metal oxides. If oxides of cobalt, copper, iron, nickel, or tin are used, the resulting semiconductor will have a negative temperature coefficient (NTC) of resistance.

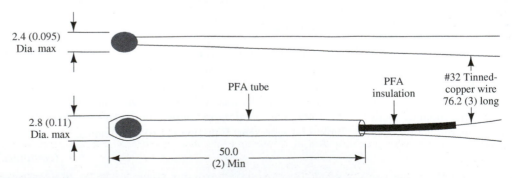

FIGURE 11.10 Precision thermistor elements, 44000 Series.

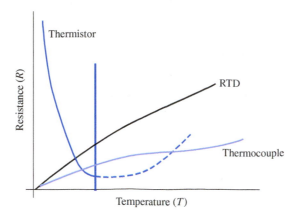

FIGURE 11.11 Characteristics of three temperature sensors.

Unlike an RTD's metal probe where the resistance increases with temperature, the thermistor responds inversely with temperature. As mentioned earlier, the relationship between resistance and temperature is nonlinear. In a thermistor, the resistance changes negatively and sharply with a positive change in temperature, as shown in Figure 11.11.

Note that in Figure 11.11, the resistance of the thermistor increases with temperature if the temperature is higher than a threshold. However, in this temperature range, the thermistor fails to give a correct temperature reading. Thermistors are fairly limited in their temperature range, working only over a nominal range of 0°C to 100°C.

The negative resistance–temperature relationship of a thermistor might be expressed by an exponential (i.e., nonlinear) function:

$$R = R_0 \times e^{\beta(\frac{1}{T} - \frac{1}{T_0})} \tag{11.7}$$

where

- T = absolute temperature (in kelvin)
- T_0 = reference temperature, usually the room temperature (25°C; 77°F; 298.15 K)
- R = resistance at the temperature T (Ω)
- R_0 = resistance at T_0
- β = calibration constant that varies with the thermistor material (usually between 3000 and 5000 K).

To solve for the temperature, T, take the logarithm (ln) of both sides of the Equation (11.7) because R and T are positive real numbers, and solve for $1/T$:

$$\frac{1}{T} = \frac{1}{T_0} + \frac{1}{\beta}\left[\ln(R) - \ln(R_0)\right] \tag{11.8}$$

Accordingly:

$$T = \frac{T_0 \times \beta}{\beta + T_0\left[\ln(R) - \ln(R_0)\right]} \tag{11.9}$$

If Equation (11.7) is differentiated and rearranged, the sensitivity of the thermistor, α, at the temperature, T, corresponds to:

$$\alpha \approx \frac{1}{R}\frac{dR}{dT} = -\frac{\beta}{T^2} \tag{11.10}$$

Because the constant β may vary slightly with temperature, several well-known temperature conditions may be used as check points, for example, ice water at 0°C (32°F) and boiling water at 100°C (212°F); other pre-calibrated thermometers may also be used to calibrate/curve-fit the value of β.

However, β may vary considerably across the temperature range of interest. In this case, a calibrated curve-fit of the R–T relationship may be used, rather than Equation (11.9). A suitable nonlinear function that can be generated via curve fitting and converts the measured resistance, R, of a thermistor to temperature, T, is:

$$\frac{1}{T} = A + B \ln R + C(\ln R)^3 \qquad (11.11)$$

where R is measured in ohms, T in kelvin, and A, B, and C are curve-fitting constants. The A, B, and C values can be obtained by substituting three pairs of values of (R,T) within the operating range into Equation (11.11) to generate three equations. Then the A, B, and C constants can be acquired after solving the three generated equations.

EXERCISE 11.1

Assume the following measurement (R,T) pairs, (100,280), (95,290), (90,310). Use Equation (11.11) to find the constants, A, B, and C. (*Hint:* write three equations for constants A, B, and C; MATLAB can be used to solve the equations.)

EXAMPLE 11.4   Applications of Temperature Sensors

Describe some applications of temperature sensors.

SOLUTION

Some applications of temperature sensors include the following:

1. *Air conditioning:* Temperature sensors are used to measure the temperature of the air to control air conditioners.
2. *Heating systems:* Temperature sensors are used to control heating systems to maintain the temperature in a room.
3. *Cooling systems in computers:* When a personal computer is running for a long time, the motherboard generates a great deal of heat. Using the temperature sensor to sense the temperature inside the computer's case, the computer can automatically turn on the fan to reduce system temperature. It may also shut down the computer when high temperatures are detected.
4. *Food transport and storage:* Temperature sensors are used to make sure food does not exceed safe temperatures to minimize harmful bacterial growth.

APPLICATION EXAMPLE 11.5   RTD in Vehicles

The resistance of an RTD over a small temperature range can be expressed as:

$$R_{RTD} = R_o\left[1 + \alpha(T - T_0)\right]$$

The RTD has the resistance of 30 Ω at a temperature of 0°C. Assume the coefficient of the resistance, α, for this RTD is 0.004225°C^{-1}. The RTD is placed on a running car engine to measure the temperature. The measured resistance of the RTD is 36.78 Ω. Determine the car engine's temperature.

SOLUTION

Here, $R_o = 30$ Ω, and $\alpha = 0.004225$°C^{-1}. Using the equation introduced in the example:

$$T = \frac{\dfrac{R_{RTD}}{R_o} - 1}{\alpha} + T_0 = \frac{\dfrac{36.78}{30} - 1}{0.004225} + 0 = 53.49°C$$

11.2.3 Accelerometers

Accelerometers (acceleration sensors) are used to measure acceleration, vibration, and mechanical shock. Acceleration is the time rate of change of velocity with respect to a reference system. Acceleration is a vector quantity, and is closely related to velocity and displacement. The linear and angular relationships for movement correspond to:

$$\text{Linear:} \quad v = \frac{dx}{dt}, \quad a = \frac{d^2x}{dt^2} \tag{11.12}$$

$$\text{Angular:} \quad \omega = \frac{d\theta}{dt}, \quad \gamma = \frac{d^2\theta}{dt^2} \tag{11.13}$$

where

- x = linear displacement
- θ = angular displacement
- v = linear velocity
- ω = angular velocity
- α = linear acceleration
- γ = angular acceleration
- t = the time.

In general, accelerometers sense acceleration by incorporating a seismic mass on which the external acceleration can be applied. They measure the displacement of the seismic mass, the force exerted by the seismic mass against the package, or the force required to keep them in place.

A typical accelerometer is illustrated in Figure 11.12. A seismic mass, m, is suspended by a spring, k, in a sensor package. When the sensor is moved with an acceleration, a, a relative displacement, x_{rel}, of the seismic mass is produced by the inertial force and is detected as an electrical signal. The motion equation corresponds to:

$$ma = m\frac{d^2x_{rel}}{dt^2} + \lambda\frac{dx_{rel}}{dt} + kx_{rel} \tag{11.14}$$

where

- k is the spring constant
- λ is the damping constant.

In the steady state, the relationship between the displacement, x_{rel}, and the acceleration, a, corresponds to:

$$F = ma = kx_{rel}$$

Accordingly:

$$\frac{x_{rel}}{a} = \frac{m}{k} \tag{11.15}$$

The displacement of the seismic mass, x_{rel}, can be detected by either monitoring the strain induced in the spring or directly measuring the position of the seismic mass.

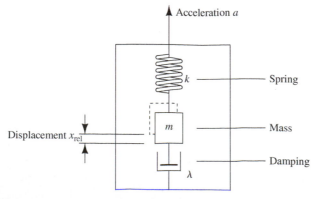

FIGURE 11.12   Accelerometer.

EXAMPLE 11.6	Applications of Accelerometers

Describe some applications of accelerometers.

SOLUTION

Some applications of accelerometers include the following:

1. *Sport watches:* When runners wear sport watches, their acceleration and speed can be determined by watches that contain accelerometers.
2. *Digital cameras:* Some digital cameras contain accelerometers to determine the orientation of the photo being taken.
3. *Rockets:* Accelerometers are used in rocket design to maintain the rocket's stability.
4. *Vehicles:* Accelerometers are used widely in vehicles to measure motion of the vehicle. A typical application is that accelerometers can help decide whether air bags should be deployed.

EXAMPLE 11.7	Accelerometer Measurement

An engineer intends to measure the acceleration of an elevator to make the elevator more comfortable for the passengers. The seismic mass, m, for the accelerometer is 50 kg and the spring constant for the accelerometer is 1.0×10^3 N/m. If the measured mass displacement of the accelerometer is 0.1 m, what is the acceleration of the elevator?

SOLUTION

According to the Equation (11.15):

$$\frac{x_{\text{rel}}}{a} = \frac{m}{k}$$

Using this equation:

$$a = \frac{k}{m}x_{\text{rel}} = \frac{1.0 \times 10^3}{50} \times 0.1 = 2 \text{ m/s}^2$$

11.2.4 Strain-Gauges/Load Cells

As discussed in Chapter 1, a load cell is a sensor that converts force into a measurable electrical output. Strain-gauge-based load cells are the most commonly used type of load cells. Strain is defined as the displacement and deformation when a force is applied to an object. Strain can be measured by a strain gauge. Strain-gauge load cells convert the load acting on them into electrical signals. The gauges used in the load cells are bonded onto a beam or structural member that deforms when weight is applied. When weight is applied, the strain changes the electrical resistance of the gauges in proportion to the load.

Load cell designs differ as the type of output signal generated (pneumatic, hydraulic, electric) changes. Load cells may also be categorized according to the technique they use to detect weight (bending, shear, compression, tension, etc.). This section examines the bonded semiconductor strain gauge as an example. The bonded semiconductor strain gauge depends on the piezo-resistive effects of silicon and measures the change in the resistance with stress as opposed to strain. In order to measure strain with a bonded resistance strain gauge, it must be connected to an electric circuit that is capable of measuring small changes in resistance corresponding to strain.

A *Wheatstone bridge* is a circuit used to measure static or dynamic electrical resistance. The output voltage of the Wheatstone bridge is expressed in millivolts output per volt input. The Wheatstone circuit is also well suited for temperature compensation. Strain gauge transducers usually employ four strain-gauge elements electrically connected to form a Wheatstone bridge circuit as shown in Figure 11.13. Here, the output voltage corresponds to:

$$V_{\text{o}} = V_{\text{i}}\left[\frac{R_3}{R_3 + R_{\text{g}}} - \frac{R_2}{R_1 + R_2}\right] \tag{11.16}$$

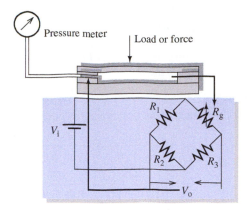

FIGURE 11.13 Strain-gauge load cell using a Wheatstone bridge.

If R_1, R_2, R_3, and R_g are equal, V_o will be zero. However, if R_g is changed to some value that does not equal R_1, R_2, and R_3, the bridge will become unbalanced and V_o will be greater than zero. The sensor may occupy one, two, or four arms of the bridge, depending on the application. The voltage, V_o, is equivalent to the difference between the voltage drop across R_1 and R_g.

In the field of measurement science, there are two ways to measure quantities: the "mass balance" (also called "null system") and the "bathroom or fish scale" methods. Like the mass balance scale, the Wheatstone bridge compares an unknown resistance against a known value until both are equal. On the other hand, a standard ohmmeter is an electronic device that converts resistance to current. The accuracy of a standard ohmmeter is dependent on a battery just like the accuracy of the bathroom or fish scale is dependent on the gravity constant.

The Wheatstone bridge circuit is almost universally used in load cells and other strain-gauge sensors because it facilitates the cancellation of unwanted temperature effects. Most strain-gauge materials are sensitive to temperature variations and their resistance tends to change resistance as they age.

Through proper flexure design and gauge placement, a linear relationship between the applied force and the sensed strain is achievable. Therefore, Equation (11.16) can be expressed by:

$$V_o = V_i \, K \, F \tag{11.17}$$

where K is the calibration factor and F is the input force.

EXERCISE 11.2

Verify Equation (11.16) for the Wheatstone bridge of Figure 11.13. Assuming, $R_1 = 1$ kΩ, $R_2 = 2$ kΩ, $R_3 = 1.5$ kΩ, and $R_g = 2$ kΩ, $V_i = 5$ V, find the voltage, V_o.

EXAMPLE 11.8 **Applications of Load Cells**

Describe some applications of load cells.

SOLUTION

Some applications of load cells include the following:

1. *Weighing of cargo:* Load cells can be used to weigh trucks, tanks, or vessels.
2. *Tension measurement:* Load cells can be used to measure underwater tension.

EXAMPLE 11.9 **Strain-Gauge Load Cell**

For a strain-gauge load cell, the relationship between the output voltage and the mass weighed is:

$$M = 3.5 V_o \text{ kg}$$

(continued)

EXAMPLE 11.9   **Continued**

The load cell has the same structure as shown in Figure 11.13. $V_i = 5$ V, and $R_1 = R_2 = R_3 = 100$ Ω. If $R_g = 80$ Ω, what is the mass?

SOLUTION

According to Equation (11.16), the output voltage corresponds to:

$$V_o = V_i \left[\frac{R_3}{R_3 + R_g} - \frac{R_2}{R_1 + R_2} \right]$$

$$= 5 \left[\frac{100}{100 + 80} - \frac{100}{100 + 100} \right]$$

$$= \frac{5}{18}$$

$$\approx 0.278 \text{ V}$$

According to the relationship between the output voltage and the mass, which is given in this example:

$$M = 3.5 \, V_o \text{ kg}$$
$$= 3.5 \times 0.278 \text{ kg}$$
$$= 0.973 \text{ kg}$$

11.2.5 Acoustic Sensors

A sound sensor is a device that converts acoustic energy into electrical energy. The frequency range of sound is extremely wide. Sound propagates in various media, such as solids, fluids, or gases. Accordingly, a wide range of sound sensors have been designed for different environments. This section concentrates on sound sensors, such as microphones, which are utilized for the detection of airborne sound in the audio-frequency range.

Generally, microphones consist of a diaphragm that vibrates by impinging waves of acoustic pressure. The motion of this diaphragm is converted into alternating electromotive forces or electric currents by a transducing mechanism. Piezoresistive microphones, capacitive microphones, and piezoelectric microphones are the most common types of microphones.

A typical capacitive microphone consists of a diaphragm, a back plate, a polarization voltage, resistors, and capacitors as depicted in Figure 11.14. The equivalent circuit of a capacitive microphone is shown in Figure 11.15. The diaphragm is placed close to the rigid metal back plate. The parallel diaphragm and back plate form a parallel plate capacitor. The sound pressure fluctuations alter the capacitor plate space and results in the capacitance changes [$\Delta C(t)$]. The output voltage across the capacitor fluctuates with change in capacitance. In Figure 11.14, the polarization voltage, E_0, is applied to generate an initial charge, Q_0, on the capacitors. If

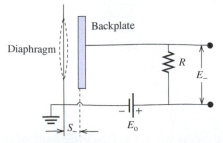

FIGURE 11.14   The basic principle of a capacitor microphone.

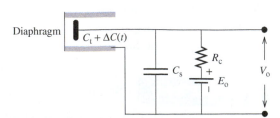

FIGURE 11.15   Equivalent circuit of a capacitive microphone.

charge is generated by the capacitor itself, sensors using such capacitors are often called electret microphones (a typical condenser microphone).

Now, considering Figure 11.15, the total capacitance, C_o, of the microphone corresponds to:

$$C_o = C_t + C_s \qquad (11.18)$$

In Figure 11.15, C_s represents the stray capacitance in the sensor, and C_t represents the capacitance of the diaphragm. If C_t is constant, the capacitor will be equivalent to an open circuit. In this case, the voltage across the capacitor C_o is:

$$V_o = E_o \qquad (11.19)$$

and the charge stored on the capacitor (C_o) is:

$$q_c = C_o E_o \qquad (11.20)$$

The resistance, R_c, is large to ensure that the time constant, $R_c C_o$, is long compared to the lowest sound frequency to be measured. Accordingly, the charge stored in the capacitor will approximately be constant. If the charge is kept constant, a change in capacitance will lead to a change in voltage. When the capacitance varies, the following equation is used:

$$q_c = (C_o + \Delta C)V_c \qquad (11.21)$$

Here, the voltage variation corresponds to:

$$\Delta V_o = V_o - V_c \qquad (11.22)$$

Using Equations (11.19) to (11.22):

$$\begin{aligned}
\Delta V_o &= V_o - V_c \\
&= E_o - \frac{C_o E_o}{C_o + \Delta C} \\
&= \frac{\Delta C}{C_o + \Delta C} E_o
\end{aligned} \qquad (11.23)$$

Now, because $C_o \gg \Delta C$, Equation (11.23) is simplified to:

$$\Delta V_o \approx \frac{\Delta C}{C_o} E_o \qquad (11.24)$$

Thus, as the capacitance of C_t changes, V_o will change. To ensure that the deflection of the diaphragm for a given sound pressure is independent of frequency and to achieve proportionality between the sound pressure and the output voltage, a stiffness-controlled operation of the membrane is required. Therefore, the fundamental resonance frequency of the membrane is placed at the upper limit of the operating frequency range.

EXAMPLE 11.10   Applications of Acoustic Sensors

Describe some examples of real-world applications of acoustic sensors.

SOLUTION

Some applications of acoustic sensors include the following:

1. *Avalanche warning:* By measuring the wind or air pressure, acoustic sensors can be used for avalanche warning and research.
2. *Oceanographic data collection:* Acoustic sensors are placed on the seafloor. Ranging measurements can be achieved by acoustic sensors that transmit and receive an acoustic signal from a near-surface projector and by noting timing and/or position measurements. These sensors help offshore exploration, reduce the impact of tsunami, aid in navigation, etc.

EXAMPLE 11.11	**Equivalent Impedance and Resistance of Capacitive Microphone**

Find the equivalent impedance of a capacitive microphone shown in Figure 11.15. Next, approximate the equivalent resistance of a capacitive microphone.

SOLUTION

The equivalent impedance is:

$$Z = \frac{\dfrac{R_c}{j\omega C_o}}{R_c + \dfrac{1}{j\omega C_o}}$$

$$= \frac{R_c}{1 + j\omega R_c C_o}$$

$$= \frac{R_c(1 - j\omega R_c C_o)}{1 + (\omega R_c C_o)^2}$$

$$= \frac{R_c}{1 + (\omega R_c C_o)^2} - j\frac{\omega R_c^2 C_o}{1 + (\omega R_c C_o)^2}$$

Because the time constant $R_c C_o$ is large we can take $\omega R_c C_o \gg 1$:

$$Z \simeq \frac{1}{\omega^2 R_c C_o^2} + \frac{1}{j\omega C_o} = R_T + \frac{1}{j\omega C_o}$$

Thus, the equivalent resistance of a microphone is:

$$R_T = \frac{1}{\omega^2 R_c C_o^2}$$

EXERCISE 11.3

Replace the capacitance in Z in Figure 11.15 with a 10-mH inductance and repeat Example 11.11.

APPLICATION EXAMPLE 11.12	**Microphone Impedance**

Most microphones are designed while their impedance is not "matched" to the load to which they are connected. Impedance matching has been discussed in detail in Chapter 6. Impedance unmatch alters their frequency response and leads to a distortion effect (hum effect), especially at high sound pressure levels.

   An engineer intends to test a new microphone and see whether it produces sound that is comfortable for listening. Based on the impedance calculated in Example 11.11, the circuit is shown in Figure 11.16. In this figure, $5\cos(3000t)$ is the output voltage of the microphone,

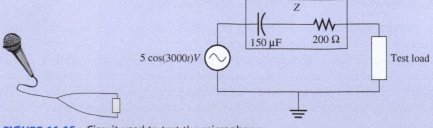

FIGURE 11.16   Circuit used to test the microphone.

and Z is its equivalent impedance. Determine the load at which the engineer needs to test the microphone to establish a "worst-case" sample.

SOLUTION

To test the relative listening comfort of the microphone's hum, the load of the microphone must match its impedance. Based on the concept of impedance matching studied in Chapter 6:

$$Z_L = Z^* = 200 - \frac{1}{j3000 \times 150 \times 10^{-6}} = 200 + j2.22 \ \Omega$$

Because the imaginary part of Z_L is positive, it must be inductive. Therefore, Z_L consists of a resistor and an inductor in series, which is shown in Figure 11.17.

$$R_L = 200 \ \Omega$$

$$L = \frac{2.22}{3000} = 0.74 \ \text{mH}$$

200 Ω   0.74 mH

FIGURE 11.17
Equivalent circuit for the load.

11.2.6 Linear Variable Differential Transformers (LVDT)

The linear variable differential transformer (LVDT) is a displacement sensor, which consists of a transformer with a single primary coil and two identical secondary coils connected in a series-opposing manner, as depicted in Figure 11.18.

An LVDT is operated based on the mutual inductance concept that is discussed in more detail in Chapter 12. The primary coil is fed from an AC excitation supply and induces voltage across the two secondary coils. The output voltage is given by:

$$v_o = v_1 - v_2 \tag{11.25}$$

The iron core between the primary and secondary coils can be displaced by an external motion. This changes the magnetic coupling between the primary and secondary coils. The induced voltages have equal magnitudes when the iron core is in the central position, and the voltage output $v_o \approx 0$. An upward displacement of the iron core from the central position increases the coupling (mutual inductance) between the primary and the top secondary coil, while decreasing the coupling for the other secondary coil, and correspondingly induces a greater voltage in the top secondary coil.

As a result, $v_o > 0$ when the iron core has an upward displacement, and $v_o < 0$ when the iron core is displaced downward. If the primary coil has the resistance, R_p, and the self-inductance, L_p, then:

$$R_p \cdot i + L_p \cdot \frac{di}{dt} = v_s \tag{11.26}$$

Note that at the left side of Equation (11.26), the total voltage induced from secondary coils is zero. This is because the current directions in secondary coils are opposite. Thus, their

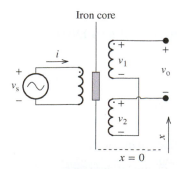

FIGURE 11.18   Linear variable differential transformer.

overall effect on the induced voltages at the primary coil would be zero. Then, the voltages induced in the secondary coils correspond to:

$$v_1 = M_1 \cdot \frac{di}{dt} \tag{11.27}$$

and

$$v_2 = M_2 \cdot \frac{di}{dt} \tag{11.28}$$

Thus:

$$v_o = (M_1 - M_2) \cdot \frac{di}{dt} \tag{11.29}$$

where M_1 and M_2 are the mutual inductances between the primary and the respective secondary coils. The position of the iron core determines M_1 and M_2. When the core is at the central position, $M_1 = M_2$, and $v_o = 0$; when the core is displaced upward away from the central position, $M_1 > M_2$, and $v_o > 0$; when the core is displaced downward away from the central position, $M_1 < M_2$, and $v_o < 0$.

Because an LVDT is usually excited by an AC signal, the current, i, can be represented by:

$$i = \cos(\omega t - \phi) \tag{11.30}$$

and v_1, v_2, and v_o correspond to:

$$v_1 = M_1 \sin(\omega t - \phi), \tag{11.31}$$
$$v_2 = M_2 \sin(\omega t - \phi), \tag{11.32}$$
$$v_o = v_1 - v_2 = (M_1 - M_2)\sin(\omega t - \phi) \tag{11.33}$$

The magnitude of the output voltage, v_o, will be the same if the magnitude of displacements $+x$ or $-x$ of the iron core away from the central position is equal. The directional information of the movement of the iron core is available in the phase of the output voltage: the movement in the two different directions differs by 180°. As a result, the amplitude and the phase of the output voltage, v_o, depends on the displacement, x.

In other words, the amplitude of v_o is modulated by the displacement. The relationship between the magnitude of the output voltage and the displacement, x, of the iron core is approximately linear over a reasonable range of movement of the iron core on either side of the central position. It is expressed by a constant of proportionality, G, as:

$$x = G \cdot v_o \tag{11.34}$$

where G is called the sensitivity (or gain) of the transformer.

EXAMPLE 11.13   Applications of LVDT

Describe some applications of LVDT.

SOLUTION

Some applications of LVDT include the following:

1. *Weighing systems:* LVDTs can be used to measure spring deformation in weighing systems. The measured displacement can be used to calculate the applied force based on the characteristics of the spring.

2. ***Displacement sensing:*** An LVDT may be mounted externally in parallel with the cylinder, thus making the core assembly free to move with the piston rod, and allowing measurement of the displacement of a piston.

3. ***Bill detector in ATMs:*** Using an LVDT, ATMs can detect the number of bills inserted. When bills pass, a motion signal is transferred to the LVDT core element and then a changing output signal is generated. As the bills pass between the rollers, the voltages vary according to the thickness of the bills.

EXAMPLE 11.14   **Displacement in LVDT**

For an LVDT shown in Figure 11.19, an engineer can obtain the displacement by measuring the voltage output according to Equation (11.34). Assume $G = 0.7282$ mm/V. Determine the displacement if the measured instant voltage $v_o = 5$ V.

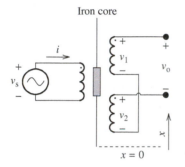

FIGURE 11.19   Circuit for the LVDT Example 11.13.

SOLUTION

Using Equation (11.34) and replacing for G and v_o:

$$x = G \times v_o = 0.7282 \times 5 = 3.641 \text{ mm}$$

11.3 SIGNAL CONDITIONING

The direct output of a sensor is generally not available in a form that computers can process. Signal conditioning is required to convert sensor output to an appropriate form. The two most important signal conditioning functions are *amplification* and *filtering*. If necessary, signal conditioning can also convert currents to voltages, or supply excitations to the sensors to convert a change in resistance, inductance, or capacitance to a change in voltage. This section describes amplifiers and filters.

11.3.1 Amplifiers

An amplifier is a device that increases the power contained within a signal. Often a low-level signal from a sensor needs to be amplified. Operational amplifier circuits have been discussed in Chapter 8.

11.3.2 Active Filters

Sensor signals may encounter interference from noise or undesired input. Therefore, filters are required to eliminate interference and noise effects. Active filters consist of resistors, capacitors, and amplifiers, whereas passive filters consist of resistors, inductors, and capacitors. Chapter 7

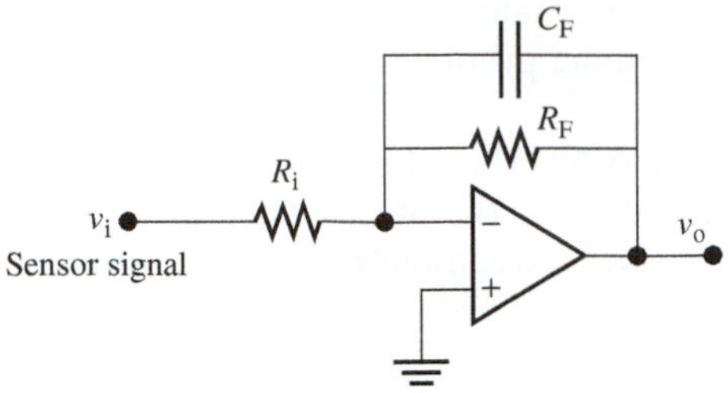

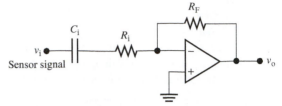

FIGURE 11.20 First-order, low-pass filter with an operational amplifier.

FIGURE 11.21 First-order, high-pass filter with an operational amplifier.

reviewed different types of passive filters in detail, including low-pass, high-pass, and band-pass filters.

Compared to active filters, passive filters have some disadvantages. First, inductors selected for the audio-frequency range, in which most mechanical measurements are made, are large and expensive. Second, passive filters restrict the sharpness of the cutoff frequency because they do not have power amplification, leading to loading in measurements. Finally, passive filters are limited to frequency selection.

Most active filters are built using operational amplifier circuits and different characteristics are defined by amplified input and feedback components that consist of resistors and capacitors. Some advanced active filters, such as Butterworth filters and Chebyshev filters, have widespread applications in instrumentation circuits.

Two examples of active filters are shown in Figures 11.20 and 11.21. A first-order, low-pass filter is shown in Figure 11.20.

According to Equation (8.54), the gain of an inverting amplifier corresponds to:

$$A_L(j\omega) = \frac{V_o}{V_i} \tag{11.35}$$

Based on Equation (8.55):

$$\frac{V_o}{V_i} = -\frac{Z_F}{Z_i} \tag{11.36}$$

where Z_F is the impedance of a paralleled capacitor, C_F, and a resistor, R_F, that is:

$$Z_F = \frac{1}{j\omega C_F} \parallel R_E = \frac{R_F}{1 + j\omega R_F C_F} \tag{11.37}$$

and

$$Z_i = R_i \tag{11.38}$$

Therefore, the gain of the low-pass, active filter is:

$$A_L(j\omega) = -\frac{\frac{1}{j\omega C_F} \parallel R_F}{R_i} = -\frac{\frac{R_F}{1 + j\omega R_F C_F}}{R_i} = -\frac{R_F}{R_i(1 + j\omega R_F C_F)} \tag{11.39}$$

A first-order, high-pass filter is shown in Figure 11.21.

According to Equation (8.55), the gain for the inverting amplifier is:

$$A_H(j\omega) = \frac{V_o}{V_i} = -\frac{Z_F}{Z_i} \tag{11.40}$$

where

$$Z_F = R_F \tag{11.41}$$

and

$$Z_i = R_i + \frac{1}{j\omega C_i} \tag{11.42}$$

Therefore, the gain of the high-pass, active filter is:

$$A_H(j\omega) = -\frac{R_F}{R_i + \dfrac{1}{j\omega C_i}} = -\frac{j\omega R_F C_i}{1 + j\omega R_i C_i} \tag{11.43}$$

The amplitude and phase of the frequency dependent gains in Equations (11.32) and (11.43) can be sketched in terms of frequency using the Bode plot or PSpice as discussed in Chapter 7.

EXERCISE 11.4

Refer to Chapter 7, sketch the Bode plot of the amplitude of the gain in Equation (11.43), assuming $C_i = 1\ \mu F$ and resistor $R_F = R_i = 1\ k\Omega$. Investigate the impact of changing R_i to $100\ \Omega$ and $10\ k\Omega$.

APPLICATION EXAMPLE 11.15   Robotic Arm

An operational amplifier is used to design controllers for mechanical applications. Assume that the transfer function of a robotic arm is $G(s) = 1/(s^2 + 1)$. The robotic arm's transfer function is an unstable system; therefore, a proportional-integral-derivative (PID) controller is added as depicted in Figure 11.22.

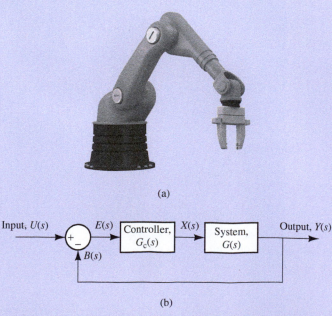

(a)

Input, $U(s)$ → $E(s)$ → [Controller, $G_c(s)$] → $X(s)$ → [System, $G(s)$] → Output, $Y(s)$

$B(s)$

(b)

FIGURE 11.22   (a) Robotic arm. (Used with permission from Nuno Andre/Shutterstock.com.) (b) Block diagram of the robotic arm mechanical system.

(continued)

APPLICATION EXAMPLE 11.15 Continued

a. Given the circuit of the PID controller shown in Figure 11.23, determine the transfer function of the controller assuming $R_1 = 2000\ \Omega$, $C_1 = 875\ \mu F$, $R_2 = 2500\ \Omega$, $C_2 = 400\ \mu F$, $R_3 = 450\ \Omega$, $R_4 = 480\ \Omega$.

b. Plot the step response of a close-loop transfer function for the robotic arm:
   i. Without a PID controller
  ii. With a PID controller

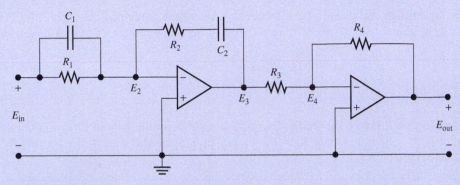

FIGURE 11.23 An operational amplifier circuit representing a PID controller.

SOLUTION

a. First, find the expression of the transfer function of the PID controller in terms of resistance and capacitance.
Using $s = j\omega$, the impedance of the capacitor can be expressed as:

$$Z_c = \frac{1}{j\omega c} = \frac{1}{CS}$$

The impedance Z_1 of R_1 and C_1 is:

$$Z_1 = \left(\frac{1}{R_1} + C_1 S \right)^{-1}$$

$$Z_1 = \frac{R_1}{R_1 C_1 s + 1}$$

The impedance Z_2 of R_2 and C_2 is:

$$Z_2 = R_2 + \frac{1}{C_2 s}$$

$$Z_2 = \frac{R_2 C_2 s + 1}{C_2 s}$$

Using the golden rules of op-amp, $E_2 = 0$ and:

$$\frac{E_{in} - E_2}{Z_1} = \frac{E_2 - E_3}{Z_2}$$

Thus,

$$\frac{E_{in}}{Z_1} = -\frac{E_3}{Z_2}$$

or

$$\frac{E_3}{E_{in}} = -\frac{Z_2}{Z_1}$$

Applying the same method to the second op-amp with $E_4 = 0$ shows:

$$\frac{E_3 - E_4}{R_3} = \frac{E_4 - E_{out}}{R_4}$$

$$\frac{E_3}{R_3} = -\frac{E_{out}}{R_4}$$

$$\frac{E_{out}}{E_3} = -\frac{R_4}{R_3}$$

Thus, the transfer function of the PID controller is:

$$\frac{E_{out}}{E_{in}} = \frac{E_{out}}{E_3} \times \frac{E_3}{E_{in}} = \frac{R_4}{R_3} \cdot \frac{Z_2}{Z_1}$$

or: $\quad \frac{E_{out}}{E_{in}} = \frac{R_4 R_2}{R_3 R_1} \frac{(R_1 C_1 s + 1)(R_2 C_2 s + 1)}{R_2 C_2 s}$

Substituting the resistance and capacitance results in:

$$G_c = \frac{2.333\,(s + 1)\,(s + 0.5714)}{s}$$

b. Define the $E(s)$, $X(s)$, and $B(s)$ as shown in Figure 11.22 with $U(s)$ as the input and $Y(s)$ as the output of the system.

From Figure 11.22, $E(s) = U(s) - B(s)$. In addition:

$$\frac{X(s)}{E(s)} = G_c(s)$$

$$\frac{Y(s)}{X(s)} = G(s)$$

$$\frac{Y(s)}{E(s)} = \frac{Y(s)}{X(s)} \times \frac{X(s)}{E(s)} = G(s)G_c(s)$$

Using this equation and representing $E(s)$ in terms of $Y(s)$ in $E(s) = U(s) - B(s)$, and knowing that $B(s) = Y(s)$ shows:

$$\frac{Y(s)}{U(s)} = \frac{G(s)G_c(s)}{1 + G(s)G_c(s)}$$

i. Without a PID controller:

For a system without a PID controller, let the $G_c(s) = 1$. Then:

$$\frac{Y(s)}{U(s)} = \frac{G(s)}{1 + G(s)}$$

where $G(s) = 1/(s^2 + 1)$. To set up the transfer function, step response, or a Bode plot in MATLAB, first determine the vector for the numerator and denominator of the transfer function. The numerator of the $G(s)$ is 1; thus, write num = [1]. The denominator of $G(s) = s^2 + 1 = (1)s^2 + (0)s + 1$; thus, write den = [1 0 1].

Now, see that the numerator and denominator vectors can be defined in terms of the highest to the lowest order of the coefficient of "s." Use MATLAB to define the transfer function using "tf(num,den)." Next, plot the step response of the system using the function "step(G,T)." T is the total sampling time. The complete

(continued)

APPLICATION EXAMPLE 11.15   Continued

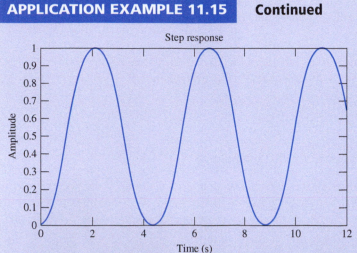

The MATLAB code is:

```
num = [1];
den = [1 0 1];
G = tf(num,den);
step(G/(1+G),12);
```

FIGURE 11.24   Step response for robotic arms.

MATLAB code is shown in Figure 11.24. Here, be aware that "step(G,T)" represents the open-loop step response. However, the system here is a closed-loop function, therefore it is "step(G/(1+G),T)."

ii. With a PID controller:

The closed-loop transfer function for the system with a PID controller is:

$$\frac{Y(s)}{U(s)} = \frac{G(s)G_c(s)}{1 + G(s)G_c(s)}$$

The given transfer function for the controller is:

$$G_c = \frac{2.333\,(s + 1)\,(s + 0.5714)}{s}$$

$$G_c = \frac{2.333\,(s^2 + 1.5714s + 0.5714)}{s}$$

Using this, define the transfer function for the controller in MATLAB, with the numerator and denominator of numc = 2.333*[1 1.5714 .5714], and denc = [1 0], respectively. The step response of the system with the PID controller is shown in Figure 11.25. Comparison of Figures 11.24 and 11.25 shows that a controller is required to let the robotic arm achieve stability during the operation.

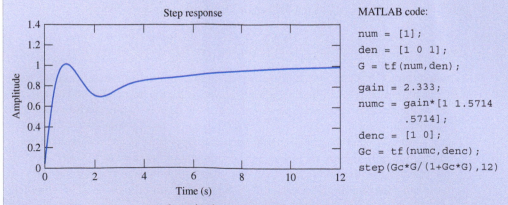

MATLAB code:

```
num = [1];
den = [1 0 1];
G = tf(num,den);

gain = 2.333;
numc = gain*[1 1.5714
        .5714];
denc = [1 0];
Gc = tf(numc,denc);
step(Gc*G/(1+Gc*G),12)
```

FIGURE 11.25   Step response for robotic arm with PID controller.

11.4 DATA ACQUISITION

A data acquisition system consists of an analog multiplexer (MUX), a buffer, a sample/hold amplifier, and an ADC as shown in Figure 11.26. As discussed in Chapter 8, a buffer made by operational amplifiers reduces the impact that two parts of a system may have on each other. A buffer has a very large (ideally infinity) input impedance and a small (ideally zero) output impedance.

11.4.1 Analog Multiplexer

A multiplexer (MUX) in a data acquisition system processes several different signals. In Figure 11.26, the channel to be monitored is selected and applied to a sampler. The sampled output is converted to digital using the multiplexer that is controlled by the control circuit, which simultaneously triggers the sampling circuit and the ADC. The multiplexer selects the next channel after the ADC completes the conversion and sends an appropriate signal to the control circuit.

11.4.2 Analog-to-Digital Conversion

Analog-to-digital converters (ADCs) are designed to transform continuous analog signals into a discrete binary code suitable for digital processing. There are two main steps in analog-to-digital conversion: *sampling* and *quantization*. First, the analog signal is sampled at regular time intervals to generate a discrete-time signal. Second, each sample of the analog voltage is converted into an equivalent digital value.

11.4.2.1 SAMPLING
The process of sampling is illustrated in Figure 11.27.

To be accurate, sampling must be conducted at a rate higher than twice the highest frequency of the analog signal. This rate is called the **Nyquist rate**. If the rate of sampling is less than the **Nyquist rate**, the reconstruction of the sample value cannot match the original analog signal. In this case, **aliasing** has occurred. Here, sampling may transmute a high-frequency signal into a lower frequency one, as illustrated in Figure 11.28. Thus, to avoid aliasing, the sampling rate, f_s, should be at least twice the highest frequency, f_H, in the sampled analog signal. That is, the Nyquist criterion corresponds to:

$$f_s \geq 2f_H \tag{11.44}$$

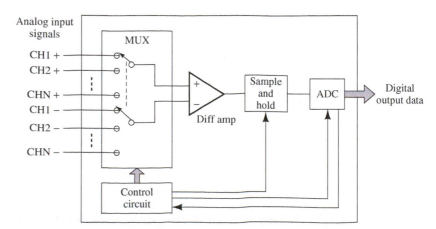

FIGURE 11.26 Data acquisition system.

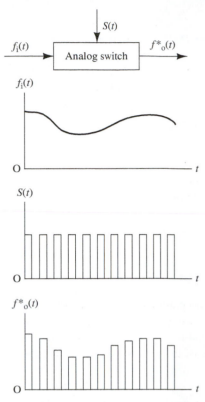

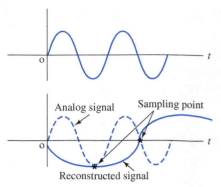

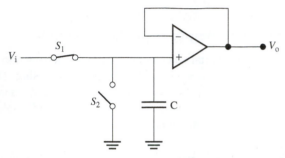

FIGURE 11.28 Conversion of an analog signal to a discrete signal if the sampling rate is not high enough.

FIGURE 11.27 Analog signal sampling.

FIGURE 11.29 A simple sample and hold circuit.

EXAMPLE 11.16   Sampling Rate

A signal has frequency components up to 60 kHz. What is the minimum sampling rate that should be used to ensure that the signal is reconstructed properly?

SOLUTION

Based on the Nyquist rate criterion:

$$f_s = 2 \times f_H = 2 \times 60 \text{ kHz} = 120 \text{ kHz}$$

11.4.2.2 SAMPLE AND HOLD

The ADC requires a certain amount of time to perform the conversion, called the conversion time. If the analog input varies within the conversion time, a conversion error may occur. To hold the input signal at a constant level within the conversion time, a sample and hold circuit is required before applying the signal to the ADC.

A typical operational amplifier circuit that performs the sample and hold function is shown in Figure 11.29. When switch S_1 is closed (S_2 is open), the capacitor, C, is charged by the input sample signal. The output, V_o, changes with the voltage of capacitor, C. When the sampled signal pulse is over (the voltage is zero), the capacitor, C, holds the sampled signal voltage until the next sampled signal pulse, and the output, V_o, keeps the same voltage. Switch S_2 is used to short-circuit the capacitor, discharging it and setting the output to 0 V, if necessary.

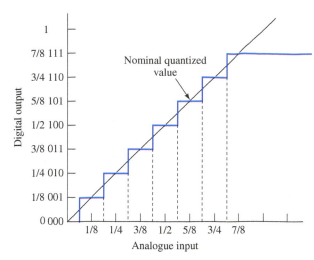

Analogue input	Binary code	Represented Analogue Signal
7/8 - 1V	111	$7\Delta = 7/8$ (V)
3/4 - 7/8V	110	$6\Delta = 3/4$ (V)
5/8 - 3/4V	101	$5\Delta = 5/8$ (V)
1/2 - 5/8V	100	$4\Delta = 1/2$ (V)
3/8 - 1/2V	011	$3\Delta = 3/8$ (V)
1/4 - 3/8V	010	$2\Delta = 1/4$ (V)
1/8 - 1/4V	001	$1\Delta = 1/8$ (V)
0 - 1/8V	000	$0\Delta = 0$ (V)

FIGURE 11.30 A/D conversion.

11.4.2.3 QUANTIZATION

Quantization describes the procedure whereby the continuous analog signal is quantized and encoded in the binary format. First, the range of the analog input signal is divided into a finite number ($2^N - 1$) of intervals, where N is the number of bits available for the corresponding binary word. Second, a binary word is assigned to each interval; the binary word is the digital representation of any voltage (current) that falls within that interval. This is illustrated in Figure 11.30.

As shown in Figure 11.30, for eight levels of quantization, only three bits are needed. As the number of quantization levels increases, the number of bits required to represent those levels increases as well. For example, for a 16-level quantizer, four bits are needed to represent those levels and for a 32-level quantizer, five bits are needed to represent all levels. In general, for an L level quantizer, the number of required bits corresponds to:

$$N = \log_2 L \tag{11.45}$$

At any particular value of the analog signal, the digital representation is either the discrete level immediately above or below this value. If the difference between two successive discrete levels is represented by the parameter Δ, then, in Figure 11.30, the maximal quantization error is Δ, that is, $V/8$. This error is known as the quantization error and is proportional to the resolution of the analog-to-digital converter, that is, the number of bits used to represent the samples in digital form.

In the quantization process, all infinite amplitude levels of analog signals are mapped into certain predetermined levels. Obviously, the reconstructed signal will not be exactly the same as the original one, because the quantization will only represent certain levels of amplitude. As smaller number of intervals is selected, a larger length binary code is created at output and the digital code more accurately represents the original signal. In other words, quantization error decreases as the number of bits per sample increases. The number of bits controls the resolution of the ADC.

Resolution, Δ, is defined as the maximum analog voltage range divided by the number of quantization levels, $L = 2^N$, where N is the number of bits [see Equation (11.45)]. In Figure 11.30, the resolution is $V/8$. If the analog voltage range is ± 5 V, an 8-bit ADC's resolution would be $\Delta = 10/2^8 = 39.0625$ mV.

Therefore, the resolution of a computer-based measurement system is limited by the word length of the ADC. The effect of finite word length can be modeled by adding quantization noise to the reconstructed signal. It can be shown that the root mean square (rms) of the quantization noise is:

$$\sigma_{\text{qrms}} = \frac{\Delta}{2\sqrt{3}} \tag{11.46}$$

where Δ is the quantization zone or resolution.

EXAMPLE 11.17   **Quantization Error**

A 10-bit ADC is designed to accept signals ranging from −5 to 5 V. Determine the width of each quantization zone (the ADC's resolution), and the rms of quantization error.

SOLUTION

The converter has 2^{10} zones = 1024 zones. Total range covered by the ADC = $5 − (−5) = 10$ V. Therefore, the width of each quantization (the ADC's resolution) is:

$$\Delta = \frac{10}{1024} = 9.8 \text{ mV}$$

The corresponding quantization rms error is:

$$\sigma_{qrms} = \frac{9.8}{2\sqrt{3}} = 2.83 \text{ mV}$$

EXAMPLE 11.18   **Signal-to-Noise Ratio Created by Quantization**

A 5-V_{peak} sinusoid is converted by a 10-bit ADC designed to accept a signal from −4 V to 4 V. Assuming quantization is the only source of error (noise), find the signal-to-noise ratio (SNR), expressed in decibels.

SOLUTION

The signal power corresponds to [see Equation (6.1)]:

$$P = \frac{V^2}{R} = \frac{\left(\frac{5}{\sqrt{2}}\right)^2}{R} = \frac{12.5}{R} \text{ W} \ (R \text{ is arbitrary})$$

The width of each quantization level is:

$$\Delta = \frac{8}{2^{10}} = 7.8125 \text{ mV}$$

Using Equation (11.46), quantization noise corresponds to:

$$\sigma_{qrms} = \frac{\Delta}{2\sqrt{3}} = \frac{7.8125}{2 \times 1.732} = 2.255 \text{ mV}$$

Accordingly, the quantization noise power is:

$$P_q = \frac{(2.255 \text{ mV})^2}{R} = \frac{5.085}{R} \text{ μW}$$

Finally:

$$SNR_{dB} = 10 \log \left(\frac{\frac{12.5}{R} \text{ W}}{\frac{5.085}{R} \text{ μW}} \right) = 10 \log (2.458 \times 10^6) = 63.91 \text{dB}$$

11.5 GROUNDING ISSUES

11.5.1 Ground Loops

A ground is a point where the voltage does not change regardless of the amount of current supplied to it or drawn from it. Ground specifies a reference voltage for a circuit or system.

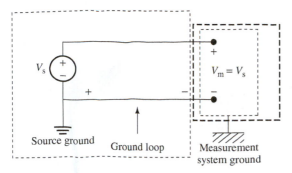

FIGURE 11.31 Ground loop.

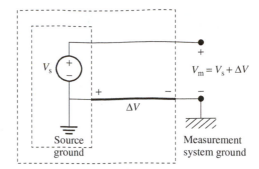

FIGURE 11.32 Differential measurement system.

For a measurement system, one important point in dealing with grounded signal sources is the *ground loop*. A ground loop is an undesired current path due to several reference voltages being connected together through different wires. A ground loop will not lead to any problem if these ground wires have zero impedance so that all of the ground points are at the same voltage. However, the resistance of some ground wires is nonzero, which changes the voltage between various ground points when current flows through them.

Figure 11.31 illustrates an example of a ground loop that is created across the measuring instrument and the instrument case grounds. Assuming that the resistance of the ground wires are not the same, a potential difference will be created across these two wires. When the measuring instrument and the instrument case grounds are connected together, a potential current will flow from one to the other through the small (but nonzero) resistance of the connecting wire. Thus, the voltage measured by the instrument will include the unpredictable voltage difference, ΔV. This leads to a measurement error.

As shown in Figure 11.32, to avoid ground loop in measurement instruments, the instrument case ground and the source ground are disconnected. Such a measurement system is called a differential measurement system. Proper grounding is also essential for safety, as discussed in Chapter 15.

APPLICATION EXAMPLE 11.19   Data Acquisition

Today, computers are widely used in research, control, testing, and measurement. Engineers can use computers to process the data collected from sensors. To obtain proper results from the rough data of sensors, using the knowledge gained through this chapter, construct a basic data acquisition system consisting of the system elements:

- Personal computer (software)
- Sensors
- Signal conditioning
- DAQ hardware

SOLUTION

Sensors are used to collect field data, and then data are applied to the *signal conditioning system* to filter or amplify the data, if necessary. To be readily accepted by a general-purpose computer, the data are then sent to the *data acquisition hardware* that periodically samples the signal and converts the sampled signals to digital words using an ADC. The digital words are then read by the *personal computer* for further processing. The system structure is shown in Figure 11.33.

(continued)

APPLICATION EXAMPLE 11.19 **Continued**

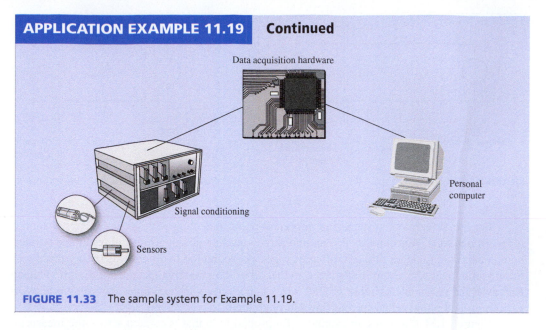

FIGURE 11.33 The sample system for Example 11.19.

11.6 USING PSPICE TO DEMONSTRATE A COMPUTER-BASED INSTRUMENT

This section introduces how to use PSpice for computer-based instrument analysis. Then, it addresses how to set up an op amp circuit, and run simulations to analyze relevant systems.

EXAMPLE 11.20 **PSpice Analysis**

For a strain-gauge load cell, the relationship between the output voltage and the mass weighed is $M = 3.5\ V_o$ kg. The load cell has the same structure as shown in Figure 11.13. $V_i = 5$ V, and $R_1 = R_2 = R_3 = 100\ \Omega$. Use PSpice to sketch the circuit and determine the mass if $R_g = 80\ \Omega$.

SOLUTION

The PSpice schematic solution is shown in Figure 11.34.

The output voltage, V_o, corresponds to:

$$V_o = 2.778 - 2.5$$
$$V_o = 0.278 \text{ V}$$

Thus, the mass of the load can be calculated, which is:

$$M = 3.5 \times V_o$$
$$M = 0.973 \text{ kg}$$

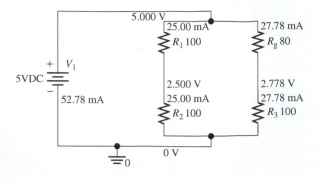

FIGURE 11.34 PSpice schematic solution for Example 11.20.

EXAMPLE 11.21   **PSpice Analysis**

Figure 11.35 shows an inverting operational amplifier. Consider the input voltage as 10 V, $R_i = 100\ \Omega$, $R_F = 200\ \Omega$, and the output load is 1000 Ω. Determine V_o using PSpice.

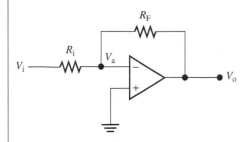

FIGURE 11.35   An inverting operational amplifier.

SOLUTION

1. First, go to "Place" > "Part" (see Figure 2.63) and type "uA741" to obtain the op amp.
2. Next, right click on the uA741 part and choose "Mirror Vertically" to invert the part upside down as shown in Figure 11.36.
3. Set up the PSpice schematic as shown in Figure 11.37, and run the simulation as a bias point analysis. The result is shown in Figure 11.37.

FIGURE 11.36   Mirror vertically.

The V_o of the circuit is −1.68 V. Notice that there is not any ideal op amp in PSpice, while $\mu A741$ is one of the op amp parts that is most commonly used. An ideal op amp model can be created through PSpice software. However, for the student/demo version, the PSpice model editing software is limited to diodes. Therefore, ideal op amps are not discussed in this section.

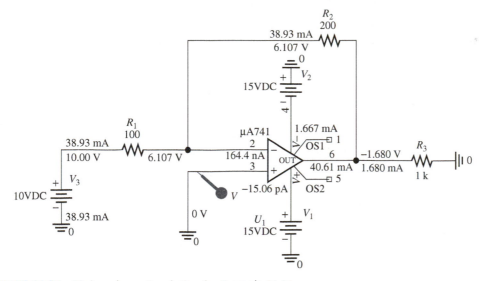

FIGURE 11.37   PSpice schematic solution for Example 11.21.

EXAMPLE 11.22   **PSpice Analysis**

Given the first-order, high-pass filter shown in Figure 11.38, plot the frequency response of the output with the load of 1 kΩ. Assume that the input voltage is 10 V 60 Hz, C is 10 μF, R_i is 100 Ω, and R_F is 200 Ω.

(continued)

EXAMPLE 11.22   **Continued**

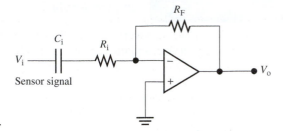

FIGURE 11.38   Circuit for Example 11.22.

SOLUTION

Follow the steps in Example 11.21 to set up an op amp in the PSpice circuit. The PSpice schematic solution is shown in Figure 11.39. Set the simulation analysis type to be AC sweep, with the logarithmic increment. The frequency plot and its phase diagram for the output are shown in Figure 11.40.

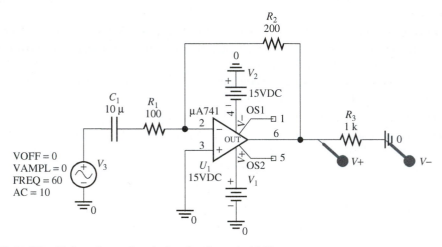

FIGURE 11.39   PSpice schematic solution for Example 11.22.

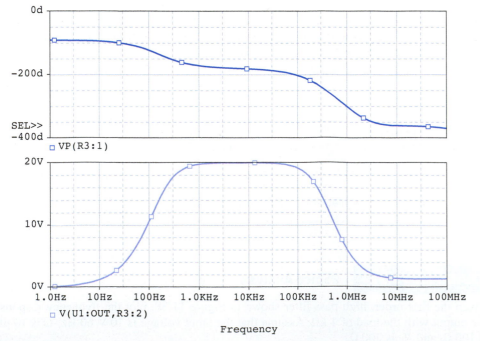

FIGURE 11.40   Bode plot for Example 11.22.

11.7 WHAT DID YOU LEARN?

- A computer-based instrumentation system consists of sensors, signal conditioning, a data-acquisition system, an analog-to-digital converter, and one or more general-purpose computer(s).
- A sensor, also called a transducer, is defined as a device that responds to a physical stimulus (e.g., heat, light, sound, pressure, magnetism, or motion) and transmits a resulting impulse for measurement or control.
- A pressure sensor is a device that responds to pressure applied to its sensing surface and converts the pressure to a measurable signal.
  Temperature sensors are used to measure temperature. For a thermocouple, the temperature difference between the two junctions of the thermocouple can be detected by measuring the change in the voltage across the dissimilar metals at the temperature measurement junction. A resistance temperature detector (RTD) is a variable-resistance device whose resistance is a function of temperature. RTDs are more accurate and stable than thermocouples.
- Accelerometers (acceleration sensors) are used to measure acceleration, vibration, and mechanical shock based on the effect of Newton's second law.
- A sound sensor is a device that converts acoustic energy into electrical energy. A microphone is one example of a typical sound sensor. The diaphragm of a microphone vibrates by impinging waves of acoustic pressure.
- A linear variable differential transformer (LVDT) is a displacement sensor that is operated based on the concept of mutual inductance.
- Signal conditioning is required to convert a sensor's output to an appropriate form. Two of the most important signal conditioning functions are amplification and filtering. An amplifier is a device that increases the power contained within a signal. A filter eliminates interference (noise or undesired input).
- A data acquisition system consists of an analog multiplexer, a buffer, a sample and hold amplifier, and an analog-to-digital converter.
- Analog-to-digital converters are designed to transform data in the form of continuous analog variables into a discrete binary code suitable for digital processing.
  There are two main steps in analog-to-digital conversion: sampling and quantization. First, the analog signal is sampled at regular time intervals. Next, quantization describes the procedure whereby the continuous analog signal is quantized and encoded in binary form.
- A ground specifies a reference voltage for a circuit or system.
- To avoid ground loop in measurement instruments, the instrument case ground and the source ground must be disconnected (see Figure 11.31).

Further Reading

Turner, J.D., and Prelove, A.J. 1991. *Acoustics for engineers.* New York: Macmillan.

Alan S. Morris. 2001. *Measurement and instrumentation principles,* 252–259. Oxford: Butterworth-Heinemann.

Chester L. Nachtigal. 1990. *Instrumentation and control: Fundamentals and applications.* New York: Wiley.

Turner, J., and Hill, M. 1999. *Instrumentation for engineers and scientists.* Oxford: Oxford University Press.

Gopel, W., Hesse, J., and Zemel, J.N. 1994. *Sensor: A comprehensive survey.* New York: VCH Verlagsgesellschaft, Weinheim (Federal Republic of Germany), VCH Publishers.

Francis, S. Tse, and Ivan E. Morse. 1989. *Measurement and instrumentation in Engineering.* New York: Marcel Dekker.

Problems

*B refers to basic, A refers to average, H refers to hard, and * refers to problems with answers.*

SECTION 11.2 SENSORS

11.1 (B) Explain the function of each element of a computer-based instrumentation system.

11.2 (B) Explain the basic concepts of sensors.

11.3 (B)* Select one: The relationship between pressure and force can be expressed as:
(a) Pressure = force × area
(b) Pressure = force/area

(c) Force = pressure/area

(d) None of above

11.4 (A) Piezoelectric sensors can measure the pressure of the object according to the equation:

$$p = \frac{V_0}{K_E x}$$

where, p is the pressure in N/m^2 and x is the displacement in meter. K_E is the voltage sensitivity of the sensor. Here, assume $K_E = 0.045$ V·m/N. If an object is placed on the sensor, and measures $V_0 = 2.75$ V, and $x = 0.85$ mm, then what is the pressure from the object?

11.5 (B) What is the Seebeck effect?

11.6 (A)* In Figure P11.6, which curve represents the characteristic of a thermistor?

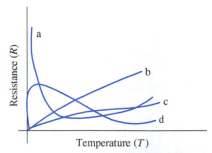

FIGURE P11.6 Resistance versus temperature curves for Problem 11.6.

11.7 (B) Which temperature sensors have a negative resistance–temperature relationship?

(a) Thermocouples

(b) All RTDs

(c) Thermistors

(d) Infrared thermometers

11.8 (H) A suitable curve fit that converts the measured resistance, R, of a thermistor to temperature, T, is given by:

$$\frac{1}{T} = A + B \ln R + C(\ln R)^3$$

where R is in ohms, T in kelvin, and A, B, and C are curve-fitting constants. An engineer has measured data pairs of resistance and temperature for the thermistor as follows:

$$(R_1, T_1),\ (R_2, T_2),\ \text{and}\ (R_3, T_3).$$

Use these measured data to express the coefficients A, B, and C.

11.9 (A)* Which equation correctly expresses the acceleration? (x = linear displacement, θ = angular displacement, v = linear velocity, ω = angular velocity, a = linear acceleration, α = angular acceleration, t = time)

(a) $a = \dfrac{dx}{dt}$ (b) $\omega = \dfrac{d^2\theta}{dt^2}$ (c) $\alpha = \dfrac{d^2\theta}{dt^2}$ (d) $\omega = \dfrac{d^2x}{dt^2}$

11.10 (A) An engineer designs a toy car. To measure the acceleration of this kind of toy car, he also designs a simple accelerometer (a spring with spring constant 100 N/m and a mass $m = 2$ kg) and binds it to the car. When the car is accelerated, he measures the displacement of the mass which is 0.01 m. What is the acceleration of this toy car?

11.11 (A)* Accelerometers are used widely in vehicles to decide whether safety air bags should be deployed. Assume that the vehicle's air bag is deployed when the seismic mass m for the accelerometer is 0.2 kg and the vehicle's acceleration is 150 m/s^2. If at this moment, the mass displacement is measured as 0.02 m, then what is the spring constant for this accelerometer?

11.12 (A) How is a Wheatstone bridge used to construct a load cell?

11.13 (A)* A load cell is constructed as shown in Figure 11.13. There is a linear relationship between the output voltage, V_0, and the mass, M, weighted: $M = kV_0 + b$, where k and b are constant. The steady input voltage to the cell is $V_i = 3$ V, and $R_1 = R_2 = R_3 = 200$ Ω. To calculate the value of k and b, two different masses ($M_1 = 1$ kg, $M_2 = 0.8$ kg) are loaded to the cell, respectively, and the R_g is measured as 100 Ω and 120 Ω, respectively. Find k and b.

11.14 (A) For an LVDT, the relationship between the magnitude of the output voltage and the displacement, x, of the iron core is approximately linear over a reasonable range of movement of the iron core and is expressed by a constant of proportionality, G, as: $x = G \cdot V_0$. An engineer uses this kind of LVDT (shown in Figure P11.14) to measure the displacement of a gear. The engineer measures the instant value of output voltage: $V_1 = 5$ V and $V_2 = 2$ V, and the sensitivity (or gain) G of the transformer is 0.6 mm/V. Determine the displacement of the gear.

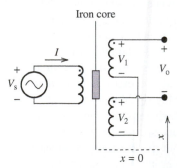

FIGURE P11.14 LVDT.

11.15 (H)* Figure P11.15(a) shows a capacitive microphone output. The voltage output of the circuit is shown in Figure P11.15(b). What is the change of capacitance due to the sound pressure on the diaphragm?

11.16 (B) Describe at least three sensors that are part of your body and explain their function.

11.17 (B) Describe at least three sensors applied in daily life and explain their function.

11.18 (A) The resistance of an RTD is calculated with $R_{RTD} = R_0[1 + \alpha(T - T_0)]$, and the coefficient $\alpha = 0.004°C^{-1}$. The room temperature is (25°C), $R_0 = 40$ Ω. An RTD is put on a running CPU and $R_{RTD} = 48$ Ω. What is the surface temperature of the CPU?

11.19 (A) Explain the basis of a thermoelectric cooler.

11.20 (H)* A set of accelerators are installed on an object as shown in Figure P11.20. Data are collected from each accelerator from time $T = 0$ s and $T = 3$ s, $a_x = 4$ m/s^2

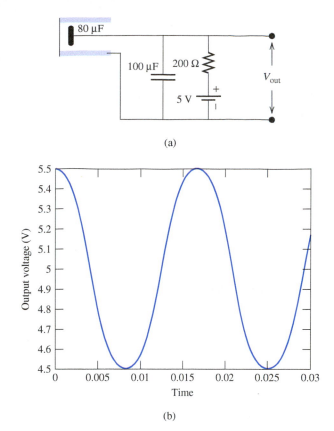

(a)

(b)

FIGURE P11.15   (a) Capacitive microphone circuit for Problem 11.15; (b) Voltage output plot for Problem 11.15.

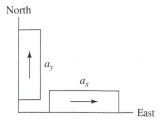

FIGURE P11.20   Accelerator set.

and $a_y = 3$ m/s^2. Assume the object is static at $T = 0$ s; calculate the object speed at $T = 3$ s.

11.21 (A) What is the advantage of applying a Wheatstone bridge in load cells compared to directly measuring the gauge resistance?

11.22 (H) If you connect a passive analog speaker to a microphone connector on a recorder, and you start to record when you speak to the passive speaker, can the recorder record your voice? If it can, what is the difference between the voices recorded using speaker and a microphone?

SECTION 11.3 SIGNAL CONDITIONING

11.23 (B) Explain why signal conditioning is mandatory in a computer-based instrumentation system.

11.24 (A)* Consider the high-pass, active filter shown in the Figure P11.24. If $v_i(t) = 10 \cos(1000\pi \cdot t)$ V, $R_i = 150$ kΩ, $R_F = 200$ kΩ, and $C_i = 200$ pF, find $v_o(t)$.

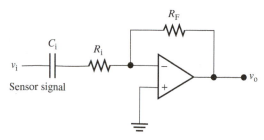

FIGURE P11.24   High-pass, active filter.

11.25 (A) Consider the low-pass, active filter shown in the Figure P11.25. If $v_i(t) = 4 \cos(2000\pi t)$ V, $R_i = 10$ kΩ, $R_{F1} = 40$ kΩ, $R_{F1} = 60$ kΩ, and $C_F = 500$ pF, find $v_o(t)$.

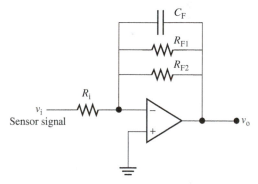

FIGURE P11.25   Low-pass, active filter.

11.26 (A) A simple band-pass filter with op amp is shown in the Figure P11.26. Derive the expression for the gain of this active filter.

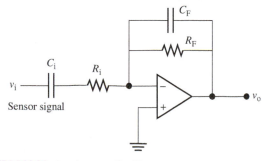

FIGURE P11.26   Band-pass, active filter.

11.27 (H) Given that both voltage sources, V_1 and V_2, are 120 V, 60 Hz what is the output voltage of the circuit shown in Figure P11.27?

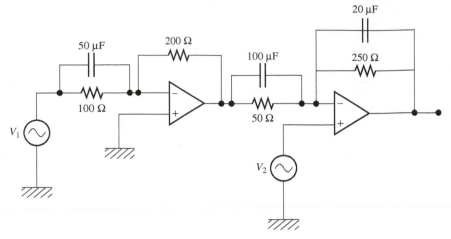

FIGURE P11.27 Circuit for Problem 11.27.

11.28 (H)* A data acquisition subsystem requires its input to be larger than 1 V and smaller than 10 V. The output of the sensor connected to the data acquisition sub-system is 1 to 5 mV. Calculate the expected voltage gain of the signal conditioner.

11.29 (A) Find the gain of the two low-pass filters shown in Figure P11.29, and analyze their pros and cons.

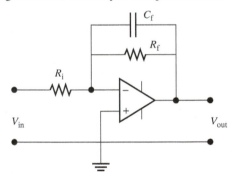

(a) Active, low-pass filter

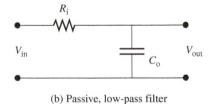

(b) Passive, low-pass filter

FIGURE P11.29 (a) Active and (b) passive low-pass filters.

11.30 (A)* An engineer wants to check a high power pulse transmitter's power supply wave form using an oscilloscope. The voltage range of the power supply is between 6 and 12 kV, and the oscilloscope's input range is 2 mV to 10 V. Which probe should the engineer select? (The parameter is the ratio of input to output).
(a) 10:1
(b) 100:1
(c) 1000:1
(d) 10,000:1

SECTION 11.4 DATA ACQUISITION

11.31 (B) What is signal sampling?

11.32 (B) A signal has frequency components up to 50 kHz. What is the minimum sampling rate that should be used to ensure that the signal is reconstructed properly?

11.33 (B) What is aliasing? Under what conditions does it occur?

11.34 (B) If the analog voltage range is ± 3 V, and a 16-bit ADC is used to convert the analog signal to digital signal, what is the resolution of this ADC?

11.35 (A)* A resolution of the sampled values of an ADC is desired to be 2%. What is the required number of bits for this ADC?

11.36 (A) Consider a voltage source, $V_S(t)$ of 5 V, 60 Hz. Use MATLAB to plot the output of sampling of this voltage source for 0.1 s if the sampling rate is:
(a) 60 Hz
(b) 120 Hz
(c) 10 times of voltage source frequency

11.37 (H) At any particular value of the analog signal, the digital representation is either the discrete level immediately above this value or the discrete level immediately below this value. For eight levels of quantization with three bits, if the difference between two successive discrete levels is represented by the parameter Δ, design a quantization method to make the maximal quantization error equal to $\Delta/2$.

11.38 (H)* A 3-V peak sinusoid is converted to a digital signal using an N-bit ADC which is designed to accept signals from −5 to 5 V. Find the smallest N to make the SNR (expressed in decibels) greater than 50 dB.

11.39 (B) Explain the necessity of ADC in a computer-based system.

11.40 (A) When the input is −5 ~ 5 V, compare the noise generated by an 8-bit and a 12-bit ADC. What are the advantages and disadvantages of using a 12-bit AD convertor compared to an 8-bit ADC?

11.41 (H) Prove that when an ADC output is increased by one bit, an extra 6 dB SNR is generated by the ADC.

11.42 (H)* Assume there is a signal between −6 and +6 V, and the required measurement RMS error is 1 mV. If a digital voltage meter is used, an ADC with how many bits is needed?

11.43 (A) The output SNR of an ADC is related to:
(a) Input SNR
(b) Input signal amplitude
(c) The number of ADC output bits
(d) The sampling rate

SECTION 11.5 GROUNDING ISSUES

11.44 (B) What is the ground loop problem?

11.45 (B) What is a data acquisition system?

11.46 (A)* Determine the voltage output, V_{out}, shown in Figure P11.46 if the ground of the voltage source and ground of the load are connected. The material of the wire that connects the devices in the circuit is copper with the resistance per length of 0.54 Ω/m.
If the connection between both grounds is cut, what is the voltage output, V_{out}?
(Provide the solution up to two decimal points.)

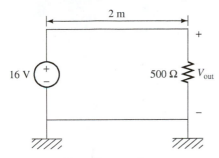

FIGURE P11.46 Circuit for Problem 11.46.

11.47 (B) How should the ground loop effect in a measurement system be minimized? Provide an example when the input is an AC signal.

11.48 (A)* A circuit provides 5 V power for an analog part and 3 V for a digital part shown in Figure P11.48. Calculate the voltage difference measured at V_{out} when $R_G = 0\ \Omega$ (no ground loop) and $R_G = 1\ \Omega$ (ground loop).

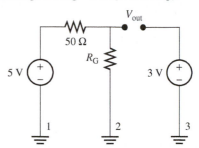

FIGURE P11.48 Ground loop.

SECTION 11.6 USING PSPICE TO DEMONSTRATE A COMPUTER-BASED INSTRUMENT

11.49 (A) Consider the high-pass, active filter shown in the Figure P11.4. If $v_i(t) = 10\cos(1000\ \pi t)$ V, $R_i = 150$ kΩ, $R_F = 200$ kΩ, and $C_i = 200$ pF. Use PSpice to plot the output of $v_o(t)$.

11.50 (A) Consider the low-pass, active filter shown in the Figure P11.5. If $v_i(t) = 4\cos(2000\ \pi t)$ V, $R_i = 10$ kΩ, $R_{F1} = 60$ kΩ, $R_{F2} = 40$ kΩ, and $C_F = 500$ pF. Use PSpice to plot the output of $v_o(t)$.

CHAPTER 12

Principles of Electromechanics

(Used with permission from oviii/Shutterstock.com.)

12.1 INTRODUCTION

In the previous chapters, you learned about electrical circuits and the concepts of voltage, current, resistance, capacitance, and inductance. You also learned about the role of resistors, capacitors, and inductors. The resistance of a resistor varies with its structure and material. Capacitance is generated by an electrical field within two (usually) parallel plates. It is also a function of the shape and the material that fills the space between the two plates. Finally, inductance is generated due to a magnetic field around a twisted wire. Inductance varies with the structure of the wire. This chapter examines magnetism, magnetic fields, and magnetic circuits.

In daily life, you see many applications of magnetic fields and magnetic circuits. Magnetic screwdriver tips and the Apple MagSafe magnetic power cord connection used on MacBooks (Macintosh laptops) shown in Figure 12.1 are examples of magnetic circuits. Electromagnetic applications in mechanical engineering include the variety of electric motors that will be addressed in Chapter 13. More specific applications include magnetic levitation (MagLev) trains and the launching and braking systems of newer roller coasters. Magnetic resonance imaging (MRI)

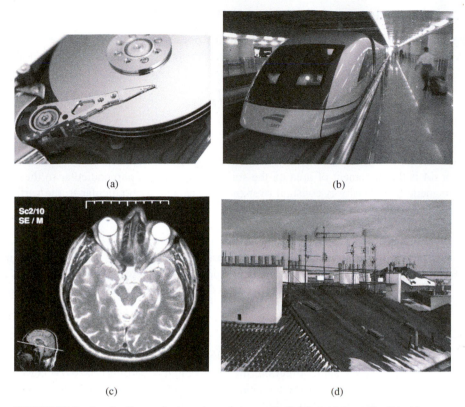

(a)   (b)

(c)   (d)

FIGURE 12.1 Applications of magnets and magnetic fields: (a) Magnetic disk drive
(Used with permission from Alias Studiot Oy/Shutterstock.com.); (b) MagLev train in
Shanghai, China (Used with permission from Lee Prince/Shutterstock.com.); (c) MRI
image cross section of a human head (Used with permission from xpixel/Shutterstock.
com.); (d) Rooftop antennas (Used with permission from vedius/Shutterstock.com.).

is a biomedical engineering application of magnetic fields. Magnetic fields are also an important
component in the operation of antenna systems. Antennas are used in all wireless communication systems (e.g., radar systems, cellular phones, radio, and TV) to transmit signals through
space and to receive signals propagated in space. Figure 12.1 shows some of these applications.
Magnetic fields can also be used to noninvasively measure current.

Other applications of magnetic fields include metal detectors, magnetic swipe cards such as
credit and debit cards, and magnetic recording tape. As discussed in Chapter 8, a transformer is
a magnetic circuit. As discussed in Chapter 9, transformers are used to increase voltage for long-distance transmission and to decrease the standard 120 V alternating current (AC) to a voltage safe
for operating low-voltage devices, such as computers. The structure of a transformer is discussed
in detail in this chapter.

12.2 MAGNETIC FIELDS

Magnetic fields are created around current-conducting wires and permanent magnets. The field
is created by electrical charges in motion. In a permanent magnet, this charge in motion is caused
by the spin of electrons in the atoms. If the direction of motion of each electron is random, the
atomic fields created by the random movement will cancel each other, and the overall structure
will not have any magnetic property. If all electrons spin in the same direction, the atomic fields
aid each other, producing a net external effect. This produces a magnetic property in the structure. In a wire, the field is created by the flow of charges (i.e., current) in the wire. The magnetic
field can be visualized as lines of magnetic flux that form closed paths. Figure 12.2 represents a
magnetic field (flux lines) created around a wire that carries current.

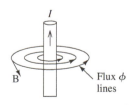

FIGURE 12.2
Magnetic field around
a wire.

12.2.1 Magnetic Flux and Flux Intensity

Magnetic flux is defined as the integral of the flux density over a given surface area. Magnetic flux is denoted by ϕ, and its unit is webers (Wb). The magnetic flux passing through the surface area, A, is given by:

$$\phi = \int_A B \cdot dA \tag{12.1}$$

Vector B represents the magnetic flux density, in webers per square meter (Wb/m^2) or teslas (T). The flux density, vector B, has a direction tangent to the flux lines as shown in Figure 12.2. Vector dA is the increment of area on the surface. Here, dA is perpendicular to the surface. If the magnetic flux density is constant over the surface and perpendicular to the surface, using Equation (12.1), the flux can be simplified to:

$$\phi = B \cdot A \tag{12.2}$$

Note that the product of the two vectors of the magnetic field and the surface in Equation (12.1) is called the dot product. The dot product of two vectors X and Y is a scalar that is equal to $|X||Y| \cos\theta$. Here, $|\cdot|$ represents the magnitude of the two vectors, and θ is the angle between the two vectors.

EXAMPLE 12.1 **Dot Product**

If two vectors have the same direction, then $\theta = 0$, and accordingly $X \cdot Y = |X||Y|$, because $\cos 0 = 1$. If two vectors are perpendicular, then, $\theta = 90°$ and $X \cdot Y = 0$. Note that $\cos 90° = 0$. As mentioned earlier, the product of the two vectors will be a scalar at all times.

EXAMPLE 12.2 **Dot Product**

Usually, a two-dimensional vector, X, is represented by $X = X_1 \cdot a_x + X_2 \cdot a_y$. Here, X_1 and X_2 are scalars, and a_x and a_y are unity magnitude vectors. In addition, a_x and a_y are perpendicular vectors; that is, the angle between these two vectors is 90°.

Assuming $X = -a_x + 2a_y$ and $Y = 2a_x + 3a_y$, find the dot product of X and Y.

SOLUTION

Because the angle between two similar vectors is zero, $a_x \cdot a_x = a_y \cdot a_y = 1$. In addition, because a_x and a_y are perpendicular vectors, $a_x \cdot a_y = a_y \cdot a_x = 0$. Next, calculate:

$$\begin{aligned}
X \cdot Y &= (-a_x + 2a_y) \cdot (2a_x + 3a_y) \\
&= -2a_x \cdot a_x - 3a_x \cdot a_y + 4a_y \cdot a_x + 6a_y \cdot a_y \\
&= -2 + 6 \\
&= 4
\end{aligned}$$

Thus, the dot product of two vectors is a scalar.

EXERCISE 12.1

Repeat Example 12.2 for the following vectors:

$$X = a_x + a_y, \text{ and } Y = a_x - a_y$$
$$X = 3a_x + 2a_y, \text{ and } Y = 2a_x + 3a_y$$

FIGURE 12.3 Two different versions of the right-hand rule.

(a)

(b)

12.2.2 Magnetic Field Intensity

The magnetic field intensity [unit: amperes per meter (A/m)] is represented by the vector *H*. The relationship of *H* (the magnetic intensity created within the coil) and the flux density, *B*, corresponds to:

$$B = \mu H \tag{12.3}$$

where μ is the magnetic permeability of the medium. For free space:

$$\mu = \mu_0 = 4\pi \times 10^{-7} \text{ (Wb/A} \cdot \text{m)} \tag{12.4}$$

Iron and some other materials have high permeability, given by their relative permeability defined as:

$$\mu_r = \frac{\mu}{\mu_0} \tag{12.5}$$

12.2.3 The Right-Hand Rule

The right-hand rule is a simple rule that represents the direction of current if the direction of the magnetic field is known, and vice versa. Two versions of this rule are discussed: (a) if a *wire* is held with the thumb in the direction of the current, the fingers will coil in the direction of the magnetic field [see Figure 12.3(a)]; (b) If a *coil* is grasped with fingers pointed in the direction of the current, the thumb will point in the direction of the magnetic field inside the coil [see Figure 12.3(b)].

EXAMPLE 12.3   Direction of Current

A wire perpendicular to the earth's surface exhibits a counterclockwise flux density, vector *B*. In which direction is the current flowing?

SOLUTION

Out of the earth.

EXAMPLE 12.4   Direction of Flux

A coil is placed pointing north and south. If the current in the coil flows clockwise (assume you are facing north), what is the direction of the flux density, vector *B*?

SOLUTION

North.

12.2.4 Forces on Charges by Magnetic Fields

When charges move within a magnetic field, a force is applied to them by the magnetic field. This force is a function of the speed of the movement, the strength of the magnetic field, and the magnitude of the electric field (that is a function of charge), which can be represented by:

$$f = qu \times B \tag{12.6}$$

where

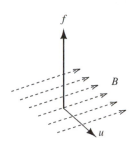

FIGURE 12.4 Force on moving a charge in a magnetic field.

- f = force vector in newtons (N)
- q = charge in coulombs (C)
- u = velocity of charge (m/s), a vector, which represents the direction of motion
- B = magnetic flux density (T), a vector, which represents the direction of B.

The product in Equation (12.6) (i.e., "$\times$") is called the cross product. The cross product of two vectors is a new vector perpendicular to the surface formed by the two vectors (see Figure 12.4). If the charge direction of movement (or the velocity vector) makes an angle θ with the magnetic field, then the magnitude of the force is:

$$f = quB \sin \theta \tag{12.7}$$

The direction of this force is at a right angle with the plane formed by B and u, as shown in Figure 12.4. The technique of finding the direction of this force is discussed in detail in Chapter 13.

12.2.5 Forces on Current-Carrying Wires

A wire carrying current includes charges flowing through it. When this conductor is inserted into a magnetic field, a force is exerted on the wire by the magnetic field. Indeed, the force applied to the moving charges leads to a force that is applied to the wire. The method of determining the direction of the force will be discussed in detail in Chapter 13. For a straight wire, the magnitude of the force is expressed as:

$$f = ilB \sin \theta \tag{12.8}$$

where

- l = length of the wire in meters
- i = current through the wire in amperes
- θ = angle between the wire and the magnetic field
- B = the magnetic field in teslas.

EXAMPLE 12.5   **Force Exerted on a Wire**

A nearly straight power line of length 2 m carrying 100 A of current is accidentally placed in a position where it is perpendicular to a magnetic field of $B = 5$ T.

  a. How much force is exerted on the wire?
  b. Should this force be of concern to the engineers?

SOLUTION

  a. Using Equation (12.8) since θ is 90°: $f = l \cdot i \cdot B = 2 \cdot 100 \cdot 5 = 1000\,\text{N}$
  b. Yes, it should be of concern if the power lines are not engineered to handle the effects of this force.

APPLICATION EXAMPLE 12.6   Power Line Hazards

High-voltage power lines are often placed near residential areas, as shown in Figure 12.5. The current through these power lines creates a magnetic field. The strength of this magnetic field corresponds to:

$$B = \frac{\mu_0 i}{2\pi r}$$

where

- B = magnitude of the magnetic field in teslas
- μ_0 = magnetic permeability of free space and $\mu_0 = 4\pi \times 10^{-7} \, (\text{Wb}/\text{A} \cdot \text{m})$; its unit is henry (H) per meter
- i = current through the wire in amperes
- r = distance from the center of the wire in meters.

Concerns have been raised that this magnetic field and the resulting electric field can cause health hazards and other undesired effects. Because of this hazard, some states limit the field to $B = 200$ mG [1 tesla = 10,000 gauss (G)]. Chapter 15 discusses power line safety issues in more detail.

a. If a house is located 50 m from a 500-kV line carrying 2.5 kA, find the magnetic field at this house.

b. Considering $B = 200$ mG as a limit to mitigate safety issues, how far should a 745-kV line bearing 3 kA be placed from a neighborhood?

c. Distribution lines carry current into neighborhoods. These lines have current levels of around 300 A. How far should these lines to be placed from the houses?

FIGURE 12.5 Power lines near houses. (Used with permission from © Diana Ninov/Alamy.)

SOLUTION

a. Using the equation for magnetic fields introduced in this example:

$$B = \frac{\mu_0 i}{2\pi r} = \frac{\mu_0 \cdot 2500}{2\pi 50} = 10^{-5} \, \text{T}$$

b. 200 mG is equivalent to 20 μT. Therefore:

$$20 \times 10^{-6} = \frac{\mu_0 3000}{2\pi r} \rightarrow r = 30 \, \text{m}$$

In other words, the residence should be located at least 30 m from the power lines.

c. In this case:

$$20 \times 10^{-6} = \frac{\mu_0 300}{2\pi r} \rightarrow r = 3 \, \text{m}$$

EXERCISE 12.2

Repeat Example 12.6(b) and determine the distance if the limit on the magnetic field is changed to 10 mG.

12.2.6 Flux Linkages

Assume that a coil has N number of turns. If the flux due to each turn is ϕ, the total flux linkage within the coil (due to N turns) corresponds to:

$$\lambda = N \cdot \phi \qquad \textbf{(12.9)}$$

Equation (12.9) assumes that each turn of the coil has the same flux link. This concept is illustrated in Figure 12.6.

12.2.7 Faraday's Law and Lenz's Law

Faraday's law of magnetic induction states that a voltage, e (also called electromotive force), is induced in a circuit whenever flux linkage changes.

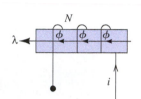

FIGURE 12.6
Relationship between N, i, ϕ, and λ.

Question: What can cause flux linkages to change?

Answer: Time varying the magnetic field or a coil moving relative to the magnetic field.

This induced voltage corresponds to:

$$e = \frac{d\lambda}{dt} \qquad \textbf{(12.10)}$$

Here, λ is the flux linkage within the coil defined in Equation (12.9). Lenz's law states that the direction of an induced voltage (electromotive force) is such that it generates a current that creates a magnetic field that is opposite in direction to the field that generates the electromotive force. In other words, an induced voltage produces a magnetic field that opposes the field that generated the induced voltage.

EXAMPLE 12.7   Induced Voltage in a Coil

A flux of $\phi(t) = 5 \cos(20\pi t)$ passes through a coil having $N = 600$ turns. Find the voltage $e(t)$ induced in the coil.

SOLUTION

First, find the flux linkage of the coil using Equation (12.9):

$$\lambda(t) = N \cdot \phi(t) = 3000 \cos(20\pi t)$$

Next, find the induced voltage using Equation (12.10):

$$e(t) = \frac{d\lambda}{dt} = -60{,}000\pi \sin(20\pi t) = -60{,}000\pi \cos\left(20\pi t - \frac{\pi}{2}\right)$$

12.3 MAGNETIC CIRCUITS

Just as electrical current can flow in electrical circuits, magnetic flux can also flow in magnetic circuits. A simple electrical circuit consists of a voltage source and a flow of current through a medium (a wire) or a resistor, which has a resistance and controls the current flowing through it. The concept of a magnetic circuit is similar to an electrical circuit. A magnetic circuit consists

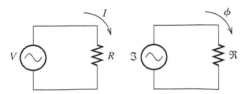

FIGURE 12.7 Comparison of electrical and magnetic circuits.

TABLE 12.1 The Analogy of Magnetic and Electrical Circuits

Electrical Circuit	Magnetic Circuit
Voltage	Magnetomotive force (mmf)
Current	Magnetic flux
Resistance	Reluctance

of a source [magnetomotive force (mmf)], a flow of magnetic flux, and a medium within which the flux is generated that controls the magnitude of the magnetic field flow (flux). For instance, the magnetic field flows better within a ferrite compared to free space. It is attenuated quickly in free space.

Thus, a circuit can be defined for the path of flow of magnetic flux. Similar to electrical resistance for electrical circuits, a magnetic reluctance can be defined for a magnetic circuit. Figure 12.7 compares a magnetic circuit and an electrical one. Table 12.1 examines this analogy further. As shown in Table 12.1, the equivalent of resistance in a magnetic field is called **reluctance**.

Some examples of magnetic circuits are shown in Figure 12.8. In Figure 12.8(a), the flux path is uniformly ferrite. Thus, the reluctance is constant. But in Figure 12.8(b), there is an air gap in the flux path. Therefore, the reluctance of part of the path is different from the other part. This section provides an introduction to magnetic circuits.

12.3.1 Magnetomotive Force

As discussed earlier, mmf forms the energy source of the magnetic circuit. To define this quantity, consider the ringed shape coil of Figure 12.9. The coil mmf, $\Im$, of an N-turn coil is defined as the number of turns multiplied by the current through the coil, that is:

$$\Im = N \cdot i \tag{12.11a}$$

It is clear that as the number of turns of current increases, the resulting magnetic field and flux increase.

In addition, the mmf of a coil is also equal to the magnetic intensity created within the coil, H, multiplied by the length of the path of flow, l. l is measured by the middle path of the coil, which is denoted by the dashed line in Figure 12.9. Therefore, the path length is the peripheral of the circular-shaped coil, that is, $2\pi r$. Accordingly, mmf corresponds to:

$$\Im = H \cdot l = H \cdot 2\pi r \tag{12.11b}$$

When an N-turn coil carrying current, i, is wound around a magnetic core, the mmf, $\Im$, produces a flux, ϕ, that is concentrated (mostly) within the core.

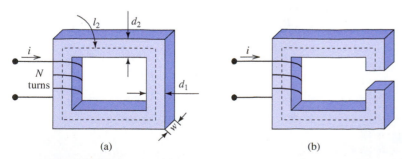

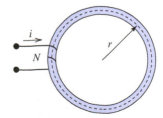

FIGURE 12.8 Magnetic circuits: (a) simple rectangular; (b) with air gap.

FIGURE 12.9 Ringed coil.

12.3.2 Reluctance

The reluctance of a magnetic material is defined as:

$$\Re = \frac{l}{\mu \cdot A} \tag{12.12}$$

Here, l is the length and A is the cross section of the magnetic material. In addition, μ is the magnetic permeability of the medium. Thus, the reluctance is a function of the medium and its structure (shape). The unit of $\Re$ is 1/H. Next, this section examines the relationship between the main components of the magnetic circuit: mmf, flux, and reluctance.

Considering a coil is wound around the core of Figure 12.10 with the cross-sectional area $A = w \cdot d$ (w is the width), the magnitude of flux density corresponds to:

$$B = \frac{\phi}{A} \tag{12.13}$$

Using Equation (12.3), the magnitude of field intensity is:

$$H = \frac{B}{\mu} = \frac{\phi}{A\mu} \tag{12.14}$$

Begin with Equation (12.11b), and use Equation (12.14) to replace H with B/μ. Thus:

$$\Im = \frac{B}{\mu} \cdot l \tag{12.15}$$

Next, use Equation (12.13) to replace B with ϕ/A in Equation (12.15). This leads to:

$$\Im = \frac{\phi}{\mu} \cdot \frac{l}{A} = \frac{l}{\mu \cdot A} \cdot \phi \tag{12.16a}$$

Finally, using the definition of reluctance in Equation (12.12), the mmf force corresponds to:

$$\Im = \Re \cdot \phi \tag{12.16b}$$

Equation (12.16) is basically Ohm's law of magnetic circuits, relating mmf, reluctance, and flux. Thus, comparing a magnetic circuit and an electrical circuit, the mmf is equivalent to voltage, the reluctance is equivalent to resistance, and the flux is equivalent to current. Accordingly, the analysis of a magnetic circuit is analogous to that of a resistive electrical circuit.

The simple magnetic circuit of Figure 12.10 has four legs: two have the mean length l_1 and the cross-sectional area $A_1 = d_1 \cdot w_1$; the other two legs have the mean length l_2 and the cross-sectional area $A_2 = d_2 \cdot w_2$. This magnetic circuit is equivalent to four series reluctances, and the total reluctance is given by:

$$\Re_{\text{series}} = 2\Re_1 + 2\Re_2 \tag{12.17}$$

with:

$$\Re_n = \frac{l_n}{\mu \cdot A_n}, \; n = 1, 2 \tag{12.18}$$

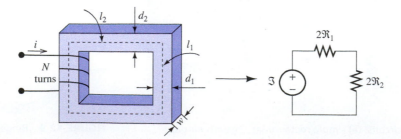

FIGURE 12.10 A simple magnetic circuit.

Note that the following assumptions are made:

1. The magnetic fluxes of all turns of the coil are linked.
2. The magnetic flux is contained exclusively within the magnetic core.
3. The flux density is uniform across the cross-sectional area of the core.

EXAMPLE 12.8   **Magnetic Circuit**

Given the current $i = 5$ A, and $\mu_r = 2000$, find the mmf, the total reluctance, ϕ, B, and H for the magnetic circuit in Figure 12.11.

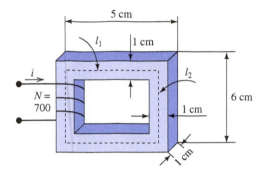

FIGURE 12.11 Figure for Example 12.8.

SOLUTION

Using Equation (12.11a), the mmf corresponds to:

$$\Im = N \cdot i = 5 \cdot 700 = 3500\,\text{A} \cdot t$$

To find the total reluctance, consider the mean length, which is calculated based on Figure 12.11. The mean path lengths are:

$$l_1 = 4\,\text{cm, and } l_2 = 5\,\text{cm}$$

The cross section of the area is:

$$A = 0.01 \cdot 0.01 = 0.0001\,\text{m}^2$$

Thus, the reluctances for the core segments are:

$$\mathfrak{R}_1 = \frac{l_1}{\mu_0 \mu_r A} = \frac{0.04}{4\pi \times 10^{-7} \times 2000 \times 0.0001} = 159.15 \times 10^3 \text{A} \cdot t/\text{Wb}$$

$$\mathfrak{R}_2 = \frac{l_2}{\mu_0 \mu_r A} = \frac{0.05}{4\pi \times 10^{-7} \times 2000 \times 0.0001} = 198.94 \times 10^3 \text{A} \cdot t/\text{Wb}$$

In addition, the total reluctance of all segments is:

$$\mathfrak{R}_{\text{total}} = 2\mathfrak{R}_1 + 2\mathfrak{R}_2 = 716.18 \times 10^3 \text{A} \cdot t/\text{Wb}$$

Using Equation (12.16b), the magnetic flux corresponds to:

$$\varphi = \frac{\Im}{\mathfrak{R}_{\text{total}}} = \frac{3500}{716.18 \times 10^3} = 4.887\,\text{mWb}$$

(*continued*)

EXAMPLE 12.8 **Continued**

Using Equations (12.13) and (12.14), the magnetic flux density and field intensity correspond to:

$$B = \frac{\phi}{A} = \frac{4.887 \times 10^{-3}}{0.0001} = 48.87 \text{ Wb/m}^2$$

And:

$$H = \frac{B}{\mu} = \frac{48.87}{4\pi \times 10^{-7} \times 2000} = 19.44 \times 10^3 \text{A} \cdot \text{t/m}$$

EXAMPLE 12.9 **Magnetic Circuit**

Find the current, i, drawn by the magnetic circuit in Figure 12.12(a). The flux is $\phi = 2\,\text{mWb}$ and $\mu_r = 1000$.

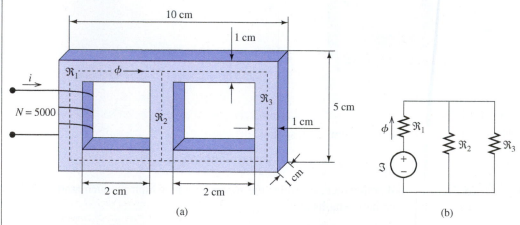

FIGURE 12.12 (a) Magnetic circuit with two loops; (b) its equivalent circuit.

SOLUTION

The equivalent magnetic circuit of Figure 12.12(a) is sketched in Figure 12.12(b). In this figure, reluctance $\Re_1$ is the equivalent reluctance for the vertical segment at the far left and the two left parallel segments of Figure 12.12(a). Reluctance $\Re_2$ is the reluctance of the middle section of Figure 12.12(a) and reluctance $\Re_3$ is the reluctance of the far right vertical section and the right two parallel sections of Figure 12.12(a). The cross-sectional area for all loops in the magnetic circuit is $A = 0.0001\,\text{m}^2$, and the mean lengths of reluctances are:

$$l_1 = 3 + 3 + 4 = 10 \text{ cm}$$
$$l_2 = 4 \text{ cm}$$
$$l_3 = 3 + 3 + 4 = 10 \text{ cm}$$

Calculating their corresponding reluctances shows that:

$$\Re_1 = \frac{l_1}{\mu_0\mu_r \cdot A} = 795.8 \times 10^3 \text{ A} \cdot \text{t/Wb}$$
$$\Re_2 = 318.3 \times 10^3 \text{ A} \cdot \text{t/Wb}$$
$$\Re_3 = 795.8 \times 10^3 \text{ A} \cdot \text{t/Wb}$$

$\mathfrak{R}_1$ and $\mathfrak{R}_2$ are in parallel, and their parallel equivalent is in series with $\mathfrak{R}_3$. Thus, the total reluctance corresponds to:

$$\mathfrak{R}_{eqv} = \left(\frac{1}{\mathfrak{R}_2} + \frac{1}{\mathfrak{R}_3} \right)^{-1} + \mathfrak{R}_1 = 1.0232 \times 10^6 \, \text{A} \cdot \text{t/Wb}$$

The current can then be calculated using Equation (12.16b), $N = 5000$ (see Figure 12.12):

$$i = \frac{\mathfrak{I}}{N} = \frac{\varphi \cdot \mathfrak{R}_{eqv}}{N} = 409.3 \, \text{mA}$$

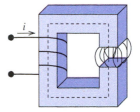

FIGURE 12.13 A magnetic circuit with an air gap.

12.3.2.1 AIR GAPS

A magnetic circuit may contain one or more air gaps, as shown in Figure 12.13. Examples include horseshoe electromagnets (see Figure 12.14) and solenoids (see Figure 12.15). With an air gap, the flux lines tend to bow out in the air gap. This effect is called the **fringing effect**. Accordingly, the effective cross-sectional area of the air gap is larger than the core. This effect is approximated by *adding the gap length to the cross-section length* for the computation of the cross-sectional area.

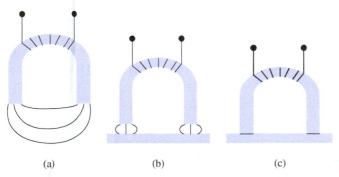

(a)   (b)   (c)

FIGURE 12.14 Air gaps: Horseshoe electromagnets: (a) with no load; (b) lifting a load; (c) load in contact with a magnet (air gap is eliminated).

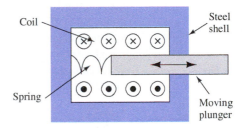

FIGURE 12.15 Solenoid; when current flows through the coil, the plunger is drawn into the shell. If current is interrupted, the spring forces the plunger out. The air gap is located between the plunger and the shell.

APPLICATION EXAMPLE 12.10   Solenoid Lock

A solenoid can be used as an electrical lock. Figure 12.16(a) shows a door utilizing a solenoid lock. Figure 12.16(b) shows the lock in the locked position. When the solenoid is energized, the moving piston is drawn back against the spring, unlocking the door as shown in Figure 12.16(c). However, many solenoid locks are energized when the lock is engaged, and de-energized when the lock is open. This allows occupants to evacuate a building if the power fails.

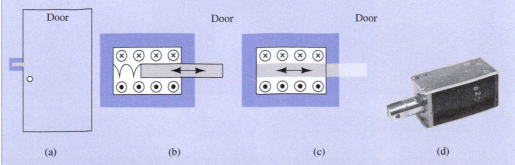

(a)   (b)   (c)   (d)

FIGURE 12.16 Solenoid lock: (a) door with solenoid lock; (b) locked; (c) unlocked; (d) real product (Used with permission from © David J. Green–electrical/Alamy.).

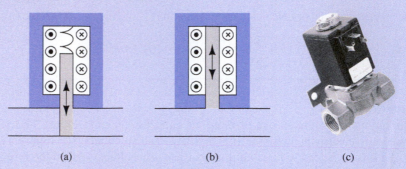

APPLICATION EXAMPLE 12.11   **Solenoid Valve**

A solenoid can be used to activate a valve, as shown in Figure 12.17. Water flow is controlled by the moving piston of the solenoid.

(a)   (b)   (c)

FIGURE 12.17   Solenoid valve: (a) closed, (b) open, (c) real product (Used with permission from David J. Green–technology/Alamy.).

12.3.2.2 RELAYS

Relays are electric switches that are controlled by solenoids. Their operation is very similar to the solenoid valve explained in Example 12.11. They have variety of applications. They can control a high-voltage (current) circuit using a low-voltage (current) signal. For example, the starter of an automobile uses a low-current signal to control a high-current signal that creates the initial rotation of the car generator. Logic functions can be created by relays, which close or open some gates based on the voltage applied to them. Accordingly, different types of relays have been designed, for example, latch relay, reed relay, and mercury-wetted relays.

EXERCISE 12.3: DIFFERENT TYPES OF RELAYS

Conduct a research and determine different types of relays available, their differences, and their applications.

APPLICATION EXAMPLE 12.12   **Speaker**

A speaker also works via a solenoid that is excited by the signal created by our voice. The signal creates a movement in the core. The core is connected to a diaphragm, which moves that based on the applied signal. The diaphragm creates the voice. Figure 12.18 presents the principles of a speaker.

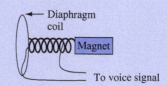

Diaphragm
coil

Magnet

To voice signal

FIGURE 12.18   Principles of speakers.

EXERCISE 12.4: MICROPHONE

Conduct a research to understand how a solenoid might be used to build microphones. Is using solenoid the only way to create a microphone?

EXAMPLE 12.13   **Magnetic Circuit with Gap**

Find the flux and the flux density in the gap for the magnetic circuit of Figure 12.19(a), given $\mu_{\text{rcore}} = 4000$ and current $i = 3$ A.

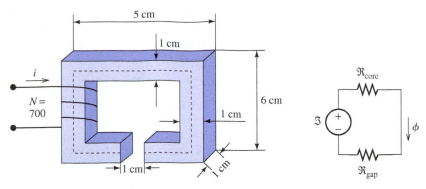

FIGURE 12.19   (a) Magnetic circuit with air gap; (b) its equivalent circuit.

SOLUTION

The equivalent circuit of Figure 12.19(a) has been sketched in Figure 12.19(b). First, find the core reluctance. To do that, first, find the length.

$$l_{\text{core}} = 5 + 5 + 4 + 3 = 17 \,\text{cm}$$

From the figure, it is clear that:

$$A_{\text{core}} = 0.0001 \,\text{m}^2$$

Thus:

$$\mathcal{R}_{\text{core}} = \frac{l_{\text{core}}}{\mu_{\text{rcore}} \cdot \mu_0 \cdot A_{\text{core}}} = 338.2 \times 10^3 \,\text{A} \cdot \text{t/Wb}$$

To find the gap reluctance, first approximate its cross section. Adding the length of the gap, $l_{\text{gap}} = 1$ cm to each side of the cross section results in:

$$A_{\text{gap}} = (0.01 + 0.01) \times (0.01 + 0.01) = 4 \times 10^{-4} \,\text{m}^2$$

Note that:

$$\mu_{\text{gap}} \approx \mu_0 = 4\pi \times 10^{-7}$$

Thus:

$$\mathcal{R}_{\text{gap}} = \frac{0.01}{4\pi \times 10^{-7} \cdot 4 \times 10^{-4}} = 19.86 \times 10^6 \,\text{A} \cdot \text{t/Wb}$$

Accordingly, the total reluctance is:

$$\mathcal{R}_{\text{total}} = \mathcal{R}_{\text{core}} + \mathcal{R}_{\text{gap}} = 20.22 \times 10^6 \,\text{A} \cdot \text{t/Wb}$$

And the total mmf is:

$$\Im = 3 \cdot 700 = 2100 \,\text{A} \cdot \text{t}$$

(continued)

| EXAMPLE 12.13 | **Continued** |

Thus, the flux and flux density correspond to:

$$\phi = \frac{\Im}{\Re_{total}} = 103.8 \times 10^{-6} \text{ Wb}$$

and

$$B_{gap} = \frac{\phi}{A_{gap}} = 0.260 \text{ T}$$

| EXAMPLE 12.14 | **Magnetic Circuit with Gap** |

For the horseshoe magnet of Figure 12.20, find the current, i, required to produce a magnetic field of 3 mT in the air gap. The core reluctance, $\Re = 500 \times 10^3$ A·t/Wb, and the core cross-sectional area, $A_{core} = 10^{-4}$ m^2.

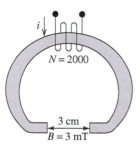

FIGURE 12.20 Horseshoe magnet.

SOLUTION

Using the cross-sectional area information provided and adding 0.03 m to consider the gap effect shows:

$$A_{gap} = (0.01 + 0.03) \times (0.01 + 0.03) = 0.016 \text{ m}^2$$

Thus, the gap reluctance corresponds to:

$$R_{gap} = \frac{l_{gap}}{\mu_{rcore} \times \mu_0 \times A_{core}} = \frac{0.03}{4\pi \times 10^{-7} \times 0.016} = 1.492 \times 10^6$$

Accordingly, the total reluctance, including the core reluctance, is $(1.492 + 0.5) \times 10^6$. In addition:

$$\phi = B \cdot A = 3 \cdot 0.016 = 0.048 \text{ mWb}$$

Now, using $i \cdot N = \phi \cdot \Re$ the result is:

$$i = \frac{0.048 \times 10^{-3} \cdot (1.492 + 0.5) \times 10^6}{2000} = 48 \text{ mA}$$

12.4 MUTUAL INDUCTANCE AND TRANSFORMERS

The inductance of a coil is related to the flux linkage, λ [defined in Equation (12.9)] and current by:

$$L = \frac{\lambda}{i} \tag{12.19}$$

Assuming that the flux is confined to the core so that all the flux goes through all turns, thus, no leakage, using Equation (12.9), the inductance corresponds to:

$$L = \frac{N \cdot \phi}{i} \tag{12.20}$$

Substituting $\phi = N \cdot i / \Re$ [see Equation (12.16b)] into Equation (12.20) shows:

$$L = \frac{N^2}{\Re} \qquad (12.21)$$

Therefore, the inductance depends on the number of coil turns, the core dimensions, and the core material.

12.4.1 Mutual Inductance

Mutual inductance occurs when two or more coils are wound on the same coil, as shown in Figure 12.21. As can be seen from the figure, this is a form of a magnetic circuit. It is clear that the magnetic flux created by current i_1 in the coil with N_1 turns is generated within the core and flows through the second coil and vice versa. The magnetic field generated by one coil that flows through the other leads to a mutual inductance. This concept is evaluated in this section. The **self-inductances** of the coils, L_1 and L_2, correspond to:

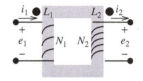

FIGURE 12.21 Simple mutual induction.

$$L_1 = \frac{\lambda_{11}}{i_1} = \frac{N_1^2}{\Re} \qquad (12.22)$$

$$L_2 = \frac{\lambda_{22}}{i_2} = \frac{N_2^2}{\Re} \qquad (12.23)$$

In Equations (12.22) and (12.23), $\Re$ refers to reluctance [see Equation (12.21)]. The **mutual inductance** between the coils corresponds to:

$$M = \frac{\lambda_{21}}{i_1} = \frac{\lambda_{12}}{i_2} \qquad (12.24)$$

In Equations (12.22)–(12.24):

- λ_{11} = flux linkage of Coil 1 caused by the current in itself
- λ_{12} = flux linkage of Coil 1 caused by the current in Coil 2
- λ_{21} = flux linkage of Coil 2 caused by the current in Coil 1
- λ_{22} = flux linkage of Coil 2 caused by the current in itself.

The flux linkage λ_{21} is due to the number of turns of Coil 2 when the flux of Coil 1 flows through it and corresponds to:

$$\lambda_{21} = N_2 \phi_1 \qquad (12.25)$$

Equivalently, the flux linkage λ_{12} is due to the number of turns of Coil 1 when the flux of Coil 2 flows through it and corresponds to:

$$\lambda_{12} = N_1 \phi_2 \qquad (12.26)$$

Now, in Equation (12.24) replacing λ_{21} with Equation (12.25) and ϕ_1 with Equation (12.16b), i.e., $\phi_1 = N_1 i_1 / R$, we have:

$$M = \frac{N_1 N_2}{R} \qquad (12.27)$$

In addition, the transformer coupling factor is defined as: $K = \dfrac{M}{\sqrt{L_1 L_2}}$. As a result, the total flux linkages of the two coils are:

$$\lambda_1 = \lambda_{11} \pm \lambda_{12} = N_1(\phi_1 \pm \phi_2) \qquad (12.28)$$

$$\lambda_2 = \lambda_{22} \pm \lambda_{21} = N_2(\phi_2 \pm \phi_1) \qquad (12.29)$$

The $+$ sign should be applied if the fluxes generated by the two coils aid each other (in the same direction), and $-$ sign should be applied if the fluxes oppose each other. Example 12.15 clarifies this concept.

| **EXAMPLE 12.15** | **Self and Mutual Inductance** |

Two coils are wound on a rectangular core having a reluctance of $\Re = 2.5 \times 10^6$ A-turns/Wb, as shown in Figure 12.22.

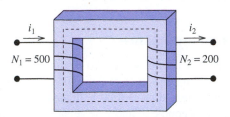

FIGURE 12.22 Figure for Example 12.15.

    a. Find the self-inductances and the mutual inductance for the coils.
    b. Find the total flux linkages given $i_1 = 20$ A, $i_2 = 50$ A.

SOLUTION

    a. To find the self-inductance and mutual coupling, use Equations (12.22)–(12.24). Using Equations (12.22) and (12.23), the self-inductances are equal to:

$$L_1 = \frac{N_1^2}{\Re} = \frac{500^2}{2.5 \times 10^6} = 100\,\text{mH}$$

$$L_2 = \frac{N_2^2}{\Re} = \frac{200^2}{2.5 \times 10^6} = 16\,\text{mH}$$

Flux due to i_1 is:

$$\phi_1 = \frac{N_1 \times i_1}{\Re} = \frac{500 i_1}{2.5 \times 10^6} = 200 \times 10^{-6} i_1$$

Flux linkage between the flux that is due to i_1 and Coil 2:

$$\lambda_{21} = N_2 \phi_1 = 40 \times 10^{-3} i_1$$

The mutual inductance between the coils is found to be:

$$M = \frac{\lambda_{21}}{i_1} = \frac{N_2 \phi_1}{i_1} = \frac{N_1 N_2 i_1}{R i_1} = \frac{N_1 N_2}{R} = 40\,\text{m}$$

    b. Equations (12.22) and (12.23) can be used to find the linkages of Coils 1 and 2 due to their own currents:

$$\lambda_{11} = L_1 i_1 = 100\,\text{mH} \times 20 = 2\,\text{Wb} - \text{turn}$$

$$\lambda_{22} = L_2 i_2 = 16\,\text{mH} \times 50 = 0.8\,\text{Wb} - \text{turn}$$

Next, find the linkages of Coils 1 and 2 due to the current of other coils. Using Equation (12.24):

$$\lambda_{21} = M i_1 = 40\,\text{mH} \times 20 = 0.8\,\text{Wb} - \text{turn}$$

$$\lambda_{12} = M i_2 = 40\,\text{mH} \times 50 = 2\,\text{Wb} - \text{turn}$$

To find the total flux, using the right-hand rule, observe that the fluxes generated by the two coils are in the same direction. Thus, using Equations (12.28) and (12.29) shows:

$$\lambda_1 = \lambda_{11} + \lambda_{12} = 4\,\text{Wb} - \text{turn}$$

$$\lambda_2 = \lambda_{22} + \lambda_{21} = 1.6\,\text{Wb} - \text{turn}$$

| EXAMPLE 12.16 | **Making an Inductance Using a Magnetic Core** |

As covered in Chapter 5, an inductor is a passive device that resists the change of current flow. Inductance is measured in **henries**. An engineer wants to make an inductor of a specific inductance value. Given a coil wrapped on a core having a reluctance of $\Re = 2.5 \times 10^6$ (see Figure 12.23), find the number of turns, N, required to produce an inductance of 100 mH.

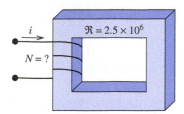

FIGURE 12.23 Magnetic core for Example 12.16.

SOLUTION

Using Equation (12.21) shows:

$$N^2 = L \cdot R = 100 \times 10^{-3} \cdot 2.5 \times 10^6$$

Thus:

$$N = 500$$

| EXAMPLE 12.17 | **Making an Inductance Using a Magnetic Core** |

An engineer desires to build an inductor with an inductance of 50 mH using the core of Figure 12.24. Find the length, l, of the core. Assume $\mu_r = 2000$.

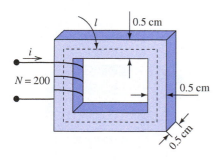

FIGURE 12.24 Magnetic circuit for Example 12.17.

SOLUTION

First, find the required reluctance:

$$L = \frac{N^2}{\Re} \rightarrow \Re = \frac{N^2}{L} \rightarrow \Re = \frac{200^2}{0.05} = 800 \times 10^3 \text{A} \cdot \text{t/Wb}$$

Now, find the length. The basic equation is:

$$\Re = \frac{l}{\mu_r \cdot \mu_0 \cdot A} \rightarrow l = \Re \cdot \mu_r \cdot \mu_0 \cdot A$$

By applying the quantities, the length is:

$$l = 800 \times 10^3 \cdot 2000 \cdot 4\pi \times 10^{-7} \cdot 0.005^2 \rightarrow l = 5.03 \,\text{cm}$$

12.4.1.1 DOT CONVENTION

Customarily, a dot is placed at the end of each coil in a schematic to show how the fluxes interact. Dots are placed such that, if both currents enter or leave at the dotted terminals, the mutual flux will be added to the self-flux, and vice versa.

According to the right-hand rule, the current entering the coil at either dot shown in Figure 12.25 creates a flux in the core in a clockwise direction. Current entering at both dots results in a constructive combining of fluxes induced by each current.

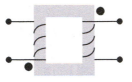

FIGURE 12.25   Dot convention.

EXAMPLE 12.18	**Dot Convention**

If a dot is placed on the primary coil as shown in Figure 12.26, where on the secondary coil should the corresponding dot be placed?

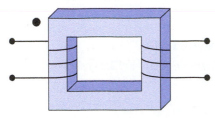

FIGURE 12.26   Magnetic circuit for Example 12.18.

SOLUTION

Here, the other dot should be placed in the right-hand side (secondary coil) such that it leads to a flux in the same direction as that of the left-hand side (primary coil).

According to the right-hand rule, if current enters the primary at the dot, a flux will be produced in the core in a counterclockwise direction. This induces a current in the secondary, which enters at the top and leaves at the bottom. Therefore, the dot should be placed at the top of the secondary coil.

EXAMPLE 12.19	**Dot Convention**

If a dot is placed on the primary coil as shown in Figure 12.27, where on the secondary coil should the corresponding dot be placed?

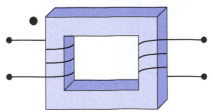

FIGURE 12.27   Magnetic circuit for Example 12.19.

SOLUTION

The other dot should be placed in the right-hand side (secondary coil) such that it leads to a flux in the same direction as that of the left-hand side (primary coil). According to the right-hand rule, if current enters the primary at the dot, a flux will be produced in the core in a counterclockwise direction. This induces a current in the secondary, which enters at the bottom and leaves at the top. Therefore, the dot should be placed at the bottom of the secondary coil.

12.4.2 Transformers

A transformer is a device used to increase or decrease voltage or current. It consists of two or more coils wound on the same core. A transformer **only** works with a **time-varying** current or **AC**. As shown by Figure 12.28(a), a transformer is a magnetic circuit with a core and two coils.

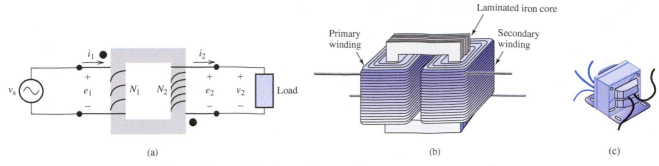

FIGURE 12.28 (a) Structure of a transformer; (b) a view of the transformer winding in an actual setting; (c) a picture of typical transformer.

If a time-varying source is connected to the input, a corresponding voltage is induced at the output. Figure 12.28(b) presents a view of an example of a transformer. Figure 12.28(c) shows a picture of a real transformer.

The analysis of a transformer is based on three stated assumptions:

1. The magnetic fluxes of all turns of the coil are linked.
2. The magnetic flux is contained exclusively within the magnetic core.
3. The flux density is uniform across the cross-sectional area of the core.

The primary coil creates a time-varying flux in the core. The flux causes a time-varying current to be induced in the secondary coil. Therefore, a transformer converts electrical energy to magnetic energy and magnetic energy back to electrical energy. Recall from Equation (12.10) that:

$$e = \frac{d\lambda}{dt}$$

Plugging Equation (12.9) into (12.10) leads to the equation for the induced voltage with respect to flux and turns ratio:

$$e = N\frac{d\phi}{dt} \tag{12.30}$$

Therefore, the electromotive forces (voltages) in Figure 12.28(a) are:

$$e_1 = N_1\frac{d\phi}{dt} = v_1 \tag{12.31}$$

$$e_2 = N_2\frac{d\phi}{dt} = v_2 \tag{12.32}$$

An ideal transformer transfers all the incoming power to the secondary coil. However, there are losses in the coils and in the magnetic core. These losses are usually converted to heat. Therefore, the power efficiency of a transformer is given by:

$$\text{Power efficency} = \eta = \frac{\text{output power}}{\text{input power}} \tag{12.33}$$

Assuming the input mmf is completely transferred to the output (without mmf loss), it is an ideal transformer, and based on Equation (12.11a):

$$\Im = N_1 \cdot i_1 = N_2 \cdot i_2$$

Accordingly:

$$\frac{i_1}{i_2} = \frac{N_2}{N_1} \tag{12.34}$$

In addition, equating $d\phi/dt$ in Equations (12.31) and (12.32) shows that:

$$\frac{v_2}{v_1} = \frac{N_2}{N_1} \tag{12.35}$$

Equations (12.34) and (12.35) are two basic equations that are used to maintain the relationship of the input and output voltages and currents in transformers. Note that based on Equations (12.34) and (12.35) in an *ideal* transformer:

$$v_1 i_1 = v_2 i_2$$

In other words, the input power (energy) in an ideal transformer is equal to the output power (energy). Note that, in general, transformers are not ideal; some energy is dissipated in the transformer, that is, in a *real* transformer:

$$v_1 i_1 > v_2 i_2$$

APPLICATION EXAMPLE 12.20   AC-to-DC Converter

Transformers are commonly used in the process of converting AC to low-voltage direct current (DC) used by electronic devices such as computers. The standard voltage available in North America is 120 V AC, 60 Hz. Computers, however, operate on 5 and 12 V DC.

In Figure 12.29, the transformer steps the AC down to a lower AC voltage. The bridge rectifier, consisting of four diodes, converts this AC into a pulsating DC. This DC is smoothed out by capacitors. In some systems, a voltage regulator is used to remove all remaining fluctuations in the voltage, resulting in a steady DC voltage. Chapter 8 discussed the fundamentals of voltage regulators—the waveforms at different stages are found in Section 8.3.3.

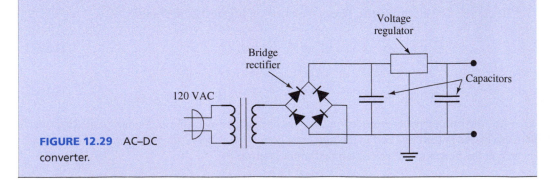

FIGURE 12.29   AC–DC converter.

EXERCISE 12.5

Based on materials learned in Chapter 8, compare the ripple created by full- and half-wave rectifiers at the output of the voltage regulator.

EXAMPLE 12.21   **Transformer Turn Ratio**

A transformer is used to step down 12 kV to 480 V. If the transformer is ideal, find the required turns ratio.

SOLUTION

Based on Equation (12.35):

$$\frac{N_1}{N_2} = \frac{v_1}{v_2} = \frac{12,000}{480} = 25.00$$

EXAMPLE 12.22   **Output Power of a Magnetic Circuit**

In the magnetic circuit shown in Figure 12.30, $i_1 = 5\,\text{A}$, and $P_1 = 100\,\text{W}$. Find the output current, i_2, and the output power, P_2.

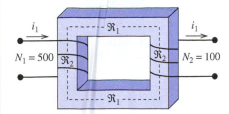

FIGURE 12.30   Magnetic circuit for Example 12.22.

SOLUTION

The secondary current can be found using Equation (12.34):

$$i_2 = \frac{N_1}{N_2} \cdot i_1 = 25\,\text{A}$$

The primary voltage corresponds to:

$$e_1 = \frac{100\,\text{W}}{5\,\text{A}} = 20\,\text{V}$$

Thus, the secondary voltage and power correspond to:

$$e_2 = e_1 \cdot \frac{N_2}{N_1} = 4\,\text{V}$$

$$p_2 = i_2 e_2 = 25\,\text{A} \cdot 4\,\text{V} = 100\,\text{W}$$

EXAMPLE 12.23   **Applications of Transformers**

Transformers are found everywhere. Figure 12.31 shows some common transformers. All electrical systems that use AC rely on transformers to deliver power.

From heavy machinery in a factory to small alarm clocks, everything that "plugs in" depends on transformers. They are used to "step up" voltages to very high levels, which make cross-country electrical transmission more efficient as verified in Chapter 9. They are also used to "step down" this high voltage to the relatively safe 120 V. Transformers are also used to step down this 120 V to the much lower levels used for power electronic devices, such as televisions, stereos, and computers. This low voltage, ranging from 3 to 20 V, is often converted from AC to DC.

(continued)

EXAMPLE 12.23   **Continued**

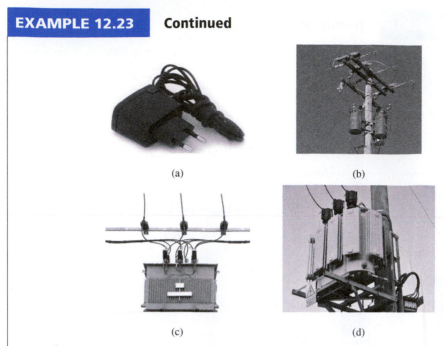

(a)   (b)

(c)   (d)

FIGURE 12.31   Various transformers. (Used with permission from (a) Ariy/Shutterstock.com; (b) Oscar C. Williams/Shutterstock.com; (c) vichie81/Shutterstock.com; and (d) Kasiap/Shutterstock.com.)

APPLICATION EXAMPLE 12.24   **Transformers in Car Ignition System**

An application of transformers is in car ignition systems (see Figure 12.32). This system is composed of a battery that is connected to a transformer of a coil. The 12-V battery output is connected to the primary of a transformer and its secondary is connected to the spark plug through a distributor. The voltage at the spark plug is selected to be in the order of 40,000 to 100,000 V to travel across the gap in the spark plugs and create a good spark. Note that, here, the voltage at the primary of the transformer eventually increases from 0 to 12 V, which increases the secondary from 0 to 40,000 V.

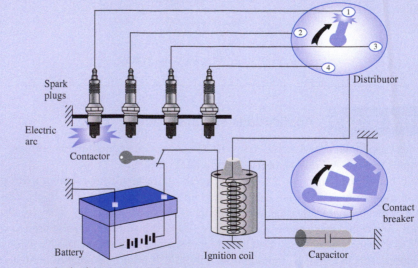

FIGURE 12.32   The ignition system of a vehicle.

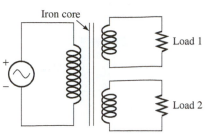

FIGURE 12.33 Multiple-winding transformers.

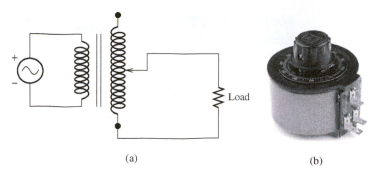

(a)   (b)

FIGURE 12.34 (a) Variable transformer; (b) Variac. (Used with permission from © David J. Green–electrical/Alamy.)

12.5 DIFFERENT TYPES OF TRANSFORMERS

Multiple Winding Transformers: A transformer may have one primary and multiple secondary. Each secondary might be used to generate different voltages or currents. Here, the secondary of the transformer can feed different devices that may need different voltages. Figure 12.33 represents the structure of a multiple-winding transformer. As shown in the figure, all windings are applied on the same iron core.

Variable Transformer: The secondary of a variable transformer is adjustable. That is, the number of secondary turns and thus the output voltage and current can be adjusted. Figure 12.34(a) presents an autotransformer. Sometimes, power companies intentionally deliver a high voltage to the houses in some rural areas where the distance of homes to the main transformer is high. This is mainly used to compensate for the transmission line losses. For example, in place of delivering 120 V, they deliver 125 V. In order to apply corrections, a transformer is used at the residential to adjust the voltage to the correct value. A variable transformer is also called *Variac* [see Figure 12.34(b)].

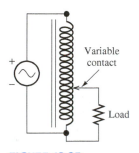

FIGURE 12.35 Autotransformer.

Autotransformer: This transformer is similar to the variable transformer; however, in this transformer, both primary and secondary windings are only on one winding as shown in Figure 12.35. The secondary has a variable contact, which determines the total voltage and current at the secondary of the transformer. An example of application of these transformers is in speed control of electric trains.

12.6 USING PSPICE TO SIMULATE MUTUAL INDUCTANCE AND TRANSFORMERS

This section introduces the steps required to simulate a simple mutual inductance and transformer system using PSpice. In order to simulate a mutual inductance and transformer, the coil ratio (or the inductance ratio) and the coupling factor must be determined. The following examples explain the necessary steps to set up the corresponding PSpice circuit.

EXAMPLE 12.25 **PSpice Analysis**

Two coils are wound on a rectangular core having a reluctance of $\Re = 2.5 \times 10^6$ A-turns/Wb, as shown in Figure 12.36. Assume an AC current source of 20 A, 60 Hz, with a load of 500 Ω connected to the output current. Use PSpice to obtain the output current, i_2.

(continued)

EXAMPLE 12.25 **Continued**

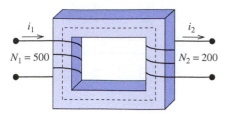

FIGURE 12.36 Figure for Example 12.25.

SOLUTION

First, determine the inductance L_1 and L_2.

$$L_1 = \frac{N_1^2}{\Re} = \frac{500^2}{2.5 \times 10^6} = 100\,\text{mH}$$

$$L_2 = \frac{N_2^2}{\Re} = \frac{200^2}{2.5 \times 10^6} = 16\,\text{mH}$$

The mutual inductance between the coils is found to be (see Equation 12.27):

$$M = \frac{N_1 N_2}{\Re} = \frac{(500)(200)}{2.5 \times 10^6} = 40\,\text{mH}$$

Next, determine the coupling factor, K, which corresponds to:

$$K = \frac{M}{\sqrt{L_1 L_2}} = \frac{40 \times 10^{-3}}{\sqrt{(100 \times 10^{-3})(16 \times 10^{-3})}} = 1$$

To draw the PSpice schematic circuit and run the simulation perform the following steps:

1. Go to "Place" > "Part" and type "ISIN" to choose the sinusoidal current source and place it as shown in Figure 12.37.
2. Follow the steps shown in Example 7.17 in Chapter 7 to set up an AC current source for ISIN.
3. Next, place the inductors as shown in Figure 12.37. However, there is no dot symbol displayed in PSpice for the mutual inductor. One solution is to assume that one pin number of the inductor represents the dot symbol. In this case, assume that pin number 2 is the dot for every inductor. There should be no wire connection between the primary inductor L_1 and the secondary inductor L_2. However, since PSpice does not allow any isolated components, a very large resistor (R_1) is used to imitate the "open circuit" between L_1 and L_2.
4. In PSpice, the pin number of an inductor will not show up by default. Go to Inductor Properties by double-clicking on it.

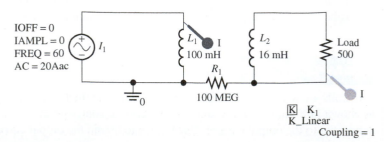

FIGURE 12.37 PSpice schematic solution for Example 12.25.

5. Next, click on the Pin Tab as shown in Figure 12.38 (shown by the dashed arrow), then highlight both Name "1" and "2" (by the solid arrow), and choose the "Display Properties." Choose "Value Only" as the Display Format, click OK, and close the properties menu.

6. Continue setting up the PSpice circuit as shown in Figure 12.37. Before running the simulation, the Coupling Factor must be set up on the PSpice circuit. Go to "Place" "Part" and type "K_LINEAR." Double-click the "COUPLING = ?" and enter the coupling factor that is calculated (enter 1 in this example).

7. In addition, determine which inductors are coupled together. Double-click the Coupling Factor, "K_LINEAR" to enter the properties menu. Under L_1 and L_2 (see arrow in Figure 12.39), enter L_1 and L_2 (or the inductor name that appeared in the circuit).

8. Set the simulation type to AC Sweep Analysis. Choose AC Sweep Type as "Logarithmic" with Start, End frequency, and Points per Decade as shown in Figure 12.40.

9. Run the simulation; the result is shown in Figure 12.41.

Figure 12.41 shows that the current flow through the load increases as frequency increases. The primary coil of a transformer draws certain amount of current to create a magnetic flux, which is known as *magnetizing current*. The magnetizing current is higher at lower

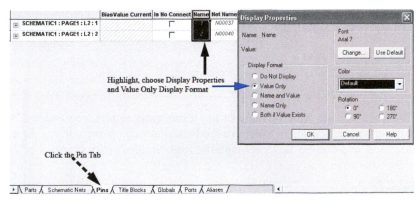

FIGURE 12.38 Steps to display the pin number of the inductor in PSpice.

FIGURE 12.39 Set up a coupled inductor.

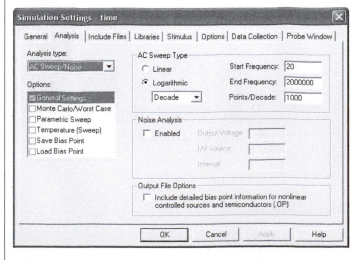

FIGURE 12.40 Setup for the simulation in Example 12.25.

(*continued*)

EXAMPLE 12.25 Continued

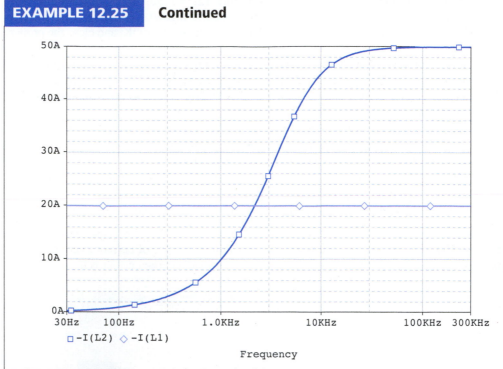

FIGURE 12.41 Simulation result for Example 12.25.

frequencies. This reduces the efficiency of the transformer at low frequencies. It should be noted that the current is divided between the load and the magnetizing current. As magnetizing current increases, less current flows through the load. Students may increase or decrease the load to observe the change of plot!

EXAMPLE 12.26 **PSpice Analysis**

Given the ideal transformer circuit in Figure 12.42, assume that the voltage source is 100 V, 60 Hz, and that resistor R_1 is 50 Ω and resistor R_2 is 100 Ω. Use PSpice to plot the input and output voltages of the transformer, if N_1 is 1000 and N_2 is 100.

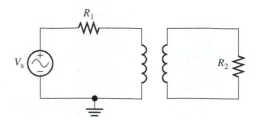

FIGURE 12.42 Circuit for Example 12.26.

SOLUTION

First, calculate the inductance ratio for L_1 and L_2:

$$\frac{L_2}{L_1} = \left(\frac{N_2}{N_1}\right)^2$$

$$\frac{L_2}{L_1} = \left(\frac{100}{1000}\right)^2 = 0.01$$

$$L_1 = 100L_2$$

This is a step-down voltage converter.

1. Set up the sinusoidal voltage source and resistor as shown in Figure 12.43. A large resistance (through a resistor) has been used as the bridge between the input and the output of the transformer.
2. To place a transformer into the circuit, go to "Place" > "Part" and type "XFRM_LINEAR." Place the transformer part into the circuit as shown in Figure 12.43.
3. Next, double-click the TX1 or the transformer to enter the properties menu.
4. For the ideal case, set the Coupling Factor to "1" (solid arrow in Figure 12.44). Then, determine the value of both inductances in the transformer. Given that L_1 is 100 times larger than L_2, choose L_1 to 1000 H while L_2 to 10 H (dashed arrow in Figure 12.44).
5. Set the simulation type as Time Analysis and run the simulation.
6. The simulation result is shown in Figure 12.45.

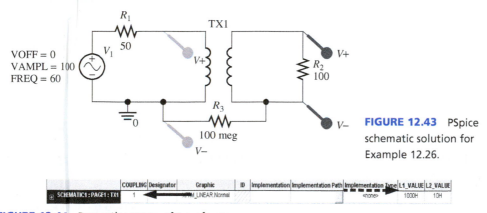

FIGURE 12.43 PSpice schematic solution for Example 12.26.

FIGURE 12.44 Properties menu of transformer.

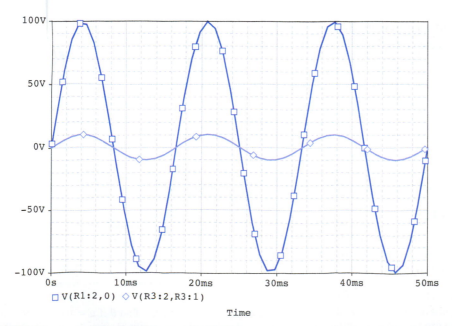

FIGURE 12.45 Simulation result for Example 12.26.

12.7 WHAT DID YOU LEARN?

- Current produces a magnetic field around it. The right-hand rule can be used to determine the direction of the field.
- Force is exerted on any charge moving through a magnetic field. Force is expressed as shown in Equation (12.6):

$$f = qu \times B$$

- Force is exerted on any current-carrying wire in a magnetic field. Its magnitude is expressed in Equation (12.8):

$$f = ilB \sin \theta$$

- The relationship between magnetic flux, ϕ, and flux density, B, is [see Equation (12.1)]:

$$\phi = \int_A B \cdot dA$$

- Magnetic flux density, B, and magnetic field intensity, H, are related as shown in Equation (12.3):

$$B = \mu H$$

- Faraday's law of magnetic induction states that a voltage, e (also called electromotive force), is induced in a circuit whenever flux linkage changes. The induced voltage can be calculated using Equation (12.10):

$$e = \frac{d\lambda}{dt}$$

- Lenz's law states that an induced voltage produces a magnetic field that opposes the field that generated the induced voltage.
- Magnetic flux flows in circuits can be analogous to electrical circuits. Magnetomotive forces are analogous to voltage sources, reluctance is analogous to resistance, and flux is analogous to current.
- The inductance of a transformer corresponds to Equation (12.21).
- Mutual inductance occurs when two or more current-carrying coils are wound on the same core. It corresponds to (see Equation (12.27):

$$M = \frac{N_1 N_2}{R}$$

- A transformer is a device used to increase or decrease voltage or current. For an ideal transformer, the voltage across each coil is proportional to its number of turns.

Problems

B refers to basic, A refers to average, H refers to hard, and * refers to problems with answers.

SECTION 12.2 MAGNETIC FIELDS

12.1 (B) A wire extending perpendicular to the face of a clock carries a current toward the clock. In which direction is the magnetic density vector B?

12.2 (B) A coil lying east and west produces a flux density vector pointing east. In which direction does the current flow?

12.3 (B)* An electron is moving in a magnetic field as shown in Figure P12.3. Find the force exerted on the electron given $u = 10^6$ m/s, $B = 10$ T, and $\theta = 120°$.

FIGURE P12.3 Force on a moving electron in a magnetic field.

12.4 (B) In Figure P12.3, find θ if $u = 10^6$ m/s, $B = 10$ T, and the exerted force on the electron is 1.6×10^{-12} N.

12.5 (A)* A power line that carries large amount of current creates a magnetic field around it and a magnetic field density that is

$$B = \frac{\mu_r \mu_0 i}{2\pi r}$$

where i is the current, r is the distance from the center of the line (measured perpendicularly), $\mu_r = 1$ for the air, and μ_0 is the magnetic permeability of the free space. Find the carrying current if $B = 10^{-3}$ T at 0.5 m from the power line.

12.6 (B) A straight wire of 1-m length carries a current of 5 A. It is placed in a position making an angle of 40° to a magnetic field of $B = 2$ T. How much force is exerted on the wire?

12.7 (A)* A straight wire of 200-cm length, carrying a current of 2 A, is placed in a magnetic field of $B = 10$ T. If a force of 2 N is exerted on the wire, find the angle of the wire with respect to the magnetic field.

12.8 (H) A magnetic field can be obtained by winding a current-carrying wire onto a cylindrical coil as shown in Figure P12.8.

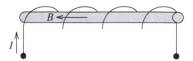

FIGURE P12.8 The magnetic field of a coil.

The length of the cylindrical core is l, μ denotes the permittivity of the core, N denotes the number of turns, and i denotes the current. If the radius of the core is r, then when $l \gg r$, it can be shown that the flux density of the interior of the core is approximated by:

$$B = \frac{\mu N}{l} i$$

(a) Derive the equation for the flux (the cross section of the core is a circle).

(b) Find the current required to produce a magnetic flux density $B = 0.5$ T, given $N = 100$, $l = 0.2$ m, and air core, that is, $\mu = \mu_0$.

12.9 (A) In Problem 12.8, how does the flux change for the following cases?

(a) Replace the core with the material with double μ.

(b) The number of turns N is doubled.

(c) The radius, r, is doubled and the current, i, is halved.

(d) The length, l, is doubled and N is doubled as well.

12.10 (A) A flux of $\phi(t) = 16 \cos(50\pi t)$ Wb passes through a coil having $N = 500$ turns. Find the voltage $e(t)$ induced in the coil.

12.11 (A)* A voltage of $V_s(t) = 30 \cos(200t)$ is connected to a coil of $N = 800$ turns. Find the flux $\phi(t)$ induced in the coil.

SECTION 12.3 MAGNETIC CIRCUITS

12.12 (B) Find the reluctance of the core given in Figure P12.12 if $\mu_r = 1000$.

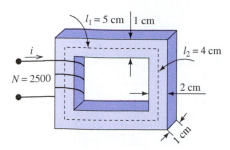

FIGURE P12.12 Magnetic core.

12.13 (H) As shown in Problem 12.8, the flux density generated by winding wire carrying current onto a coil is given by:

$$B = \frac{\mu N}{l} i$$

It is also known that the mmf is defined as:

$$\Im = N \cdot i$$

where N denotes the number of turns, i denotes the coil current, l denotes the length of magnetic path, and μ denotes the permittivity of the core.
The magnetic resistance is defined as:

$$\Re = \frac{\Im}{\phi}$$

where ϕ is the flux through the magnetic path.

Prove that $\Re = \dfrac{l}{\mu A}$ where A is the cross-sectional area.

12.14 (H) A closed magnetic core is given in Figure P12.14.

(a) Let ϕ denotes the flux in the core and μ denotes the permittivity of the core, find H_{12}, H_{23}, H_{34}, and H_{41} (H_{12} denotes the magnetic field intensity in the branch $P_1 P_2$ and others are similarly defined).

(b) Verify that $\Im = H_{12}l_2 + H_{23}l_1 + H_{34}l_2 + H_{41}l_1$.

A more general equation is $\Im = \sum_m H_m l_m$ and it is applicable to any closed path.

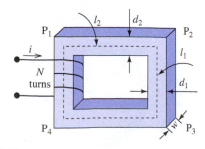

FIGURE P12.14 Magnetic core.

12.15 (A) In practice, circuits must often be designed to produce a certain amount of magnetic flux in an air gap.

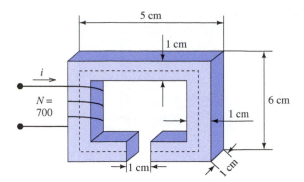

FIGURE P12.15 Magnetic circuit with an air gap.

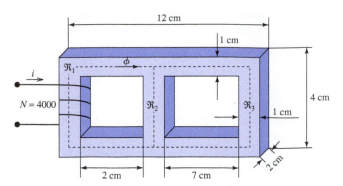

FIGURE P12.19 Magnetic circuit with two loops.

Consider the circuit in Figure P12.15. Assume that $\mu = 4000\mu_0$ for the core.

    To produce a magnetic flux density of 0.26 T in the air gap, follow the steps below to find the current, i, in the coil:

(a) Compute H_c (the magnetic field intensity of the core) and H_a (the magnetic field intensity of the air).

(b) Compute the total mmf, $\mathfrak{F}$.

(c) Compute the current, i.

12.16 (A) Repeat Problem 12.15 if the material of the core is iron, that is, $\mu = 1000\mu_0$.

12.17 (A)* Given $\mu_r = 3000$ and $i = 7$ A, find the mmf and the flux ϕ for the magnetic circuit in Figure P12.17.

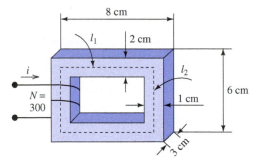

FIGURE P12.17 Magnetic core.

12.18 (A) Given $\mu_r = 4000$, $r = 2$ cm, depth $= 0.5$ cm, and $i = 7$ A, find the mmf, the flux ϕ, the flux density B, and the flux intensity H for the magnetic circuit in Figure P12.18. (Assume that the ring has a square cross-sectional area.)

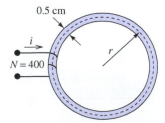

FIGURE P12.18 Magnetic circuit.

12.19 (H)* Find the current, i, drawn by the magnetic circuit in Figure P12.19 if the flux $\phi = 15$ mWb, and $\mu_r = 4000$.

12.20 (H) Repeat Problem 12.19 if there is an air gap at the center of the $\mathfrak{R}_2$ branch shown in Figure 12.19. The length of the gap is 1 cm.

12.21 (H)* If $i = 10$ A and $\mu_r = 2000$, find the flux, ϕ_B, for the circuit shown in Figure P12.21. The depth of the core is 1 cm and the radius, r, is 4 cm. Assume a square-shaped cross-sectional area.

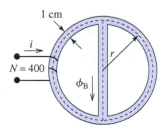

FIGURE P12.21 Magnetic circuit with two loops.

12.22 (A) Find the flux and the flux density in the gap for the magnetic circuit shown in Figure P12.22, given $\mu_{rcore} = 3000$ and $i = 5$ A.

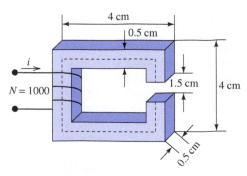

FIGURE P12.22 Magnetic circuit with air gap.

12.23 (H) It is desired that the gap in Figure P12.23 has a flux density of $B = 5$ T. Given $\mu_{rcore} = 2000$ and $i = 3$ A, find the required turns ratio, N.

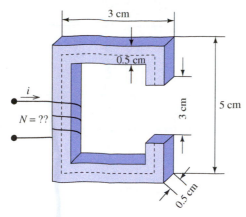

FIGURE P12.23   Magnetic circuit with air gap.

SECTION 12.4 MUTUAL INDUCTANCE AND TRANSFORMERS

12.24 (B)* Two coils are wound on a core as shown in Figure P12.24. The core reluctance is $\Re = 6.3 \times 10^6$ A-turns/Wb. Find the self-inductances of the coils.

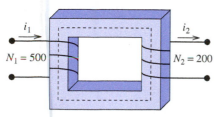

FIGURE P12.24   Magnetic circuit.

12.25 (B) A coil with 300 turns shown in Figure P12.25 has a flux of 3.2×10^{-4} Wb when $i = 0.5$ A. Compute the inductance of this coil.

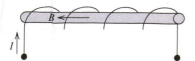

FIGURE P12.25   Magnetic circuit.

12.26 (H) The mutual inductance between the coils shown in Figure P12.26 is denoted by M. Prove that $M = N_1 N_2 / \Re$, where N_1 and N_2 are the number of turns of the primary and secondary coils, respectively, and $\Re$ is the magnetic reluctance of the core.

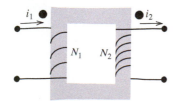

FIGURE P12.26   Mutual induction.

12.27 (B) Two coils are wound on a rectangular core having a reluctance of $\Re = 4.8 \times 10^6$ A-turns/Wb, as shown in

Figure P12.27. Find the self-inductances and the mutual inductance for the coils.

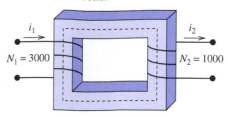

FIGURE P12.27   Magnetic circuit.

12.28 (A)* Given a coil wrapped on a core having a reluctance of $\Re = 50 \times 10^6$ (see Figure P12.28), find the number of turns, N, required to produce an inductance of 20 mH. Note: use 1/H for the unit of reluctance.

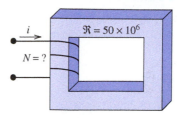

FIGURE P12.28   The structure for Problem 12.28.

12.29 (A) An inductor is desired with an inductance of 25 mH. In Figure P12.29, find the length, l, of the core that produces this inductance given $\mu_r = 1000$.

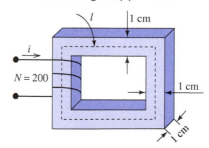

FIGURE P12.29   Magnetic circuit.

12.30 (A) If a dot is placed on the primary coil as shown in Figure P12.30, where on the secondary coil should the corresponding dot be placed?

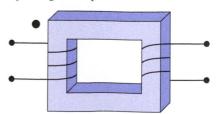

FIGURE P12.30   Magnetic circuit with a dot.

12.31 (A)* A nonideal transformer has two kinds of power loss: the copper loss and core loss. The copper loss of the windings is due to the wire resistance, which converts electrical power to heat and the core loss is due to leakage flux.

Assume that a transformer has 300-W core loss. The wire resistance of the windings is 1.4 Ω and the current through it is 15 A. The output power of this transformer is 20,000 W.

(a) Compute the copper loss power.

(b) Compute the power efficiency of the transformer.

12.32 (B) An ideal transformer has a turns ratio of $N_2/N_1 = 1/4$. Find the output AC voltage if the input voltage (AC) is

(a) 2 V

(b) −5 V

(c) 10 V

12.33 (A)* A transformer is used to step up 560 V to 8.5 kV. If the transformer is ideal, find the required turns ratio.

SECTION 12.6 USING PSPICE TO SIMULATE A MUTUAL INDUCTANCE AND TRANSFORMERS

12.34 (B) In Figure P12.34, the voltage source is 10 V, 60 Hz, the resistor, R, is 10 Ω, and the load has a resistance of 100 Ω. Use PSpice to plot the voltage input to the transformer and across the load if N_1 is 100 and N_2 is 400. Assume an ideal-case transformer.

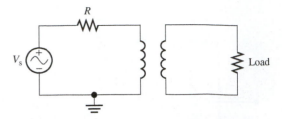

FIGURE P12.34 Circuit for Problems 12.34 and 12.35.

12.35 (B) In Figure P12.34, the voltage source is 10 V, 60 Hz, the resistor, R, is 500 Ω, and the load has a resistance of 200 Ω. Use PSpice to plot the voltage input to the transformer and across the load if N_1 is 600 and N_2 is 200. Assume an ideal-case transformer.

12.36 (A) Assume that a mutual inductance circuit has an AC current source of 15 A, 60 Hz, and that its output is connected to a load with 100 Ω. Also assume that the dots of both inductors are placed at the same side. Use PSpice to plot the input and output of the current if N_1 is 400, N_2 is 250, and reluctance is $\Re = 2.5 \times 10^6$.

CHAPTER 13

Electric Machines

(Used with permission from © BL Images Ltd/Alamy.)

13.1 INTRODUCTION

In Chapter 12 you studied magnetic circuits, transformers, and the relationship between electricity and magnetism. In this chapter, you will learn about devices whose operation can be explained based on the relationship between electricity and magnetism. These devices are called *electric machines*. Do you know what central air conditioners, vacuum cleaners, and CD players all have in common? Each operates using an electric machine of one kind or another.

Electric machines are devices that convert energy from one form (e.g., potential energy of the stored water in the dam) to another, that is kinetic energy of the machine. Electric machines are classified into two types. Devices that generate electrical energy through the input of some other form of energy are called *generators*. Devices that transform electrical energy into mechanical energy are called *motors*. *Alternators* are generators that generate alternating current (AC). An example is a car alternator, which supplies the vehicle's equipment such as air conditioning, radio, lights, and speakers. You may be amazed at the wide range of applications in which electric motors are used. Electric motors are

TABLE 13.1 Examples of Various Applications of Electric Motors

Domestic and Office	Automobile	Medicine	Industry
Central air conditioners	Power windows	Dentist drills	Drilling machines
Food processors	Sunroof operation	Electric wheelchairs	Grinders
Vacuum cleaners	Radiator fans	Artificial hearts	Lathe machines
Refrigerators	Fuel pumps		Milling machines
Hair dryers	Starter motors		Compressors
Sewing machines	Windshield wipers		Industrial fans, blowers
CD players, cameras			Cranes, lifts
Ceiling fans			Elevators
Photocopiers			Motors in robots

found in applications as diverse as domestic appliances, automobiles, and industrial equipment. Table 13.1 lists various applications for electric motors. This leads us to an obvious question: Why have electric machines become so widespread across so many diverse fields?

13.1.1 Features of Electric Machines

- Wide range of torque–speed characteristics make them suitable for a variety of applications
- Smooth speed control can be attained via semiconductor devices, such as thyristors
- Compatible with different loads as operating characteristics can be easily modified
- High efficiency, high-load capacity, and long life
- Low noise and low maintenance
- Smooth acceleration and deceleration
- Can be operated in diverse environments (e.g., radioactive, submerged in liquids, etc.)
- Use a clean source of energy
- Fast response, that is, no warm-up time is required; thus, instant loading is achievable
- Compact, reliable, and economical
- Remote control is possible, if required

13.1.2 Classification of Motors

Now that you are familiar with various features of electric machines, we can investigate the governing principles of motors and generators. Motors are classified into two main categories: *alternating current (AC) motors* and *direct current (DC) motors*. AC and DC designations refer to the type of current supply required for the motor to operate. The tree diagram shown in Figure 13.1 depicts

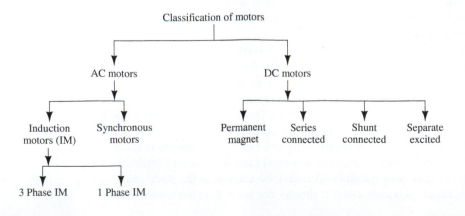

FIGURE 13.1

Classifications of motors.

the broad classifications of motors. The next sections of this chapter explain the different classes of motors revealed in Figure 13.1.

13.2 DC MOTORS

13.2.1 Principle of Operation

"When a current-carrying conductor is placed in a magnetic field, it experiences a mechanical force, the direction of which is given by the right-hand rule (or Fleming's left-hand rule)."

Based on this principle, the force applied to the current corresponds to:

$$\mathbf{F} = l\mathbf{I} \times \mathbf{B}$$

where $\mathbf{F}$ is the electromagnetic force vector, l is the length of the current-carrying wire, $\mathbf{I}$ is the current vector, and $\mathbf{B}$ is the magnetic field vector measured in tesla. Sometimes, the current, $\mathbf{I}$, is replaced with the current density, $\mathbf{J}$, measured in terms of amperes per square meter (A/m^2). In that case, the $\mathbf{F}$ would be the electromagnetic force density measured in terms of newton per square meter (N/m^2). The multiplication shown by "×" refers to the fact that the direction of the force is perpendicular to the plane formed by the direction of the current and the magnetic field vectors. The direction of this force can be determined by what is called the *right-hand rule*. This rule is shown visually in Figure 13.2(a). In this figure, you can see that the direction of all fingers (except the thumb) are first pointed toward the direction of current $\mathbf{I}$. Then, when the vector formed by $\mathbf{I}$ is rotated toward vector $\mathbf{B}$, the direction of the thumb indicates the direction of force applied to current $\mathbf{I}$.

The direction of the force can be equivalently explained by Fleming's "left-hand rule" as shown in Figure 13.2(b). Here, the thumb, the first, and the second fingers on the left hand are held so that they are at right angles to each other. If the first finger points in the direction of the magnetic field and the second finger in the direction of the current in the wire, then the thumb will point toward the direction of the force on the conductor.

13.2.2 Assembly of a Typical DC Motor

To function, a DC motor needs a DC supply. A DC supply can be provided via rectifiers (AC to DC converters) as explained in Chapter 8. The DC supply can also be provided by batteries. The main components of a DC motor include the stator, rotor, windings, and commutator as explained below.

- *Stator:* A stator is usually the static component of the motor. Most electric motors employ a cylindrical stator that has an even number of magnetic poles. The stator can be formed either by permanent magnets (PMs) [as shown in Figure 13.3(a)] or by field windings [see Figure 13.3(b)]. Field windings are applied to the slots cut within the surface of the stator structure.

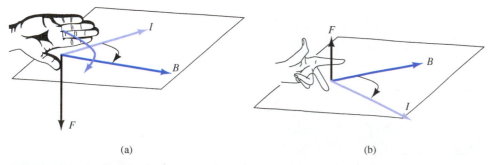

(a)　　　　　　　　　　　　　　(b)

FIGURE 13.2　Evaluating the direction of the force applied to a current, I, using (a) right-hand rule; (b) Fleming's left-hand rule.

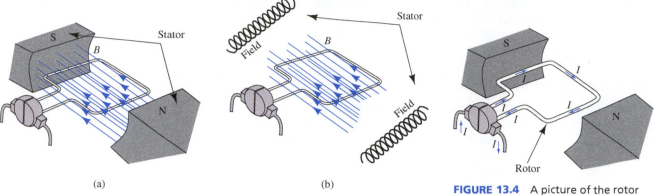

(a)   (b)

FIGURE 13.3 DC motor: (a) Permanent magnet stator; (b) field winding stator.

FIGURE 13.4 A picture of the rotor with slots.

- ***Rotor:*** A rotor is the rotating component of the motor. It is located inside the stator and is basically a laminated iron cylinder mounted on a shaft. The slots are cut lengthwise on the surface of the rotor to accommodate the armature conductors. Figure 13.4 represents one winding loop of the rotor.
- ***Windings:*** There are two types of windings: field and armature. Both types carry DCs. Field winding sets up the magnetic field of the stator, whereas armature winding does the same for the rotor. Figure 13.3 shows the two types of windings.
- ***Commutator:*** A commutator reverses the direction of the current in each conductor through the armature as it passes from one pole to another and helps to develop continuous unidirectional torque. Figure 13.5 is an actual photograph of a commutator.
- ***Brushes:*** The electrical connections to the commutator are made by brushes as illustrated in Figure 13.6. In contrast to the slip rings found in AC motors, brushes are more prone to wear.

13.2.3 Operation of a DC Motor

In many DC motors, field windings (in the stator) create the required magnetic field, while armature conductors (rotor conductors) carry the current. Thus, the force experienced by the current-carrying conductors placed in a magnetic field is transferred to the armature conductors. The direction of this force is given by the right-hand rule or Fleming's left-hand rule as shown in Figure 13.2. The armature conductors are placed in slots on the periphery of the rotor. The force applied to each conductor acts as a twisting force on the rotor. The cross product of twisting force and the radius vector of rotation is called *torque*. Figure 13.7 explains the operation of a DC motor.

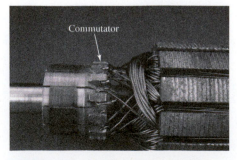

FIGURE 13.5 Commutator. (Used with permission from © sciencephotos/Alamy.)

FIGURE 13.6 Brushes.

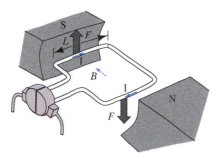

FIGURE 13.7 Operation of a DC motor: force applied to the conductors.

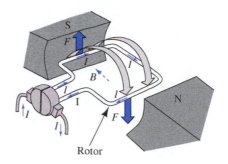

FIGURE 13.8 Rotation created by the torque produced.

In Figure 13.7, the direction of the magnetic field is depicted by the dashed arrow. The direction of current through the armature conductors is shown by the colored solid arrows. The direction of force experienced by each of the rotor conductors is denoted by the gray arrow. The curved gray arrows in Figure 13.8 depict the direction of the resultant torque produced by the force experienced by the individual rotor conductors.

EXERCISE 13.1

Apply the right-hand rule (or Fleming's left-hand rule) to Figures 13.7 and 13.8. Is the direction of the resultant force shown correctly?

EXAMPLE 13.1   Force Direction

In Figures 13.7 and 13.8, what happens when the direction of the field is reversed? What happens when the direction of the field and the direction in which the current is flowing reverse simultaneously?

SOLUTION

If the field direction reverses, the force will reverse. However, if the current and the field reverse their directions simultaneously, then the direction of the resultant force will remain unchanged. This can be easily verified using the right-hand rule.

13.2.4 Losses in DC Machines

In both generators and motors, a certain amount of energy is always dissipated during the energy conversion process. This dissipated energy appears in the form of heat and therefore increases the temperature of the machine. This phenomenon limits the performance and the lifetime of the components of the machine and reduces its power output. In addition, the dissipated energy is the detrimental factor in determining the kilowatt (kW) rating of the motor. Different losses that occur within a DC machine are classified into three main types:

   i. *Copper* losses
  ii. *Iron* or *core* losses
 iii. *Mechanical* losses.

Each type of loss is explained in detail:

i. *Copper losses:* These losses are due to the current flowing through the resistance of various windings. These losses take place in armature as well as field circuits. Copper losses in armature are load dependent and are therefore named *variable losses.*

ii. *Iron or core losses:*

Following are the sources of these losses:

- *Eddy current losses:* Eddy currents are created when there is a relative motion between a conductor and the magnetic field produced by the stationary stator. According to Lenz's law, the generated electric field is directed in such a way that it opposes the cause producing it. Thus, there is a loss of energy. Eddy currents also produce heat, which results in losses. These losses can be minimized by using materials with low electrical conductivity, such as ferrites, or by using laminations.

- *Hysteresis losses:* Hysteresis loss is a heat loss, which is caused by the magnetic properties of the armature. Armature rotation leads to the change of direction of magnetic flux, and thus continuous movement of magnetic particles. Magnetic particles are particles within a magnetic element. These particles create the magnetic property of the material. Indeed, a magnet is formed by the integration of many magnetic particles. Figure 13.9 represents the schematics of magnetic particles that form a magnet.

    A molecular friction is produced as magnetic particles try to align themselves with the changing direction of the magnetic field. This leads to an increase in temperature, which increases armature resistance. This loss can be minimized by proper selection of materials for armature. Examples of proper materials include steel or laminated iron.

- *Mechanical losses:* These losses are due to the rotation of the armature and include bearing friction loss, brush friction loss, and windage or air friction loss. These losses vary with speed and are considered constant for machines having fairly constant speed.

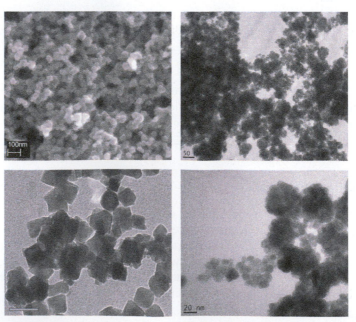

FIGURE 13.9 Magnetic particle under the microscope. (Photos courtesy of Silvio Dutz, IPHT Jena.)

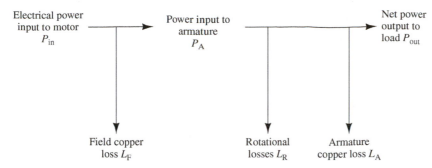

FIGURE 13.10 Power stages in a DC motor.

iii. *Iron and copper losses* are often grouped together and called *rotational losses*. The lost power is converted into heat and it is independent of the load; thus, it is called a *constant loss*.

Some losses in a machine are not exactly deterministic and are therefore called *stray losses*. These losses could be due to the magnetic flux leakages and short-circuit currents in the coils undergoing commutation. In small machines, these losses are negligible.

Figure 13.10 provides a summary of all losses. As depicted in this figure, the net output power is calculated after deducting the losses explained in this section.

EXAMPLE 13.2 **Drawbacks of DC Motors**

What is the most defining drawback of DC motors in general?

SOLUTION

The biggest drawback in any DC motor is the use of the commutator and brush arrangement and the relative motion between them. As a result of this arrangement, they:

- Require additional maintenance
- Have a shorter life span
- Possess a higher chance of sparking at the commutator and brush assembly

---

13.3 DIFFERENT TYPES OF DC MOTORS

As summarized in Figure 13.1, DC motors are divided into four main categories: (1) shunt-connected, (2) series-connected, (3) permanent magnet, and (4) separately excited motors. This section examines the equations that maintain the relationship of the voltage across armature, applied torque, generated power, and its angular speed (see Section 13.3.1). You will learn to analyze the characteristics of DC motors in other sections.

13.3.1 Analysis of a DC Motor

The emf, E_A, that is created across the rotor (see Figure 13.11) represents the average voltage induced in the armature due to the motion of conductors relative to the magnetic field. The concept of back-emf is discussed in greater detail later in the chapter. In general, in motors, E_A is referred to as back-emf because it opposes the externally applied voltage (i.e., the source that produces it). This is explained by Lenz's law.

Lenz's law states that an electromagnetic field interacting with a conductor generates electrical current which induces a magnetic field that opposes the magnetic field generating the current. In other words, the induced emf and the change in flux have opposite signs.

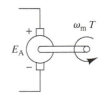

FIGURE 13.11
Armature back-emf and its relationship with torque and angular speed.

An expression for the induced emf is defined using the following parameters:

P = the number of stator poles

ϕ = flux per stator pole in weber

$\omega_m = 2\pi N$ is the angular velocity, with the unit of radians per minute

N = speed of armature in revolutions per minute (rpm)

Z = total number of armature conductors

According to Faraday's law of electromagnetic induction, the average emf induced in each conductor is proportional to the rate at which rotating parts cut the flux. It corresponds to:

$$e = \frac{d\phi}{dt} \tag{13.1}$$

Now, the total flux produced is given as:

$$d\phi = \phi \cdot P \tag{13.2a}$$

The time required for one revolution is:

$$dt = \frac{60}{N} \tag{13.2b}$$

Because the rate change of flux as described in Equation (13.1) is defined as total flux in one revolution, by replacing $d\phi$ and dt in Equation (13.1) with those in Equations (13.2a) and (13.2b), for the emf per conductor, we see that:

$$e = \frac{\phi \cdot P \cdot N}{60} \tag{13.3}$$

The total emf for Z conductors in the DC motor corresponds to:

$$E_A = \frac{\phi \cdot P \cdot N \cdot Z}{60} \tag{13.4}$$

Using Equation (13.5), we see that the induced voltage (emf) measured in volts in the armature is:

$$E_A = K \cdot \phi \cdot \omega \tag{13.5}$$

In Equation (13.5), $K = P \cdot Z/2\pi$ is called the machine constant and varies with the machine design. In addition,

$$\omega = \frac{2\pi N}{60} = \frac{\omega_m}{60} \tag{13.6}$$

is the angular velocity of the motor in radians per second. This equation represents that the induced voltage is directly proportional to the speed of the motor as well as the flux produced by each stator pole.

The torque (in newtons meter) produced by a conductor is:

$$T = B \cdot I_A \cdot l \cdot r \tag{13.7}$$

where:

- I_A = armature current in the conductor (in amperes)
- l = length of each conductor (in meters)
- r = radius at which the conductor is placed (in meters)

Because the motor contains Z conductors, the total torque corresponds to:

$$T = B \cdot I_A \cdot l \cdot r \cdot Z \tag{13.8}$$

We know that the relationship between the magnetic flux, magnetic field, and the cross-sectional area of the magnetic flux corresponds to:

$$B = \frac{\phi}{A} \tag{13.9}$$

This torque acting on the rotor is directly proportional to the magnetic flux and armature current. Replacing for B in Equation (13.8) using Equation (13.9), we see that:

$$T = \frac{\phi}{A} \cdot I_A \cdot l \cdot r \cdot Z \tag{13.10}$$

Thus, we can represent the torque as:

$$T = K_A \cdot \phi \cdot I_A \tag{13.11}$$

where:

$$K_A = \frac{lrZ}{A} \tag{13.12}$$

K_A is a constant that varies with the specifications of the motor such as its diameter, length, and the number of conductors. The generated power (mechanical power) by the motor corresponds to:

$$P = E_A \cdot I_A \tag{13.13}$$

Now, the relationship between generated power, torque, and its angular velocity corresponds to:

$$P = T \cdot \omega \tag{13.14}$$

Equating Equations (13.13) and (13.14) then replacing E_A with Equation (13.5) and T with Equation (13.11) results in $K = K_A$.

EXAMPLE 13.3   Induced emf

Consider a two-stator pole motor where each pole has a flux of 15 mWb. Assume the motor rotates at 600 rpm. Calculate the induced emf if the motor has two armature conductors.

SOLUTION

First, the angular velocity of the motor is:

$$\omega = 2\pi N/60$$
$$\omega = 20\pi$$

And the machine constant, K (defined in Equation (13.6)), is:

$$K = \frac{2 \times 2}{2\pi} = \frac{2}{n}$$

Therefore, the induced emf, E_A, is:

$$E_A = K \cdot \phi \cdot \omega$$
$$E_A = \frac{2}{\pi}(15 \times 10^{-3})(20\pi)$$
$$E_A = 0.6 \text{ V}$$

EXAMPLE 13.4   Induced emf

A DC motor that supplies a maximum torque of 50 N m rotates at the speed of 200 rpm. Assume that the induced current of the motor is 20 A. Determine the induced emf.

SOLUTION

Considering Equations (13.13) and (13.14), the induced emf is:

$$E_A = \frac{T\omega}{I_A}$$
$$E_A = \frac{50(200)(2\pi)/60}{20}$$
$$E_A = 52.3 \text{ V}$$

13.3.2 Shunt-Connected DC Motor

The shunt-connected DC motor derives its name from the fact that the armature and field windings are connected in parallel (shunt). The inductance effects of these windings are ignored because they behave as a short circuit for DC. Figure 13.12 represents an equivalent circuit of a DC shunt motor. The rotor and armature circuits are connected in parallel.

In Figure 13.12, R_A represents the effect of the resistance of the armature windings plus the brush resistance. The resistance, R_f, represents the field resistance, and I_f and V are the field current and voltage across the field resistance, respectively. Normally, in a DC motor, R_f is selected to be large; thus, the field current is small: it should be mainly enough to produce the required flux. Thus, almost the entire current from the supply is applied to the armature. The voltage across the armature is partially dissipated in the armature windings and partially converted to the required emf that leads to the rotation.

We can now analyze the equivalent circuit of a shunt DC motor (Figure 13.12). Considering steady-state status:

$$V = R_f \cdot I_F \tag{13.15}$$

Now, examine the torque and speed relationship for a shunt motor. Applying Kirchhoff's voltage law (KVL) to the circuit of Figure 13.12 results in:

$$V = R_A I_A + E_A \tag{13.16}$$

Next, replacing for I_A using Equation (13.11), and for E_A using Equation (13.5):

$$V = \frac{R_A T}{K\phi} + \frac{K\phi\omega_m}{60} \tag{13.17}$$

Applying some mathematical manipulations to Equation (13.17), the torque can be computed as:

$$T = \frac{K\phi}{R_A}\left(V - \frac{K\phi\omega_m}{60}\right) \tag{13.18}$$

The relationship of the torque and the angular frequency, ω_m, is derived using Equation (13.18). Observe that the speed at no load condition ($T = 0$) is maximum and the maximum torque would lead to zero speed ($\omega_m = 0$). Here, the load refers to mechanical load. Thus, when there is no load, the torque required to move the load is zero.

Equation (13.11) depicts that torque, T, is directly proportional to the armature current. At zero current, the torque is zero. This fact has been shown in Figure 13.13(a): torque increases when the armature current increases and vice versa. Now, consider the speed of a shunt motor when it is operating at a no load condition.

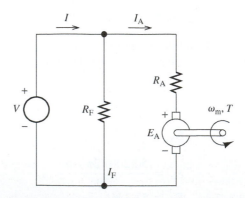

FIGURE 13.12 Equivalent circuit of a DC shunt motor.

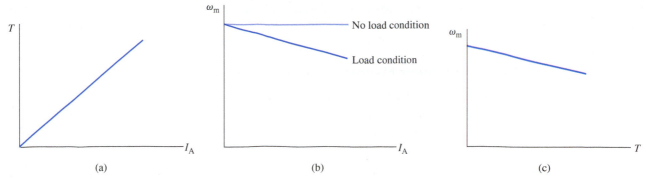

FIGURE 13.13 Torque and speed characteristics.

As shown in Figure 13.13(b), because flux and back-emf are constants, at a no load condition, using Equation (13.5), the speed would be constant and does not change with the armature current. However, under load conditions, with increase in armature current, using Equation (13.16), the emf decreases and thus using Equation (13.6) we can see that as emf decreases, the speed decreases as well.

Now consider Figure 13.13(c), which illustrates that when torque increases, the speed of the motor will decrease. This fact can be easily determined from Equation (13.17): For a constant V, as T increases, the motor speed decreases.

Applications: DC shunt motors are normally preferred for constant speed applications, such as automotive applications. However, because AC motors are simpler and less expensive, they are used in wide range of applications compared to DC motors. See Section 13.7 for more details.

EXAMPLE 13.5 **Back-emf**

A 220-V supply provides a current of 20 A to a shunt motor. The armature resistance is 0.3 Ω. Find back-emf and power created in the motor when the field resistance is 150 Ω.

SOLUTION

Using Figure 13.12 and Equation (13.15) the field current corresponds to:

$$I_F = \frac{\text{Supply voltage}}{R_F} = \frac{220}{150} = 1.46 \text{ A}$$

The value of the armature current is:

$$I_A = I_S - I_F = 20 - 1.46 = 18.53 \text{ A}$$

Thus:

$$E_A = V_S - I_A \cdot R_A = 220 - 18.53 \times 0.3 = 214.44 \text{ V}$$

As a result, power corresponds to:

$$P = E_A I_A = 214.44 \times 18.53 = 3973.57 \text{ W}$$

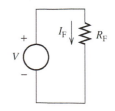

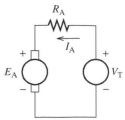

FIGURE 13.14

Equivalent circuit of a separately excited DC motor.

13.3.3 Separately Excited DC Motors

As shown in Figure 13.14, two sources of power are used in these motors: one for the stator windings and the other for rotor. This facilitates speed control by varying either of the two voltage sources. Characteristics of a separately excited DC motor are the same as those of the DC shunt motor. The main advantage of using two separate sources is to be able to control motor speed by varying either of the two sources. The disadvantage of this type is the use of a secondary DC voltage source.

13.3.4 Permanent Magnet (PM) DC Motor

The operation and characteristics of permanent magnet DC motors are the same as DC shunt motors with a wound stator. The only difference in architecture is that in a permanent magnet (PM) DC motor, the field is created by permanent magnets on the stator instead of a field winding. Figure 13.15 shows the components of a permanent magnet DC motor with all the components distinctly marked.

Figure 13.16 represents the equivalent circuit of a permanent magnet DC motor. Here, R represents the armature resistance. Remember that the field winding does not exist in a PM DC motor. Instead, a permanent magnet is used. As described in Section 13.3.1, the torque is related to the armature current by Equation (13.11). Now, because the flux is constant in PM DC motors, Equation (13.11) becomes:

$$T = K_T \cdot I_A \tag{13.19}$$

where $K_T = K_A \phi$ is the torque constant. Due to the relationship of the constant, K_A, and the motor dimensions and structure, [see Equation (13.12)] K_T is determined by the dimensions and structure of the motors. In addition, using Equation (13.5), the back-emf of PM motors corresponds to:

$$E_A = K_A \cdot \omega \tag{13.20}$$

where $K_A = K \cdot \phi$ is a constant.

Now, using the circuit model of Figure 13.5, note that:

$$V_S = I_A R + E_A \tag{13.21}$$

Replacing for E_A and I_A using Equations (13.20) and (13.19), after some simple mathematical manipulations:

$$\omega = \frac{V_S}{K_A} - \frac{TR}{K_A K_T} \tag{13.22}$$

Equation (13.22) and Figure 13.17 illustrate the torque–speed characteristics of a permanent motor. From Figure 13.17, we can conclude that the torque developed by a PM DC motor is directly proportional to the applied voltage and inversely proportional to the motor speed.

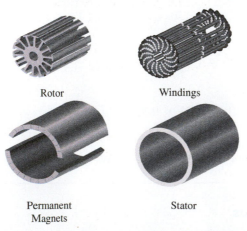

Rotor   Windings

Permanent   Stator
Magnets

FIGURE 13.15 Permanent magnet DC motor. (From www. cst.com. Reprinted with permission of Computer Simulation Technology.)

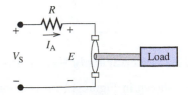

FIGURE 13.16 Equivalent circuit of a PM DC motor.

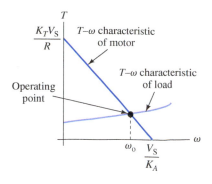

FIGURE 13.17
Torque–speed
characteristics of the
PM DC motor.

Figure 13.13(b) shows that when a load is applied to a motor, the torque and angular velocity of the motor decreases. In Figure 13.17, the T–ω characteristic of load represents the required torque with respect to the angular velocity to lift a load. The T–ω characteristic of motor and load intersect at the operating point, ω_0. At the operating point, the torque that is produced by the motor equals the torque required to move the load. If the torque supplied by the motor exceeds the torque required (i.e., in the left region of operating point), it will allow more load onto the motor until the motor speed is slowed down at the operating point. If the load increases after reaching the operating point, the motor will fail to function.

EXERCISE 13.2

What peculiar and important characteristic of DC series motors can we infer from Figure 13.17?

Advantages

- Higher efficiency as no power is needed to produce the stator flux.
- It is smaller in size and has a simpler architecture; thus, it can be used for simple applications such as toys.
- Motors can produce high torque with low heat loss.

Disadvantages

- Magnets can become demagnetized due to excessive use and overheating.
- The PM flux density is smaller compared to that of motors with stator winding. Thus, the torque produced is lower compared to a wound stator motor with the same power rating. This limits the use of PM DC motors to low-power applications such as motors used in toys, or low-power home applications, such as washing machines.
- PM motors have more variable characteristics than other types of motors. This can add to manufacturing cost and make PM motors less desirable if little variability is required.

EXAMPLE 13.6   Back-emf

A certain PM motor is used in a grass cutter (see Figure 13.18). The motor produces a back-emf of $E = 12$ V at the speed of 4000 rpm. Assuming the field current remains constant, find the back-emf for the speed of 1000 and 4500 rpm.

SOLUTION

In PM DC motors $K_A = K \cdot \phi$ is constant; therefore, using Equation (13.20):

$$\frac{E_1}{\omega_{m1}} = \frac{E_2}{\omega_{m2}}$$

(continued)

FIGURE 13.18 PM DC motor used in a grass cutter. (Used with permission from © David J. Green–technology/Alamy.)

EXAMPLE 13.6 **Continued**

Replacing for E_1 and ω_{m1} and ω_{m2}:

$$\frac{12}{2\pi \times 4000} = \frac{E_2}{2\pi \times 1000}$$

or

$$E_2 = 3 \text{ V}$$

Similarly, for $N = 4500$:

$$E_2 = 13.5 \text{ V}$$

EXAMPLE 13.7 **Armature Current and Voltage**

Find the armature current and armature voltage (without load) of a PM DC motor when the source current is 20 A, field current is 1.2 A, source voltage is 225 V, and the armature resistance is 0.3 Ω.

SOLUTION

Using Figure 13.12, the armature current corresponds to:

$$I_A = I_S - I_F = 20 - 1.2 = 18.8 \text{ A}$$

In addition, the voltage can be found using:

$$E_A = V_S - I_A \cdot R_A = 225 - 18.8 \times 0.3 = 219.36 \text{ V}$$

EXAMPLE 13.8 **DC Motor Torque and Speed**

Using the result of Example 13.7, find the speed and the torque of a DC motor with the power of 2300 W, $K = 1$, and $\phi = 3.2$ Wb.

SOLUTION

The speed of a motor is given by Equation (13.5):

$$\omega = \frac{E_A}{K \cdot \phi}$$

$$\omega = \frac{219.36}{3.2} = 68.55 \text{ rad/s}$$

In order to find the torque, use Equation (13.14):

$$T = \frac{P}{\omega}$$

$$T = \frac{2300}{68.55} = 33.55 \text{ N m}$$

EXAMPLE 13.9 **DC Shunt versus Permanent Magnet Motor**

Considering the similarity in the construction of the DC shunt motor and the permanent magnet DC motor, what are the differences between the two in terms of characteristics and applications?

SOLUTION

Permanent magnet (PM) DC motors support lower torques compared to the DC shunt motor because of their use of permanent magnets. PM DC motors have limited power range up to 10 hp while DC shunt motors offer power ranges up to 200 hp.

13.3.5 Series-Connected DC Motor

DC motors exhibit considerable differences in characteristics when the connection between the armature and field is changed from parallel to series. Series DC motors have higher starting torque compared to parallel motors; however, their speed may have variations, in other words their speed stability (or speed regulation) is poor. Speed regulation is defined as the difference between full-load and no-load speed. Series-connected motors are used for applications such as cranes, where a high starting load is required.

The equivalent circuit diagram of a series-connected DC motor is shown in Figure 13.19. The inductance of the field winding is neglected because inductance behaves like a short circuit for DC under steady-state conditions.

Let's analyze the equivalent circuit of the DC series motor. As shown in Figure 13.19, in this case, the armature and field windings will be in series and thus their currents will be equal. Accordingly, the relationship between the magnetic flux and the armature current corresponds to:

$$\phi = K_1 \cdot I_A \tag{13.23}$$

Here, K_1 is a constant that varies with the physical parameters of the field winding. The relationship between force and current was discussed in Equation [12.11(a)] (see Chapter 12). This relationship corresponds to magnetic motive force (mmf).

Replacing for the flux in Equations (13.5) and (13.11) with the flux in Equation (13.23), the created emf, E_A, and torque, T, correspond to:

$$E_A = K \cdot K_1 \cdot \omega I_A \tag{13.24}$$

and

$$T = K \cdot K_1 \cdot I_A^2 \tag{13.25}$$

Here, I_A is armature current which is proportional to field current in series motors, and ω_m is the angular speed. Applying Kirchhoff's voltage law to the equivalent circuit of Figure 13.19:

$$V = (R_F + R_A) \cdot I_A + E_A \tag{13.26}$$

Replacing Equation (13.5) for E_A in Equation (13.26), and substituting for the flux using Equation (13.23), the expression for I_A can then be shown as:

$$I_A = \frac{V}{R_A + R_F + K \cdot K_1 \cdot \omega} \tag{13.27}$$

Now, squaring both sides of Equation (13.27) and using Equation (13.25) results in:

$$T = \frac{K \cdot K_1 \cdot V^2}{(R_A + R_F + K \cdot K_1 \cdot \omega)^2} \tag{13.28}$$

Equation (13.28) shows that the torque is inversely related to the speed: As the speed increases, the torque decreases. This fact is depicted in Figure 13.20.

Applications: DC series motors can develop very high torque at low speed and therefore are preferred for traction type loads. This is due to the high armature current in Equation (13.27). Thus, they are used in cranes, electric locomotives, conveyors, and so on.

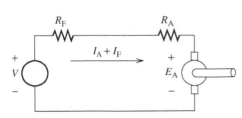

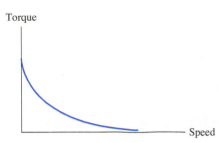

FIGURE 13.19   Equivalent circuit of the DC series motor.

FIGURE 13.20   Torque and speed characteristics of a DC series motor.

EXAMPLE 13.10   DC Motor Speed

A DC series motor running at the speed of 800 rpm draws 18 A from the supply. If the load is changed such that the current drawn by the motor is 55 A, calculate the speed of the motor on the new load. Armature and field winding resistances are 0.2 and 0.4 Ω, respectively. Assume a supply voltage of 230 V and also that the flux produced is proportional to the current.

SOLUTION

Using Equation (13.27), calculate the constant $K \cdot K_1$:

$$K \cdot K_1 = \frac{V - I_A(R_A + R_F)}{I_A \cdot \omega}$$
$$= 0.1454$$

Next, calculate the speed of the motor at the current of $I_A = 55$ A. Using Equation (13.27):

$$\omega = \frac{230 - 33}{55 \cdot 0.1454} = 24.6 \, \frac{\text{rad}}{\text{s}}.$$

EXAMPLE 13.11   DC Motor Voltage and Torque

A DC motor is used in a conveyor belt. The belt has been wrapped around the rollers, and one of the rollers is connected to the motor. The series-connected motor rotates at 1200 rpm with the torque of 12 N m and the field current of 40 A. Find (a) output voltage, and (b) torque, assuming $R_A = R_F = 0$.

SOLUTION

a. In order to find the output voltage, equate Equations (13.13) and (13.14), which leads to:

$$E_A = \frac{T \cdot \omega}{I_A}$$

Before solving, first find ω. We know (Equation (13.6)):

$$\omega = \frac{2\pi N}{60} = \frac{2\pi \cdot 1200}{60} = 1215.66 \text{ rad/s.}$$

Thus, we can solve for the voltage:

$$V = \frac{12 \times 125.66}{40} = 37.69 \text{ V}$$

b. The field current is given and we know that we are using a series DC motor; thus, $I_A = I_F$ and $I_A = 40$ A. Using Equation (13.25):

$$K \cdot K_1 = \frac{T}{I_A^2} = \frac{12}{1600}$$

Now, to find the torque when R_A and $R_F = 0$, use Equation (13.28):

$$T = \frac{K \cdot K_1 \cdot V^2}{(R_A + R_F + K \cdot K_1 \cdot \omega)^2}$$

or:

$$T = \frac{37.69}{125.66} = 0.3 \text{ N m}$$

13.3.6 Summary of DC Motors

Table 13.2 summarizes the different types of DC motors, their torque characteristics, power, and applications.

13.4 SPEED CONTROL METHODS

Speed control in motors is required in many applications, such as conveyors, extruders, surface winders, machine tool spindles of lathe machines, pumps, some fans, and so on. In these applications, the motor speed must be able to be adjusted either to very high or low values. Several methods can be used to control the speed the of DC motors.

13.4.1 Speed Control by Varying the Field Current

Consider a shunt-connected DC motor. Equation (13.5) shows that keeping E_A constant, the motor speed and the flux created by the field winding are inversely proportional. E_A can be kept constant if the armature resistance, R_A, is low and thus the voltage drop across this resistance is negligible. In this case, E_A will be equal to the supply voltage which is assumed constant. Therefore, a resistance in series with the field winding (see Figure 13.21) can be used to adjust the field and, accordingly, the motor speed. As the resistance increases, the field current and, therefore, the flux decreases. This leads to an increase in the speed of the motor.

Accordingly, this method of speed control is only applicable to shunt-connected DC motors. It should be noted that in the case of a PM DC motor, field windings are not available. In

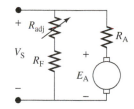

FIGURE 13.21 Speed control by varying the field current.

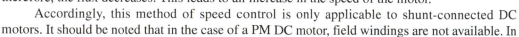

TABLE 13.2 Summary of all DC Motors Studied

Motor Types	Characteristics	Power Range	Applications
DC shunt motor	Constant flux constant speed motor creates moderate torque at start	Up to 200 hp	Machine tools like lathes, milling machines, grinding machines, centrifugal and reciprocating pumps, blowers, and fans
DC series motor	Creates dangerously high torques at low speeds and should be always connected to the load; speed can be varied	Up to 200 hp	Preferred for traction-type loads; employed in electric locomotives, conveyors, cranes, elevators, trolleys
Permanent magnet DC motor	Higher efficiency, smaller size and simpler architecture; magnets can become demagnetized due to excessive use and overheating; produces lower torque	Up to 10 hp	Power windows in automobiles, computer peripherals
Separately excited motors	Can be controlled either by varying the voltage applied to the field winding or by varying the voltage applied to the armature; can produce high torques	Up to 100 hp	Traction applications, to control the speed and torque of the motor by changing both armature voltage and stator current

addition, in series-connected motors, the current through the armature is equal to the field current [see Equation (13.23)]. Thus, it is not possible to control the speed using field current in either PM DC motors or series-connected motors.

Characteristics of the field current speed control system:

- Speed control is smooth and easy.
- Speed control above the rated value is also possible by making I_A very large (rated value is the speed that the motor rotates without causing overheating).
- The method is efficient as the power loss in the field circuit is minimal due to the lower value of the field current.
- As a disadvantage: if the speed increases to very high values, the commutation operation might be affected—an undesirable result.

| **EXAMPLE 13.12** | **Supply Voltage and Field Impact on Shunt-Connected Motor** |

What would happen if supply voltage varies in a shunt-connected motor? What would happen if the field circuit gets opened?

SOLUTION

I_A and I_F depend on supply voltage. Thus, if supply voltage decreases, both currents will decrease and, accordingly, the produced flux will decrease. Because the voltage drop across armature resistance is very small, $V_S \approx E_A = K \phi \omega$. Therefore, reducing V_S decreases the flux. In this case, there will only be a very small reduction in speed.

If the field circuit is open-circuited, the field current and therefore electromagnetic flux will be zero. Now assuming the motor has an initial speed, its speed increases infinitely while its torque decreases to zero. Note that as a practical matter, the speed cannot increase infinitely. Mechanical losses would overheat the motor and cause damage that would ultimately stop its rotation.

| **EXAMPLE 13.13** | **DC Shunt Motor Current and Speed** |

A 250-V DC shunt motor has a shunt field resistance of 150 Ω and an armature resistance of 0.3 Ω. For a given load, the motor runs at 1400 rpm, and drawing 20 A current. If a resistance of 75 Ω is added in series with the field, find the new armature current and speed. Assume load torque and flux are constant.

SOLUTION

Figure 13.12 represents the equivalent circuit of a DC shunt motor. The field current corresponds to:

$$I_f = \frac{V_f}{R_f} = \frac{250}{150} = 1.66 \text{ A}$$

Now:

$$I_A = I_L - I_f = 20 - 1.66 = 18.34 \text{ Amps}$$

Observe that the current of the field circuit is much smaller than the armature current. Thus, adding the speed control resistance leads to minor power dissipation. Now, applying KVL to the loop of Figure 13.12:

$$E_A = V_S - I_A R_A = 250 - 5.502 = 244.5 \text{ V}$$

Using Equation (13.5):

$$E_A = K\phi\omega$$

$$K\phi = \frac{E_A}{\omega} = \frac{244.5}{2 \times \pi \times 1400} = 0.0278$$

Considering Figure 13.12, note that:

$$I_f = \frac{V_S}{R_F + R_{adj}} = \frac{250}{225} = 1.11 \text{ Amps}$$

The new armature current corresponds to:

$$I_A = I_L - I_f = 20 - 1.11 = 18.89 \text{ Amps}$$

Thus, the new E_A is:

$$E_A = V_S - I_A R_A = 250 - 5.667 = 244.33 \text{ V}$$

Again, using Equation (13.6):

$$\omega = \frac{244.33}{0.0278} = 8789 \text{ rad/s}$$

13.4.2 Speed Control by Varying the Armature Current

This method is applicable to all types of DC motors. The voltage across the armature is controlled when a resistance is inserted in series with the armature. Speed has a direct relationship with the voltage applied across the armature. For a certain load, the value of the armature current remains constant. Thus, when a resistance is added in series with the armature, the new resistance is called R_A as illustrated in Figure 13.22. In this case, the same relationship as Equation (13.18) can be calculated for the torque, that is:

$$T = \frac{K\phi}{R_A}(V_T - K\phi\omega)$$

The torque–speed relationship for different armature resistances is sketched in Figure 13.23.

The main disadvantage of this method of speed control is the fact that it needs the addition of an extra resistor as shown in Figure 13.22. This resistor consumes energy. Specifically, the speed can be adjusted to lower values if the series resistance is adjusted to higher values. This leads to the waste of considerable energy in the form of the heat dissipated in the series resistance. It should be noted that this problem is not too severe when the speed control resistance is located in the shunt resistor. As discussed earlier, the current through the shunt resistor is very minimal and so is the dissipated power.

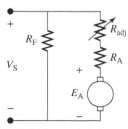

FIGURE 13.22 Speed control by the armature current.

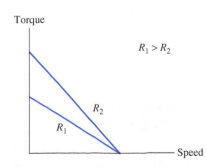

FIGURE 13.23 Torque–speed characteristics.

13.5 DC GENERATORS

A DC generator converts mechanical energy into electrical energy. Examples of the sources of mechanical energy include a steam turbine, diesel engine, or wind power. DC generators have applications in the production of industrial materials, vehicle battery charging, and street lights. Small generators produce power in the range of 1 to 10 kVA. Large industrial generators produce power in the range of 8 to 30 kVA for homes or small shops.

Generators have a standard *rating*. Rating defines the values for voltage and current of a machine at which the machine works with minimal copper losses. These standard rating definitions are designed to allow correct selection of machines. This section details the principles of DC generators and describes different types of DC generators, their features, and characteristics.

13.5.1 The Architecture and Principle of Operation of a DC Generator

The architecture of DC generators is very similar to that of DC motors, which have already been discussed. A generator consists of a stationary stator. The rotor receives the input from the prime mover. An example of a prime mover is a water turbine. The main function of the stator is to create the required flux via windings. The DC generator also consists of a commutator, whose purpose is to convert the alternating emf generated into unidirectional emf. It is made up of copper segments insulated from each other by a layer of mica (Figure 13.24).

Accordingly, the fundamental principle underlying the operation of a DC generator is Faraday's law of electromagnetic induction: when the flux linkage of a coil or conductor changes, electromagnetic force (emf) is created across the coil. Changes in flux linkage occur only when there is a relative motion between the flux and the coil. The direction of this emf can be found via Fleming's left-hand rule or the right-hand rule explained in Figure 13.2.

A relative motion between flux and coil is created by rotating the conductors—with respect to the flux—using a prime mover. To achieve higher voltages, a large number of conductors needs to be connected together in a specific manner to form a winding. This winding is placed on a rotor and is called an armature winding. There are two types of armature windings: *lap winding* and *wave winding*, which are shown in Figures 13.25 and 13.26.

> *Lap winding:*   Here, the connections start from the conductor in slot 1 and subsequently overlap each other as the winding proceeds until the starting point is reached. As illustrated

FIGURE 13.24
Commutator in a DC generator. (Photo courtesy P. D. Simpson & Company.)

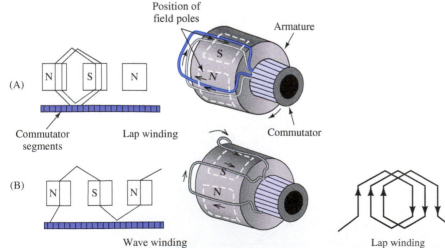

FIGURE 13.25   (a) Lap and (b) wave winding.

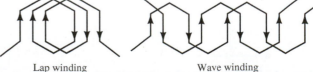

FIGURE 13.26   Types of armature winding.

in Figure 13.25, the end of one coil is connected to a commutator and the end of other coil is placed just under the same pole and in this way all coils are connected.

Accordingly, the total number of conductors is divided into a number of parallel paths, which is equal to the number of magnetic poles in the machine. The current-carrying capacity of this type of winding is higher because of the presence of several parallel paths. This type of winding (lap winding) is used in DC generators designed for high-current applications. DC generators need many pairs of poles and brushes as shown in Figure 13.25.

Wave winding: In this type, windings are organized in a way to avoid overlapping. They are organized like a progressive wave and hence the name. The winding is divided into two parallel paths irrespective of the number of poles on the machine. The lower number of parallel paths results in lower carrying current capacity. This type of winding is used in high-voltage applications. As shown in Figure 13.26, the two ends of each coil are connected to a commutator separated by the distance between poles. This configuration makes the series addition of the voltages in all the windings between brushes. This type of winding only requires one pair of brushes to provide only two paths regardless of the number of poles.

13.5.2 emf Equation

Let us now find an expression for the induced emf in terms of the parameters of DC generators. The expression is similar to Equation (13.5). The only difference is that here there are R parallel paths within which the armature conductors are divided. R equals the number of poles for lap winding, and $R = 2$ for wave winding.

There are Z conductors and R parallel paths; therefore, Z/R series-connected conductors will be available and the same emf is induced in each of them. The total induced emf defined in Equation (13.5) for motors can then be replaced by the following for DC generators:

$$E_A = e \cdot \frac{Z}{R} = \frac{\phi \cdot P \cdot N \cdot Z}{60 \cdot R} \tag{13.29}$$

Here, $N/60$ can be replaced by $\omega/2\pi$ and $K = (P \cdot N \cdot Z)/(2\pi R)$. Thus, the result is an equation similar to Equation (13.6) in Section 13.2.

$$E_A = K \cdot \phi \cdot \omega \tag{13.30}$$

Equation (13.30) is known as the emf equation for a DC generator. Equation (13.6) for the emf of DC motors represents the average voltage induced in the armature due to the motion of conductors relative to the magnetic field. In motors, E_A is sometimes referred to as back-emf because it opposes the externally applied voltage (the cause producing it). This voltage is exactly the same as what is found using Equation (13.30).

EXAMPLE 13.14 **DC Shunt Generator**

A four-pole DC shunt generator has the speed of 1200 rpm. The armature has 700 conductors and the flux per pole is 30 mWb. Find the emf voltage. Assume the number of parallel paths equals the number of poles.

SOLUTION

Using Equation (13.29):

$$E = \frac{\phi \cdot P \cdot N \cdot Z}{60R} = \frac{30 \times 10^{-3} \times 4 \times 1200 \times 700}{60 \times 4} = 420 \text{ V}$$

Vehicle Battery Charging

Explain the source of energy for charging the battery of an automobile.

SOLUTION

The battery is charged by the car itself. In the engine, the chemical energy of the fuel is converted into mechanical energy (rotary motion) available at the crankshaft. This is delivered to the input of the generator using a fan belt. The shaft of the generator has a cylindrical structure and has electric conductors on its periphery attached to it. Thus, it forms the rotor of the car's generator. The rotary movement at the input causes the rotor to move, and therefore, the conductors cut the magnetic flux present inside the generator, and electricity is generated. This electrical energy is fed to rectifiers that convert alternating (time varying) voltage at the generator to constant (DC) voltage required to charge the battery. Rectifiers were studied in Chapter 8. Figure 13.27 represents the car generator structure and the circuit for charging the battery.

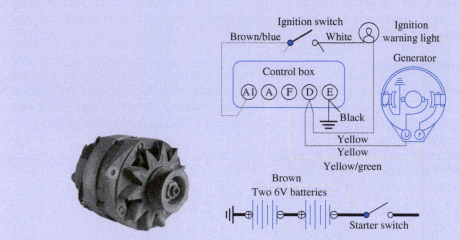

FIGURE 13.27 Car generator (Used with permission from © INSADCO Photography/Alamy.) and battery charging circuit.

13.6 DIFFERENT TYPES OF DC GENERATORS

This section examines different types of DC generators, including separately excited DC generators and shunt-connected DC generators. The section also investigates load regulation characteristics of DC generators.

A DC generator is basically an energy converter. In DC generators, the field windings must be excited with DC. The two main types of DC generator are called *shunt* and *series*. When a shunt field configuration is used, it is called a *shunt generator*. As the name indicates, the shunt field is connected in parallel with the armature. When separate voltage sources are used, the generator is called a *separately excited generator*. These types of generators are examined further in the rest of this section.

13.6.1 Load Regulation Characteristics of DC Generators

In any generator, the induced emf generates current that flows through the load. As shown in the equivalent circuits in Figure 13.28, as the load increases from zero, the current increases and the voltage drop across the armature resistance, R_A, increases. Therefore, the terminal voltage (load voltage) of any generator reduces as the load increases. As shown in Figure 13.28, the armature current increases if the load increases. This in turn decreases the back-emf.

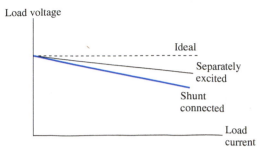

FIGURE 13.28 Load regulation characteristics for different types of DC generators.

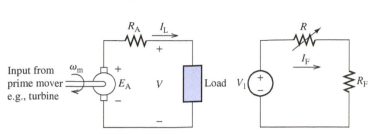

FIGURE 13.29 Equivalent circuit for a separately excited DC generator.

Voltage regulation of a generator is defined as the change in its output voltage from no load to full load. It is usually expressed as a percentage.

$$\% \text{ Voltage regulation} = \frac{(V_{\text{NL}} - V_{\text{FL}})}{V_{\text{FL}}} \times 100 \qquad \textbf{(13.31)}$$

where:

- V_{NL} = terminal voltage of the generator at no load ($I_L = 0$)
- V_{FL} = terminal voltage of the generator at full load

Figure 13.28 compares the load (voltage) regulation characteristics of different types of generators. For a shunt-connected DC generator, as the load on the generator increases (load current increases), the load voltage drops considerably. On the other hand, the load regulation characteristics for the separately excited DC generator show a much lower drop in load voltage with increasing load current. Thus, the separately excited DC generator has better performance than its shunt counterpart.

13.6.2 Separately Excited DC Generator

As its name suggests, with this generator there are two separate sources of voltage, one for the armature circuit and the other for the field circuit. As shown in Figure 13.29, in a separately excited DC generator, a prime mover creates the rotational movement of the generator at an angular velocity of ω_m. Induced armature voltage (i.e., the back-emf E_A) generates current flow through the load. Here, I_L and I_F are the load and field currents, respectively. Figure 13.29 (right side) shows the field circuit in which V_1 is the field voltage, and R_A, and R_F are armature and field resistances, respectively. The voltage across load is represented as V. E_A is the emf generated by the prime mover and R is the variable resistance inserted to control the flux generated, and therefore the load voltage.

EXAMPLE 13.16   **Voltage Regulation of Generator**

A separately excited generator is rated for a load voltage of 100 V with a full-load current of 23 A at 1800 rpm. The no-load voltage is 120 V. Find:

a. The armature resistance and the developed torque at full load
b. The voltage regulation

SOLUTION

a. Considering Figure 13.29:

$$R_A = (V_{\text{noloud}} - V_{\text{fullload}})/I_L = (120 - 100)/23 = 0.87 \text{ ohms}$$

(continued)

EXAMPLE 13.16   **Continued**

Using Equation (13.30):

$$E_A = K \cdot \phi \cdot \omega$$

$$K \cdot \phi = \frac{E_A}{\omega} = \frac{120}{2\pi \times 1800} = 0.0106$$

Using Equation (13.11):

$$T = K \cdot \phi \cdot I_A = 0.0106 \times 23 = 0.2438 \text{ N m}$$

b. Voltage regulation is given by Equation (13.31):

$$\text{Voltage regulation} = \frac{V_{noLoad} - V_{fullLoad}}{V_{fullLoad}} \times 100 = \frac{120 - 100}{120} \times 100 = 16.67\%$$

Similar to the case of DC motors, the disadvantage of a separately connected machine is that two power sources are required. In the case of a shunt-connected DC generator, only one power source is required as discussed in the next section.

13.6.3 Shunt-Connected DC Generator

In these generators, the field is in shunt with the armature. Here, the load voltage can be controlled by changing the resistor, R. The load regulation of a shunt-connected DC generator is not as good as its separately excited counterpart because as the load current increases, the voltage drop across the armature resistance increases and as a result the field current reduces as described in Figure 13.30. Here, the armature current, I_A, is a combination of field and load currents, I_L and I_F. Thus, the voltage drop across R_A will be larger due to the high resistance of R_A. Shunt-connected generators have large load regulation as compared to separately excited generators because the field current falls as the load current increases due to the drop across the armature resistance.

13.7 AC MOTORS

This section outlines AC machines such as AC motors and generators. Here, AC stands for *alternating current*. The section covers three-phase as well as single-phase motors.

- An AC motor takes in AC supply as an input and converts it to rotational motion: conversion of electrical energy to mechanical energy.
- An AC generator accepts mechanical (rotational motion) as input and produces alternating supply at its output: conversion of mechanical energy to electrical energy.

AC motors are discussed first. According to the classification of motors discussed earlier, AC motors are basically divided into two types:

1. Synchronous motors
2. Induction motors

It should be noted that when comparing an AC and DC motor, the speed of a DC motor is controlled by its voltage, whereas an AC motor speed is a function of supply frequency. That

FIGURE 13.30

Equivalent circuit for a shunt-connected DC generator.

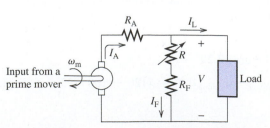

is why conventional analog electric clocks with motors are so accurate. In addition, in an AC motor, the commutator that is required in DC motors no longer needed because the AC itself causes a reversal of the magnetic force. Therefore, no change of brushes needed and no cleaning is required. In other words, AC motors are mechanically simpler. Thus they are less expensive. Finally, in many AC motors, the power consumption is fixed and is not purely a function of voltage, lowering the voltage, the current increases.

13.7.1 Three-Phase Synchronous Motors

These motors operate at a constant speed and hence the name *synchronous speed*. Section 13.7.1.3 details the principles of synchronous speed. These motors need a rotating magnetic field (RMF) for their operation. Let's see how that can be created.

13.7.1.1 CREATING A ROTATING MAGNETIC FIELD (RMF)

The generation of synchronous speed motors involves the use of polyphase (multiple phase) electric supply and stationary windings in the stator (and rotor) placed physically apart in space. Each of these stationary windings is called a *phase*. The physical separation between these phases should be equal to the electrical phase difference between the currents from the polyphase supply. Typically, a three-phase supply is used for these motors.

As discussed in Chapter 9, the phase difference between the output currents from a three-phase supply is 120°. Hence, it is also mandatory that the physical separation between the stationary windings (phases) is 120°. The number of stationary windings (phases) has to be equal to the number of supply phases.

When all of these conditions are satisfied, a rotating magnetic field is generated. Because three-phase supply is applied to the three stationary windings, three fluxes are produced. The resultant interaction of these fluxes is a flux of constant magnitude with its axis rotating in space. This corresponds to the fact that windings don't need to be physically rotated, which is definitely an advantage as it avoids any possible mechanical damage in the windings.

13.7.1.2 RMF IN A SYNCHRONOUS MOTOR

A three-phase synchronous motor consists of three-phase windings, placed in lengthwise slots cut in the stator. The three phases (stationary windings) are separated from each other by 120° and connected in either a star (having a common point called neutral) or delta (windings connected end-to-end to form a triangle) arrangement and powered by a balanced, three-phase electric supply. The same concept has been discussed in Chapter 9 for power systems.

Figure 13.31 shows a star-connected, balanced, three-phase winding and the direction of rotation of RMF. In reality, each winding consists of a large number of turns that spread over the entire circumference of the stator. A balanced supply corresponds to the production of equal magnitude of sinusoidal flux across all windings.

Assume the phase sequence of the windings is denoted as R–Y–B. The fluxes generated in these three phases are depicted by ϕ_R, ϕ_Y, ϕ_B. These are sinusoidal as shown:

$$\phi_R = k\, i_R(t) \cos\theta \qquad (13.32)$$
$$\phi_Y = k\, i_Y(t) \cos(\theta - 120°) \qquad (13.33)$$
$$\phi_B = k\, i_B(t) \cos(\theta - 240°) \qquad (13.34)$$

where k is a constant that depends on the geometry and materials of the stator and rotor, and $i_R(t)$, $i_Y(t)$, and $i_B(t)$ are currents of windings. As we have learned, applying a balanced three-phase source to the windings results in currents with the same magnitude, which are 120° apart in phase. This fact is shown in Figure 13.32.

The currents are given by:

$$i_R(t) = I_m \cos(\omega t) \qquad (13.35)$$
$$i_Y(t) = I_m \cos(\omega t - 120°) \qquad (13.36)$$
$$i_B(t) = I_m \cos(\omega t - 240°) \qquad (13.37)$$

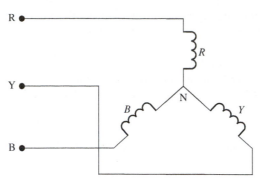

FIGURE 13.31 Star-connected three-phase winding.

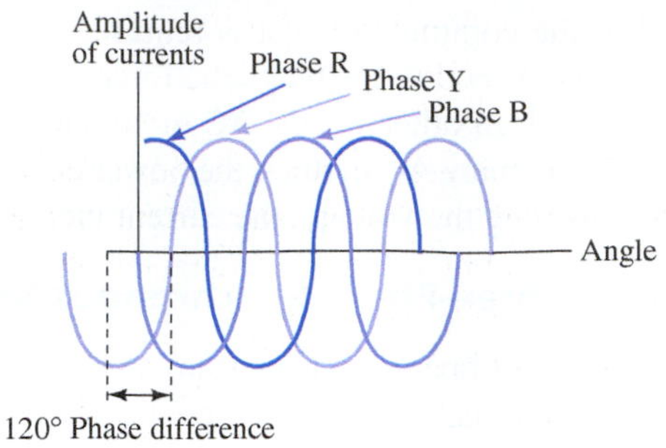

FIGURE 13.32 Waveform of the three currents.

Here, I_m represents the maximum current. From Equations (13.32) to (13.34), the magnitude of the flux in each phase can be inferred to be the same.

Figure 13.33 represents the phasor diagram for the three fluxes and that results for the condition of $\theta = 0$. In addition, $\phi_T = \phi_R + \phi_Y + \phi_B$ is the total or the resultant of the three fluxes obtained by phasor addition.

In the phasor diagram of Figure 13.33, the magnitudes of ϕ_R, ϕ_Y, ϕ_B can be found using Equations (13.32) to (13.34). Assuming $\theta = 0$, $\phi_Y = 0.25\,\phi_R$, $\phi_B = 0.25\,\phi_R$. Using these values, the total amplitude of $\phi_T = 1.5\phi_R$. This can be easily proven by applying some basic geometric principles.

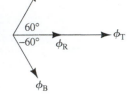

FIGURE 13.33
Directions of fluxes and the calculation of resultant flux when $\theta = 0$.

EXERCISE 13.3

Calculate the total flux when $\theta = 90$. $\phi_Y = 0.25\,\phi_R$, and $\phi_B = 0.25\,\phi_R$

13.7.1.3 SYNCHRONOUS SPEED AND TORQUE
The speed of the rotation of the resultant flux, that is, the rotating magnetic field in rpm is N rpm. Accordingly, the revolutions per second correspond to $N/60$. P refers to the number of poles. The speed of a synchronous motor is determined by the number of poles and the frequency of signal, not line voltage. In a three-phase system $P = 6$, and for a two-phase system $P = 2$. As illustrated in Figure 13.34, a three-phase induction motor has six poles.

Equation (13.6) was presented for the angular speed of a two phase system. In general, for a P pole system the frequency of signal at the output of a generator (or the frequency of signal applied to an AC motor) corresponds to the equation:

$$f = \frac{N}{60} \cdot \frac{P}{2} \tag{13.38}$$

Accordingly:

$$N = \frac{120 \cdot f}{P} \tag{13.39}$$

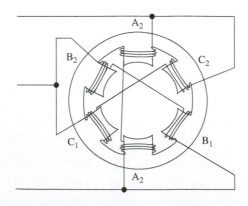

FIGURE 13.34 A three-phase synchronous motor.

Thus, the rotation speed, N, varies with the supply frequency as well as the number of poles. The torque developed by a synchronous motor can be written in terms of stator and rotor currents. It corresponds to:

$$T = \phi_s \cdot \phi_r \cdot \sin(\gamma) = K \cdot I_s \cdot I_r \sin(\gamma) \qquad \textbf{(13.40)}$$

where γ is the angle between the stator and rotor fields, and $I_s(\phi_s)$ and $I_r(\phi_r)$ are the stator and rotor currents (fields). In addition, similar to K_1 in Equation (13.23), K is a constant that varies with the physical parameters of the field winding.

EXAMPLE 13.17 **Synchronous Speed**

A six-pole, three-phase synchronous motor is supplied by a 60-Hz supply. Determine its synchronous speed.

SOLUTION

Using Equation (13.39), the synchronous speed can be found to be:

$$N = \frac{120 \cdot f}{P} = \frac{120 \times 60}{6} = 1200 \text{ rpm}$$

13.7.1.4 STRUCTURE OF SYNCHRONOUS MOTORS

The earlier sections of this chapter examined the structure of DC machines. There is a minor constructional difference between AC and DC machines. The synchronous motor basically consists of a stator and a rotor. The stator is powered by a three-phase AC supply while a rotor might be powered by a DC or an AC supply to create poles and support the process of rotation. Figure 13.35 shows the architecture of a synchronous motor with a salient pole rotor.

Stator: The stator contains a set of windings (also called an *armature*) that creates the stator rotating magnetic field (RMF). These windings might be star- or delta-connected (see Chapter 9). These fields consist of P number of magnetic poles. The number P is an even number, as for every north pole there is a south pole.

Rotor: The rotor of a synchronous machine can be either a salient pole (also called a projected pole) or non-salient pole (also called a cylindrical pole) as shown in Figure 13.35(a) and (b), respectively. Many manufacturers use a salient pole type of construction because it has a large starting torque. Salient poles are divided into two types: permanent magnet or electromagnet. Non-salient pole rotors are also called *drum* or *wound* rotors. Salient rotors are designed for lower speeds (less than 1500 rpm) while non-salient rotors are designed for higher speeds (higher than 1500 rpm).

The field current can be supplied by an external DC source; however, in most cases an AC generator output is rectified and applied to the field. This AC generator is called an *exciter*. The input energy to this generator is created by the synchronous motor rotation. The exciter avoids the requirement of using a DC source and its related maintenance issues.

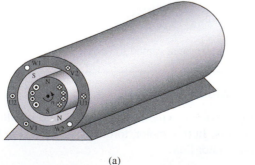

(a)

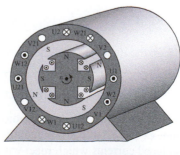

(b)

FIGURE 13.35 The architecture of a synchronous motor. Adapted with permission from Hermann Merz, Electrical Machines and Drives: Fundamentals and Calculation Examples for Beginners. Berlin: VDE Verlag, 2002.

13.7.1.5 OPERATION

The excitation provided to the stator generates a magnetic field (RMF) in the air gap between the stator and rotor. The principle of operation can be summarized as follows: "When a current-carrying conductor (rotor) is placed in a magnetic field (RMF) it experiences a force."

The rotating magnetic field is equivalent to a physically rotating magnet. This rotation occurs at the synchronous speed of N_s. For simplicity, let us consider that the stator consists of two poles N_1 and S_1 rotating at a synchronous speed. The rotor field current also creates two rotor poles N_2 and S_2. Consider that at a given instant, the stator and rotor poles are aligned such that like poles are near each other. Like poles repel each other and so if the stator poles are rotating clockwise, then the rotor poles will experience a torque in the anticlockwise direction. But due to inertia, it takes some time for the rotor to start rotating. This time period might be in the order of one half-cycle of the electric supply of the stator.

At the start of the next half-cycle, the positions of the two poles of the stator would be exactly opposite to the start of first half-cycle. Thus, now the unlike poles are near each other and as a result the rotor poles will experience torque in the clockwise direction. Hence, in one cycle of electric supply to the stator, the rotor experiences zero net torque. Even if it is assumed that at the start, the unlike stator and rotor poles are facing each other, still—because of the inertia of the rotor—the net torque experienced by the rotor would be zero. Accordingly, a synchronous motor is not self-starting.

This leads to the question: How can a synchronous motor be started?

The answer lies in rotating the rotor of the synchronous motor using an external drive at a speed almost equal to the synchronous speed. At the beginning of this process, the rotor does not have any excitation. After the rotor starts rotating at speeds almost equal to N_s, it is then supplied with electric current to generate poles. At a certain instant, unlike stator and rotor poles may face each other with magnetic axes almost aligned toward each other. In this case, they experience a force of attraction between the two and it is as if the stator and rotor are locked together magnetically and from now on they continue to occupy same relative positions.

The rotor now experiences a unidirectional torque in the direction of stator field and is said to be *synchronized* with the stator. Now, the external drive can be removed and the synchronous motor is now up and running at the speed of N_s and will continue to run. As a result, synchronous motors are inherently constant speed motors. Even though mechanical load on the motor may be increased (up to a certain limit which is determined by other parameters of the motor) it continues to rotate at a constant speed. This is an advantage of synchronous motors. The speed may change if there are fluctuations in the supply frequency.

Synchronous motors have the following drawbacks:

- Because three-phase constant voltage and constant frequency are supplied, speed variation/control is not possible unless a supply with variable frequency is provided.
- They require an external drive to start and bring them up to their synchronous speed.
- Two separate drives, one AC and one DC are needed.

However, synchronous motors still find applications in many devices. Examples include fans and blowers, rolling mills, cement mills, textile mills, and motor generator sets.

EXERCISE 13.4

Can the impedance of the rotor conductors be assumed to be purely resistive?

13.7.2 Three-Phase Induction Motor

The three-phase induction motor is used in a wide range of applications. More than 80% of motors used today are induction motors. In this motor, torque is developed in the rotor due to induced currents which react with the stator flux.

The *three-phase induction* motor derives its name from the fact that currents are "induced" in the rotor due to the RMF produced in the stator. Thus, no separate excitation is required for the rotor.

13.7.2.1 STRUCTURE

The generation of RMF and the architecture of the stator are exactly the same as the synchronous motor discussed in the previous section. The difference between the two motors from an architectural point of view lies in the rotor. Figure 13.36 shows a photograph of an induction motor, normally used in a home for water pumping purposes.

FIGURE 13.36 Three-phase induction motor. (Used with permission from © Scott Bowman/Alamy.)

Rotors.   Two kinds of rotor architectures are used in induction motors:

1. Squirrel cage or short-circuited rotors [Figure 13.37(a)]
2. Slip ring or wound rotors [Figure 13.37(b)].

Squirrel Cage Rotors.   This rotor is characterized by simplicity and ruggedness as shown in Figure 13.37. It consists of aluminum bars called rotor conductors placed in lengthwise slots cut along the circumference of a cylindrical rotor made up of iron. These aluminum bars are shorted at the end by rings and the entire structure looks like a cage, and hence the name (see Figure 13.38). A slip ring and brush arrangement is not needed which helps to reduce the required maintenance and keeps the architecture simple. This type of rotor is most commonly used. When rotor and stator fields are stationary with respect to each other, starting torque is generated. If this starting torque is sufficient to spin the rotor, then it will move up to its operating speed. Moderate starting torque is achievable using this rotor design. Starting torque is an important feature of an electric machine. It is the torque provided by the motor at the zero speed.

Slip Ring Rotors.   In this case, the rotor is exactly identical to the stator. It contains a set of three-phase coils placed in slots and configured in such a way that the number of poles on the stator and rotor are equal. Three terminals of the windings are brought out to the external terminals through the use of a slip ring and brush arrangement. Advantages of slip ring rotors include: (1) good speed and torque control can be achieved with variable resistances connected to its terminals and (2) supporting very high-starting torques.

Disadvantages of this type of rotor include:

- High cost of construction.
- Less rugged than alternatives.
- Maintenance is required due to the brush and slip ring arrangement.

13.7.2.2 PRINCIPLE OF OPERATION

An earlier section outlined how a rotating magnetic field produces the same effect as rotating magnets. Suppose that the direction of rotation is clockwise. Now, because the rotor is stationary, there is a relative motion between the RMF and the rotor. The RMF that is cut by the rotor conductors leads to the induction of electromotive force (emf) in the rotor conductors. This makes the current flow within the rotor.

(a)

(b)

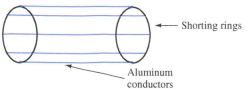

— Shorting rings

Aluminum conductors

FIGURE 13.37   (a) Squirrel cage (Used with permission from © David J. Green–electrical/Alamy.) (b) wound rotor. (Used with permission from © Lyroky/Alamy.)

FIGURE 13.38   Diagram of a squirrel cage rotor.

According to Lenz's law, this current should be in a direction such that it opposes the cause producing it. The cause producing the current is the induced emf, and an emf exists in the first place because of the relative motion between the stator and the rotor conductors. Therefore, to oppose or reduce the relative motion, the rotor starts rotating in the same direction as the stator and also tries to catch up with the stator in terms of speed.

EXAMPLE 13.18 **Synchronous Speed**

Does the rotor ever catch up with the synchronous speed of the stator rotating field? What would happen in such a scenario?

SOLUTION

If the rotor does manage to rotate at the same speed as that of the stator rotating field then there would be no relative motion between those two and consequently there would be no motoring action. In this case, the motor speed would reduce to zero, but as soon as the motor speed began to decline immediately a relative motion would be created again and the motor would start. In practice, however, this does not happen owing to the large inertia of the motor. As a result: *in the steady state, the rotor always rotates at a speed less than the synchronous speed.*

13.7.2.3 THE CONCEPT OF SLIP

The rotor's induced voltage depends on (1) the relative speed of the stator field with respect to the rotor and (2) on the number of poles. N denotes the rotor speed, and N_s denotes the synchronous speed. The rotor speed ranges from zero to synchronous speed. The difference between these two speeds is called the *slip speed*. The slip speed is defined as a fraction of N_s and corresponds to:

$$s = \frac{N_s - N}{N_s} \tag{13.41}$$

The maximum value of slip is $s = 1$, which occurs when the motor starts as $N = 0$ or when the rotor is stationary. The $s = 0$ condition occurs when rotor runs at synchronous speed as can be verified using Equation (13.41).

EXAMPLE 13.19 **Induction Motor Speed**

A six-pole, three-phase induction motor has a full-load slip of 2% with the frequency of 60 Hz. Calculate the full-load speed of the motor.

SOLUTION

Using Equation (13.39), the synchronous speed of motor N_s is:

$$N_s = \frac{120 \times \text{frequency}}{\text{No. of poles}} = \frac{60 \text{ s/min} \times 60 \text{ r/s}}{6/2} = 1200 \text{ rpm}$$

Using Equation (13.41):

$$N = N_s(1 - s) = 0.98 \times 1200 = 1176 \ r/min$$

In radians per second (see Equation (13.6)):

$$\omega = 123.15 \text{ rad/s}.$$

Full-load speed is 1176 rpm.

13.7.2.4 TORQUE EQUATION

The torque developed inside the induction motor depends on the following factors:

- The amount of RMF, which interacts with rotor conductors and the induced emf
- The magnitude of the rotor current when the rotor runs
- The power factor of the rotor when it runs

Stator and rotor circuits of an induction motor are shown in Figure 13.39. Important elements include:

I_{rr} = magnitude of the rotor current when it runs

ϕ = flux that induces emf in the rotor

ϕ_{rr} = power factor of the rotor when it runs

E_r = rotor's voltage

E_s = stator's voltage

R_c = core loss resistance

X_m = mutual reactance

X_r, X_s = reactance of the stator and rotor, respectively

R_r, R_s = resistance of the stator and rotor, respectively

Based on the definition of slip speed in Equation (13.41) and the corresponding discussion, when a load is applied to a motor, the actual speed, N, is different from the synchronous speed, N_s. The slip denoted by s is defined as a fraction of N_s: $s = (N_s - N)/N_s$. Using the relationship between the rotor voltage and its angular speed in Equation (13.30), the rotor-induced voltage at the slip, s, E_{rr} can be related to the rotor voltage at synchronous speed, E_r, by:

$$E_{rr} = E_r \cdot s \qquad (13.42)$$

The reactance, X_r, is computed using the synchronous speed which is proportional to the AC signal frequency. An applied load decreases that frequency by ratio s. Therefore, the actual reactance is $s \cdot X_r$. Accordingly, applying KVL to the circuit of Figure 13.39(b), and assuming that an applied load creates a slip, the rotor current with slip, s, corresponds to:

$$I_{rr} = \frac{E_{rr}}{\sqrt{R_r^2 + (s \cdot X_r)^2}} \qquad (13.43)$$

$$I_{rr} = \frac{sE_r}{\sqrt{R_r^2 + (s \cdot X_r)^2}} \qquad (13.44)$$

The parameter s that appears in the denominator of Equation (13.44) represents the change in speed when a load is applied.

Equations (13.13) and (13.14) show that power can be expressed in terms of torque, angular velocity, voltage, and current. Assuming resistor R_r represents the applied load, the power output at R_r can be expressed as:

$$P_{out} = I_{rr}^2 R_r \qquad (13.45)$$

Combining Equations (13.14) and (13.45) and knowing that the angular frequency is reduced to $S\omega$:

$$ST\omega = I_{rr}^2 R_r \qquad (13.46)$$

Substituting Equation (13.44) into Equation (13.46), the torque can be expressed in terms of voltage, resistance, and angular velocity.

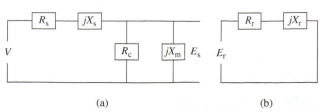

(a)   (b)

FIGURE 13.39 The (a) stator and (b) rotor circuits of an induction motor.

$$T = \frac{E_r^2}{\omega} \frac{sR_r}{R_r^2 + (sX_r)^2} \tag{13.47}$$

The maximum torque that corresponds to the slip s_{max} can be obtained by taking its derivative with respect to slip and setting it to zero, which corresponds to:

$$\frac{dT}{ds} = \left(\frac{E_r^2}{\omega}\right)\left[\frac{R_r}{R_r^2 + (sX_r)^2} - \frac{2s^2X_r^2R_r}{(R_r^2 + (sX_r)^2)^2}\right] \tag{13.48}$$

By taking $dT/ds = 0$, the slip value that corresponds to the maximum torque, s_{max} is:

$$s_{max} = \frac{R_r}{X_r} \tag{13.49}$$

By substituting the s_{max} obtained in Equation (13.49) into Equation (13.47), the maximum torque (T_m) is represented as:

$$T_m = \frac{E_r^2}{2\omega X_r} \tag{13.50}$$

The maximum torque is also known as *breakdown torque* and the speed at this torque is called *breakdown speed*.

13.7.2.5 TORQUE–SPEED CHARACTERISTICS

The region near the synchronous speed is called the *stable region* and usually represents the operating range of the induction motor. In this region, the speed ranges between 90% and 95% of the synchronous speed. In this region, up to the point of maximum torque, the slip is directly proportional to the torque (considering small slip) as illustrated in Figure 13.40. When the torque of the motor decreases, the magnitude of inductive reactance in Figure 13.39 will increase and the magnitude of the current will become almost independent of slip.

As the slip increases beyond the point of maximum torque, the inductive component in Equation (13.44) dominates the resistive one and the equation reduces to satisfy the condition in which the torque varies inversely with slip. In any machine, when load increases the speed reduces. In this case, the reduction in speed refers to an increase in slip, which in turn reduces the torque. This is undesirable as one would want the torque to increase as a response to higher loads. Accordingly, higher loads lead to further decrease in speed and torque and eventually the motor will stop. As depicted in Equation (13.48), at any speed, the developed torque is proportional to the square of the applied voltage. Figure 13.40 also illustrates that motor speed does not change significantly around the no-load condition. Thus, practically, the induction motor is called a*constant speed motor*.

13.7.2.6 ROTOR INPUT POWER VERSUS ROTOR COPPER LOSS AND MECHANICAL POWER

According to Equation (13.14), the mechanical power of a motor is related to the torque by $P = T \cdot \omega$, in which $\omega = 2 \cdot \pi \cdot N/60$, and N is the speed in rpm. P_s is defined as the total input power applied to the induction motor at the stator. In addition, assume the percentage of this

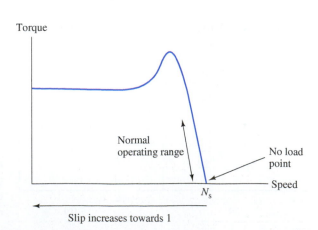

FIGURE 13.40

Torque–speed characteristics of an induction motor.

power that is the power P_r is applied to the rotor. P_r is coming from the stator side through the rotating magnetic field that has the synchronous speed, N_s. Assuming that the rotor rotates with synchronous speed, the power then corresponds to:

$$P_r = T \times \frac{2 \cdot \pi \cdot N_s}{60} \qquad (13.51)$$

Now, the rotor tries to apply all of this power to the mechanical load on the motor. But the rotor has the speed of N, which is slightly lower than the synchronous speed. If the synchronous speed N_s is replaced with the rotor speed N, the equation of P_m is obtained—which is the power provided by the rotor to the mechanical load—and is shown as:

$$P_m = T \times \frac{2 \cdot \pi \cdot N}{60} \qquad (13.52)$$

The difference between Equations (13.51) and (13.52) is the value of the rotor copper losses which corresponds to:

$$P_C = P_r - P_m = T \times \frac{2 \cdot \pi}{60} \times (N_s - N) \qquad (13.53)$$

Dividing Equation (13.53) by Equation (13.51) results in:

$$\frac{P_C}{P_r} = \frac{N_s - N}{N_s} \qquad (13.54)$$

However, based on the definition in Equation (13.41):

$$s = \frac{N_s - N}{N_s}$$

Therefore:

$$\frac{P_C}{P_r} = s \qquad (13.55)$$

Using a similar approach, the ratio is:

$$P_r : P_C : P_m \text{ is equal to } 1 : s : 1 - s \qquad (13.56)$$

This demonstrates a very important identity relating the power at various stages inside the motor to slip s. This equation shows the role of slip, s, in getting good performance in an induction motor.

EXERCISE 13.5

Verify Equation (13.56).

EXAMPLE 13.20   Slip and Speed of Induction Motor

An eight-pole, three-phase induction motor is supplied by a 60-Hz source. At full load, the frequency of the emf induced in the rotor is 5 Hz. Find the full-load slip and the full-load speed.

SOLUTION

First, calculate the synchronous speed using Equation (13.39):

$$N_s = \frac{120 \times f}{P} = \frac{120 \times 60}{8} = 900 \text{ rpm}$$

The full-load slip corresponds to:

$$s = \frac{f_R}{f} = \frac{5}{60} = 0.083$$

Using Equation (13.41) to compute the full-load speed:

$$N = (1 - s) N_s = 825.3 \text{ rpm}$$

| **EXAMPLE 13.21** | **Total Power Transferred to the Rotor from Stator** |

The torque at the load of a three-phase, 60-Hz, six-pole induction motor is 150 N m. The frequency of the emf induced in the rotor is 4 Hz. Mechanical losses are 500 W. Calculate the total power available to the rotor from the stator.

SOLUTION

To calculate the synchronous speed, use Equation (13.39):

$$N_s = \frac{120 \times f}{P} = \frac{120 \times 60}{6} = 1200 \text{ rpm}$$

Next, the full-load slip corresponds to:

$$s = \frac{f_R}{f} = \frac{4}{60} = 0.067$$

Thus:

$$N = (1 - s) N_s = 1119.6 \text{ rpm}$$

Now, using Equation (13.52):

$$P_m = T \times \frac{2\pi N}{60} = 150 \times 117.71 = 17{,}656 \text{ W}$$

Therefore, the total power available to the rotor is:

$$P_r = P_m + P_c = 17{,}656 + 500 = 18.16 \text{ KkW}$$

13.7.2.7 EFFECT OF EXTERNAL RESISTANCE ON TORQUE

An external resistance might be applied to a motor to control the speed of the motor. Adding an external resistance is only possible for slip ring or wound rotor induction motors.

Recall the analysis of Section 13.7.2.4 to compute the equation of the torque of induction motors. From Equation (13.49), observe that the maximum value of the torque is independent of the rotor resistance at standstill. Therefore, adding an external resistance does not alter the maximum torque value but can definitely impact the value of the slip at which that maximum value occurs. As observed in Section 13.7.2.3, s_{max} is the value of slip at which the maximum torque is attainable. Based on Equation (13.49), s_{max} corresponds to:

$$s_{max} = \frac{R_r}{X_r}$$

Therefore, the addition of external resistance to the rotor changes the value of s_{max}. Accordingly, the maximum torque will occur at higher values of slip. Now, by referring to Equation (13.48), the expression for the torque corresponds to:

$$T = \frac{s \cdot E_r^2 \cdot R_r / \omega}{R_r^2 + (s \cdot X_r)^2} \tag{13.57}$$

When the motor starts, the value of slip is 1, that is, $s = 1$ in Equation (13.41). Thus, the starting torque can be controlled by adding an external resistance to the increasing rotor resistance.

If the maximum torque is desired at the motor's start, then the value of slip should be equal to 1 at maximum torque. Equation (13.49) shows that this is only possible when $R_r = X_r$. However, such a high value of resistance—if kept permanently in the circuit—could lead to very high copper losses. Therefore, this level of resistance is undesirable. In practice, once the motor starts, this high resistance is gradually reduced to zero. Reducing the resistance after start ensures good performance at the time of start and during running conditions.

13.7.2.8 APPLICATIONS OF THE INDUCTION MOTOR

Squirrel cage induction motors are used in applications that need moderate starting torque, such as lathe machines, water pumps, grinders, and printing machines.

Slip ring induction motors are used in applications that need higher starting torque such as lifts, cranes, elevators, and compressors.

Induction motors are considered more reliable than synchronous or DC machines because they do not require slip rings or brushes.

13.7.3 Losses in AC Machines

The losses in AC machines (motors as well as generators) are almost the same as those discussed in Section 13.2.5 for DC machines. However, in the case of AC machines, iron losses or core losses that occur in the stator and rotor core are frequency dependent. Stator iron losses constitute the major part of these losses because stator frequency is the supply frequency. In the case of the rotor (for induction motors) the rotor frequency is directly proportional to the slip. Therefore, it is much smaller compared to supply frequency. Accordingly, rotor iron losses are quite minimal.

When a conductor is rotated in a magnetic field, currents are induced in it. These induced currents are called *eddy currents*. These currents create some power dissipation (loss) in the form of heat. *Hysteresis loss* is a heat loss caused by the magnetic properties of the armature. *Mechanical losses* are due to the rotation of the armature and include bearing friction loss, brush friction loss, and windage or air friction loss.

13.7.4 Power Flow Diagram for an AC Motor

Figure 13.41 represents the power flow diagram for an AC motor. In summary, there are basically three stages of losses during power flow of an AC motor. The first stage is stator losses that occur at the stator part of motor when input power is applied to the stator. The second stage is rotor losses that occur at the rotor part of the motor. The third stage of losses is mechanical losses. Power is reduced because of these losses.

Besides the machine type, that is, motor or generator, an electric machine nameplate usually includes some information that represents its nominal characteristics such as its rated voltage, its rated current, and its voltage or current regulation. The rated voltage is the terminal root mean square (rms) voltage for which the machine is designed and the rated current is the terminal rms current that does not cause overheating. Voltage or current regulation is the maximum voltage or current that maintains a constant speed or torque when the load varies.

Table 13.3 summarizes different types of AC motors and compares their characteristics.

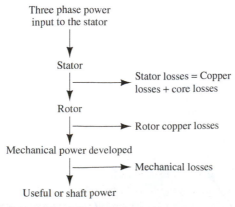

FIGURE 13.41 Power flow diagram for AC motors.

TABLE 13.3 Summary of All AC Motors Studied

Motor Types	Characteristics	Power Range	Applications
Three-phase induction motor (Squirrel cage rotor)	Moderate starting torque. Simple robust and maintenance free construction. Speed control by addition of rotor resistance is not possible	1–5000 hp	Lathe machines, water pumps, grinders, printing machines, large refrigeration and air-conditioning units, small compressors
Three-phase induction motor (Slip ring rotor)	Very high starting torque. Speed control by rotor resistance possible. High cost. Complicated construction which requires maintenance due to use of slip rings and brush	1–5000 hp	Cranes, hoists, elevators, large compressors, industrial fans and blowers
Three-phase synchronous motor	Constant speed irrespective of load, need of a starting device, need for two different excitation sources, variable frequency drives required for speed control	Up to 50,000 hp	Motor generator sets, timing devices, centrifugal pumps, textile mills, cement mills, rolling mills

13.8 AC GENERATORS

This section examines synchronous generators. Recall that synchronous machines refer to those that operate at their synchronous speed, N_s. Close to 98% of the world's power generators are synchronous generators. Induction generators are not often employed because of their lower performance. The principle of operation of synchronous generators is similar to that of DC generators.

Section 13.5 on DC generators explained that the electromotive force (emf) generated in the armature winding of DC generators has an alternating nature. The commutator is a device, which converts the alternating emf into unidirectional emf. As a result, theoretically, replacing the commutator with a slip ring and brush assembly (to tap the generated AC voltages) and making the armature rotate should convert a DC generator to an AC generator (see Figure 13.42). However, this concept cannot be implemented practically. Why not?

As we discussed previously, the relative motion between a conductor and the magnetic flux is the basis of producing emf in the conductor. Thus, both the rotation of the conductor in a stationary magnetic field and the rotation of the magnetic field with stationary conductors lead to generation of emf. In DC generators, the preferred architecture involves the conductors rotating in a stationary magnetic field. In AC generators, it is more common for the conductors to be stationary in a rotating magnetic field. Thus, the stator, also called the armature (stationary part), carries the conductors while the rotor is the rotating part, which carries the field winding (the source of flux).

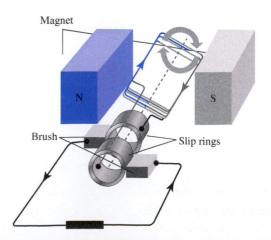

FIGURE 13.42 The structure of AC generators.

AC generators are configured differently than DC generators because of the following reasons:

- AC generators are usually of much higher capacity than their DC counterparts. Higher demands for power require the current-carrying conductor to be much thicker. Thicker conductors require deeper slots to house them. This results in an increase in the overall size of the stator. Rotating a large stator calls for more output from the prime mover. In addition, it is not easy or recommended to tap high voltages from rotating armatures in spite of the slip ring and brush arrangement.
- We can completely eliminate the necessity of a slip ring and brush assembly for the stator if it is kept stationary. This is because an emf is generated in the armature conductors, which can be directly connected to the transmission line.
- A cooling arrangement is an important consideration for the stator of an AC generator because very high voltage levels are generated. Efficient cooling is possible when the component to be cooled is stationary.
- A stationary stator winding avoids mechanical damage due to rotation and protects the armature from the centrifugal forces due to rotation.

13.8.1 Construction and Working

The construction of the synchronous generator is not very different from the synchronous motor studied earlier. A stator consists of slots to hold windings. Windings are separated from each other by 120° in space. The voltages generated by the windings have 120° phase difference with respect to each other. Steel is used as the material for construction to minimize hysteresis losses.

The rotor can be either salient pole or cylindrical in type, as shown in Figure 13.43:

- The *salient pole* has poles projecting out and field windings are located on them.
- The *cylindrical* style houses the winding in slots and the remaining un-slotted portion forms the magnetic poles. The cylindrical style is preferred for many applications that demand high speeds. Higher speeds are achievable because of the mechanical strength of the cylindrical style, which creates the ability to bear the heavy centrifugal forces due to the rotation at high speeds. A rotor has to be driven by a prime mover like an internal combustion (IC) engine (an engine that converts chemical energy into useful mechanical energy by burning fuel) or a steam turbine.

The rotor is supplied by a DC excitation and also it is rotated via a prime mover. This creates the rotation of the magnetic field. Now, the armature conductors (stator windings) cut this flux and the emf is induced in them. The synchronous speed for a rotor is given in Equation (13.38).

13.8.2 Winding Terminologies for the Alternator

Before deriving the emf equation of the alternator, a few important winding terms must be defined. A familiarity with the terminology of winding is needed because the winding impacts the amount of emf generated.

In the stator of synchronous generators, there are six terminals (ports) for the three windings (two per winding). Three out of these six terminals are connected in star or delta and the

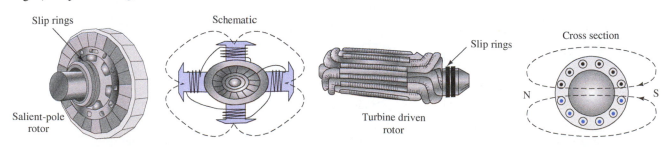

(a) Salient pole rotor and its cross-section   (b) An example of a cylindrical rotor – turbine driven rotor, and its cross-section.

FIGURE 13.43 Salient pole and cylindrical rotor. (a) Salient pole rotor and its cross section; (b) an example of a cylindrical rotor— turbine driven rotor, and its cross section.

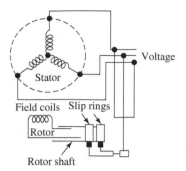

FIGURE 13.44 Circuit of an AC generator.

remaining three are used as outputs. Output terminal voltage is tapped as shown in Figure 13.44. Each of the three windings are called a *phase* and the emf generated in each is called *emf per phase*, E_{ph}. All the emfs generated should be added together.

- A *conductor* is a metallic wire under the influence of the magnetic field.
- A *turn* is formed when two conductors from two different slots are connected.
- When a number of turns are grouped together, a *coil* is formed (e.g., the field coil shown in Figure 13.44).
- The number of slots per pole or the distance between two adjacent poles is called *pole pitch*.

Accordingly, one pole corresponds to 180° of emf. This 180° is called *one pole pitch*. The number of slots per pole is referred to as *n*; therefore, these *n* slots are responsible for producing 180° phase difference.

- *Slot angle* is the phase difference created as a result of each slot on the armature of the alternator and it is denoted by *β* which corresponds to:

$$\beta = \frac{180}{n} \tag{13.58}$$

- *Single and double-layered winding* is distinguished by the number of coil sides (one or two) in each slot. Double-layered winding is preferred to save space.
- When the coil side in one slot is connected to another coil side, that is, one pole pitch away the winding is said to be *full-pitched winding*. Anything shorter is called *short-pitched winding*.

Figure 13.45(a) and (b) show a one-slot-per-pole alternator. Each coil in the figures is aligned with each slot. The coil group in Figure 13.45(a) is also called half-coil winding, because only one side of the slot has a coil. In Figure 13.45(b), both sides of the slot contain a coil. Therefore, the arrangement in Figure 13.45(b) is known as a whole-coil winding. Figure 13.45(c)

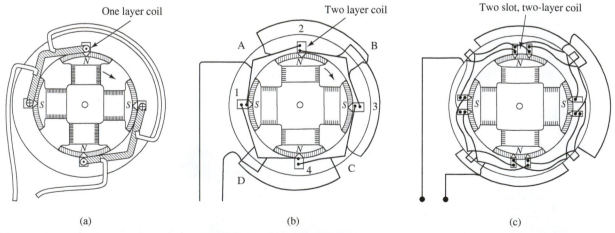

(a)   (b)   (c)

FIGURE 13.45 Alternator: (a) One-layer, one slot per pole; (b) double-layer, one slot per pole; (c) double-layer, two slots per pole. (From "A Course in Electrical Engineering, Volume 2, from ALTERNATING CURRENTS" by Chester Laurens Dawes. Copyright © 1952 The McGraw-Hill Companies, Inc.)

shows a two-slot-per-pole, two-layer alternator. Two whole-coil winding slots are aligned with each pole in the alternator.

For example, suppose an alternator has eight slots per pole. If the coil side in slot 1 is connected to the coil side in slot 9, such that the two slots are one pole pitch apart, then the winding is called full-pitched winding. If the coil side in slot 1 is connected to the coil side in slots 7 or 8, then it is called short-pitched winding.

In practice, short-pitched windings are employed more frequently because:

1. shorter copper conductors may be used for end connections;
2. high frequency harmonics are eliminated, which in turn results in reduced eddy current and hysteresis losses.

In Figure 13.46, α denotes the angle by which the coils are short pitched. The factor by which the generated emf is used is called the **pitch factor** and it corresponds to:

$$K_{\text{coil}} = \cos\left(\frac{\alpha}{2}\right) \tag{13.59}$$

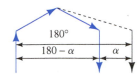

FIGURE 13.46 Short-pitched winding.

- In a *concentrated winding*, conductors that belong to a particular phase are placed in a single slot whereas in a *distributed winding* the conductors belonging to a particular phase are distributed across all slots. Distributed windings reduce harmonics heat dissipation and perform better than concentrated windings.

In practice, double-layered, short-pitched, and distributed winding is used for alternators.

In concentrated windings, because the entire winding is kept in one slot, the induced emf in each coil has the same phase. In contrast, in distributed windings, each slot adds a phase difference of β degrees to the generated emf. Consequently, there is a reduction in the emf. The factor by which the emf is reduced is called the *distribution factor* and corresponds to:

$$K_{\text{dist}} = \frac{\sin\left(\dfrac{m\beta}{2}\right)}{m \cdot \sin\left(\dfrac{\beta}{2}\right)} \tag{13.60}$$

where m is the number of slots per pole, per phase. For a three-phase machine, $m = n/3$ (n is the number of slots per pole).

13.8.3 The emf Equation of an Alternator

For this example, assume a full-pitch concentrated winding. The average induced emf for a single conductor is $d\phi/dt$, which is the ratio of the change of flux per unit time. For one revolution of the rotor, the emf induced in a single conductor is the ratio of the total cut flux and the time required for each revolution. If ϕ is the flux per pole, and P is the number of poles:

$$d\phi = \phi \cdot P \tag{13.61}$$

$$dt = \frac{60}{N_{\text{s}}} \tag{13.62}$$

Using Equations (13.61) and (13.62):

$$E_{\text{av}} = \frac{\phi \cdot P}{\dfrac{60}{N_{\text{s}}}} \tag{13.63}$$

In the Equation (13.63), $60/N_{\text{s}}$ is the time of one revolution. Here, refer to Equation (13.39), N_{s} is the synchronous speed and corresponds to:

$$N_{\text{s}} = \frac{120 \cdot f}{P} \tag{13.64}$$

Substituting Equation (13.64) into (13.63):

$$E_{\text{av}} = \phi \cdot 2f \tag{13.65}$$

Equation (13.65) represents the average value of emf induced in a single conductor. For the induced emf per turn (a turn consists of two conductors as defined previously), Equation (13.65) should be revised to:

$$E_{av} = 2 \cdot 2f\phi \qquad \qquad \textbf{(13.66)}$$

A factor of 2 in Equation (13.66) can only come into the picture when the winding is full pitch and the emfs produced by the two conductors (separated by 180° for full pitch) assist each other.

Assume there are T turns per phase and these turns are connected in series. Considering a concentrated winding, and that all these turns are within the same slot and they are in phase, then the average value of the induced emf per phase corresponds to:

$$E_{phase} = T \cdot E_{av} = T \cdot 4f\phi \qquad \qquad \textbf{(13.67)}$$

APPLICATION EXAMPLE 13.22 — Hydroelectric Power Generation

Potential energy of stored water in dams is converted to kinetic energy when it flows through pipes onto the blades of a turbine. The turbine blades are connected to the rotor inside the generator by a common shaft. The rotation of turbine blades creates the rotation of the rotor conductors. Accordingly, the magnetic flux is cut inside the generator, and an emf and, finally, electricity is generated. This method is called hydroelectric power generation.

Other techniques like nuclear power generation, electricity from wind energy, and electricity from solar energy are slowly gaining momentum and their generation capacities are increasing manifold every year. For example, in 2003, 85% of the total power generation in France happened via nuclear power generation.

EXAMPLE 13.23 — AC Generator Speed

Find the speed of a six-pole AC generator when its frequency is 60 Hz.

SOLUTION

To find the speed of a generator, the frequency and the number of poles must be known; in this case, 60 Hz and six (6), respectively. Using Equation (13.64):

$$N_s = \frac{120 \cdot f}{P} = \frac{120 \cdot 60}{6} = 1200 \text{ rpm}$$

EXAMPLE 13.24 — Induced emf of AC Generator

Calculate the induced emf per phase of a four-pole AC generator which has a torque of 6 N·m. The speed of the AC generator is 1500 rpm and the flux of the AC generator is 0.6 Wb. The number of turns per phase is six.

SOLUTION

In order to calculate the induced emf per phase of an AC generator, first find the frequency of the generator. Because the number of poles and the speed of the generator are given, using Equation (13.64):

$$f = \frac{N_s \cdot P}{120} = \frac{1500 \cdot 4}{120} = \frac{150}{3} = 50 \text{ Hz}$$

Now, calculate the induced emf per phase. Using Equation (13.67):

$$E_{phase} = T \cdot 4 \cdot f \cdot \phi = 6 \cdot 4 \cdot 50 \cdot 0.6 = 720 \text{ V}$$

13.9 SPECIAL TYPES OF MOTORS

13.9.1 Single-Phase Induction Motors

The first single-phase induction motor was built by Nikola Tesla in the USA in 1943. He was a Yugoslav–American scientist. The SI unit of magnetic flux density is named after Tesla. The stator of a single-phase induction motor has a winding called the *main winding* and the rotor is squirrel cage just like the three-phase induction motor discussed earlier. As described for the three-phase motor, the flux (RMF) rotates in a given direction and its direction can be reversed by interchanging the Y and B phases.

However, in single-phase motors, the total flux can be expressed as the sum of two equal flux components rotating in opposite directions. One of these components rotates in the same direction as the rotor and is called the forward component, while the other is called the reverse component. Thus, it can be inferred that because the torques produced by these two flux components are equal and opposite they will cancel each other and, as a result, net starting torque will be zero. Therefore, this motor cannot deliver power to a load from a standing start.

The mechanism of a single-phase induction motor is shown in Figure 13.47. Figure 13.47(a) shows the rotor at its initial stationary position. Here, the forces in each pole are in equilibrium states. In Figure 13.47(b), when an initial torque, T_o, is applied, it makes the rotor begin to rotate. At this point, the forces are not in equilibrium anymore, and the south pole of the rotor, S_R, is pulled to the left, toward the north pole of the stator, N_S. Then, the north pole of the rotor, N_R, will be pulled toward to the right, toward the south pole of the stator, S_S. The nonequilibrium force causes the rotor to rotate after the initial torque has been applied. In order to allow the rotor to rotate continuously, the stator is connected to an AC source. The AC source switches the pole of the stator continuously, which prevents the equilibrium state from being achieved.

13.9.2 Stepper Motors

A stepper motor is a motor that maps electrical pulses into a series of discrete angular movements called steps. There is a one-to-one relationship between the input pulse and the step movement of the shaft. These motors are normally divided into two types, *variable reluctance* and *permanent magnet* stepper motors, as explained below.

Applications of these motors include the following:

- Precise position control such as controlling the position of the magnetic head for reading or writing information onto computer discs
- Odometers in automobiles
- Controlling fuel-to-air ratio before injection into a cylinder in a car
- Sound and vision equipment
- Driving the arms of a watch

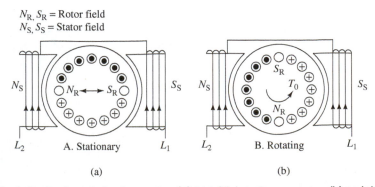

FIGURE 13.47 A single-phase induction motor. (a) At initial stationary status; (b) an initial torque is applied to begin the rotation.

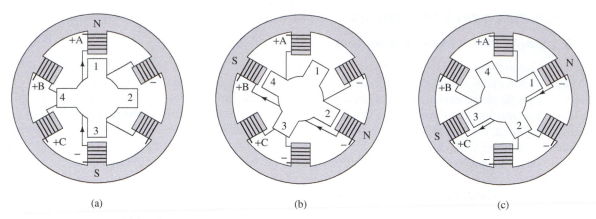

13.9.2.1 VARIABLE RELUCTANCE STEPPER MOTORS

In a variable reluctance stepper motor, the stator usually consists of six poles and phase winding as shown in Figure 13.47. One stator phase includes windings on two diametrically opposite poles that are serially connected. The rotor has poles with no excitation winding.

Figure 13.48 shows how a variable reluctance stepper motor works. Three poles in the stator are labeled as A, B, and C, and the poles in the rotor are labeled as 1, 2, 3, and 4 [see Figure 13.48(a)]. The magnetic field is activated across A, B, and C sequentially. Based on the status of the poles shown in Figure 13.48(a), when current passes through A (ON), while it does not pass through B and C (OFF), the rotor will be in stationary. Next, poles A and C are turned OFF and pole B is turned ON. Then the rotor is turned and its poles (2 and 4) are aligned with pole B in the stator that is shown in Figure 13.48(b). Again, when poles B and A are turned OFF, with pole C turned ON, the rotor will rotate and its poles (1 and 3) will be aligned with pole C in the stator. At this status, the rotor rotates in the clockwise direction. However, it can be carefully designed to ensure its rotation in the desired direction.

13.9.2.2 PERMANENT MAGNET STEPPER MOTOR

In a permanent magnet stepper motor, the stator has four poles around which excitation coils are wound. These are spaced 90° apart as shown in Figure 13.49(a). These four excitation coils form the four phases of the stator.

The rotor is cylindrical in nature and permanently magnetized (made of materials such as ferrite). Now, if the phases are excited in a particular sequence there will be an interaction across the two fluxes: (a) the stator flux due to the current passing through the poles, and (b) the permanent rotor flux. Due to this interaction, a torque (angular movement or step) will be produced.

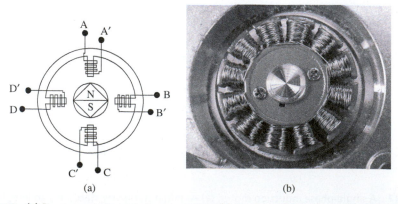

FIGURE 13.49 (a) Permanent magnet stepper motor. (b) Brushless DC motor. (Used with permission from © Lyroky/Alamy.)

The rotor position is stepped (angular movement) by applying a sequence of pulses to the stator windings. The step size will be 90°.

An increase in the number of stator poles guarantees a smaller step size but increases the difficulties in construction. As a result, hybrid stepper motors have been developed to combine the properties of variable reluctance and permanent magnet stepper motors.

13.9.3 Brushless DC Motors

Do you know which motor is used in the fan present in your computers for ventilation?
Do you know which motor is used in the hard drive in your computer?

The answer to both questions is the *brushless DC motor* as shown in Figure 13.49(b).

Learning the operation of a brushless DC motor is a logical extension of the principles of the previously studies permanent magnet stepper motor. The only difference in the architecture of brushless DC motors is the use of position sensors and control devices such as a thyristor that is an electronic gate. The position sensors can be based on Hall effect principles or can be based on optical position sensors. The Hall effect was discovered by Edwin Hall in 1879. Based on the *Hall effect*, a potential difference (*Hall voltage*) is created across an electric conductor located in a magnetic field when an electric current flows through the conductor.

This motor is equipped with a sensor that senses the magnetic field which is called Hall IC. The magnets of the motor are labeled 1, 2, 3, and 4. As one magnet gets closer to the Hall IC in Figure 13.50(a), the magnetic field of magnet 1 is detected by the Hall IC sensor. Then, the Hall IC sensor sends a signal to the transistor which produces a current in the electromagnet that creates a force capable of rotating the motor. Figure 13.50(b) shows that when a motor rotates and there is no magnet close to the Hall IC, there will be no signal in the transistor and as a result no force is created by the electromagnet. However, the inertia created by the initial force maintains the rotation. Thus, magnet 2 in Figure 13.50(c) will go into a position similar to the magnet 1 in Figure 13.50(a). In this case, the magnetic field is detected by the Hall IC sensor and a signal is sent to rotate the motor. As this process continues, it maintains the rotation of the rotor.

Figure 13.51 represents a more complex brushless motor. Here, H_1, H_2, and H_3 are sensors. The most important element in the operation of these motors is the timing of the switching of the power between the three stator windings which is determined by the rotor position. The rotor position is an input to the position sensor whose output is a digital signal.

The speed of the brushless DC motor is a function of the amplitude and the duration of the firing pulses applied to each of the stator windings. The brushless DC motor is gaining considerable popularity in industry due to the following characteristics:

- The absence of brushes and the commutator means no maintenance is required.
- Longer life.
- Greater reliability.

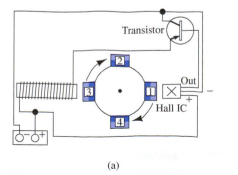
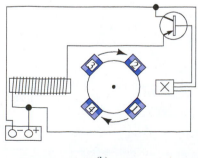
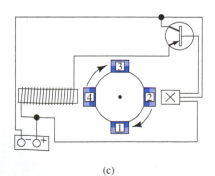

(a)   (b)   (c)

FIGURE 13.50 A simple illustration of the Hall effect switch in a motor. (Reprinted with permission. http://www.simplemotor.com.)

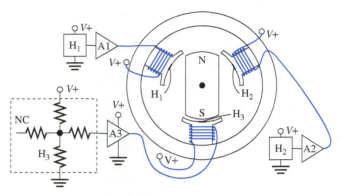

FIGURE 13.51   A sensor-based brushless DC motor.

- Greater efficiency.
- Lower inertia and friction which means they can accelerate faster and move to higher speeds quicker.
- The armature winding is located on the stator which supports effective cooling.
- They have low starting torque and high cost; however, their cost reduces as the cost of thyristorized circuits decreases.

Applications

- Video recorders
- Computer peripherals
- Biomedical instruments
- Commercial ovens
- Film processing equipment
- Printing technology
- Material handling equipment
- Centrifuges, grinders, and fans
- Hybrid vehicles

13.9.4 Universal Motors

DC series motors operated using single-phase AC sources are called *universal motors*. Universal motors are possible only because DC series motors can be operated using either AC or DC sources. This flexibility occurs because the commutator arrangement permits the field and armature currents to remain in phase. An AC motor with a brush and commutator arrangement is classified as a universal motor. Other names for the universal motor include *AC commutator motor* or *AC series motor*. The name AC series motor is derived from the fact that the stator and rotor are in series and are designed such that they operate on a 50 or 60 Hz supply. Different components of a universal motor are shown in Figure 13.52.

FIGURE 13.52

Universal motor. (Reproduced with permission from Blackstone Industries.)

Field

Brush holders

Commutator

Armature

Motor shaft

Carbon brush (Fits in holders)

From the discussion on DC series motors, you learned that the torque produced is proportional to the square of the applied voltage [see Equation (13.28)]. This leads to the conclusion that the torque produced is unidirectional and it is independent of the polarity of the applied voltage. The only restriction in powering a DC series motor using a single-phase AC source is that the stator must use laminated construction in order to reduce the eddy current losses. This is a constructional difference between DC series motors and universal motors.

In the equivalent circuit of a DC series motor, we ignored the effect of the inductance of the armature and field windings. As we explained in Chapter 4, an inductance is equivalent to a short circuit when a DC source is applied. However, when the same motor is connected to a single-phase AC source, the inductance cannot be ignored. If it were ignored, there would be a finite voltage drop across this inductance. This voltage drop would then reduce the current in the circuit. To cope with this problem, the inductance of the windings must be reduced. This is accomplished by using fewer turns on the winding.

However, fewer turns corresponds to lower flux. This can be addressed by incorporating a larger number of armature conductors. A larger number of conductors increases the armature reaction which can be offset using a compensatory winding. This is defined as a winding connected in series with armature winding to produce mmf.

Characteristics of the universal motor include the following:

- The ratio of the developed power to the motor weight is very high.
- The value of starting torque is very high without causing the current to rise too high.
- The universal motor suits applications in which load torque varies over a wide range.
- The universal motor can be operated at very high speeds.
- One disadvantage is the short lifespan of brushes and commutators used.
- As a result of the short lifespan, in applications which require the motor to run for a long time, induction motors could replace universal motors.

Applications

- Small appliances like mixers, blenders, drills, saws
- Vacuum cleaners
- Sewing machines

Table 13.4 represents a summary of all motors discussed in this section.

TABLE 13.4 Summary of all Special Types of Motors Studied

Motor Types	Characteristics	Power Range	Applications
Single-phase induction motor	Simple and rugged construction	Up to 5 hp	Fans, water pumps, and refrigerators
Stepper motor	Smooth rotation, linear relation between electrical input and step rotation	Sub-fractional horsepower rating	Used for precise position control in printers, floppy disk drives, robotics, process control, machine tools
Brushless DC motor	No maintenance required, long life, high reliability and efficiency, low inertia and friction, low starting torque and high costs, ability to run at speeds as high as 50,000 rpm	Up to 300 hp	Video recorders, computer peripherals, biomedical instruments, commercial ovens, film processing equipment, printing technology, and material handling equipment
Universal motor	High-power-to-weight ratio, can produce high starting torque, can operate at very high speeds, suits wide variety of applications demanding different torques, shorter lifespan	Used exclusively in fractional horsepower range	Small appliances like mixers, blenders, drills, saws, vacuum cleaners, sewing machines

13.10 HOW IS THE MOST SUITABLE MOTOR SELECTED?

The beginning of the chapter outlined how each motor suits a wide variety of applications. Because of this variety, more than one motor or more than one design of the same motor could be used for a specific application. However, engineers are often called upon to select the best possible design for each situation. Proper motor selection can help achieve the greatest possible efficiency and productivity while putting the lowest load on resources like power and cost.

This section covers a few points that should be taken into account before finalizing the selection of a motor for a specific application:

- Starting as well as running characteristics (operating voltages and currents)
- Life span
- Possible speed control and maximum operating speed
- Size and weight considerations
- Level of noise present in the operating environment
- Suitability of the motor for constant, intermittent, and frequently variable loads
- Capacity in case of overloads and torque characteristics
- Available power supply
- Installation as well as running cost
- Frictional characteristics

To explain these points more clearly, consider an application-based problem and design modifications that can be applied in induction motors to make them suitable for different torque requirements and therefore different applications.

EXAMPLE 13.25 **Motor Design**

What are possible modifications in the design of a squirrel cage induction motor so that it can serve different applications with a variety of torque–speed characteristics?

SOLUTION

Here, to understand the different torques present in the motor, it is helpful to review the torque–speed characteristics of a squirrel cage induction motor as shown in Figure 13.53.

Starting torque, also called locked-rotor torque, is the torque produced by the induction motor at start-up of the motor. Therefore, *starting current* of the motor is also called the *locked-rotor current.*

Pull-up torque is also called the minimum torque developed by the induction motor.

Breakdown torque is the maximum torque generated by the induction motor without stalling. Sometimes breakdown torque is called *pull-out torque.*

Full-load torque is the torque which the motor produces at rated load.

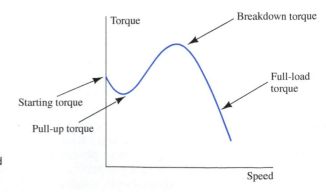

FIGURE 13.53 Torque–speed characteristics.

Torque–speed characteristics of a squirrel cage induction motor can be altered to suit different applications by modifying the construction and design. Different torque–speed characteristics can be generated to suit different applications as illustrated in Figure 13.54.

The National Electrical Manufacturers Association (NEMA) has identified four such designs as part of their standard. Each is considered, along with its corresponding performance curve.

Design A   This design is characterized by a starting current close to 10 times that of the full-load current, high breakdown torque, and low slip. Starting torque is close to 150% of full-load torque and breakdown torque is close to 200% of full-load torque. Motors having this design are characterized by high efficiency.

Design B   This design has a high reactance obtained by narrower rotor bars, which limits the starting current to five times the full-load current. Starting torque and slip are the same as in the design A motor. Breakdown torque is lower than design A motors, which is a drawback of this design and consequently the design B motor cannot be used for loads which have high peak values. Motors of this design are used in applications with relatively low starting torque requirements like fans, blowers, centrifugal pumps, and so on.

Design C   This motor has the highest starting torque of any of the designs discussed so far, with values going up to 200% of full-load torque. However, the breakdown torque is lower than either design A or B. Full-load torque is, however, the same as in designs A and B. At rated load, the values of slip are higher and, as a result, these motors are less efficient at rated load. This motor is particularly used for accelerating heavy loads like conveyors, agitators, reciprocating pumps, and so on.

Design D   This design has the highest starting torque of any of the designs discussed with values approximately 280% to 300% of full-load torque. However, in this design, torque is inversely proportional to speed. Starting current is the same as in the motor of design B and slip is high. This design has the highest slip among any of the designs and is primarily used where the starting torque requirement is very high but the load is not continuous, such as in hoists and elevators.

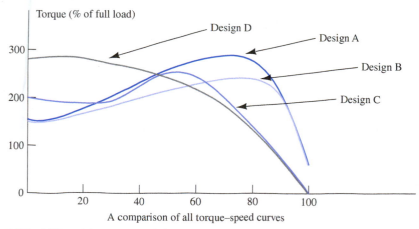

FIGURE 13.54   Different torque–speed characteristics.

13.11 SETUP OF A SIMPLE DC MOTOR CIRCUIT USING PSPICE

In this section, a few methods to simulate a DC motor circuit are introduced using a simple PSpice schematic. In addition, you will learn how to set up a global parameter sweep simulation. In some cases during PSpice simulation setup, you may require multiple electronic parts of the same kind, for example, impedance with the same value. Instead of changing the value of each part one by one, a part can be defined called "PARAMETERS" for all electronic parts. Then, the value assigned to PARAMETERS can be changed and the value of all associated components will change.

The part defined by PARAMETERS has a few functions in PSpice. First, it serves as a model in PSpice to connect the electronic parts you may wish to simultaneously control to the global parameter sweep option in the simulation setting. In addition, it works as a global parameter where it can assign a single value to multiple parts at once. For example, if there are 10 resistors and you wish to divide into two sets, each consisting of five resistors, and allocate the same magnitude to each five resistors, you can use PARAMETERS to configure such that R_1 and R_2 represent the value for each set of five resistors. You only need to change the value of R_1 and R_2 in the PARAMETERS part, instead of changing the resistance of each resistor one by one.

It should be mentioned that for many electronic parts such as variable resistors, "R_var" cannot be referred directly by a parameter model in the simulation. Thus, parameter sweep is not available for them. The solution to this problem is to use the part, PARAMETERS, which serves as an intermediate part that connects the electronic part to the parameter model in the parameter sweep simulation. The impedance/parameters of the electronic parts are connected and assigned through the PARAMETERS. Then, the PARAMETERS passes the impedance/parameters that are assigned into the simulation to run.

EXAMPLE 13.26   **PSpice Analysis**

A DC shunt motor is designed to drive a load in the range of 20 to 50 N. The armature resistance, R_A, and the field resistance, R_F, are 0.5 and 80 Ω, respectively. In addition, assume the torque–emf relationship is $T = K_A E_A$ where $K_A = 10$; note that this relationship can be maintained using Equations (13.24) and (13.25), in which $K_A = I_A / \omega_m$. Use PSpice to simulate and plot the power consumed by the motor assuming the voltage supply is 10 V.

SOLUTION

First, determine the induced emf by the motor, which is:

$$E_A = \frac{T}{K_A}$$

Using this equation and given the range of torque is 20 to 50 N, the range of induced emf is calculated as 2 to 5 V. Assume that the induced emf is a voltage source. The PSpice circuit is shown in Figure 13.55.

Here, the label of voltage source, "VM" (in the colored box) represents the induced emf produced by the motor. The label of any components in PSpice can be renamed by double-clicking on the label (such as the "VM" inside the colored box).

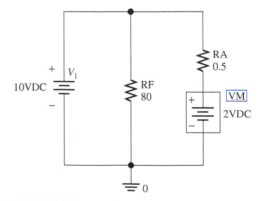

FIGURE 13.55 PSpice schematic for Example 13.26.

Next, simulate the circuit for a given range of induced emf. Set the analysis type of the simulation to "DC sweep" (colored solid arrow in Figure 13.56). Then, determine which voltage source needs to be swept in the simulation, that is, the induced emf, "VM" (dashed arrow).

The minimum and maximum induced emf is set to 2 and 5 V, respectively. Using an appropriate increment, the sweep type is selected to be linear (black arrow), with the minimum and maximum induced emf obtained from the calculation (dotted arrow).

The PSpice Add Trace menu allows us to plot various expressions for the circuit using the basic expression of mathematic functions. Then, to plot the power consumed by motor, first, go to the Add Trace menu. The power consumed by an electronic device corresponds to $P = V \cdot I$. Here, V is the voltage across the motor (also known as induced emf), and I is the current flow through the motor. Then, inside the Trace Expression column (see the arrow in Figure 13.57), type input "V(VM:+,VM,−)*I(VM)." Here, "V(VM:+,VM,−)" is the voltage across the motor, and I(VM) is the current flow through the motor and the symbol "*" represents the multiplication between two variables. The simulation plot is shown in Figure 13.58.

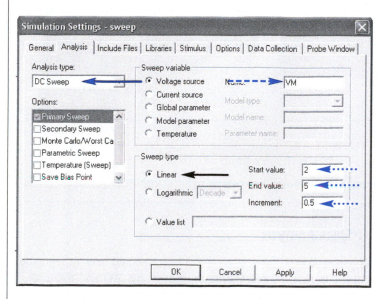

FIGURE 13.56 Simulation setting for Example 13.26.

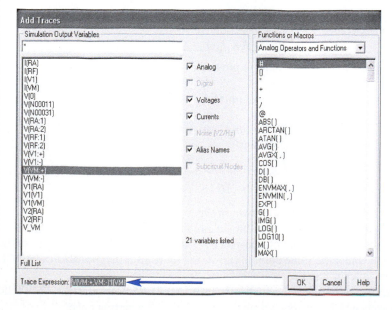

FIGURE 13.57 Add trace.

(continued)

EXAMPLE 13.26 **Continued**

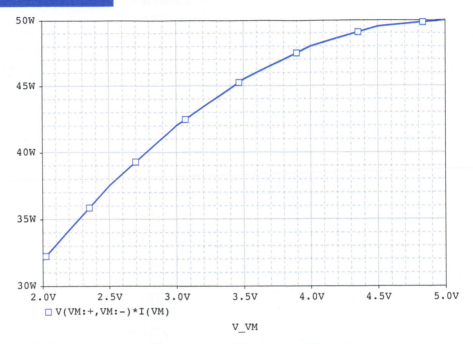

□ V(VM:+,VM:-)*I(VM)

V_VM

FIGURE 13.58 Power consumed by the motor with respect to different torque.

EXAMPLE 13.27 **PSpice Analysis**

A DC series motor with constant $K = 0.008$ and $K_1 = 2.5$ is connected to a voltage source of 40 V. Given that the motor's minimum and maximum speeds are 500 and 800 rpm, use PSpice to simulate and plot the power consumed by the motor if the motor's field resistance, R_F, is 0.4 Ω and the armature resistance, R_A, is 0.2 Ω.

SOLUTION

Based on Equation (13.24):

$$E_A = KK_1 I_A \omega$$

The resistance of the motor, R_M, can be expressed as:

$$\frac{E_A}{I_A} = KK_1 \omega = R_m$$

Then, based on the minimum and maximum speed of 500 and 800 rpm, the minimum and maximum resistance of the motor corresponds to:

$$R_{m, min} = 10 \ \Omega$$
$$R_{m, max} = 16 \ \Omega$$

Next, set up the PSpice schematic as shown in Figure 13.59.

To obtain the variable resistance in the PSpice circuit, go to "Part" and type "R_var." To simulate the motor with different values of resistance in "R_var," go to the "R_var" properties by double clicking on it. Under the value column (arrow in Figure 13.60), type "RESISTANCE" instead of inserting any integer number.

Next, another new part is introduced, which is the PARAMETERS. Here, the PARAMETERS work as global parameters which can assign a single value to multiple parts at once.

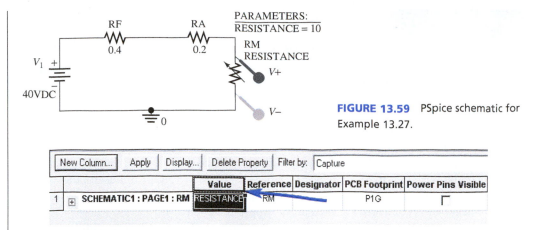

FIGURE 13.59 PSpice schematic for Example 13.27.

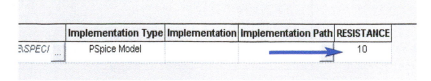

FIGURE 13.60 R_VAR's properties menu.

To place the PARAMETERS into the PSpice circuit, go to "Parts" and type "PARAM." It can be found under the "Special" library. Next, go into the PARAMETERS properties menu, and set the RESISTANCE value to minimum resistance, that is 10 Ω (arrow in Figure 13.61). If the "RESISTANCE" does not appear in the properties menu, click the "New Column" (solid arrow in Figure 13.62), then type "RESISTANCE" under the NAME column. Here, the number 10 represents 10 Ω under the VALUE column (dashed arrow in Figure 13.62).

There are three types of simulations that provide the parameter sweep option, which are DC Sweep, AC Sweep, and Time Transient. Here, the circuit is DC, thus, the DC option is used. Note that AC Sweep is used for AC circuits. In addition, the Time Transient option is used if you desire to study transient analysis of the circuits (e.g., RC, RL, and RLC circuit) with different sets of parameters.

	Implementation Type	Implementation	Implementation Path	RESISTANCE
&SPECI	PSpice Model			10

FIGURE 13.61 Properties menu of PARAMETERS.

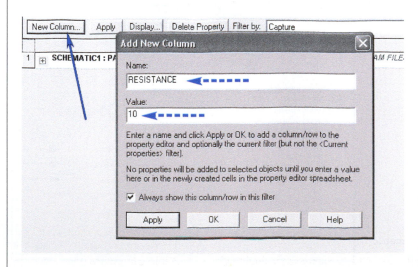

FIGURE 13.62
Insert a new column in the properties menu.

(*continued*)

EXAMPLE 13.27 **Continued**

Next, set the simulation type to be DC Sweep. Under the "Primary Sweep" (solid arrow in Figure 13.63) options, set the voltage source's name (dashed arrow) as the voltage supply in the PSpice circuit in Figure 13.63 (i.e., V1). Because the magnitude of voltage source is constant, then the start and end values of Voltage Sweep are the same (dotted arrows). Be aware that PSpice cannot run the Voltage Sweep simulation with zero increment; therefore, you must enter a number inside the increment column, such as 1.

Then, click and tick on the "Parametric Sweep" option (solid arrow in Figure 13.64). Select the "Global parameter" as the sweep variable (dashed arrow). Then, insert the parameter name that you wish to control to be swept, which is "RESISTANCE." Enter the start and end value of the sweep as the minimum and maximum resistance obtained from calculation, with an increment of 1 (dotted arrow). Then press "Apply" and "OK."

Run the simulation, and follow the steps as shown in Figure 13.57 to add the trace that represents the power consumed by the motor. The simulation plot is shown in Figure 13.65.

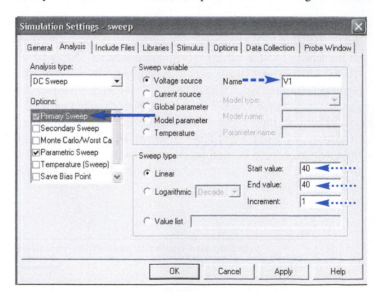

FIGURE 13.63 Primary sweep configuration.

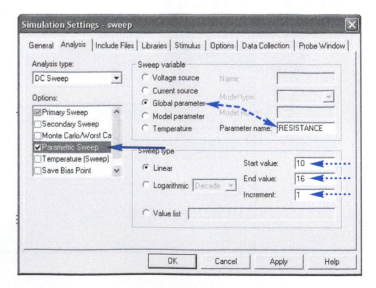

FIGURE 13.64 Parametric sweep configuration.

Note that the resistance of the motor is directly proportional to the angular velocity [see Equation (13.6)]. Then, you can conclude that the plot of Figure 13.65 also represents the power consumed by the motor with respect to the angular velocity of the motor.

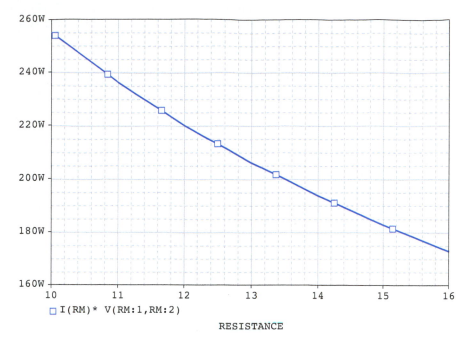

□ I(RM)* V(RM:1,RM:2)

RESISTANCE

FIGURE 13.65 Power consumed by the motor with respect to resistance (or angular velocity).

EXAMPLE 13.28   **PSpice Analysis**

Consider a speed control motor in Figure 13.66 that has an armature resistance, R_A, of 0.4 Ω and a field resistance, R_F, of 100 Ω. Assume that the motor has an impedance of 80 Ω and the circuit is connected to a voltage source of 20 V. Use PSpice to simulate the induced emf produced by the motor if the variable resistor has a resistance that varies within 5 to 50 Ω.

SOLUTION

Following the same setup in Example 13.26, the PSpice schematic is given in Figure 13.67.

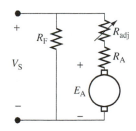

FIGURE 13.66 Speed control motor with varying current.

PARAMETERS:
RESISTANCE = 5

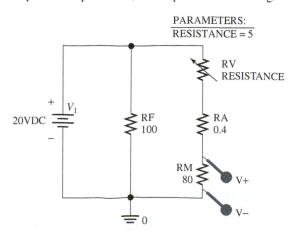

FIGURE 13.67 PSpice schematic for Example 13.28.

(continued)

EXAMPLE 13.28   **Continued**

Here, RM represents the motor's impedance, 80 Ω. Then, set up the PARAMETERS part for the circuit in PSpice by following the procedure in Figures 13.60 to 13.62. Next, refer to Figures 13.63 and 13.64. The simulation setup is configured as shown in Figure 13.68.

Then, run the simulation. The output of induced emf is given in Figure 13.69.

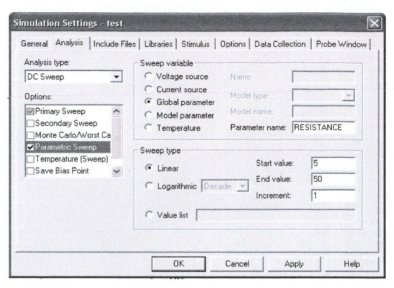

FIGURE 13.68   Global parameter configuration for Example 13.28.

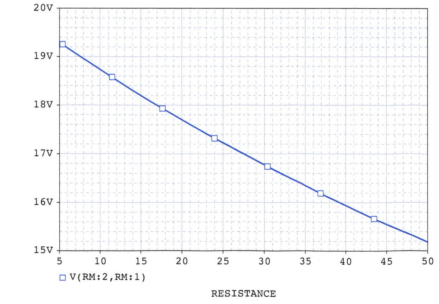

FIGURE 13.69   Output of induced emf by the motor.

13.12 WHAT DID YOU LEARN?

• Motors are classified into two main categories: AC motors and DC motors.
• DC motors have the following components: rotor, stator, windings, commutators, and brushes.
• Losses in a DC machine include: copper losses, iron or core losses, and mechanical losses.

- DC motors are divided into three main categories: (1) shunt-connected, (2) series-connected, (3) permanent magnet, and (4) separately excited motors.
- The following equations provide the basis for analyzing DC machines [Equations (13.6), (13.11), and (13.14)]:

$$E_A = K \cdot \phi \cdot \omega$$
$$T = K_A \cdot \phi \cdot I_A$$
$$P = T \cdot \omega$$

- DC generators include separately excited DC generators and shunt-connected DC generators.
- AC motors include synchronous motors and induction motors.
- Special types of motors include single-phase induction motors, stepper motors, brushless motors, and universal motors. A summary of the characteristics of each is given in Table 13.4.
- Factors to consider when selecting an electrical motor for a specific application include the following:
  - Starting as well as running characteristics (operating voltages and currents)
  - Life span
  - Speed control characteristics and maximum operating speed
  - Size and weight
  - Level of noise present in the operating environment
  - Suitability of the motor for constant, intermittent, frequently variable loads
  - Capacity in case of overloads
  - Torque characteristics
  - Available power supply
  - Installation and operation costs
  - Frictional characteristics

Further Reading

Hambley, A. 2005. *Electrical Engineering Principles and Applications.* Upper Saddle River, NJ: Prentice Hall.

Rizzoni, G. 2004. *Principles and Applications of Electrical Engineering.* New York: McGraw Hill.

Zorbas, D. 1989. *Electric Machines Principles, Applications, and Control Schematics.* St. Paul, MN: West Publishing Company.

Problems

*B refers to basic, A refers to average, H refers to hard, and * refers to problems with answers.*

SECTION 13.2 DC MOTORS

13.1 (B) If the current always flows from A to B (see Figure P13.1), briefly explain the mechanism of this motor.

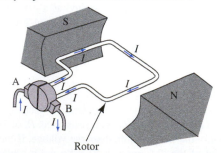

FIGURE P13.1   Rotor for Problem 13.1.

13.2 (B) Given the specfic properties in each case below, determine the result from the setup in Figure P13.2.
  (a) Determine the moving direction of the electric rod if the current flows from A to B; C is the north pole and D is the south pole.
  (b) What is the pole of the magnetic bars if current flows from B to A, and the electric rod moves in the X-direction?
  (c) Determine the current flow direction if C is the south pole and D is the north pole, but the electric rod does not move.

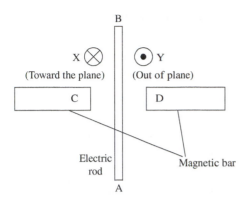

FIGURE P13.2 Diagrammatic representation for Problem 13.2.

13.3 (B) For a simple DC motor setup (see Figure P13.1), explain one way to configure the motor such that the rotor will always rotate in the same direction.

SECTION 13.3 DIFFERENT TYPES OF DC MOTORS

13.4 (B)* Consider a series-connected motor where the field and armature resistance are connected in series. The series current is 60 A, the rotating speed is 1800 rpm, and the power is 8000 W. Find the torque and the armature emf.

13.5 (B) Consider a DC motor that has 540 conductors, four poles, and the flux of 5 mWb per pole. What is the emf if its rotational speed is 1000 rpm?

13.6 (B) Equation (13.11) shows the relationship between the torque, flux, and the armature current. In many cases, the magnetic flux is considered to be constant. What will happen to the armature current if the torque is reduced by half the original torque? Assume that the number of conductors and poles remain the same.

13.7 (B)* A PM motor that drives a load at 500 N m with motor speed of 1200 rpm is connected to a voltage source of 50 V at 10 A. What is the torque constant, k_T, and armature constant, k_A?

13.8 (B) A motor is connected to a 110-V voltage source running at the speed of 500 rpm. The motor has a magnetic flux of 0.5 Wb and the constant K_1 is 4. Determine the current flow through the circuit.

13.9 (A)* Find the speed for a shunt DC motor that takes 35 A current with a supply of 240 V and a torque of 72 N m. The armature and field resistances are 0.30 and 200 Ω, respectively.

13.10 (A) A series-connected DC motor has 400 conductors and six poles. It produces a 120-N m torque when the armature current is 50 A. What is the magnetic flux per pole for the motor?

13.11 (A)* A separately excited DC motor has a supply voltage of 220 V to the field resistance. If the supply voltage to the armature is 75% of the supply voltage to field resistance, what is the back-emf of a motor with armature resistance of 0.6 Ω and armature current of 40 A?

13.12 (A) A 230-V DC shunt motor supplies the current at 93.5 A, with field resistance at 120 Ω and armature resistance at 0.3 Ω. What is the field current if the back-emf of the motor is 203 V?

13.13 (A)* The T–ω characteristics of a load is represented by $\omega = 0.02T + 900$ Write the expression of the speed at the operating point of a PM motor.

13.14 (A) What is the resistance required for a PM motor to drive a load at a torque of 2000 N m given that the T–ω relationship is $\omega = 0.04T + 120$ rad/s, the voltage source, V_S, is 100 V, k_a is 0.02, and k_T is 100.

13.15 (A)* A 5-A series motor that consumes 400 W is connected to a 110-V voltage source. What is the total resistance of R_A and R_F required if the motor is running at 400 rad/s? Also, what is the maximum torque that the motor can drive?

13.16 (A) A series-connected motor in Figure P13.16 is driving a load of 2.5 N m at 100 rad/s. The constant K is 2 and K_1 is 0.05. Determine the minimum voltage of the source and the armature current.

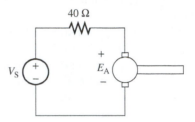

FIGURE P13.16 Circuit for Problem 13.16.

13.17 (H)* The DC shunt motor is running at full load with a supply voltage of 440 V and a supply current of 70 A. The rotation speed in full load is 1000 rpm, the field resistance is 200 Ω and the armature resistance is 0.2 Ω. Calculate its speed if the load torque is reduced to three-fourth of the full-load torque, and the armature resistance is increased by 0.5 Ω. Neglect the voltage drop due to the brushes and assume constant flux.

13.18 (H) A 230-V DC shunt motor runs at 1200 rpm on full load and the output power is 18.5 kW. Its efficiency is 85%, the armature resistance is 0.5 Ω, and field resistance is 50 Ω. If load torque is reduced to 100 N m and an additional 1.5 Ω resistance is connected in series with the armature, what will be the back-emf generated in the machine? Assume load torque to be constant.

13.19 (H) A 230-V DC shunt motor has an armature resistance of 0.05 Ω. When connected to a 230 V direct current supply it develops a back-emf of 225 V at 1200 rpm. Calculate armature current, armature current at start, and speed of the machine if operated as a generator in order to deliver an armature current of 80 A at 230 V. Assume a constant flux.

13.20 (H)* A resistor is connected in a circuit to control the current flow through a PM motor. The motor has an armature resistance of 0.5 Ω and is driving a load of 700 N m at 4 A with a 200 V supply. What is the maximum load that the motor is able to drive if the resistance is changed to 80 Ω?

13.21 (H) Given a PM motor with resistance, R, of 10 Ω and k_a of 4, what is the required source voltage, if the operating point of the PM motor is 40 rad/s and it drives a load of 25 N m?

SECTION 13.4 SPEED CONTROL METHODS

13.22 (A) For a speed control motor (see circuit in Figure P13.22), given $E_A = K \phi \, \omega_m$, express the speed of the motor in terms of K, ϕ, I_A, I_F, R_{adj}, R_F, and R_A.

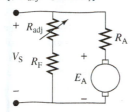

FIGURE P13.22 Circuit for Problem 13.22.

13.23 (A) Explain the mechanism of a speed control motor during the change of resistance of an adjustable resistor, for the cases that the adjustable resistor is connected in series with
(a) Armature resistor
(b) Field resistor

13.24 (A)* A speed control motor with flux constant, K_1 of 0.04, K of 0.125, is running at 1000 rpm 10 A when the resistance is set to 50 Ω. The R_A is 0.5 Ω and R_F is 150 Ω, respectively, and the adjustable resistor has a range of 50 to 100 Ω and is connected in series with the field resistor. What is the current flow through the field resistor?

13.25 (H) A speed control motor is input with a voltage source of 200 V at 25 A. The armature resistance, R_A, and field resistance, R_F, are 0.5 and 20 Ω, respectively. The adjustable resistor has a range of 20 to 180 Ω, which is connected in series with the field resistor. If the motor constant K is 50 and flux constant K_1 is 0.5, write the function of the torque produced by motor T in terms of R_{adj}, and sketch the T–R_{adj} plot.

13.26 (H) A speed control motor is input with a voltage source of 120 V at 30 A, with R_A and R_F are 0.2 and 40 Ω, respectively. The adjustable resistor has a range of 80 to 120 Ω and it is connected in series with the field resistor. If the motor constant, K, is 30, and flux constant, K_1, is 0.04, is it possible to drive a load that requires 1525 N m?

13.27 (H)* A student bought a 24 V, 4 A speed control DC motor, and he wants to know an estimate of the field and armature resistance of this motor. He connects the motor to a power source and runs two different speed and torque tests, which are (1) lowest resistance, and (2) highest resistance. The results he attained are:
1. $\omega = 32.80$ rad/s, $T = 2.312$ N m
2. $\omega = 29.93$ rad/s, $T = 2.738$ N m

Assume that the adjustable resistor is connected in series with the field resistor, which is stated in the manual. Also, assume that the lowest resistance of the adjustable resistor is 0 Ω. Find the field and armature resistance of this motor.

SECTION 13.5 DC GENERATORS

13.28 (B) A lap-winding generator has 40 slots and eight conductors per slot with four poles. The flux is 0.02 Wb per pole. Find the emf of the generator with a speed 1200 rpm.

13.29 (B) Consider a wave-winding generator with 540 conductors with eight poles. If the flux is 5 mWb per pole and the rotation speed is 1000 rpm, calculate the emf of the generator.

13.30 (B)* The terminal voltage of a DC generator at no load is 400 V. When full load is applied to the generator, the terminal voltage is read as 325 V. What is the voltage regulation of this DC generator?

13.31 (A) Calculate the output voltage when the supply voltage is 220 V and the current is 150 A. The shunt resistance is 50 Ω and the armature resistance is 0.02 Ω.

13.32 (A) A DC generator outputs a voltage of 220 V and a current of 150 A. The shunt field resistance is 50 Ω and armature resistance is 0.02 Ω. Find the power generated through the armature current.

13.33 (A)* A six-pole DC generator with a wave-wound armature has 50 slots and 25 conductors per slot. Find the generated emf if it is driven at 30 revolutions per second and the total useful flux in the machine is 0.05 Wb.

13.34 (A) A four-pole, wave-winding DC generator has 400 conductors on its armature with flux of 0.02 Wb per pole. The speed of the generator is 1000 rpm and the current is 100A. Calculate the output voltage and torque of the generator.

13.35 (A) A four-pole, wave-winding DC generator has 400 conductors. The input voltage is 220 V and the power is 12 kW. The generator runs at the speed of 1100 rpm. Calculate the flux and armature resistance.

13.36 (H)* A four-pole, wave-winding DC generator has 540 conductors. Its speed is 1000 rpm and flux per pole is 25 mWb. The armature resistance is 0.8 Ω and field resistance is 50 Ω. At no load, the terminal voltage is measured as 376 V. Calculate the load current for full load, when the voltage regulation is 22% and the power generated across the field resistance is 924.5 W.

13.37 (H) The armature of a four-pole, lap-wound DC shunt generator has 150 conductors. The resistance of the shunt field is 85 Ω and armature resistance is 0.02 Ω. If the flux per pole is 0.07 Wb, find the speed of the machine when supplying 100 kW at a terminal voltage of 220 V.

13.38 (H) A four-pole, lap-wound 800 rpm DC shunt generator has an armature resistance of 0.5 Ω and field resistance of 250 Ω. The armature has 720 conductors and flux per pole is 30 mWb. If the load resistance is 25 Ω, determine its terminal voltage.

SECTION 13.7 AC MOTORS

13.39 (B) For a star-connected, three-phase winding AC motor, what can you conclude from the resultant flux?

13.40 (B)* An AC induction motor has a synchronous speed of 2000 rpm. A load has been applied to the motor and the rotor speed is 1900 rpm. What is the slip of the induction motor?

13.41 (B) An AC motor outputs the power to the rotor at 13 kW. If the slip of the motor is 0.2 when a load is applied to it, what is the mechanical power and copper loss power?

13.42 (B) Determine the number of poles of an induction motor with a synchronous speed of 1800 rpm and frequency of 30 Hz.

13.43 (A) A two-pole induction motor supplies 12 kW power at 50 Hz to a load. Find the motor slip and torque when the rotor speed is 2800 rpm.

13.44 (A) Consider a six-pole induction motor rotating at synchronous speed of 1700 rpm. It produces the torque at 100 N m with frequency of 13 Hz. Find the power loss due to the copper loss effect.

13.45 (A) An eight-pole, 8 hp three-phase induction motor of 50 Hz runs at speed of 1200 rpm. Calculate the slip and frequency of the rotor currents.

13.46 (A)* Consider a two-pole 400 V induction motor with 60 Hz and power of 50 hp. It is operating at a speed of 1700 rpm. Calculate the slip rotor and frequency of the rotor current.

13.47 (A) A four-pole induction motor is operating at 80 Hz. The rotor resistance is 0.02 Ω and rotor reactance is 0.9 Ω. Find the rotor speed at the maximum slip.

13.48 (H) The torque expression of a six-pole induction motor is given as:

$$T = \frac{3sE_r^2 R_1}{\omega\left[(R_1 + R_2)^2 + (sX)^2\right]}$$

(a) Find the expression of the breakdown torque.
(b) Given that $R_1 = 0.8$ Ω, $R_2 = 0.3$ Ω, $E_r = 550$ V, $X = 0.15$ Ω, $N_s = 1800$ rpm, and $N = 1700$ rpm. What is the torque produced by the motor?

13.49 (H) A mechanical lifting arm with length of 5 m is designed to lift a load of 200 kg. It is equipped with a four-pole, 60 Hz AC induction motor with 1% copper losses. Determine the required power output for this mechanical load. (Assume gravitation acceleration is 9.8 m/s².)

13.50 (H)* A six-pole, 60 Hz AC induction motor in Section 13.7.2.4 has the rotor resistance and reactance at 0.3 and 1.3 Ω, respectively. The rotor-induced voltage at breakdown torque is measured as 237 V. Determine the breakdown torque and the percentage of power losses due to copper losses.

SECTION 13.8 AC GENERATORS

13.51 (B) Why is a short-pitched winding alternator preferred in many applications as compared to one that has full-pitched winding?

13.52 (B) Determine the slot angles for the following configuration:
(a) Single phase, three slots per pole
(b) Single phase, eight slots, two poles
(c) Three phases, six slots per pole

13.53 (B) What is the distribution factor for a three-phase machine that has 12 slots and three poles?

13.54 (A) Does an eight-pole alternator generate more average induced emf than a four-pole alternator? Why? Assume both alternators have the same synchronous speed and magnetic flux.

13.55 (A) An AC generator is designed to generate an average 100 V voltage using an eight-pole, two-conductor configuration. What is the synchronous speed the generator should run at if the flux of the AC generator is 0.5 Wb?

13.56 (A)* Calculate the flux of a six-pole AC generator that generates 120 V per phase. The synchronous speed of the AC generator is 1200 rpm and there are six turns per phase.

13.57 (H) Using a simple two pole (a north pole and a south pole) AC generator, sketch the configuration diagram and plot to explain how the voltage/current is generated for one cycle.

SECTION 13.11 SETUP OF A SIMPLE DC MOTOR CIRCUIT USING PSPICE

13.58 (B) A DC shunt motor is designed to be such that its minimum and maximum induced emf are 2 and 5 V, respectively. Given that the voltage source is 12 V, the armature resistance, R_A, and field resistance, R_F, are 0.2 and 75 Ω, respectively. Use PSpice to simulate and plot the power consumed by the motor.

13.59 (H) Consider a DC motor that varies the armature current to control its speed has an armature resistance, R_A, of 2 Ω and a field resistance, R_F, of 80 Ω, respectively. Assume that the motor has an impedance of 50 Ω with the coefficient $K\phi$ of 0.04. Let the circuit connect to a voltage source of 20 V. Use PSpice to simulate and plot the angular velocity of the motor if the resistor controller has a resistance range from 0 to 40 Ω. Then, determine the approximate minimum and maximum angular velocity of the motor from the simulation plot.

13.60 (H) An electronic device is connected to a shunt DC generator where its armature resistance, R_A, and field resistance, R_F, are 0.5 and 100 Ω, respectively. The DC generator has a range of angular velocity between 1000 and 1800 rpm with the coefficient $K\phi$ of 0.02. Assume that the electronic device has an impedance of 25 Ω and requires a minimum current of 1 A. Use PSpice to determine if the DC generator is capable of supplying the power to the electronic device or not.

Electrical Measurement Instruments

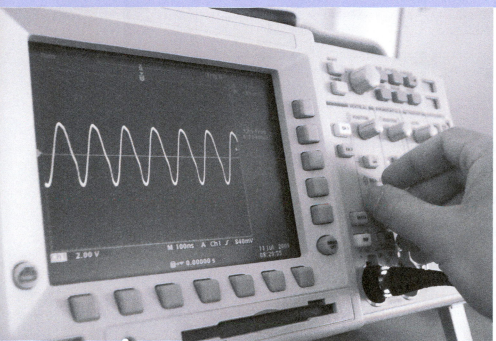

(Used with permission from Rannev/Shutterstock.com.)

14.1 INTRODUCTION

For each physical quantity outlined in this text, a unit has been defined, which describes the nature of the quantity. The previous chapters introduced many quantities related to electrical engineering and their corresponding units. Examples include current (unit: amperes), voltage (unit: volts), resistance (unit: ohms), capacitance (unit: farads), and frequency (unit: hertz). This chapter introduces you to the many measurement devices available, which can specify the magnitude of many of these quantities.

The purpose of electrical and electronic instruments is to measure physical and electrical quantities. The precision of these measurements depends on factors such as the sensitivity and the accuracy of instruments. To accurately measure these quantities, in this chapter, you will learn how to use each instrument properly and how to interpret measured values. In addition, you will learn to be aware of the sources of error in the measurement process, and techniques used to predict and address these errors. Specifically, this chapter discusses the effect of loading on measurement error. This chapter studies the application of widely used electrical and electronic instruments. Instruments presented in this chapter include the voltmeter, ohmmeter, ammeter, oscilloscope, spectrum analyzer, and function generator.

14.2 MEASUREMENT ERRORS

The *error* in a measurement is defined as the deviation of the measured value from the actual, true, or expected value. If x_m and x_{true} are the measured and true values, respectively, the error is:

$$e = x_{true} - x_m \qquad (14.1)$$

The percentage of error corresponds to:

$$\text{Error percentage} = \left| \frac{e}{x_{true}} \right| \times 100 = \left| \frac{x_{true} - x_m}{x_{true}} \right| \times 100 \qquad (14.2)$$

Accuracy is an important specification of any measurement instrument. Basically, it represents the degree of the reliability of an instrument. The accuracy is defined as:

$$A = 1 - \left| \frac{x_{true} - x_m}{x_{true}} \right| \qquad (14.3)$$

The percentage of accuracy corresponds to:

$$a = A \times 100 \qquad (14.4)$$

Many environmental and systematic factors affect measurement. For example, suppose you want to measure the resistance of a resistor. The temperature of the environment changes the length and the cross-sectional area of the resistor. Thus, if you measure the resistance while holding the resistor in your hand versus placing the resistor on a table, the results might be different. The systematic and environmental factors in a measurement are random and cannot be predicted.

In addition, if you measure a quantity repeatedly, you may obtain different values. Assuming a large number of independent observations of a quantity, that is, considering measurements as independent random variables, then the average of the measurements taken is most likely closest to the true value. Therefore, if $x_1, x_2, ..., xN$ (N, the number of instances, must be large enough, for example, $N > 25$) are the recorded measurement values, their average or arithmetic mean is calculated as:

$$\bar{x} = \frac{x_1 + x_2 + \cdots + x_N}{N} = \frac{1}{N}\sum_{i=1}^{N} x_i \qquad (14.5)$$

When the sample size (N) is large enough, and measurement values vary over the numerous samples, the average is considered the most accurate measured value. A quantitative measure that indicates the closeness of each individual measurement to the average of the set of the measurements is called *precision*. Mathematically, precision of a measurement, x_i, is expressed as:

$$\text{Precision} = 1 - \left| \frac{x_i - \bar{x}}{\bar{x}} \right|, i \in \{1, 2, \ldots, N\} \qquad (14.6)$$

The *standard deviation* is a measure of variability (which can be due to error) in a set of measurements. The standard deviation can be computed as:

$$\sigma = \sqrt{\frac{1}{N-1}\sum_{i=1}^{N} (x_i - \bar{x})^2} \qquad (14.7)$$

The following examples show how to apply these definitions to a problem.

EXAMPLE 14.1   Resistor Tolerance

A value called *tolerance* is allocated to all standard resistors. It specifies a limit on the maximum error that can be observed in the value allocated to a given quantity. If the manufacturer's specification of a resistor shows its resistance as $100\ \Omega \pm 2\%$, the value of the resistance would be $100\ \Omega$ with a maximum of $2\% \times 100\ \Omega = 2\ \Omega$ tolerance.

Now, suppose you have two resistors, R_1 and R_2, with corresponding resistances of $R_1 = 100\ \Omega \pm 10\%$, $R_2 = 200\ \Omega \pm 10\%$. What would be the tolerance percentage of the series and parallel configurations of these resistors?

SOLUTION

Based on the specified tolerance, the maximum and minimum values for the two resistors, R_1 and R_2, correspond to:

$$90 < R_1 < 110$$
$$180 < R_2 < 220$$

The equivalent series resistance without error is:

$$100\ \Omega + 200\ \Omega = 300\ \Omega.$$

The minimum and maximum series resistances are:

$$R_{\text{series, min}} = 90\ \Omega + 180\ \Omega = 270\ \Omega$$
$$R_{\text{series, max}} = 110\ \Omega + 220\ \Omega = 330\ \Omega$$

Thus, the total tolerance of the series connection of resistors corresponds to $\pm 30\ \Omega$ and the tolerance percentage is:

$$30/300 \times 100 = 10\%$$

The equivalent parallel resistance without error is:

$$\left(\frac{1}{100} + \frac{1}{200}\right)^{-1} = 66.6\ \Omega$$

The minimum and maximum parallel resistances are:

$$R_{\text{parallel, min}} = \left(\frac{1}{90} + \frac{1}{180}\right)^{-1} = 60.0\ \Omega$$

$$R_{\text{parallel, max}} = \left(\frac{1}{110} + \frac{1}{220}\right)^{-1} = 73.3\ \Omega$$

Therefore, the tolerance is about $6.6\ \Omega$ and the tolerance percentage is $6.6/66.6 \times 100 = 9.9\%$.

EXAMPLE 14.2   Thermistor

A chemical engineer uses a thermistor (temperature-varying resistor) to measure temperature. Engineers also use ohmmeters to measure resistance. The ohmmeter applies a known voltage to the thermistor and measures the current drawn, I. By applying Ohm's law, the resistance is then calculated internally from this voltage and current.

(continued)

EXAMPLE 14.2   **Continued**

To prevent short circuit between the leads of the ohmmeter, which creates an extremely high current that may damage the ohmmeter, a small resistor, R_{int}, is placed in series with the leads, as indicated in Figure 14.1(a). This example studies the impact of this resistance on measurements.

   If the short-circuit current is 12 A, then what is the value of R_{int}? Use this value of R_{int} and find the error percentage if the actual value of R is:

a. $R = 0.2\ \Omega$	**f.** $R = 100\ \Omega$
b. $R = 1\ \Omega$	**g.** $R = 1\ k\Omega$
c. $R = 5\ \Omega$	**h.** $R = 10\ k\Omega$
d. $R = 10\ \Omega$	**i.** $R = 100\ k\Omega$
e. $R = 50\ \Omega$	**j.** $R = 1\ M\Omega$

The error percentage is defined by Equation (14.2) as:

$$\text{Error percentage} = \left| \frac{\text{measured value} - \text{actual value}}{\text{actual value}} \right| \times 100\%.$$

SOLUTION

R_{int} can be found in a way similar to finding a Thévenin resistance: as seen in Figure 14.1(b), the short-circuit current, $I_{SC} = 12$ A, and the open-circuit voltage is $V_{OC} = 6$; thus,

$$R_{int} = \frac{6}{12} = 0.5\ \Omega$$

Now, based on the circuit of Figure 14.1(a), the measured current corresponds to:

$$I = \frac{6}{R + 0.5}$$

The measured resistance, R_{meas}, will then be:

$$R_{meas} = \frac{V_s}{I} = \frac{6}{\dfrac{6}{R + 0.5}} = R + 0.5$$

Thus, the measured resistance is not exactly equal to the connected resistance, R. Therefore, the measurement error percentage is:

$$\text{Error percentage} = \left| \frac{R + 0.5 - R}{R} \right| \times 100\% = |(1 + 0.5/R - 1)| \times 100\%$$
$$= \frac{0.5}{R} \times 100\%$$

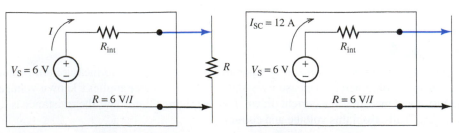

FIGURE 14.1   (a) Ohmmeter circuit; (b) calculating internal resistance.

Now, if you change the resistance, the error percentage will also change. The errors corresponding to the errors in the measuring the resistance, R, are calculated to be:

Error = 250%	Error = 0.5%
Error = 50%	Error = 0.05%
Error = 10%	Error = 0.005%
Error = 5%	Error = 0.0005%
Error = 1%	Error = 0.00005%

You can observe that for small thermistor resistances, the error percentage is considerable. Therefore, if the thermistor resistance is known to be small, this internal resistance needs to be considered while taking readings. However, this error can be ignored if the thermistor resistance is high.

14.3 BASIC MEASUREMENT INSTRUMENTS

Today's instruments are mainly based on digital circuits. As discussed in Chapter 11, these circuits convert the voltage or current into digital using analog-to-digital converters, and then use digital circuits, such as counters, to calculate the voltage or current. This chapter examines the fundamentals of analog measurement devices. The main goal is to practice an application of the basic circuit theory you have learned in Chapters 2 and 3, while learning measurement fundamentals.

The most basic analog electrical meter is a *galvanometer*. A galvanometer consists of a coil of wire suspended in a magnetic field, and a spring. When current passes through a coil in a magnetic field, the coil experiences a torque proportional to the current. This torque rotates the coil until the equilibrium is maintained between the opposite field and spring torques. The angle of rotation depends on the amount of current passing through the coil. The needle attached to the coil must be calibrated to indicate the current. The scale of the needle can also be calibrated to show other units such as voltage and resistance. Therefore, the galvanometer can be used as a multipurpose measurement instrument. A galvanometer is shown in Figure 14.2.

In order to use the galvanometer for measurement purposes, there must be a linear relationship between the current and the galvanometer's deflection angle. Assuming this linearity is maintained for currents between 0 and i_g, the current, i_g, is usually in the order of microamperes. In this sense, it is sometimes said that the sensitivity of galvanometers is high, that is, a small current may lead to a large deflection of the galvanometer needle.

FIGURE 14.2
A galvanometer.

14.3.1 An Ammeter Built Using a Galvanometer

An *ammeter* measures the current of a circuit. Assume that a galvanometer is used to construct an ammeter to measure currents between 0 and i amperes. A typical ammeter consists of a galvanometer in parallel with a resistance [see Figure 14.3(a)]. This resistance is called a shunt resistance. The galvanometer is usually not capable of tolerating high-level currents. Thus, the shunt resistance needs to be low enough to allow most of the current to flow through the resistance.

Assume R_g is the internal resistance of the galvanometer and R_{sh} is the shunt resistance. As seen in Figure 14.3, the voltage across the galvanometer wire's internal resistance and the shunt resistance is the same; thus:

$$R_g i_g = R_{sh} i_{sh} \tag{14.8}$$

As seen in Figure 14.3, $i_{sh} = i - i_g$. Thus, using Equation (14.8), the shunt resistance can be obtained using the following equation:

$$R_{sh} = \frac{i_g}{i - i_g} \times R_g \tag{14.9}$$

The shunt circuit allows us to measure a current, i, that is n times larger than the current passing through the galvanometer, i_g. The number, n, is called a multiplying factor. The measured current, i, and the current passing through the galvanometer, i_g, are related as:

$$i = n i_g \tag{14.10}$$

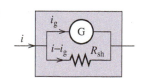

(a)

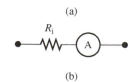

(b)

FIGURE 14.3 (a) Ammeter circuit; (b) the symbol of an ammeter.

Substituting Equation (14.10) into Equation (14.9):

$$R_{sh} = \frac{R_g}{n - 1}$$ (14.11)

EXAMPLE 14.3   **Galvanometer**

A galvanometer is used to construct an ammeter that measures currents in the range of 0–5 A. The resistance of the wiring in the galvanometer is 500 Ω and a maximum current of 5 mA can flow through the device without harming it. Determine the resistance of the shunt resistor connected in parallel with the galvanometer.

SOLUTION

Referring to Figure 14.3, the voltages across parallel resistors are the same. Thus:

$$5 \text{ mA} \times 500 \text{ Ω} = (5 \text{ A} - 5 \text{ mA}) \times R_{sh}$$

Solving for the shunt resistance, R_{sh}:

$$R_{sh} = 0.5 \text{ Ω}$$

Several shunt resistors can be incorporated to increase the range of current measurement.

The ammeter must be placed in *series* with the circuit for which the current is measured. Thus, its resistance should be kept low to maintain its minimal effect on the circuit's actual current. An ideal ammeter has an internal resistance of zero. The symbol of an ammeter is shown in Figure 14.3(b), where R_i represents the internal resistance of the ammeter. Given this equivalent circuit, it is clear that applying the ammeter in a circuit to measure the current would impact the actual current (see Examples 14.4 to 14.6).

EXERCISE 14.1

Use Figure 14.3(a) and show that the internal resistance, R_i, of an ammeter is $R_{sh} \| R_g$. Calculate R_i given $R_g = 1 \text{ Ω}$ and $R_{sh} = 1 \text{ kΩ}$ and compare it with R_g. In this exercise, you will find that R_i will be very small because $R_{sh} \| R_g < R_g$; and R_g itself is usually very small.

Current Measurement Via Ammeter is an Invasive Measurement: It should be noted that measuring current is an *invasive* process that is required to open the circuit in order to place the ammeter in the circuit. A *noninvasive* approach for measuring the current is to measure the voltage across a known resistance. By dividing the voltage and the resistance, the current can be measured.

Another noninvasive approach to current measurement is to measure the magnetic field around the wire. A typical sensor that is used to measure the magnetic effects is called Hall sensor. The details of these sensors are out of the scope of this chapter. Hall sensors are used for the measurement of current in some applications such as automotive.

14.3.2 A Voltmeter Built Using a Galvanometer

This chapter already outlined that an ammeter should be placed in series with the circuit because it measures the current through the circuit. What about the *voltmeter*?

A voltmeter should be placed in *parallel* with the element whose voltage is being measured. Why?

Because voltage is a relative quantity: in other words, the voltage of one point is measured with respect to another point.

A galvanometer can be used to construct a voltmeter as shown in Figure 14.4(a). Here, a large resistor is located in series with the galvanometer (R_s). The reasons are twofold:

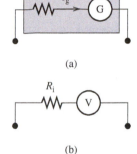

(a)

a. To keep the current through the voltmeter small in order to reduce the impact of the voltmeter on the circuit.

b. To ensure that the galvanometer operates within its linear operation by reducing the current flowing through the galvanometer.

Thus, if the maximum current that maintains linear operation of the galvanometer is i_g, the maximum measurable voltage of the voltmeter would be $V = (R_s + R_g) \times I_g$.

Note that unlike the ammeter, the circuit branch does not need to be broken in order to measure the voltage. The voltage can be measured by simply connecting the voltmeter probes across the desired nodes. Because the voltmeter's resistance is high, it does not change the current in the circuit. Thus, like the ammeter, when voltmeter is applied in a circuit, there will be no change in the circuit variables of voltage, current, and resistance. The symbol of a voltmeter with its internal resistance is shown in Figure 14.4(b). An ideal voltmeter has an infinite resistance.

(b)

FIGURE 14.4
(a) Voltmeter circuit;
(b) the symbol of a voltmeter.

14.3.3 An Ohmmeter Built Using a Galvanometer

The galvanometer can also be used to construct an *ohmmeter*. The principle of an ohmmeter is based on the Wheatstone bridge, introduced first in Chapter 1. Figure 14.5 shows an unknown resistance, R_2, that can be measured and three adjustable resistances R_1, R_3, R_4. The three adjustable resistances can be varied until the current passing through the galvanometer reduces to zero. In this case, resistors R_1 and R_3 will be in parallel. Therefore, the voltage across these two resistors will be equivalent, and:

$$I_1 R_1 = I_3 R_3 \tag{14.12}$$

Similarly, the resistors R_2 and R_4 are also in parallel and:

$$I_2 R_2 = I_4 R_4 \tag{14.13}$$

Dividing Equations (14.12) and (14.13), we obtain the unknown resistance using the following equation:

$$R_2 = \frac{R_4}{R_3} R_1 \tag{14.14}$$

Based on the procedure explained in this subsection, to measure the resistance of a resistor, the resistor should be disconnected from the circuit and connected in the circuit of Figure 14.5. Measuring a resistor while it is connected to a circuit is equivalent to connecting the Wheatstone circuit (Figure 14.5) to a circuit where the resistor is located. This leads to totally different and incorrect reading.

14.3.4 Multi-Meters

A multi-meter is an instrument that measures voltage, current, and resistance. Modern multi-meters use digital instruments to measure voltage, current, and resistance. These instruments are usually more accurate than galvanometer-based multi-meters. Figure 14.6 shows a typical digital multi-meter.

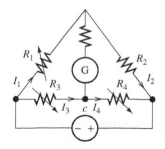

FIGURE 14.5 Wheatstone bridge.

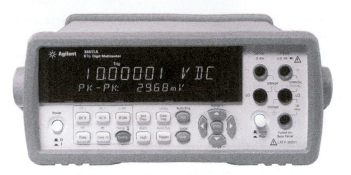

FIGURE 14.6 A digital multi-meter. (© Agilent Technologies, Inc. 2012. Reproduced with permission. Courtesy of Agilent Technologies, Inc.)

EXAMPLE 14.4   **Instrument Reading Error**

In Figure 14.7, determine the error percentage of a current reading due to the insertion of an ammeter. $R_3 = R_2 = 100\ \Omega$, $R_1 = 200\ \Omega$, and the internal resistance of the ammeter is $R_{in} = 5\ \Omega$.

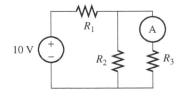

FIGURE 14.7 Circuit for Example 14.4.

SOLUTION

Before inserting the ammeter, the total current flowing through the 10 V source is calculated to be:

$$I = \frac{10}{R_1 + R_2 \| R_3} = 0.04\ \text{A}$$

Because the two parallel resistors, R_3 and R_2, are equivalent, the current passing through R_3 is $0.04/2 = 0.02$ A.

After inserting the ammeter, due to the resistance of the ammeter, the final resistance of the two parallel branches will not be equivalent. In this case, the total current flowing through the 10 V resistor will be:

$$I = 10/(R_1 + R_2) \| (R_3 + 5\ \Omega) = 0.0398\ \text{A}$$

In this case, using the current division concept, the current flowing through R_3 is:

$$\frac{100}{100 + 105} \times 0.0398\ \text{A} = 0.0194\ \text{A}$$

Therefore, the error percentage is:

$$\text{Error percentage} = \frac{0.02 - 0.0194}{0.02} \times 100 = 3\%$$

Thus, the error percentage due to the current reading corresponds to 3%.

EXERCISE 14.2

In Example 14.4, how will the percentage of error change if the ammeter's internal resistance increases to $R_g = 15\ \Omega$?

EXAMPLE 14.5   Instrument Reading Error

In Figure 14.8, determine the error percentage of the voltage reading due to the insertion of a voltmeter. $R_1 = 3\ k\Omega$, $R_2 = 2\ k\Omega$, and the internal resistance of the voltmeter is 50 $k\Omega$.

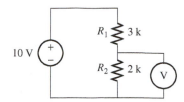

FIGURE 14.8   Circuit for Example 14.5.

SOLUTION

The actual voltage of R_2 is:

$$\frac{2\ k\Omega}{2\ k\Omega + 3\ k\Omega} \times 10\ V = 4\ V$$

The voltage after inserting the voltmeter will then be:

$$\frac{2\ k\Omega \parallel 50\ k\Omega}{2\ k\Omega \parallel 50\ k\Omega + 3\ k\Omega} \times 10\ V = 3.906\ V$$

Thus, the error percentage is:

$$\text{Error percentage} = \frac{4 - 3.906}{4} \times 100 = 2.35\%$$

EXAMPLE 14.6   Instrument Measurement Error

Figure 14.9(a) represents a simple voltage divider circuit. A voltmeter with an internal resistance of 1 $M\Omega$ is connected to measure V_A, the voltage across R_2. Calculate the voltage measured for V_A. Next, calculate the measured V_A if the same voltage divider is constructed using 1 $M\Omega$ resistors, as shown in Figure 14.9(b).

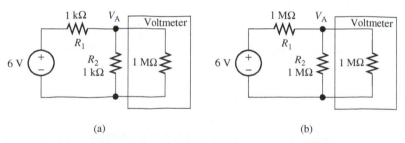

(a)   (b)

FIGURE 14.9   (a) Voltage divider with voltmeter; (b) voltage divider using high-value resistors.

(continued)

EXAMPLE 14.6   **Continued**

SOLUTION

Without applying a voltmeter and using a simple voltage dividing, observe that the actual voltage of V_A is 3 V. For Figure 14.9(a):

$$\frac{V_A - 6}{1\ k} + \frac{V_A}{\left(\dfrac{1}{1\ k} + \dfrac{1}{1\ M}\right)^{-1}} = 0$$

Factorizing V_A and rearranging:

$$2.001 \times V_A = 6 \rightarrow V_A = 2.9985\ \text{V}$$

Also observe that the voltage, V_A, is ~3 V in this case, because the resistors available in this circuit are much smaller than the voltmeter's resistor. Replacing R_1 with a 1 MΩ resistor as shown in Figure 14.9(b):

$$\frac{V_A - 6}{1\ M} + \frac{V_A}{\left(\dfrac{1}{1\ M} + \dfrac{1}{1\ M}\right)^{-1}} = 0$$

This leads to:

$$(1.5 \times 10^6) \times V_A = 6 \times 0.5 \times 10^6 \rightarrow V_A = 2.00\ \text{V}$$

The voltage measured from Figure 14.9(b) is much different than the actual voltage. This example shows that a voltmeter will load a circuit if its internal resistance is low, or if the resistance of the circuit is in the range of the voltmeter's internal resistance. Usually, voltmeters are equipped with buffers at their output (circuits made by operational amplifiers as discussed in Chapter 8). The output resistance of these circuits is high enough to minimize the error for many real-world measurements.

Resistance Measurement: In general when a resistor is connected in the circuit, the resistance of the resistor cannot be measured: The resistor should be disconnected prior to the measurement. An ohmmeter includes an internal voltage source. Thus, if it is connected to a resistor while the resistor is not disconnected from the circuit, current and voltages could be measured in different components of the circuit.

Summary:

a. An ideal current meter has zero internal resistance: an ammeter should be applied in series with the element whose current we intend to measure. Thus, this current measurement approach is invasive.

b. An ideal voltmeter has an infinity internal resistance. It is applied in parallel to the element that we intend to measure its voltage.

c. An alternative noninvasive method of measuring current is to measure the voltage across a known resistor and then calculate the current using the Ohms law.

d. Another noninvasive current measuring method is to use a Hall sensor whereby the magnetic field caused by the current is measured. This is commonly used in automotive testing.

e. A resistor should be disconnected from the circuit prior to the measurement of its resistance.

14.4 TIME DOMAIN AND FREQUENCY DOMAIN

This section explains the two instruments used to measure the characteristics of signals in the time and frequency domains, that is, the *oscilloscope* and the *spectrum analyzer*, respectively. In order to explain these two instruments, a short review of the representation of signals in the time and frequency domains is provided. A sinusoid is a simple but an important example of a signal. This example helps to better explain a signal in the frequency domain.

Sinusoids play an important role in science and engineering. A sinusoid is a sine function, that is, a function of time. The general form of a sinusoid corresponds to:

$$f(t) = A \cos (2\pi f t + \theta) + B \qquad \textbf{(14.15)}$$

Here:

- A = amplitude
- f = frequency (Hz)
- θ = phase angle
- B = DC offset; if $B = 0$, then the sinusoid will be balanced with respect to the *x*-axis (see Figure 14.10), if $B > 0$, the sinusoid will be shifted upward (see Figure 14.11).

The parameters introduced in Equation (14.15) are illustrated in Figures 14.10 to 14.12. Typically, in the sinusoid of Figure 14.10, the amplitude of the sinusoid is 1, the period is 4 s; thus, the frequency is 0.25 Hz, the phase is zero, and the DC offset is zero. In Figure 14.11, V_{peak} is the amplitude, the phase is 0°, and there is a nonzero DC offset.

Sinusoids describe the characteristics of many physical processes. Examples of sinusoids are ocean waves, radio/television signals, alternating current (AC) voltages and currents, and acoustic pressure variations of a single musical note. Some processes include single sinusoid waves, such as the output of an AC power supply. Other processes such as ocean and sound waves are a combination of many sinusoids with different frequencies.

14.4.1 The Time Domain

Sound is a practical means of explaining the time domain and the frequency domain. Sound is perceived when fluctuations in air pressure vibrate your eardrum. The amplitude of pressure determines the sound volume. Air pressure fluctuates cyclically, and the number of cycles per

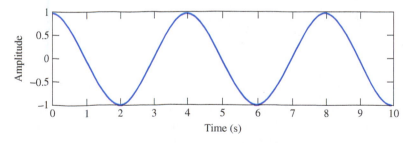

FIGURE 14.10 A sinusoid with $B = 0$.

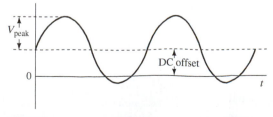

FIGURE 14.11   DC offset of a sinusoid ($B > 0$).

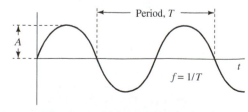

FIGURE 14.12   Voltages and period of a sinusoid.

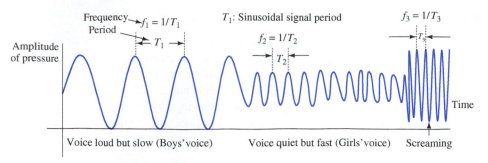

FIGURE 14.13 Several frequencies shown in the time domain.

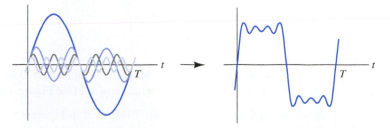

FIGURE 14.14 Combining sinusoids creates different periodic signals.

second is the *frequency* of the sound. Frequency has a unit of *hertz* (Hz), where 1 Hz equals one cycle per second. The amplitude of pressure corresponds to the loudness of the sound and the frequency corresponds to the pitch of the sound. Figure 14.13 shows the amplitude of air pressure vibrations with respect to time for different frequencies.

Complex sound waveforms are the summation of multiple sinusoids of various frequencies, amplitudes, and phase. Complex sound waves can also be realized by adding sinusoids of appropriate amplitudes, frequencies, and phases. Figure 14.14 shows how multiple sinusoids can be combined in the time domain to create a waveform similar to a square wave. In general, based on the Fourier transform of periodic signals, any periodic signal can be represented by adding a number of sinusoids with different amplitudes and frequencies. Fourier transform was discussed in Chapter 7.

14.4.2 The Frequency Domain

As discussed, all periodic signals (including sound) consist of multiple sinusoids with different frequencies. What if an engineer desires to know the frequency components of the sound? Plotting in the *frequency domain* provides an alternate representation of the sound. When plotting in the frequency domain, the *x*-axis is the frequency and the *y*-axis is the amplitude. The plot consists of vertical lines. As seen in Figure 14.15, the position of a vertical line on the *x*-axis represents its frequency, and the height of the line represents the amplitude at that frequency. A *spectrum analyzer* is used to plot signals in the frequency domain (see Section 14.5).

Note that in Figure 14.15 each frequency is represented by a line. Here, each line represents a tone. However, in general, the transmitted signals are not a tone alone. For example your

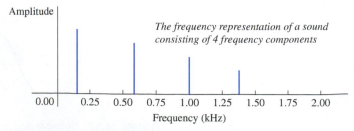

FIGURE 14.15 A frequency domain plot.

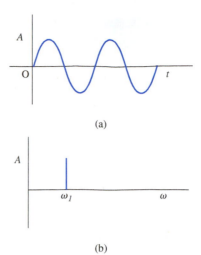

FIGURE 14.17 (a) A sinusoid in the time domain; (b) a sinusoid in the frequency domain.

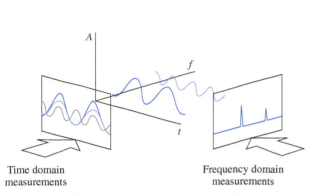

Time domain measurements

Frequency domain measurements

FIGURE 14.16 A time domain plot versus a frequency domain plot.

voice consists of a large number of signals. Therefore, in reality, in the frequency domain, we generally observe a signal as a cluster of lines. Usually, we observe a signal at a cluster of frequencies centered around a given frequency. In communication systems, this frequency is called the *carrier*. For example, when you tune your radio on the frequency of 200 MHz, the signal indeed is transmitted over 200 MHz $\pm$ 20 kHz. In this case, 40 kHz is the bandwidth (BW) of the signal (see Section 14.6).

14.4.3 Time Domain Versus Frequency Domain

Figure 14.16 illustrates the difference between the time and frequency domains. All time-varying signals can be presented in both time and frequency domains. In the time domain, the amplitude is sketched as a function of time, and in the frequency domain, the amplitude is sketched as a function of frequency.

For example, given a single sinusoid that is defined as $f(t) = A \sin \omega_1 t$ with a frequency of $f = \omega_1/2\pi$, its time and frequency domain representations are sketched in Figure 14.17.

EXAMPLE 14.7 **Sum of Sinusoids**

Evaluate the sum of the sinusoids in Figure 14.18(a) and (b).

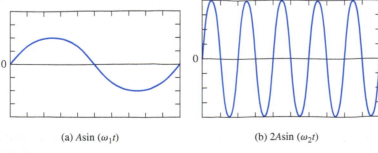

(a) $A\sin(\omega_1 t)$

(b) $2A\sin(\omega_2 t)$

FIGURE 14.18 (a) Sinusoid A; (b) sinusoid B.

(*continued*)

EXAMPLE 14.7 **Continued**

The combination of sinusoids A and B in the time and frequency domains is shown in Figure 14.19(a) and (b), respectively. You are encouraged to use MATLAB to sketch Figures 14.18 and 14.19(a).

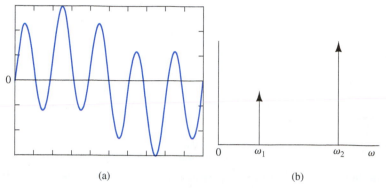

(a)   (b)

FIGURE 14.19 A combined signal in (a) the time domain and (b) the frequency domain.

EXERCISE 14.3

Use MATLAB to sketch the time and frequency domain of the following signals:

a. $f(t) = \sin(100\,t)$
b. $g(t) = \cos(150\,t)$
c. $h(t) = \cos(90\,t + 20°)$
d. $x(t) = f(t) + g(t)$
e. $y(t) = f(t) + g(t) + h(t)$

14.5 THE OSCILLOSCOPE

An *oscilloscope* displays signals in the time domain. An analog oscilloscope consists of a cathode-ray tube with a vacuum inside. An electron beam emitted by the heated cathode is accelerated and focused by one or more anodes, and strikes the front of the tube, producing a bright spot on the phosphorescent screen. The electron beam is deflected by the signal voltages applied to two sets of plates fixed in the tube. Many advanced oscilloscopes used today are digital and do not possess a cathode-ray tube. The Agilent 54621D, a typical oscilloscope, is shown in Figure 14.20. The oscilloscope screen functions similar to a television screen. Figure 14.21 shows examples of signals that appear on an oscilloscope screen.

An oscilloscope is capable of measuring the period, T, frequency, f, amplitude, A, and phase, θ, of a periodic signal. It is also used to measure the rise and fall times of transients. This concept is explained in detail later in this section. An oscilloscope screen is divided by a grid of horizontal and vertical lines into *divisions*, as shown in Figure 14.22. The horizontal lines denote voltage and the vertical lines denote time. A division is the difference in value between two adjacent lines. For example, the space between V_1 and V_2 in Figure 14.22 represents one voltage division, and the space between t_1 and t_2 is one time division. The *voltage per division* is given by:

$$\Delta V = V_2 - V_1 \tag{14.16}$$

and the *time per division* is given by:

$$\Delta t = t_2 - t_1 \tag{14.17}$$

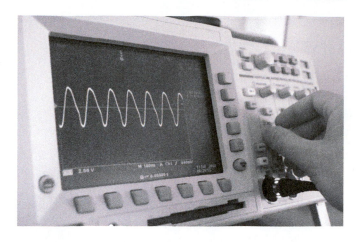

FIGURE 14.20 An oscilloscope. (Used with permission from Rannev/Shutterstock.com.)

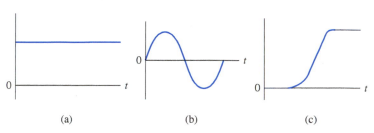

(a) (b) (c)

FIGURE 14.21 Sample signals as displayed by an oscilloscope: (a) DC; (b) AC sinusoid; (c) transient.

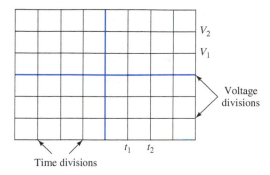

FIGURE 14.22 The oscilloscope grid.

These two quantities are adjusted by the controls of the oscilloscope. These quantities are required to calculate some measurements from the oscilloscope, such as period and amplitude. An oscilloscope typically has the following controls:

Run/Stop: This allows the oscilloscope to continually acquire and display data or stop acquiring data, respectively. When the "run" mode is enabled, the oscilloscope screen will be continually updated with new data. When "stop" is engaged, no data is acquired, and the last acquisition is continuously displayed.

Single: Push of the "single" button makes the oscilloscope acquire only enough data to refresh its screen. When data acquisition is complete, the oscilloscope switches to "stop" mode.

Horizontal Scale: The horizontal scale (sometimes labeled timescale) adjusts the horizontal timescale, in seconds/division. This allows zooming in/out on the signal horizontally.

Horizontal Position: This controls the position of the displayed signal along the horizontal axis. This allows aligning the signal plot with the timescale and taking measurements.

Vertical Scale: Each channel has a knob to adjust the voltage per division, which allows zooming in/out on the signal vertically.

Vertical Position: Each channel also has a knob to control the vertical position of the signal on the screen. This allows overlapping or separating signals when displaying multiple signals simultaneously.

X–Y mode: For oscilloscopes with two analog channels, the scope can be configured to use one as the x-axis and the other as the y-axis.

Cursors: The cursors are an important tool for measuring signal values in an oscilloscope. Cursors are lines that can be moved about the oscilloscope screen to measure a time interval or a voltage difference. Typically, an oscilloscope has four cursors, two horizontal, and two vertical, labeled X_1, X_2, Y_1, and Y_2 as shown in Figure 14.23. The oscilloscope

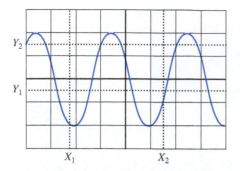

FIGURE 14.23
Oscilloscope cursors.

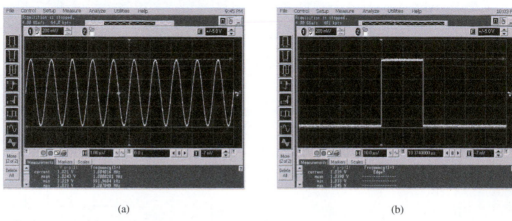

(a)   (b)

FIGURE 14.24 (a) A single sinusoid; (b) a single pulse displayed on an oscilloscope.

screen displays the time values of the vertical cursors, and the voltage values of the horizontal cursors. These values are usually placed below the signal plot on the screen. The cursors can be used to measure signal period, frequency, and amplitude.

Trigger: The trigger is used to synchronize the time base of the oscilloscope to the input signal. Otherwise, the signal would drift horizontally. It is desirable to have the trigger enabled for many applications of the oscilloscope. When the trigger is activated, it sends a pulse to the oscilloscope, which refreshes the screen (equivalent to the "single" function described previously).

Figure 14.24(a) and (b) shows two popular signals, a sinusoid and a pulse, displayed on an oscilloscope. You should note that the generated pulse has a rise time, that is, it does jump from zero to the desired voltage within a short (nonzero) time period.

EXAMPLE 14.8   **Measuring Amplitude, Period, Frequency, and Phase**

Use the oscilloscope screen of Figure of 14.25 to determine the amplitude, A, period, T, frequency, f, and phase, θ, of the sinusoid $f(t) = A \cos\left(2\pi f t + \theta\right)$.

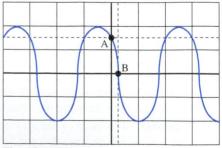

Time/div = 0.25 ms
Volt/div = 0.2 V

FIGURE 14.25 An oscilloscope screen for Example 14.8.

SOLUTION

Amplitude: The waveform spans four voltage divisions. Therefore, the peak-to-peak voltage is 0.8 V. The amplitude, A, is given by:

$$V_{p-p}/2 = 0.4 \text{ V}$$

Period: The figure shows that one period spans three time divisions. Thus, the period is:

$$T = 0.25 \text{ ms} \times 3 = 0.75 \text{ ms}$$

Frequency: The frequency, f, is the reciprocal of the period, therefore:

$$f = 1/T = 1.33 \text{ kHz}$$

Phase: To find the phase, θ, plug in and solve for it. Assume that the center of the oscilloscope display is the origin (0,0), and point B represents $t = 0.25$ ms, $f(t) = -0.4$ V. Note that there are different options to find t and $f(t)$. For example, the horizontal and vertical cursors may be adjusted at point A [$f(t)$] or B of Figure 14.25.

Now, using $f(t) = A \cos(2\pi f t + \theta)$, and knowing that at $t = 0.25$ ms, $f(t) = -0.4$ V, and applying the calculated parameters of amplitude and frequency:

$$-0.4 = 0.4 \cos(2\pi \times 1.33 \times 10^3 \times 0.25 \times 10^{-3} + \theta)$$

Solving for θ:

$$\theta = 1.052433 \ rad = 60.3°$$

Therefore, the sinusoid function corresponds to:

$$f(t) = 0.4 \cos(2.33 \times 10^3 \pi t + 1.0524)$$

APPLICATION EXAMPLE 14.9   **Displaying Transients**

The trigger function is used to display circuit transients. A transient is a change in voltage or current due to the variations in system states. For example, a transient occurs when the voltage of a resistive–capacitive (RC) circuit eventually changes as a result of the switch movement as discussed in Chapter 4. A transient can be created by toggling a switch, pushing a button, or turning off a knob. The voltage or current jump is altered due to the circuit transient response for a short time period and then it is adjusted to a new (steady) state.

Transients are important for mechanical engineers. When a motor is switched on, the voltage across the motor spikes before it settles down. This is due to the inherent inductance in the coils of the motor. This "overshoot" is shown in Figure 14.26(a) and (b). Figure 14.26 shows the situation when a motor is switched from 0 to 5 V. Due to the transients, the voltage across the motor jumps to nearly 7 V, then it oscillates before it finally settles on 5 V. Due to Ohm's law, this spike in voltage results in a spike in current.

The oscilloscope "trigger" function allows us to capture the signal at the moment of the transient, and therefore, to study the signal behavior at that point. If an increase in voltage is expected, the trigger would be set to "rising edge," otherwise it would be set to "falling edge." To see this, first set the trigger threshold to a reasonable level. (If the current voltage is 0 V, a threshold of 0.5 or 1.0 V is desirable.) Next, press the "single" button. The oscilloscope will wait for the transient. When it occurs, it will display the signal at the time of the transient.

The rise or fall time of a transient can be measured using the oscilloscope's cursors. In practice, the rise time is often defined as the time period between the signal at 5% of its maximum value and the signal at 95% of its maximum value, as shown in Figure 4.27(a). Likewise, the fall time is often defined as the time period between the signal at 95% of its

(continued)

APPLICATION EXAMPLE 14.9   Continued

maximum value to 5% of its maximum value, as shown in Figure 14.27(b). The cursors can be positioned over 95% and 5% values, to measure the time difference $t_2 - t_1$.

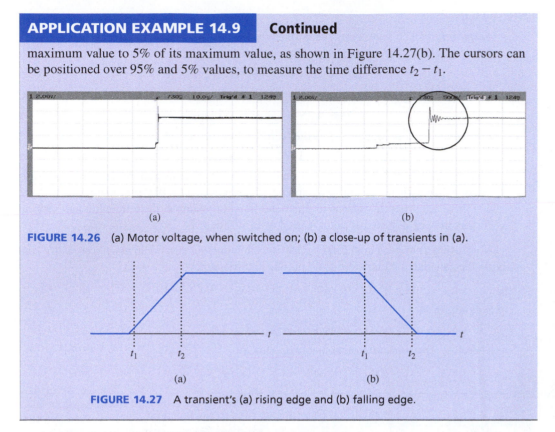

FIGURE 14.26   (a) Motor voltage, when switched on; (b) a close-up of transients in (a).

FIGURE 14.27   A transient's (a) rising edge and (b) falling edge.

APPLICATION EXAMPLE 14.10   De-Bouncing a Circuit

One practical application of an oscilloscope is the measurement of switch bounce. Switch bounce is a form of transient (see Chapters 5, 10, and 15). When a switch is turned ON or OFF, it does not reach its steady-state position immediately. It may bounce between OFF and ON for a fraction of a second. This is due to the mechanical structure of the switch. Figure 14.28(a) shows the output voltage of a switch when it is turned ON.

    To measure this transient, the trigger conditions are set on its rising edge. When the switch is turned ON, the voltage rises from 0 V to the source voltage, V_s. The oscilloscope should be set to its "normal" mode. "Automatic" mode is NOT desired because it may trigger the oscilloscope at undesired times. This ensures that the waveform is not displayed until the trigger conditions are met (i.e., the switch is toggled). The trigger voltage level

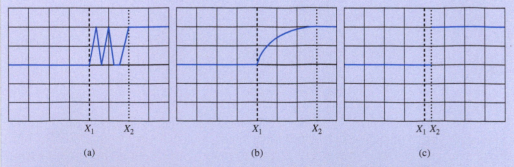

FIGURE 14.28   Measuring rise time of a switch (a) exhibiting switch bounce; (b) de-bounced with an RC circuit; and (c) de-bounced with an RS flip-flop.

should be selected somewhere between 0 V and V_s. The oscilloscope should be then run for a "single" acquisition (as opposed to constantly running) because the signal only appears once. Figure 14.28(b) and (c) shows the output of the de-bounced circuits when an RC circuit (see Example 5.2) and an RS flip flop (see Example 10.7), respectively, are used.

EXAMPLE 14.11   **Transient Fall Time**

What is the fall time of the transient in Figure 14.29?

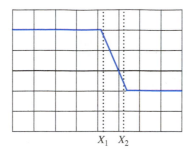

Time/div = 5 μs, Volt/div = 1 V
$X_1 = 1.35$ μs, $X_2 = 6.89$ μs

FIGURE 14.29 Transient for Example 14.9.

SOLUTION

The signal drops by three divisions. Each division is equivalent to 1 V, thus it drops by 3 V. Therefore, the 95% and 5% voltages are:

$$V_{95\%} = 3 \times 0.95 = 2.85 \text{ V}$$
$$V_{5\%} = 3 \times 0.05 = 0.15 \text{ V}$$

Note that the cursors in Figure 14.29 mark these values. Thus, the fall time is:

$$X_2 - X_1 = 6.89 - 1.35 = 5.54 \text{ μs.}$$

Note: when using an oscilloscope, you will need to adjust the cursors yourself.

14.6 THE SPECTRUM ANALYZER

Spectrum analyzers are used to analyze the distribution of signal powers in the *frequency domain*. A spectrum analyzer generates the frequency components of a signal, the power of each frequency component, and the bandwidth occupied by the signal. Signal bandwidth represents the frequencies over which the power of the signal is higher than the noise power level. A spectrum analyzer consists of a filter whose center frequency is swept and the output amplitude over each center frequency is measured as shown in Figure 14.30(a). A typical spectrum analyzer is shown in Figure 14.30(b).

14.6.1 Adjusting the Spectrum Analyzer's Display Window

Similar to an oscilloscope, a spectrum analyzer's display can be adjusted both horizontally and vertically. Remember, that for an oscilloscope, the display range changes by varying the time/div and the voltage/div. In a spectrum analyzer, the display is adjusted by adjusting the center

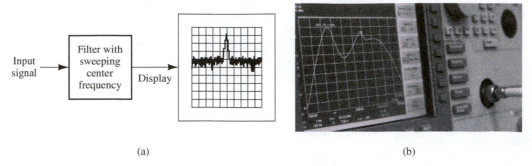

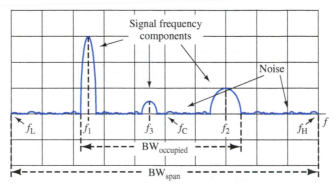

(a)   (b)

FIGURE 14.30 (a) A spectrum analyzer's structure. (b) An Agilent E4408B spectrum analyzer. (Used with permission from © GIPhotoStock/Alamy.)

FIGURE 14.31 A spectrum analyzer screen with various items labeled.

frequency and the span bandwidth. The center frequency, f_C, is displayed at the center of the spectrum analyzer. The span bandwidth, $\mathrm{BW}_{span} = f_H - f_L$, is the bandwidth displayed on the screen: f_L is the lowest frequency, and f_H is the highest frequency (see Figure 14.31). Here, f_L is given by:

$$f_L = f_C - \frac{\mathrm{BW}_{span}}{2} \tag{14.18}$$

and f_H is given by:

$$f_H = f_C + \frac{\mathrm{BW}_{span}}{2} \tag{14.19}$$

The "scale type" and "scale/div" are used to adjust the vertical scale. The scale type determines the vertical scale as linear or logarithmic. The scale/div specifies the difference between grid markings. Its units depend on the selected scale type. For a linear scale, unit is watts (W), and for a logarithmic scale, unit is decibel meter (dB m). Note that dB m refers to 10 times the logarithm of a quantity measured in milliscale. For example, if $P = 10$ mW, then it is equivalent to $P_{dB} = 10 \log 10 = 10$ dB m. The center frequency and span bandwidth can be adjusted to display different frequency ranges of the spectrum.

Peak Search: The frequency components displayed can be easily found using the peak search. It locates the highest peak first. For this peak, the spectrum analyzer gives its frequency and power (in dB m). The "next peak" button allows you to locate the second highest peak. Pushing the "next peak" button repeatedly locates each subsequent next highest peak.

Sweep Time: The sweep time is the time the spectrum analyzer takes to scan through the span bandwidth. A larger sweep time makes the spectrum analyzer refresh its screen less often, while a smaller sweep time results in a higher refresh rate. Other values that a

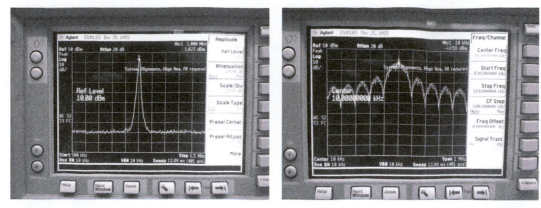

FIGURE 14.32   (a) A single sinusoid; and (b) a single pulse displayed on a spectrum analyzer.

spectrum analyzer can find include the occupied bandwidth, $BW_{occupied}$, the bandwidth that the signal spans, the channel power, the average power across a given channel bandwidth, and the spectral density of the signal (the power per hertz). Two common signals, a sinusoid and a pulse, displayed on a spectrum analyzer, are shown in Figures 14.32(a) and 14.32(b), respectively.

EXAMPLE 14.12   Spectrum Analyzer

An engineer intends to resolve the source of noise coming from an engine. The engineer connects a microphone to the engine and an oscilloscope. The oscilloscope generates the plot as shown in Figure 14.33(a).

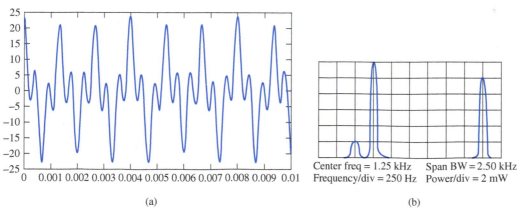

Center freq = 1.25 kHz   Span BW = 2.50 kHz
Frequency/div = 250 Hz   Power/div = 2 mW

(a)   (b)

FIGURE 14.33   Engine noise in (a) the time domain and (b) the frequency domain.

The frequency components cannot be easily determined from the time domain plot. The engineer is interested in studying each frequency component separately to see if they are coming from different parts of the engine. To display the plot in the frequency domain, the engineer connects a microphone to a spectrum analyzer. The plot is shown in Figure 14.33(b). Given the frequency domain plot of Figure 14.33(b), what are the frequency components of the signal?

SOLUTION

The center frequency is 1.25 kHz, the span bandwidth is 2.50 kHz, the leftmost frequency is 0 Hz, and the rightmost frequency is 2.5 kHz. The component frequencies are 500, 750, and 2,250 Hz. The component powers are 2, 12, and 10 mW, respectively.

APPLICATION EXAMPLE 14.13   Communication Systems

An oscilloscope and a spectrum analyzer can be used together in order to simultaneously investigate the behavior of signals (with applications in communications) in the time and frequency domains. The output of a function generator can be connected to an oscilloscope

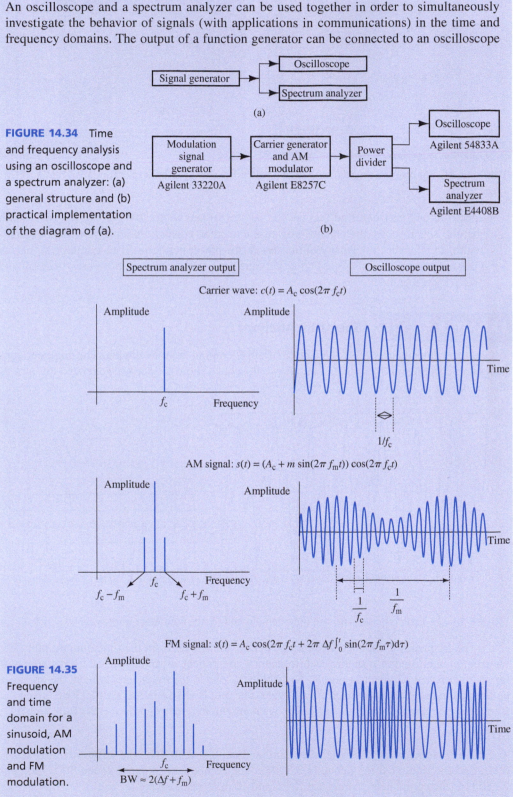

FIGURE 14.34 Time and frequency analysis using an oscilloscope and a spectrum analyzer: (a) general structure and (b) practical implementation of the diagram of (a).

FIGURE 14.35 Frequency and time domain for a sinusoid, AM modulation and FM modulation.

and spectrum analyzer, as shown in Figure 14.34, and the time and frequency domains can then be observed and compared. Note that many signal generators can generate amplitude modulated (AM) and frequency modulated (FM) signals (see Section 14.7).

The time and frequency domains of these signals are shown and compared in Figure 14.35. Using a spectrum analyzer, it is possible to determine the bandwidth (BW) of these signals as shown in the figure. The bandwidth of a communication signal represents the frequency band required to transmit the signal as discussed in Equations (14.18) and (14.19). Comparing the AM and FM signal in Figure 14.35, it is observed that the FM signal needs a higher bandwidth.

EXAMPLE 14.14	**Signal Bandwidth Reading Using Spectrum Analyzer**

Figure 14.36 represents actual oscilloscope and spectrum analyzer outputs for a single sinusoid, AM- and FM-modulated waves. The carrier (center) frequency in all figures is 500 kHz. The modulating wave for both AM and FM signals is the square wave. It is clear that the bandwidth of the FM signal is higher than that of the AM signal.

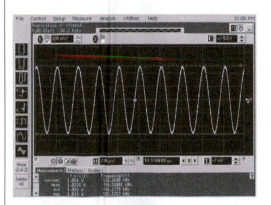

(a)

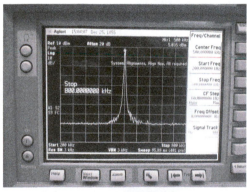

(b)

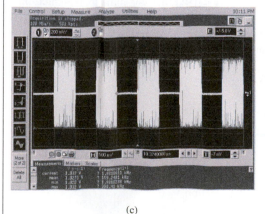

(c)

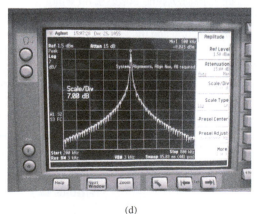

(d)

FIGURE 14.36 Actual oscilloscope and spectrum analyzer outputs: (a) Time domain, single frequency; (b) frequency domain, single frequency; (c) time domain, modulated AM; (d) frequency domain, modulated AM; (e) time domain, modulated am; and (f) frequency domain, modulated AM.

(continued)

| EXAMPLE 14.14 | **Continued** |

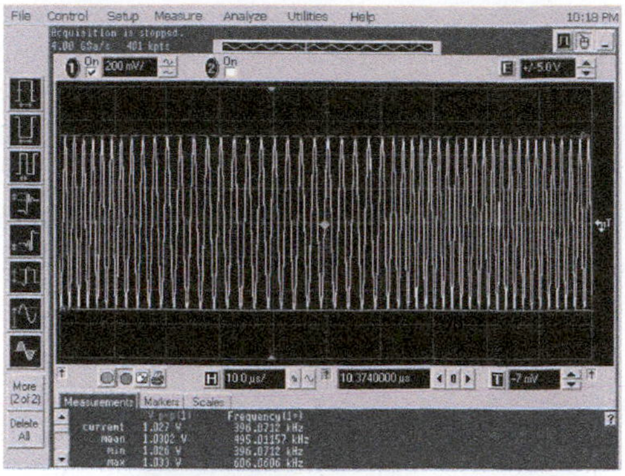

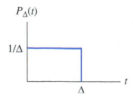

(e) 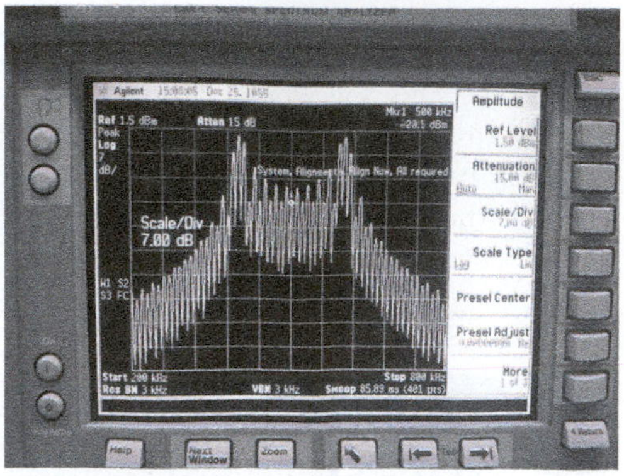   (f)

FIGURE 14.36   (continued)

| EXAMPLE 14.15 | **Spectrum Analyzer Observations** |

In Figure 14.15, we mentioned that the spectrum of sine waves is represented by impulses. However, when a sine wave is applied to a spectrum analyzer, the spectrum shown on the analyzer is not quite impulse, but it is consistent with Figures 14.32(a), 14.33(b), or 14.36(b). Explain the reason.

SOLUTION

Theoretically a sine wave that is stretched from $-\infty$ to $+\infty$ can be represented via an impulse. That is, the Fourier transform of $\sin(2\pi f_0 t)$, $-\infty < t < \infty$ has the $\delta(f\text{-}f_0)$ component. However, in reality, we observe a small portion of a sine wave and not from $-\infty$ to $+\infty$, that is, we observe a sine wave multiplied by a pulse. Mathematically, a small window of observation is represented by $\sin(2\pi f_0 t)\, P_\Delta(t)$ in which $P_\Delta(t)$ corresponds to a pulse with the duration of Δ and the height of $1/\Delta$ (see Figure 14.37). Note that here, Δ represents the observation window of the signal $\sin(2\pi f_0 t)$. As the observation window Δ increases, we expect that the signal represented on the spectrum analyzer becomes closer to the ideal impulse, $\delta(f\text{-}f_0)$. The Fourier transform of $P_\Delta(t)$ is a sinc function and the Fourier transform of $\sin(2\pi f_0 t)\, P_\Delta(t)$ would be the convolution of a sinc function and $\delta(f\text{-}f_0)$, which corresponds to a sinc function centered at f_0 as shown in Figure 14.36(b).

FIGURE 14.37   The pulse function, $P_\Delta(t)$.

For interested readers, it should be mentioned that a delta function is a limit of a pulse with the duration Δ and the height of $1/\Delta$ as Δ approaches zero, that is:

$$\delta(t) = \lim_{\Delta \to 0} P_\Delta(t)$$

Important properties of $\delta(t)$ are:

$$\int_{-\infty}^{+\infty} \delta(t)\, dt = 1$$

And for any function $f(t)$:

$$\int_{-\infty}^{+\infty} f(t)\, \delta(t - \tau)\, dt = f(\tau)$$

EXERCISE 14.4

Calculate the spectrum of $\sin(2\pi f_0 t)\, P_\Delta(t)$. Show that as Δ, the duration of the pulse $P_\Delta(t)$, increases, the width of the spectrum observed on the spectrum analyzer decreases. In other words, the spectrum approaches its ideal shape that is an impulse.

EXERCISE 14.5 RADIO SPECTRUM EXPLORATION

Today, there are many signal sources that transmit or broadcast signals over different frequencies. To explore available frequencies in the air, a dipole antenna can be connected to a spectrum analyzer and the output signal can be observed in the frequency domain. It is interesting to know that when tuning a radio station via a spectrum analyzer via connecting an antenna to its input port, the received signal is not a pure sinusoid. The main reason (see Example 14.13) is the fact that the transmitted signals are modulated. Thus, they are a combination of multiple sinusoids. If you observe the signal using a spectrum analyzer, you will see a broad range of frequencies.

14.7 THE FUNCTION GENERATOR

A *function generator* is used to produce electric waveforms and to test the response of circuits to a specific signal. These can be periodic (repetitive), or once-only signals. Virtually, all function generators can generate periodic waveforms. For periodic waveforms, the operator can choose the waveform shape, amplitude, frequency, and DC offset. An example of a function generator (Agilent 33220A) is shown in Figure 14.38.

A function generator usually has the following controls:

Function: This control selects the desired output waveform. Usual choices include sinusoid, square wave, triangle wave, and sawtooth wave. These basic waveforms are shown in Figure 14.39(a) to (d).

Frequency: This control specifies the frequency of the output waveform. Some function generators allow the operator to choose the period $T = 1/f$ instead. The period of a sinusoid is shown in Figure 14.12.

Offset: This control specifies the DC offset of the periodic signal. This offset is illustrated in Figure 14.11.

Frequency Sweep: This control generates a waveform of the specified type and amplitude, but varies the frequency over time. The operator specifies a start and stop frequency. The function generator starts at the start frequency, and progresses to the stop frequency. This feature is useful for examining the frequency response of a circuit over a range of frequencies.

Modulation: As mentioned in Example 14.13, some function generators also allow modulation of a signal with the generated frequency. Modulation is the process of

FIGURE 14.38 A function generator. (© Agilent Technologies, Inc. 2012. Reproduced with permission. Courtesy of Agilent Technologies, Inc.)

(a)　　　　　　(b)　　　　　　(c)　　　　　　(d)

FIGURE 14.39 Typical function generator waveforms: (a) sinusoid waveform; (b) square wave; (c) triangle wave; and (d) sawtooth wave.

varying the amplitude or the phase of a known sinusoidal signal via a waveform that carries a message (information). This waveform is called baseband signal in the era of communication systems. A special input is allocated to the baseband signal. The modulation may be either amplitude modulation (AM) or frequency modulation (FM). AM is applied to the amplitude of the sinusoidal signal, while FM is applied to the instantaneous frequency of the sinusoidal signal. AM is the most common modulation offered by function generators.

14.8 WHAT DID YOU LEARN?

- Measurement error is defined as the absolute value of the difference between the actual value and the measured value.

- Error percentage $= \left| \dfrac{e}{x_{\text{true}}} \right| \times 100 = \left| \dfrac{x_{\text{true}} - x_{\text{m}}}{x_{\text{true}}} \right| \times 100$ [see Equation (14.2)].

- To calculate the accuracy of a measurement, use Equation (14.3):

$$A = 1 - \left| \frac{x_{\text{true}} - x_{\text{m}}}{x_{\text{true}}} \right|$$

- The arithmetic mean (average) of multiple independent measurements is defined as [Equation (14.5)]:

$$\bar{x} = \frac{x_1 + x_2 + \cdots + x_N}{N} = \frac{1}{N} \sum_{i=1}^{N} x_i$$

  An average of several measurements is considered to be more precise than any single measurement.

- The *standard deviation* is a measure of variation (often due to error) in a set of measurements.

  The standard deviation can be computed as shown in Equation (14.7):

$$\sigma = \sqrt{\frac{1}{N-1} \sum_{i=1}^{N} (x_i - \bar{x})^2}$$

- The galvanometer is the most basic analog electrical meter. It is based on the fact that a magnetic field exerts a torque on the current-bearing coil (see Figure 14.2). A galvanometer can be used to construct analog voltmeters, ammeters, and ohmmeters.

- A *voltmeter* is connected in parallel with the voltage to be measured. An *ammeter* is connected in series with the current to be measured. An *ohmmeter* is connected in parallel with the resistance to be measured, while the resistor is disconnected from the circuit.
- Modern multimeters use digital instruments to measure voltage, current, and resistance.
- The time domain expresses a signal as a function of time. An oscilloscope is used to display a signal in the time domain.
- The frequency domain expresses a signal as a function of frequency. A spectrum analyzer is used to display a signal in the frequency domain.
- A function generator is used to produce waveforms as a source for circuit testing.

Problems

*B refers to basic, A refers to average, H refers to hard, and * refers to problems with answers.*

SECTION 14.2 MEASUREMENT ERRORS

14.1 (B)* Suppose an engineer has two resistors, R_1 and R_2, with corresponding resistances:

$$R_1 = 1000 \ \Omega \pm 5\%$$
$$R_2 = 5000 \ \Omega \pm 10\%$$

What is the minimum and maximum tolerable value of the series and parallel configurations of these resistors?

14.2 (A) For the resistor network of Figure P14.2, find the minimum and maximum tolerable equivalent resistance (R_{eqv}).

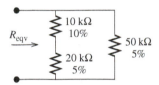

FIGURE P14.2 Resistor network.

14.3 (H) An ohmmeter applies a known voltage to the thermistor and measures the current drawn, I. The resistance is then calculated internally from this voltage and current. To prevent a short circuit between the leads from causing an extremely high current, a small resistor, R_{int}, is placed in series with the leads, as indicated in Figure P14.3.

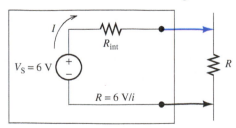

FIGURE P14.3 Ohmmeter circuit.

If the short circuit current is 10 A, then what is the value of R_{int}?
Using the value of R_{int}, find the error percentage given the actual value of R is:
a. $R = 0.2 \ \Omega$

b. $R = 5 \ \Omega$
c. $R = 10 \ \Omega$
d. $R = 50 \ \Omega$
e. $R = 100 \ \Omega$

14.4 (H) The ammeter with a 250 Ω resistance in Figure P14.4 reads a minimum current at 17.7 mA and a maximum current at 18.7 mA with a voltage source of 10 V.
a. Determine the resistance of the resistor, R, with its tolerance percentage and measurement of error percentage.
b. If more resistors, R, are supplied, what is the setup that is able to reduce the reading error?

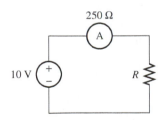

FIGURE P14.4 Circuit for Problem 14.4.

14.5 (H)* A resistor network is shown in Figure P14.5, $R_1 = 5$ kΩ $\pm 10\%$, $R_2 = 10$ k$\Omega \pm 20\%$, $R_3 = 2$ k$\Omega \pm 10\%$, and $R_4 = 8$ kΩ $\pm 5\%$. Calculate the maximum and minimum value of V_4.

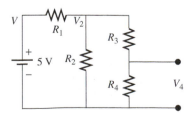

FIGURE P14.5 Resistor network.

14.6 (B) A resistor's resistance is measured resulting in the resistance values in Table P14.6. Assume that the multimeter is unbiased and measurements are independent. Using the values listed in Table P14.6, estimate the true value of the resistor's resistance, R_{true}, and calculate the measurement's standard deviation, σ_R.

TABLE P14.6 Measured Resistance Value (Ω)

No. of meas. (i)	1	2	3	4	5	6	7	8	9	10
Resistance value (R_i)	97.63	97.89	100.11	97.56	99.92	97.74	97.30	99.48	101.91	102.94
No. of meas. (i)	11	12	13	14	15	16	17	18	19	20
Resistance value (R_i)	100.26	101.31	97.66	99.08	99.48	97.57	97.36	101.86	100.02	98.71

14.7 (A)* A square wave has a 50% duty cycle, 1 V peak-to-peak amplitude ($V_{\text{p--p}}$) and 0.5 V offset. What is its RMS amplitude (V_{RMS})? If its offset is 0, what is the corresponding RMS amplitude?

14.8 (B) Assume that you have an ohmmeter that includes a series resistor, R_{int}, to avoid extremely high current. If you increase the value of R_{int}, does the measurement precision increase or decrease?

SECTION 14.3 BASIC MEASUREMENT INSTRUMENTS

14.9 (A) Discuss the suitable cases where the shunt resistor must be connected to the galvanometer to meet the design requirements in:
a. Parallel
b. Series

14.10 (B)* Assume you use a galvanometer to construct an ammeter that measures currents in the range of 0–10 A. The wire resistance of the galvanometer is 200 Ω and a maximum current of 8 mA can flow through the device. Determine the resistance of the shunt resistor connected in parallel with the galvanometer.

14.11 (B) Assume you use a galvanometer to construct a voltmeter that measures voltages from 0 to 10 V. If $i_{\text{gmax}} = 5$ mA and $R_G = 500\Omega$, find the value of shunt resistor, R_{sh}.

14.12 (A) Figure P14.12 shows a simple ohmmeter, made using a galvanometer. If $i_{\text{gmax}} = 2$ mA and $R_G = 1$ kΩ, find the value of the shunt resistor. (Hint: the maximum current occurs when the measured resistance is 0 Ω.)

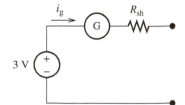

FIGURE P14.12
Ohmmeter.

14.13 (A)* In Figure P14.13, determine the error percentage of the current reading due to the insertion of an ammeter. $R_1 = 3$ kΩ, $R_2 = 5$ kΩ, and the internal resistance of the ammeter is $R_g = 50$ Ω.

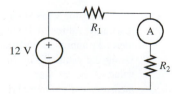

FIGURE P14.13 Circuit for Problem 14.13.

14.14 (A) For the circuit shown in Figure P14.14, determine the error percentage of the voltage reading due to the insertion of a voltmeter. $R_1 = 5$ kΩ, $R_2 = 10$ kΩ, and the internal resistance of the voltmeter is 50 kΩ.

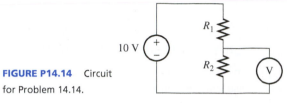

FIGURE P14.14 Circuit for Problem 14.14.

14.15 (A) Figure P14.15a represents a simple voltage divider circuit. A voltmeter with an internal resistance of 1 MΩ is connected to measure V_A, the voltage across R_2. Calculate the voltage measured for V_A and the error from the actual value. If the same voltage divider is constructed using 200 kΩ resistors, as shown in Figure P14.15(b), calculate the measured V_A and the error from the actual value.

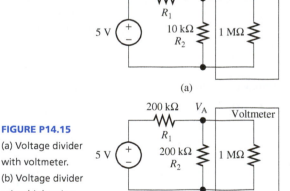

FIGURE P14.15
(a) Voltage divider with voltmeter.
(b) Voltage divider using high-value resistors.

14.16 (H)* A circuit is shown in Figure P14.16; $R_1 = 100$ Ω, $R_2 = R_3 = 150$ Ω. Assume you have an ammeter with 5 Ω internal resistor (R_{int}), and want to calculate the source voltage

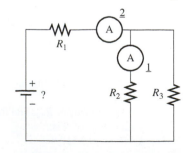

FIGURE P14.16
A complex voltage divider.

by measuring the current at point 1 or 2. To which point should you apply the ammeter to get better accuracy?

14.17 (A) A 5 V source (V_s) with a 5 Ω internal resistor ($R_{int,s}$) is providing power for a circuit whose equivalent resistance (R_{eqv}) is 500 Ω (shown in Figure P14.17). Assume that you use a voltmeter with a 1 MΩ internal resistor ($R_{int,vm}$) to check the voltage on the circuit. When you begin to check the voltage, how will the voltage applied on the circuit change? How much will it change? If the voltmeter's internal resistor is 500 Ω, how much would the change be?

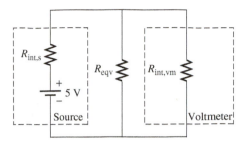

FIGURE P14.17   Using a voltmeter to check circuit voltage.

14.18 (B) Assume you have a galvanometer, its internal resister is 99 Ω, maximum current is 10 mA, and resolution ($I_{re,ga}$) is 0.1 mA (the current numbers that can be read from the galvanometer are 0, 0.1, 0.2, ... , 10 mA). Using the galvanometer to construct an ammeter with current range 0 ~ 1 A, what is the ammeter's resolution ($I_{re,am}$)?

14.19 (B)* Assume you have a galvanometer, its internal resister (R_{int}) is 100 Ω, maximum current ($I_{max,ga}$) is 10 mA, and resolution ($I_{re,ga}$) is 0.1 mA. If you use the galvanometer to construct a voltmeter with voltage range 0 ~ 100 V, what is the voltmeter's resolution ($V_{re,vm}$)?

14.20 (B) If you have an ideal ammeter and an ideal voltmeter, what is the internal resistance of each? Can you construct an ideal ammeter with a galvanometer having 1 Ω resistance?

14.21 (A) A circuit is shown in Figure P14.21, $R_1 = 100$ Ω, $R_2 = 400$ Ω. When you use a voltmeter with 10 kΩ internal resistance to measure the voltage applied on R_1 and R_2, where will you get a larger error percentage?

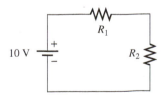

FIGURE P14.21   Simple volt-dividing circuit.

SECTION 14.4 TIME DOMAIN AND FREQUENCY DOMAIN

14.22 (H) The summation of two sine waves is shown on the oscilloscope screen (Figure P14.22). Find the frequency and RMS amplitude of the two sine waves. If you apply the signal to a spectrum analyzer, how much power will you get at each frequency?

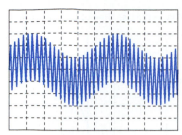

Time/div = 20 ms   Time delay = 0 s
Volt/div = 1 V   Voltage offset = 0 V

FIGURE P14.22   Oscilloscope screen for Problem 14.22.

14.23 (H)* A sine wave $v(t) = 2 \sin(2\pi \times 100t)$ is generated using a function generator, and is put into an oscilloscope. The function generator's internal resistance, R_{int}, is 50 Ω. If you set the oscilloscope's voltage/div to 1 V, and the input resistance, R, to 1 MΩ, what is the peak-to-peak amplitude you will see on the oscilloscope? If you change the oscilloscope's input resistance, R, to 50 Ω, what is the peak-to-peak amplitude of the signal displayed on the oscilloscope screen?

SECTION 14.5 THE OSCILLOSCOPE

14.24 (B) Use the oscilloscope screen in Figure P14.24 to determine the amplitude, A, period, T, frequency, f, and phase, θ, of this sinusoidal signal:

$$f(t) = A \cos(2\pi ft + \theta)$$

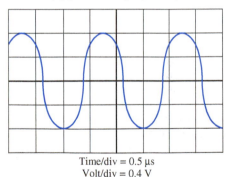

Time/div = 0.5 μs
Volt/div = 0.4 V

FIGURE P14.24   Oscilloscope screen for Problem 14.24.

14.25 (A)* Find the peak-to-peak and offset voltages of the signal displayed on the oscilloscope screen in Figure P14.25.

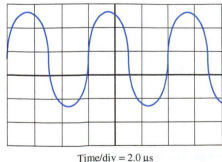

Time/div = 2.0 μs
Volt/div = 100 mV

FIGURE P14.25   Oscilloscope screen for Problem 14.25.

14.26 (A) What is the rise time of the transient in Figure P14.26?

Time/div = 2 μs, Volt/div = 0.5 V

FIGURE P14.26 Transient.

14.27 (A) Write the function for the sine signal shown in Figure P14.27. If it is a cosine signal, what is its function?

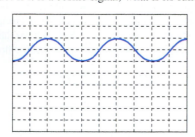

Time/div = 1.0 μs   Time delay = 0 s
Volt/div = 1 V   Voltage offset = 0 V

FIGURE P14.27 Oscilloscope screen for Problem 14.27.

14.28 (H) If you set the oscilloscope's run control to single, time/div to 1 s, delay to 0 s, voltage/div to 1 V, trigger to DC, 1.5 V, how would the signal given in the following equation be displayed on the oscilloscope screen? If you set the trigger to DC, 2.5 V, what signal would be shown on the oscilloscope screen? $v(t) = \sin(\pi t) + 2\delta(t - 1) + 3\delta(t - 2)$.

14.29 (H) Set an oscilloscope to its X–Y mode. Let $x_1 = \sin(\pi t)$ and $y_1 = \cos(\pi t)$, what would be shown on the oscilloscope screen? If $x_2 = \sin(\pi t)$ and $y_2 = 2\cos(\pi t)$, what would be shown on the oscilloscope screen?

14.30 (A)* Assume a load of 50 Ω. Find the power and the frequency of the signal shown in Figure P14.30. If there is a 1 V DC offset in the signal, what will the signal power be?

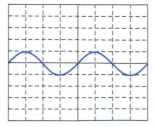

Time/div = 0.1 μs   Time delay = 0 s
Volt/div = 1 V   Voltage offset = 0 V

FIGURE P14.30 Oscilloscope screen for Problem 14.30.

14.31 (B) Assume you have a short time period (18 ms) square wave signal. Its duty cycle is 50%, period is 4 ms, peak-to-peak amplitude is 1 V, and offset is 0.5 V. The entire signal on the oscilloscope screen as shown in Figure P14.31. What is the oscilloscope's setup?

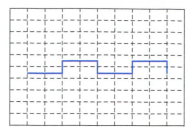

FIGURE P14.31 Square wave on oscilloscope screen.

SECTION 14.6 THE SPECTRUM ANALYZER

14.32 (B) Given the frequency domain plot of Figure P14.32, what are the frequency components of the signal?

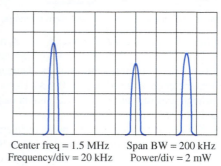

Center freq = 1.5 MHz   Span BW = 200 kHz
Frequency/div = 20 kHz   Power/div = 2 mW

FIGURE P14.32 Frequency domain plot for Problem 14.32.

14.33 (A) Find the frequencies shown in Figure P14.33. Call them $f1$, $f2$, and $f3$ respectively from left to right. In addition, find the power levels of these frequencies.

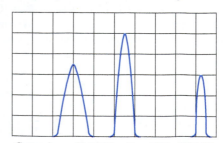

Center freq = 10.0 MHz   Span BW = 2.0 MHz
Frequency/div = 0.2 MHz   Power/div = 5 mW

FIGURE P14.33 Frequency domain plot for Problem 14.33.

14.34 (B)* Is a spectrum analyzer able to measure the phase of the input signal?

14.35 (A) Find the frequency and power (in dBm) of the signals shown in Figure P14.35.

14.36 (A) Find the 3 dB and 6 dB bandwidths of the signal shown in Figure P14.36. The signal XdB bandwidth is defined as $BW_{xdB} = f_{H,xdB} - f_{L,xdB}$ where $f_{H,xdB}$ is the signal's higher frequency, at which the signal power is XdB lower than the signal's largest power; and, $f_{L,xdB}$ is

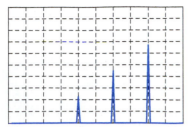

Center freq = 20 MHz   Span BW = 10 MHz
Frequency/div = 1 MHz   Power/div = 2 mW

FIGURE P14.35 Signals in the frequency domain for Problem 14.35.

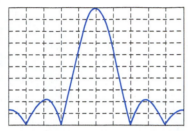

Center freq = 2.5 GHz   Span BW = 1 GHz
Frequency/div = 100 kHz   Power/div = 3 dB

FIGURE P14.36 Signals in the frequency domain for Problem 14.36.

the signal's lower frequency, at which the signal's power is XdB lower than the signal's largest power.

14.37 (H)* Write the function for the signal shown in Figure P14.37, assuming the load of the spectrum analyzer is 50 Ω.

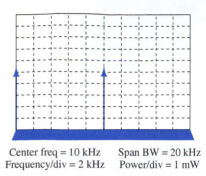

Center freq = 10 kHz   Span BW = 20 kHz
Frequency/div = 2 kHz   Power/div = 1 mW

FIGURE P14.37 Signals in the frequency domain for Problem 14.37.

SECTION 14.7 THE FUNCTION GENERATOR

14.38 (B) Use a function generator to generate a square wave with 2 V high level, −1 V low level, 50% duty cycle, and 1,000 Hz frequency. What is the function generator's setup and what is the offset of the square wave?

14.39 (A) Use a function generator to generate a sine wave with 2 V_{RMS}, 2000 Hz frequency, and zero offset. What is the function generator's setup and what is the peak-to-peak amplitude of the generated signal?

14.40 (A)* A function generator's output range is 20 mV_{p-p}–20 V_{p-p} in an open circuit and there is a series output impedance (R_{int}) of 50 Ω in the function generator. What is the output amplitude range of the function generator when the output load (R_{load}) is 150 Ω?

CHAPTER 15

Electrical Safety

(Used with permission from Jhaz Photography/Shutterstock.com.)

15.1 INTRODUCTION

The preceding chapters covered the various applications of electricity. This chapter discusses safety issues regarding electricity. In addition to electrical shock, this chapter addresses hazards relating to electromagnetic (radio) waves, electrical arcs, and explosive atmospheres. Moreover, the chapter presents the National Electric Code (NEC) and defines its role and regulations that promote electrical safety. The entire chapter can be considered as an application chapter for many concepts presented in this book.

15.2 ELECTRIC SHOCK

Electrical shock is caused by bodily connection to two points that have different electrical potentials. In other words, electric shock occurs when a person simultaneously touches two points with a voltage drop across them. The *ground* or the *Earth* can be one of the contacts. As a result, touching a live wire and the ground at the same time can deliver an electrical shock. Electric shock can be lethal when it passes through the heart or the head. Electric current passing through the head affects the brain, while current passing through the chest can stop the heart. For these reasons, if a worker needs to test a live circuit, one hand can be placed in the worker's pocket so that if something goes wrong, there is less likelihood the current will pass through the worker's heart.

15.2.1 Shock Effects

Alternating current (AC) is the main source of most electrocutions. AC current can interfere with the natural electrical pulse of the heart. This interference happens even at low voltage levels. The AC causes ventricular fibrillation, which can quickly lead to death. However, direct current can also be fatal at high current levels. With either AC or DC, the electrical *current* passing through the body is the main cause of death. Whether a shock victim lives or dies depends on how much current flows through his or her body. It only takes 100 mA or 0.1 A to disrupt a victim's heart. Low currents can send the victim's heart into ventricular fibrillation.

Higher currents can stop a victim's heart. However, it is more difficult to treat ventricular fibrillation than restarting a stopped heart. Therefore, low-current shocks can be even more dangerous than high-current shocks. Table 15.1 summarizes the effects of various current levels on the human body.

Electric shock is not the only hazard posed by electrical equipment. Electric energy generates electromagnetic radiation proportional to the magnitude of the current. This radiation can be hazardous, as covered in Section 15.3.2.

According to Ohm's law, the amount of current flowing through the body depends on the body's resistance. A low body resistance and a moderate voltage will produce a dangerously high current through the body. Factors that can lower bodily resistance include wetness or moisture on the skin, bare feet on bare earth (dirt), and broken or bleeding skin. Normal hand-to-hand resistance is about 300 kΩ. But the resistance with wet skin is less than 20 kΩ, which means the current is 15 times greater. Having bare feet on bare ground also gives a better electrical contact with ground, lowering resistance even further.

However, electrical contact to broken or bleeding skin yields the highest, most dangerous current levels. Most of the body's resistance to electricity is found in the skin. Due to the highly conductive blood, water, and nervous system, the internal resistance behind the skin is very low. As a result, the body's resistance is lower at any point of broken skin. Touching electrical contacts with bleeding wounds or stabbing electrified leads into the skin can cause death, even when the contact is with low-voltage DC because the resulting current is so high.

TABLE 15.1 Electric Shock Effects on the Human Body

Cadick, J. 1994. *Electrical Safety Handbook*, 1.4. New York: McGraw-Hill.

Current	Effect on Body	Potential for Fatality
<1 mA	None	None
1 mA	Perception threshold, mild sensation	None
1–10 mA	Mild to painful sensation	None
10 mA	Paralysis threshold, cannot release hand grip	Minimum level that has a slight risk of fatality
30 mA	Respiratory paralysis (cannot breathe)	Frequently fatal
75 mA	Fibrillation threshold 0.5%	Heart action disrupted, possibly fatal
250 mA	Fibrillation threshold 99.5%	Usually fatal (very high risk)
4 A	Heart paralysis threshold (heart stops)	Possible to revive victim; fatal if not intervened quickly
5 A	Tissue burning	Yes, if vital organs are burned

| EXAMPLE 15.1 | **Fatal Voltage** |

If a person's body resistance is 100 kΩ, what voltage can potentially be fatal?

SOLUTION

The minimal current for a fatality is in the order of 0.1 A. According to Ohm's law:

$$V = 100 \text{ k}\Omega \times 0.1 \text{ A} = 10 \text{ kV}$$

| EXAMPLE 15.2 | **Electric Current in a Swimming Pool** |

An electrical power cord falls into a swimming pool, as shown in Figure 15.1. If the pool water has the resistivity of 175 kΩ m, find the current, I, that flows through the water. Assume that each longitude cross section of the pool acts as an insulator. In this case, we can consider the simplified model shown in Figure 15.2 for the problem. Here, the two dashed planes represent the two sides of the water resistor. The separation of these two planes is assumed to be 2 cm.

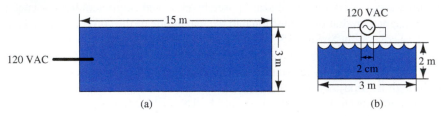

FIGURE 15.1   Swimming pool configuration for Example 15.2. (a) Overhead view; (b) cross-sectional view.

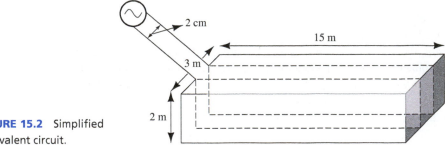

FIGURE 15.2   Simplified equivalent circuit.

SOLUTION

The electrical current flows through a conductor consisting of water. This conductor is 2 cm long and has a cross-sectional area of $15 \times 2 = 30 \text{ m}^2$.

$$R = \frac{175,000 \times 0.02}{30} = 116.7 \ \Omega$$

Therefore,

$$I = \frac{120}{116.7} = 1.029 \text{ A}$$

Note that this current is 10 times the lethal limit that is in the order of 0.1 A (see Table 15.1). Also, note that the pool wall can act as a ground, causing current to flow through the entire body of water, and not just the "conductor" section.

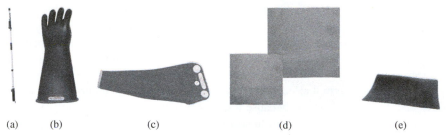

(a) (b) (c) (d) (e)

FIGURE 15.3 Safety tools: (a) hot stick; (b) insulating gloves; (c) insulating sleeve; (d) insulating blanket; and (e) insulating mat. (Photos courtesy of Salisbury by Honeywell.)

15.2.2 Shock Prevention

To prevent shock, prior to working on equipment, all power should be turned off if at all possible. If de-energizing the circuit is not practical, there are several things that can be done to significantly reduce the risk of shock. One typical safety measure is to wear rubber gloves and rubber sleeves to prevent electrical contact with the body. A rubber sleeve covers a worker's arms and shoulders. Use of rubber mats and rubber boots, specifically designed to serve as insulation, can also prevent a path to a ground. Also rubber blankets and covers can be used to temporarily cover energized equipment. Rubber line hose can be used as a temporary insulator to cover exposed power lines. Hot sticks (long, insulated sticks) can be used to handle energized wires and other equipment (see Figure 15.3).

Insulated tools are also available for use with energized equipment. Figure 15.3 shows some of this safety equipment. These are a few of the safety products available that are designed to reduce the risk of shock from working on live equipment.

15.3 ELECTROMAGNETIC HAZARDS

Radio frequency (RF) hazards are due to electromagnetic waves or radio waves. Electromagnetic waves are invisible (with the exception of visible light) and are used for all wireless communication applications, including radio, TV, cellular phones, wi-fi, and RADAR. RF hazards result from the emission of radio frequency radiation. A few common devices that emit radio frequency radiation include cell phones, microwave ovens, wi-fi, and two-way radios. Radio frequency energy can cause thermal effects such as heating and burns, and, at certain frequencies, the radiation can even cause cancer. There are two primary types of RF hazards—low-frequency and high-frequency RF hazards. High-frequency RF energy is much more dangerous than low-frequency RF energy.

15.3.1 High-Frequency Hazards

High-frequency RF radiation can cause cancer. One example of a source of high-frequency hazard is a radar transmitter. RADAR stands for RAdio Detection And Ranging. The RADAR acronym has become so common it has now simply become its own commonly recognized word, *radar*. These systems usually consist of a side-by-side transmitter and receiver. A radar transmitter transmits a high-power burst of RF energy. The reflection of this signal from targets is processed by the receiver for detection and localization purposes. Extended exposure to radar signals has been linked to a number of health problems, including leukemia and brain tumors. A 1953 study at Hughes Aircraft Corporation found "excessive amounts of internal bleeding, leukemia, cataracts, headaches, brain tumors, heart conditions, and jaundice in those employees working with radar [1]."

Exposure to high-frequency RF signals, above 10^{17} Hz, is another cause of health problems. These frequencies include gamma, beta, and alpha radiation, as well as some X-ray frequencies. Ionization causes electrons to be stripped from atoms. This can produce molecular changes, potentially leading to DNA and biological tissue damage. Ionizing radiation is well known to cause cancer, but exposure to this radiation is not likely in most workplaces.

RF energy can also cause thermal effects, the heating of tissue, and even burns. Exposure to high levels of RF energy can be harmful because it can heat biological tissue rapidly. Power density levels of 1 to 10 mW/cm^2 can result in measurable heating of tissue. Very high power levels, 100 mW/cm^2 or higher, can easily cause heating of tissue and increasing of body temperature. This is especially of concern with the eyes and testes, as these areas have less blood flow, and thus less ability to dissipate the extra heat.

There has been much debate over the health effects of using cell phones. Cell phones use frequencies of 850, 900, or 1850 MHz, which are in the microwave (RF) range for wireless communication. Cellular systems consist of two main components: (1) base stations (cell towers), which communicate with a number of mobile users that are located in their coverage area simultaneously and (2) mobile stations (users). The distance of base stations and their output power (which is typically less than 100 W) avoids any considerable effect on users. However, there is a debate on the relationship of cell phones and cancer.

Is there a link between cell phones and cancer? The transmitting power of a cell phone is often in the order of 1 W. However, when you hold the cell phone close to your head, it can affect your body. Some studies show that extensive use can cause change in a user's DNA, a precursor to cancer [2]. The Austrian Chamber of Physicians has claimed that long-term use can cause cancer and recommends that children should never use it [3]. Studies in 1999 and 2001 showed that long-term use could disrupt brain wave patterns [4] and even cause eye cancer [5]. There are several class-action lawsuits against the cell-phone industry, claiming that the devices can cause brain tumors.

However, other studies conclude that there is no cell phone–cancer link. The American Cancer Society denies any link. "The link between cell-phone use and cancer, especially brain cancer is a myth" says Ted Gansler, Director of Medical Content, American Cancer Society [6]. Anthony Swerdlow of the Institute of Cancer Research concludes that there is no substantial risk of cancer after the first decade of cell-phone use [7]. The general consensus between both sides is that more research is needed for a definite answer.

Cell phones give off radiation at all times when turned on, even when not in use. However, more radiation is emitted when talking or text messaging. Also, when far from a base station, more power is required for communication even when not using the phone. To determine how much radiation your phone emits, check its specific absorption rate (SAR). The SAR for each specific phone model can be found at http://www.fcc.gov/cgb/sar/.

Research has also suggested a link between carrying a cell phone in one's pocket and male infertility. Ashok Agarwal, Ph.D., Director of the Center for Reproductive Medicine at the Cleveland Clinic, recently completed a study in which cell phones were set down for 1 h in talk mode, next to sperm samples in test tubes. He found that the sperm's motility and viability were significantly reduced, and levels of harmful free radicals increased after exposure [8].

Radio and TV transmitters pose similar risks to cell phones. These use frequencies from 500 kHz to 1 GHz.

EXAMPLE 15.3   **Power Density and High-Frequency RF Hazard**

For a radio transmitter, the power transmitted is 100 kW and the antenna gain is 50. (a) If an employee is standing 2 m from the antenna, what is the power density of the radio frequency radiation passing through his body? (b) What if the employee stands 20 m from the antenna?

The power density is given by:

$$D = \frac{P_t \cdot G_t}{4\pi d^2}$$

where D is the power density (W/m^2); P_t is the transmitted power; G_t is the antenna gain; d is the distance (m).

SOLUTION

a. Considering a 2-m separation, the power density corresponds to

$$D = \frac{100{,}000 \times 50}{4\pi 2^2} = 99.472 \text{ kW/m}^2$$

b. Considering a 20-m separation, the power density is:

$$D = \frac{100{,}000 \times 50}{4\pi 20^2} = 994.72 \text{ W/m}^2$$

Notice that the extra distance in this scenario greatly reduces the power density. If the distance becomes 10 times greater, the power density would reduce by 100 times.

EXAMPLE 15.4   **Power Density and RF Hazard**

A person holds a cell phone 5 mm away from his head. If the phone is transmitting 3 W, and the antenna gain is 10, what is the power density at the surface of his head? Should this power density level be of concern?

SOLUTION

Using the equation introduced in Example 15.3:

$$D = \frac{P_t \cdot G_t}{4\pi d^2} = \frac{3 \times 10}{4\pi (0.005)^2} = 95.49 \text{ kW/m}^2$$

A power density of 100 mW/cm^2 is considered very high. This value must first be converted into mW/cm^2.

$$95{,}490 \text{ W/m}^2 \times \frac{1 \text{ m}^2}{100^2 \text{ cm}^2} \times \frac{1000 \text{ mW}}{1 \text{ W}} = 9549 \text{ mW/cm}^2$$

9549 mW/cm^2 is much greater than 100 mW/cm^2. So this radiation should be of great concern to the user. Holding a cell phone about 5 cm away from head could reduce this density to 95.4 mW/cm^2.

15.3.2 Low-Frequency Hazards

High-tension power lines emit RF radiation at a frequency of 60 Hz. Is this radiation dangerous? A 2001 California health department review found that living near high-voltage power lines resulted in "Added risk of miscarriage, childhood leukemia, brain cancer, and greater incidence of suicide [9]." But, ultimately, they added that risk was a "very small increased lifetime risk."

How close can people safely live to power lines? It depends on the voltage and current through the power lines. How can the voltage of a power line be determined? The "High Voltage" signs near power lines do not typically give the voltage. One way of determining the voltage is by observing the size and other characteristics of the transmission towers carrying the lines.

The values in Table 15.2 indicate that, for higher voltages, the towers are generally taller; however, this is not always the case. Note the overlapping as evidenced by the minimum and

TABLE 15.2 Typical Transmission Tower Specifications [10]

Voltage (kV)	Conductor Spacing (m)			Transmission Tower Height (m)		
	Median	*Min*	*Max*	*Median*	*Min*	*Max*
345	7	5	10	32	18	43
500	9	6	13	39	30	52
765–800	13	12	16	42	35	53

maximum heights. The conductors (wires) for higher voltages are usually spaced farther apart. The conductor spacing is the minimum distance between any two conductors on a tower. Therefore, the voltage level of a power line can be roughly determined by (1) the distance between the conductors and (2) the height of the tower. Higher voltage levels utilize larger transmission towers and wider conductor spacing. Figure 15.4 shows transmission towers bearing different voltages, and Figure 15.5 shows close-up views of the power lines.

As can be seen from Figure 15.5, some conductors are combined in multiple conductors. Figure 15.5(b) shows two-conductor pairs, and Figure 15.5(c) and (d) shows four conductors. Paired conductor lines carry more power as compared to single conductor lines, for example, four conductors can carry four times more current. As discussed in Chapter 9, current flows on the surface of the conductor. To maximize current, it is desired to maximize conductor surface area while minimizing the conductor size. Thus, more current can flow through four individual conductors than can flow through a single large one with a diameter equivalent to the combination of four individual conductors.

(a)

(b)

(c)

(d)

FIGURE 15.4 (a) 12.47-kV line (Courtesy of Shah & Associates, Inc., www.shahpe.com.); (b) 345-kV line (Courtesy of Argonne National Laboratory.); (c) 500-kV line (Courtesy of © TebNad/Fotolia.); (d) 765-kV line (Courtesy of American Electrical Power.).

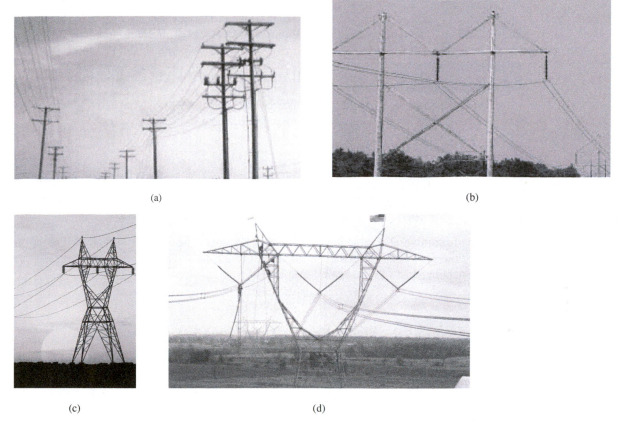

FIGURE 15.5 Close-up views of (a) 12.47-kV line (Courtesy of Shah & Associates, Inc., www.shahpe.com.); (b) 345-kV line (Courtesy of Argonne National Laboratory.); (c) 500-kV line (Courtesy of © TebNad/Fotolia.); (d) 765-kV line (Courtesy of American Electrical Power.).

The electromagnetic radiation is directly proportional to the current level through the power lines. Table 15.3 represents typical current ratings for different voltage power lines. Please note that these are the rated currents; the current at a given time more or less depends on the load on the wire at that time.

The magnitude of the magnetic field produced by a (straight) power line is given by:

$$B = \frac{\mu_0 \times i}{2\pi r} \tag{15.1}$$

where B is the magnitude of the magnetic field, in tesla (T); μ_0 is the magnetic permeability of free space; i is the current (A); r is the distance from the conductor (m).

The strength of the magnetic field is measured by a device called a magnetometer (also called a guassmeter). A typical magnetometer is shown in Figure 15.6.

How much radiation is safe? In the United States, no national limit has been set. The US Federal government does not acknowledge electromagnetic radiation as dangerous and, therefore,

FIGURE 15.6 Typical magnetometer. (Photo courtesy AlphaLab, Inc.)

TABLE 15.3 Current Levels Through Different Power Lines

Voltage (kV)	Current (A)	Safe Distance (m)
12.47	100	~5
345	1000–2500	25
500	2000–3000	30
745	2500–4000	40

has not issued any regulations concerning building near power lines. However, some US states limit radiation to 200 mG. One tesla equals 10,000 Gauss. The government of Sweden has stricter limits on health issues and has set the maximum magnetic field exposure limit to only 2.5 mG.

EXAMPLE 15.5	Power Line Hazard; Magnetic Field Calculation

If a house is placed 50 m from a 500-kV power line bearing 3000 A, what is the strength of the magnetic field? Is this field safe (according to the US state standards)?

SOLUTION

Using Equation (15.1):

$$B = \frac{4\pi \times 10^{-7} \times 3000}{2 \times \pi \times 50} \times 10{,}000 = 120 \text{ mG}$$

The maximum allowable in some states is 200 mG. This value is safe according to this specification. However, the Swedish maximum is 2.5 mG. This value greatly exceeds the Swedish maximum.

EXAMPLE 15.6	Power Line Hazard; Magnetic Field Calculation

Find the maximum magnetic field magnitude directly under a 745-kV line.

SOLUTION

From Table 15.2, the minimum height of a 745-kV transmission tower is 35 m. From Table 15.3, the maximum current through this line is 4 kA. Therefore,

$$B = \frac{4\pi \times 10^{-7} \times 4000}{2 \times \pi \times 35} \times 10{,}000 = 228.6 \text{ mG}$$

EXAMPLE 15.7	Power Line Hazard; Magnetic Field Calculation

According to the Swedish standard, what is the minimum distance homes should be placed from a 12.47-kV line rated at 500 A? As 12.47-kV lines are often located in or near neighborhoods, is this distance practical?

SOLUTION

Using Equation (15.1) and the Swedish standard of 2.5 mG:

$$B = \frac{4\pi \times 10^{-7} \times 500}{2\pi r} \times 10{,}000 = 2.5 \text{ mG}$$

The safe distance, r, can then be calculated from this equation:

$$r = 400 \text{ m}$$

The safe distance, 400 m, is not practical because city lots are generally less than 100 m in length. Even if a house is located in the center of the lot and the power lines run along the property line at the rear of the lot, there is not enough room to implement this standard. Accordingly, the lines are placed under the ground. The ground reduces the RF intensity.

EXERCISE 15.1

It is notable that in some cities, it is required that the power lines are located under the ground. This improves the sight by removing the towers; it may save installation costs and reduce the radiation effects. You are encouraged to investigate how installing the cables under the ground would impact their radiation. Typically, considering the lines are located 1 m under the ground, calculate the magnetic field at the distance of 10 to 50 m.

15.3.3 Avoiding Radio Frequency Hazards

Here are some tips to avoid RF hazards:

- Use a "hands-free" device instead of holding a cell phone directly to your head.
- Whenever possible, store your cell phone away from your body to minimize exposure.
- Avoid standing near radio antennas and transmitters.
- Do not live in close proximity to high-tension power lines.

15.4 ARCS AND EXPLOSIONS

Explosions in any context can be a serious safety concern. Safety precautions can help to reduce or eliminate the risk of explosions. Because this chapter covers electrical safety, it will focus mostly on electrical causes of explosions and how to prevent them. Explosions are caused by a mixture of explosive materials and an ignition source. The most common electrical cause of explosion ignition is arcs. To reduce explosion hazards, both the amount of explosive materials in the atmosphere and the risk of arcs can be reduced.

Factors contributing to an explosive atmosphere include flammable gasses and organic compounds. These materials, when released into the air (above certain concentrations) may pose an explosion hazard if exposed to an ignition source. These materials include (but are not limited to) acetylene, ethyl ether, gasoline, alcohols, acetone, methane, propane, and hydrogen. It is also important to note here that high concentrations of oxygen increase the risk of fire and explosion. While oxygen itself cannot explode, it is required for any combustion to occur. Increased oxygen levels result in increased potential for combustion and explosions.

15.4.1 Arcs

An explosive atmosphere does not result in an explosion unless an ignition source is available. Arcs and sparks are common ignition sources. A spark is basically a short-lived arc. An arc occurs when a substantial amount of current flows through superheated air. The energy flowing through the air causes the air to break down and forms plasma, which is an ionized gas, that is, an electrical conductor. The current then flows through this plasma mixture consisting of a vaporized electrical conductor and ionized air particles.

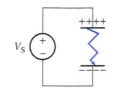

FIGURE 15.7 Electrical causes of arcs.

An electric arc occurs when two conductors having a large potential difference are located in close proximity, as shown in Figure 15.7.

High voltage causes the gas between the conductors to ionize, allowing the current to arc through it, resulting in a spark. Lightning (Figure 15.8) is an example of a large arc.

Arcs release extremely high levels of thermal energy. An arc can generate temperatures as high as 50,000 K at its terminals. These high temperatures can create fatal burns at distances of more than 8 ft from the arc. The amount of energy (and heat) in an arc is proportional to the maximum short-circuit power at that point of the circuit. Low-voltage systems are just as susceptible to arcing as high-voltage systems.

FIGURE 15.8 Lighting is a large arc. (Used with permission from Jhaz Photography/Shutterstock.com.)

When ionized, oxygen (O_2) is converted to ozone (O_3). The unique smell caused by sparks is the ozone, that is, the result of the spark. While ozone is toxic, minimal amounts are produced by most sparks, and it quickly decays back to oxygen (O_2). So, ozone generated by sparks does not pose any health hazard. Ionization is also caused by the electric field around power lines. This effect is called *corona discharge*. The crackling or buzzing sound typical of power lines is caused by corona discharge. While this effect does not present a health hazard (corona discharge is used to generate ozone by air purifiers), it is undesirable in power lines because it results in energy loss.

EXAMPLE 15.8   Arc Power

Suppose an arc occurs across the circuit breaker between points A and B in Figure 15.9. (Assume the breaker is open.) (a) What is the maximum power available to the arc? (b) What is the maximum power of an arc between A and the ground?

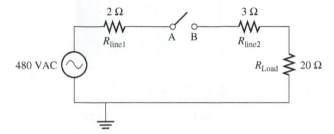

FIGURE 15.9 Circuit for Example 15.8.

SOLUTION

a. An arc across points A–B acts as a short circuit across these points. The maximum current through this short circuit is simply the current through the series circuit.

$$I_{SC} = \frac{480}{2 + 3 + 20} = 19.2 \text{ A}$$

   The maximum voltage occurs when the switch is open, that is, 480 V. As a result, the maximum power is:

$$P_{max} = 480 \times 19.2 = 9.22 \text{ kW}$$

b. Since an arc between point A and the ground acts as a short circuit between these two points, the current only flows through the 2 Ω line resistance.

$$I_{SC} = \frac{480}{2} = 240 \text{ A}$$

   The maximum voltage is 480 V. Therefore,

$$P_{max} = 480 \times 240 = 115.2 \text{ kW}$$

A very common electrical cause of arcs is transients in high-voltage circuits. A transient can occur during the changing of the state of a circuit, such as by flipping a switch or adding a resistor. A simple switch and light bulb circuit (such as a ceiling light or a flashlight) has small inductances contained in both the connecting wires and the bulbs (see Figure 15.10). Recall from Chapter 4 that an inductor resists changes in current. If the current changes, the inductor tries to maintain the current at the previous level.

When a switch is turned off, the inductances try to maintain the previous current level. However, because the circuit is open, this is not possible. Ohm's law states that $V = I \times R$. Opening the switch changes the circuit resistance to infinity ($R = \infty$). Because the inductor

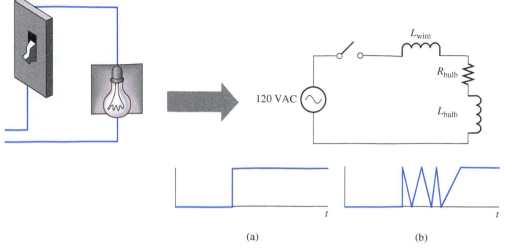

FIGURE 15.10
Equivalent model for a simple circuit shown to the left.

FIGURE 15.11 (a) Ideal switch response; (b) switch exhibiting switch bounce.

tries to maintain the current at the previous level, it produces a brief voltage across the circuit of $I \times R = I \times \infty$. This voltage appears across the break in the circuit, that is, across the switch. Therefore, the inductances in the circuit cause a brief spike in the voltage across the switch contacts as they open, which can generate sparks.

A switch may also create arcs when it is closed. This is due to the switch bounce effect. Because a switch has mechanical contacts, it doesn't turn on immediately, but fluctuates between off and on for a brief period of time (on the order of microseconds). For more details on switch bounce, see Examples 5.2, 10.18, and 14.10. This rapid fluctuation, in combination with the closeness of the contacts, can create sparks. Figure 15.11 compares the ideal switch response with what is observed in switch bounce.

A switch can also create a spark when it closes. However, the sparks when it opens are usually larger than those occurring when it closes due to the high voltage involved. In high-voltage systems, the narrowing gap between the switch contacts may allow current to arc across the gap, due to the high voltage. For this reason, when power is reconnected to a large city, it is not connected to the entire city all at once. Rather, it is reconnected sequentially to one small subset of the city at a time. This reduces the current load handled by each reconnect switch and, therefore, the maximum arcing power.

Switches are not the only cause of electric sparks. Sparks can occur at any moving contact and even in fixed contacts. Fixed contacts include motor contacts, screw terminals, wire connections, and contacts in outlets. Welding equipment by its very nature is a great cause of sparks and should not be used anywhere near explosive atmospheres. Another cause of sparks is friction between moving surfaces.

15.4.2 Blasts

Arcs can cause *blasts* or electrical explosions. An arc superheats air instantaneously. This causes a rapid expansion of air, resulting in air pressures of 100 to 200 lb/ft^2. Note that an explosive atmosphere is not required for a blast. A blast can occur without any outside fuel. However, an atmosphere containing explosive elements will make the blast more severe. Also note that not all arcs are accompanied by a blast.

While sparks are a common trigger of explosions, explosions can also be ignited by heat and fire sources, such as cigarettes, matches, candles, and pilot lights.

15.4.3 Explosion Prevention

To prevent explosions, an explosive atmosphere must be avoided. Proper ventilation can be used to remove explosive material from the air. Also, it is important to keep flammable materials

FIGURE 15.12 Safety switches (© tarczas/ Alamy).

tightly sealed to prevent the release of fumes resulting in an explosion hazard. This reduces or eliminates the fuel required for an explosion. To reduce the explosion risk, it is also necessary to prevent sparks or keep sparks that can ignite an explosion and the explosive environment separate. Because it is impossible to prevent the sparks from occurring across switch contacts, the sparks must be isolated from the surrounding environment. To do this, safety switches are used to isolate the switch contacts from any nearby explosive environment.

A *safety switch* is a special spark-resistant switch that must be used where explosion hazards exist (see Figure 15.12). The switch has completely enclosed contacts, which isolate any sparks from the outside environment. This prevents transient sparks from igniting explosions.

In conclusion, colocated explosive materials and ignition sources of explosions should be avoided. The ignition source is usually a spark. To reduce this explosion hazard, both the explosive material and the spark hazard must be reduced.

15.5 THE NATIONAL ELECTRIC CODE

The National Electric Code (NEC) provides standards and requirements regarding electrical wiring and safety in the United States. It is published by the National Fire Prevention Association (NFPA). The NEC is not a law, but its use is commonly mandated by state and local codes. It is intended to make use of electricity safer by reducing or eliminating most electrical safety hazards. Electricians are required to follow the NEC when wiring a building. When they are finished, a building inspector ensures that the wiring is "up to code." If it is not, it must be fixed before the building is occupied.

The NEC was established in 1897. The NEC focuses on electric utilization systems and does not address generation, transmission, or distribution systems. The NEC also does not apply to ships, trains, mines, or communications equipment that are under the control of a communication utility.

The NEC covers all industrial, commercial, and residential electric systems. The NEC covers four basic installation types: electrical conductors within or on public or private buildings or structures, including recreational vehicles and mobile homes, conductors and equipment that connect to electric supplies, any other outside conductors and equipment, and fiber optic cable [12].

15.5.1 Shock Prevention

The NEC offers guidelines aimed at preventing shocks by requiring proper polarization, grounding, and requiring special outlets called *ground fault circuit interrupters* (GFCIs) or *arc fault circuit interrupters* (AFCI) (discussed in Section 15.5.2.2) in certain areas.

15.5.1.1 POLARIZATION

Figure 15.13 shows a typical outlet used in the United States. You can see that the neutral slot is longer than the hot slot. This is called *polarization*. It is intended to allow a plug to be inserted into the outlet only in one way. By its nature, AC is not polarized, because it is constantly reversing direction. With this in mind, what is the purpose of polarization? Some electrical devices, if they do not have a grounding prong, use the neutral or return as the ground. The neutral wire is grounded at the circuit breaker panel. Polarization ensures that the internal ground is not connected to the hot conductor to better avoid any shock hazard.

FIGURE 15.13 Slots on a standard outlet in the United States. (Used with permission from © B.A.E. Inc./Alamy.)

Hot

Neutral

Ground

APPLICATION EXAMPLE 15.9   Polarization

Suppose a man touches a metal toaster and is shocked. The toaster, having a polarized plug, is inspected, and nothing is found amiss. The inspector then looks at the outlet. He measures 120 V between the long slot and ground, and 0 V between the short slot and ground. What is the problem?

SOLUTION

The long slot is the neutral connection. This slot should be connected to ground at the circuit breaker panel, thus having a voltage of 0 V with respect to ground. The short slot is the hot connection and should have a voltage of 120 V with respect to ground. The inspection has revealed that the measurements are the opposite of what they should be. Therefore, the outlet is wired backward, with the hot and neutral wires swapped.

15.5.1.2 EQUIPMENT GROUNDING

Equipment grounding connects the circuit of a piece of equipment electrically to the ground, or the earth. The NEC details the proper methods to achieve this ground connection. Some of these ways include putting a rod into the ground or clamping a wire to a cold water pipe, which ultimately makes its way back to the ground. The NEC must be followed to ensure that the ground connection works properly. The outlets must be wired properly so that their ground slot is actually connected to a ground. Figure 15.13 shows the grounding slot on a standard outlet. This slot is used to connect metal cases and chassis of electrical devices such as computers, refrigerators, and microwaves to the ground.

The electrical wiring is grounded at one point, typically at the main circuit breaker panel. This serves to prevent ground loops (see Section 11.4). The return or neutral conductor is also grounded along with the ground conductor. It is typical to select the return path conductor as the reference node (ground) in electronic circuits (see Section 3.5).

Grounding prevents electric shock by redirecting loose current (e.g., from a broken wire resting on a metal case) to the ground. Because the resistance through the ground connection is much lower than the connection through your body to ground, the current will take the path of least resistance and travel through the ground wire to ground, and not through you.

APPLICATION EXAMPLE 15.10   Grounding

Figure 15.14 presents an example of how grounding works. The black wire (which is the incoming or hot wire) has broken loose and is touching the metal case. When proper grounding is used, this stray current is redirected to the ground (as shown by the dashed arrows) and the shock hazard is eliminated. When there is no grounding or grounding is not properly implemented, the stray current is not redirected, the housing is energized, and an unseen shock hazard is created. An example of this is using a grounding adapter to plug into a two-slot outlet and not connecting the grounding tab to the screw (see Example 15.12).

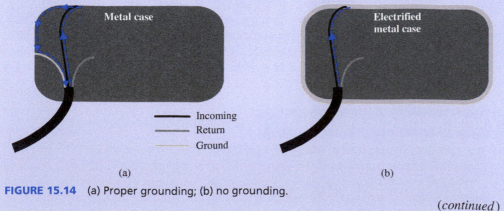

Metal case   Electrified metal case

——— Incoming
——— Return
——— Ground

(a)   (b)

FIGURE 15.14   (a) Proper grounding; (b) no grounding.

(continued)

APPLICATION EXAMPLE 15.10   Continued

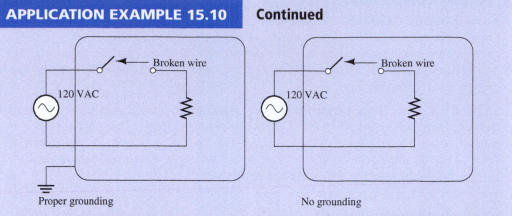

FIGURE 15.15 Schematic diagram of proper grounding.

Figure 15.15 shows the equivalent circuit for Figure 15.14. The stray current from the broken wire is redirected to the ground. Since the ground is at a zero potential (0 V), the loose current flows through it. Normally, this loose current is powerful enough to blow a fuse or trip a circuit breaker. This interrupts the current and alerts the user to the problem. However, when there is no ground, the power stays connected, leaving the operator unaware of a potentially fatal shock hazard. Proper grounding is used to eliminate this situation.

APPLICATION EXAMPLE 15.11   Checking for a Proper Ground

A voltmeter can be used to verify a proper ground. First, measure the line voltage by inserting the probes in the two vertical slots [see Figure 15.16(a)]. This serves two purposes. First, power to the outlet is verified and, second, the voltage can be observed for reference purposes. Now, insert a voltmeter probe in the incoming slot (the shorter slot), and in the ground slot [Figure 15.16(b)]. If the ground is working properly, the measured voltage should be very close to the line voltage.

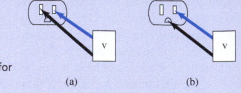

FIGURE 15.16   Testing an outlet: (a) checking for power; (b) checking for proper ground.

(a)   (b)

In older buildings, outlets may not have a ground slot. However, these outlets may have an internal ground connection. To test for this connection, measure the voltage between the incoming slot and the faceplate screw (see Figure 15.17). If this voltage matches the line voltage, an internal ground is available. Example 15.12 explains how to safely connect a grounded plug to a two-slot outlet.

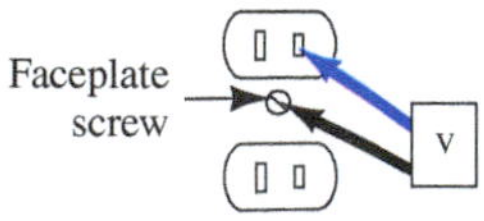

Faceplate screw

FIGURE 15.17
Checking for ground on older outlets.

APPLICATION EXAMPLE 15.12   Grounding Older Systems

If it is necessary to plug a three-prong, grounded plug into a two-slot outlet, a device called a *grounding adapter* can be used. This adapter has a three-slotted outlet on one side, and a two-prong plug with a grounding tab on the other side (see Figure 15.18). Before using this adapter, it is necessary to verify the existence of an internal ground (see Example 15.11).

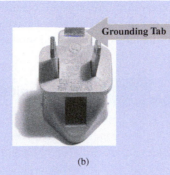

Grounding Tab

(a)   (b)

FIGURE 15.18 Grounding adapter: (a) front; (b) rear. (Used with permission from Alexander Remy Levine/Shutterstock.com.)

To use the adapter, first remove the faceplate screw from the outlet. Plug in the adapter and replace the screw, using it to connect the grounding tab to the outlet. The outlet must have an internal ground, and the tab must be connected to the screw to ensure proper grounding. If either condition is not fulfilled, the grounding has been defeated, and a shock hazard can be present. If the building wiring lacks an internal ground, it is highly recommended that the wiring be completely replaced.

Note that some devices such as CMOS (see Chapter 8) need careful grounding protection. They are very sensitive to the static electricity of the body. Therefore, when working with these devices, it is highly suggested to connect the body (e.g., using a wristband) to the ground.

Note that similar electronic devices are put and sold in conductive bags in order to prevent electric shock from body. A conductive bag keeps the electric charges in the outside layers of the bag and avoids harming the electric device inside. Similarly, in some workshops, conductive mats are used to conduct the electric charges of the body to pass through to the ground and avoid hazards for the devices.

EXERCISE 15.2

Airplanes are subject to high-voltage electric shocks due to lightening or passing through clouds that carry electric charges [11]. Explain why passengers are safe inside of the airplane.

15.5.1.3 THE GROUND FAULT CIRCUIT INTERRUPTER

A *ground fault circuit interrupter* (GFCI) is a special kind of circuit breaker that the NEC requires in certain locales to prevent shocks. This device is most commonly available as an electrical outlet, which is readily distinguished by its "Test" and "Reset" buttons. Figure 15.19 shows a standard GFCI outlet with its "Test" and "Reset" buttons.

A GFCI compares the current levels in both the incoming and return conductors. If there is even a minor difference between these levels, the GFCI shuts off the current immediately. It can shut off current in a fraction of a second. This can increase the safety of the outlet. For example, if you come into contact with electricity, some of the current flows through your body. The current tries to find some other path to make the potential equal to zero rather than going through the return wire in the circuit. Kirchhoff's current law says that the current flowing into and out of a node must sum to zero. If the incoming and return currents do not sum to zero, the electricity is taking an additional path (i.e., through your body). Thus, a properly functioning GFCI will shut off the current fast enough to prevent tissue damage from the shock. Example 15.13 shows how a GFCI works.

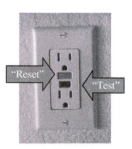

"Reset"   "Test"

FIGURE 15.19 GFCI outlet. (Used with permission from © Ted Foxx/Alamy.)

Even though it is a valuable safety device, the GFCI is not foolproof. While it prevents shocks between the incoming and ground connections, it doesn't prevent shocks between the incoming and returning conductors (unless there is a path to ground). For example, if you touch the incoming conductor with one hand and the returning with the other, while wearing rubber boots, the GFCI will not trip and you will be shocked (see Example 15.14).

The NEC requires GFCI protection in areas where electricity is used close to water or dampness. Water on skin reduces its resistance, increasing the conductivity and the potential for fatal shocks. Some of these locations include outdoors, bathrooms, garages, kitchens, unfinished basements, crawlspaces, and near pools and hot tubs.

APPLICATION EXAMPLE 15.13   Ground Fault Circuit Interrupter

A GFCI monitors the currents I_{in} and I_{out} measured by ammeters A_1 and A_2, respectively. Under normal circumstances, $I_{in} = I_{out}$. If the current is redirected through a body to a ground, the current in will not be equal to the current out. When this happens, the GFCI will "trip," and shut off current to the circuit. For the circuit shown in Figure 15.20, will the GFCI trip?

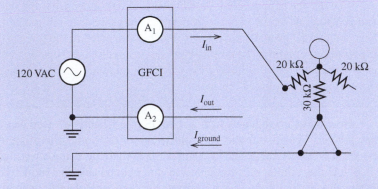

FIGURE 15.20 GFCI for Example 15.13.

SOLUTION

By inspection, it is observed that the entire current, I_{in}, passes through the person's body, and none of it passes through the return wire. A_1 measures a current, while A_2 measures no current (0 A). Therefore, the GFCI will trip.

Note:

If the person in Figure 15.20 does not touch the wire, there would not be any current to the ground, and both I_{in} and I_{out} will be equal to 0 A. Therefore, in this case, the GFCI would not trip.

APPLICATION EXAMPLE 15.14   GFCI

An individual comes into contact with the incoming and return conductors of a GFCI protected outlet, as shown in Figure 15.21. He is wearing rubber boots, which isolate his feet from the ground. (a) Find the currents I_{in}, I_{out}, and I_{ground}. (b) Does the GFCI trip?

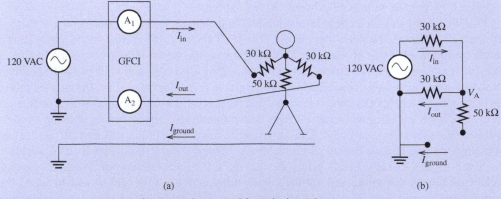

(a)   (b)

FIGURE 15.21 (a) Figure for Example 15.14; (b) equivalent circuit.

SOLUTION

a. The circuit can be modeled as shown in Figure 15.21(b). In this situation, because the person is wearing rubber shoes, and the resistance of those shoes is almost infinity, no current flows through the 50-kΩ resistor. Applying KVL:

$$\frac{V_A - 120}{30,000} + \frac{V_A}{30,000} = 0$$

Using this equation:

$$V_A = 60 \text{ V}$$

Now:

$$I_{in} = I_{out} = \frac{V_A}{30,000} = 2 \text{ mA}$$

b. Because I_{out} equals I_{in}, the GFCI does NOT trip; thus, the person in this example gets shocked as the current flows through his body.

EXERCISE 15.3

Repeat Example 15.14 assuming the resistance of the shoes is not infinity, rather assume it is about 100 kΩ.

15.5.2 Fire Prevention

The NEC also seeks to reduce electrical fire hazards [13]. This is mainly achieved by preventing over-current and arcs. These topics are discussed in the following subsections.

15.5.2.1 OVER-CURRENT PROTECTION

Excessive current causes conductors to dissipate more power. This leads to extreme heat in the conductors, which can ignite surrounding material. The NEC regulations, therefore, attempt to reduce or eliminate the risk of this over-current. To protect from over-current, the NEC stipulates the proper wire gauge that must be used in specific situations. The wires must have a diameter capable of handling the current and avoiding overheating and, therefore, fire. Common wire gauges and their current capacities are shown in Table 15.4. Note that in the American Wire Gauge system, a smaller number means a larger wire diameter.

Extension cords can overheat when the wrong gage of wire is used. Recall that:

$$R = \rho \times \frac{l}{A} \tag{15.2}$$

When the ratio of length to cross-sectional area is unfavorable, R increases. The dissipated power corresponds to $P = I^2 \times R$. Thus, an increase in R is equivalent to higher voltage drop and higher dissipated power in the extension cord.

Chapter 15 • Electrical Safety

TABLE 15.4 Common Wire Gauges

Wire Gauge	Uses	Current Capacity
12	In-wall wiring	20 A
14	In-wall wiring	15 A
16	Extension cords, power strips	10 A
18	Most power cords	7 A

EXAMPLE 15.15   **Over-Current**

Three devices rated at 1000 W, 5 A, and 200 W, respectively, are connected to a 120-V, 14-gauge line on a 15-A circuit breaker. (a) Does this make the circuit breaker trip? (b) Does the wire overheat?

SOLUTION

It is clear that devices are connected in parallel in buildings. Note that if devices were connected in series, a disconnection in one device would lead to the disconnection of the circuit and loss of power to all devices. In addition, connecting in series would make it impossible to serve all loads with the same voltage.

a. Due to the parallel connection of devices, the total current is a sum of all currents. The currents through the 1000-W and 200-W devices respectively correspond to:

$$\frac{1000}{120} = 8.33 \text{ A}$$

and

$$\frac{200}{120} = 1.67 \text{ A}$$

Due to the parallel connection of devices, the total current is:

$$i_{\text{total}} = 8.33 + 1.67 + 5 = 15 \text{ A}$$

By comparing 15 A with the circuit breaker and wire ratings, it can be determined that the circuit breaker will trip.

b. Because the circuit breaker trips, the wire does not overheat, and no fire hazard is present.

To minimize over-current, the NEC mandates specific requirements on the number of outlets in a room. More specifically, it defines a minimum number of outlets per wall. If a wall is more than 4 ft wide, it must have an outlet. Outlets must be spaced a maximum of 10 ft from each other along each wall. This is intended to eliminate the need for extension cords, because most electrical devices have 6-ft cords and—with this spacing requirement—can reach an outlet regardless of their position along the wall. Extension cords can overheat, get frayed and damaged, and ignite electrical fires.

The NEC also specifies a maximum number of outlets per circuit. A *circuit* includes everything connected to a certain circuit breaker in the circuit breaker box. Too many outlets on one circuit can encourage the user to overload the circuit with too many electrical devices. The NEC restricts the number of outlets to around eight on a standard circuit.

Circuit breakers and fuses are over-current prevention devices. They disconnect current when it exceeds the maximum rated value. A fuse is a strip of metal enclosed in a glass case that melts or *blows* when the maximum current level is exceeded. A circuit breaker is a special kind of switch that *trips* when high currents flow through it. Fuses are destroyed when their maximum level is exceeded and, therefore, must be replaced. On the other hand, circuit breakers trip (move to an "off" position) and are simply reset. There is no need to replace a circuit breaker when it trips.

A *short circuit* occurs when the line conductor makes contact with the return conductor. Because the resistance of the wires is very low (nearly 0 Ω), the current may spike to dangerously high levels. An overload occurs when too many devices are plugged into one circuit, drawing more current than the circuit can handle. Either of these situations can generate a fire. Therefore, when the current exceeds the maximum level of the over-current device, it shuts off, preventing overheating and fire.

Important Note: Contrary to popular belief, circuit breakers and fuses do not prevent electric shock. This is because they are rated in the ampere range (commonly 15 or 20 A). Recall that only 0.1 A is required to be fatal.

15.5.2.2 THE ARC FAULT CIRCUIT INTERRUPTER

The arc fault circuit interrupter (AFCI) is a special circuit breaker that detects arcs in electrical circuits. Recall from Section 15.4.1 that an arc occurs when a substantial amount of current flows through superheated air. The 2005 NEC requires AFCI's in circuits feeding bedroom outlets. The 2008 NEC code requires AFCIs in circuits feeding outlets in "family rooms, dining rooms, living rooms, parlors, libraries, dens, bedrooms, sunrooms, recreation rooms, closets, hallways, or similar rooms or areas…". They are not required in areas where GFCI protection is required.

15.6 WHAT DID YOU LEARN?

- Electric shock is caused by the contact of two points with different electric potential.
- The level of electrical *current* passing through the human body is the main cause of fatality in the case of shock.
- High-frequency RF radiation can cause cancer. High RF energy can burn biological tissue.
- There is currently a debate over the potential level of radiation hazard presented by cell phones.
- High-tension power lines produce a magnetic field in their vicinity. Exposure to strong magnetic fields (>200 mG) is not safe.
- Explosive atmospheres are caused by flammable substances in the air.
- Electric sparks can cause explosions. Sparks need to be isolated from dangerous atmospheres.
- A *safety switch* is a special spark-resistant switch that can prevent explosion.
- The NEC is a set of regulations that intends to make the use of electricity safer.
- An outlet with *polarization* has a neutral slot that is longer than the hot slot. By doing this, the internal ground (with a longer lead) is not connected to the hot wire and a shock is avoided.
- All equipment should be electrically connected to a ground.
- A ground fault circuit interrupter (GFCI) is required in areas where electricity is used close to water or dampness.
- Fuses and arc fault circuit interrupters (AFCIs) are used to prevent over-current, which results in fire hazards.

References

1. Maisch, D. 2001. Mobile phone use: it's time to take precautions. *Journal of the Australasian College of Nutritional and Environmental Medicine* 20(1): 4.
2. Wired News. 20 Dec 2004. Lab tests show mobile-phone risk. http://www.wired.com/gadgets/wireless/news/2004/12/66097. Accessed 8 July 2008.
3. IOL. 30 Aug 2005. Physicians warn of cellphone cancer risk. http://www.int.iol.co.za/index.php?set_id=1&click_id=31&art_id=qw112540806283B223. Accessed 8 July 2008.
4. van Grinsven, L. 2001. Cellphones may harm by speeding up the brain, *IOL.* 21 Sept 2001. http://www.int.iol.co.za/index.php?set_id=1&click_id=31&art_id=qw1001086502683B243. Accessed 8 July 2008.
5. Stang, Andreas, A. et al. 2001. The possible role of radiofrequency radiation in the development of uveal melanoma. *Epidemiology* 12(1). http://www.epidem.com/pt/re/epidemiology/pdfhandler.00001648-200101000-00003.pdf;jsessionid=DDKXjIV2hVRM74W17Nr71fnL7FkpEbC9POtSlBM23dGj9A2EaL6l!-1660146838!-949856145!9001!-1. Accessed 8 July 2008.
6. Brown, K. 2005. Residents worry, but companies say cellphone antennas are safe. *Queens Chronicle* 6 Oct 2005. http://www.zwire.com/site/index.cfm?newsid=15340673&BRD=2731&PAG=461&dept_id=574995&rfi=8. Accessed 8 July 2008.
7. Rediff News. 1 Aug 2005. Cell phones don't cause cancer: study. http://in.rediff.com/news/2005/aug/31cellphone.htm. Accessed 8 July 2008.
8. Evans, J.A. nd. The cell tolls for thee. MSN Health and Wellness. http://health.msn.com/health-topics/cancer/articlepage.aspx?cp-documentid=100211877&page=1. Accessed 5 Aug 2008.
9. ENS. 16 July 2001. Power lines, wiring pose health risks. http://www.feb.se/EMFguru/Elf/calif-power.html. Accessed 8 July 2008.
10. Electric Power Research Institute. 1982. *Transmission Line Reference Book 345 kV and Above*, 2 edn, pp. 45–61. Palo Alto: Electric Power Research Institute.
11. DetectorTechnologies.com. nd. DC magnetometer (Gauss). http://www.detectortechnologies.com/store/detail.aspx?ID=7. Accessed 5 Aug 2008.
12. Cadick, J. 1994. *Electrical Safety Handbook*, 4.15. New York: McGraw-Hill.
13. National Fire Prevention Association. 2007. *National Electric Code*, 2008 edn, Quincy: National Fire Prevention Association, Section 210.12.

Problems

*B refers to basic, A refers to average, H refers to hard, and * refers to problems with answers.*

SECTION 15.2 ELECTRIC SHOCK

15.1 (B)* If a person comes into contact with 120 V, what body resistance can potentially result in a fatal shock?

15.2 (A) An electrical power cord falls into a swimming pool, as shown in Figure P15.2. If the pool water has a resistivity of 175 kΩ m, find the current, I, that flows through the water. Assume that the pool bottom acts as a ground (return conductor), and the pool walls act as insulators.

15.3 (B)* A current is going through your body and you do not feel it, what is the possible value of the current?
(a) 100 mA
(b) 10 mA
(c) 5 mA
(d) 0.1 mA

15.4 (H) When shoes are scuffed on a rug on a dry winter day, our bodies can be easily charged up to a potential

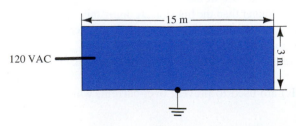
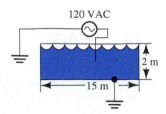

FIGURE P15.2 Swimming pool diagram for Problem 15.2.

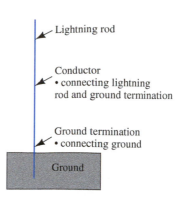

FIGURE P15.12   Building with a lightning rod and the basic structure of it. (Used with permission from © Michael Hawkridge/Alamy.)

of several thousand volts (about 1000 to 3000 V) static electricity with respect to the ground. A person shocked by this high-voltage static will not die. However, if the same person were shocked by a power supply with 200 V, the person would die. Explain the reason.

15.5 (B) In China and Europe, the voltage of the power supply for houses and apartments is 50 Hz, 220 to 240 V, while in the United States the power supply voltage is 60 Hz, 100 to 127 V. Assume the resistance of a body is 500 to 1000 Ω; compare the two power supply voltages' impact if someone is shocked.

15.6 (A)* The actual resistance of the body varies depending on the points of contact and the skin condition (moist or dry). Usually, the resistance from hand to foot is close to 500 Ω. Assume a person stands on the ground with bare feet and one hand touches a 120-V power supply cable. Is it fatal?

15.7 (B) Which of the following cannot help prevent electric shock?
(a) Rubber gloves
(b) Wet wood chair
(c) Metal workbench
(d) Dry wood workbench

15.8 (B) You need to replace a broken fuse in an instrument. To avoid shock, what will you do?

15.9 (H) Explain the reason that a high-power transmitter design lab is always equipped with wooden bar stools, and why the bar stools should be kept dry.

15.10 (H)* Assume that the density of a flash of lightning hitting the ground is 4 lightning strikes per km^2 per year, on average. Also assume that the lightning-attractive area of one house is 1250 m^2 (which means if lightning hits a spot within 1250 m^2 of a house, it can be attracted by the house and hit it). Calculate the probability of a lightning strike on the house. On average, how many years will it take for the house to be hit once?

15.11 (A)* What should you not do during a lightning storm?
(a) Go underneath trees or rain/sun shelters
(b) Touch metal objects or water
(c) Use the telephone

(d) Go into fully enclosed all-metal vehicles or substantial buildings

15.12 (B) Why can a lightning rod installed on the top of a building protect the building from lightning strikes (see Figure P15.12)?

15.13 (A) Describe the method used to protect an airplane in a storm from lightning.

15.14 (H) A safety fuse has been installed in a surge protector to prevent large currents from destroying electronic equipment. The surge protector is a safety fuse that opens the circuit automatically if the current within the circuit increases beyond a threshold. Sketch the circuit for the case when a lightning strike occurs and induces large current into a circuit that has radio equipment connected in series with a surge protector. Then, explain how the safety fuse works in this case.

SECTION 15.3 ELECTROMAGNETIC HAZARDS

15.15 (A)* For a radio transmitter, the power transmitted is 50 kW, and the antenna gain is 20.
(a) If an employee is standing 10 m from the antenna, what is the power density of the radio frequency radiation passing through his body?
(b) At what distance is the power density of the radiation 1 kW/m^2?

15.16 (A) A cell phone, held 5 mm from the head, transmits 0.3 W at an antenna gain of 1.5. What is the power density on the skin? Is this radiation level of concern?

15.17 (A)* If a house is placed 30 m from a 745-kV power line bearing 4000 A, what is the strength of the magnetic field? Is this field safe (according to the US state standards)?

15.18 (B) Find the maximum magnetic field magnitude directly under a 12.47-kV line. The line is 5 m above the ground and is rated at 800 A.

15.19 (B) If the maximum acceptable magnetic field magnitude is 200 mG, how far should houses be built from a 345-kV line rated at 2000 A based on the US state standards?

15.20 (B) List four kinds of radiations that can cause cancer.

15.21 (H) A cell phone's transmitting power is 1 W and its antenna power gain is 1.5. You can either use the cell phone close to your ear (0.5 in. to your head) by hand holding it or use a hands-free headset that allows the phone to be 3 ft away from your body. Compare the power density and select the method from the two that will best minimize the high-frequency impact on your body.

15.22 (A)* A hot pot consumes 2200 W and the power supply voltage is 120 V. Assume you are sitting close to the hot pot cable at a distance of 1 ft. Is this safe, according to US state standards?

15.23 (H) Why are overhead high-voltage electric power transmission lines more commonly used rather than underground high-voltage electric power transmission lines? *Hint:* compare them in terms of cost, heat dissipation, and environmental impacts.

15.24 (A) If someone's bone is broken, the doctor will use an X-ray machine to check the condition of the bone. Explain why a person should not be exposed to the X-ray for an extended period of time.

SECTION 15.4 ARCS AND EXPLOSIONS

15.25 (A)* What is the maximum power available for an arc across the load (between point C and the ground) in Figure P15.25? Also, find the maximum power for an arc across the circuit breaker between points A and B.

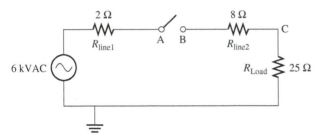

FIGURE P15.25   Circuit for Problem 15.25.

15.26 (B) Figure P15.26 shows a basic flashlight circuit. When the switch is opened, a spark occurs across its contacts. What is the maximum power in this spark?

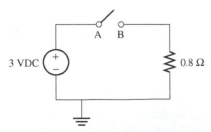

FIGURE P15.26   Circuit for Problem 15.26.

15.27 (H) Why is a spark-resistance switch needed in a milling plant?

15.28 (H) Why on a dry day in the winter is it easy to be shocked by an arc when you touch a metal door handle after you walk on a carpet? How can you avoid this kind of shock?

15.29 (A) Explain the reason why there are no walls around gas stations.

15.30 (A)* Calculate the energy dissipated in an arc generated on the switch in the lamp circuit shown in Figure P15.30. Assume the arc period is 1 ms.

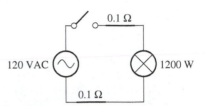

FIGURE P15.30   Lamp circuit.

15.31 (H) Explain the basics of explosion prevention.

SECTION 15.5 THE NATIONAL ELECTRIC CODE

15.32 (B) Based on the measurements in Figure P15.32, does this outlet have a proper ground?

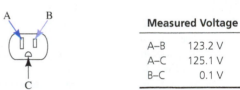

Measured Voltage	
A–B	123.2 V
A–C	125.1 V
B–C	0.1 V

FIGURE P15.32   Outlet for Problem 15.32.

15.33 (B) Can a grounding adapter be safely used with this older outlet?

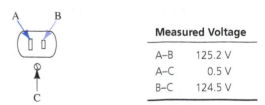

Measured Voltage	
A–B	125.2 V
A–C	0.5 V
B–C	124.5 V

FIGURE P15.33   Outlet for Problem 15.33.

15.34 (A) For each of the scenarios described below, state when current will flow through the person's body when he is in contact with incoming and return conductors.
(a) With a rubber glove and a rubber boot
(b) With a rubber glove
(c) With a rubber boot

15.35 (A)* An individual comes into contact with the incoming and return conductors of a GFCI-protected circuit, as shown in Figure P15.35.
(a) Find the currents I_{in}, I_{out}, and I_{ground}.
(b) Does the GFCI trip? If not, how much current flows through his body?

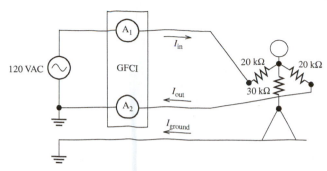

FIGURE P15.35 Circuit diagram for Problem 15.35.

15.36 (A) An individual comes into contact with the incoming and return conductors of a GFCI-protected circuit, as shown in Figure P15.36. He is wearing rubber boots, which isolate his feet from the ground.
  (a) Find the currents I_{in}, I_{out}, and I_{ground}.
  (b) Does the GFCI trip? If not, how much current flows through his body?

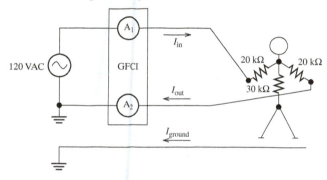

FIGURE P15.36 Circuit diagram for Problem 15.36.

15.37 (A) A 120-V circuit has devices rated at 200 W, 1500 W, 2 A, and 300 W, respectively. The 12-gauge line is connected to a 20-A circuit breaker.
  (a) Does the circuit breaker trip?
  (b) Is a fire hazard present?

15.38 (A)* A 120-V circuit on a 20-A circuit breaker has devices rated at 1, 200 W, 3 A, 60 W, 100 W, and 300 W, respectively. The line gauge is 14.
  (a) Does the circuit breaker trip?
  (b) Is a fire hazard present?

15.39 (H) The following devices, rated at 440, 520, 360, 25, and 110 W, respectively, are connected to a socket with a 120-V voltage source. The line gauge is 16.

  (a) Is a fire hazard present?
  (b) If there is a fire hazard, what kind of advice can be given to prevent it?

15.40 (B) What is the purpose of NEC?

15.41 (A)* There is an old outlet shown in Figure P15.41. Which of the following equipment cannot be connected to the old outlet?
  (a) Cell-phone charger transformer
  (b) Hot-water pot with plastic case
  (c) Personal computer
  (d) Electric fan

FIGURE P15.41 Old outlet.

15.42 (A) Compare the advantages and disadvantages of fuses and circuit breakers.

15.43 (A) As an instrument ages, the connection between the ground and the instrument's metal case may decay and generate a resistance, for example $R_1 = 100 \ \Omega$ in Figure P15.43. Assume that the incoming line voltage is 120 V and that there is a broken point that connects to the instrument's metal case, and $R_2 = 100 \ \Omega$. Please use the figure to explain the importance of dependable grounding in the building's wiring system, assuming a body's resistance to be $R_3 = 500 \ \Omega$.

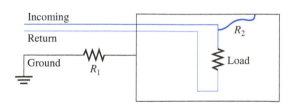

FIGURE P15.43 Example of a fault ground.

15.44 (B) What will happen if a line is carrying a current greater than the rated value? For example, what happens when a 10-A rated line carries 15-A current?

15.45 (A) Compare the functions of AFCI and GFCI protected circuits.

APPENDIX A

Solving Linear Equations

A set of linear equations will have a unique solution if the determinant of the system matrix is nonzero. The solution to the linear system $A\,x = b$ is $x = A^{-1}b$. Finding the inverse of a matrix is difficult, especially in cases where the matrix is large. Cramer's rule provides a method that makes it easier to determine the solution to the linear system.

CRAMER'S RULE

Suppose A is an $n \times n$ invertible matrix. The solution to the system $A\,x = b$ is given by:

$$x_1 = \frac{\det(A_1)}{\det(A)},\ x_2 = \frac{\det(A_2)}{\det(A)},\ \cdots,\ x_n = \frac{\det(A_n)}{\det(A)}$$

where A_i is the matrix found by replacing the ith column of A with b.

EXAMPLE A.1   **Cramer's Rule**

Use Cramer's rule to determine the solution to the following linear equations:

$$4x_1 + 5x_2 = 3$$
$$2x_1 - x_2 = 1$$

SOLUTION

From the problem, develop matrix:

$$A = \begin{bmatrix} 4 & 5 \\ 2 & -1 \end{bmatrix}$$

and:

$$b = \begin{bmatrix} 3 \\ 1 \end{bmatrix}$$

Calculating the determinant of matrix A results in $\det(A) = -14 \neq 0$. Thus, this method results in a unique solution for this linear system.

The matrices A_1 and A_2 are found by respectively replacing the first and the second columns of A with b. Therefore:

$$A_1 = \begin{bmatrix} 3 & 5 \\ 1 & -1 \end{bmatrix}$$

$$A_2 = \begin{bmatrix} 4 & 3 \\ 2 & 1 \end{bmatrix}$$

(continued)

EXAMPLE A.1   **Continued**

Now, according to Cramer's rule:

$$x_1 = \frac{\det(A_1)}{\det(A)} = \frac{-8}{-14} = \frac{4}{7}$$

$$x_2 = \frac{\det(A_2)}{\det(A)} = \frac{-2}{-14} = \frac{1}{7}$$

MATLAB software can also be used to determine the solution to a linear system. The MATLAB function "inv" determines the inverse of a matrix. Using this function, the command $x = \text{inv}(A)*b$ can be used in order to find the system solution.

APPENDIX B

Laplace Transform

The Laplace transform or simply *L-transform* is defined as the transformation from the time domain to the frequency domain (here, called the *s*-domain). It can be expressed mathematically as:

$$L(x(t)) = X(s) = \int_0^\infty x(t)e^{-st}dt \qquad \textbf{(B.1)}$$

where $x(t)$ is the *t*-domain signal and $X(s)$ is the *s*-domain signal. The complex angular frequency, *s*, is given by:

$$s = \sigma + j\omega \qquad \textbf{(B.2)}$$

Notice that the time domain signal is expressed in lowercase (e.g., $x(t)$); however, uppercase notation (e.g., $X(s)$) is used for *s*-domain signals.

EXAMPLE B.1 **Laplace Transform**

Find the Laplace transform for the step voltage defined by:

$$v(t) = \begin{cases} E & t \geq 0 \\ 0 & \text{else} \end{cases}$$

SOLUTION

$$V(s) = \int_0^\infty v(t)e^{-st}dt$$

$$= \int_0^\infty Ee^{-st}dt$$

$$= E\int_0^\infty e^{-st}dt$$

$$= E\frac{-1}{s}e^{-st}\Big|_0^\infty$$

$$= \frac{E}{s}$$

Note: The last equality is true when the real part of *s*, $R(s) > 0$ so that $e^{-st} \to 0$ as $t \to \infty$.

EXAMPLE B.2 **Laplace Transform**

Find the Laplace transform for the decaying exponential function defined by:

$$x(t) = \begin{cases} e^{-at} & t \geq 0 \\ 0 & \text{else} \end{cases}$$

SOLUTION

$$X(s) = \int_0^\infty x(t)e^{-st}dt$$

$$= \int_0^\infty e^{-at}e^{-st}dt$$

(continued)

EXAMPLE B.2	Continued

$$= \int_0^\infty e^{-(s+a)t}dt$$

$$= \frac{-1}{s+a}e^{-(s+a)t}\Big|_0^\infty$$

$$= \frac{1}{s+a}$$

Note: The last equality is true when $R(s) > -a$ so that $e^{-(s+a)t} \to 0$ as $t \to \infty$.

EXAMPLE B.3	Laplace Transform

Find the Laplace transform of:

$$x(t) = 5 + e^{-2t}$$

SOLUTION

From the previous examples, it is known that $L(5) = 5/s$, and $L(e^{-2t}) = 1/(s+2)$. Because the Laplace transform is an integration process it possesses the linearity property, therefore:

$$X(s) = L(5) + L(e^{-2t}) = \frac{5}{s} + \frac{1}{s+2} = \frac{6s+10}{s(s+2)}$$

Table B.1 represents the Laplace transforms of some important functions. This table can be used to find the Laplace transform as well as the inverse Laplace transform. The following two examples help to explain how to find the inverse Laplace transform and how to solve a differential equation.

TABLE B.1 The Laplace Transform of Basic Functions and Some of its Properties

$x(t)$	$X(s)$
Basic Functions	
$\delta(t)$	1
1	$\frac{1}{s}$
t	$\frac{1}{s^2}$
t^n	$\frac{n!}{s^{n+1}}$
e^{at}	$\frac{1}{s-a}$
$\cos(\omega t)$	$\frac{s}{s^2+\omega^2}$
$\sin(\omega t)$	$\frac{\omega}{s^2+\omega^2}$
s-domain shift	
$x(t)e^{at}$	$X(s-a)$
te^{at}	$\frac{1}{(s-a)^2}$
$\cos(\omega t)e^{at}$	$\frac{s-a}{(s-a)^2+\omega^2}$

(continued)

TABLE B.1   Continued

x(t)	X(s)
$\sin(\omega t)e^{at}$	$\dfrac{\omega}{(s-a)^2 + \omega^2}$
t-domain shift	
$x(t-a)$	$X(s)e^{-as}$
Differentiation	
$x'(t)$	$sX(s) - x(0)$
$x''(t)$	$s^2X(s) - sx(0) - x'(0)$
$x^{(n)}(t)$	$s^nX(s) - s^{n-1}x(0) - s^{n-2}x'(0) - \cdots - x^{(n-1)}(0)$
Integration	
$\displaystyle\int_0^t x(\tau)d\tau$	$\dfrac{1}{s}X(s)$
Convolution	
$x_1(t) * x_2(t)$	$X_1(s) \cdot X_2(s)$
$x_1(t) \cdot x_2(t)$	$X_1(s) * X_2(s)$

EXAMPLE B.4   **Inverse Laplace Transform**

Find the inverse Laplace transform of $Y(s)$ given by:

$$Y(s) = \frac{8}{s^3 + 4s^2 + 3s}$$

SOLUTION

To solve this problem, use the partial fraction procedure:

$$Y(s) = \frac{8}{s(s^2 + 4s + 3)} = \frac{8}{s(s+1)(s+3)} = \frac{k_0}{s} + \frac{k_1}{s+1} + \frac{k_2}{s+3}$$

where k_0, k_1, and k_2 are constants (called the residues) of the poles 0, 1, and 3, respectively. Note that the poles of $Y(s)$ are the zeros of its denominator. These residues can be calculated as follows:

$$k_0 = sY(s)\big|_{s=0} = \frac{8}{(0+1)(0+3)} = \frac{8}{3} = 2.67$$

$$k_1 = (s+1)Y(s)\big|_{s=-1} = \frac{8}{-1 \times (-1+3)} = \frac{8}{-2} = -4$$

$$k_1 = (s+3)Y(s)\big|_{s=-3} = \frac{8}{-3 \times (-3+1)} = \frac{8}{+6} = +1.33$$

and; therefore:

$$Y(s) = \frac{2.67}{s} - \frac{4}{s+1} + \frac{1.33}{s+3}$$

Using Table B.1 and the linearity property of Laplace transform, $y(t)$ corresponds to:

$$y(t) = 2.67 - 4e^{-t} + 1.33e^{-3t}, \qquad t > 0$$

EXAMPLE B.5   **Solving Differnetial Equations Using Laplace Transform**

Find the solution of the first-order differential equation:

$$x'(t) + 3x(t) - 24 = 0$$

  a. Assuming zero initial conditions
  b. $x(0) = 2$

SOLUTION

Applying Laplace transform:

$$sX(s) - x(0) + 3X(s) - \frac{24}{s} = 0$$

  a. $x(0) = 0$, thus:

$$(s + 3)X(s) = 0 + \frac{24}{s} = \frac{24}{s}$$

  or:

$$X(s) = \frac{24}{s(s + 3)} = \frac{8}{s} - \frac{8}{s + 3} = 8\left(\frac{1}{s} - \frac{1}{s + 3}\right)$$

  Its inverse Laplace transform corresponds to:

$$x(t) = 8(1 - e^{-3t})$$

  b. $x(0) = 2$, thus:

$$(s + 3)X(s) = 2 + \frac{24}{s} = \frac{2s + 24}{s}$$

  or:

$$X(s) = \frac{2s + 24}{s(s + 3)} = \frac{8}{s} - \frac{6}{s + 3}$$

  Here, the second equality is based on partial fraction procedure explained in Example B.4.
  Its inverse Laplace transform corresponds to:

$$x(t) = 8 - 6e^{-3t}, \qquad t > 0$$

EXAMPLE B.6   **Solving Differnetial Equations Using Laplace Transform**

Find the solution of the second-order differential equation, assuming all initial conditions are zeros:

$$x''(t) + 5x'(t) + 6x(t) - 50 = 0$$

SOLUTION

Applying Laplace transform:

$$s^2X(s) - sx(0) - x'(0) + 5sX(s) - 5x(0) + 6X(s) - \frac{50}{s} = 0$$

Setting $x(0) = x'(0) = 0$:

$$(s^2 + 5s + 6)X(s) = \frac{50}{s}$$

$$X(s) = \frac{50}{s(s^2 + 5s + 6)} = \frac{50}{s(s + 2)(s + 3)} = \frac{8.33}{s} - \frac{25}{s + 2} + \frac{16.67}{s + 3}$$

Using Table B.1 to find inverse Laplace transform results in:

$$x(t) = 8.33 - 25e^{-2t} + 16.67e^{-3t}, \qquad t > 0$$

APPENDIX C

Complex Numbers

Any sinusoidal signal, like the one shown in Figure C.1, can be represented by its amplitude and its phase. These signals can be expressed using complex numbers.

FIGURE C.1 A sinusoidal signal.

The signal shown in Figure C.1 can also be expressed in exponential form, that is, $A_1 \cos(\omega t + \phi_1) = \text{Re}\left\{A_1 e^{j\omega t} e^{j\phi_1}\right\}$. Here, $A_1 e^{j\phi_1}$ is the complex number in exponential form.

From the 13th century to the early 17th century, numbers were only used to represent real things, for example, ONE piece of cake or TWO cars, as seen below.

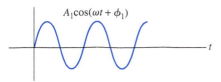

At that time, a mathematician solving $X^2 = 4$ would have arrived at a single solution, $X = 2$.

In the middle of the 17th century, negative numbers were added to the number system to represent things like debt, height below sea level, for example, $-2\$$, -30 m, and so on.

Once negative numbers appeared, solving $X^2 = 4$ results in the solutions $X_1 = 2$ and $X_2 = -2$.

By the end of 18th century, the notion of complex numbers had been developed in the mind of scientists. However, the scientists could not intuitively explain the concept using real identities.

Now that complex numbers were available, solving $X^2 = -4$ resulted in the solution $X = \pm j2$, that is, $(j2)^2 = -4$, $(-j2)^2 = -4$, where $j = \sqrt{-1}$.

DEFINITION OF A COMPLEX NUMBER

First, $j^2 = -1$; correspondingly $j = \sqrt{-1}$. Here, j represents the square root of -1. Sometimes, symbol i is used instead of j.

Second, an *imaginary number* is defined as the product of a real number and the imaginary operator j; for example, $j5$ is an imaginary number, or, a pure imaginary number.

A *complex number* is defined as the sum of a real number and an imaginary number, that is, $a + jb$, where a and b are real numbers. $2 + j3$ and $1.1 + j5.3$ are examples of complex numbers. A complex number, for example, $Z = a + jb$ is said to have a real part a and an imaginary part b.

Actually, a *real number* is simply a complex number with an imaginary part equal to zero, that is, $a + j0$. For example, $1.1 + j0 = 1.1$, $2.4 + j0 = 2.4 \ldots$. A pure imaginary number is a complex number with a real part equal to zero.

A *complex number* can be expressed in rectangular form $a + bj$, in exponential form $re^{j\theta}$, or in polar form $r\angle\theta$.

The addition, subtraction, multiplication, and division of complex numbers in rectangular form can be accomplished in a way similar to algebraic arithmetic.

OPERATIONS OF COMPLEX NUMBERS IN RECTANGULAR FORM

Assume: $Z_1 = a + jb$ and $Z_2 = c + jd$.

- The conjugate of the complex number, Z_1, is (changing the sign of the imaginary part):

$$Z_1^* = a - jb, \text{ where * represents the conjugate.}$$

- If $Z_1 = Z_2$, then $a = c$ and $b = d$.

If two complex numbers are equal, then each number's real part and imaginary part must be equal to the real part and imaginary part of the other number, respectively.

- Addition: $Z_1 + Z_2 = (a + c) + j(b + d)$.

To add two complex numbers, add the real and the imaginary parts of them, respectively.

- Subtraction: $Z_1 - Z_2 = (a - c) + j(b - d)$.

To subtract two complex numbers, subtract the real part of Z_2 from the real part of Z_1, and subtract the imaginary part of Z_2 from the imaginary part of Z_1.

- Multiplication: $Z_1 \times Z_2 = (ac - bd) + j(bc + ad)$.

To multiply two complex numbers, use the same rules as in the algebraic arithmetic.

$$Z_1 \times Z_2 = (a + jb)(c + jd) = ac + jad + jbc - bd = (ac - bd) + j(bc + ad).$$

- Division:

$$\frac{Z_1}{Z_2} = \frac{a + jb}{c + jd} = \frac{(a + jb)(c - jd)}{(c + jd)(c - jd)} = \frac{(ac + bd) + j(bc - ad)}{c^2 + d^2}$$

To divide two complex numbers, multiply the numerator and the denominator by the complex conjugate of the denominator.

EXAMPLE C.1 — Operations of Complex Numbers

For the given complex numbers: $Z_1 = 3 + j4$; $Z_2 = 9 + j12$.

- Addition: $Z_1 + Z_2 = (3 + 9) + j(4 + 12) = 12 + j16$
- Subtraction: $Z_1 - Z_2 = (3 - 9) + j(4 - 12) = -6 - j8$
- Multiplication: $Z_1 \times Z_2 = 27 + j36 + j36 - 48 = -21 + j72$
- Division: $\dfrac{Z_1}{Z_2} = \dfrac{(3 + j4)(9 - j12)}{(9 + j12)(9 - j12)} = \dfrac{75}{81 + 144} = \dfrac{75}{225} = \dfrac{1}{3}$
- Conjugate of Z_1: $Z_1^* = 3 - j4$

COMPLEX PLANE

A complex number can be represented in a *complex plane*, like the one shown in Figure C.2 (*x*-axis: real part; *y*-axis: imaginary part).

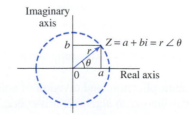

FIGURE C.2 A Complex Plane.

A complex number has three forms:

Rectangular Form: $Z_1 = a + jb$

Exponential Form: $Z_1 = re^{j\theta}$

$$\begin{cases} r = \sqrt{a^2 + b^2} \\ \theta = \tan^{-1}\dfrac{b}{a} \\ a = r \cdot \cos\theta \\ b = r \cdot \sin\theta \end{cases} \qquad \text{(C.1)}$$

Polar Form: $Z_1 = r\angle\theta$

In the complex plane, the length, r, of the arrow represents the magnitude of the complex number, and θ is the angle between the arrow and the positive real axis.

Using Equation (C.1), a complex number can be converted from rectangular form to exponential form, or vice versa. It is easy to convert the exponential form to polar form, because both forms use magnitude and angle to express the complex number.

EXAMPLE C.2   **Rectangular Form to Exponential and Polar Form Conversion**

Convert the complex number $Z_1 = 4 + j3$ from rectangular form to exponential and polar forms.

SOLUTION

Using Equation (C.1):

$$r = \sqrt{3^2 + 4^2} = 5, \quad \text{and}$$

$$\tan\theta = \frac{3}{4} = 0.75, \quad \theta = \tan^{-1}0.75 = 36.87°$$

Therefore:

$$Z_1 = 5e^{j36.87°}$$

The corresponding polar form is:

$$Z_1 = 5\angle 36.87°$$

EXAMPLE C.3   **Exponential to Rectangular and Polar Form Conversion**

Convert the complex number $Z_1 = 2.828e^{j45°}$ from exponential form to rectangular form and polar forms.

SOLUTION

The polar form for $Z_1 = 2.828e^{j45°}$ is:

$$Z_1 = 2.828\angle 45°$$

Using Equation (C.1):

$$a = r\cos\theta = 2.828 \times \cos(45°) = 2$$
$$b = r\sin\theta = 2.828 \times \sin(45°) = 2$$

Therefore, the rectangular form is:

$$Z_1 = a + jb = 2 + j2$$

| EXAMPLE C.4 | Polar to Exponential and Rectangular Form Conversion |

Convert the complex number $Z_1 = 10\angle 60°$ from polar form to exponential and rectangular forms.

SOLUTION

The exponential form for $Z_1 = 10\angle 60°$ is:

$$Z_1 = 10e^{j60°}$$

Using Equation (C.1):

$$a = r\cos\theta = 10 \times \cos(60°) = 5$$
$$b = r\sin\theta = 10 \times \sin(60°) = 8.66$$

Therefore, the rectangular form is:

$$Z_1 = a + jb = 5 + j8.66$$

OPERATIONS IN EXPONENTIAL AND POLAR FORMS

Assume that complex numbers Z_1 and Z_2 are expressed in exponential form: $Z_1 = r_1e^{j\theta}$, $Z_2 = r_2e^{j\phi}$.

- Complex conjugates are:

$$Z_1^* = r_1e^{-j\theta} = r_1\angle(-\theta) \quad \text{and} \quad Z_2^* = r_2e^{-j\phi} = r_2\angle(-\phi)$$

- Addition and subtraction:

If the complex numbers are expressed in exponential or polar form, they need to be converted into rectangular form for addition and subtraction.

- Multiplication:

$$Z_1 \times Z_2 = r_1e^{j\theta} \cdot r_2e^{j\phi} = r_1r_2e^{j(\theta+\phi)} = r_1r_2\angle(\theta + \phi)$$

To multiply two complex numbers in exponential or polar form, multiply the magnitude of them and add the angles.

- Division:

$$\frac{Z_1}{Z_2} = \frac{r_1e^{j\theta}}{r_2e^{j\phi}} = \frac{r_1\angle\theta}{r_2\angle\phi} = \frac{r_1}{r_2}\angle(\theta - \phi)$$

To divide two complex numbers in exponential or polar form, divide the magnitude of them and subtract the angle of the divisor from the angle of the dividend.

- Power, n, of a complex number:

$$Z_1^n = (r_1e^{j\theta})^n = r_1^ne^{jn\theta} = r_1^n\angle n\theta,$$

The magnitude is powered to n and the n of the angles are added together.

- n root of a complex number:

$$Z_1^{\frac{1}{n}} = (r_1e^{j\theta})^{\frac{1}{n}} = r_1^{\frac{1}{n}}e^{j\theta \cdot \frac{1}{n}} = r_1^{\frac{1}{n}}\angle\left(\frac{\theta + 2k\pi}{n}\right), \quad k = 0, \pm 1, \pm 2, \ldots$$

| EXAMPLE C.5 | Multiplicaton and Division in Polar Form |

Assume $Z_1 = 2.828\angle 45°$ and $Z_2 = 3\angle 30°$, calculate Z_1Z_2 and Z_1/Z_2.

SOLUTION

$$Z_1 \times Z_2 = (2.828 \times 3)\angle(45° + 30°) = 8.484\angle75°$$

$$\frac{Z_1}{Z_2} = \frac{2.828}{3}\angle(45° - 30°) = \frac{2.828}{3}\angle15° = 0.9427\angle15°$$

EXAMPLE C.6 **Complex Operations**

Assume $Z_1 = 3\angle45°$, $Z_2 = 4\angle45°$, calculate $Z_1 + Z_2$, Z_1Z_2, and Z_1/Z_2.

SOLUTION

First, convert Z_1 and Z_2 to rectangular form, then add them.

Using Equation (B.1):

The real part a of the Z_1 is:

$$a = 3 \times \cos(45°) = 2.1213$$

The imaginary part b of the Z_1 is:

$$b = 3 \times \sin(45°) = 2.1213$$

The real part c of the Z_2 is:

$$c = 4 \times \cos(45°) = 2.8284$$

The imaginary part d of the Z_2 is:

$$d = 4 \times \sin(45°) = 2.8284$$

Therefore:

$$Z_1 + Z_2 = 2.1213 + j2.1213 + 2.8284 + j2.8284 = 4.9497 + j4.9497$$

Convert the result to polar form:

$$Z_1 + Z_2 = 7\angle45°$$

$$Z_1 \times Z_2 = (3 \times 4)\angle(45° + 45°) = 12\angle90°$$

$$\frac{Z_1}{Z_2} = \frac{3}{4}\angle(45° - 45°) = 0.75\angle0°$$

EULER'S IDENTITY

Euler's identities state that:

$$e^{j\theta} = \cos\theta + j\sin\theta \tag{C.2}$$

$$e^{-j\theta} = \cos\theta - j\sin\theta \tag{C.3}$$

$$\cos\theta = \frac{e^{j\theta} + e^{-j\theta}}{2} \tag{C.4}$$

$$\sin\theta = \frac{e^{j\theta} - e^{-j\theta}}{j2} \tag{C.5}$$

Using the Euler's identities, the cosine and sinusoidal functions can be expressed as complex numbers, and the calculation can be simplified.

EXAMPLE C.7	Application of Euler's Identity

Calculate e^{1-j1}.

SOLUTION

Method 1: Angle in radians:

$$e^{1-j1} = e^1 \cdot e^{-j1} = e^1(\cos 1 - j \sin 1) = 2.71828(0.5403 - j0.8415) = 1.469 - j2.29$$

Method 2: Angle in degrees:

$$e^{1-j1} = e^1 \cdot e^{-j1} = e^1(\cos 1 - j \sin 1) = e^1(\cos 57.3° - j \sin 57.3°) = 2.71828(0.5403 - j0.8415)$$
$$= 1.469 - j2.29$$

Here, "1" is the angle in radians. To convert radians into degrees or vice versa, use the following equation:

$$\frac{\theta_{rad}}{\theta_{deg}} = \frac{\pi}{180}$$

where, $\pi = 3.1415926$.

SUMMARY

A complex number can be expressed as:

$$Z_1 = a + jb = \text{Re}[Z_1] + j\,\text{Im}[Z_1] = re^{j\theta} = \sqrt{a^2 + b^2}\,e^{j\tan^{-1}(b/a)} = \sqrt{a^2 + b^2}\angle\tan^{-1}(b/a).$$

SELECTED SOLUTIONS FOR PROBLEMS

Chapter 2

2.9: $5t$	**2.10:** 25 C	**2.11:** −3 A	**2.14:** 15 J	**2.16:** 2.122 J
2.22: −1 A, −2 A, 8 A	**2.23:** 13 V	**2.26:** 25 mA	**2.28:** $7R/5$	**2.30:** 0.2 A
2.36: (a) 0.25 A, (b) $R_1/R_2 = 1/2$	**2.37:** $7R/5$	**2.41:** 70 V	**2.45:** 120 mA, 133.3 kΩ	**2.48:** 0.001284, 2.364×10^{-4}, 9.304×10^{-8}
2.52: 500 W	**2.58:** (a) 30 W, (b) −2 A	**2.62:** 2.5 V	**2.66:** 6.002 V, 4.8 W	

Chapter 3

3.6: 60 Ω	**3.8:** $10R/3$	**3.11:** (a) 0 (b) 1.760R	**3.13:** 5 Ω
3.15: $15R/8$	**3.19:** 8 A, 4 A, 4 A, 2 A	**3.22:** 10 mA, 7 mA	**3.28:** 26.32 V
3.30: A	**3.35:** 25 V, 75 V	**3.38:** −22.72 V, −6.22 V, 3.33 V	**3.44:** 64.84 V, 44.55 V, 38.55 V
3.48: 0	**3.53:** 11 V, 16.1 mA, 683 Ω	**3.57:** 1 A	**3.64:** 1.4 A
3.69: 5 Ω	**3.71:** 683.3 Ω	**3.74:** 989 μA, 1.8 mA, 6 mA	**3.75:** 12.51 mV

Chapter 4

4.1: (d)	**4.5:** 60 μC	**4.7:** 42.4 nm	**4.14:** (a) 33.3 V (b) 667 μJ	**4.19:** 4 V
4.22: 2 A	**4.24:** 1.36×10^{-4} J	**4.28:** 41.69 μF	**4.32:** (b) 5 Gal	**4.35:** 27.8 V
4.43: 30.2 mH	**4.45:** 55.49 mH	**4.48:** 89.54 mH		

Chapter 5

5.3: $10 - 10e^{-100t}$	**5.5:** $5e^{-t/1.5 \times 10^{-3}}, t \geq 0$	**5.8:** 8.63×10^{-3} s	**5.14:** $T = RC(V_f/V_i)$
5.15: R	**5.16:** R and V_f RV_f = constant	**5.22:** $6.667e^{-10t} - 666.67e^{-100t}$	**5.25:** 0 A; 0.05 A
5.28: $3750e^{-3125t}$ V	**5.32:** $\dfrac{40}{3} - \dfrac{40}{3}e^{-37.5}$ V	**5.35:** 0.0594 A; 0 A; 0.286 A	**5.38:** 1 A; 5 V

5.43: Doubled; three multiple; halved   **5.46:** Over-damped

5.49: $0.396e^{-1000t} - 0.396 \cos(100t) + 3.96 \sin(100t)$ A

5.56: $v_c(t) = 10\cos(50t) + 10\sin(50t) - 5e^{-50t}$

Chapter 6

6.1: $-7.32 + j1.04$

6.3: 50 V

6.8: 110.31 V

6.11: $\sqrt{22}$ V

6.13: $4\cos(4\omega t + 60°) + 2.5$ V

6.17: $340.3\cos(120\pi t + 51.6°)$ V

6.20: $177.58 + j68.97\,\Omega$

6.24: $50 - j1970\ \Omega$

6.28: $450 + j8.75\Omega$

6.30: $R_1C_1 = R_2C_2$

6.33: $35.8\cos(200t + 56.57°)$ mA

6.36: $7.032\angle124.86°$

6.39: $1.961\cos(200t - 11.31)$ A; $0.392\angle78.69$

6.43: \$0.37

6.46: $Z_N = 6.24 + j39\Omega$; $I_N = 4\cos(200t + 30°)$.

6.50: $93.84 - 2.25j\,\Omega$

6.53: $\dfrac{j3RZ_L + 2R^2}{2R + jZ_L}$

6.58: 127.85 W; 0; 127.85 W

6.62: $0.657\angle53.2°$

6.64: (1) 0.9629; (2) 1 W.

6.67: 91.75 μF

6.69: $47.14\angle0°A$; $166.7\angle66.42°A$

6.71: 0.2; 16.5 μF

Chapter 7

7.1: $\dfrac{1}{1 + j(f/20 \times 10^6)}$

7.4: $\dfrac{1}{1 + j(f/34)}$

7.6: $\dfrac{3}{2\pi}$ Hz, $\dfrac{2}{\pi}$ Hz

7.9: $\dfrac{1}{1000\sqrt{\pi}}$ F

7.13: 500 Hz, 0.1

7.14: $\dfrac{1}{1 + j(f/400)}$

7.16: $3\cos(2\times10^5\pi\cdot t - 120°)$

7.20: $8\cos(2 \times 10^5\pi\cdot t + 150°)$

7.24: $\dfrac{j\dfrac{f}{4}}{1 + j\dfrac{f}{4}}$

7.28: 200 Hz, $\dfrac{j(f/200)}{1 + j(f/200)}$

7.32: 17.68 to 19.89 nF

7.35: 20

7.38: A series resonance band pass filter; 45 Hz

7.40: 9.9 to 86.9 nF

7.46: $\dfrac{1}{1 + j(f/500)}$

Chapter 8

8.2: (c)

8.6: Two

8.11: (a) 3.65 mA; (b) −1.71 mA

8.14: 0.0229 A

8.18: (a) 3.3 V, 5.3 V; (b) −0.2 A

8.20: If $v_s < 21$ V, $v_o = 0$; else $(93.75v_S(t) - 328.125)$ mV.

8.21: 41.67 mA; 41.67 mA; 0

8.27: 6.4 V

8.29: $R < 25.5\ \Omega$

8.31: 516.67 μF; 5.24

8.35: (c)

8.37: 99.6

8.40: 27.4 kΩ

8.44: 1 V

8.48: −0.857; −18.57; 75 kΩ; 3 kΩ

8.53: 2.93 V

8.64: NAND gate

8.65: AND gate

8.70: (a) 4; (b) 16 V

8.71: 8 V

8.73: $\dfrac{V_O}{V_S} = -\left[\dfrac{R_2}{R_1} + \dfrac{R_3}{R_1} + \dfrac{R_3}{R_1} \times \dfrac{R_2}{R_4}\right]$

8.75: 328.8 mA

8.78: $V_{out}(t) = -\dfrac{1}{RC}\displaystyle\int_0^t V_{in}(\tau)d\tau$

Chapter 9

9.3: 165.4 μF; 55.14 μF

9.5: 9.68 kW; 29.04 kW

9.9: 8 μF

9.12: 960 W

9.14: $I_A = 12$ A, $I_B = 24$ A, $I_C = 22.28$ A

9.18: 30 Ω

9.20: 14 mH

9.25: (a) 15248.11 V; (b) add a capacitive load to make phase current 50.2 A

9.27: $P_2 = P_1$, resistive; $P_2 > P_1$, inductive; $P_1 > P_2$, capacitive

9.30: $325.5 \times 10^6 \Omega.m$; 1618.25 Ω

9.34: 152410 ∠ −49.41° V

9.38: 9.55 Ω

9.42: 3.37 m

9.45: 1.13 Ω

9.48: 11.33 pF/m; 10.89 pF/m

9.52: 15

Chapter 10

10.3: 139

10.6: 13.8125

10.13: 28D.D3

10.19: 8130

10.23: E9.D

10.24: 1333.0302

10.32: 1

10.35: $\overline{A} + (\overline{B} \cdot C)$

10.38: 1

10.41: A

10.42: $\overline{A} \cdot B$

10.45: (d)

10.48: $A \cdot B + C$

10.51: $(A + B)(\overline{CD})$

10.54: $\overline{A} \cdot \overline{B} + \overline{C \cdot D} + C \cdot D \cdot B \cdot C + C \cdot D \cdot \overline{C} \cdot \overline{B}$

10.58: $D = C + A.B$

10.60: 4

10.69: 1

10.70: 0

10.71: (c)

Chapter 11 Partial Answers

11.3: (b)

11.6: (a)

11.9: (c)

11.11: 1,500

11.13: $k = 1.6$; $b = 0.2$

11.15: 18 μF

11.20: 22.5 m/s

11.24: $v_o(t) = 1.252 \cos(1000\pi \cdot t + 264.61°)$

11.28: from 1,000 to 2,000

11.30: (d)

11.35: $N \geq 6$

11.38: $N = 9$

11.42: $N >= 12$

11.46: 15.97 V

11.48: 0.1 V

Chapter 12

12.3: 1.39×10^{-12} N

12.5: 2.5 kA

12.7: 30°

12.11: $0.1875 \sin(200t)$ mWb

12.17: 2,100 A ·t; 15.834 mWb

12.19: 0.7927 A

12.21: 3.519 mWb

12.24: 39.68 mH; 6.35 mH

12.28: 1,000

12.31: (a) 315 W; (b) 97%

12.33: 15.178

Chapter 13

13.4: 42.44 Nm; 133.3 V	**13.7:** 50; 1.25	**13.9:** 1,030 rad/min	**13.11:** 141 V
13.13: $\omega = \dfrac{0.02k_T V_S + 900R}{(0.02k_a k_T + R)}$.	**13.15:** 6 Ω; 1 Nm	**13.17:** 948.33 rpm	**13.20:** 437.5 Nm
13.24: 0.275 A	**13.27:** 40 Ω; 0.5 Ω	**13.30:** 23.08%	**13.33:** 375.5 V
13.36: 182.18 A	**13.40:** 0.05	**13.46:** 0.528; 28.83 Hz	**13.50:** 4,197 Nm; 23.08%
13.56: $\phi = \dfrac{5}{2\pi}$ Wb			

Chapter 14

14.1: Series: 5,450– 6,550 Ω; Parallel: 784–882 Ω.	**14.5:** Min: 1.72; max: 2.27	**14.7:** 0.71 V; 0.5 V	**14.10:** 0.16 Ω
14.13: 6.25%	**14.16:** Point 1	**14.19:** 1 V	**14.23:** 4 V; 2 V.
14.25: 400 mV; 75 mV	**14.30:** 0.03 W	**14.34:** No	
14.37: $v(t) = 0.5 + 0.7\sin(2\pi \times 10000 \times t + \theta)$		**14.40:** 15 mV$_{pp}$ ~ 15 V$_{pp}$	

Chapter 15

15.1: 1.2 kΩ	**15.3:** (d)	**15.6:** It is fatal	**15.10:** 1/200 years
15.11: A(a), B, C(b), (c)	**15.15:** 795.77 W/m^2; 8.92 m	**15.17:** 266.7 mG; not safe	**15.22:** Safe
15.25: 3.6 MW; 102.9 kW	**15.30:** 1.18 J	**15.35:** 3.75 mA, 2.25 mA, 1.5 mA; it does trip	**15.38:** No; yes
15.41: (c)			

INDEX